Computational Fluid Dynamics
in INDUSTRIAL COMBUSTION

INDUSTRIAL COMBUSTION SERIES
Edited by Charles E. Baukal, Jr.

PUBLISHED TITLES

Oxygen-Enhanced Combustion
Charles E. Baukal, Jr.

Heat Transfer in Industrial Combustion
Charles E. Baukal, Jr.

Computational Fluid Dynamics in Industrial Combustion
Charles E. Baukal, Jr., Vladimir Y. Gershtein, and Xianming Li

FORTHCOMING TITLES

The John Zink Combustion Handbook
Charles E. Baukal, Jr.

Computational Fluid Dynamics *in* INDUSTRIAL COMBUSTION

Edited by

Charles E. Baukal, Jr., Ph.D.

John Zink Co. LLC,
Tulsa, Oklahoma

Vladimir Y. Gershtein, Ph.D.

Air Products and Chemicals, Inc,
Allentown, Pennsylvania

Xianming Li, Ph.D.

Air Products and Chemicals, Inc,
Allentown, Pennsylvania

CRC Press
Taylor & Francis Group
Boca Raton London New York

CRC Press is an imprint of the
Taylor & Francis Group, an **informa** business

CRC Press
Taylor & Francis Group
6000 Broken Sound Parkway NW, Suite 300
Boca Raton, FL 33487-2742

First issued in paperback 2019

© 2001 by Taylor & Francis Group, LLC
CRC Press is an imprint of Taylor & Francis Group, an Informa business

No claim to original U.S. Government works

ISBN-13: 978-0-8493-2000-2 (hbk)
ISBN-13: 978-0-367-39798-2 (pbk)

Library of Congress Card Number 00-041353

Library of Congress Cataloging-in-Publication Data
Computational fluid dynamics in industrial combustion / edited by Charles E. Baukal,
Jr.., Vladimir Y. Gershtein, Xianming Li
    p. cm. — (Industrial combustion series)
Includes bibliographical references and index.
ISBN 0-8493-2000-3
    1. Fluid dynamics—Mathematics. 2. Combustion engineering—Industrial applications.
3. Numerical analysis. I. Baukal, Charles E. II. Gershtein, Vladimir Y. III. Li, Xianming

TA357.C58778 2000
621.402′3—dc21   00-041353

Visit the Taylor & Francis Web site at
http://www.taylorandfrancis.com

and the CRC Press Web site at
http://www.crcpress.com

Preface

This book is intended to fill a gap in the literature for books on computational fluid dynamics (CFD) modeling in industrial combustion, written primarily for the practicing engineer. Many textbooks have been written both on CFD and on combustion, but both types of books generally have only a limited amount of information concerning the combination of CFD and industrial combustion. The main purpose of this book is to codify the many relevant books, papers, and reports that have been written on this subject into a single coherent reference source. The key difference for this book compared to others is that it looks at each topic from a somewhat narrow scope to see how that topic affects modeling in industrial combustion.

The book is basically organized into three parts. The first part deals with the basics of CFD in combustion and includes two introductory chapters and four chapters on some of the popular techniques and methodologies used to solve combustion problems. The second part of the book deals with specific applications of CFD modeling in industrial combustion and includes chapters on modeling to design burners, impinging flame jets, and modeling in the steel, aluminum, glass, gas-turbine, and petrochemical industries. The third part of the book deals with advanced optimization and visualization techniques for enhancing the usefulness of CFD modeling. The book has numerous discussions on the use of oxygen to enhance combustion, something that has received very little coverage in previous references on CFD in industrial combustion.

This book attempts to focus on those topics of interest to the practicing engineer. It does not claim to be exhaustive, but is comprehensive and representative of the techniques used and the problems encountered in industry. There are certainly many niche applications like packed bed combustion, material synthesis in flames, and flare applications that are not discussed here because they are fairly narrow in scope. References are provided for the interested reader who would like more information on a particular subject. It is always a struggle to know what to include and what not to include in a book. Here, the guideline that has been used is to try and touch on the topics in CFD that are relevant to a wide range of industrial combustion applications.

About the Editors

Charles E. Baukal, Jr., Ph.D., P.E., is the Director of the John Zink Company LLC R&D Test Center in Tulsa, OK. He previously worked for 13 years at Air Products and Chemicals, Inc. (Allentown, PA) in the area of oxygen-enhanced combustion. He has 20 years of experience in the fields of heat transfer and industrial combustion and has authored more than 50 publications in those fields. He is the editor of the book *Oxygen-Enhanced Combustion*, the author of the book *Heat Transfer in Industrial Combustion*, and the general editor of the *Industrial Combustion* series, all with CRC Press (Boca Raton, FL). He has a Ph.D. in mechanical engineering from the University of Pennsylvania, is a licensed Professional Engineer in the state of Pennsylvania, has been an adjunct instructor at several colleges, and has eight U.S. patents. He can be reached at baukalc@kochind.com.

    Vladimir Y. Gershtein, Ph.D., is an industry specialist at Air Products and Chemicals, Inc. He is involved in the development and commercialization of new technologies for different processes including high temperature systems. He has done extensive research in the areas of heat and mass transfer, and oxygen-based combustion. He has developed numerical codes for commercial modeling in the glass and aluminum industries. From 1974 to 1989, he worked at the Research Center in St. Petersburg, Russia where he was involved in various research projects including: studying thermophysical and electrical properties of different materials, development of new materials, and application of these materials for different industries; heat and mass transfer for industrial boilers; development of new compact heat exchangers for diesel and chemical industries; development, testing and application of industrial "thermogeneraters" (combustion devices) for chemical industry. Dr. Gershtein received his Ph.D. in Physics and Mathematics from the Russian Academy of Science, Institute of Thermal Physics. He has written 24 technical publications in the area of thermal science. He can be reached at gershtvy@apci.com.

    Xianming (Jimmy) Li, Ph.D., is currently a Lead Research Engineer at Air Products and Chemicals, Inc., in Allentown, PA. Dr. Li is a specialist in the field of combustion and energy systems. His field experience includes utility boilers, glass melters, hydrogen plants, aluminum furnaces and steel facilities. His expertise includes fluid flow, heat transfer, emission control, and process optimization. From 1990 through 1995, Dr. Li was a CFD applications specialist for Fluent Inc. of Lebanon, NH. He provided consulting on CFD applications in the areas of fluid flow, heat transfer, mass transfer and combustion. His Ph.D. thesis work at the Georgia Institute of Technology in Atlanta involved analytical and numerical research on ignition of solid materials. Dr. Li completed the two-year Wharton Management Program at the University of Pennsylvania and is a member of ASME, ASHRAE, Phi Kappa Phi, and The Combustion Institute. Dr. Li has written 53 technical publications in the area of thermal science and holds three U.S. patents. He can be reached at lixm@apci.com.

Contributors

Beth Anne V. Bennett, Ph.D., is an Associate Research Scientist and Lecturer in the Department of Mechanical Engineering at Yale University. She received her Ph.D. in Mechanical Engineering from Yale University in 1997. Her research interests include the development of time- and memory-efficient finite-difference solution-adaptive gridding techniques, with applications in the areas of multidimensional laminar combustion phenomena and fluid dynamics. She is the author of six refereed journal articles and is currently treasurer of the New Haven Section of ASME. She can be reached at beth.bennett@yale.edu.

D. Scott Crocker, Ph.D., is a group leader at CFD Research Corporation where he has worked in the areas of gas turbine fuel nozzle and combustor design, particularly for low emissions applications. He has also been responsible for software development in the areas of combustion and spray. He has a Ph.D. in Mechanical Engineering from the University of Tennessee, Knoxville. He has 13 publications on combustion related topics and is a member of the IGTI Combustion & Fuels Committee. He can be reached at dsc@cfdrc.com.

Werner J.A. Dahm, Ph.D., is Professor of Aerospace Engineering at The University of Michigan, where he is Head of the Laboratory for Turbulence & Combustion (LTC). He has authored over 90 journal articles, conference papers, and technical publications and given over 70 invited and plenary lectures in the areas of turbulence and combustion. Professor Dahm has served on numerous technical, advisory, and organizational committees, and as a consultant for industry. He has graduate degrees in Mechanical Engineering and in Aeronautics, and in 1998 was made a Fellow of the American Physical Society (Division of Fluid Dynamics). He can be reached at wdahm@umich.edu.

Andrew M. Eaton, Ph.D., is a Research Scientist at Morton Thiokol in Brigham City, UT. Prior to this position, he worked as a Research Associate in the Advanced Combustion Engineering Research Center (ACERC) at Brigham Young University. His research interests include computer modeling of flow and combustion systems. He has worked in computational modeling of turbulent flow, heat transfer, and combustion systems. He can be reached at eatonam@thiokol.com.

Richard D. Frederiksen, Ph.D., is a Lead Technical Specialist in Combustion Systems at NGB Technologies, Inc. in Ann Arbor, MI. He has worked in the field of fluid dynamics, turbulence, and combustion, and holds a Ph.D. in Aerospace Engineering from The University of Michigan.

Lori Freitag, Ph.D., is a computer scientist in the Mathematics and Computer Science Division at Argonne National Laboratory (Argonne, IL). She received her Ph.D. in applied mathematics and now works primarily in the areas of adaptive mesh computations and interactive visualization environments. She is the author of more than 30 technical publications and received the Presidential Early Career Award for Scientists and Engineers for her visualization work in 1997. She can be reached at freitag@mcs.anl.gov.

Michael Henneke, Ph.D., is currently a CFD Engineer at John Zink Company in Tulsa, Oklahoma. His academic background is in the area of reacting flow modeling and radiative transport. He holds a Ph.D. in Mechanical Engineering from The University of Texas at Austin. He has published three refereed journal papers as well as many non-refereed articles and has given a number of presentations on computational fluid dynamic modeling of industrial combustion systems. He can be reached at hennekem@kochinc.com.

Scott C. Hill, Ph.D., is a Research Associate in the Advanced Combustion Engineering Research center (ACERC) at Brigham Young University. Prior to this position, he was a scientist at Los Alamos National Laboratory. He has worked on computational modeling of flow and combustion codes and graphics libraries for both workstations and PCs. He can be reached at sch13@juno.com.

Bryan C. Hoke, Jr., Ph.D., is a Senior Principal Research Engineer at Air Products and Chemicals, Inc. (Allentown, PA) where he is a member of the Combustion Center of Excellence and specializes in CFD modeling and combustion related technology development for the glass industry. Bryan has a Ph.D. in Chemical Engineering from Lehigh University (1992). He has more than 25 publications and 1 U.S. patent. Bryan is currently the Treasurer of the Heat Transfer and Energy Conversion Division of AIChE and was elected for two terms as a Director of the division. He can be reached at hokebc@apci.com.

Richard W. Johnson, Ph.D., is a Consulting Engineer at the Idaho National Engineering & Environmental Laboratory (INEEL) in Idaho Falls, Idaho. His Ph.D. is in the area of turbulence modeling. He has experience in the fields of fluid dynamics, turbulence modeling, heat transfer, CFD and design optimization. He is Editor-in-Chief of the CRC Press (Boca Raton, FL) *Handbook of Fluid Dynamics* and is a member of the Fluids Engineering Division of the American Society of Mechanical Engineers (ASME). He can be reached at rwj@inel.com.

Mark D. Landon, Ph.D., is an Advisory Engineer the Idaho National Engineering and Environmental Laboratory. He has worked in the field of design optimization and computational dynamics for eighteen years and has a Ph.D. in Structural Mechanics–Civil Engineering. He has over fifteen publications and one patent. he can be reached at m19@inel.gov.

Sanjay R. Mathur, Ph.D., is Manager of R&D at Fluent, Inc. He has worked in the field of computational fluid dynamics and unstructured grid methods. He has been actively involved in the development of commercial CFD software for industrial applications. He has a Ph.D. in Aerospace Engineering from Iowa State University. He has over 30 technical publications, and is a member of the ASME and the AIAA. He can be reached at sm@fluent.com.

Jayathi Y. Murthy, Ph.D., is an Associate Professor in the Department of Mechanical Engineering at Carnegie Mellon University. She has worked in the field of computational fluid dynamics both in academia and in industry, and has a Ph.D. in Mechanical Engineering from the University of Minnesota. She has over over 50 technical publications, and is a member of the ASME and the AIAA. She can be reached at murthy@andrew.cmu.edu.

Ernest C. Perry, Brigham Young University, Provo, UT. He can be reached at perrye@et.byu.edu.

Petr Schill, Ph.D., is a Senior Research Associate at Glass Service, Inc., Vasetin, Czech Republic. He is working in the field of mathematical modeling for glass furnaces and has developed an original code that has been applied to many industrial furnaces. He has a Ph.D. in solid state physics from Charles University, Prague, and has published approximately 50 papers related to mathematical modeling of glass furnaces. He is a member of TC21 ICG, the American Ceramic Society, and the Czech Glass Society. He can be reached at Glass Service, Inc. [research@gsl.cz].

Clifford E. Smith, M.S., is Vice President/Engineering at CFD Research Corporation (CFDRC). He has worked in the field of gas turbine combustion for over 27 years, specializing in fuel nozzle innovation/design and combustion instability modeling. Mr. Smith has an M.S. degree from California Institute of Technology, and worked at Pratt & Whitney for 13 years before joining CFDRC. He has over 25 publications in combustion, and is the chairman of the Combustion and Fuels Committee of the International Gas Turbine Institute (IGTI). He can be reached at ces@crdrc.com.

Mitchell D. Smooke, Ph.D., is the chairman and the Strathcona Professor of Mechanical Engineering at Yale University. He earned a Ph.D. in Applied Mathematics from Harvard University in 1978 and an M.B.A. in Management and Finance from the University of California at Berkeley in 1983. His primary research interests lie in the areas of computational combustion, chemical vapor deposition, and the numerical solution of ordinary and partial differential equations. He has published numerous papers on the computational structures of flames. He can be reached at smooke–mitchell@yale.edu.

L. Douglas Smoot, Ph.D., is a Professor of Chemical Engineering at Brigham Young University. He is Dean Emeritus of the College of Engineering at Technology and the Founding Director of the Advanced Combustion Engineering Research Center (ACERC). He has done research and published extensively for over four decades in combustion, explosions, and fossil fuels. He can be reached at lds@byu.com.

Mark J. Stock, M.S., is a Technical Specialist in Numerical Modeling at NGB Technologies, Inc. in Ann Arbor, MI. He has worked in computer systems and numerical modeling, and holds an M.S. in Aerospace Engineering from The University of Michigan.

Gretar Tryggvason, Ph.D., is a Professor and Head of Mechanical Engineering at the Worcester Polytechnic Institute. Previously, he was a Professor of Mechanical Engineering at the University of Michigan in Ann Arbor. He has published papers on multiphase and free surface flows, vortex dynamics, and combustion, boiling, solidification, and numerical methods. He has also consulted for private industry and government agencies. He is Associate Editor of the *Journal of Computational Physics.* He can be reached at gretar@wpi.edu.

Acknowledgments

Chuck Baukal would like to thank his wife Beth and his daughters Christine, Caitlyn, and Courtney for their patience and help during the writing of this book. He would like to thank David Koch and Roberto Ruiz of John Zink Company LLC (Tulsa, OK) for their support in the writing of this book. He would also like to thank the good Lord above, without whom this would not have been possible. Scott Crocker would like to thank Marni Kent, Denise Rynders, and Sharon Corbin for their team effort and patience in assembling the text and figures for Chapter 12. Lori Freitag thanks the engineers at Fuel Tech, Inc. (particularly William Michels) and Air Products and Chemicals, Inc. (particularly Vladimir Gershtein and Mark D'Agostini) for their collaborative work on the case studies presented. The work of the author on those projects was originally supported in part by the Mathematical, Information, and Computational Sciences Division subprogram of the Office of Computational and Technology Research, U.S. Department of Energy under contract W-31-109-Eng-38. Vladimir Gershtein would like to thank his daughter Marianna and his wife Nellie for their understanding and patience during the many long hours required for the preparation of his materials. Bryan Hoke and Petr Schill would like to dedicate Chapter 10, "CFD Modeling for the Glass Industry," to the memory of Alexandr (Sasa) Franek, who was a great contributor to the advancement of glass furnace modeling. Those active in the field of glass furnace modeling will miss Sasa and his valuable contributions. Bryan and Petr would like to thank their families for their patience and understanding during the preparation of this chapter. Jimmy Li would like to dedicate this book to the newest member of his family, Anna, who came into the world just in time for his first draft of the manuscript and who has given much meaning to his work. He thanks his wife, Dr. Pingping Ma, for her understanding and support and the following individuals at Air Products and Chemicals, Inc.: Mr. Michael Lanyi for valuable discussions and mentoring; Mr. P. Buddy Eleazer III for introducing the application that resulted in the case study in Chapter 8; and Dr. Vladimir Gershtein for encouragement at times of difficulty.

Contents

Section I

Modeling Techniques

Section

1 Introduction

Charles E. Baukal, Jr.

CONTENTS

1.1 INDUSTRIAL COMBUSTION

The subject of this book is computational fluid dynamics (CFD), specifically in industrial combustion systems. It is important to define what is and what is not meant by industrial combustion. Here, it refers to fossil fuel combustion for processing materials in the following industries:

- Ferrous metals production including ironmaking and steelmaking
- Non-ferrous metals production including aluminum, copper, brass, lead, zinc, etc.

- Heat treating including annealing, brazing, carburizing, normalizing, sintering, and tempering
- Paper and allied products
- Chemical, petrochemical, and hydrocarbon industries
- Minerals including glass, cement, lime, bricks, ceramics, and refractories
- Waste incineration

It also includes power generation (boilers and stationary gas turbines) but does not include propulsion (internal combustion, gas turbine, or rocket engines).

The U.S. Dept. of Energy Office of Industrial Technologies has developed an Industrial Combustion Technology Roadmap for the industrial combustion community.[1] One of the highest priority recommendations for process improvements in that roadmap to be reached by the year 2020 is an improved robust design tool for new burners. This would include the development of a unified code that accounts for burner-furnace geometry and includes emissivity and multi-flame interactions. Other recommendations related to CFD and burners include: adaptation of existing complex CFD codes to burners, use of existing validation facilities to test the robustness of burner design tools, and improved understanding of the combustion process within and outside the burner. These recommendations underscore the importance of CFD for modeling industrial combustion processes. Three of the major components of industrial combustion systems — burners, combustors, and the load — are briefly discussed next.

1.2 COMBUSTION SYSTEM COMPONENTS

There are four components that are important in most industrial combustion processes: the burner(s), furnace, load, and heat recovery system if present (see Figure 1.1). Because the heat recovery system is not usually modeled, it is not discussed here. Although there are other important components in a combustion system, such as the flow control system, they are not normally modeled and are also not considered here.

1.2.1 BURNERS

The burner is the device that is used to combust the fuel with an oxidizer to convert the chemical energy in the fuel into thermal energy. A given combustion system may have a single burner or many burners, depending on the size and type of the application. There are many types of burners designs that exist due to the wide variety of fuels, oxidizers, combustion chamber geometries, environmental regulations, thermal input sizes, and heat transfer requirements, which includes

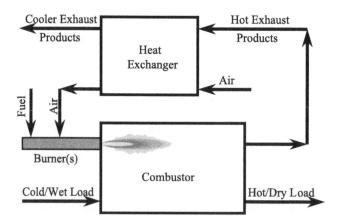

FIGURE 1.1 Schematic of the major components in a combustion system.

things like flame temperature, flame momentum, and heat distribution. Some of these design factors are briefly discussed next.

1.2.1.1 Fuel Effects

Depending on many factors, certain types of fuels are preferred for certain geographic locations due to cost and availability considerations. Gaseous fuels, particularly natural gas, are commonly used in most industrial heating applications in the United States. In Europe, natural gas is also commonly used along with light fuel oil. In Asia and South America, heavy fuel oils are generally preferred although the use of gaseous fuels is on the rise. Fuels also vary depending on the application. For example, in incineration processes, waste fuels are commonly used either by themselves or with other fuels like natural gas. In the petrochemical industry, fuel gases often consist of a blend of several fuels, including gases like hydrogen, methane, propane, butane, and propylene.

The fuel choice has an important influence on modeling the flame. In general, solid fuels like coal are more complicated to model because there are numerous processes that take place during solid fuel combustion, including processes like vaporization, devolatilization, and char oxidation. This is discussed further in Chapter 4. Liquid fuels like oil are somewhat simpler to model than solid fuels but are more difficult to simulate than gaseous fuels. Both solid and liquid fuels tend to produce very luminous flames that contain soot particles that radiate like blackbodies to the heat load. Gaseous fuels like natural gas often produce non-luminous flames because they burn so cleanly and completely without producing soot particles. Flames produced by a fuel like hydrogen are completely nonluminous as there is no carbon available to produce soot. Highly luminous flames may be more difficult to simulate because of the soot production processes and the resulting radiation. However, nonluminous flame radiation is also complicated by the spectral nature of the radiation. In most cases, the fuel choice is dictated by the customer as part of the specifications for the system and is not chosen by the modeler who must simulate whatever fuel has been selected.

In some cases, the burner may have more than one type of fuel. An example is shown in Figure 1.2.[2] Dual-fuel burners are typically designed to operate on either gaseous or liquid fuels. These burners are used where the customer may need to switch between a gaseous fuel like natural gas and a liquid fuel like oil, usually for economic reasons. These burners normally operate on one fuel or the other, and occasionally on both fuels. Another application where multiple fuels may be used is in waste incineration. One method of disposing of waste liquids contaminated with hydro-carbons is to combust them by direct injection through a burner. The waste liquids are fed through the burner, which is powered by a traditional fuel such as natural gas or oil. The waste liquids often have very low heating values and are difficult to combust without an auxiliary fuel. This further complicates the burner design where the waste liquid must be vaporized and combusted concurrently with the normal fuel used in the burner. Multiple fuel burners further complicate the modeling process as well.

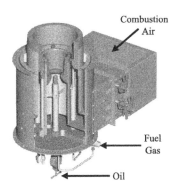

FIGURE 1.2 Typical combination oil and gas burner. (Courtesy of John Zink Co. LLC, Tulsa, OK.)

1.2.1.2   Oxidizer Effects

The predominant oxidizer used in most industrial heating processes is atmospheric air. This can present challenges in some applications, where highly accurate control is required, due to the daily variations in the barometric pressure and humidity of ambient air. The combustion air is sometimes preheated and sometimes blended with some of the products of combustion which is usually referred to as flue gas recirculation (FlGR). In certain cases, preheated air is used to increase the overall thermal efficiency of a process. FlGR is often used to both increase thermal efficiency and to reduce NOx emissions. The thermal efficiency is increased by capturing some of the energy in the exhaust gases that are used to preheat the incoming combustion oxidizer. NOx emissions may also be reduced because the peak flame temperatures are reduced, which can reduce the NOx emissions that are highly temperature dependent. There are also many high-temperature combustion processes that use an oxidizer containing a higher proportion of oxygen than the 21% (by volume) found in normal atmospheric air. This is referred to as oxygen-enhanced combustion (OEC) and has many benefits, including increased productivity and thermal efficiency while reducing the exhaust gas volume and pollutant emissions.[3] A simplified global chemical reaction for the stoichiometric combustion of methane with air is given as follows:

$$CH_4 + 2O_2 + 7.52N_2 \rightarrow CO_2 + 2H_2O + 7.52N_2 + \text{trace species} \qquad (1.1)$$

This compares to the same reaction where the oxidizer is pure O_2 instead of air:

$$CH_4 + 2O_2 \rightarrow CO_2 + 2H_2O + \text{trace species} \qquad (1.2)$$

The volume of exhaust gases is significantly reduced by the elimination of N_2. In general, a stoichiometric oxygen-enhanced methane combustion process can be represented by:

$$CH_4 + 2O_2 + xN_2 \rightarrow CO_2 + 2H_2O + xN_2 + \text{trace species} \qquad (1.3)$$

where $0 \leq x \leq 7.52$, depending on the oxidizer. The N_2 contained in air acts as a ballast that can inhibit the combustion process and have negative consequences. The benefits of using OEC must be weighed against the added cost of the oxidizer, which in the case of air is essentially free except for the minor cost of the air-handling equipment. The use of a higher purity oxidizer has many conse-quences with regard to modeling; these are considered elsewhere in the book (see Chapters 7 to 11).

1.2.1.3   Gas Recirculation

A common technique used in combustion systems is to design the burner to induce furnace gases to be drawn into the burner to dilute the flame, usually referred to as furnace gas recirculation (FuGR). Although the furnace gases are hot, they are still much cooler than the flame itself. This dilution can accomplish several purposes. One is to minimize NOx emissions by reducing the peak temperatures in the flame, as in flue gas recirculation. However, FuGR may be preferred to flue gas recirculation because no external high-temperature ductwork or fans are needed to bring the product gases into the flame zone. Another reason to use FuGR may be to increase the convective heating from the flame because of the added gas volume and momentum. An example of FuGR into the burner is shown in Figure 1.3.[4]

1.2.1.4   General Burner Types

There are numerous ways to classify burners. Some of the common ones are discussed here. One common method for classifying burners is according to how the fuel and the oxidizer are mixed. In premixed burners, shown in a cartoon in Figure 1.4a, the fuel and the oxidizer are completely

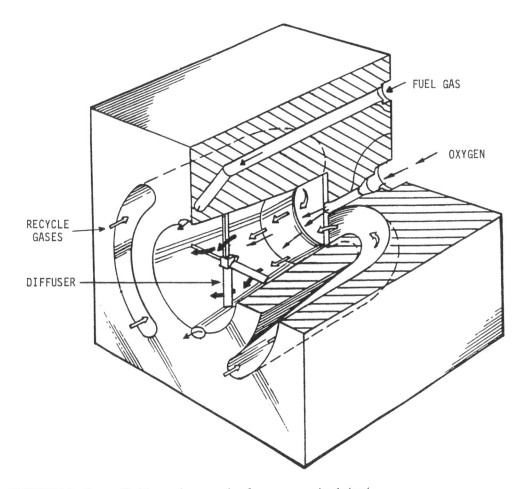

FUEL GAS

OXYGEN

RECYCLE
GASES

DIFFUSER

FIGURE 1.3   Oxygen/fuel burner incorporating furnace gas recirculation.[4]

mixed before combustion begins. Porous radiant burners are usually of the premixed type. Premixed burners often produce shorter and more intense flames, compared to diffusion flames. This can produce high-temperature regions in the flame, leading to nonuniform heating of the load and higher NOx emissions. In diffusion-mixed burners, shown schematically in Figure 1.4b, the fuel and the oxidizer are separated and unmixed prior to combustion, which begins where the oxidizer/fuel mixture is within the flammability range. Oxygen/fuel burners (see Chapter 7) are usually diffusion burners, primarily for safety reasons, to prevent flashback and explosion in a potentially dangerous system. Diffusion gas burners are sometimes referred to as "raw gas" burners as the fuel gas exits the burner essentially intact, with no air mixed with it. Diffusion burners typically have longer flames than premixed burners, do not have as high temperature a hot spot, and usually have a more uniform temperature and heat flux distribution. It is also possible to have partially premixed burners, shown schematically in Figure 1.4c, where a portion of the fuel is mixed with the oxidizer. This is often done for stability and safety reasons where the partial premixing helps anchor the flame, but not fully premixing lessons the chance for flashback. This type of burner often has a flame length and temperature and heat flux distribution that are between the fully premixed and diffusion flames.

Another burner classification based on mixing is known as staging: staged air and staged fuel. Schematics of staged air and staged fuel burners are shown in Figure 1.5. Examples of staged air and staged fuel burners are shown in Figures 1.6 and 1.7. Secondary and sometimes tertiary injectors

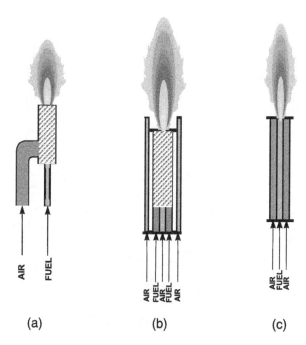

FIGURE 1.4 Schematics of (a) premixed burner, (b) diffusion burner, and (c) partially premixed burner.

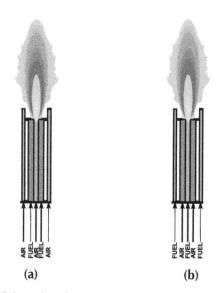

FIGURE 1.5 Schematics of (a) staged air burner and (b) staged fuel burner.

in the burner are used to inject a portion of the fuel and/or oxidizer into the flame, downstream of the root of the flame. Staging is often done to reduce NOx emissions and to produce longer flames. These longer flames typically have a lower peak flame temperature and more uniform heat flux distribution than nonstaged flames.

Most industrial burners are known as forced-draft burners. This means that the oxidizer is supplied to the burner under pressure. For example, in a forced-draft air burner, the air used for combustion is supplied to the burner by a blower. In natural-draft burners, the air used for combustion is induced into the burner by the negative draft produced in the combustor. A sample schematic is shown in Figure 1.8 and an example is shown in Figure 1.9. In this type of burner, the pressure drop

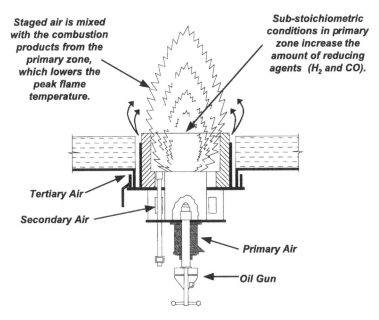

Staged air is mixed with the combustion products from the primary zone, which lowers the peak flame temperature.

Sub-stoichiometric conditions in primary zone increase the amount of reducing agents (H_2 and CO).

Tertiary Air

Secondary Air

Primary Air

Oil Gun

FIGURE 1.6   Example of a staged-air oil burner. (Courtesy of John Zink Co. LLC, Tulsa, OK. With permission.)

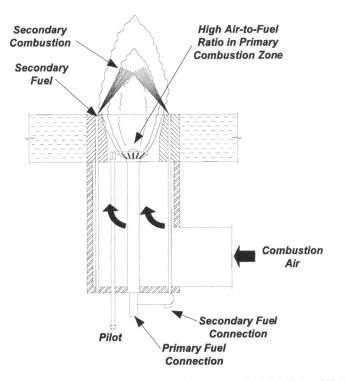

Secondary Combustion

Secondary Fuel

High Air-to-Fuel Ratio in Primary Combustion Zone

Combustion Air

Secondary Fuel Connection

Pilot

Primary Fuel Connection

FIGURE 1.7   Example of a staged fuel burner. (Courtesy of John Zink Co. LLC, Tulsa, OK. With permission.)

and combustor stack height are critical in producing enough suction to induce sufficient combustion air into the burners. This type of burner is commonly used in the chemical and petrochemical industries in fluid heaters. The main consequence of the draft type on heat transfer is that the natural-draft flames are usually longer than the forced-draft flames so that the heat flux from the flame is distributed over a longer distance and the peak temperature in the flame is often lower.

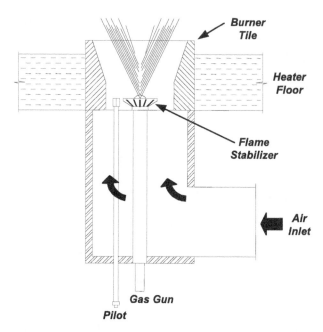

Burner Tile

Heater Floor

Flame Stabilizer

Air Inlet

Gas Gun

Pilot

FIGURE 1.8   Example of a natural draft burner. (Courtesy of John Zink Co. LLC, Tulsa, OK. With permission.)

FIGURE 1.9   Example of a natural draft burner. (Courtesy of John Zink Co. LLC, Tulsa, OK. With permission.)

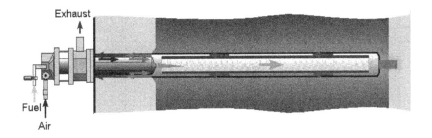

FIGURE 1.10 Example of a single-ended recuperative radiant tube burner. (Courtesy of WS Thermal Process Technology, Inc., Elyria, OH.)

The burners discussed above are all examples of direct-firing where the flame is not confined by an intermediate surface. There are also some burners that have such an intermediate surface surrounding the flame to separate the combustion products from the load. This is often done in metal-treating applications where those products would contaminate the load and reduce its quality. The main example of this type of burner is referred to as a radiant tube burner where the flame is contained by a ceramic or high-temperature metal alloy tube. The flame heats the tube, which then radiates to the load. The three primary designs for radiant tube burners are single-ended, U-tube, and straight-through, depending on how the combustion products flow through the tube. An example of a single-ended radiant tube burner is shown in Figure 1.10. The modeling analysis may be more complicated because of the intermediate surface.

1.2.2 COMBUSTORS

This chapter section briefly introduces some of the combustors commonly used in industrial heating and melting applications. There are many important factors that must be considered when designing a combustor. A primary consideration for any combustor is the type of material that will be processed. One obvious factor of importance in handling the load and transporting it through the combustor is its physical state — whether it is a solid, liquid, or gas. Another factor is the transport properties of the load. For example, the solid may be granular or it might be in the form of a sheet (web). Related to that is how the solid will be fed into the combustor. A granular solid could be fed into a combustor continuously with a screw conveyor or it could be fed in with discrete charges from a front-end loader. The shape of the furnace will vary according to how the material will be transported through it. For example, limestone is fed continuously into a rotating and slightly downwardly inclined cylinder (see Figure 1.13 for example).

1.2.2.1 Furnace Types

Furnaces are often classified as to whether they are batch or continuous. In a batch furnace, the load is charged into the furnace at discrete time intervals where it is heated. There may be multiple load charges, depending on the application. Normally, the firing rate of the burners is reduced or turned off during the charging cycle. On some furnaces, a door may also need to be opened during charging. These significantly impact the modeling of the system because of the time dependence of the process, which is dynamic and constantly changing as a result of the cyclical nature of the load charging. This makes analysis of these systems more complicated.

In a continuous furnace, the load is fed into and out of the combustor constantly. The feed rate may change sometimes due to conditions upstream or downstream of the combustor or due to the production needs of the plant, but the process is nearly steady state. This makes continuous processes

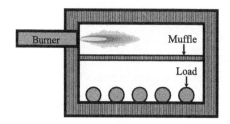

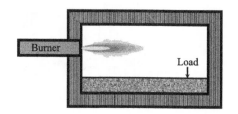

FIGURE 1.11 Elevation view of an indirect-fired furnace. (Courtesy of CRC Press, Boca Raton, FL.)

FIGURE 1.12 Elevation view of a direct-fired furnace. (Courtesy of CRC Press, Boca Raton, FL.)

simpler to analyze as there is no need to include time in the computations. It is often easier to make meaningful measurements in continuous processes due to their steady-state nature.

There are some furnaces that are semicontinuous, where the load can be charged in a nearly continuous fashion but the finished product can be removed from the furnace at discrete intervals. An example is an aluminum reverberatory furnace that is charged using an automatic side-well feed mechanism (see Chapter 9). In that process, shredded scrap is continuously added to a circulating bath of molten aluminum. When the correct alloy composition has been reached and the furnace has a full load, some or all of that load is then tapped out of the furnace. The effect on modeling is somewhere between that for batch and continuous furnaces.

Combustors are often classified as indirect or direct heating. In indirect heating, there is some type of intermediate heat transfer medium, between the flames and the load, that keeps the combustion products separate from the load. One example is a muffle furnace in which there is a high-temperature ceramic muffle between the flames and the load (see Figure 1.11). The flames transfer their heat to the muffle, which then radiates to the load, which is usually some type of metal. The limitation of indirect heating processes is the temperature limit of the intermediate material. Although ceramic materials have fairly high temperature limits, other issues such as structural integrity over long distance spans and thermal cycling can still reduce the recommended operating temperatures. Another example of indirect heating is in process heaters where fluids are transported through metal tubes that are heated by flames. As a result of the temperature limits of the heat exchange materials, most higher temperature processes are of the direct-heating type where the flames can directly radiate heat to the load. In direct heating, there is no intermediate surface to separate the combustion products from the load (see Figure 1.12).

1.2.2.2 Furnace Geometry

Another common way of classifying combustors is according to their geometry, which includes their shape and orientation. The two most common shapes are rectangular and cylindrical. The two most common orientations are horizontal and vertical, although inclined furnaces are commonly used in certain applications such as rotary furnaces (see Figure 1.13).[5] An example of using the shape and orientation of the furnace as a means of classification would be a vertical cylindrical heater (sometimes referred to as a VC) used to heat fluids in the petrochemical industry. Both the furnace shape and orientation have important effects on the modeling of the system. They also determine the type of analysis that will be used. For example, in a VC heater, it is often possible to model only a slice of the heater due to its angular symmetry, in which case cylindrical coordinates would be used. On the other hand, it is usually not reasonable to model a horizontal rectangular furnace using cylindrical coordinates, especially if buoyancy effects are important.

Some furnaces are classified by what they look like. One example is a shaft furnace used to make iron (see Figure 1.14).[5] The raw materials are loaded into the top of a tall, thin, vertically oriented cylinder. Hot combustion gases generated at the bottom through the combustion of coke

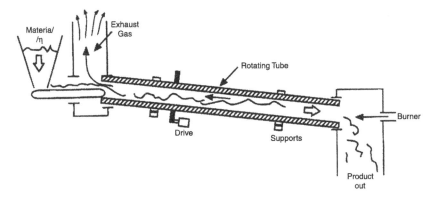

FIGURE 1.13   Elevation view of an inclined rotary kiln. (Courtesy of CRC Press, Boca Raton, FL.)

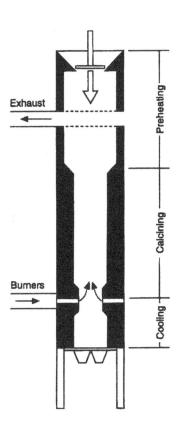

FIGURE 1.14   Elevation view of a vertical shaft
kiln. (Courtesy of CRC Press, Boca Raton, FL.)

flow up through the raw materials that get heated. The melted final product is tapped out of the
bottom. The furnace looks and acts almost like a shaft because of the way the raw materials are
fed in through the top and exit at the bottom. A transfer vessel used to move molten metal around
in a steel mill is often referred to as a ladle because of its function and appearance. These ladles
are preheated using burners before the molten metal is poured into them to prevent the refractory-
lined vessels from thermally shocking.

Another aspect of the geometry, which is important in some applications, is whether the furnace
is moving or not. For example, in a rotary furnace for melting scrap aluminum, the furnace rotates
to enhance mixing and heat transfer distribution (see Figure 1.15). This again affects the type of
analysis that would be appropriate for that system and can add some complexity to the computations.

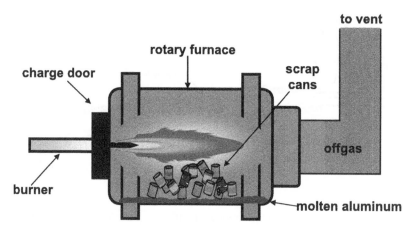

FIGURE 1.15 Elevation view of a rotary scrap aluminum melting furnace. (Courtesy of CRC Press, Boca Raton, FL.)

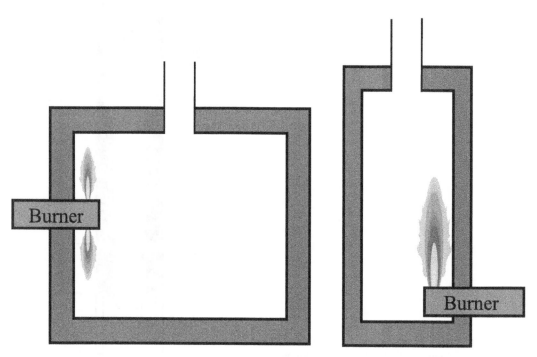

FIGURE 1.16 Elevation views of different wall-fired furnace configurations. (Courtesy of CRC Press, Boca, Raton, FL.)

The burner orientation with respect to the combustor is also sometimes used to classify the combustor. For example, a wall-fired furnace has burners located in and firing along the wall. There are many possible configurations for how the burners fire along the wall, as shown in Figure 1.16. Some furnaces fire with burners located in the middle of the floor (hearth) as shown in Figure 1.17a. Other furnaces, often referred to as down-fired furnaces, fire with burners located in the ceiling as shown in Figure 1.17b. Some furnaces fire a combination of hearth and wall burners (see Figure 1.18).

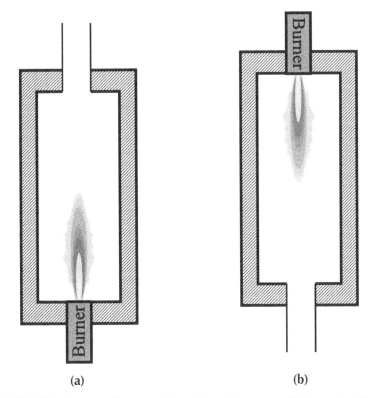

FIGURE 1.17   Elevation views of (a) hearth-fired furnace and (b) down-fired furnace.

1.2.3   HEAT LOAD

This chapter section provides a brief introduction to some of the important issues concerning the heat load in a furnace or combustor.

1.2.3.1   Process Tubes

In petrochemical production processes, process heaters are used to heat petroleum products up to operating temperatures. The fluids are transported through the process heaters in process tubes. These heaters often have a radiant section and a convection section. In the radiant section, radiation from burners heats the process tubes. In the convection section, the combustion products heat the tubes by flowing over the tubes. The design of the radiant section is especially important as flame impingement on the tubes can cause premature failure of the tubes or cause the hydrocarbon fluids to coke inside the tubes, which reduces the heat transfer to the fluids. Examples of different types of process heaters are shown in Figure 1.19.[6] These types of heaters can be more complicated to model because they are fully three-dimensional, and they have an intermediate heat transfer surface (the process tubes) and a load consisting of a fluid flowing through pipes.

1.2.3.2   Moving Substrate

In some applications, heaters and burners are used to heat or dry moving substrates or webs. An example is shown in Figure 1.20. One common application is the use of gas-fired infrared burners to remove moisture from paper during the forming process.[7] These paper webs can travel at speeds over 300 m/s (1000 ft/s) and are normally dried by traveling over and contacting steam-heated cylinders. Convection dryers are also used to heat and dry substrates. Typically, high-velocity heated air is blown at the substrate from both sides so that the substrate is elevated between the nozzles.

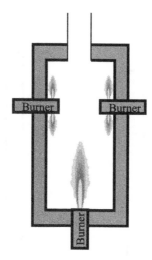

FIGURE 1.18 Elevation view of a furnace fired with hearth and wall-fired burners. (Courtesy of CRC Press, Boca Raton, FL.)

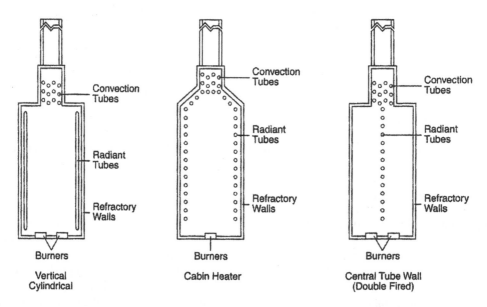

FIGURE 1.19 Elevation view of different types of process heaters. (Courtesy of CRC Press, Boca Raton, FL.)

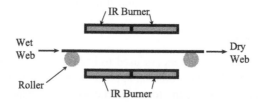

FIGURE 1.20 Elevation view of gas-fired infrared burners heating a moving substrate. (Courtesy of CRC Press, Boca Raton, FL.)

In many cases, the heated air is used for both heat and mass transfer, to volatilize any liquids on or in the substrate (such as water) and then carry the vapor away from the substrate. An important aspect of heating webs is how energy is transferred into the material and vapors are transported out of the material. Therefore, modeling these systems is complicated by the importance of both heat and mass transport.

1.2.3.3 Opaque Materials

This type of load encompasses a wide range of materials, including granular solids like limestone and liquids like molten metal. For this type of load, the heat transfers to the surface of the load and must conduct down into the material. This process can be enhanced by proper mixing of the materials so that new material is constantly exposed to the surface, as in rotary kilns or in aluminum reverberatory furnaces that have molten metal pumps to continuously recirculate the metal through the heating zone. The potential problems with this type of load include overheating the surface materials or having lower thermal efficiencies by limiting the heat transfer to the surface to prevent overheating. Again, the modeling can be complicated by the heat transport through the solids that may be moving in a complicated manner.

1.2.3.4 Transparent Materials

The primary example of this type of load is glass, which has selective radiant transmission properties. In glass-melting processes (see Chapter 10), the primary mode of heat transfer is by radiation. Flames have specific types of radiant outputs that vary as a function of wavelength.[8] If the flame is nonluminous, the flame usually has higher radiant outputs in the preferred wavelengths for water and carbon dioxide bands. If the flame is luminous, it has a broader, more graybody-type spectral radiant profile. Luminous flames are preferred in melting glass because of the selective transmission properties of molten glass.[9] This allows a significant portion of the radiation received at the surface of the glass to penetrate into the glass which enhances heat transfer rates and reduces the chances of overheating the surface (which would reduce product quality). This can greatly complicate the modeling process as spectral radiation may need to be included.

1.3 LITERATURE DISCUSSION

This chapter section briefly considers some of the relevant literature on the subjects of CFD, industrial combustion, and the combination of CFD modeling of industrial combustion processes. Many textbooks have been written on both CFD and combustion, but both book types generally have only a limited amount of information concerning the combination of CFD modeling of industrial combustion. Many of the CFD books are written at a highly technical level for use in upper-level undergraduate or graduate-level courses. The books typically have a broad coverage with less emphasis on practical applications due to the nature of their target audience. This chapter section briefly surveys books related to CFD, industrial combustion, and CFD modeling of industrial combustion.

1.3.1 COMPUTATIONAL FLUID DYNAMICS

Numerous excellent books have been written on the subject of CFD. However, almost none of them has any significant discussion of industrial combustion.[10-26] This is not surprising as there are numerous applications of CFD, which makes it very difficult to be exhaustively comprehensive. Many of the CFD books have no specific discussion of any type of combustion. Oran and Boris (1987) have written a significant work on modeling reactive flows and have a brief section specifically on one-dimensional flame modeling.[27] Weicheng (1991) wrote a book on the use of a software package called GM80 in combustion applications, but nothing specifically for industrial combustion processes.[28] Oliemans (1991) edited a book on CFD for the petrochemical industry; but surprisingly, there was nothing on combustion, which is an important part of most petrochemical processes.[29] Gosman (1998) has given a review of recent developments in the use of CFD for modeling general industrial processes.[30]

1.3.2   INDUSTRIAL COMBUSTION

Many theoretical books have been written on the subject of combustion but most have little if anything on the CFD modeling of the combustion process.[31-35] Barnard and Bradley (1985) have a brief chapter on industrial applications, but have little on modeling of those processes or from flames.[36] A more recent book by Turns (1996), which is designed for undergraduate and graduate combustion courses, contains more discussions of practical combustion equipment than most similar books.[37] Warnatz et al. (1996) have written a brief treatise on various aspects of combustion, including modeling and simulation, but have no discussion of industrial combustion processes.[173]

There have also been several books written on the more practical aspects of combustion. Griswold's (1946) book has a substantial treatment of the theory of combustion, but is also very practically oriented and includes chapters on gas burners, oil burners, stokers and pulverized-coal burners, heat transfer, furnace refractories, tube heaters, process furnaces, and kilns.[38] Stambuleanu's (1976) book on industrial combustion has information on actual furnaces and on aerospace applications, particularly rockets.[39] There is much data in the book on flame lengths, flame shapes, velocity profiles, species concentrations, and liquid and solid fuel combustion. A book on industrial combustion has significant discussions on flame chemistry.[40] Keating's (1993) book on applied combustion is aimed at engines and has no treatment of industrial combustion processes.[41] A recent book by Borman and Ragland (1998) attempts to bridge the gap between the theoretical and practical books on combustion.[42] However, the book has little discussion about most typical industrial applications. Even handbooks on combustion applications have little if anything on industrial combustion systems.[43-47]

1.3.3   CFD MODELING OF INDUSTRIAL COMBUSTION

A number of books have been written on the subject of modeling combustion processes,[48-55] but relatively few about industrial combustion, and even fewer have specifically concerned large-scale industrial combustion systems. Khalil (1982) presented modeling results for six large-scale industrial furnaces that had published experimental data for comparison.[56] These six studies had burners with and without quarls (burner tiles); methane, natural gas and propane fuels; firing rates ranging from 0.74 to 13 MW (2.5 to 44 × 10^6 Btu/h); furnace lengths ranging from 4.5 to 11 m (15 to 36 ft); and swirl numbers ranging from 0 to 5.0. The modeling results using the k-ε turbulence model were in good agreement with the published experimental data. Brauner and Schmidt-Lainé (1988) edited a book on combustion modeling containing papers presented at a NATO workshop.[57] The papers were generally more fundamental in nature, with little discussion of industrial applications. Oran and Boris (1991) have edited a large book on combustion modeling.[58] Part 1 of the book concerns modeling the chemistry of combustion. Part 2 contains information on flames and flame structure. Part 3 is on high-speed reacting flows. Part 4 is humorously entitled "(Even More) Complex Combustion Systems" and has chapters on liquid and solid fuel combustion, as well as on pulse combustion. This book is more theoretical in nature and is intended for aerospace combustion. However, it does have some useful information pertinent to industrial combustion, which is referred to later in this book. A book edited by Larrouturou (1991) looks at modeling of some fundamental processes in combustion science, but does not specifically consider large-scale industrial flames.[59] The papers have significant discussions of flame chemistry and fluid dynamics, but very little on heat transfer from the flames. Rhine and Tucker (1991) have written a book on modeling of natural gas-fired industrial furnaces and boilers.[60] Chung (1993) has edited a book that contains chapters on the various techniques used to model combustion processes, but without any specific applications to industrial combustion problems.[61] In a handbook on fluid dynamics, Lilley (1998) has a brief treatment of combustion modeling with only very brief discussions of industrial applications.[62] Many papers have also been written on the subject of modeling the heat transfer in combustion systems. An example is a recent paper by Eaton et al. (1999) that presents an extensive review of comprehensive combustion models and includes discussions of submodels, data needed to validate codes, and comparison of the model results against experimental data for three cases.[63] Many other papers are given next and in other chapters throughout this book.

1.4 CFD MODELING

1.4.1 INCREASING POPULARITY

Computational fluid dynamics (CFD) is a numerical tool for simulating the complicated fluid flow, heat transfer, and chemical reactions in a combustor. This tool has gained in popularity in recent years due to a number of factors. One obvious change is the dramatic increase in computer power that is available at a cost that is affordable for virtually all businesses. Each new generation of computer hardware continues to have more speed and more memory at a lower cost. The personal computers of today are more powerful than the workstations of only a few years ago. Another reason for the growing popularity of CFD codes is that they are now available at a reasonable cost that usually includes some type of support to aid the user both in the general use of the code as well as the application of the code to the specific needs of the user. Another change is that the CFD computer codes now have very easy-to-use front ends for setting up problems and viewing the results. Before the days of the commercial CFD codes, data had to be input in a certain format that generally required detailed knowledge of both the code and the computer. The results were generally only available in tabular form, which made it difficult and unwieldy to visualize the predictions. Now the data can be input without any knowledge of the details inside the code and without regard for the specific computer operating system. The graphical user interfaces are much more powerful and can be used to quickly set up very complicated flow geometries, including smoothly contoured walls that had to modeled using a stairstep type of approach in the past. The results of the modeling can be quickly and easily viewed in visually appealing and useful formats. One researcher has renamed "CFD" to be "colorized fluid dynamics" because of the explosion of color and pretty pictures that can be generated with the codes that are available today.[64] Another important factor in the adoption of CFD is the improvements in the physics and number of submodels that are available to the user. It is now possible to model compressible and incompressible, viscous and nonviscous, laminar and turbulent, high and low pressure, reacting and nonreacting, multiphase, and many other types of flows. The user often has a choice of several submodels for thermal radiation, turbulence, soot, and pollution chemistry, for example. Another important factor that has accelerated the use of CFD is experimental validation where experimental data has been compared against numerical predictions to show the validity of the predictions. Where the correlation between the predictions and the experimental data was poor, changes have been made to the code to improve its weaknesses. As with most tools of this type, increased adoption of the codes has led to greater acceptance by the engineering community. This growth in popularity is expected to continue to increase as more and more engineers are trained in its use.

1.4.2 OBJECTIVES

There are many reasons to model a combustion system. The most obvious is to gain insight into a particular configuration in order to optimize it. Optimization means different things to different people. It may mean maximizing thermal efficiency, minimizing pollutant emissions, maximizing throughput, minimizing operating costs, or some combination of these. Another reason to model is in the development of new technologies. New geometries can be tested relatively quickly compared to building an entire combustion system. Ideally, modeling is done in conjunction with experimentation to validate a particular new design. Doing modeling first can save considerably on prototype development time and costs by eliminating particular designs without having to actually test them. However, in most cases, it is not possible to completely eliminate prototype testing because of the uncertainty and limitations of combustion modeling, especially when it comes to new configurations that may never have been tried before. Another reason for modeling is to aid in scaling systems to either larger or smaller throughputs. Simple velocity or residence time scaling laws often do not apply to complicated combustion problems.[65] Modeling can be used for predictive purposes to test different scenarios that may be too risky or expensive to try in an existing operational industrial combustion system. For example, a glass producer may want to evaluate the impact of replacing an existing air preheat system with pure oxygen. Another reason to do combustion modeling is to help determine the location for instrumentation. For example, models can be used to help decide where to locate thermocouples in a furnace wall at potential hot spots in order to prevent refractory damage. Although experiments are normally used to validate modeling results, the opposite may also be true. In industrial combustion systems, large-scale probes may be necessary due to water-cooling requirements for survivability. These large probes can cause significant disturbances in the process, which can be simulated with models. The model results can then be used to determine the relevance of experimental measurements.

Models can also be used to simulate potentially dangerous conditions to assess the consequences in order to design the proper safety equipment and procedures. A more recent use of computer modeling is for control of processes where the models are used to predict the results under the given conditions and then adjust the operating parameters to produce the desired results. This includes the use of artificial intelligence where the control system has a large database of past operating conditions and the associated results so that the system can then predict and adjust itself to meet new operating conditions. Examples include making adjustments as equipment ages and deteriorates as well as for new materials being processed. In the past, these adjustments would have been based on the knowledge and experience of the operators and were often trial-and-error. Newer control systems promise more sophisticated and systematic evaluation of the given operating conditions and desired results.

1.4.3 RISKS AND CHALLENGES

One of the risks of computer modeling is that too much faith might be placed in the results. Some tend to believe anything generated by a computer. However, if the computer models have not been properly validated, then the results may be highly suspect. For the foreseeable future, it is likely that models will continue to use various approximations (e.g., turbulence) in order to get solutions in a reasonable amount of time. Therefore, the user must exercise good judgment and not try to overextend the results beyond what is warranted. For example, in many cases, models are very useful in predicting pollutant emission trends but are often very inaccurate in predicting the actual emissions. Knowledge of the model's capabilities helps one understand which results are more reliable and which ones are less reliable. The bulk fluid flow and heat transfer in a combustion system can usually be predicted with a high degree of accuracy, while the small-scale turbulence and trace species predictions may be less reliable. Therefore, it is recommended that computer modeling of combustion systems only be done by those who have been properly trained in that area.

Patankar and Spalding (1974) noted some of the important aspects of the problem statement for industrial combustion modeling problems:[66]

- Geometry of the combustion chamber
- Fuel and air input conditions
- Thermal boundary conditions
- Thermodynamic, transport, radiative, and chemical-kinetic properties

as well as the desired outputs of models:

- Velocity, temperature, composition, etc. throughout the chamber
- Heat flux and temperature at the wall

Kotake et al. (1993) have a chapter in their book on numerical simulations on combustion modeling.[67] They noted some specific challenges of modeling combustion, including:

- Insufficient understanding of combustion mechanisms
- Unclarified correlations between these mechanisms and the turbulent structure
- Stiffness (a measure of the difficulty in solving the equations; the more stiff the equations the more difficult they are to solve) in conjunction with computations of chemical reactions

1.4.4 POTENTIAL PROBLEMS

As with most things, CFD modeling is not a panacea that can be used to solve all problems. As the saying goes, "Garbage in = garbage out," or the results are only as good as the input data. It is incumbent upon the user to input the relevant geometry and boundary conditions into the simulation and to select the appropriate submodels for a given problem. Because of the ease-of-use of the codes available today, it is possible for someone with little or no training in fluid dynamics, heat transfer, and chemical reactions to make predictions that may or may not be credible. It is often observed that anything predicted by a computer must, by definition, be correct. Those who are skilled in this art know that this is far from the truth and tend to use the codes to predict trends, rather than to guarantee absolute numbers, unless they have a great deal of experience with a particular type of problem and have experimental data to validate the code under those conditions. It is very easy for the CFD codes to be misapplied to problems that are beyond the range of their validity. Therefore, it is appropriate to use the caveat, "let the user beware."

There are still many limitations of the physical models in the codes. One example of considerable importance in most combustion problems is turbulence. The empirical k-ε model has been around for many years and has been widely used despite the many known limitations. Two other submodels of great importance in combustion modeling are radiation and chemical reactions. These are all discussed in some detail elsewhere in this book. Suffice it to say that further research is required and is ongoing to improve those submodels.

Combustion problems are among the most complex that CFD codes are used to solve because they usually involve complicated geometries and fluid dynamics, heat transfer (including nonlinear thermal radiation), and chemical reactions (that may include many species and literally hundreds of individual chemical reactions). Many industrial combustors have two very different length scales because the combustor itself may be relatively large, while the length scale required to properly simulate the individual flames in the combustor may be several orders of magnitude smaller. Therefore, large-scale problems may require hundreds of thousands — if not millions — of gridpoints. As the complexity of the problem increases, so do the number of iterations required to get a converged solution. The large number of gridpoints and complicated physics often equate to long computation times, depending on the available hardware. A calculation of a large industrial furnace with multiple burners and the full set of physics, modeled on a typical workstation, may take literally weeks to get a converged solution. Normally, multiple simulations of a given problem are required to find the optimum set of operating conditions. However, it is rare that weeks are available to get such solutions. This means that some simplification of the problem is required. This usually involves fairly intimate knowledge of the physics of the problem and preferably some prior knowledge of typical results. It is usually desirable to have some base case that the model results can be compared against to determine the validity of the numerical predictions and simplifying assumptions. Sometimes, the time, experience with related problems, and base case data are lacking so that the modeler is left to use his or her best judgment as to how to simplify a given problem. Experienced users know the inherent dangers in blindly simplifying a problem in order to get a "solution" within the time available.

It is tempting to use only CFD to design new combustion equipment because it is often much cheaper and faster than building prototypes that are usually tested first under controlled laboratory conditions before trying them out in actual field installations. Further, it is also tempting to use CFD to guarantee the performance of combustion equipment because it may be difficult — if not impossible — to test the equipment in every conceivable type of application. Too much confidence in CFD codes is potentially dangerous without proper experimental validation. A more logical approach is to use a combination of numerical modeling in conjunction with experimental measurements. CFD modeling can be used to dramatically reduce the cycle times for developing new products by rapidly simulating and wide range of configurations that would be both time-consuming and expensive to do with prototypes (see Chapter 7 for example). CFD modeling can also be used to scale-up lab results or field results from one specific application to another type of application. Only proper experimental validation can ensure the user of the usefulness of the modeling results. Unfortunately, this step is often overlooked or ignored and can result in spurious predictions.

1.5   INDUSTRIAL COMBUSTION EXAMPLES

The two major parts of industrial combustion problems consist of the burners and the combustors. Often, these are modeled separately for a variety of reasons as discussed throughout this book. Examples of modeling different types of burners and combustors are given next.

1.5.1   MODELING BURNERS

There have been numerous papers on modeling industrial burners. A sampling of references for modeling industrial burners include:

- Radiant tube burners[68-70]
- Swirl burners[71]
- Oil burners[72]
- Pulse combustion burner[73]
- Porous radiant burners[74-80]
- Industrial hydrogen sulfide burner[81]

Butler et al. (1986) gave a general discussion of modeling burners using the finite volume techniques.[82] Schmücker and Leyens (1998) described the use of CFD to design a new nozzle-mix burner referred to as the Delta Burner.[83] Schmidt et al. (1998) described the use of CFD to redesign a burner, originally firing on coal, to fire on natural gas for use in a rotary kiln.[84]

Modeling radiant burners poses the additional challenge of simulating a porous medium.[85] Perrin et al. (1986) discussed the use of a numerical model for the design of a single-ended radiant tube for immersion in and heating of a bath of molten zinc.[86] Most of the heat transfer is by radiation. Hackert et al. (1998) simulated the combustion and heat transfer in two-dimensional porous burners.[87] Two different porous geometries were simulated: (1) a honeycomb consisting of parallel, non-connecting passages, and (2) a separated plates geometry consisting of parallel but broken walls where there was no continuous solid path which minimizes the importance of solid conduction. Spatial calculated temperatures compared favorably with measured values. The radiant efficiency in the stable region of the flame ranged from 15 to 17% and 22% for the two geometries, respectively. They determined a volumetric Nu of 5.4 ± 0.3, regardless of burning rate or pore size, for the separated plates geometry, which compares favorably with other values reported in the literature. Fu et al. (1998) used a one-dimensional model to simulate the performance of a porous radiant burner.[88] The model accounted for the interaction of convection, conduction, radiation, and chemical reaction in the burner, which consisted of two layers of reticulated ceramics having different porosities. The model showed that the radiant efficiency increased with the volumetric heat transfer coefficient and the effective thermal conductivity of the solid matrix and decreased when the firing rate increased and the equivalence ratio decreased.

1.5.2 MODELING COMBUSTORS

Numerous papers have been presented over the past few decades on modeling of industrial combustors. Some of the earliest work was done at Imperial College (U.K.).[66,89,90] Lilley (1979) reviewed the approaches, equations, and solution methods employed for using CFD to model practical combustors, including industrial furnaces.[91] A sampling of references are given for modeling industrial combustors:

- General combustors and boilers[92-108]
- Glass furnaces[109-116]
- Aluminum reverberatory furnaces[117]
- Metal reheat furnaces[118-122]
- Radiant tube batch furnace[123]
- Flash smelting furnaces[124,125]
- Industrial coal combustors[126-130]
- Steam reformers for liquefied petroleum gas conversion[131]
- Fluid catalytic crackers[132,133]
- Petrochemical cylindrical heater[134]
- Kraft recovery furnace for paper manufacturing[135]
- Pulse combustors[136-139]
- Vortex combustor[140]
- Rotary hearth calciners[141]
- Rotary kilns[142,143]
- Cement rotary kiln[144]
- Vertical lime kilns[145]
- Generic oxy/fuel-fired furnace[146]
- Rotary kiln incinerator[147-150]
- Municipal solid waste incinerator ash-melting furnace[151]
- Hazardous waste incineration furnace[152]
- Baking ovens[153]

Song and Viskanta (1986) did a parametric study of the thermal performance of a generic natural gas-fired furnace.[154] Carvalho and Nogueira (1997) modeled glass-melting furnaces, cement kilns, and baking ovens.[155] Some papers have specifically focused on radiation modeling in combustors,[156] while others have considered swirling flows in furnaces.[157,158]

Hutchinson et al. (1975) numerically studied the effect of the presence or absence of a quarl (burner block or tile).[159] A two-dimensional axisymmetric finite difference model was used with a four flux radiation model. The chemical reactions were indirectly incorporated through a mixture fraction equation and by varying the gas properties. Numerical predictions and experimental measurements showed that the presence of a quarl shortened the flame on an air/natural gas burner with swirl, firing into a water-cooled cylindrical furnace. The recirculation zone in the furnace was pulled closer to the burner with the quarl, as compared to the same burner without a quarl. Higher temperatures and heat fluxes were measured closer to the burner with the quarl, than without the quarl. The quarl shortened the flame and moved the heat release closer to the burner. The calculations showed that the convective heat transfer and the radiative heat transfer to the furnace wall were of the same order with or without the quarl.

Nelson (1979) considered modeling of turbulent reacting flows in large gas-fired furnaces.[160] The models were classified into three categories: (1) empirical, (2) modular, and (3) fundamental. Empirical models predict overall characteristics but are usually only valid for a specific furnace and cannot be easily adapted to other configurations. Modular models predict stability and residence-time distributions (macromixing) based on average gas velocities, but do not accurately predict the temperature-time distribution in the flow (micromixing). Fundamental models are the CFD models considered in this book that predict the detailed micromixing characteristics of the flow.

Viskanta and co-workers performed parametric computational studies of both direct- and indirect-fired furnaces.[161-164] Chapman et al. (1989) presented the results of parametric studies of a direct-fired continuous reheating furnace.[161] They developed a simplified mathematical model that accurately calculated the heat balance throughout the furnace. The combustion space was divided into zones that were considered to be well-stirred reactors. The load in each zone was further subdivided into smaller control volumes. Radiation was modeled using Hottel's zone method. A primary objective of the study was to compute the furnace thermal efficiency as a function of a variety of parameters, including the load velocity, load emissivity, furnace combustion space height, and refractory emissivity. The furnace efficiency increased rapidly with the initial increase in load velocity and then leveled off with further increases in velocity. The load heat flux was relatively insensitive to the load velocity except for the lowest speed, where the flux was considerably lower. The heat flux to the load was most sensitive to the load emissivity. The heat flux to the load was relatively insensitive to the height of the combustion space. Although the model used in the study was fairly simple, with no detailed fluid flow calculations, it was useful for studying a wide range of values for different parameters, which makes it a valuable design tool for studying other configurations. In a companion study, Chapman et al. (1989) also did a parametric modeling study of direct-fired batch reheating furnaces.[162] Again it was shown that the heat flux to the load was very sensitive to the emissivity of the load. Chapman et al. (1994) modeled a direct-fired metal reheat furnace with impinging flame jets.[163] The model simulated an actual steel reheat furnace at an Inland Steel Co. plant in East Chicago, Indiana. The heat transfer to the load was affected by the fuel firing rate.

Carvalho and Nogueira (1993) reviewed the modeling of industrial glass-melting processes.[165] They noted that modeling glass melting is especially important because of the difficulty in making measurements due to the very high temperatures and corrosive environment. The models are useful for improving glass quality, increasing thermal efficiency, reducing pollution emissions, and improving equipment reliability. The authors recommended further research in simulating flow modeling, batch melting, fining and refining, foam formation/elimination, homogenizing, refractories, radiative transfer, and the mass transfer between the glass melt and the combustion chamber. Glass modeling is particularly difficult if the combustion space and molten glass are coupled together because of the large disparity in flow types (see Chapter 10). The combustion space may have turbulent gas flow while the molten glass is very low-speed flow of a highly viscous material. The system is further complicated by the transparent radiative characteristic of the glass and by the chemical reactions occurring in both the gas space and liquid glass phase. In many cases, electrodes are located in the liquid glass and gases may be bubbled through the glass to stimulate stirring and circulation patterns in the glass. The load consists of fine solid materials at the inlet of the tank and liquid glass at the outlet of the tank. Despite this complexity, models have been successfully used to further the understanding of and improve the glass production process.

1.6 FUTURE CHALLENGES

There have been important advances in the modeling of combustion in recent years.[166] However, Liñán and Williams (1993) believe that "full description with complete chemistry of a turbulent combustion process is more than a generation away."[167] Many areas of combustion chemistry need considerable amounts of further research.[168] Kuo (1986) writes:[169]

> Although many idealized problems of combustion have been attacked and solved in simplified forms in the past, there are still many practical problems that defy exact solution today. Real life is indeed much more complicated than the idealized situations considered in the past. The investigation of real-life problems in the combustion field has really just begun.

Dervieux and Larrouturou (1989) believe that modeling turbulent combustion remains a major challenge.[170] Moss (1994) writes that "The direct numerical simulation (DNS) of complex, three-dimensional, reacting flows still lies far beyond presently envisaged computing capability."[171] Despite the exponential growth in computing power, reliable predictions of combustor performance are not yet on the horizon.[172] Despite the length of time it has been around, despite its importance to man, and despite vast amounts of research, combustion is still far from being completely understood.

REFERENCES

1. U.S. Dept. of Energy Office of Industrial Technologies, Industrial Combustion Roadmap, U.S. DOE, Washington, D.C., April 1999.
2. API Publication 535: Burner for Fired Heaters in General Refinery Services, 1st edition, American Petroleum Institute, Washington, D.C., July 1995.
3. C.E. Baukal, Ed., *Oxygen-Enhanced Combustion*, CRC Press, Boca Raton, FL, 1998.
4. K. J. Fioravanti, L. S. Zelson, and C. E. Baukal, Flame Stabilized Oxy-Fuel Recirculating Burner, U.S. Patent 4,954,076 issued 04 September 1990.
5. G.L. Shires, Kilns, in *International Encyclopedia of Heat & Mass Transfer*, G.F. Hewitt, G.L. Shires, and Y.V. Polezhaev, Eds., CRC Press, Boca Raton, FL, 1997, 651-653.
6. G.L. Shires, Furnaces, in *The International Encyclopedia of Heat & Mass Transfer*, G.F. Hewitt, G.L. Shires, and Y.V. Polezhaev, Eds., CRC Press, Boca Raton, FL, 1997, 493-497.
7. S. Longacre, Using infrared to dry paper and its coatings, *Process Heating*, 4(2), 45-49, 1997.
8. C.E. Baukal, *Heat Transfer in Industrial Combustion*, CRC Press, Boca Raton, FL, 2000.
9. B. Ji and C.E. Baukal, Spectral radiation properties of oxygen-enhanced/natural gas flames, *Proceedings of 1998 International Gas Research Conference*, November 8-11, 1998, San Diego, CA, Dan Dolenc, Ed., 5, 422-433, 1998.
10. P.J. Roach, *Computational Fluid Dynamics*, revised printing, Hermosa Publishers, Albuquerque, NM, 1976.
11. H.J. Wirz and J.J. Smolderen, *Numerical Methods in Fluid Dynamics*, Hemisphere, Washington, D.C., 1978.
12. K.W. Morton and M.J. Baines, Eds., *Numerical Methods for Fluid Dynamics*, Academic Press, London, 1982.
13. R. Peyret and T.D. Taylor, *Computational Methods for Fluid Flow*, Springer-Verlag, New York, 1983.
14. M. Holt, *Numerical Methods in Fluid Dynamics*, Springer, Berlin, 1984.
15. T.K. Bose, *Computational Fluid Dynamics*, Wiley, New York, 1988.
16. C.A.J. Fletcher, *Computational Techniques for Fluid Dynamics*, 2 volumes, Springer, Berlin, 1991.
17. C.T. Shaw, *Using Computational Fluid Dynamics*, Prentice-Hall, New York, 1992.
18. H.K. Versteeg and W. Malalasekera, *An Introduction to Computational Fluid Dynamics: The Finite Volume Method*, Longman Scientific & Technical, New York, 1995.
19. R.L. Batra, *Role of Finite Elements and Boundary Elements in Computational Fluid Dynamics*, Wiley, New York, 1996.
20. R. Peyret, Ed., *Handbook of Computational Fluid Mechanics*, Academic Press, London, 1996.
21. J.F. Wendt, Ed., *Computational Fluid Dynamics: An Introduction*, Springer, Berlin, 1996.
22. C. Pozrikidis, *Introduction to Theoretical and Computational Fluid Dynamics*, Oxford University Press, New York, 1997.

23. V.K. Garg, *Applied Computational Fluid Dynamics*, Marcel Dekker, New York, 1998.
24. M.Griebel, T. Dornseifer, and T. Neunhoeffer, *Numerical Simulation in Fluid Dynamics*, Society for Industrial and Applied Mathematics, Philadelphia, 1998.
25. W. Shyy, *Computational Techniques for Complex Transport Phenomena*, Cambridge University Press, Cambridge, 1997.
26. J.H. Ferziger and M. Peri'c, *Computational Methods for Fluid Dynamics*, Springer, Berlin, 1999.
27. E.S. Oran and J.P. Boris, *Numerical Simulation of Reactive Flow*, Elsevier, New York, 1987.
28. F. Weicheng, *Computer Modelling of Combustion Processes*, Pergamon Press, Oxford, 1991.
29. R.V.A. Oliemans, *Computational Fluid Dynamics for the Petrochemical Process Industry*, Kluwer Academic, Dordrecht, The Netherlands, 1991.
30. A.D. Gosman, Developments in industrial computational fluid dynamics, *Trans. IChemE*, 76(A), 153-161, 1998.
31. R.A. Strehlow, *Fundamentals of Combustion*, Inter. Textbook Co., Scranton, PA, 1968.
32. F.A. Williams, *Combustion Theory*, Benjamin/Cummings, Menlo Park, CA, 1985.
33. B. Lewis and G. von Elbe, *Combustion, Flames and Explosions of Gases*, 3rd ed., Academic Press, New York, 1987.
34. W. Bartok and A.F. Sarofim, Eds., *Fossil Fuel Combustion*, Wiley, New York, 1991.
35. I. Glassman, *Combustion*, 3rd ed., Academic Press, New York, 1996.
36. J.A. Barnard and J.N. Bradley, *Flame and Combustion*, 2nd ed., Chapman and Hall, London, 1985.
37. S.R. Turns, *An Introduction to Combustion*, McGraw-Hill, New York, 1996.
38. J. Griswold, *Fuels, Combustion and Furnaces*, McGraw-Hill, New York, 1946.
39. A. Stambuleanu, *Flame Combustion Processes in Industry*, Abacus Press, Tunbridge Wells, U.K., 1976.
40. E. Perthuis, *La Combustion Industrielle*, Éditions Technip, Paris, 1983.
41. E.L. Keating, *Applied Combustion*, Marcel Dekker, New York, 1993.
42. G. Borman and K. Ragland, *Combustion Engineering*, McGraw-Hill, New York, 1998.
43. C.G. Segeler, Ed., *Gas Engineers Handbook*, Industrial Press, New York, 1965.
44. R. D. Reed, *Furnace Operations*, 3rd ed., Gulf Publishing, Houston, TX, 1981.
45. R. Pritchard, J. J. Guy, and N. E. Connor, *Handbook of Industrial Gas Utilization*, Van Nostrand Reinhold, New York, 1977.
46. Reed, R. J., *North American Combustion Handbook*, 3rd ed., Volume I, North American Mfg. Co., Cleveland, OH, 1986.
47. IHEA, *Combustion Technology Manual*, 5th ed., Industrial Heating Equipment Assoc., Arlington, VA, 1994.
48. R. Glowinski, B. Larrouturou, and R. Temam, Eds., *Numerical Simulation of Combustion Phenomena: Proceedings of the Symposium,* held at INRIA, Sophia-Antipolis, France, May 21–24, 1985, Springer-Verlag, Berlin, 1985.
49. J. Buckmaster and T. Takeno, Eds., *Mathematical Modeling in Combustion Science: Proceedings of a Conference Held in Juneau, Alaska, August 17–21, 1987,* Springer-Verlag, Berlin, 1988.
50. B. Engquist, M. Luskin and A. Majda, Eds., *Computational Fluid Dynamics and Reacting Gas Flows*, Springer-Verlag, New York, 1988.
51. J. Bebernes and D. Eberly, *Mathematical Problems from Combustion Theory*, Springer-Verlag, New York, 1989.
52. A. Dervieux and B. Larrouturou, Eds., *Numerical Combustion: Proceedings of the Third International Conference on Numerical Combustion,* held in Juan les Pins, Antibes, May 23–26, 1989, Springer-Verlag, Berlin, 1989.
53. M. Onofri and A. Tesei, *Fluid Dynamical Aspects of Combustion Theory*, Halsted Press, New York, 1992.
54. J. Buckmaster and T. Takeno, Eds., *Modeling in Combustion Science*, Springer-Verlag, Berlin, 1995.
55. B.N. Chetverushkin, J.A. Désidéri, Y.A. Kuznetsov, J. Périaux, Kh.A. Muzafariv, and O. Pironneau, *Experimentation, Modelling, Computation in Flow, Turbulence and Combustion*, John Wiley, New York, 1996.
56. E.E. Khalil, *Modelling of Furnaces and Combustors*, Abacus Press, Kent, U.K., 1982.
57. C.-M. Brauner and C. Schmidt-Lainé, *Mathematical Modeling in Combustion and Related Topics*, Kluwer Academic, Dordrecht, The Netherlands, 1988.
58. E.S. Oran and J.P. Boris, Eds., *Numerical Approaches to Combustion Modeling*, Vol. 135: Progress in Astronautics and Aeronautics, American Institute of Aeronautics and Astronautics, Washington, D.C., 1991.

59. B. Larrouturou, *Recent Advances in Combustion Modelling*, World Scientific, Singapore, 1991.

60. J. M. Rhine and R. J. Tucker, *Modelling of Gas-Fired Furnaces and Boilers*, McGraw-Hill, London, 1991.

61. T.J. Chung, *Numerical Modeling in Combustion*, Taylor & Francis, Washington, D.C., 1993.

62. D.G. Lilley, Chemically reacting flows (combustion), *The Handbook of Fluid Dynamics*, R.W. Johnson, Ed., CRC Press, Boca Raton, FL, 1998, chap. 16.

63. A.M. Eaton, L.D. Smoot, S.C. Hill, and C.N. Eatough, Components, formulations, solutions, evaluation, and application of comprehensive combustion models, *Prog. Energy Comb. Sci.*, 25(4), 387-436, 1999.

64. R. Dibble, University of California at Berkeley, private communication, 1997.

65. R. Weber, Scaling characteristics of aerodynamics, heat transfer, and pollutant emissions in industrial flames, *Twenty-Sixth Symposium (International) on Combustion*, The Combustion Institute, Pittsburgh, PA, 1996, 3343-3354.

66. S. Patankar and B. Spalding, Simultaneous predictions of flow patterns and radiation for three-dimensional flames, *Heat Transfer in Flames*, N.H. Afgan and J.M. Beer, Eds., Scripta Book Co., Washington, D.C., 1974, 73-94.

67. S. Kotake, K. Hijikata, and T. Fusegi, *Numerical Simulations of Heat Transfer and Fluid Flow on a Personal Computer*, Elsevier, Amsterdam, 1993.

68. A.M. Lankhorst and J.F.M. Velthuis, Ceramic recuperative radiant tube burners: simulations and experiments, in *Tranport Phenomena in Combustion*, Vol. 2, S.H. Chan, Ed., Taylor & Francis, Washington, D.C., 1996, 1330-1341.

69. F. Mei and H. Meunier, Numerical and experimental investigation of a single ended radiant tube, in *ASME Proceedings of the 32nd National Heat Transfer Conf.*, Vol. 3: Fire and Combustion, L. Gritzo and J.-P. Delplanque, Eds., ASME, New York, 1997, 109-118.

70. H. Ramamurthy, S. Ramadhyani, and R. Viskanta, Development of fuel burn-up and wall heat transfer correlations for flows in radiant tubes, *Num. Heat Transfer, Part A*, 31, 563-584, 1997.

71. S. Bortz and A. Hagiwara, Inviscid model for the prediction of the near field region of swirl burners, in *Industrial Combustion Technologies*, M.A. Lukasiewicz, Ed., American Society of Metals, Materials Park, OH, 1986, 89-97.

72. T.A. Butcher, Computational fluid dynamics in oil burner design, *Proc. 1997 Oil Heat Tech. Conf. & Workshop*, Paper 97-01, 1997, 7-15.

73. Y. Tsujimoto and N. Machii, Numerical analysis of pulse combustion burner, *Twenty-First Symposium (International) on Combustion*, The Combustion Institute, Pittsburgh, PA, 1986, 539-546.

74. T.W. Tong, S.B. Sathe, and R.E. Peck, Improving the performance of porous radiant burners through use of sub-micron size fibers, in *Heat Transfer Phenomena in Radiation, Combustion, and Fires*, R.K. Shah, Ed., ASME HTD-Vol. 106, pp. 257-264, New York, 1989.

75. S.B. Sathe, R.E. Peck, and T.W. Tong, A numerical analysis of combustion and heat transfer in porous radiant burners, *Int. J. Heat Mass Transfer*, 33(6), 1331-1338, 1990.

76. S.H. Chan and K. Kumar, Analytical investigation of SER recuperator performance, in *Fossil Fuel Combustion Symposium 1990*, S. Singh, Ed., ASME PD-Vol. 30, pp. 161-168, New York, 1990.

77. P.-F. Hsu, J.R. Howell, and R.D. Matthews, A numerical investigation of premixed combustion within porous inert media, *J. Heat Transfer*, 115(3), 744-750, 1993.

78. R. Mital, J.P. Gore, R. Viskanta, and S. Singh, Radiation efficiency and structure of flames stabilized inside radiant porous ceramic burners, in *Combustion and Fire*, M.Q. McQuay, W. Schreiber, E. Bigzadeh, K. Annamalai, D. Choudhury, and A. Runchal, Eds., *ASME Proceedings of the 31st National Heat Transfer Conf.*, Vol. 6, ASME HTD-Vol. 328, pp. 131-137, New York, 1996.

79. C.L. Hackert, J.L. Ellzey, and O.A. Ezekoye, Numerical simulation of a porous honeycomb burner, in *ASME Proceedings of the 32nd National Heat Transfer Conf.*, Vol. 3: Fire and Combustion, L. Gritzo and J.-P. Delplanque, Eds., ASME, New York, 1997, 147-153.

80. P.H. Bouma and L.P.H. De Goey, Premixed combustion on ceramic foam burners, *Comb. Flame*, 119, 133-143, 1999.

81. M.M. Sidawi, B. Farouk, and U. Parekh, A numerical study of an industrial hydrogen sulfide burner with air- and oxygen-based operations, in *Heat Transfer in Fire and Combustion Systems — 1993*, B. Farouk, M.P. Menguc, R. Viskanta, C. Presser, and S. Chellaiah, Eds., ASME HTD-Vol. 250, 227-234, New York, 1993.

82. G.W. Butler, J. Lee, K. Ushimaru, S. Bernstein, and A.D. Gosman, A Numerical Simulation Methodology and its Application in Natural Gas Burner Design, in *Industrial Combustion Technologies*, M.A. Lukasiewicz, Ed., American Society of Metals, Warren, PA, 1986, 109-116.

83. A. Schmücker and R,E, Leyens, Development of the Delta Burner Using Computational Fluid Dynamics, *Proc. 1998 International Gas Research Conf.*, Vol. V: Industrial Utilization, D.A. Dolenc, Ed., Gas Research Institute, Chicago, 1998, 516-526.

84. B. Schmidt, B. Spiegelhauer, N.B. Kampp Rasmussen, and F. Giversen, Development of a Process Adapted Gas Burner Through Mathematical Modelling and Practical Experience, *Proc. 1998 International Gas Research Conf.*, Vol. V: Industrial Utilization, D.A. Dolenc, Ed., Gas Research Institute, Chicago, 1998, 578-584.

85. J.R. Howell, M.J. Hall, and J.L. Ellzey, Combustion within porous media, in *Heat Transfer in Porous Media*, Y. Bayazitoglu and U.B. Sathuvalli, Eds., ASME, HTD-Vol. 302, pp. 1-27, New York, 1995.

86. M. Perrin, P. Lievoux, R. Borghi, and M. Gonzalez, Utilization of a numerical model for the design of a gas immersion tube, in *Industrial Combustion Technologies*, M.A. Lukasiewicz, Ed., Amer. Soc. Metals, Materials Park, OH, pp. 127-134, 1986.

87. C.L. Hackert, J.L. Ellzey, and O.A. Ezekoye, Combustion and heat transfer in model two-dimensional porous burners, *Comb. Flame*, 116, 177-191, 1999.

88. X. Fu, R. Viskanta, and J.P. Gore, Modeling of thermal performance of a porous radiant burner, in *Combustion and Radiation Heat Transfer*, R.A. Nelson, K.S. Ball, and Z.M. Zhang, Eds., Proceedings of the ASME Heat Transfer Division — 1998, Vol. 2, ASME HTD-Vol. 361-2, pp. 11-19, New York, 1998.

89. S.V. Patankar and D.B. Spalding, A computer model for three-dimensional flows in furnaces, *Fourteenth Symposium (Int.) on Combustion*, The Combustion Institute, Pittsburgh, PA, 1973, 605-614.

90. L.S. Caretto, A.D. Gosman, S.V. Patankar, and D.B. Spalding, Two calculation procedures for steady, three-dimensional flows with recirculation, in *Proc. 3rd Int. Conf. on Numerical Methods in Fluid Dynamics*, Springer, Berlin, 1972, 60-68.

91. D.G. Lilley, Flowfield modeling in practical combustors, *J. Energy*, 3(4), 193-210, 1979.

92. S.V. Patankar and D.B. Spalding, A computer model for three-dimensional flow in furnaces, *Fourteenth Symposium (International) on Combustion*, The Combustion Institute, Pittsburgh, PA, 1972, 605-614.

93. T.M. Lowes, M.P. Heap, S. Michelfelder, and B.R. Pai, Paper 5. Mathematical modelling of combustion chamber performance, *J. Institute Fuel*, 46(38), 343-351, 1973.

94. W. Richter, Prediction of heat and mass transfer in a pulverised fuel furnace, *Letters in Heat & Mass Transfer*, 1, 83-94, 1978.

95. M.M.M. Abou Ellail, A.D. Gosman, F.C. Lockwood, and I.E.A. Megahed, Description and validation of a three-dimensional procedure for combustion chamber flows, *J. Energy*, 2(2), 71-80, 1978.

96. A.D. Gosman, F.C. Lockwood, and A.P. Salooja, The prediction of cylindrical furnaces gaseous fueled with premixed and diffusion burners, *Seventeenth Symposium (International) on Combustion*, The Combustion Institute, Pittsburgh, PA, 1978, 747-760.

97. E.E. Khalil, On the modelling of turbulent reacting flows in furnaces and combustion chambers, *Acta Astronautica*, 6(3-4), 449-465, 1979.

98. E.E. Khalil, P. Hutchinson, and J.H. Whitelaw, The calculation of the flow and heat-transfer characteristics of gas-fired furnaces, *Eighteenth Symposium (International) on Combustion*, The Combustion Institute, Pittsburgh, PA, 1980, 1927-1938.

99. E.E. Khalil, Numerical computations of turbulent swirling flames in axisymmetric combustors, in *Flow, Mixing and Heat Transfer in Furnaces*, K.H. Khalil, F.M. El-Mahallawy, E.E. Khalil, and D.B. Spalding, Eds., Pergamon Press, Oxford, 1978, 231-246.

100. K. Görner, Prediction of the turbulent flow, heat release and heat transfer in utility boiler furnaces, in *Coal Combustion*, J. Feng, Ed., Hemisphere Publishing, New York, 1988, 273-282.

101. R. Görner and W. Zinser, Prediction of three-dimensional flows in utility boiler furnaces and comparison with experiments, *Comb. Sci. Tech.*, 58, 43-58, 1988.

102. P.A. Gillis and P.J. Smith, An evaluation of three-dimensional computational combustion and fluid dynamics in industrial furnace geometries, in *Twenty-Third Symp. (Int.) on Combustion*, The Combustion Institute, Pittsburgh, PA, 1990, 981-991.

103. M.G. Carvalho, J.B. Lopes, and M. Nogueira, A three-dimensional procedure for combustion and heat transfer in industrial furnaces, in *Advanced Computational Methods in Heat Transfer*, Vol. 3: Phase Change and Combustion Simulation, L.C. Wrobel, C.A. Brebbia, and A.J. Nowak, Eds., Springer-Verlag, Berlin, 1990, 171-183.

104. H. Meunier, Modelling of industrial furnaces, *Proc. of 2nd European Conf. on Industrial Furnaces and Boilers,* Portugal, April, R. Collin, W. Leuckel, A. Reis, and J. Ward, Eds., 1991, 1-21.

105. J.M. Rhine and R.J. Tucker, *Modelling of Gas-fired Furnaces and Boilers and Other Industrial Heating Processes*, British Gas and McGraw-Hill, New York, 1991.

106. V. Sidlauskas and M. Tamonis, Mathematical modeling of the thermal process in industrial combustion chambers, *Heat Transfer — Soviet Research*, 23(7), 897-914, 1991.

107. M. Matsumura, S. Ito, Y. Ichiraku, and T. Saeki, Heat transfer simulation in industrial gas furnaces, *Proc. of 1992 International Gas Research Conf.*, H.A. Thompson, Ed., Government Institutes, Rockville, MD, 1993, 2195-2204.

108. C.J. Hoogendoorn, Full modelling of industrial furnaces and boilers, in *Tranport Phenomena in Combustion*, Vol. 2, S.H. Chan, Ed., Taylor & Francis, Washington, D.C., 1996, 1177-1188.

109. A.D. Gosman, F.C. Lockwood, I.E.A. Megahed, and N.G. Shah, The prediction of the flow, reaction and heat transfer in the combustion chamber of a glass furnace, in *AIAA 18th Aerospace Sciences Meeting*, January, Pasadena, CA, 1980, 14-46.

110. M.D.G. Carvalho and F.C. Lockwood, Mathematical simulation of an end-port regenerative glass furnace, *Proc. Instn. Mech. Engrs.*, 199(C2), 113-120, 1985.

111. M.G. Carvalho, D.F.G. Durão, and J.C.F. Pereira, Prediction of the flow, reaction and heat transfer in an oxy-fuel glass furnace, *Eng. Comput.*, 4(1), 23-34, 1987.

112. L. Post and C.J. Hoogendoorn, Heat Transfer in Gas-Fired Glass Furnaces, *VDI, Berichte*, No. 645, 457-466, 1987.

113. C.J. Hoogendoorn, L. Post, and J.A. Wieringa, Modelling of combustion and heat transfer in glass furnaces, *Glastech. Ber.* 63(1), 7-12, 1990.

114. V.B. Kut'in, S.N. Gushchin, and V.G. Lisienko, Heat transfer in the cross-fired glass furnace, *Glass & Ceramics*, 54(5-6), 135-138, 1997.

115. M.G. Carvalho, M. Nogueira, and J. Wang, Mathematical modelling of the glass melting industrial process, in *Proc. XVII Int. Congress on Glass*, Beijing, China, International Academic Publishers, 6, 69-74, 1995.

116. V.B. Kut'in, S.N. Gushchin, and V.G. Lisienko, Heat Exchange in the Cross-Fired Glass Furnace, *Glass & Ceramics*, 54(5-6), 172-174, 1997.

117. V.Y. Gershtein and C.E. Baukal, Model Prediction Comparison for Aluminum Reverberatory Furnace Firing on Air-, Air-Oxy-, and Oxy/Fuel, presented at *1999 Minerals, Metals & Materials Society Annual Meeting & Exhibition,* February 28–March 4, 1999, San Diego, CA.

118. R.M. Davies, D.M. Lucas, B.E. Moppett, and R.A. Galsworthy, Isothermal model studies of rapid heating furnaces, *J. Inst. Fuel*, 44, 453-461, 1971.

119. Y.K. Lee, H.S. Park, and K.W. Cho, Effect of fuel gas preheating on combustion and heat transfer in reheating furnace, *Proceedings of the 13th Energy Engineering World Congress, Energy & Environmental Strategies for the 1990s,* Atlanta, GA, October 1990, 1991, chap. 78, 461-466.

120. R. Klima, Improved knowledge of gas flow and heat transfer in reheating furnaces, *Scandinavian J. Metallurgy (Suppl.)*, 26, 25-32, 1997.

121. C. Zhang, T. Ishii, and S. Sugiyama, Numerical modeling of the thermal performance of regenerative slab reheat furnaces, *Num. Heat Trans.*, 32A(6), 613-631, 1997.

122. J.M. Blanco and J.M. Sala, Improvement of the efficiency and working conditions for reheating furnaces through computational fluid dynamics, *Industrial Heating*, Vol. LXVI(5), 63-67, 1999.

123. H. Ramamurthy, S. Ramadhyani, and R. Viskanta, Thermal system model for a radiant tube batch reheating furnace, *Proc. of 1992 International Gas Research Conf.*, H.A. Thompson, Ed., Government Institutes, Rockville, MD, 1993, 2205-2216.

124. N.D.H. Munroe, Experimental and numerical modeling of transport phenomena in a particulate reacting system, in *Heat Transfer in Fire and Combustion Systems — 1993*, B. Farouk, M.P. Menguc, R. Viskanta, C. Presser, and S. Chellaiah, Eds., ASME HTD, New York, 250, 69-78, 1993.

125. T. Ahokainen, A. Jokilaakso, O. Teppo, and Y. Yang, Flow and Heat Transfer Simulation in a Flash Smelting Furnace, NTIS Report DE95779247, U.S. Dept. of Commerce, Springfield, VA, 1994.

126. D.L. Smoot, Pulverized coal diffusion flames: a perspective through modelling, in *Eighteenth Symp. (Int.) on Combustion*, The Combustion Institute, Pittsburgh, PA, 1981, 1185-1202.

127. W.A. Fiveland and R.A. Wessel, Numerical model for predicting performance of three-dimensional pulverized-fuel fired furnaces, *ASME J. Engng. Gas Turbines and Power*, 110, 117-126, 1988.

128. S. Li, B. Yu, W. Yao, and W. Song, Numerical modelling for pulverised coal combustion in large furnace, in *Proc. 2nd Int. Symp. on Coal Combustion*, Beijing, China, Report, X. Xu, L. Zhou, and W. Fu, Eds., China Machine Press, Beijing, China, 1991, 167-173.

129. R. Boyd and A. Lowe, Three-dimensional modelling of a pulverised coal fired utility furnace, in *Coal Combustion*, Hemisphere, New York, 1988, 165-172.

130. B.S. Brewster, S.C. Hill, P.T. Radulovic, and L.D. Smoot, Comprehensive Modeling, in *Fundamentals of Coal Combustion for Clean and Efficient Use*, L.D. Smoot, Ed., Elsevier, Amsterdam, 1993, chap. 8.

131. K. Kudo, H. Taniguchi, and K. Guo, Heat-transfer simulation in a furnace for steam reformer, *Heat Transfer — Japanese Research*, 20(8), 750-764, 1992.

132. K.N. Theologos and N.C. Markatos, Advanced modeling of fluid catalytic cracking riser-type reactors, *AIChE J.*, 36(6), 1007-1017, 1993.

133. S.L. Chang, C.Q. Zhou, S.A. Lottes, B. Golchert, and M. Petrick, A numerical investigation of the scaled-up effects on flow, heat transfer, and kinetics processes of FCC units, in *Combustion and Radiation Heat Transfer*, R.A. Nelson, K.S. Ball, and Z.M. Zhang, Eds., *Proceedings of the ASME Heat Transfer Division — 1998*, Vol. 2, ASME, New York, HTD-Vol. 361-2, pp. 73-81, 1998.

134. H. Zuqi, Y. Kuangjoing, and Q. Jialin, The mathematical modelling for heat transfer in the radiative chamber of a petro-chemical tubular furnace, *Proc. of Interpec'91: Petroleum Refining & Petroleum*, Bejing, Sept., 1991, 1655-1661.

135. A.K. Jones and P.J. Chapman, Computational fluid dynamics combustion modelling — a comparison of secondary air system designs, *Tappi J.*, 76(7), 195-202, 1993.

136. B. Ponizy and S. Wojcicki, On modeling of pulse combustors, *Twentieth Symp. (Int.) on Combustion*, The Combustion Institute, Pittsburgh, PA, 1984, 2019-2024.

137. P.K. Barr and H.A. Dwyer, Pulse combustor dynamics: a numerical study, in *Numerical Approaches to Combustion Modeling*, E.S. Oran and J.P. Boris, Ed., Vol. 135, Progress in Astronautics and Aeronautics, American Institute of Aeronautics and Astronautics, Washington, D.C., 1991, 673-710.

138. P.K. Barr, J.O. Keller, and J.A. Kezerle, SPCDC: a user-friendly computation tool for the design and refinement of practical pulse combustion systems, *Proc. of 1995 International Gas Research Conf.*, D.A. Dolenc, Ed., Government Institutes, Rockville, MD, 1996, 2150-2159.

139. E. Lundgren, U. Marksten, and S.-I. Möller, The enhancement of heat transfer in the tail pipe of a pulse combustor, *Twenty-Seventh Symp. (Int.) on Combustion*, The Combustion Institute, Pittsburgh, PA, 1998, 3215-3220.

140. J. Zhang and S. Nieh, Simulation of gaseous combustion and heat transfer in a vortex combustor, *Num. Heat Trans., Part A*, 32(7), 697-713, 1997.

141. H.C. Meisingset, J.G. Balchen and R. Fernandez, Mathematical modelling of a rotary hearth calciner, *Light Metals 1996*, W. Hale, Ed., The Minerals, Metals & Materials Society, Warren, PA, 1996, 491-497.

142. J.R. Ferron and D.K. Singh, Rotary Kiln Transport Processes, *AIChE J.*, 37(5), 747-758, 1991.

143. A.A. Boateng, On flow induced kinetic diffusion and rotary kiln bed burden heat transport, in *ASME Proceedings of the 32nd National Heat Transfer Conf.*, Vol. 3: Fire and Combustion, L. Gritzo and J.-P. Delplanque, Ed., ASME, New York, 1997, 183-191.

144. P.S. Ghoshdastidar and V.K. Anandan Unni, Heat transfer in the non-reacting zone of a cement rotary kiln, in *Heat Transfer Phenomena in Radiation, Combustion, and Fires*, R.K. Shah, Ed., ASME, New York, HTD-Vol. 106, 113-122, 1989.

145. E.E. Khalil, Flow and combustion modeling of vertical lime kiln chambers, in *Industrial Combustion Technologies*, M.A. Lukasiewicz, Ed., American Society of Metals, Warren, PA, 1986, 99-107.

146. B. Farouk and M.M. Sidawi, Effects of nitrogen removal in a natural gas fired industrial furnace: a three dimensional study, in *Heat Transfer in Fire and Combustion Systems — 1993*, B. Farouk, M.P. Menguc, R. Viskanta, C. Presser, and S. Chellaiah, Eds., ASME, New York, HTD-250, 173-183, 1993.

147. W.D. Owens, G.D. Silcox, J.S. Lighty, X.X. Deng, D.W. Pershing, V.A. Cundy, C.B. Leger, and A.L. Jakway, Thermal analysis of rotary kiln incineration: comparison of theory and experiment, *Combust. Flame*, 86, 101-114, 1991.

148. D. Pal, J.A. Khan, and J.S. Morse, Computational Modelling of an Industrial Rotary Kiln Incinerator, in *Heat Transfer in Fire and Combustion Systems*, A.M. Kanury and M.Q. Brewster, Eds., ASME, New York, HTD-199, 167-173, 1992.

149. J.A. Khan, D. Pal, and J.S. Morse, Numerical modeling of a rotary kiln incinerator, *Hazardous Waste & Hazardous Materials*, 10(1), 81-95, 1993.

150. F.C. Chang and C.A. Rhodes, Computer Modeling of Radiation and Combustion in a Rotary Solid-Waste Incinerator, Argonne National Lab, U.S. Dept. of Energy Report ANL/ET/CP-85778, 1995.

151. N. Machii, K. Nishimura, D. Liu, and K. Shibata, Development of CAE for MSWI ash melting furnace, *Proc. 1998 Int. Gas Research Conf.*, Vol. V: Industrial Utilization, D.A. Dolenc, Ed., Gas Research Institute, Chicago, 1998, 696-704.

152. S.E. Bayley, R.T. Bailey, and D.C. Smith, Heat Transfer Analysis of Hazardous Waste Containers Within a Furnace, in *Combustion and Fire*, M.Q. McQuay, W. Schreiber, E. Bigzadeh, K. Annamalai, D. Choudhury, and A. Runchal, Eds., *ASME Proc. 31st National Heat Transfer Conf.*, Vol. 6, ASME, New York, HTD-328, 61-69, 1996.

153. M.G. Carvalho and N. Martins, Mathematical modelling of heat and mass transfer phenomena in baking ovens, *5th Int. Conf. on Computational Methods and Experimental Measurements*, in *Computational Methods and Experimental Measurements V*, A. Sousa, C.A. Brebbia, and G.M. Carlomagno, Eds., Computational Mechanics Publications, 1991, 359-370.

154. T.H. Song and R. Viskanta, Parametric Study of the Thermal Performance of a Natural Gas-Fired Furnace, in *Fossil Fuel Combustion 1991*, R. Ruiz, Ed., ASME, New York, PD-Vol. 33, 135-141, 1991.

155. M. Carvalho and M. Nogueira, Improvement of energy efficiency in glass-melting furnaces, cement kilns and baking ovens, *Applied Therm. Eng.*, 17(8-10), 921-933, 1997.

156. J.A. Wieringa, J.J. Elich, and C.J. Hoogendoorn, Spectral radiation modelling of gas-fired furnaces, in *Proc. 2nd European Conf. on Industrial Furnaces and Boilers*, Portugal, April, R. Collin, W. Leuckel, A. Reis, and J. Ward, Eds., 1991, 36-53.

157. M.J.S. de Lemos, Computation of heated swirling flows with a fully-coupled numerical scheme, in combustion and fire, M.Q. McQuay, W. Schreiber, E. Bigzadeh, K. Annamalai, D. Choudhury, and A. Runchal, Eds., *ASME Proceedings of the 31st National Heat Transfer Conf.*, Vol. 6, ASME, New York, HTD-328, 139-145, 1996.

158. M.J.S. de Lemos, Simulation of Vertical Swirling Flows in a Model Furnace with a High Performance Numerical Method, in *Combustion and Radiation Heat Transfer*, R.A. Nelson, K.S. Ball, and Z.M. Zhang, Eds., *Proc. ASME Heat Transfer Division — 1998*, Vol. 2, ASME, New York, HTD-361-2, 21-28, 1998.

159. P. Hutchinson, E.E. Khalil, J.H. Whitelaw, and G. Wigley, Influence of burner geometry on the performance of small furnaces, *2nd Eur. Symp. on Combustion*, 1975, 659-665.

160. H.F. Nelson, Combustion modeling in large gas-fired furnaces, *Letters in Heat & Mass Trans.*, 6(1), 23-33, 1979.

161. K.S. Chapman, S. Ramadhyani, and R. Viskanta, Modeling and analysis of heat transfer in a direct-fired continuous reheating furnace, in *Heat Transfer in Combustion Systems*, N. Ashgriz, J.G. Quintiere, H.G. Semerjian, and S.E. Slezak, Eds., ASME, New York, HTD-122, 35-43, 1989.

162. K.S. Chapman, S. Ramadhyani, and R. Viskanta, Modeling and analysis of heat transfer in a direct-fired batch reheating furnace, in *Heat Transfer Phenomena in Radiation, Combustion, and Fires*, R.K. Shah, Ed., ASME, New York, HTD-106, 265-274, 1989.

163. R. Viskanta, K.S. Chapman, and S. Ramadhyani, Mathematical modeling of heat transfer in high-temperature industrial furnaces, in *Advanced Computational Methods in Heat Transfer*, Vol. 3: Phase Change and Combustion Simulation, L.C. Wrobel, C.A. Brebbia, and A.J. Nowak, Eds., Springer-Verlag, Berlin, 1990, 117-131.

164. K.S. Chapman, S. Ramadhyani, and R. Viskanta, Two-dimensional modeling and parametric studies in a direct-fired furnace with impinging jets, *Comb. Sci. Tech.*, 97, 99-120, 1994.

165. M.G. Carvalho and M. Nogueira, Modelling of glass melting industrial process, *J. de Physique*, 3(7.2), 1357-1366, 1993.

166. J.D. Buckmaster and T. Takeno, *Mathematical Modeling in Combustion Science*, Springer-Verlag, New York, 1988.

167. A. Liñán and F.A. Williams, *Fundamental Aspects of Combustion*, Oxford University Press, Oxford, 1993.

168. W.C. Gardiner, Jr., *Combustion Chemistry*, Springer-Verlag, New York, 1984.

169. K.K. Kuo, *Principles of Combustion*, John Wiley & Sons, New York, 1986.

170. A. Dervieux and B. Larrouturou, *Numerical Combustion*, Springer-Verlag, New York, 1989.

171. J.B. Moss, Combustion Research — 25 Years On, *Comb. Sci. Tech.*, 98, 337-340, 1994.

172. F. Weinberg, Combustion Research for the 21st Century — Some Speculative Extrapolations, *Comb. Sci. Tech.*, 98, 349-359, 1994.

173. J. Warnatz, U. Maas, and R.W. Dibble, *Combustion: Physical and Chemical Fundamentals, Modeling and Simulation, Experiments, Pollutant Formation,* Springer, Berlin, 1996.

2 CFD Modeling

Charles E. Baukal, Jr.

CONTENTS

2.1 MODELING APPROACHES

A complete combustion system may be extremely complex and can include a wide range of physical processes that are often highly interactive and interdependent. A given combustion system may include:

- Turbulent fluid dynamics in the flame with laminar fluid dynamics in the bulk of the combustor
- Multi-dimensional flows that could include swirl
- Multiple phases that could include gases, liquids, and solids, depending on the fuel composition

- Combustion of solid fuels that may include processes like vaporization, devolatilization, and char production and destruction
- Very high temperature, velocity, and species gradients in the flame region with much lower gradients in the bulk of the combustor
- Large material property variations caused by the wide range of temperatures, species, and solids present in the system
- Multiple modes of heat transfer, especially radiation that is highly nonlinear and may include wavelength dependence
- Complex chemistry involving numerous reactions and many species, most of which are in trace amounts
- Porous media
- Catalytic chemical reactions in some limited applications
- Complex, nonsymmetrical furnace geometries that may include an intermediate surface to separate the combustion products from the load
- Multiple flame zones produced by burners that may be operated at different conditions and whose flames interact with each other
- A heat load that may be moving in a complicated manner and interacting with the combustion space above it in a nonlinear manner
- A heat load that may produce volatile species during the heating process
- A heat load whose properties may vary greatly with temperature, physical state, and even wavelength (for radiation)
- A transient heating and melting process that may include discrete material additions and withdrawals

There are many challenges caused by this complexity, including inadequate physics to properly model the problem, large numbers of gridpoints requiring large amounts of computer memory, and long computation times. The simulation results may be difficult to validate as many of the experimental measurements are difficult, time-consuming, and costly to make in industrial combustors. Therefore, in most combustion simulations, simplifying assumptions must be made to get cost-effective solutions in the amount of time available for a given problem. The actual simplifications depend on many factors, including the level of accuracy required, the available amount of computing power, the skill and knowledge of the modeler, the experience with the given system being simulated, and the time available to get a solution. These simplifying approaches are briefly discussed here. More detailed information on each aspect of modeling is given later in this chapter. Spalding (1963) discussed simplifying approaches to combustion modeling and noted that the main concern is which modeling rules can be ignored to simplify the problem and then to estimate the errors in the resulting predictions.[1] Also noted was the difficulty in matching all the dimensionless groups in a large-scale problem with small-scale experiments. Weber et al. (1993) classified models for designing industrial burners into three categories.[2] First-order methods give rough qualitative estimates of heat fluxes and flame shapes. Second-order methods give higher accuracy results than first-order methods for temperature, oxygen concentration, and heat flux. Third-order methods further improve accuracy over second-order methods and give detailed species predictions in the flame that are useful for pollutant formation rates. The order used will in large part depend on the information and accuracy needed.

2.1.1 FLUID DYNAMICS

There are a variety of methods available to simulate the fluid flow in a combustion system. The Navier-Stokes equations are generally accepted as providing an "exact" model for turbulent fluid flow systems.[3] Unfortunately, these equations for systems of practical interest are too complicated to solve exactly either analytically or numerically. Therefore, different types of approximations have been suggested for solving these equations. These are very briefly discussed next with appropriate references for the reader interested in more detail.

2.1.1.1 Moment Averaging

This has been by far the most popular method used in simulating large-scale industrial combustion problems, primarily because of the ready availability of commercial software programs like PHOE-NICS, FLUENT (see Chapter 3), FLOW-3D, TEACH, PCGC-3 (see Chapter 4), Harwell-3D, GENMIX, and others to solve these problems. In this method, the turbulent velocity components are decomposed into average and fluctuating terms and solved using the famous k-ε closure equations.[4] Despite the well-known limitations of this approach, it remains the most popular choice for solving practical combustion problems. This may be due to the fact that it has been around for decades and therefore the software has been highly developed. Finite-difference,[5] finite element,[6] and finite volume[7] techniques have been used to simulate turbulent flows. The commercial codes today are very user friendly and have excellent pre- and post-processing packages to make setting up the problem and viewing the results relatively simple and straightforward. Because of the popularity and widespread use of this method, it is discussed throughout this book.

2.1.1.2 Vortex Methods

Most numerical approaches for solving fluid flow problems use an Eulerian scheme with a fixed coordinate system that is discretized into small parts. One problem with this approach is that there may often be areas in the flow where the gradients are very high and require very fine discretization, while in nearby areas the gradients may be much lower and need much less discretization. To further complicate this disparity, these areas may be moving. Finite difference solution convergence problems result from having fine cells next to coarse cells. Therefore, the choice is to either use finer or coarser cells for both areas. If finer cells are used, then accuracy is improved, but with a significant penalty in solution times. If coarser cells are used, then solution times are improved, but accuracy is sacrificed. An alternative approach is to use a Lagrangian system with a moving coordinate system that can keep track of the finer details of high gradient areas, without the burden of unnecessary detail in areas that do not require it. Some Lagrangian methods use gridpoints that are transported along flow trajectories while other Lagrangian methods are grid-free.[8] The Navier-Stokes equations are set up and solved in terms of vorticity:

$$\frac{\partial \vec{\omega}}{\partial t} + \vec{u} \cdot \nabla \vec{\omega} = \vec{\omega} \cdot \nabla \vec{u} \qquad (2.1)$$

where $\omega = \nabla \times \vec{u}$, $\nabla = (\partial/\partial x, \partial/\partial y, \partial/\partial z)$, $\vec{u} = (u,v,w)$ and $\vec{x} = (x,y,z)$. Velocities are then calculated from the vorticity solutions. This method has been applied to industrial combustion simulations.[9] Variations of this method have also been referred to as large-eddy simulations (LES).[10-12] Dahm and co-workers have developed a method known as the Local Integral Moment or LIM (see Chapter 5) that is based on large-eddy simulation concepts.[13-16]

2.1.1.3 Spectral

This is an approximation method where the solutions for the scalar variables in the partial differential equations are simulated as a truncated series expansion[17]:

$$C(x,t) = \sum_{k=0}^{N} c_k(t)\phi(x) \qquad (2.2)$$

where $C(x,t)$ is a scalar variable like temperature, N is the finite wavenumber truncation cutoff, $c_k(t)$ are the expansion coefficients, and $\phi_k(x)$ are the basis functions chosen to best represent the flow. This solution approach is more global than finite difference discretization approaches, which tend to be more local. Therefore, spectral methods can provide more accurate approximations of the solution compared to moment methods, although this is not always the case. Solution times may be longer and the selection of the proper basis functions is critical to the success of this approach that has been used in combustion problems.[18,19] This has not been a popular method for solving industrial combustion problems. This method could become more popular if the appropriate user-friendly software were developed and commercialized. Karniadakis and Sherwin (1999) have written a general book on the use of spectral elements in CFD.[20]

2.1.1.4   Direct Numerical Simulation

In this method, usually referred to as DNS,[21-27] no assumptions are made regarding the turbulent behavior of the flow. The exact Navier-Stokes equations are solved at small enough length and time scales that the complete physics of the problem can be captured. This approach obviously requires tremendous computing power and is not currently used for solving industrial combustion problems. However, as rapid advancements in computers continue, including parallel processing, large memories, and fast computer speeds, this method may become more prevalent in the future. This method is currently being used to solve fundamental combustion[28] and aerospace propulsion problems using supercomputers where the simulation costs are not a significant portion of the overall cost of new developments. At this time, the economics of DNS are not justified for most industrial combustion equipment manufacturers and end users where the cost of these calculations could dwarf the actual cost of the combustion system itself.

2.1.2   Geometry

There are several different levels of complexity concerning the geometry of a given combustion system, ranging from zero-dimensional up to fully three-dimensional. These levels are briefly discussed here and have been discussed in more detail by Khalil (1982).[29]

2.1.2.1   Zero-dimensional

Numerous modeling approaches to handling the complexity of large-scale combustors are possible and have been used. Before the advent of CFD codes, a common modeling approach was to do an overall heat and material balance on the system. This is often referred to as zero-dimensional modeling because it does not give any spatial resolution. This type of zero-dimensional modeling does not involve any analysis of the fluid dynamics. It can, however, include detailed analysis of the chemical reactions and is often referred to as a stirred reactor or stirred vessel. This type of modeling was made easier with the advent of electronic spreadsheets, but still requires numerous assumptions and simplifications. A more recent type of zero-dimensional model may include very detailed chemistry but still no fluid flow. In that type, the reactor is assumed to be typically either constant pressure or constant volume (see Figure 2.1). The main variable then becomes time, which may be finite or infinite (equilibrium).

Zero-dimensional models give a reasonable approximation of the overall performance of the system, but give very little information on the detailed performance, such as where potential hot spots on a furnace wall might be, for example. Despite the obvious disadvantages, there are some advantages of zero-dimensional modeling. One is that solutions can be obtained very quickly. This is important in parameter studies where a large number of variables are to be investigated and where fast results are needed. Another advantage is that these models can be very helpful in developing an understanding of the system performance that can sometimes be lost when detailed analyses are done. One can see the forest before looking at the individual trees in the forest. Another

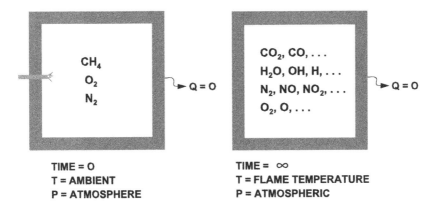

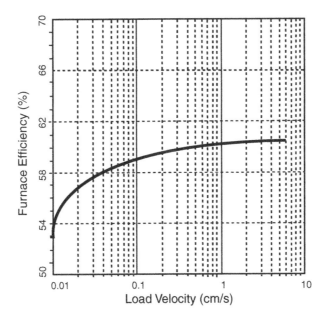

FIGURE 2.1 Adiabatic equilibrium reaction process.

advantage is that this type of modeling does not require the same level of training as with complicated modeling, so it can be done by a wider group of personnel.

As an example of zero-dimensional modeling, Viskanta and co-workers did parametric computational studies of both direct and indirect-fired furnaces.[30-33] Chapman et al. (1989) presented the results of parametric studies of a direct-fired continuous reheating furnace.[30] A simplified mathematical model was developed that accurately calculated the heat balance throughout the furnace. The combustion space was divided into zones that were considered to be well-stirred reactors. The load in each zone was further subdivided into smaller control volumes. Radiation was modeled using Hottel's zone method. A primary objective of the study was to compute the furnace thermal efficiency as a function of a variety of parameters, including the load velocity, load emissivity, furnace combustion space height, and refractory emissivity. The furnace efficiency increased rapidly with the initial increase in load velocity and then leveled off with further increases in velocity (see Figure 2.2). Although the model used in the study was fairly simple, with no detailed

FIGURE 2.2 Predicted furnace efficiency as a function of load velocity. (Courtesy of ASME, New York.[30] With permission.)

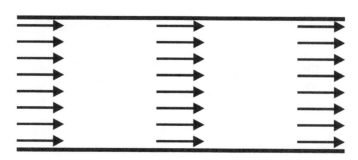

FIGURE 2.3   Plug-flow reactor.

fluid flow calculations, it was useful for studying a wide range of values for different parameters, which makes it a valuable design tool for studying other configurations. Other examples of zero-dimensional models of furnaces include rapidly heating cylindrical metal billets[34] and simulating a batch-fed solid waste incineration process.[35]

2.1.2.2   One-dimensional

The next level of complexity involves one-dimensional modeling. This is where only one spatial dimension is considered. Although this greatly simplifies the number of equations, these models may still be fairly complicated and provide many details into the spatial changes of a given parameter. One-dimensional modeling is often used to examine the detailed chemistry in a combustion process that may be simulated as a plug-flow reactor (see Figure 2.3).

Despite the limitations, there are advantages to using this type of geometrical simplification. In certain applications, these models are particularly relevant, with little or no sacrifice in accuracy and resolution. An example is in porous radiant burners and flat flames that are both essentially one-dimensional in nature. Another obvious advantage is that faster results are possible, compared to multi-dimensional modeling. One-dimensional models also greatly simplify the task of radiation modeling, which can become very complicated in multi-dimensional geometries. However, it should be noted that one-dimensional models may still be fairly complicated and may include very detailed chemistry, multiple phases, porous media, and radiation. As an example, Singh et al. (1991) reported on a one-dimensional model used to simulate ceramic radiant burners.[36] For that type of burner, the one-dimensional model is generally very adequate. El-Mahallawy et al. (1982) developed a one-dimensional empirical/analytical model to calculate the heat transfer from swirling oil flames to the walls of a water-cooled cylindrical furnace.[37] A sample result is shown in Figure 2.4.

2.1.2.3   Multi-dimensional

The highest level of geometrical complexity involves multi-dimensional modeling, both two- and three-dimensional geometries. Geometry simplifications are often used to reduce the computing requirements for simulating combustion systems. Wherever possible, three-dimensional (3-D) problems are simulated by two-dimensional (2-D) models or by axisymmetric geometries which are three-dimensional problems that can be solved in two spatial variables. In the early days of CFD, it was not uncommon to simulate a rectangular furnace as a cylindrical axisymmetric geometry to reduce the problem from 3-D to 2-D, where angular symmetry was assumed (see Figure 2.5). A related simplification is modeling certain types of cylindrical problems as angular slices instead of modeling the entire cylinder. For example, if a burner has four injectors equally spaced angularly and radially from the center point, then this can be modeled as a 90° slice of a cylinder by using symmetric boundary conditions (see Figure 2.6). This is still a three-dimensional problem, but only one quarter of the geometry needs to be solved.

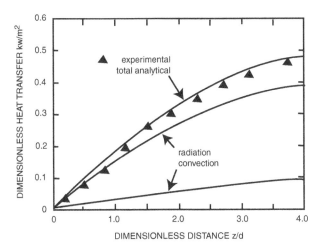

FIGURE 2.4   Comparison of the predicted and measured total heat flux from a swirling oil flame to the walls of a water-cooled cylindrical furnace. (Modified from Reference 37.)

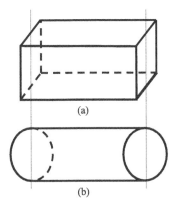

FIGURE 2.5   (a) Rectangular furnace (3-D) represented by (b) cylindrical geometry (2-D axisymmetric).

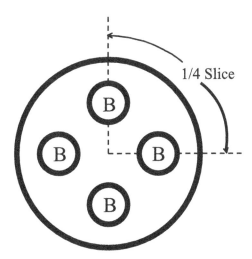

FIGURE 2.6   Slice modeling of a cylindrical geometry with angular symmetry.

Another type of geometric simplification in multi-dimensional modeling involves limiting the number of gridpoints due to the limitations of the computer speed and memory. It may not always be possible to model the entire combustion system, so an approach that has often been taken is to separately simulate the flame region where small-scale effects are important and the combustor where large-scale effects are predominant. The results of the flame simulation can then be used as inputs to the large-scale modeling of the combustor itself. For example, a single flame can be more accurately input as a heat source using the detailed modeling results for that flame. Another common method for minimizing the number of gridpoints is to model only a small portion or section of a combustor. For example, most glass furnaces have multiple burners symmetrically firing parallel to the molten glass bath. Often, only a single slice of the furnace containing one burner is modeled. Although this precludes simulating the flame-to-flame interactions, it is a reasonable assumption to make in order to get timely and cost-effective solutions of acceptable accuracy.

The obvious advantage of multi-dimensional modeling is that much higher spatial resolution is possible. This can provide important insight into the problem that is not possible with simpler geometrical models. This resolution is particularly important in simulating burner performance because the burner geometry is normally too complicated to model as a one-dimensional problem. However, there are some obvious disadvantages to multi-dimensional modeling, including longer computational times, difficulties in visualizing and interpreting the results, and more difficulty in separating the effects of individual parameters. Gillis and Smith (1990) evaluated a 3-D model for industrial furnaces and compared modeling results against experimental data for two pilot-scale furnaces.[38]

2.1.3 REACTION CHEMISTRY

The reaction chemistry is another important aspect of most industrial combustion problems. Modeling approaches for this chemistry range from nonreacting up to multiple reactions with multiple species, and finite rate kinetics. The different approaches commonly used in modeling combustion problems are briefly discussed next.

2.1.3.1 Nonreacting Flows

When CFD codes first became commercially available, the chemistry submodels were very primitive and greatly increased the computation time, often beyond the capability of the available hardware. Therefore, a common approach to simulating combustion problems was to model them as nonreacting flows. This has sometimes been referred to as "cold-flow" modeling, which is really a misnomer as the flame was often simulated as a flow input of hot inert gases to the combustor. A variation of this approach is to use a nonreacting gas that has the thermophysical properties, like viscosity, thermal conductivity, and specific heat, of the combustion products as a function of temperature. Those properties are separately calculated, typically using some type of equilibrium chemistry calculation. The properties are then curvefit with temperature and included in the CFD codes. This type of nonreacting flow model can give fairly accurate predictions in many cases for the overall energy transfer in a large-scale combustor.

Nonreacting flow modeling may grossly oversimplify a problem, but it can give considerable insight into the flow patterns inside the combustor. The flame can also be simplified to be a heat source, in order to avoid modeling the chemical reactions in the flame zone. The difficulty is how to specify the heat release profile of the flame, especially because that is something usually desired of the modeling itself. Although there are some advantages to using nonreacting chemistry, such as simplicity and speed, this approach is rarely used in most types of combustion modeling today because it is too limited and unrealistic.

2.1.3.2 Simplified Chemistry

The term "simplified" chemistry is a somewhat relative term, but generally refers to reducing the number of chemical equations used to represent a system, reducing the complexity of the reaction mechanism, or a combination of both. In the first approach, a very limited number of reactions and species are used to represent the actual combustion reaction system that may involve hundreds of reactions and dozens of species. In this approach, a greatly reduced set of reactions is used.[39] Often the goal of this approach is to predict flow and heat transfer information, but not detailed species such as pollutant emissions like NOx. Simplified chemistry was often used in the early days of CFD modeling because of the limitations of the submodels and computer memory and because the main interests were things like the heat transfer to the load and the walls and the bulk gas flow in the system. For example, the earliest models for simulating the combustion of methane used a single-step reaction such as the following (unbalanced equation shown for simplicity):

$$CH_4 + O_2 + N_2 \rightarrow CO_2 + H_2O + O_2 + N_2 \tag{2.3}$$

Infinite rate kinetics were used and no minor species were included. This simplified chemistry could obviously not be used to predict pollutant emissions like NOx, but was useful for simulating the flow patterns and heat transfer in the combustor. An example of a slightly more complicated reaction set is given by Westbrook and Dryer (1981)[40]:

$$CH_4 + 1.5O_2 \rightarrow CO + 2H_2O \tag{2.4a}$$

$$CH_4 + 2H + H_2O \rightarrow CO + 4H_2 \tag{2.4b}$$

A more popular approach in recent years is to use slightly more complicated reduced sets. An example of a four-step reduced mechanism set for methane flames is given by[41]:

$$CH_4 + 2H + H_2O \rightarrow CO + 4H_2 \tag{2.5a}$$

$$CO + H_2O \rightarrow CO_2 + H_2 \tag{2.5b}$$

$$H + H + M \rightarrow H_2 + M \tag{2.5c}$$

$$O_2 + 3H_2 \rightarrow 2H + 2H_2O \tag{2.5d}$$

Smooke and Bilger (1991) have edited a volume considering reduced kinetic mechanisms specifically for air/methane flames.[42] Peters and Rogg (1993) have also edited a book containing papers on reduced kinetic mechanisms for combustion modeling.[43]

Another aspect to simplified chemistry models involves not only the number of equations used, but the type of chemical kinetics that are being simulated. This second approach to simplifying the chemistry is sometimes referred to as reaction mechanism simplification, or mechanism reduction. Infinite rate kinetics, or equilibrium chemistry, is an example of this type of approach that is often used in combustion modeling. This means that the chemical reactions are assumed to be infinitely fast and therefore independent of time. This is often a reasonable assumption to make but is again dependent on the specific problem and required level of accuracy. Another variation of this approach is an empirical correlation for the chemistry of a given system.

2.1.3.3 Complex Chemistry

Another approach to modeling of combustion systems is to use very detailed chemistry. This approach is commonly used if detailed information on gas species is required, such as when, for example, NOx emissions need to be predicted. Again, "complex" chemistry is a somewhat vague and relative term, but here refers to many multi-step reactions with many species. The actual numbers of reactions and species depend on the given problem and the level of detail required. Complex chemistry also concerns finite rate kinetics where the reaction rates are time dependent.

2.1.4 Radiation

Kocaefe et al. (1987) gave a brief review of some of the methods used for radiation modeling.[44] They concluded that the imaginary planes and discrete transfer methods have good accuracy and low computation times, while the zone method has the lowest computation time if the interchange factors are known or only calculated once. Baukal (2000) discusses in detail the importance of radiation in industrial combustion.[45]

2.1.4.1 Nonradiating

Another type of simplification involves using known empirical relationships for the problem at hand. These empirical correlations normally only apply to a specific set of conditions and problems but can be very useful for reducing the size and complexity of the problem. For example, it may be possible to simulate the nonlinear radiation from the flame to the load and combustor walls as a type of radiation heat transfer coefficient in order to make the radiation linear with temperature and therefore much easier to solve:

$$h_{rad} = f\left(T_{source}^4 - T_{sink}^4\right) \approx f\left(T_{source} - T_{sink}\right) \tag{2.6}$$

This approach should be used with caution only after careful examination and understanding of the system under investigation. Although this may limit the generality of the problem, this type of simplification may greatly reduce the time to get solutions. This makes it possible to do more simulations of the problem and may be especially useful for finding optimized conditions.

The key to using any simplifications is to understand the resulting inaccuracies they introduce. Therefore, it is usually prudent to have experimental data to compare against any simplified numerical simulations. It is also advisable to use the most complicated possible model for at least a base-case problem, which can then be used to compare against the simplified results. If the simplified results compare favorably to the full-blown simulation, there is some justification for using the simplifications. However, if the simplifications do not compare favorably to the comprehensive model results, then further analysis is warranted to understand the discrepancies.

As computer power continues to improve, fewer and fewer simplifications will be necessary. Eventually it will be possible to do direct numerical simulations (DNS) so that even the turbulent fluid flow will not need to be approximated because it will be possible and practical to model the small length scales present in turbulent fluid flows.

2.1.4.2 Participating Media

Participating media includes nonluminous gaseous radiation and luminous radiation from particle-laden (e.g., soot) gaseous flows. Bhattacharjee and Grosshandler (1989) noted three factors that complicate gaseous radiation modeling: (1) the spectral variation of the properties requires calculations over the spectrum; (2) the gaseous composition is not homogeneous over the entire space, which means that integrations must be done over every line of sight; and (3) the asymmetry of most real problems means integration over all solid angles.[46] Sivathanu et al. (1990) noted that there are accurate methods for calculating weakly radiating turbulent diffusion flames, but that it is much more difficult to model the strongly radiating turbulent diffusion flames that are used in many industrial combustion systems.[47] Turbulence can significantly increase the mean radiation levels from diffusion flames.[48-53]

Hoogendoorn et al. (1990) used a 15-band gas radiation model in a well-stirred furnace zone method to simulate the nonluminous radiation in a natural gas-fired regenerative glass melter.[54] The results showed that as much as 99% of the heat transfer to the melt was by radiation. They compared the axial heat flux distribution using both a simplified plug-flow model and a more complete model. The plug-flow model gave both unrealistically high fluxes in the middle of the melter and low fluxes near the ends of the melter. The addition of additives to the combustion products was shown to slightly increase the heat flux to the glass. However, this would lead to lower flue gas outlet temperatures that reduces the performance of the regenerative air preheater so that the overall effect of the additive in this process would be minimal. Increasing the roof emissivity from 0.4 to 1 was shown to increase the heat transfer to the glass melt by only 8%.

Yan and Holmstedt (1997) presented a fast narrow-band (FASTNB) computer model for predicting the radiation intensity in a general nonisothermal and nonhomogeneous combustion environment.[55] The model is used to calculate the spectral absorption coefficients for CO_2, H_2O, and soot. It is claimed to be as much as 20 times faster than a benchmark model called RADCAL,[56-58] with only a 1% deviation from that model. Further development was proposed for inclusion of other gases like carbon monoxide, methane, propylene, and acetylene.

Liu et al. (1998) presented a new approximate method for non-gray gas radiative heat transfer using a statistical narrow-band model that utilizes a local absorption coefficient that is calculated using local rather than global properties.[59] The main advantage of the proposed approximate method is considerable savings in computational time, up to a two orders of magnitude reduction. The method also improves the accuracy of the calculations compared to methods using global properties.

Gritzo and Strickland (1999) presented a gridless, integral method for solving the radiative transport equations for use in combustion calculations using Lagrangian techniques to solve the fluid dynamics.[60] Their approach is particularly compatible to parallel computing. It is shown that this method compares favorably against other popular methods used in grid-based solution techniques, which can have significant errors when adapted to gridless solution schemes. Previous methods to solve for radiation have relied on grid-based calculations and are not optimal for transport element methods (see Chapter 1).

For some combustion processes, primarily those involving solid and liquid fuels, spectral radiation from particulates may be significant. Ahluwalia and Im (1990) presented a three-dimensional spectral radiation model that they used to model the burning of deeply cleaned coals in a pulverized coal furnace.[61] Spectroscopic data were used to calculate the absorption coefficients of the gases. The extinction and scattering efficiencies of the particulates were calculated using Mie theory. The optical properties of the char, ash, and soot were determined from reflectivity, transmissivity, and extinction measurements. The radiation from the char was as much as 30% of the nonluminous gaseous radiation. The heat transfer in the furnace ranged from 168 to 221 MW (5.73 to 7.54 × 10⁸ Btu/hr), depending on the specific fuel used. It is noted that ashes rich in iron enhance radiative heat transfer and fine grinding of the coal improves furnace heat absorption. In a later paper, Ahluwalia and Im (1994) used a hybrid technique to solve spectral radiation involving gases and particulates in coal furnaces.[62] To optimize computational speed and accuracy, the discrete ordinate method (S_4), modified differential approximation (MDA), and P_1 approximation were combined and used in different ranges of optical thicknesses. The MDA method has been shown to be sufficiently accurate for all optical thicknesses but computationally slow for the optically thin and thick limits.[63] There were significant discrepancies between the predicted and calculated heat fluxes. This was explained by the difficulty in making heat flux measurements in industrial furnaces.

The soot, char, and ash contributions to heat transfer were approximately 15%, 3%, and up to 14%, respectively.

2.1.5 TIME DEPENDENCE

Another critical aspect of combustion modeling is whether or not the solution is time dependent. Nearly all industrial combustion processes are time dependent at small length scales due to turbulence. However, these processes are normally modeled as steady-state systems because of the limitations of the turbulent submodels, the large increase in computer time required to simulate transient combustion, and the lack of need for such detailed information in most industrial combustion systems.

2.1.5.1 Steady State

In steady-state simulations, there is no time dependence of the solution. The problem with turbulent combustion problems is that the time scale, especially in the near flame region, is very small. To accurately simulate an entire system using such a small time scale is normally computationally prohibitive. However, this is not usually a problem in most cases because that type of detail is not required. The standard approach has been to simulate the average properties. There is considerable debate about averaging turbulent properties, but the reality is that virtually all commercial codes have some type of turbulence averaging (discussed in more detail below). For many continuous industrial applications, the heating process is essentially steady state, with fixed and usually known combustion chamber wall temperatures, fuel firing rates, and material feed rates. Processes that are truly varying with time, such as batch heating processes, may be simulated using average conditions over the entire cycle or by making a series of steady-state calculations to simulate various steps in the process.

2.1.5.2 Transient

Transient or time-varying calculations are rarely made for large-scale industrial combustion processes due to the large computational time required, which usually exceeds the amount of time allowed for the needed simulations. As computer speeds continue to increase, this type of computation will increase in popularity for those applications that have significant variations during a given cycle. A good example of such an application is scrap metal melting. Initially, a charge of cold solid scrap metal is charged into a cold furnace. Then the burners begin to heat up and melt down the metal, as well as heat up the furnace. At any given time there may be a mixture of solid and liquid metal. When the charge is at or nearly fully molten, a second charge of cold scrap may be added to the first melted charge. Several more charges are possible, depending on the application. An accurate simulation of this process should include a fully transient computation.

2.2 PHYSICS

The purpose of this chapter section is not to present the equations used in CFD modeling that represent the physics, but rather to discuss some of the important issues. The equations are presented elsewhere in the book.

2.2.1 FLUID FLOW

In industrial combustion, it is normally assumed that the flows are low Mach number, which simplifies the fluid dynamics where the flow is incompressible.[64] One of the earliest and still widely used algorithms was developed by Patankar and Spalding.[65,66] The algorithm is known by the acronym SIMPLE, which stands for semi-implicit pressure-linked equation.

An aspect of fluid dynamics that is important in most industrial combustion problems is turbulence. Many good general references on turbulence modeling are available.[67-91] Those including the effects of combustion are briefly discussed next. Spalding (1976) discussed the eddy-break-up model for turbulent combustion.[92] Bilger (1976) modeled turbulent diffusion flames using cylindrical coordinates.[93] Prudnikov (1959),[94] Spalding (1976),[95] Bray (1978, 1996),[96,97] Pope (1990),[98] and Ashurst (1994)[99] have given reviews of combustion in turbulent flames. Jones and Whitelaw (1984) compared some of the available turbulence models against experimental data.[100] Faeth (1986) reviewed the interactions between turbulence and heat and mass transfer processes in flames.[101] Arpaci (1993) discussed a method for including the interaction between turbulence and radiation.[102] Hussaini et al. (1994) have edited a book on turbulence and combustion.[103] Yoshimoto et al. (1996) gave a typical example of modeling a furnace using the k-ε turbulence model.[104] Hanifi et al. (1999) have edited a book with chapters on modeling turbulence in combustion.[105] Lindstedt and Váos (1999) gave a good discussion of the closure problem using the Reynolds stress equations to solve turbulent flame models.[106] Swaminathan and Bilger (1999) discussed the stationary laminar flamelet and conditional moment closure submodels used for simulating turbulent combustion.[107] Vervisch and Veynante (1999) have edited a lecture series book on turbulent combustion that presents a state-of-the-art review of the ongoing activities in this area and gives current research directions.[108] Peters (2000) has written a book on turbulent combustion that was not yet available as of the time of this writing.[109]

Turbulence is very important in most industrial combustion applications that involve high-speed flows. Therefore, this phenomenon must be included in most types of models if representative results are expected. Some examples of modeling swirling flows, which may be important in certain types of combustors, are given elsewhere in this chapter.

2.2.2 RADIATION

The neglect of radiation cannot be justified in combustion system modeling.[52] Nonluminous and luminous radiation can greatly complicate a problem because of the spectral dependence of the solution.[45] Cess and Tiwari (1972)[110] and Ludwig et al. (1973)[111] gave a very extensive treatment of gaseous radiation, including the methods available at that time for analyzing those types of problems. They gave different techniques used for computing nonluminous radiation and both experimental and computational data on a wide variety of gases, including CO, CO_2, H_2O, HCl, HF, NO, and OH. Beér and co-workers (1972, 1974) presented a discussion of some of the early methods used for radiation analysis, including the zone method and some of the flux methods.[112,113] Lowes et al. (1974) reviewed some of the methods used to analyze radiation in furnaces, including two flux and multi-flux models.[114] Buckius and Tien (1977) showed that computations using non-gray homogeneous and nonhomogeneous radiation models for infrared flame radiation compared favorably with experimental measurements.[115] Crosbie and Dougherty (1981) gave an extensive review of exact methods for solving the radiative transport equation.[116] However, it is noted that exact solutions are not practical for engineering problems.[52] Wall et al. (1982) used a simple zoned model using the Monte Carlo technique for the radiative heat transfer in a pilot-scale furnace for oil and gas flames.[117] They used a predetermined convective coefficient of 5.8 W/m^2-K (1.0 Btu/hr-ft^2-°F) for the transfer from the gas to the furnace walls. Predictions showed good agreement with experimental measurements. Hayasaka (1987) described a method called radiative heat ray (RHR) that is intended to model the actual radiation phenomenon from an atom and is claimed to be more computationally efficient than either the Hottel zone or Monte Carlo methods.[118] Bhattacharjee and Grosshandler (1989, 1990) developed a model termed the effective angle model (EAM), which promises computer storage and computational time savings compared to other models.[46,119] The EAM should be effective for calculating radiation in two-dimensional or cylindrical combustors with black walls that have either a specified temperature or heat flux condition. Viskanta and Mengüç (1990) gave an extensive review of radiative heat transfer in combustion systems, including some simple examples.[120] They noted that the following characteristics are needed from a radiation model:

- Capability of handling inhomogeneous and spectrally dependent properties
- Capability of handling highly anisotropic radiation fields due to large temperature gradients and anisotropically scattering particles present in the medium
- Compatibility with finite-difference/finite-element algorithms for solving transport equations

They also compared the different techniques for modeling radiative heat transfer as shown in Table 2.1. Komornicki and Tomeczek (1992) developed a modification of the wide-band gas model for use in calculating flame radiation.[121] Their model compared well against experimental data and both a narrow-band model and an unmodified wide-band model. Soufiani and Djavdan (1994) compared the weighted-sum-of-gray-gases (WSGG) and the statistical narrow-band (SNB) radiation models.[122] The WSGG model is much less computationally intensive than the SNB model. They found that the WSGG only introduced small errors when the gas mixture was nearly isothermal and surrounded by cold walls. However, significant inaccuracies were found when using the WSGG where large temperature gradients existed. Lallemant et al. (1996)[123] compared nine popular total emissivity models used in CFD modeling for H_2O-CO_2 homogenous mixtures with the exponential wide-band model (EWBM).[124,125] They recommended the use of the EWBM in conjunction with WSGG models. A more recent review by Carvalho and Farias (1998) presented the various models that have been used to simulate radiation in combustion systems.[126] These methods include:

1. Zone method usually referred to as Hottel's zonal method[127]
2. Monte Carlo method which is a statistical method[128]
3. Schuster-Hamaker-type flux models[129-131]
4. Schuster-Schwarzschild-type flux models[132-135]
5. Spherical harmonic flux models (*P-N* approximations)[136]
6. Discrete ordinates approximations[137-141]
7. Finite volume method[142,143]
8. Discrete transfer method[144-146]

A schematic of some of the popular radiation models is shown in Figure 2.7.[51] The discrete exchange factor (DEF) method has been used by Naraghi and co-workers.[147-149] Denison and Webb (1993) presented a spectral radiation approach for generating WSGG models.[150] Unlike some other methods, the absorption coefficient is the modeled radiative property that permits arbitrary solution of the radiative transfer equation.

Some examples of the application of various radiation models to industrial combustion processes are given next. Siddall and Selcuk (1974) described the application of the two-flux method to a process gas heater.[151] Docherty and Fairweather (1988) showed that their predictions using the discrete transfer method for radiation from nonhomogeneous combustion products compared favorably to narrow-band calculations.[146] Cloutman and Brookshaw (1993) described a numerical algorithm for solving radiative heat losses from an experimental burner.[152] Abdullin and Vafin (1994) modeled the radiative properties of a waterwall combustor to determine their effects on the heat transfer in a tube furnace.[153] They modeled down-fired burners bounded by rows of vertical tubes with the exhaust at the bottom of the furnace. The results showed peak heat fluxes at about 20% of the distance from the ceiling and the floor, with radiation far exceeding convection. As expected, the tube emissivity was an important parameter in the heat flux in the combustor. The partial pressure of the combustion products also had an interesting effect because it affected the gas

TABLE 2.1

Comparison of Techniques for Modeling Radiative Heat Transfer

Method	Remarks	Advantages	Disadvantages
Mean-beam length	Approximation of radiation heat flux using concept of gas emissivity	Simple, possible to include detailed spectral information	Isothermal system; uncertain accuracy; insufficient detail; difficult to generalize
Zone	Approximation of system by finite size zones containing uniform temperature and composition gases	Nonhomogeneities in temperature and concentration of gases can be accounted for	Cumbersome; restricted to relatively simple geometries; difficult to account for scattering and spectral information of gases; not compatible with numerical algorithms for solving transport equations
Differential spherical harmonics, moment	Approximation of RTE in terms of the moments of intensity	RTE is recast into a system of differential equations; absorption and scattering can be accounted for; compatible with numerical algorithms for solving transport equations	Unknown accuracy as the relationship between RT and the flux equations is not always explicit
Flux and discrete ordinates	Approximation of angular intensity distribution along discrete directions and solutions of these equations numerically	Flexible; higher order approximations are accurate; can account for spectral absorption by gases and scattering by particles; compatible with numerical algorithms for solving transport equations	Time-consuming; requires iterative solution of finite-difference equations; simple flux approximations are not accurate
Discrete transfer, ray tracing, numerical	Solves RTE approximately along a line-of-sight	Can use spectral information; flexible; compatible with numerical algorithms	Time-consuming if scattering by particles is to be acounted for; accuracy is poor if few rays are considered in scattering media
Monte-Carlo	Simulation of physical process using purely statistical techniques and following individual photons	Flexibility for application to complex geometries; absorption and scattering by particles can be accounted for	Can be time-consuming; not compatible with numerical algorithms for solving transport equations
Hybrid	New procedures, relatively untested, which use a combinatino of two or more methods	Different methods can be developed to account for geometric effects; flexible; may be compatible with the numerical algorithm	Relatively untested; cannot be generalized to all systems

From R. Viskanta and M.P. Mengüç, *Handbook of Heat and Mass Transfer*, N. Cheremisinoff, Ed., Gulf Publishing, Houston, TX, 1990, 970-971. With permission.

radiation and absorptivity. The peak radiation to the waterwall was predicted for a gas partial pressure of the combination of CO_2 and H_2O of 0.27 atm (0.27 barg).

Ahluwalia and Im (1992) presented an improved technique for modeling the radiative heat transfer in coal furnaces.[154] Coal furnaces differ from gas-fired combustion processes because of the presence of char and ash, which produce significant quantities of luminous radiation. This improved technique was developed to help solve three-dimensional spectral radiation transport equations for the case of absorbing, emitting, and anisotropically scattering media that are present in coal systems. The incorporation of spectral radiation can significantly increase the computational

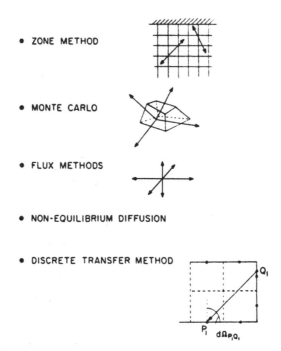

- ZONE METHOD

- MONTE CARLO

- FLUX METHODS

- NON-EQUILIBRIUM DIFFUSION

- DISCRETE TRANSFER METHOD

FIGURE 2.7   Common approaches to radiation modeling.

time and complexity, depending on how the spectra is discretized. The technique is a hybrid combination of the discrete ordinate method (S_4),[155] modified differential approximation (MDA),[156] and P_1 approximation for use in different ranges of optical thicknesses. It combines a char burnout model and spectroscopic data for H_2O, CO_2, CO, char, soot, and ash. It is used to determine the influence of ash composition, ash content, and coal preparation on heat absorption by the furnace. In the simulation of an 80 MW_e corner-fired coal boiler, predicted wall heat fluxes ranging from approximately 100 to 600 kW/m² (32,000 to 190,000 Btu/hr-ft²) compared favorably with experimental measurements.

Song and Viskanta (1987) discussed the modeling of radiation and turbulence as applied to combustion.[157] Köylü and Faeth (1993) discussed modeling the properties of flame-generated soot.[158] They evaluated approximate methods for calculating the following properties for both individual aggregates and polydisperse aggregate populations: the Rayleigh scattering approximation,[159] Mie scattering for an equivalent sphere,[160] and Rayleigh-Debye-Gans (R-D-G) scattering[161] for both given and fractal aggregates. Available measurements and computer simulations were not adequate to properly evaluate the approximate prediction methods. Given those limitations, Rayleigh scattering generally underestimated scattering, Mie scattering for an equivalent sphere was unreliable, and R-D-G scattering gave the most reliable results. Bockhorn (1994) edited a book on soot formation in combustion, including several sections on modeling, although nothing specifically on industrial combustion.[162] Bressloff et al. (1996) presented a coupled strategy for predicting soot and gas species concentrations, and radiative exchange in turbulent combustion.[163] Good agreement was found with experimental data on temperature, mixture fraction, and soot volume fraction. Bai et al. (1998) discussed soot modeling in turbulent jet diffusion flames.[164] Brookes and Moss (1999) showed the intimate connection between soot production and flame radiation, which must be accurately accounted for in modeling.[165]

Numerous methods exist for modeling the nonluminous spectral radiation from combustion products like H_2O, CO_2, and CO. Wide-band models were the first to be used because they are the simplest to implement.[166,167] More advanced models incorporated narrow-band approximations. Taine (1983)[168] and Hartmann et al. (1984)[169] have computed line-by-line calculations for single absorption bands of CO_2 and H_2O, respectively. Goody (1964) developed a statistical narrow-band model.[170] Properties used in the band models are often taken from Ludwig and co-workers (1973).[111] Song et al. (1997) used a statistical narrow-band hybrid model to calculated the gaseous radiation heat transfer in a side-port-fired glass furnace firing on natural gas.[171] This study showed the need to include spectral calculations because a gray-medium assumption overestimates the heat transfer and produces inconsistent results.[172]

2.2.3 CHEMISTRY

Many schemes have been proposed for the number of equations and reaction rates that may be used to simulate combustion reactions. One that has commonly been used is known as CHEMKIN, which was developed at Sandia National Labs.[173] Another source of chemical kinetic data is a database, formed through funding by Gas Research Institute (Chicago, IL), known as GRI-Mech.[174] The National Institute of Standards and Technology or NIST (Washington, D.C.) has also assembled and maintains a very extensive database (over 37,000 separate reactions for over 11,400 distinct reactant pairs) of chemical kinetic data taken from over 11,000 papers.[175] The CEC Group on Evaluation of Kinetic Data for Combustion Modeling was established by the European Energy Research and Development Programme to compile a database of critically evaluated chemical kinetic data.[176,177] Gaz de France has sponsored research toward improving the chemical kinetic modeling of natural gas combustion.[178] Gardiner (1984),[179] Sloane (1984),[180] Hucknall (1985),[181] and Libby and Williams (1994)[182] have written or edited books concerning combustion chemistry. Golden (1991) reviewed the rate parameters used in combustion chemistry modeling.[183] A number of papers have been written that discuss chemistry in combustion processes.[184-193] Some papers specifically consider the interaction of turbulence and the chemical reactions.[194]

One approach that has been used to simplify the chemistry in combustion modeling is to use a statistical approach referred to as the probability-density function (PDF) approach.[195,196] This is coupled with the solution of the energy and momentum equations along with the species equations. The PDF approach is most suited to turbulent, reactive flows because the complex chemical reactions can be treated without modeling assumptions.[197] However, some simplifications are usually required because of the excess computer time requirement for complete PDF modeling.

Modeling soot formation in flames is also a challenging aspect of a simulation. Coelho and Carvalho (1994) compared different soot formation models for turbulent diffusion propane flames with 500°C (930°F) air preheat.[198] They used the soot formation models given by Khan and Greeves (1974)[199] and Stewart et al. (1991).[200] They used soot oxidation models given by Magnussen and Hjertager (1977),[201] Lee et al. (1962),[202] and Nagle and Strickland-Constable (1961).[203] By comparing with available soot data, it was found that the Stewart et al. model gave better predictions than the Khan and Greeves model, once the constants were properly tuned. There was not enough data to determine which soot oxidation model gave the best predictions. Delichatsios and Orloff (1988) studied the interaction between luminous flame radiation and turbulence.[204] They concluded that soot formation was determined by the straining rate of the small (Kolmogorov) scales. Boerstoel et al. (1996) found that experimental data compared favorably with several different soot formation and oxidation models for a high-temperature furnace.[205] Said et al. (1997) proposed a simple two-equation model for soot formation and oxidation in turbulent flames.[206] Xu et al. (1998) studied the soot produced by fuel-rich, oxygen/methane, atmospheric pressure, laminar premixed flames.[207] Their measurements showed good agreement with the soot computational models proposed by Frenklach and Wang (1990)[208] and Leung and Lindstedt (1995).[209]

In certain industrial heating processes, there may be additional chemical reactions besides those involved in the combustion. The additional reactions may come from the material processing in the combustor or they may also come from downstream processing of the exhaust gases, especially in the case of many of the post-treatment pollutant reduction technologies such as selective noncatalytic reduction (SNCR) or methane re-burn for NOx reduction. In the latter case, it may be argued that the downstream treatment technologies involving combustion are a part of the overall combustion system. Several examples will suffice to illustrate these "other" chemical reactions. In the glass-melting process, there are many chemical reactions produced during the melt-in of the incoming batch materials, including the production of CO_2 and some corrosive species (see Chapter 10). In the flash smelting process used for the processing of copper, there are many chemical reactions involving copper, sulfur, and iron that are separate from the combustion reactions. In the methane re-burn NOx reduction technology, methane is injected downstream of the main combustion zone where it chemically reduces much of the NO generated in the flame region back to N_2 and O_2. It is not the purpose here to detail the non-combustion reactions, but merely to point out that they may need to be included in the model as they are directly or indirectly coupled to the combustion system.

2.2.4 MULTIPLE PHASES

In some combustion systems, multiple phases are present. The most notable involve the combustion of liquid and solid fuels. In the case of a liquid fuel, the fuel is atomized or vaporized into a fine spray, which can then be combusted. In the case of a solid fuel, the fuel normally must be finely ground so that complete combustion can be achieved. In both cases, the modeling effort is significantly complicated. In the U.S., the vast majority of industrial combustion processes use gaseous fuels. Therefore, modeling the combustion of liquid or solid fuels will not be treated here. It is recommended that the interested reader consult some of the numerous references that are available for liquid and solid fuel combustion modeling and given throughout this chapter and in the general references cited in Chapter 1. Chapter 4 discusses modeling coal combustion.

2.3 MODEL VALIDATION

Roache (1998) has written an entire book about verification and validation of computational fluid dynamics modeling.[210] He noted the distinction between verification ("solving the equations right") and validation ("solving the right equations"). Here, no discussion will be given on verification as this is available in any good CFD book (including the book by Roache), which as Roache notes is a more mathematical issue. For the purposes of this chapter section, it will be assumed that the mathematical models in a given code have been properly debugged and produce reliable results within a given accuracy range. Of more concern in combustion modeling is validation — to make sure the appropriate physics are being used for the problem under consideration and that those physics are properly simulated. Validation is a much more difficult problem than verification, especially in combustion simulations, due to the difficulties of making relevant and accurate measurements in harsh environments.

The American Institute of Aeronautics and Astronautics (Reston, VA) has written a booklet with guidelines for assessing the validity and credibility of computational fluid dynamics simulations.[211] The booklet uses the definitions for verification and validation as defined by the U.S. Department of Defense.[212,213] Verification is defined as "the process of determining that a model implementation accurately represents the developer's conceptual description of the model and the solution to the model." Validation is defined as "the process of determining the degree to which a model is an accurate representation of the real world from the perspective of the intended uses of the model." The booklet distinguishes between uncertainty or potential deficiency and error or known deficiency in a model. The requirements for good experimental data are discussed to ensure the validity of the comparison between the numerical and experimental results.

One of the seductive aspects of computer modeling is that virtually any type of problem can be simulated. How a problem is modeled depends on many things; but if enough assumptions are made, it is possible to generate computational "results." For the naïve and inexperienced, the tendency may be to believe anything that is generated on a computer, because how can a computer be wrong? The caveat, "garbage in, garbage out" definitely applies to computer modeling of complex industrial combustion problems. Any given problem may have many assumptions that need to be made, so the results are only as good as the model and the accompanying assumptions. Paraphrasing an anonymous researcher, "Everyone believes a computer analysis except the one who did it, and no one believes experimental results except the one who made them." This is to say that most people inherently realize the difficulties of making experimental measurements in complex geometries, but most naturally believe the results generated by a computer.

As any good modeler knows, a model is only as good as its validation. Models must constantly be tested against experimental measurements when they are applied to new problems. Model validation is particularly difficult for industrial combustion problems because of the difficulty in making measurements in harsh environments and because of the high cost involved in making those measurements. Most measurements made in industrial combustors are with intrusive water-cooled probes because many of the nonintrusive laser-based techniques have not been developed yet for large scales, are not rugged enough for the environments, or are too costly to use outside the lab. These intrusive probes are often larger than those used in labs because more water cooling is required in high-temperature combustors. Therefore, the flow is disturbed by the probes, which makes it more difficult to compare the measurements with the modeling results. In general, there is relatively little experimental data available for industrial combustors that has enough information to do a complete model validation. This is an important research need for the future to generate comprehensive data sets in a wide range of industrial combustion systems that can be used for model validation. Some typical model validation cases are given next and are representative of those available to date.

Fiveland et al. (1996) presented four validation cases for comparison against codes developed by Babcock and Wilcox.[214] Although the codes are primarily directed at large-scale boilers, the validations were done for a broader range of cases, including flow in a curved duct, nonreacting flow and natural gas combustion in a swirl stabilized flame, and swirling flow coal combustion in a one-sixth scale model of a utility boiler. Kaufman and Fiveland (1995) generated a large set of experimental data for the swirl stabilized natural gas flame case in the Burner Engineering Research Lab at Sandia National Labs (Livermore, CA) as part of program partially funded by the Gas Research Institute (Chicago, IL).[215] This data was used for the model validation. The model results were generally very good, except in the recirculation zones. Further work was recommended to improve the chemistry and turbulence models.

REFERENCES

1. D.B. Spalding, The art of partial modeling, *Ninth Symposium (International) on Combustion*, Academic Press, New York, 1963, 833-843.
2. R. Weber, A.A. Peters, P.P. Breithaupt, and B.M.V. Visser, Mathematical modeling of swirling pulverized coal flames: what can combustion engineers expect from modeling?, *Amer. Soc. of Mech. Eng. (ASME) FACT*, 17, 71-86, 1993.
3. E.S. Oran and J.P. Boris, Detailed modeling of combustion systems, *Prog. Energy Comb. Science*, 7(1), 1-72, 1981.
4. B.E. Launder and D.B. Spalding, The numerical computation of turbulent flows, *Lectures in Mathematical Modeling of Turbulence*, Academic Press, London, 1972.
5. J.I. Ramos, Finite-difference methods in turbulent combustion, in *Numerical Modeling in Combustion*, T.J. Chung, Ed., Taylor & Francis, Washington, D.C., 1993, 281-373.
6. T.J. Chung, Finite element methods in turbulent combustion, in *Numerical Modeling in Combustion*, T.J. Chung, Ed., Taylor & Francis, Washington, D.C., 1993, 375-397.
7. H.A. Dwyer, Finite-volume methods in turbulent combustion, in *Numerical Modeling in Combustion*, T.J. Chung, Ed., Taylor & Francis, Washington, D.C., 1993, 399-408.
8. A.F. Ghoniem, Vortex simulation of reacting shear flow, in *Numerical Approaches to Combustion Modeling*, E.S. Oran and J.P. Boris, Eds., Vol. 135, Progress in Astronautics and Aeronautics, American Institute of Aeronautics and Astronautics, Washington, D.C., 1991, 305-348.

9. L.-F. Martins and A.F. Ghonien, Simulation of the nonreacting flow in a bluff-body burner — effect of the diameter ratio, in *Heat and Mass Transfer in Fires and Combustion Systems*, W.L. Grosshandler and H.G. Semerjian, Eds., ASME, New York, HTD-148, 33-44, 1990.

10. C. Fureby, E. Lundgren, and S.-I. Möller, Large eddy simulation of combustion, in *Tranport Phenomena in Combustion*, Vol. 2, S.H. Chan, Ed., Taylor & Francis, Washington, D.C., 1996, 1083-1094.

11. H.G. Weller, G. Tabor, A.D. Gosman, and C. Fureby, Application of a flame-wrinkling LES combustion model to a turbulent mixing layer, *Twenty-Seventh Symp. (Int.) on Combustion*, The Combustion Institute, Pittsburgh, PA, 1998, 899-907.

12. P.E. Desjardin and S.H. Frankel, Two-dimensional large eddy simulation of soot formation in the near-field of a strongly radiating nonpremixed acetylene-air turbulent jet flame, *Comb. Flame*, 119, 121-132, 1999.

13. G. Tryggvason and W.J.A. Dahm, An Integral Method for Mixing, Chemical reactions, and extinction in unsteady strained diffusion layers, *Comb. Flame*, 83(3-4), 207-220, 1990.

14. C.H.H. Chang, W.J.A. Dahm, and G. Tryggvason, Lagrangian model simulations of molecular mixing, including finite rate chemical reactions, in temporally developing shear layer, *Phys. Fluids A*, 3(5), 1300-1311, 1991.

15. W.J.A. Dahm, G. Tryggvason, and M. Zhuang, Intergral method solution of time-dependent strained diffusion-reaction layers with multistep kinetics, *SIAM J. Appl. Math.*, 56(4), 1039-1059, 1996.

16. W.J.A. Dahm, G. Tryggvason, J.A. Kezerle, and R.V. Serauskas, Simulation of turbulent flow and complex chemistry by local integral moment (LIM) modeling, *Proc. of 1995 Int. Gas Research Conf.*, D.A. Dolenc, Ed., Govt. Institutes, Rockville, MD, 1996, 2169-2178.

17. M.Y. Hussaini and T.A. Zang, Spectral methods in fluid dynamics, *Annu. Rev. of Fluid Mechanics*, 19, 339-367, 1987.

18. P.A. McMurtry and P. Givi, Spectral simulations of reacting turbulent flows, in *Numerical Approaches to Combustion Modeling*, E.S. Oran and J.P. Boris, Eds., Vol. 135, Progress in Astronautics and Aeronautics, American Institute of Aeronautics and Astronautics, Washington, D.C., 1991, 257-303.

19. P. Givi and C.K. Madnia, Spectral Methods in Combustion, in *Numerical Modeling in Combustion*, T.J. Chung, Ed., Taylor & Francis, Washington, D.C., 1993, 409-452.

20. G.E. Karniadakis and S.J. Sherwin, *Spectral/hp Element Methods for CFD*, Oxford University Press, New York, 1999.

21. V. Eswaran and S.B. Pope, Direct numerical simulations of the turbulent mixing of a passive scalar, *Physics of Fluids*, 31(3), 506-520, 1988.

22. T. Baritaud, T. Poinsot and M. Baum, Eds., *Direct Numerical Simulation for Turbulent Reacting Flows*, Editions Technip, Paris, 1996.

23. J.H. Chen, J.M. Card, M. Day, and S. Mahalingam, Direct numerical simulation of turbulent non-premixed methane-air flames, in *Tranport Phenomena in Combustion*, Vol. 2, S.H. Chan, Ed., Taylor & Francis, Washington, D.C., 1996, 1049-1060.

24. C. Hartel, Turbulent flows: direct numerical simulation and large-eddy simulation, in *Handbook of Computational Fluid Mechanics*, R. Peyret, Ed., Academic Press, London, 1996, 283-338.

25. T. Poinsot, Using direct numerical simulations to understand premixed turbulent combustion, *Twenty-Sixth Symp. (Int.) on Combustion*, The Combustion Institute, Pittsburgh, PA, 1996, 219-232.

26. M. Boger, D. Veynante, H. Boughanem, and A. Trouvé, Direct numerical simulation analysis of flame surface density concept for large eddy simulation of turbulent premixed combustion, *Twenty-Seventh Symp. (Int.) on Combustion*, The Combustion Institute, Pittsburgh, PA, 1998, 917-925.

27. B. Bédat, F.N. Egolfopoulos, and T. Poinsot, Direct numerical simulation of heat release and NOx formation in turbulent nonpremixed flames, *Comb. Flame*, 119, 1999, 69-83.

28. P.A. McMurtry and P. Givi, Direct numerical simulations of mixing and reaction in non-premixed homogeneous turbulent flows, *Comb. Flame*, 77, 171-185, 1989.

29. E.E. Khalil, *Modelling of Furnaces and Combustors*, Abacus Press, Kent, U.K., 1982.

30. K.S. Chapman, S. Ramadhyani, and R. Viskanta, Modeling and analysis of heat transfer in a direct-fired continuous reheating furnace, in *Heat Transfer in Combustion Systems*, N. Ashgriz, J.G. Quintiere, H.G. Semerjian, and S.E. Slezak, Eds., ASME, New York, HTD-122, 35-43, 1989.

31. K.S. Chapman, S. Ramadhyani, and R. Viskanta, Modeling and analysis of heat transfer in a direct-fired batch reheating furnace, in *Heat Transfer Phenomena in Radiation, Combustion, and Fires*, R.K. Shah, Ed., ASME, New York, HTD-106, 1989, 265-274.

32. R. Viskanta, K.S. Chapman, and S. Ramadhyani, Mathematical modeling of heat transfer in high-temperature industrial furnaces, in *Advanced Computational Methods in Heat Transfer*, Vol. 3: Phase Change and Combustion Simulation, L.C. Wrobel, C.A. Brebbia, and A.J. Nowak, Eds., Springer-Verlag, Berlin, 1990, 117-131.

33. K.S. Chapman, S. Ramadhyani, and R. Viskanta, Two-dimensional modeling and parametric studies in a direct-fired furnace with impinging jets, *Comb. Sci. Tech.*, 97, 1994, 99-120.

34. R.M. Davies, D.M. Lucas, B.E. Moppett, and R.A. Galsworthy, Isothermal model studies of rapid heating furnaces, *J. Inst. Fuel*, 44, 453-461, 1971.

35. J.T. Kuo, System simulation and control of batch-fed solid waste incinerators, in *Heat Transfer in Fire and Combustion Systems*, W.W. Yuen and K.S. Ball, Eds., ASME, New York, HTD-272, 55-62, 1994.

36. S. Singh, M. Ziolkowski, J. Sultzbaugh, and R. Viskanta, Mathematical model of a ceramic burner radiant heater, in *Fossil Fuel Combustion 1991*, R. Ruiz, Ed., ASME, New York, PD-33, 111-116, 1991.

37. F.M. El-Mahallawy, A.S. Elasfouri, and E. Mahdi Ali, Heat liberation and heat transfer in flame tubes, in *Heat Transfer 1982*, Vol. 2, U. Grigull, E. Hahne, K. Stephan, and J. Straub, Eds., Hemisphere, Washington, D.C., 1982, 529-534.

38. P.A. Gillis and P.J. Smith, An evaluation of three-dimensional computational combustion and fluid dynamics for industrial furnace geometries, *Twenty-Third Symp. (Int.) on Combustion*, The Combustion Institute, Pittsburgh, PA, 1990, 981-991.

39. M. Frenklach, Reduction of chemical reaction models, in *Numerical Approaches to Combustion Modeling*, E.S. Oran and J.P. Boris, Eds., Vol. 135, Progress in Astronautics and Aeronautics, American Institute of Aeronautics and Astronautics, Washington, D.C., 1991, 129-154.

40. C.K. Westbrook and F.L. Dryer, Simplified reaction mechanisms for the oxidation of hydrocarbon fuels in flames, *Comb. Sci. Tech.*, 27, 31-43, 1981.

41. N. Peters, Systematic reduction of flame kinetics: principles and details, in *Dynamics of Reactive Systems*, A.L. Kuhl, J.R. Bowen, J.-C. Leyer, and A. Borisov, Eds., Vol. 113, Progress in Astronautics and Aeronautics, American Institute of Aeronautics and Astronautics, Washington, D.C., 1988, 67-86.

42. M.D. Smooke and R.W. Bilger, *Reduced Kinetic Mechanisms and Asymptotic Approximations for Methane-Air Flames: A Topical Volume*, Springer-Verlag, Berlin, 1991.

43. N. Peters and B. Rogg, *Reduced Kinetic Mechanisms for Application in Combustion Systems*, Springer-Verlag, Berlin, 1993.

44. Y.S. Kocaefe, A. Charette, and M. Munger, Comparison of the various methods for analysing the radiative heat transfer in furnaces, *Proc. Combustion Institute Canadian Section Spring Technical Meeting*, Vancouver, Canada, May 1987, 15-17.

45. C.E. Baukal, *Heat Transfer in Industrial Combustion*, CRC Press, Boca Raton, FL, 2000.

46. S. Bhattacharjee and W.L. Grosshandler, Effect of radiative heat transfer on combustion chamber flows, *Combust. Flame*, 77, 347-357, 1989.

47. Y.R. Sivathanu, J.P. Gore, and J. Dolinar, Transient scalar properties of strongly radiating flames, in *Heat and Mass Transfer in Fires and Combustion Systems*, W.L. Grosshandler and H.G. Semerjian, Eds., ASME, New York, HTD-148, 45-56, 1990.

48. G. Cox, On radiant heat transfer in diffusion flames, *Combust. Sci. Tech.*, 17, 75-78, 1977.

49. V.P. Kabashnikov and G.I. Kmit, Influence of turbulent fluctuations on thermal radiation, *Appl. Spect.*, 31, 963-967, 1979.

50. W.L. Grosshandler and P. Joulain, The Effect of Large Scale Fluctuations on Flame Radiation, *Prog. Astro. and Aero.*, Vol. 105, Part II, AIAA, Washington, D.C., 1986, 123-152.

51. A.F. Sarofim, Radiative heat transfer in combustion: friend or foe, Hoyt C. Hottel Plenary Lecture, *Twenty-First Symp. (Int.) on Combustion*, The Combustion Institute, Pittsburgh, PA, 1986, 1-23.

52. R. Viskanta and M.P. Menguc, Radiation heat transfer in combustion systems, *Prog. Energy Combust. Sci.*, 8, 97-160, 1987.

53. G.M. Faeth, J.P. Gore, S.G. Chuech, and S.M. Jeng, Radiation from turbulent diffusion flames, *Annu. Rev. Numerical Fluid Mech. & Heat Trans.*, C.L. Tien and T.C. Chawla, Eds., Hemisphere, New York, 2, 1-38, 1989.

54. C.J. Hoogendoorn, L. Post, and J.A. Wieringa, Modelling of combustion and heat transfer in glass furnaces, *Glastech. Ber.*, 63(1), 7-12, 1990.

55. Y. Zhenghua and G. Holmstedt, Fast, narrow-band computer model for radiation calculations, *Num. Heat Transfer, Part B*, 31, 61-71, 1997.

56. W.L. Grosshandler, Radiation from Nonhomogeneous Fires, Technical Report FMRC, Sept. 1979.
57. W.L. Grosshandler, Radiative heat transfer in nonhomogeneous gases: a simplified approach, *Int. J. Heat Mass Transfer*, 23, 1447-1459, 1980.
58. W.L. Grosshandler, RADCAL: A Narrow-Band Model for Radiation Calculations in a Combustion Environment, NIST Technical Note 1402, April 1993.
59. F. Liu, Ö.L. Gülder, and G.J. Smallwood, Non-grey gas radiative transfer analyses using the statistical narrow-band model, *Int. J. Heat Mass Transfer*, 41(14), 2227-2236, 1998.
60. L.A. Gritzo and J.H. Strickland, A gridless solution of the radiative transfer equation for fire and combustion calculations, *Combust. Theory Modelling*, 3, 159-175, 1999.
61. R.K. Ahluwalia and K.H. Im, Radiative Heat Transfer in PC Furnaces Burning Deeply Cleaned Coals, U.S. Dept. of Energy Report DE91 006022, Argonne National Laboratory, Argonne, IL, 1990.
62. R.K. Ahluwalia and K.H. Im, Spectral radiative heat-transfer in coal furnaces using a hybrid technique, *J. Inst. Energy*, 67, 23-29, 1994.
63. H.M. Park, R.K. Ahluwalia, and K.H. Im, Three-dimensional radiation in absorbing-emitting-scattering media using the modified differential approximation, *Int. J. Heat Mass Trans.*, 36(5), 1181-1189, 1993.
64. R.B. Pember, A.S. Almgren, W.Y. Crutchfield, L.H. Howell, J.B. Bell, P. Colella, and V.E. Beckner, An Embedded Boundary Method for the Modeling of Unsteady Combustion in an Industrial Gas-Fired Furnace, U.S. Dept. of Commerce Report No. DE96004155, Springfield, VA, 1995.
65. S.V. Patankar and D.B. Spalding, *Heat and Mass Transfer in Boundary Layers: A General Calculation Procedure*, 2nd ed., Intertext Books, London, 1970.
66. S.V. Patankar, *Numerical Heat Transfer and Fluid Flow*, McGraw-Hill, New York, 1980.
67. G.K. Batchelor, *The Theory of Homogeneous Turbulence*, Cambridge University Press, New York, 1953.
68. J.T. Davies, *Turbulence Phenomena*, Academic Press, New York, 1972.
69. B.E. Launder and D.B. Spalding, *Mathematical Models of Turbulence*, Academic Press, New York, 1972.
70. J.O. Hinze, *Turbulence*, 2nd ed., McGraw-Hill, New York, 1975.
71. P. Bradshaw, Ed., *Turbulence*, Springer-Verlag, New York, 1978.
72. W. Kollmann, *Prediction Methods for Turbulent Flows*, Hemisphere, Washington, D.C., 1980.
73. P. Bradshaw, *Engineering Calculation Methods for Turbulent Flow*, Academic Press, New York, 1981.
74. T. Tatsumi, Ed., *Turbulence and Chaotic Phenomena in Fluids*, North Holland, New York, 1984.
75. M.M. Stanisic, *The Mathematical Theory of Turbulence*, Springer-Verlag, New York, 1985.
76. M. Lesieur, *Turbulence in Fluids*, Dordrecht, Boston, 1987.
77. H.C. Mongia, R.M.C. So, and J.H. Whitelaw, Eds., *Turbulent Reactive Flow Calculations*, Gordon and Breach, New York, 1988.
78. B.S. Petukhov and A.F. Polyakov, *Heat Transfer in Turbulent Mixed Convection*, Hemisphere, New York, 1988.
79. V.R. Kuznetsov, *Turbulence and Combustion*, Hemisphere, New York, 1990.
80. W.D. McComb, *The Physics of Fluid Turbulence*, Oxford University Press, Oxford, U.K., 1990.
81. W. Rodi and E.N. Ganic, Eds., *Engineering Turbulence Modeling and Experiments*, Elsevier Science, Amsterdam, The Netherlands, 1990.
82. M. Landahl, *Turbulence and Random Processes in Fluid Mechanics*, Cambridge University Press, Cambridge, U.K., 1992.
83. S.W. Churchill, *Turbulent Flows*, Butterworth-Heinemann, Boston, 1993.
84. D.C. Wilcox, *Turbulence Modeling for CFD*, 2nd ed., DCW Industries, La Cãnada, CA, 1998.
85. A.J. Chorin, *Vorticity and Turbulence*, Springer-Verlag, New York, 1994.
86. R.J. Garde, *Turbulent Flow*, Wiley, New York, 1994.
87. U. Frisch, *Turbulence*, Cambridge University Press, Cambridge, U.K., 1995.
88. K. Hanjalic and J.C.F. Pereira, Eds., *Turbulence, Heat, and Mass Transfer*, Begell House, New York, 1995.
89. C.J. Chen and S.-Y. Jaw, *Fundamentals of Turbulence Modeling*, Taylor & Francis, Washington, D.C., 1998.
90. C.G. Speziale and R.M.C. So, Turbulence Modeling and Simulation, in *The Handbook of Fluid Dynamics*, R.W. Johnson, Ed., CRC Press, Boca Raton, FL, 1998.

91. J. Baldyga, *Turbulent Mixing and Chemical Reactions*, Wiley, New York, 1999.

92. D.B. Spalding, Development of the eddy-break-up model of turbulent combustion, *Twentieth Symp. (International) on Combustion*, The Combustion Institute, Pittsburgh, PA, 1976, 1657-1663.

93. R.W. Bilger, Turbulent jet diffusion flames, *Prog. Energy Combust. Sci.*, 1, 87-109, 1976.

94. A.G. Prudnikov, Flame turbulence, *Seventh Symp. (International) on Combustion*, Butterworths Scientific, London, 1959, 575-582.

95. D.B. Spalding, Mathematical models of turbulent flames: a review, *Combust. Sci. Tech.*, 13(1-6), 3-25, 1976.

96. K.N.C. Bray, The interaction between turbulence and combustion, *Seventeenth Symp. (International) on Combustion*, The Combustion Institute, Pittsburgh, PA, 1978, 223-233.

97. K.N.C. Bray, The Challenge of Turbulent Combustion, *Twenty-Sixth Symp. (International) on Combustion*, The Combustion Institute, Pittsburgh, PA, 1996, 1-26.

98. S.B. Pope, Computations of turbulent combustion: progress and challenges, *Twenty-Third Symp. (International) on Combustion*, The Combustion Institute, Pittsburgh, PA, 1990, 591-612.

99. W.T. Ashurst, Modeling Turbulent Flame Propagation, *Twenty-Fifth Symp. (International) on Combustion*, The Combustion Institute, Pittsburgh, PA, 1994, 1075-1089.

100. W.P. Jones and J.H. Whitelaw, Modelling and measurements in turbulent combustion, *Twentieth Symp. (International) on Combustion*, The Combustion Institute, Pittsburgh, PA, 1984, 233-249.

101. G.M. Faeth, Heat and Mass Transfer in Flames, in *Heat Transfer 1986*, Vol. 1, C.L. Tien, V.P. Carey, and J.K. Ferrell, Eds., *Proc. 8th Int. Heat Transfer Conf.,* San Francisco, 1986, 151-160.

102. V.S. Arpaci, Radiative turbulence: radiation affected turbulent forced convection, in *Heat Transfer in Fire and Combustion Systems — 1993*, B. Farouk, M.P. Menguc, R. Viskanta, C. Presser, and S. Chellaiah, Eds., ASME, New York, HTD-250, 155-160, 1993.

103. M.Y. Hussaini, T.B. Gatski, and T.L. Jackson, Eds., *Transition, Turbulence and Combustion — Vol. II: Turbulence and Combustion*, Kluwer Academic, Dordrecht, 1994.

104. T. Yoshimoto, T. Okamoto, and T. Takagi, Numerical simulation of combustion and heat transfer in a furnace and its comparison with experiments, in *Tranport Phenomena in Combustion*, Vol. 2, S.H. Chan, Ed., Taylor & Francis, Washington, D.C., 1996, 1153-1164.

105. A. Hanifi, P.H. Alfredsson, A.V. Johansson, and D.S. Hennigson, *Transition, Turbulence and Combustion Modelling*, Kluwer Academic, Dordrecht, 1999.

106. R.P. Lindstedt and E.M. Váos, Modeling of premixed turbulent flames with second moment methods, *Combust. Flame*, 116, 461-485, 1999.

107. N. Swaminathan and R.W. Bilger, Assessment of combustion submodels for turbulent nonpremixed hydrocarbon flames, *Combust. Flame*, 116, 519-545, 1999.

108. L. Vervisch and D. Veynante, Eds., *Introduction to Turbulent Combustion*, Von Karman Institute for Fluid Dynamics, Rhode Saint Genese, Belgium, 1999.

109. N. Peters, *Turbulent Combustion*, Cambridge University Press, New York, 2000.

110. R.D. Cess and S.N. Tiwari, Infrared Radiative Energy Transfer in Gases, in *Advances in Heat Transfer*, Vol. 8, J.P. Hartnett and T.F. Irvine, Eds., Academic Press, New York, 1972, 229-283.

111. C.B. Ludwig, W. Malkmus, J.E. Reardon, and J.A.L. Thomson, *Handbook of Infrared Radiation*, National Aeronautics and Space Administration report SP-3080, Washington, D.C., 1973.

112. T.R. Johnson and J.M. Beer, Radiative Heat Transfer in Furnaces: Further Development of the Zone Method of Analysis, *Fourteenth Symp. (International) on Combustion*, The Combustion Institute, Pittsburgh, PA, 1972, 639-649.

113. J.M. Beér, Methods for Calculating Radiative Heat Transfer from Flames in Combustors and Furnaces, pp. 29-45, in *Heat Transfer in Flames*, N.H. Afgan and J.M. Beer, Eds., Scripta, Washington, D.C., 1974.

114. T.M. Lowes, H. Bartelds, M.P. Heap, S. Michelfelder, and B.R. Pai, Prediction of Radiant Heat Flux Distribution, in *Heat Transfer in Flames*, N.H. Afgan and J.M. Beer, Eds., Scripta, Washington, D.C., 1974, chap. 10, 179-190.

115. R.O. Buckius and C.L. Tien, Infrared flame radiation, *Int. J. Heat Mass Transfer*, 20, 93-106, 1977.

116. A.L. Crosbie and R.L. Dougherty, Two-dimensional radiative transfer in cylindrical geometry with anisotropic scattering, *J. Quant. Spectrosc. Radiat. Transfer*, 25(6), 551-569, 1981.

117. T.F. Wall, H.T. Duong, I.M. Stewart, and J.S. Truelove, Radiative Heat Transfer in Furnaces: Flame and Furnace Models of the IFRF M1- and M2-Trials, *Nineteenth Symp. (International) on Combustion*, The Combustion Institute, Pittsburgh, PA, 1982, 537-547.

118. H. Hayasaka, A direct simulation method for the analysis of radiative heat transfer in furnaces, in *Heat Transfer in Furnaces*, C. Presser and D.G. Lilley, Eds., ASME, New York, HTD-74, 59-63, 1987.

119. S. Bhattacharjee and W.L. Grosshandler, A simplified model for radiative source term in combusting flows, *Int. J. Heat Mass Trans.*, 33(3), 507-516, 1990.

120. R. Viskanta and M.P. Mengüç, Principles of radiative heat transfer in combustion systems, in *Handbook of Heat and Mass Transfer*, N. Cheremisinoff, Ed., Vol. 4, Gulf Publishing, Houston, TX, chap. 22, 925-978, 1990.

121. W. Komornicki and J. Tomeczek, Modification of the wide-band gas radiation model for flame calculation, *Int. J. Heat Mass Transfer*, 35(7), 1667-1672, 1992.

122. A. Soufiani and E. Djavdan, A comparison between weighted sum of gray gases and statistical narrow-band radiation models for combustion applications, *Comb. Flame*, 97, 240-250, 1994.

123. N. Lallemant, A. Sayre, and R. Weber, Evaluation of emissivity correlations for H_2O-CO_2-N_2/air mixtures and coupling with solution methods of the radiative transfer equation, *Prog. Energy Combust. Sci.*, 22, 543-574, 1996.

124. D.K. Edwards, Molecular gas band radiation, in *Advances in Heat Transfer*, T.F. Irvine and J.P. Hartnett, Eds., Vol. 12, Academic Press, New York, 1976, 115-193.

125. A.T. Modak, Exponential wide band parameters for the pure rotational band of water vapor, *J. Quant. Spectosc. Radiat. Transfer*, 21(2), 131-142, 1979.

126. M.G. Carvalho and T.L. Farias, Modelling of heat transfer in radiating and combusting systems, *Trans. IChemE*, 76A, 175-184, 1998.

127. H.C. Hottel and A.F. Sarofim, *Radiative Transfer*, McGraw-Hill, New York, 1967.

128. J.R. Howell and M. Perlmutter, Monte Carlo solution of thermal transfer through radiant media between gray walls, *J. Heat Transfer*, 86(1), 116-122, 1964.

129. S. Patankar and B. Spalding, Simultaneous predictions of flow patterns and radiation for three-dimensional flames, in *Heat Transfer in Flames*, N.H. Afgan and J.M. Beer, Eds., Scripta, Washington, D.C., 1974, 73-94.

130. H.C. Hamaker, Philips Research Reports 3, 103, 112, and 142, 1947.

131. A.D. Gosman and F.C. Lockwood, Incorporation of a flux model for radiation into a finite difference procedure for furnace calculations, *Fourteenth Symp. (International) on Combustion*, The Combustion Institute, Pittsburgh, PA, 1973, 661-671.

132. T.M. Lowes, H. Bartelds, M.P. Heap, S. Michelfelder and B.R. Pai, Prediction of radiant heat flux distributions, Int. Flame Research Foundation Report GO2/A/26, IJmuiden, The Netherlands, 1973.

133. W. Richter and R. Quack, A mathematical model of a low-volatile pulverised fuel flame, in *Heat Transfer in Flames*, N.H. Afgan and J.M. Beer, Eds., Scripta, Washington, D.C., 1974, 95-110.

134. R.G. Siddall and N. Selçuk, Two-flux modelling of two-dimensional radiative transfer in axi-symmetrical furnaces, *J. Inst. Fuel*, 49, 10-20, 1976.

135. R.G. Siddall and N. Selçuk, Evaluation of a new six-flux model for radiative transfer in rectangular enclosures, *Trans. IChem*, 57, 163-169, 1979.

136. R. Viskanta, Radiative transfer and interaction of convection with radiation heat transfer, *Advances in Heat Transfer*, T.F. Irvine and J.P. Hartnett, Eds., 3, 175-252, 1966.

137. S. Chandrasekhar, *Radiative Transfer*, Dover Publications, New York, 1960.

138. W.A. Fiveland, Discrete-ordinates solutions of the radiative transport equation for rectangular enclosures, *J. Heat Transfer*, 106, 699-706, 1984.

139. A.S. Jamaluddin and P.J. Smith, Predicting radiative transfer in rectangular enclosures using the discrete ordinates method, *Comb. Sci. Tech.*, 59(4-6), 321-340, 1988.

140. A.S. Jamaluddin and P.J. Smith, Predicting radiative transfer in axisymmetric cylindrical enclosures using the discrete ordinates method, *Comb. Sci. Tech.*, 62(4-6), 173-186, 1988.

141. W.A. Fiveland and A.S. Jamaluddin, Three-dimensional spectral radiative heat transfer solutions by the discrete-ordinates method, in *Heat Transfer Phenomena in Radiation, Combustion, and Fires*, R.K. Shah, Ed., ASME, New York, HTD-106, 43-48, 1989.

142. G.D. Raithby and E.H. Chui, A finite-volume method for predicting radiant heat transfer in enclosures with participating media, *J. Heat Transfer*, 112, 414-423, 1990.

143. J.C. Chai, H.S. Lee, and S.V. Patankar, Finite volume method for radiation heat transfer, *J. Thermophysics & Heat Transfer*, 8(3), 419-425, 1994.

144. N.G. Shah, New Method of Computation of Radiative Heat Transfer in Combustion Chambers, Ph.D. thesis, Imperial College, London, 1979.

145. F.C. Lockwood and N.G. Shah, A new radiation solution method for incorporation in general combustion prediction procedures, *Eighteenth Symp. (International) on Combustion*, the Combustion Institute, Pittsburgh, PA, 1981, 1405-1414.

146. P. Docherty and M. Fairweather, Predictions of radiative transfer from nonhomogeneous combustion products using the discrete transfer method, *Combust. Flame*, 71, 79-87, 1988.

147. M.H.N. Naraghi and M. Kassemi, Radiative transfer in rectangular enclosures: a discretized exchange factor solution, *ASME Proc. 1988 National Heat Transfer Conf.*, H.R. Jacobs, Ed., Vol. 1, 259-267, 1988.

148. M.H.N. Naraghi, Radiative heat transfer in non-rectangular enclosures, in *Heat Transfer Phenomena in Radiation, Combustion, and Fires*, R.K. Shah, Ed., ASME, New York, HTD-106, 17-25, 1989.

149. M.H.N. Naraghi and B. Litkouhi, Discrete Exchange Factor Solution of Radiative Heat Transfer in Three-Dimensional Enclosures, in *Heat Transfer Phenomena in Radiation, Combustion, and Fires*, R.K. Shah, Ed., ASME, New York, HTD-106, 221-229, 1989.

150. M.K. Denison and B.W. Webb, A spectral line-based weighted-sum-of-gray-gases model for arbitrary RTE solvers, *J. Heat Transfer*, 115, 1004-1012, 1993.

151. R.G. Siddall and N. Selcuk, The application of flux methods to prediction of the behavior of a process gas heater, in *Heat Transfer in Flames*, N.H. Afgan and J.M. Beer, Eds., Scripta, Washington, D.C., 1974, chap. 11, 191.

152. L.D. Cloutman and L. Brookshaw, Numerical Simulation of Radiative Heat Loss in an Experimental Burner, Lawrence Livermore National Lab, U.S. Dept. of Energy Report UCRL-JC-115048, 1993.

153. A.M. Abdullin and D.V. Vafin, Numerical investigation of the effect of the radiative properties of a tube waterwall and combustion products on heat transfer in tube furnaces, *J. Eng. Physics and Thermophysics*, 65(2), 752-757, 1994.

154. R.K. Ahluwalia and K.H. Im, Radiative Heat Transfer in Coal Furnaces, U.S. Dept. of Energy Report DE92018770, Argonne National Lab, Argonne, IL, 1992.

155. W.A. Fiveland and A.S. Jamaluddin, Three-dimensional spectral radiative heat transfer solutions by the discrete-ordinate method, *J. Thermophysics*, 5(3), 335-339, 1991.

156. M.F. Modest, Modified differential approximation for radiative transfer in general three-dimensional media, *J. Thermophysics*, 3(3), 283-288, 1989.

157. T.H. Song and R. Viskanta, Interaction of radiation with turbulence: application to a combustion system, *J. Thermophysics*, 1(1), 56-62, 1987.

158. Ü.Ö. Köylü and G.M. Faeth, Radiative properties of flame-generated soot, *J. Heat Transfer*, 409-417, 1993.

159. C.L. Tien and S.C. Lee, Flame radiation, *Prog. Energy Combust. Sci.*, 8, 41-59, 1982.

160. W.H. Dalzell, G.C. Williams, and H.C. Hottel, A light scattering method for soot concentration measurements, *Combust. Flame*, 14, 161-170, 1970.

161. J.E. Martin and A.J. Hurd, Scattering from fractals, *J. Appl. Cryst.*, 20, 61-78, 1987.

162. H. Bockhorn, Ed., *Soot Formation in Combustion*, Springer-Verlag, Berlin, 1994.

163. N.W. Bressloff, J.B. Moss, and P.A. Rubini, CFD Prediction of Couple Radiation Heat Transfer and Soot Production in Turbulent Flames, *Twenty-Sixth Symp. (International) on Combustion*, The Combustion Institute, Pittsburgh, PA, 1996, 2379-2386.

164. X.S. Bai, M. Balthasar, F. Mauss and L. Fuchs, Detailed soot modeling in turbulent jet diffusion flames, *Twenty-Seventh Symp. (International) on Combustion*, The Combustion Institute, 1998, 1623-1630, Pittsburgh, PA.

165. S.J. Brookes and J.B. Moss, Predictions of soot and thermal radiation properties in confined turbulent jet diffusion flames, *Combust. Flame*, 116, 486-503, 1999.

166. D.K. Edwards, L.K. Glassen, W.S. Hauser, and J.S. Tuchscher, Radiation heat transfer in nonisothermal nongray gases, *J. Heat Transfer*, 86C, 219-229, 1967.

167. B. Leckner, Spectral and total emissivity of water vapor and carbon dioxide, *Comb. Flame*, 19, 33-48, 1972.

168. J. Taine, A line-by-line calculation of low-resolution radiative properties of CO_2-CO transparent nonisothermal gaseous mixtures up to 3000K, *J. Quant. Spectroscopic and Radiative Transfer*, 30, 371-379, 1983.

169. J.M. Hartmann, L. Leon, and J. Taine, Line-by-line and narrow-band statistical model calculations for H2O, *J. Quant. Spectroscopic and Radiative Transfer*, 32(2), 119-127, 1984.

170. R.M. Goody, *Atmospheric Radiation*, Vol. I, Oxford University Press, Oxford, U.K., 1964.

171. G. Song, T. Bjørge, J. Holen, and B.F. Magnussen, Simulation of fluid flow and gaseous radiation heat transfer in a natural gas-fired furnace, *Int. J. Num. Methods for Heat & Fluid Flow*, 7(2/3), 169-180, 1997.

172. R.K. Ahluwalia and K.H. Im, Spectral radiative heat-transfer in coal furnaces using a hybrid technique, *J. Institute of Energy*, 67, 23-29, 1994.

173. R.J. Kee, R.M. Rupley, and J.A. Miller, CHEMKIN-II: A Fortran Chemical Kinetics Package for the Analysis of Gas Phase Chemical Kinetics, Sandia National Laboratory Report SAND89-8009B, Livermore, CA, 1989.

174. C.T. Bowman, R.K. Hanson, D.F. Davidson, W.C. Gardiner, V. Lissianski, G.P. Smith, D.M. Golden, M. Frenklach, and M. Goldberg, http://www.me.berkeley.edu/gri_mech/, 1999.

175. W.G. Mallard, F. Westley, J.T. Herron, R.F. Hampson, and D.H. Frizzell, *NIST Chemical Kinetics Database User's Guide — Windows Version 2Q98*, National Institute of Standards and Technology, Washington, D.C., 1998.

176. D.L. Baulch, C.J. Cobos, R.A. Cox, C. Esser, P. Frank, et al., Evaluated kinetic data from combustion modelling, *J. of Phys. Chem. Ref. Data*, 21(3), 411-734, 1992.

177. D.L. Baulch, C.J. Cobos, R.A. Cox, P. Frank, G. Hayman, Th. Just, J.A. Kerr, T. Murrells, M.J. Pilling, J. Troe, R.W. Walker, and J. Warnatz, Summary Table of Evaluated Kinetic Data for Combustion Modeling: Supplement 1, *Combust. Flame*, 98, 59-79, 1994.

178. A. Turbiez, P. Desgroux, J.F. Pauwels, L.R. Sochet, S. Poitou, and M. Perrin, GDF.kin®: a new step towards a detailed kinetic mechanism for natural gas combustion modeling, *Proc. 1998 Int. Gas Research Conf.*, Vol. V: Industrial Utilization, D.A. Dolenc, Ed., Gas Research Institute, Chicago, 1998, 210-221.

179. W.C. Gardiner, Ed., *Combustion Chemistry*, Springer-Verlag, New York, 1984.

180. T.M. Sloane, *The Chemistry of Combustion Processes*, American Chemical Society, Washington, D.C., 1984.

181. D.J. Hucknall, *Chemistry of Hydrocarbon Combustion*, Chapman & Hall, London, 1985.

182. P.A. Libbey and F.A. Williams, Eds., *Turbulent Reacting Flows*, Academic Press, London, 1994.

183. D.M. Golden, Evaluation of chemical thermodynamics and rate parameters for use in combustion modeling, in *Fossil Fuel Combustion*, W. Bartok and A.F. Sarofim, Eds., Wiley, New York, 1991, chap. 2.

184. C.K. Westbrook and F.L. Dryer, Chemical Kinetics and Modeling of Combustion Processes, *Eighteenth Symp. (International) on Combustion*, The Combustion Institute, Pittsburgh, PA, 1980, 749-767.

185. F. Kaufman, Chemical Kinetics and Combustion: Intricate Paths and Simple Steps, *Nineteenth Symp. (International) on Combustion*, The Combustion Institute, Pittsburgh, PA, 1982, 1-10.

186. J. Wofrum, Chemical Kinetics in Combustion Systems: The Specific Effect of Energy, Collisions, and Transport Processes, *Twentieth Symp. (International) on Combustion*, The Combustion Institute, Pittsburgh, PA, 1984, 559-573.

187. S.W. Benson, Combustion, A Chemical and Kinetic View, *Twenty-First Symp. (International) on Combustion*, The Combustion Institute, Pittsburgh, PA, 1986, 703-711.

188. V.Y. Basevich, Chemical kinetics in the combustion process, in *Handbook of Heat and Mass Transfer*, N. Cheremisinoff, Ed., Vol. 4, Gulf Publishing, Houston, 1990, chap. 18.

189. P. Gray, Chemistry and Combustion, *Twenty-Third Symp. (International) on Combustion*, The Combustion Institute, Pittsburgh, PA, 1990, 1-19.

190. F.L. Dryer, The Phenomenology of Modeling Combustion Chemistry, in *Fossil Fuel Combustion*, W. Bartok and A.F. Sarofim, Eds., Wiley, New York, 1991, chap. 3.

191. E. Ranzi, A. Sogaro, P. Gaffuri, G. Pennati, C.K. Westbrook, and W.J. Pitz, A new comprehensive reaction mechanism for combustion of hydrocarbon fuels, *Combust. Flame*, 99, 201-211, 1994.

192. H.C. Magel, U. Schnell, and K.R.G. Hein, Simulation of detailed chemistry in a turbulent combustor flow, *Twenty-Sixth Symp. (International) on Combustion*, The Combustion Institute, Pittsburgh, PA, 1996, 67-74.

193. J.A. Miller, Theory and modeling in combustion chemistry, *Twenty-Sixth Symp. (International) on Combustion*, The Combustion Institute, Pittsburgh, PA, 1996, 461-480.

194. A.Y. Federov, V.A. Frost, and V.A. Kaminsky, Turbulent transfer modeling in flows with chemical reactions, in *Tranport Phenomena in Combustion*, Vol. 2, S.H. Chan, Ed., Taylor & Francis, Washington, D.C., 1996, 933-944.

195. S.B. Pope, PDF methods for turbulent reactive flows, *Prog. Energy Combust. Sci.*, 11(2), 119-192, 1985.

196. W.P. Jones and M. Kakhi, PDF modeling of finite-rate chemistry effects in turbulent nonpremixed jet flames, *Comb. Flame*, 115, 210-229, 1998.

197. V. Saxena and S.B. Pope, PDF simulations of turbulent combustion incorporating detailed chemistry, *Comb. Flame*, 117, 340-350, 1999.

198. P.J. Coelho and M.G. Carvalho, Modelling of soot formation in turbulent diffusion flames, in *Heat Transfer in Fire and Combustion Systems*, W.W. Yuen and K.S. Ball, Eds., ASME, New York, HTD-272, 29-39, 1994.

199. I.M. Khan and G. Greeves, A method for calculating the formation and combustion of soot in diesel engines, in *Heat Transfer in Flames*, N.H. Afgan and J.M. Beer, Eds., Scripta, 1974, chap. 25.

200. C.D. Stewart, K.J. Syed, and J.B. Moss, Modelling soot formation in non-premixed kerosene-air flames, *Combust. Sci. Tech.*, 75, 211-266, 1991

201. B.F. Magnussen and B.H. Hjertager, On mathematical modelling of turbulent combustion with special emphasis on soot formation and combustion, *Sixteenth Symp. (International) on Combustion*, The Combustion Institute, Pittsburgh, PA, 719-728, 1977.

202. K.B. Lee, M.W. Thring, and J.M. Beer, On the rate of combustion of soot in a laminar soot flame, *Comb. Flame*, 6, 137-145, 1962.

203. J. Nagle and R.F. Strickland-Constable, Oxidation of carbon between 1000-2000C, *Proc. 5th Conf. on Carbon*, Pergamon Press, 1, 154-, 1961.

204. M.A. Delichatsios and L. Orloff, Effects of turbulence on flame radiation from diffusion flames, *Twenty-Second Symp. (International) on Combustion*, The Combustion Institute, Pittsburgh, PA, 1271-1279, 1988.

205. P. Boerstoel, T.H. van der Meer, and C.J. Hoogendoorn, Numerical simulation of soot-formation and -oxidation in high temperature furnaces, in *Tranport Phenomena in Combustion*, Vol. 2, S.H. Chan, Ed., Taylor & Francis, Washington, D.C., 1996, 1025-1036.

206. R. Said, A. Garo, and R. Borghi, Soot formation modeling for turbulent flames, *Combust. Flame*, 108, 71-86, 1997.

207. F. Xu, K.-C. Lin, and G.M. Faeth, Soot formation in laminar premixed methane/oxygen flames at atmospheric pressure, *Combust. Flame*, 115, 195-209, 1998.

208. M. Frenklach and H. Wang, Detailed modeling of soot particle nucleation and growth, *Twenty-Third Symp. (International) on Combustion*, The Combustion Institute, Pittsburgh, PA, 1990, 1559-1566.

209. K.M. Leung and R.P. Lindstedt, Detailed kinetic modeling of C_1C_3 alkane diffusion flames, *Combust. Flame*, 102(1-2), 129-160, 1995.

210. P.J. Roache, *Verification and Validation in Computational Science and Engineering*, Hermosa Publishers, Albuquerque, NM, 1998.

211. Amer. Inst. Aero. and Astronautics (AIAA), Guide for the Verification and Validation of Computational Fluid Dynamics Simulations, AIAA Report G-077-1998, Reston, VA, 1998.

212. DoD Instruction 5000.61: Modeling and Simulation (M&S) Verification, Validation, and Accredization (VV&A), Defense Modeling and Simulation Office, Office of the Director of Defense Research and Engr., 1996.

213. Verification, Validation, and Accreditation (VV&A) Recommended Practices Guide, Defense Modeling and Simulation Office, Office of the Director of Defense Research and Engr., 1996.

214. W.A. Fiveland, K.C. Kaufman, and J.P. Jessee, Validation of an Industrial Flow and Combustion Model, in *Computational Heat Transfer in Combustion Systems*, M.Q. McQuay, W. Schreiber, E. Bigzadeh, K. Annamalai, D. Choudhury, and A. Runchal, Eds., *ASME Proc. 31st Annu. National Heat Transfer Conf.*, Vol. 6, ASME, New York, HTD-328, 147-157, 1996.

215. K.C. Kaufman and W.A. Fiveland, Pilot Scale Data Collection and Burner Model Numerical Code Validation, Topical Report for Gas Research Institute Contract 5093-260-2729, 1995.

3 Unstructured Mesh Methods for Combustion Problems

Jayathi Y. Murthy and Sanjay R. Mathur

CONTENTS

3.1 INTRODUCTION

During the last 2 decades, there have been enormous strides in the application of computational fluid dynamics (CFD) to combustion. Driven by the widespread availability of inexpensive computing power, CFD is today being used to analyze applications as diverse as boilers and furnaces, fluidized beds, gas-turbine combustors, gasifiers, and many others. Increasingly, there is interest in simulating evermore complex geometries and to aggregate many complex subsystems to create very large-scale, system-level simulations. Unstructured CFD methods have played a central role in facilitating this expansion. Unstructured meshes reduce mesh generation time from months to weeks or even days. Coupled with advances such as solution-adaptive meshing, unstructured mesh CFD methods promise to greatly reduce human time in the simulation process, and integrate simulation and analysis in the design cycle.

Nevertheless, simulation of industrial combustion phenomena continues to pose severe challenges to any CFD methodology. Industrial combustion problems involve not only complex geometries, but complex physics as well. Nearly all combusting flows of industrial interest are turbulent, and the interaction of turbulence and chemistry remains an enduring research challenge. Industrial combustion simulations almost always involve multi-mode heat transfer, with participating radiation playing an important role; combustion products such as carbon dioxide and water vapor, as well as particulates such as soot, act to make wavelength-dependent absorption, emission, and scattering important. In specialized applications, such as glass furnaces, radiation in semitransparent media is encountered, with large optical thicknesses in the infrared range of the radiation spectrum. When spray and coal combustion are involved, coupling to a dispersed particulate phase must also be taken into account.

Over the last decade, a variety of unstructured mesh methods have emerged to address industrial combustion. These include finite-element and finite-volume methods, as well as control-volume-based finite-element methods.[1-7] In this chapter, the focus is on finite volume methods, particularly on a class of cell-based methods developed during the last few years.[8-10] Section 3.2 presents the governing equations commonly used for modeling combusting flows, followed by a finite volume numerical procedure for discretizing these governing equations and a methodology for solving them. No attempt is made to address the plethora of physical models being used in combustion simulations today. Rather, the authors' aim is to identify general classes of physics and the corresponding equations to be solved, and to examine the issues encountered in applying unstructured mesh methodologies to them. A particular focus of this chapter is the development of fast unstructured mesh methods for participating radiation. The chapter concludes with a presentation of validation and demonstration problems using the unstructured mesh methodology.

3.2 GOVERNING EQUATIONS

Industrial combustion simulations typically involve the solution of turbulent flows with heat transfer, species transport, and chemical reactions. It is common to use the Reynolds-averaged form of the governing equations in conjunction with a suitable turbulence model. Additional equations, such as for radiative transport, or for specialized combustion models, are also used. The intent here is to identify typical equations and to devise numerical methods to solve them. Typical Reynolds-averaged governing equations for combusting flows are given below.

3.2.1 CONTINUITY EQUATION

The Reynolds-averaged mixture continuity equation for the gas phase is:

$$\frac{\partial}{\partial t}(\rho) + \nabla \cdot (\rho \mathbf{V}) = S_m \tag{3.1}$$

where t is time, ρ is the Reynolds-averaged mixture density, $\mathbf{V}$ is the Reynolds-averaged velocity vector, and S_m represents external mass sources. Typically, these would result from mass transfer interactions from a dispersed phase such as spray droplets or coal particles.

3.2.2 MOMENTUM EQUATION

The Reynolds-averaged gas-phase momentum equation is given by:

$$\frac{\partial}{\partial t}(\rho\mathbf{V}) + \nabla\cdot(\rho\mathbf{V}\mathbf{V}) = \nabla\cdot\left((\mu + \mu_t)\nabla\mathbf{V}\right) + \mathbf{F} \tag{3.2}$$

where μ_t is the turbulent viscosity, obtained from a turbulence model. $\mathbf{F}$ contains those parts of the stress term not shown explicitly as well as other momentum sources, such as drag from the dispersed phase.

3.2.3 ENERGY EQUATION

Heat transfer is governed by the energy conservation equation:

$$\frac{\partial}{\partial t}(\rho E) + \nabla\cdot(\rho\mathbf{V}E) = \nabla\cdot\left((k + k_t)\nabla T\right) + \nabla\cdot(\tau\cdot\mathbf{V}) - \nabla\cdot(p\mathbf{V}) + S_r + S_h \tag{3.3}$$

where k is the thermal conductivity, k_t is the turbulent thermal conductivity resulting from the turbulence model, τ is the stress tensor, p is the pressure, and E is the total energy defined as

$$E = e(T) + \frac{\mathbf{V}\cdot\mathbf{V}}{2} \tag{3.4}$$

and e is the internal energy per unit mass. The terms on the LHS of Equation (3.3) describe the temporal evolution and the convective transfer of total energy. The first three terms on the RHS represent the conductive transfer, viscous dissipation, and pressure work, respectively. S_r is the radiative heat source (see below). In the present form of the energy equation, reaction source terms are included in S_h, which also contains all other volumetric heat sources, including those from any dispersed phase.

3.2.4 RADIATIVE TRANSFER

In the absence of radiation–turbulence interaction, one can write the radiative transfer equation (RTE) for a gray absorbing, emitting, and scattering medium in the direction $\mathbf{s}$ as:[11]

$$\nabla\cdot(I(\mathbf{s})\mathbf{s}) = -(\kappa + \sigma_s)I(\mathbf{s}) + B(\mathbf{s}) \tag{3.5}$$

where

$$B(\mathbf{s}) = \kappa I_B + \frac{\sigma_s}{4\pi}\int_{4\pi} I(\mathbf{s}')\Phi(\mathbf{s}',\mathbf{s})d\Omega' \tag{3.6}$$

Here, $I(\mathbf{s})$ is the radiant intensity in the direction $\mathbf{s}$, κ is the absorption coefficient, σ_s is the scattering coefficient, I_B is the blackbody intensity and Φ is the scattering phase function. The radiative source term S_r in the energy equation [Equation (3.3)] is given by:

$$S_r = \kappa \int_{4\pi} \left[I(s) - I_B \right] d\Omega \tag{3.7}$$

A variety of methods for solving the RTE have been developed in the literature, including the zonal and Monte Carlo methods,[11] ray tracing methods,[12] the class of P-N approximations,[13] and the discrete ordinates[14] and finite-volume methods.[15] Although unstructured mesh analogues for many of these methods have begun to appear in the literature, this chapter focuses on the extension of the unstructured mesh methodology to the finite-volume method for participating radiation.[10] This methodology allows the solution of mixed-mode radiation problems over a wide range of optical thicknesses. For problems with relatively large optical thicknesses, as for many combustion problems, the P_1 approximation[11] has been found very useful. It involves the solution of the following transport equation for the radiation intensity G:

$$\nabla \cdot (\Lambda \nabla G) - \kappa (G - I_B) = 0 \tag{3.8}$$

where

$$\Lambda = \frac{1}{3(\kappa + \sigma_s) - C\sigma_s} \tag{3.9}$$

and C is the linear anisotropic phase function coefficient. The radiative source term S_r in the energy equation [Equation (3.3)] is then given by:

$$S_r = \kappa (G - I_B) \tag{3.10}$$

3.2.5 TURBULENCE MODELING

Although a variety of turbulence models exist in the literature, for the purposes of developing the numerical technique, the focus here is on the k-ε model,[16] which involves solution of the following transport equations for the turbulent kinetic energy k and its dissipation ε.

$$\frac{\partial}{\partial t}(\rho k) + \nabla \cdot (\rho \mathbf{V} k) = \nabla \cdot \left(\frac{(\mu + \mu_t)}{\sigma_k} \nabla k \right) + G_k - \rho \varepsilon \tag{3.11}$$

$$\frac{\partial}{\partial t}(\rho \varepsilon) + \nabla \cdot (\rho \mathbf{V} \varepsilon) = \nabla \cdot \left(\frac{(\mu + \mu_t)}{\sigma_\varepsilon} \nabla \varepsilon \right) + C_{1\varepsilon} \frac{\varepsilon}{k} G_k - C_{2\varepsilon} \rho \frac{\varepsilon^2}{k} \tag{3.12}$$

where G_k is the turbulence production term, σ_k and σ_ε are the turbulence Prandtl numbers, and $C_{1\varepsilon}$ and $C_{2\varepsilon}$ are model constants.

3.2.6 SPECIES TRANSPORT

Under the dilute approximation, the Reynolds-averaged conservation equations for the mass fraction, m_l, of species l can be written as:

$$\frac{\partial}{\partial t}(\rho m_l) + \nabla \cdot (\rho \mathbf{V} m_l) = \nabla \cdot \left(\left(\rho D + \frac{\mu_t}{\sigma_m} \right) \nabla m_l \right) + R_l \tag{3.13}$$

where D is the diffusion coefficient of species l in the mixture, and σ_m is the turbulent Schmidt number. For turbulent reacting flows, the eddy dissipation model of Magnussen and Hjertager[17] is sometimes used. This involves the solution of species transport equations of the type shown above, with a reaction rate term R_l, which is computed as the minimum of the Arrhenius rate, and that due to turbulent mixing.[17]

3.2.7 MIXTURE FRACTION/VARIANCE

For turbulent diffusion flames, the mixture fraction/probability-density-function (PDF) approach is frequently used. Because the kinetic reaction rates greatly exceed the turbulent mixing rate, the instantaneous thermochemical state of the fluid (i.e., the composition, density, temperature, etc.) can be computed from the instantaneous values of the mixture fraction and total enthalpy by assuming chemical equilibrium.[18] Time-averaged values are then obtained by weighting the instantaneous values with a probability-density function (PDF), based on the mean mixture fraction and its variance. The shape of the PDF is assumed; typically, the double-delta or the β functions are employed.[19] Nonequilibrium effects can be incorporated within this framework by using laminar flamelet models.[20] These models include the effect of strain rate on chemistry and are typically based on calculations/experiments of canonical flame structures, such as a counterflow diffusion flame.[21] To speed up computations, the equilibrium calculations and the PDF integrations are performed in a preprocessing step. The results are stored in lookup tables of time-averaged density, temperature, and species mole fractions as functions of the mean and variance of mixture fraction and the mean total enthalpy (for nonadiabatic cases) or a scalar dissipation rate parameter (for flamelet models).

The following conservation equations are solved for the time-averaged value of mixture fraction, $\bar{f}$, and its variance, $\overline{f'^2}$ (see Reference 19):

$$\frac{\partial}{\partial t}\left(\rho\bar{f}\right)+\nabla\cdot\left(\rho\mathbf{V}\bar{f}\right)=\nabla\cdot\left(\frac{\mu_t}{\sigma_h}\nabla\bar{f}\right) \tag{3.14}$$

$$\frac{\partial}{\partial t}\left(\rho\overline{f'^2}\right)+\nabla\cdot\left(\rho\mathbf{V}\overline{f'^2}\right)=\nabla\cdot\left(\frac{\mu_t}{\sigma_h}\nabla\overline{f'^2}\right)+C_g\mu_t\left(\nabla\bar{f}\cdot\nabla\bar{f}\right)-C_d\rho\frac{\varepsilon}{k}\overline{f'^2} \tag{3.15}$$

where C_g and C_d are appropriate model constants. The thermochemical state at each cell can be interpolated from the precomputed tables using the local values of $\bar{f}$, $\overline{f'^2}$, and enthalpy.

3.2.8 DISPERSED PHASE MODELS

Many combustion problems, such as those involving coal particles or fuel sprays, can be efficiently modeled using a mixed Eulerian-Lagrangian approach rather than a full Eulerian approach. Such methods are appropriate for dilute cases when the total volume of the particulate phase is small compared to that of the primary continuous phase. The secondary phase is convected with the same velocity field as the primary phase and interaction between particles is usually ignored.

Lagrangian modeling is suitable when the entrance and exits of the discrete phase are well-defined. Each entrance point can be characterized by a single particle (or a family of particles) whose trajectory through the computational domain is tracked until it exits. The mass, momentum, energy, composition, and other properties of the particle are computed along the trajectory using the Lagrangian form of the conservation laws. For example, the momentum balance equation is given by

$$\frac{d\mathbf{V}_p}{dt}=F_D\left(\mathbf{V}_c-\mathbf{V}_p\right)+\mathbf{G}\left(\rho_p-\rho_c\right)+\mathbf{F} \tag{3.16}$$

where $\mathbf{V}_p$ and ρ_p are the particle velocity and density, respectively, and $\mathbf{V}_c$ and ρ_c and the values of the velocity and density in the continuous phase. The latter are obtained from the mesh cell in which the particle is currently located. The term $F_D(\mathbf{V}_c - \mathbf{V}_p)$ represents the drag force per unit mass of the particle and is computed using empirical relations. The term $\mathbf{G}(\rho_p - \rho_c)$ denotes the buoyancy force while $\mathbf{F}$ represents all other forces acting on the particle, such as "virtual mass" force, Coriolis forces, thermophoretic forces, etc.

For combustion applications, it is also necessary to track the mass, energy, and composition of the particles. There are models for computing the energy transfer between the particle and the gas phase, including radiative exchange, and for vaporization and boiling of droplets.[18,22,23] In case of coal particles, devolatilization[24,25] and surface combustion[26,27] are also modeled during trajectory calculations. These exchange terms appear as source terms in the corresponding continuous-phase conservation equations.

3.3   NUMERICAL METHODS

3.3.1   DISCRETE SOLUTION OF GOVERNING EQUATIONS

In numerical simulations, one tries to obtain the values of independent variables such as pressure, velocity, temperature, and composition at a finite number of locations in the domain rather than seeking closed-form solutions of the governing equations. To this end, one *discretizes* the domain, that is, one divides it into smaller volumes known as cells. In the finite-volume method, the conservation laws are applied to each of these individual cells to obtain a set of simultaneous equations that relate their associated variables. Because these equations are typically nonlinear and mutually coupled, strategies must be devised for their efficient solution. Two basic types of solution techniques are possible. In explicit methods, the values of the variables are updated by visiting each cell in turn and the entire process repeated until the values stop changing. In implicit schemes, the nonlinear terms are linearized to obtain a set of algebraic linear equations, which are then solved simultaneously. Iterations are still necessary because of the nonlinearities in the governing equations, but implicit methods usually require fewer iterations than explicit methods.

3.3.2   STRUCTURED VERSUS UNSTRUCTURED DISCRETIZATION

For relatively simple geometries, it is usually convenient to discretize the domain using a logically rectangular grid, as shown in Figure 3.1(a). Such a *structured* grid arrangement makes it much easier to discretize the terms in the governing equations. Values at neighboring cells can be obtained by simple indexing, and the linear systems that result have regular banded structures that enable quick solution.

For complex geometries, however, it is not easy to discretize the domain using structured grids. In some cases it might be possible to divide the domain into several rectangular-like regions and thus use a multi-block structured grid arrangement, as in Figure 3.1(b). However, such a process is a very time-consuming and not amenable to easy automation. It is therefore highly desirable to permit division of the domain into an arbitrary arrangement of cells, as in Figure 3.1(c). Such an *unstructured* grid usually consists of tetrahedral cells, although hexahedral, prismatic, and pyramid cells are also possible. Along with allowing for a high degree of automation in the mesh generation process, unstructured grids also enable the use of solution adaption (i.e., the optimal distribution of cells depending on the solution characteristics).

This increased mesh flexibility comes at the cost of increased algorithmic and programming complexity. Discretization methods must now account for factors such as mesh non-orthogonality and arbitrary connectivity. The linear systems also have arbitrary sparse patterns requiring the use of more sophisticated solution techniques. These aspects are discussed in Section 3.4.

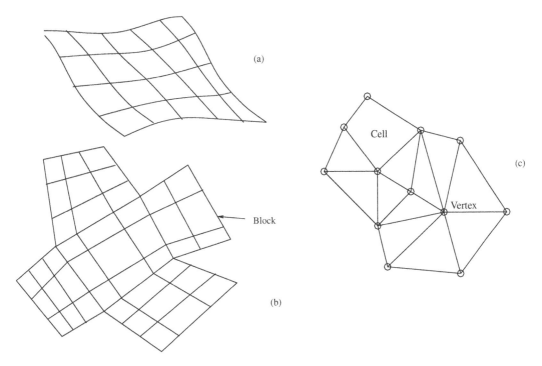

FIGURE 3.1   Mesh connectivity: (a) structured, (b) block-structured, and (c) unstructured meshes.

3.4   DISCRETIZATION OF SCALAR TRANSPORT EQUATION

The governing equations presented in previous sections can be cast into the form of a generic scalar transport equation:

$$\frac{\partial}{\partial t}(\rho\phi) + \nabla\cdot(\rho\mathbf{V}\phi) = \nabla\cdot(\Gamma\nabla\phi) + S_\phi \tag{3.17}$$

where ϕ is the transport variable, Γ is the diffusion coefficient, and S_ϕ is the source term. The equations governing the transport of mass, momentum, and energy each have the above form, and represent an appropriate choice of ϕ, Γ and S_ϕ.

The domain is discretized into arbitrary unstructured convex polyhedra called cells. The boundaries surrounding the cells are called faces, and the vertices of the polyhedra are referred to as nodes. All transport variables are stored at cell centroids. In addition to cell centroids, transport variables are also stored at the centroids of all boundary faces to aid in the imposition of nonlinear boundary conditions.

Integration and discretization of Equation (3.17) about the control volume C0 in Figure 3.2 yields:

$$\left[\frac{\partial}{\partial t}(\rho\phi)\right]_0 \Delta V_0 + \sum_f J_f \phi_f = \sum_f D_f + \left(S_\phi \Delta V\right)_0 \tag{3.18}$$

where J_f is the mass flow rate, ϕ_f is the face value of ϕ, ΔV_0 is the volume of the cell C0, D_f is the transport due to diffusion through the face f, and the summations are over the faces of the control volume. To obtain a set of linear equations, the convective and diffusive fluxes as well as volume integrals in Equation (3.18) must be written in terms of the unknowns, that is, values of ϕ at cell and boundary face centroids, as discussed next.

3.4.1 DIFFUSION TERM

The diffusion term at a face is given by:

$$D_f = \Gamma_f \nabla \phi \cdot \mathbf{A} \tag{3.19}$$

It can be written in terms of discrete values of ϕ at the two cell centers and the two vertices of the face as:[8]

$$\nabla \phi \cdot \mathbf{A} = \frac{\phi_1 - \phi_0}{ds} \frac{\mathbf{A} \cdot \mathbf{A}}{\mathbf{A} \cdot \hat{\mathbf{e}}_s} + \frac{\phi_e - \phi_a}{|\mathbf{A}|} \frac{\mathbf{A} \cdot \mathbf{A}}{\mathbf{A} \cdot \hat{\mathbf{e}}_s} \hat{\mathbf{e}}_t \cdot \hat{\mathbf{e}}_s \tag{3.20}$$

Here, $\hat{\mathbf{e}}_s$ is the unit vector in the direction joining the two cell centroids and $\hat{\mathbf{e}}_t$ is parallel to the face. This expression does not make any assumptions about mesh structure. The diffusion term has been split into two parts: a primary component representing the diffusion component in the direction of the line joining the two cell centroids and a secondary component that arises because of grid nonorthogonality. The primary component is written purely in terms of the difference of two cell values on either side of the face f, (i.e., ϕ_0 and ϕ_1). It has the same form as that obtained with a central difference scheme on a structured mesh and is treated implicitly in the discrete equations for the two cells.

In Equation (3.20), the secondary component is expressed in terms of ϕ_a and ϕ_e, the values at face vertices. In three dimensions, the secondary term would involve gradients along two directions perpendicular to the face normal. This representation is convenient for structured meshes. For unstructured meshes in three dimensions, however, the face f is a polyhedron with an arbitrary number of vertices, and these directions cannot be uniquely identified. One therefore writes the secondary component as the difference between the total diffusion term and the primary component. Thus,

$$D_f = \Gamma_f \frac{\left(\phi_1 - \phi_0\right)}{ds} \frac{\mathbf{A} \cdot \mathbf{A}}{\mathbf{A} \cdot \hat{\mathbf{e}}_s} + \Gamma_f \left(\overline{\nabla \phi} \cdot \mathbf{A} - \overline{\nabla \phi} \cdot \hat{\mathbf{e}}_s \frac{\mathbf{A} \cdot \mathbf{A}}{\mathbf{A} \cdot \hat{\mathbf{e}}_s} \right) \tag{3.21}$$

where $\overline{\nabla \phi}$ at the face is taken to be the average of the derivatives at the two adjacent cells. Methods for determining gradients of ϕ at cell centers are discussed in Section 3.4.4.

3.4.2 CONVECTION TERM

For the purposes of discretizing the scalar transport equation, the mass flow rate J_f is assumed to be known from the solution of the flow equations. As a first-order approximation, the convective flux at the face f can then be evaluated using the value of ϕ in the upwind cell:

$$J_f \phi_f = J_f \phi_{upwind} \tag{3.22}$$

Higher order schemes on structured grids typically interpolate ϕ_f along the appropriate grid line using a larger stencil. Of course, on unstructured grids this is not possible, and the face flux must be obtained using multi-dimensional *reconstruction*. Thus, for a second-order approximation, assuming a constant gradient $\nabla\phi_r$ in the cell, one can write ϕ in the upwind cell:

$$\phi_f = \phi_{upwind} + \nabla\phi_{r_{upwind}} \cdot \vec{dr} \qquad (3.23)$$

3.4.3 UNSTEADY TERM

The unsteady term is discretized using backward differences. The first-order scheme is:

$$\left[\frac{\partial}{\partial t}(\rho\phi)\right]_0 = \frac{(\rho\phi)_0^{n+1} - (\rho\phi)_0^n}{\Delta t} \qquad (3.24)$$

Higher order representations of the unsteady term can be written using more levels of storage.

3.4.4 GRADIENT CALCULATION

Gradients of the scalar variable ϕ are required to compute the secondary diffusion terms and for the linear reconstruction of face values of ϕ in the convection terms. The gradient is required at cell centroids. For structured grids, simple finite difference formulae can be used; for unstructured meshes, however, there is no straightforward extension of this practice. Node-based formulations such as finite element schemes and CVFEM[2] find gradients by using shape functions. Another approach suitable for both node and cell-based formulations is to use least-squares interpolation.[28] However, many of these techniques are cell-shape specific. A more general approach is to use the discrete form of the divergence theorem. One can define the *reconstruction gradient* as:

$$\nabla\phi_r = \frac{\alpha}{\Delta v} \sum_f \left(\bar{\phi}_f A\right) \qquad (3.25)$$

where the summation is over all the faces of the cell. The face value of ϕ is obtained by averaging the values at the neighboring cells, so that for the face f in Figure 3.2

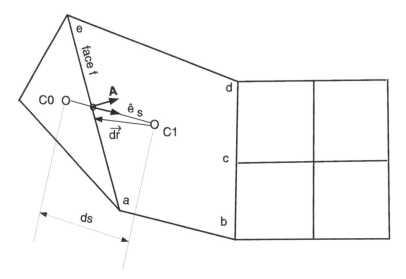

FIGURE 3.2 Control volume.

$$\overline{\phi}_f = \frac{\phi_0 + \phi_1}{2} \tag{3.26}$$

α is a factor used to ensure that the reconstruction does not introduce local extrema.[8] It is easy to see that this procedure can be used for arbitrary control volume shapes, including non-conformal grids.

Using the reconstruction gradient, the value at any face of the cell can be reconstructed as:

$$\phi_{f,c} = \phi_c + \nabla\phi_r \cdot \vec{dr} \tag{3.27}$$

The cell derivatives used for the secondary diffusion terms in Equation (3.21) are computed by again applying the divergence theorem over the control volume and using the averaged reconstructed values at the faces:

$$\nabla\phi = \frac{1}{\Delta v} \sum_f \left(\tilde{\phi}_f \mathbf{A} \right) \tag{3.28}$$

where for face f in Figure 3.2, $\tilde{\phi}_f$ is given by:

$$\tilde{\phi}_f = \frac{\phi_{f,0} + \phi_{f,1}}{2} \tag{3.29}$$

3.4.5 FINAL FORM OF DISCRETE EQUATIONS

Combining the discretized diffusion, convection, source, and unsteady terms yields linear equations of the following form at each cell i:

$$M_{ij}\phi_j + S_i = 0 \tag{3.30}$$

The matrix M is a sparse matrix because the only non-zero entries in each row are those corresponding to the neighboring cells. Thus, for a hexahedral grid of N cells, approximately $7N$ words are required for the storage of M. This system of equations is solved using an algebraic multigrid scheme (Section 3.5) to obtain the values of ϕ at all cells. In the case of nonlinearities (due to either secondary diffusion terms, second-order interpolation, or nonlinear sources), the coefficients and source terms are updated using the current values of ϕ, and the entire process is repeated until convergence. In the case of transient problems, the converged solution represents the values at the next step, and the iterations are repeated for the desired duration.

3.4.6 DISCUSSION

The scheme presented here represents the logical extension of conventional Cartesian or body-fitted structured grid cell-based schemes to unstructured meshes. Indeed, on Cartesian meshes, the discrete equations obtained are identical to those in Reference 29; similarly, for orthogonal body-fitted meshes, the decomposition of the diffusion terms shown here results in the same discrete equations used in standard structured mesh schemes.[30] The main difference lies in the evaluation of gradients; secondary diffusion terms and second-order upwind terms, which use cell gradients, result in a different S_i term in Equation (3.30) from those in References 29 and 30. Like other finite-volume methods, the scheme is *conservative* in that the net fluxes always balance. Unlike some finite-difference or finite-element based methods, this conservative property is retained even on coarse meshes.

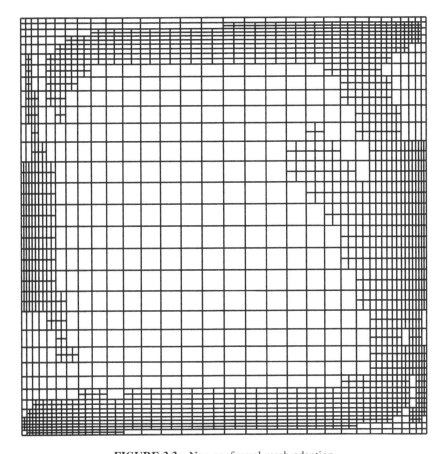

FIGURE 3.3 Non-conformal mesh adaption.

Because the scheme can accommodate arbitrary convex polyhedral cells as well as arbitrary mesh connectivity, solution adaption, both conformal and non-conformal, can be easily incorporated. In case of conformal adaption, the region marked for adaption is refined by splitting the longest edges. Additional non-marked cells may also need refinement, so as to not create "hanging" nodes, such as node *c* in Figure 3.2. With non-conformal adaption, each marked cell is isotropically subdivided, and "hanging" nodes are allowed. Non-conformal adaption facilitates highly local adaption and preserves mesh quality for cell shapes such as quadrilaterals, hexahedra, and triangles. Because the discretization supports arbitrary polyhedra, cells such as *C*1 in Figure 3.2, created by non-conformal adaption, can be handled easily. Figure 3.3 shows a quadrilateral mesh used to compute a mixed-mode heat transfer problem involving buoyancy and participating radiation.[10] The mesh was adapted to velocity magnitude using non-conformal adaption, and reflects underlying velocity field.

For unstructured meshes, the matrix M in Equation (3.30) is sparse but does not have a specific band structure. Consequently, familiar line-iterative solvers cannot be used, and more general methods are required. Point-iterative relaxation methods, such as Gauss-Seidel, used in conjunction with multigrid cycling, are usually the most effective approaches for solution of the linear systems resulting from the discretization method presented here.

3.5 MULTIGRID METHODS FOR UNSTRUCTURED GRIDS

It is a common observation that convergence rates for numerical methods tend to decrease as the problem size (i.e., the number of cells) increases. For practical industrial problems, where several components are modeled together resulting in meshes on the order of millions of cells, this becomes a limiting factor. The multigrid method is a powerful tool that helps overcome these limitations and enables efficient solution of large-scale problems.

Nonlinear problems are usually solved in an iterative manner. Iterative schemes are also used for solution of large linear systems like Equation (3.30). Starting with an initial guess, a fixed sequence of steps (called the "relaxation sweep") is applied in order to obtain a better guess, and the entire process is repeated until convergence (i.e., when the solution stops changing appreciably). If one decomposes the error (i.e., the difference between the current and the final solution) into Fourier modes, in general the starting error is composed of all modes. The fundamental reason for the decrease in convergence rate with increasing grid size is that most iterative relaxation methods are only effective in removing the higher frequency components of the error. After a few iterations, all the higher frequency errors are removed and only the low frequency errors remain, thus stalling convergence.

Multigrid methods work by transferring the problem to a coarser grid where these errors appear as comparatively higher frequency errors and thus can be removed effectively by using the same relaxation methods. This process can be continued recursively on increasingly coarser grids. Typically, the problem at all coarse levels is formulated in terms of corrections. The solution obtained on any coarse grid level is used only to correct the current guess at the next finer level; therefore, the final converged solution at the finest level depends only on the discretization at the finest level. In other words, the use of the multigrid algorithm only serves to accelerate convergence but has no effect on the final answer.

A variety of cycling strategies can be devised to sequence the relaxation sweeps at the different grid levels. Some of the most commonly used ones are shown in Figure 3.4.

The general idea of multigrid acceleration can be applied at many stages of the numerical algorithm. One common use is for the solution of linear systems of the type in Equation (3.30). In such *algebraic* multigrid methods, the coarse-level problems are formulated purely from the coefficients of the fine level (i.e., the matrix M), without reference to either the geometry or the governing equations at the coarse level. Relatively simple relaxation sweeps, such as Gauss-Seidel iteration, are usually sufficient at each of the levels. These methods can be used for the sparse linear system resulting from a single PDE or for a block-sparse system resulting from the simultaneous discretization of several PDEs. The latter approach usually results in better convergence for highly coupled systems but it does have significantly higher memory requirements.

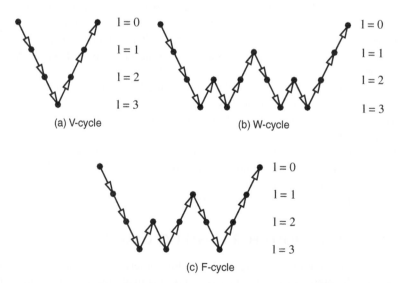

(a) V-cycle   (b) W-cycle

(c) F-cycle

FIGURE 3.4   Relaxation and grid transfer sequence for some fixed cycles.

In *geometric* multigrid methods, on the other hand, the discretization process is applied at coarse grids level as well. These methods are usually applied to a group of equations. Because they require creation of all the geometry information at the coarse levels, the implementation is more complex than that for algebraic multigrid methods. Both nonlinearity and inter-equation coupling are handled at the coarse levels. Geometric multigrid methods, therefore, have the potential of providing the best convergence acceleration.

3.5.1 MULTIGRID COARSENING

Creation of coarse levels is a common requirement for all multigrid methods. On structured meshes, coarse-level grids are usually created by combining cells along all three directions; the coarse-level grid consequently has the same topological structure as the fine-level grid and all the cells are still rectangular. Therefore, fine-level solution procedures can be easily applied at the coarse level. This property can be achieved with unstructured meshes by employing a sequence of increasingly coarse, independently created meshes. However, this greatly complicates the inter-grid transfer operations because general multi-dimensional interpolations must be used. To keep such operations simple, coarse grids are created by agglomeration, that is, each coarse-level cell is obtained by combining two or more cells from the previous fine level. For algebraic multigrid methods, the best agglomeration strategy is to group every cell with those of its n neighboring ungrouped cells for which the coefficient M_{ij} is the highest. Combining the largest coefficients allows any boundary information associated with the weaker coefficients of a stiff matrix to propagate faster and results in convergence acceleration.

An example of such a coarsening is shown in Figure 3.5 where a low-conductivity outer region, with temperatures specified at the boundaries, surrounds a highly conducting inner square domain. The ratio of conductivities is 1000; a ratio of this order would occur for a copper block in air. Coefficients resulting from the diffusion term are $O(kA/\Delta x)$, where A is the face area and Δx is a cell length scale. For interface cells in the highly conducting region, coefficients to interior cells are approximately three orders of magnitude bigger than coefficients to cells in the low-conducting

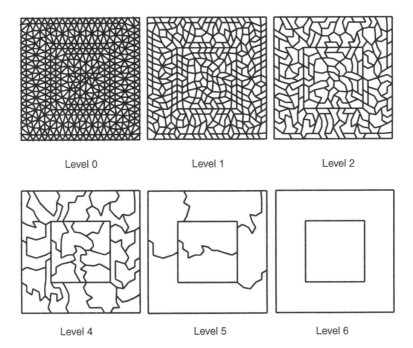

Level 0   Level 1   Level 2

Level 4   Level 5   Level 6

FIGURE 3.5 Nested multigrid coarsening.

region. However, Dirichlet boundary conditions, which set the level of the temperature field, are only available at the outer boundaries of the domain, adjacent to the low-conducting region. Information transfer from the outer boundary to the interior region is inhibited because the large-coefficient terms overwhelm the boundary information transferred through the small-coefficient terms. The coefficient-based agglomeration strategy results in the coarse levels shown in Figure 3.5. At the coarsest level, the domain consists of a single cell in the high-conducting region, and another in the low-conducting region. The associated coefficient matrix has coefficients of the same order. The mean temperature of the inner region is primarily set by the multigrid corrections at this level and results in very fast convergence. Also note that nested coarsening results in arbitrary cell shapes, not necessarily convex — and might require specialized discretization procedures if used in the context of a geometric multigrid scheme.

3.6   SOLUTION METHODOLOGY FOR COMBUSTION PROBLEMS

The discretization procedure outlined above can, in principle, be applied to the governing equations that are of interest in combustion applications. However, combustion simulations are special in a number of respects. First, the number of equations to be solved is frequently very large. This occurs, for example, when species mass fractions are transported, as in eddy dissipation models. This is also true when using radiation models such as the discrete ordinates or finite-volume methods; here, even a modest resolution of the angular space leads to a doubling of the number of unknowns. Combustion simulations are strongly nonlinear due to the dependence of properties on temperature. When mixture-fraction PDF approaches are used, the nonlinearity may be difficult to address explicitly because it is hidden in equilibrium computations and a convolution with a probability-density function. Furthermore, some of these operations are optimized through the use of table lookups, making nonlinear approaches such as Newton-Raphson iteration difficult to implement. Combustion problems involving dispersed-phase calculations typically employ a loose iteration between a Lagrangian particle tracking simulation and an Eulerian field calculation; a more implicit, tightly coupled methodology is difficult to implement in this context. Similar issues arise when using ray-tracing schemes such as the discrete transfer model[12] or the Monte Carlo method for participating radiation. Consequently, the method of choice has been a sequential technique, solving for each governing equation one after another, using Picard iteration for nonlinearities.

3.6.1   PRESSURE-VELOCITY COUPLING

One major difficulty in using a sequential procedure for the solution of the incompressible continuity and momentum equations is that pressure does not explicitly appear in the former. The SIMPLE algorithm[29] is among the most widely used approaches for overcoming this difficulty. It involves the sequential solution of the momentum equations for velocity components, using a guessed value of the pressure field and face mass fluxes. The mass fluxes are then updated using the newly obtained velocity components. Because these fluxes will not, in general, satisfy the continuity equations, a correction to these fluxes is proposed. Using the discrete momentum equations, this flux correction is then expressed in terms of a pressure correction field and an algebraic equation set for the pressure correction in all cells is formed by enforcing mass flux balance in each cell. This algorithm can be easily extended to unstructured meshes. The discrete pressure correction equation has the same form as Equation (3.30) and can be solved using the techniques discussed in Section 3.5.

Another difficulty that arises in solving incompressible flow problems is that co-located storage (i.e., pressure and velocity components being stored at the same locations) can lead to "checker-boarding." This comes about because, in the finite volume scheme, pressure is required at the faces of the cell, whereas they are stored at cell centers. Straightforward averaging decouples the variables in adjacent cells and thus permit oscillatory patterns. On Cartesian grids, this problem can be remedied by a "staggered" storage strategy, whereby pressure and velocity components are stored at different locations.[29] On body-fitted grids, however, such arrangements are difficult to implement; several alternate strategies have been proposed. The most common approach, due to Rhie and Chow,[31] involves the addition of a pressure dissipation term, obtained as the difference between the pressure derivatives averaged from the two cells and one evaluated at the face, in the face mass flux expression. This procedure can also be easily extended to unstructured meshes; details may be found in Reference 8.

3.6.2 ITERATIVE SCHEME FOR ENERGY EQUATION

For combustion problems, the energy equation [Equation (3.3)] is complicated by the presence of reaction terms, and by variable properties, especially C_p. Application of the scalar transport discretization procedure for the energy equation is not straightforward because the convective transfer is described in terms of total energy E while diffusion is governed by gradients of temperature T. Therefore it is usually cast into the temperature form:

$$\frac{\partial}{\partial t}\left(\rho C_p T\right) + \nabla \cdot \left(\rho \mathbf{V} C_p T\right) = \nabla \cdot \left(k \nabla T\right) + \tau : \mathbf{V} + \rho T \left(\frac{\partial C_p}{\partial t} + \mathbf{V} \cdot \nabla C_p\right)$$

$$+ S_r + S_h + \left(\frac{\partial \ln(1/\rho)}{\partial \ln T}\right)_p \left(\frac{\partial p}{\partial t} + \mathbf{V} \cdot \nabla p\right)$$

(3.31)

When C_p is not constant, the temperature form of the energy equation is not conservative; the non-conservative form in Equation (3.31) would destroy the attractive conservative property of the finite volume scheme. One alternative is to write the diffusion terms in terms of energy — but this is not desirable for conjugate heat transfer applications where step changes in C_p make energy double-valued at fluid–solid boundaries. Also, boundary conditions are most naturally written in terms of temperature; conversion to the energy form of complex boundary conditions, such as for radiation or surface reaction, is not convenient.

To retain temperature as the dependent variable while still ensuring that the discrete analogue of Equation (3.3) is satisfied at convergence, one can rewrite Equation (3.3) as:

$$\frac{\partial}{\partial t}\left(\rho C_p T\right) + \nabla \cdot \left(\rho \mathbf{V} C_p T\right) = \nabla \cdot \left(k \nabla T\right) + \nabla \cdot \left(\tau \cdot \mathbf{V}\right) - \nabla \cdot \left(p \mathbf{V}\right) + S_r + S_h$$

$$+ \frac{\partial}{\partial t}\left(\rho \left[C_p T^* - E^*\right]\right) + \nabla \cdot \left(\rho \mathbf{V}\left[C_p T^* - E^*\right]\right)$$

(3.32)

where T^* and E^* are the values of T and E at the current iteration. In the discretization process, one can treat the terms in T implicitly, while absorbing the T^* and E^* terms explicitly. It is easy to see that the procedure described above for the scalar transport equation can now be applied. The coefficients of the algebraic set — as well as the nonlinear source terms — are updated iteratively using prevailing values. At convergence, $T = T^*$; the unsteady and convective terms in T and T^* in Equation (3.32) cancel, and Equation (3.3) is satisfied identically. This type of iterative procedure has been shown to work well for combustion and other problems (see Reference 32).

Although temperature is used as the dependent variable, this approach ensures energy conservation regardless of the mesh size. Because temperature is continuous at fluid-solid boundaries, no special treatment is required for the diffusion terms. Another advantage of this approach over formulations using energy as the dependent variable is that in the latter, temperature must be recovered from energy. For temperature-dependent C_p, this requires finding the root of a nonlinear equation for each computational cell in the domain, and can be quite expensive. In the present approach, T is computed directly from Equation (3.32); an evaluation of E is required using Equation (3.4), but no root-finding.

For combustion calculations using the mixture-fraction/PDF approach, adiabatic combustion is sometimes assumed; but more frequently, it is necessary to solve for the total enthalpy, which is required for table lookups of temperature and composition. Because turbulent diffusion domi-nates, an assumption of unity Lewis number is usually appropriate. This simplifies the species diffusion and conduction terms of the energy equation so that it can be written in terms of the total enthalpy H as:

$$\frac{\partial}{\partial t}(\rho E) + \nabla \cdot (\rho \mathbf{V} H) = \nabla \cdot \left(\frac{k_t}{C_p} \nabla H \right) + \nabla \cdot (\rho \cdot \mathbf{V}) - \nabla \cdot (p \mathbf{V}) + S_r + S_h \qquad (3.33)$$

Note that there are no source terms due to reactions, because the conservation statement is for the total enthalpy. Now the energy equation has the same form as the scalar transport equations with $\phi = H$ and it therefore is convenient to solve directly for H. However, conjugate interfaces and boundary conditions do require special handling.

3.6.3 PARTICIPATING RADIATION

Participating radiation plays an important role in most combustion calculations. In many situations, the optical thickness involved is relatively large, and lower-order approximations, such as the P_1 model, can be used. The equations governing the class of P-N models can be cast in the form of diffusion equations, and their solution follows the basic structure described in previous sections. As computational power has become available, however, there has been interest in resolving the radiant field with greater fidelity. This is particularly true when large-scale simulations are done, which not only involve a flame calculation, but other peripheral equipment as well. Hence, one encounters a range of optical thicknesses that must be resolved. Recently, Raithby and Chui[15,33] developed a CVFEM framework for radiative heat transfer which is suitable for this type of problem. A similar approach was taken by Chai and co-workers,[34,35] who applied it to a cell-centered formulation on a structured mesh. Murthy and Mathur[10] developed an implementation for unstruc-tured meshes using the framework outlined here. Extensions for axisymmetric and periodic geom-etries,[36,37] as well as for semitransparent media with Fresnel boundaries[38] have been published. The basic methodology is presented below; details can be found in Reference 10.

3.6.3.1 Discretization

The spatial discretization is done as in Section 3.4. In addition, the angular space 4π at any spatial location is discretized into discrete non-overlapping control angles ω_i, the centroids of which are denoted by the direction vector $\mathbf{s}_i$ and the polar and azimuthal angles θ_i and ϕ_i. Each octant is discretized into $N_\theta \times N_\phi$ solid angles. The angles θ and ϕ are measured with respect to the global Cartesian system (x, y, z); θ is measured from the z-axis, and ϕ is measured from the y-axis. The angular discretization is uniform; the control angle extents are given by $\Delta\theta$ and $\Delta\phi$. For each discrete direction i, Equation (3.5) is integrated over the control volume $C0$ in Figure 3.6 and the solid angle ω_i to yield:

$$\sum_f J_f I_{if} |\mathbf{A}| = \left[-(\kappa + \sigma_s) I_{i0} + B_i \right] \omega_i \Delta V_0 \qquad (3.34)$$

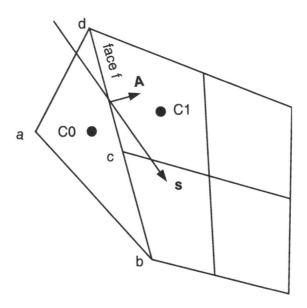

FIGURE 3.6   Control volume for radiant energy conservation in direction **s**.

where I_{if} is the intensity associated with the direction i at the face f of the control volume, and I_{i0} is the intensity at the cell $C0$ in the direction i. J_f is a geometric factor defined below. The solid angle ω_i is given by:

$$\omega_i = \int_{\Delta\phi}\int_{\Delta\theta} \sin\theta \, d\theta \, d\phi \qquad (3.35)$$

The integration may be performed exactly so that $\sum_i \omega_i = 4\pi$ is identically satisfied. The source term B_i is given by:

$$B_i = \kappa I_{B0} + \frac{\sigma_s}{4\pi}\sum_j I_{j0}\,\gamma_{ij} \qquad (3.36)$$

where

$$\gamma_{ij} = \frac{1}{\omega_i}\int \Phi\left(\mathbf{s}_{i'}, \mathbf{s}_{j'}\right) d\omega_{j'}\, d\omega_{i'} \qquad (3.37)$$

The blackbody intensity I_{B0} is based on the temperature of the cell $C0$, and I_{j0} is the cell intensity in the directions j.

3.6.3.2   Control Angle Overhang

One sees that for each direction **s**, Equation (3.5) has the form of a convection equation with a source term. It would seem that the discretization would therefore be a straightforward implementation of the numerical scheme outlined in previous sections. For example, it should be possible to "upwind" I_{if} using the schemes outlined in Section 3.4.2 for the convection term. However, for unstructured and body-fitted meshes, the computation of I_{if} and J_f is complicated by the possibility

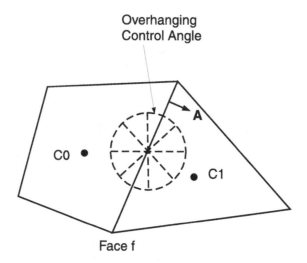

FIGURE 3.7 Control angle overhang in 2-D.

of control angle overhang, a problem specific to radiation. The situation in 2-D is shown in
Figure 3.7. Because the directions s_i are defined with respect to a global coordinate system (x, y, z)
the boundaries of the discrete solid angles ω_i do not necessarily align with control volume faces
and a portion of the control angle "overhangs" the face as shown. In 3-D, the intersection of the
control volume face with the angular discretization is more complex, involving the intersection of
an arbitrary great circle with the control angle. The energy associated with the overhanging control
angle must be taken into account.

For directions with no overhang on the face f, one writes

$$J_f = \hat{\mathbf{e}}_n \cdot \int_{\Delta\theta}\int_{\Delta\phi} \mathbf{s} \sin\theta\, d\theta\, d\phi \tag{3.38}$$

Using a standard "step" approximation for I_{if}[11]

$$I_{if} = I_{i,\text{upwind}} \tag{3.39}$$

where $I_{i,\text{upwind}}$ is the value of I_i in the "upwind" cell. A second-order approximation can also be
written. The angular integrations are done analytically.

When there is control angle overhang, it is necessary to account for the degree of overhang in
writing I_{if} and J_f. Reference 10 proposed to interpolate the face intensity in Equation (3.38) by
accounting for the fraction of overhanging control angle, using

$$J_f I_{if} = I_{if,\text{out}}\alpha_{i,\text{out}} + I_{if,\text{in}}\alpha_{i,\text{in}} \tag{3.40}$$

where

$$\alpha_{i,\text{out}} = \hat{\mathbf{e}}_n \cdot \int_{\Delta\theta}\int_{\Delta\phi} \mathbf{s} \sin\theta\, d\theta\, d\phi,\ \mathbf{s}\cdot\hat{\mathbf{e}}_n > 0$$

$$\tag{3.41}$$

$$\alpha_{i,\text{in}} = \hat{\mathbf{e}}_n \cdot \int_{\Delta\theta}\int_{\Delta\phi} \mathbf{s} \sin\theta\, d\theta\, d\phi,\ \mathbf{s}\cdot\hat{\mathbf{e}}_n \leq 0$$

The incoming and outgoing face intensities are then written as in Equation (3.39). Thus, referring to Figure 3.6,

$$J_f I_{if} = I_{i0} \alpha_{i,\text{out}} + I_{i1} \alpha_{i,\text{in}} \tag{3.42}$$

where I_{i0} and I_{i1} refer to the intensities in cells $C0$ and $C1$ in the direction i, and the terms "incoming" and "outgoing" are with respect to $C0$.

In 3-D, the overhang fractions $\alpha_{i,\text{in}}$ and $\alpha_{i,\text{out}}$ require integrations over control angle domains which are complex in shape. A discrete pixelation method was proposed to compute these fractions.[10] Here, each control angle was discretized into pixels, and each pixel was tested to determine whether it was incoming or outgoing to the face under consideration. Murthy and Mathur have found[10] that for problems with gray-diffuse boundaries, accounting for control angle overhang did not make much of a difference. For other types of problems, however, it was very important. These included rotationally periodic domains[37] and when reflection at arbitrarily oriented symmetry, Fresnel, or other specular surfaces was involved. A detailed treatment of various boundary conditions can be found in References 10, 37, and 39, as well as a description of the methodology for coupling between the energy equation and the radiative transfer equation.

3.6.4 SOLUTION METHODOLOGY FOR RADIATIVE HEAT TRANSFER

Because the overall methodology for solution is a segregated one, it is convenient to apply the same idea to the discrete radiation equations. It is customary to solve the discrete radiative transfer equations sequentially, each direction in turn; typically, the radiation calculation is appended to the end of the solution loop for flow and heat transfer. The solution of the nominally linear algebraic equation set resulting from the discretization is the same as for other equations. For unstructured meshes, an algebraic multigrid procedure has been used in the published literature.[10]

Although this type of sequential methodology is adequate for problems with moderate optical thickness, it fails for large optical thicknesses. The source of difficulty is inter-equation coupling. When the absorption coefficient is large, the energy and radiation equations are tightly coupled through the emission and absorption terms. When these tightly coupled equations are solved sequentially, oscillation and poor convergence are often encountered. Similar problems occur when the scattering coefficient is large. Here, the coupling is between the intensities in different directions in the in-scattering terms. The same issues are encountered in other radiation models as well, for example, in the P-N class of models. In Equation (3.8), the irradiation G is tightly coupled to temperature through the blackbody emission term I_B; the degree of coupling is determined by the optical thickness.

Mathur and Murthy have proposed a coupled multigrid method called COMET for accelerating the convergence of these tightly coupled problems.[40] The method is an extension of coupled multigrid methods for fluid flow. The idea here is to solve the energy equation and all the discrete radiative transfer equations in a given cell simultaneously, assuming spatial neighbors known. The procedure is applied to each cell in the domain, and the domain is swept until convergence. To eliminate long wavelength error, the point-coupled solution is used as a relaxation procedure in a multigrid scheme.

The multigrid scheme itself consists of three steps. First, coarse-level meshes must be created, as shown in Figure 3.5. It is convenient to start with the finest mesh, that is, the mesh on which we desire the final solution, and create nested coarse meshes by agglomeration based on coefficient strength. For unstructured meshes, the agglomerated coarse-level meshes do not retain the shape of the fine-level cells; in the case of triangles, for example, the coarse-level cells are arbitrary non-convex polyhedra. Once these coarse meshes are available, it is necessary to discretize the governing equations on each level of mesh. In a *geometric* multigrid procedure, the discretization process described in previous sections is applied at each mesh level and discrete equations obtained.

Alternatively, *algebraic* multigrid strategies create coarse mesh discrete equations by algebraic manipulation of fine mesh discrete equations. This procedure allows us to do the discretization only once on the fine mesh, and the discretization procedures need not be general enough to work on the arbitrary polyhedral coarse-level meshes. Once discrete equation sets are available at each mesh level, the solution strategy requires cycling between the different levels, applying the relaxation procedure at each level until appropriate criteria are met. In the case of COMET, the relaxation sweep involves one or more traverses of the computational domain, with a point-coupled solution of radiation and temperature in each cell.

Applying a multigrid method of this type to radiative heat transfer on unstructured meshes poses special challenges. First, the number of radiative transfer equations to be solved is usually large; for a 2 × 2 resolution of the octant, a total of 32 equations must be solved for 3-D problems. Coefficient storage for so many equations is not feasible and, therefore, one must be able to compute coefficients for each equation on the fly. This makes the use of algebraic multigrid methods for the radiation equations inconvenient because one cannot afford to store all the fine mesh radiation equations for manipulation at coarse levels. Thus, a geometric multigrid idea is best for the radiative transfer equations. Fortunately, the unstructured mesh discretization procedure for the radiative transfer equation is easily applied at coarse levels because it works with arbitrary, unstructured polyhedral cells. The same, however, is not true for the energy equation. Here, although arbitrary polyhedra are admitted, non-convexity poses a problem. For non-convex polyhedra, the cell centroid may fall outside the cell volume. For such a case, the diffusion coefficient

$$\frac{\mathbf{A} \cdot \mathbf{A}}{\mathbf{A} \cdot \hat{\mathbf{e}}_s} \qquad\qquad (3.43)$$

in Equation (3.21) may become negative, leading to a loss of diagonal dominance and difficulty in using iterative solvers. Thus, for the energy equation, it is difficult to discretize at coarse levels directly. An algebraic multigrid procedure that manipulates fine-level discrete equations to obtain coarse-level equations is the most suitable. Because the fine-level equations are well-behaved, coarse-level equations derived by algebraic manipulation are also well-behaved.

Mathur and Murthy applied this type of mixed geometric-algebraic multigrid idea to absorption, emission, and scattering problems across a range of optical thicknesses and obtained substantial acceleration.[40] Figure 3.8 shows a quadrilateral cavity in which absorbing and emitting radiation was computed using both COMET and the sequential procedure. All walls are black and have specified temperatures. The medium has a Planck number ($Pl = k/4L\sigma T_c^3$) of 1.4×10^{-5}; here, k is the thermal conductivity, L is the length of the bottom wall, and T_c is its temperature. There is no scattering. The optical thickness κL is varied from 0 to 220.

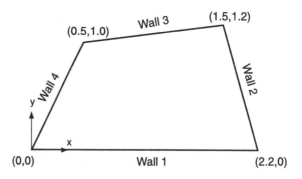

FIGURE 3.8   Radiation in quadrilateral enclosure: schematic.

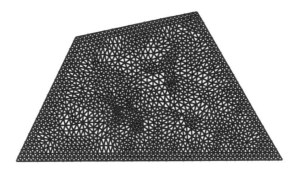

FIGURE 3.9 Radiation in quadrilateral enclosure: baseline mesh.

Computations are performed on a series of meshes, starting with the baseline unstructured mesh of 640 triangles shown in Figure 3.9, and obtaining subsequent meshes by subdividing each triangle into four. An angular discretization of $N_\theta = N_\phi = 2$ is used. Control angle overhang occurs in the ϕ direction because some of the walls are not orthogonal and therefore a 1×10 pixelation is used at all boundaries.

A comparison of COMET and the sequential solution procedure is presented in Table 3.1. Substantial acceleration over the sequential procedure is obtained for moderate to high optical thicknesses. For moderate optical thicknesses, the coupled approach takes 15 to 60% less time.

TABLE 3.1
Radiation in a Quadrilateral Cavity

	Sequential		COMET	
	CPU		CPUs	
κL	(s)	Iter	(s)	Iter
	650 cells			
0	2.41	11	3.14	10
0.1L	3.14	12	2.65	5
1.0L	5.78	25	2.73	5
10.0L	45.98	247	3.38	5
100.0L	775.61	4127	5.65	5
	2600 cells			
0	13.78	10	15.32	13
0.1L	21.23	11	13.24	5
1.0L	30.31	25	11.89	5
10.0L	186.80	262	11.83	5
100.0L	3655.93	5952	15.04	5
	10400 cells			
0	47.27	9	71.99	15
0.1L	86.42	12	62.92	5
1.0L	155.81	24	74.66	5
10.0L	998.80	271	49.19	5
100.0L	>1e5	>6000	89.55	5

Note: CPU time and iterations for convergence for a variety of grid sizes and absorption coefficients at Pl = 1.4×10^{-5} and $N_\theta = N_\phi = 2$

TABLE 3.2

Radiation in a Quadrilateral Cavity

κL	Sequential CPU (s)	Iter	COMET CPUs (s)	Iter
	650 cells			
0.1L	11.25	12	12.4	5
1.0L	18.66	25	12.82	5
10.0L	144.06	239	16.45	5
	2600 cells			
0.1L	98.53	12	81.85	5
1.0L	112.46	25	73.23	5
10.0L	690.99	252	72.65	5
	10400 cells			
0.1L	488.57	12	376.99	5
1.0L	714.97	24	344.63	5
10.0L	3209.37	266	278.73	5

Note: CPU time and iterations for convergence for a variety of grid sizes and absorption coefficients at Pl = 1.4×10^{-5} and $N_\theta = N_\phi = 4$

The speedup for higher optical thicknesses is dramatic; factors of $O(10^3)$ are obtained. Here, the speedup improves with increased grid size. Similar speedup behavior is observed for a finer angular discretization of $N_\theta = N_\phi = 4$ and at a higher Planck number of 1.4×10^{-3}, results for which are tabulated in Tables 3.2 and 3.3, respectively. For low optical thicknesses, the energy and radiation equations are nearly independent of each another and therefore there is no advantage to using COMET; one sees that the COMET procedure results in a penalty in the transparent limit. The results for this limit are included only to provide an idea of the overhead associated with the multigrid procedure. However, the overall solution times are small, so that COMET remains a useful algorithm for general-purpose solvers. In keeping with the multigrid formulation, one sees that the number of iterations to convergence required by COMET remains the same with increasing optical thickness and for all grid sizes.

3.6.5   RADIATIVE PROPERTIES

The numerical methodology for participating radiation described in this chapter is applicable for both gray and non-gray radiation. When the absorption spectrum is banded, as for glasses and ceramics, a band approach for discretizing the spectral space is easily admitted.[39] For spectrally dependent gas-phase radiation, any of the narrow- and wide-band models used with structured-mesh solutions are directly admissible. These include the weighted-sum-of-gray-gases (WSGGM) approach,[11,41] the k-distribution approach,[42] and the recently published combination of the two such as that suggested by Denison and Webb.[43]

3.7   EXAMPLES

This chapter section presents three problems designed to demonstrate the use of the finite-volume method outlined to problems of industrial interest. The first of these problems uses the finite-volume method for computing multi-mode heat transfer in a rod bed. The problem illustrates how modern CFD

TABLE 3.3
Radiation in a Quadrilateral Cavity

	Sequential		COMET	
	CPU		CPUs	
κL	(s)	Iter	(s)	Iter
	650 cells			
0.1L	3.17	11	3.10	5
1.0L	5.14	22	2.61	5
10.0L	46.28	244	3.44	5
	2600 cells			
0.1L	21.27	11	17.02	7
1.0L	27.20	22	15.52	7
10.0L	186.12	254	11.83	5
	10400 cells			
0.1L	131.42	11	83.40	8
1.0L	119.92	19	78.34	8
10.0L	947.28	271	66.40	7

Note: CPU time and iterations for convergence for a variety of grid sizes and absorption coefficients at Pl = 1.4×10^{-3} $N_\theta = N_\phi = 2$

techniques can now be used across the board in industrial problems, covering problems involving fully coupled heat transfer and radiation in complex geometries. The other two problems involve combustion and NOx predictions using the flow, turbulence and combustion models outlined previously. Local and global temperature and concentration measurements are available for detailed comparisons.

3.7.1 MULTI-MODE HEAT TRANSFER BETWEEN CYLINDRICAL RODS

Many problems of interest involve participating radiation coupled to other modes of heat transfer. In furnaces, for example, heat exchanger tubes are heated by the participating radiation from the flame, and also exchange heat by conduction, convection, and radiation with each other. This problem considers a bed of opaque conducting rods arranged as shown in Figure 3.10. The rods extend infinitely in the z-direction, and are arranged in-line in the x-y plane as shown. The bed is irradiated from the left with a blackbody emission E_{b1} and from the right by an irradiation E_{b2}. E_{b2}/E_{b1} is chosen to be 0.5. The rods are conducting, with a dimensionless conductivity k_s^*. The surface emissivity of the rods is ε. The radio l/d is chosen to be 5.333, leading to a bed porosity of 0.56. The intervening fluid is assumed to be stationary and conducting, and can absorb and emit radiation. The dimensionless conductivity and optical thickness are chosen to be $k_s^* = 0.01$ and $\tau_g = \kappa_g d = 1.0$, respectively. Further details about the problem may be found in Reference 39.

The computational domain is chosen to be one periodic module, as shown in Figure 3.10. The objective is to compute the total exchange factor defined as:

$$F_{tot} = \frac{q_{tot}(1 + l/d)}{E_{b1} - E_{b2}} \tag{3.44}$$

where q_{tot} is the total heat flux from the left boundary. Because the top and bottom walls are periodic boundaries, this also represents the total heat transfer through any cross-section in the bed.

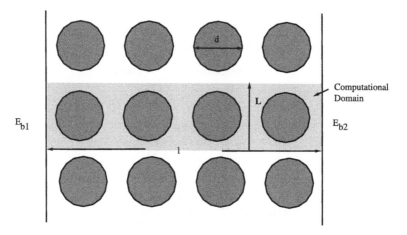

FIGURE 3.10   Multi-mode heat transfer between cylindrical rods: schematic.

TABLE 3.4
**Multi-Mode Heat Transfer
Between Cylindrical Rods**

τ_g	DTRM	Finite Volume[a]
0.1	1.045	1.031
		(1.3%)
1.0	0.756	0.747
		(1.19%)
10.0	0.282	0.288
		(2.1%)

Note: Comparison of F_{tot} between DTRM
and Finite Volume (FV) Calculations

[a] difference with respect to DTRM in
parentheses.

There being no published results available for comparison, calculations using the finite-volume scheme were compared to those obtained using the discrete transfer radiation model (DTRM).[44] A range of fluid optical thicknesses were investigated for $k_s^* = 0.01$, $k_s^* = 1.0$. The DTRM solutions were obtained with an octant angular discretization of 4×4 rays; all rays originated at the boundaries. The finite volume calculations were carried out with a $4 \times 4 \times 1 \times 10$ angular discretization. The comparison of F_{tot} between the two calculations is shown in Table 3.4. Although the underlying methodologies are very different, and it is difficult to do calculations of exactly equal resolution, the predicted values of F_{tot} are within 2.1% of each other.

Calculations were done with the 11,440-cell triangular mesh shown in Figure 3.11a; a detail of one module is shown in Figure 3.11b. Figure 3.12 shows the variation of F_{tot} for a bed of opaque conducting rods in an absorbing and emitting fluid. The case corresponds to $\tau_g = 1.0$ and $k_s^* = 0.01$. For low values of the solid conductivity, thermal radiation reaching the rods is essentially re-radiated back into the interstitial space; some of the radiation may be transmitted from the hot boundary to the cold boundary, but some of it will be reradiated back to the hot boundary and

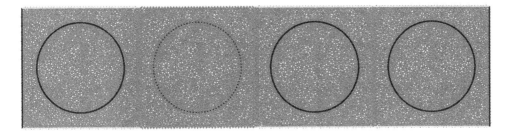

(a) Mesh

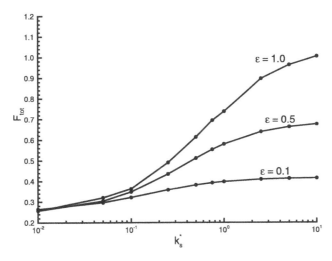

(b) Detail

FIGURE 3.11   Multi-mode heat transfer between cylindrical rods: triangular mesh.

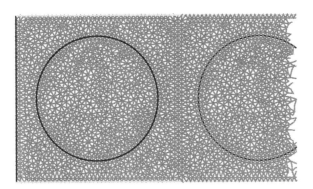

FIGURE 3.12   Multi-mode heat transfer between cylindrical rods: variation of exchange factor with rod conductivity for $\tau_g = 1.0$, $k_g^* = 0.01$.

absorbed. Consequently, the rate of transfer across the bed is small. As expected, this variation is independent of rod emissivity if its conductivity is low enough. For a given rod emissivity, the effective exchange factor of the bed increases as k_s^* increases because conduction through the rod enhances heat transfer. As k_s^* exceeds unity, conduction through the rod becomes the dominant mechanism for energy transport across the bed. However, once k_s^* becomes large enough, the limiting heat transfer rate becomes the radiant transfer rate from the boundaries to the rod (and from other rods), and increasing rod conductivity does not increase the net heat transfer through the bed any further. For a given rod conductivity k_s^*, increasing rod emissivity increases the exchange factors. This is expected because the rods are assumed to be gray. Consequently, the absorptivity of the surface also increases as ε increases. This implies that the conduction mechanism for transferring energy through the rods is activated to a greater degree for higher ε.

3.7.2 COMBUSTION IN IFRF PULVERIZED COAL BURNER

Presented here is a simulation of the International Flame Research Foundation (IFRF) 2.4-MW pulverized coal flame.[45] The combustor has a single swirl burner, and is unstaged, thus generating high NOx. In-flame measurements of temperature, velocities, chemical species, and coal burnout are reported in Reference 45.

The geometry of the burner is shown in Figure 3.13; only one quarter of the burner is simulated due to symmetry. The simulation is performed on an unstructured hexahedral mesh of 52,000 cells. Local adaption in a cylindrical volume about the burner can help resolve the local flame structure; an adapted mesh is shown in Figure 3.14. Turbulence, chemistry, and radiation are modeled with the standard k-ε, the equilibrium mixture fraction PDF, and P_1 radiation models, respectively. Coal is modeled as a discrete phase, with a Rosin-Rammler distribution of particle sizes injected at the inlet. Stochastic tracking is employed to include the effects of turbulent dispersion. Coal devolatilization is modeled with a single rate kinetic model,[24] and char combustion is modeled with a weighted kinetics/diffusion model.[26] NOx formation is computed using the model in Reference 46.

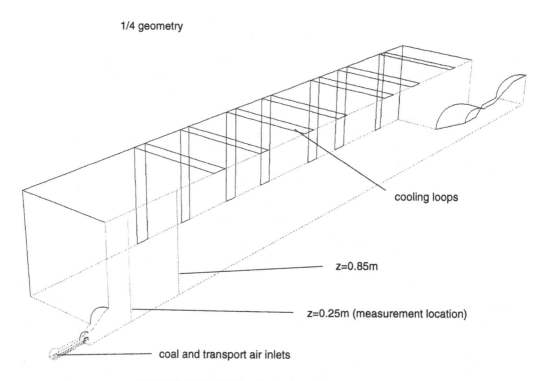

1/4 geometry

cooling loops

z=0.85m

z=0.25m (measurement location)

coal and transport air inlets

FIGURE 3.13   IFRF pulverized coal burner: geometry.

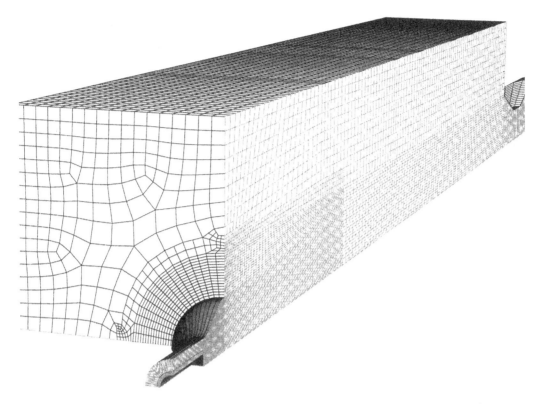

FIGURE 3.14   IFRF pulverized coal burner: non-conformal mesh adaption in burner region.

Comparisons of computed and measured CO_2, O_2, and NO concentrations as well as temperature profiles are shown in Figures 3.15 through 3.18. Two different axial locations, $z = 0.25$ m and $z = 0.85$ m shown in Figure 3.13, are used for comparison. The comparison in all cases is reasonable.

3.7.3   NOx Emission from Gas-Turbine Combustor

NOx emissions from a GE LM1600 gas turbine at part load operation (courtesy of Nova Research and Technology Center, Calgary, Alberta, Canada) is simulated here. A close-up of the geometry of the injection ports is shown in Figure 3.19. The combustor burns natural gas in a non-premixed mode, producing high NOx as a consequence of the high flame temperatures.

The domain is meshed with 286,000 hexahedral cells. Initial calculations with the equilibrium mixture fraction model underpredicted NO concentration due to its inability to capture the super-equilibrium atomic oxygen radicals. Good agreement with experiment is obtained with the laminar flamelet model, which accounts for moderate chemical non-equilibrium. Figure 3.20 shows a comparison of the predicted NO concentration at the combustor exit as a function of load. The laminar flamelet model captures the trends well, while the equilibrium PDF model substantially underpredicts the NO concentration.[46]

3.8   CLOSURE

This chapter has outlined an unstructured solution-adaptive finite volume scheme for the computation of flow, heat transfer, chemical reaction, and radiation, and demonstrated its use in practical industrial problems. The method is versatile and lends itself to efficient shared and distributed

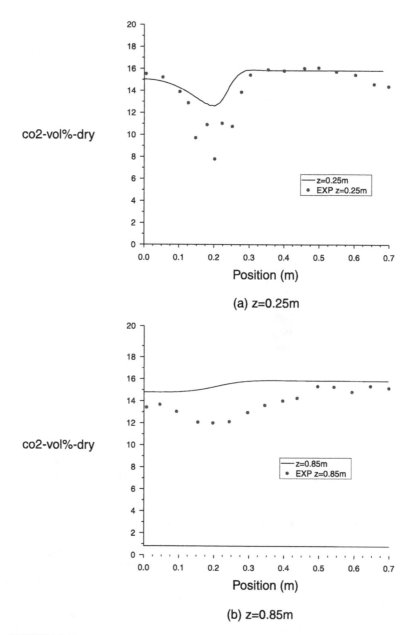

FIGURE 3.15 IFRF pulverized coal burner: profile of CO_2 concentration (dry %).

memory parallel processing. It can be used as a general-purpose engine for industrial combustion computations, and forms a viable basis for the inclusion of more complex turbulence and combustion models. A number of interesting research efforts are underway to improve the speed and convergence of this finite volume methodology. These include the development of efficient coupled multigrid methods for coupling pressure and velocity, the development of accurate large-eddy simulation (LES) implementations for unstructured meshes, and further improvements to participating radiation and multi-mode heat transfer computational procedures.

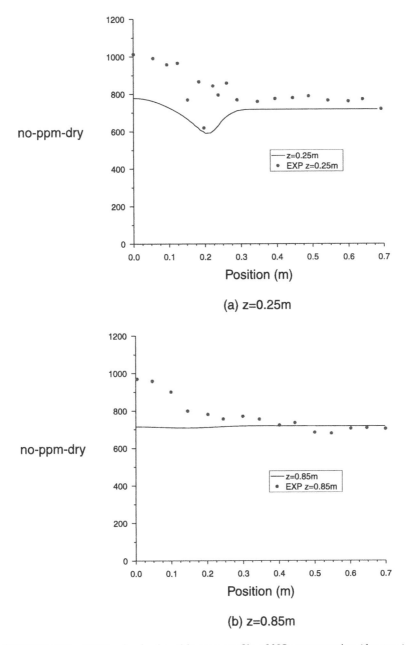

no-ppm-dry

(a) z=0.25m

no-ppm-dry

(b) z=0.85m

FIGURE 3.16 IFRF pulverized coal burner: profile of NO concentration (dry ppm).

ACKNOWLEDGMENTS

The authors wish to acknowledge the use of Fluent, Inc.'s, solver FLUENT/UNS, and its mesh generators PreBFC and TGrid, in the results presented here. Thanks also go to Dr. Graham Goldin of Fluent, Inc., for help with the example problems.

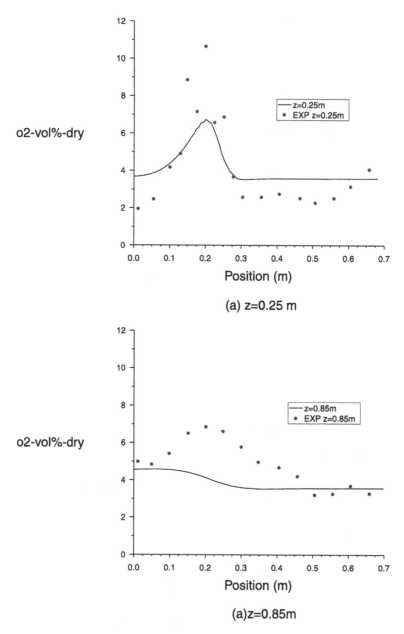

FIGURE 3.17 IFRF pulverized coal burner: profile of O_2 concentration (dry %).

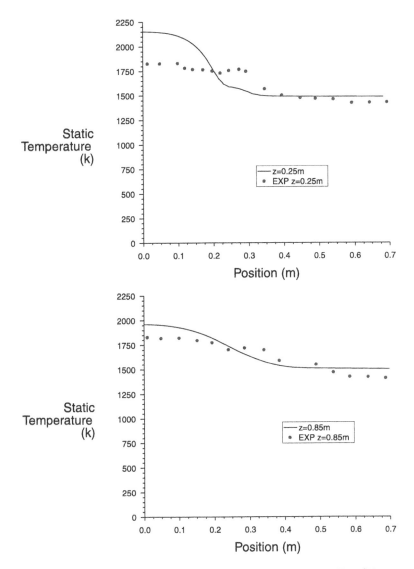

FIGURE 3.18 IFRF pulverized coal burner: temperature profiles (°K).

FIGURE 3.19   Gas-turbine combustor: inlet geometry.

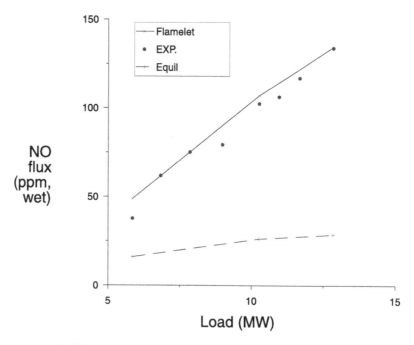

FIGURE 3.20   Gas-turbine combustor: exit NO concentration.

REFERENCES

1. O.C. Zienkewicz, *The Finite Element Method,* McGraw-Hill, New York, 1980.
2. B.R. Baliga and S.V. Patankar, A control-volume finite element method for two-dimensional fluid flow and heat transfer, *Numer. Heat Transfer,* 6, 245–261, 1983.
3. Y. Jiang and A.J. Przekwas, Implicit, pressure-based incompressible Navier-Stokes equations solver for unstructured meshes, AIAA-94-0305, 1994.
4. L. Davidson. A pressure correction method for unstructured meshes with arbitrary control volumes, *Int. J. Numer. Meth. Fluids,* 22, 265–281, 1996.
5. I. Demirdzic and S. Muzaferija, Numerical method for coupled fluid flow, heat transfer and stress analysis using unstructured moving meshes with cells of arbitrary topology, *Comput. Meth. Appl. Mech. Eng.,* 125, 235–255, 1995.
6. R. Löhner, K. Morgan, J. Peraire, and M. Vahdati, Finite element flux corrected transport (FEM-FCT) for the Euler and Navier-Stokes equations, *Int. J. Numer. Meth. Fluids,* 7, 1093–1109, 1987.
7. D. Mavriplis and A. Jameson, Solution of the Navier-Stokes equations on triangular meshes, *AIAA J.,* 28(8), 1415–1425, 1990.
8. S.R. Mathur and J.Y. Murthy, A pressure based method for unstructured meshes, *Numer. Heat Transfer,* 31(2), 195–216, 1997.
9. S.R. Mathur, J.Y. Murthy, M. Missaghi, and O.S. Faltsi-Saravelou, Computation of combusting flows using unstructured solution adaptive meshes, presented at *IMECE 96,* Atlanta, GA 1996.
10. J.Y. Murthy and S.R. Mathur, Finite volume method for radiative heat transfer using unstructured meshes, *J. Thermophys. Heat Transfer,* 12(3), 313–321, 1998.
11. M.F. Modest, *Radiative Heat Transfer,* Series in Mechanical Engineering. McGraw-Hill, New York, 1993.
12. F.C. Lockwood and N.P. Shah, A new radiation solution method for incorporation in general combustion prediction procedures, in *Proceedings of the 18th Symposium (Int.) on Combustion,* The Combustion Institute, 1981, 1405–1414.
13. M.P. Menguc and R. Viskanta, Radiative heat transfer in combustion systems. *J. Quant. Spectrosc. Radiat. Transfer,* 33, 533–549, 1985.
14. W.A. Fiveland and J.P. Jessee, Finite element formulation of the discrete ordinates method for multidimensional geometries, *J. Thermophys. Heat Transfer,* 8(3), 426–433, 1994.
15. G.D. Raithby and E.H. Chui, A finite-volume method for predicting radiant heat transfer in enclosures with participating media, *J. Heat Transfer,* 112, 415–423, 1990.
16. B.E. Launder and D.B. Spalding, *Lectures in Mathematical Models of Turbulence.* Academic Press, London, 1972.
17. B.F. Magnussen and B.H. Hjertager, Mathematical models of turbulent combustion with special emphasis on soot formation and combustion, *16th Symp. (Int.) on Combustion,* Cambridge, MA, August 1976.
18. K.K. Kuo, *Principles of Combustion,* John Wiley & Sons, New York, 1986.
19. W.P. Jones and J.H. Whitelaw, Calculation methods for reacting turbulent flows: a review, *Combust. Flame,* 48, 1–26, 1982.
20. K.N. Bray and N. Peters, Laminar flamelets in turbulent flames, in P.A. Libby and F.A. Williams, Eds., *Chemically Reacting Flows,* Academic Press, 1994.
21. A.E. Lutz, R.J. Kee, J.F. Grcar, and F. M. Rupley, OPDIF: a FORTRAN program for computing opposed-flow diffusion flames, Technical Report SAND96-8243, Sandia National Laboratories, 1997.
22. W.E. Ranz and W.R. Marshall, Evaporation from drops, part i, *Chem. Eng. Prog.,* 48(3), 141–146, 1952.
23. W.E. Ranz and W.R. Marshall, Evaporation from drops, part ii, *Chem. Eng. Prog.,* 48(4), 173–180, 1952.
24. S. Badzioch and P.G.W. Hawksley, Kinetics of thermal decomposition of pulverized coal particles, *Ind. Eng. Chem. Process Design and Development,* 9, 521–530, 1970.
25. H. Kobayashi, J.B. Howard, and A.F. Sarofim, Coal devolatalization at high temperatures, in *16th Symp. (Int.) on Combustion,* The Combustion Institute, 1976.

26. M.M. Braum and P.J. Street, Predicting the combustion behaviour of coal particles, *Combust. Sci. Tech.,* 3(5), 231–243, 1971.

27. I.W. Smith, The combustion rates of coal chars: a review, in *19th Symp. (Int.) on Combustion,* The Combustion Institute, 1982.

28. T.J. Barth, *Aspects of Unstructured Grids and Finite-Volume Solvers for the Euler and Navier-Stokes Equations,* Special Course on Unstructured Grid Methods for Advection Dominated Flows, AGARD Report 787, 1992.

29. S.V. Patankar, *Numerical Heat Transfer and Fluid Flow,* McGraw-Hill, New York, 1980.

30. K.C. Karki and S.V. Patankar, Pressure based calculation procedure for viscous flows at all speeds in arbitrary configurations, *AIAA J.,* 27(9), 1167–1174, 1989.

31. C.M. Rhie and W.L. Chow, Numerical study of the turbulent flow past an airfoil with trailing edge separation. *AIAA J.,* 21(11), 1523–1532, 1983.

32. J.Y. Murthy and S.R. Mathur, A conservative numerical scheme for the energy equation, *J. Heat Transfer,* 120(4), 1081–1085, 1998.

33. E.H. Chui and G.D. Raithby, Computation of radiant heat transfer on a non-orthogonal mesh using the finite-volume method, *Numer. Heat Transfer,* 23, 269–288, 1993.

34. J.C. Chai, G. Parthasarathy, S.V. Patankar, and H.S. Lee, A finite-volume radiation heat transfer procedure for irregular geometries, AIAA 94-2095, June 1994.

35. J.C. Chai and J.P. Moder, Spatial-multiblock procedure for radiation heat transfer, in A. Gopinath, P.D. Jones, J. Syed-Yagoobi, and K.A. Woodbury, Eds., *HTD-Vol. 332, Proceedings of the ASME Heat Transfer Division,* Vol. 1, 119–128, ASME, 1996.

36. J.Y. Murthy and S.R. Mathur, Radiative heat transfer in axisymmetric geometries using an unstructured finite volume method, *Numer. Heat Transfer,* 33(4), 397–416, 1998.

37. S.R. Mathur and J.Y. Murthy, Radiative heat transfer in periodic geometries using a finite volume scheme, *J. Heat Transfer,* 121(2), 357–364, 1998.

38. J.Y. Murthy and S.R. Mathur, A finite volume scheme for radiative heat transfer in semi-transparent media, AJTE99:6293, presented at the *ASME-JSME Thermal Engineering Joint Conference,* San Diego, CA, March 1999.

39. S.R. Mathur and J.Y. Murthy, Computation of multi-mode heat transfer using an unstructured finite volume method, presented at *ASME National Heat Transfer Conference,* Albuquerque, NM, 1999.

40. S.R. Mathur and J.Y. Murthy, Coupled ordinates method for multigrid acceleration of radiation calculations, *J. Thermophys. Heat Transfer,* 13(4), 467–473, 1999.

41. H.C. Hottel and A.F. Sarofim, *Radiative Transfer,* McGraw-Hill, New York, 1967.

42. R.M. Goody and Y.K. Yung, *Atmospheric Radiation,* Clarendon Press, Oxford, U.K., 1989.

43. M.K. Denison and B.W. Webb, K-distribution and weighted-sum-of-gray-gases: a hybrid model, in *Heat Transfer 1994,* Vol. 2, G. F. Hewitt, Ed., 1994, 19–24.

44. N.G. Shah, A New Method of Computation of Radiation Heat Transfer in Combustion Chambers, Ph.D. thesis, Imperial College of Science and Technology, London, U.K., 1979.

45. R. Weber, J. Dugue, A. Sayre, A.A. Peters, and B.M. Visser, Measurements and computations of quarl zone fluid flow and chemistry in a swirling pulverized coal flame, Technical Report IFRF Doc No. F36/y/20, International Flame Research Foundation, 1992.

46. R.K. Hanson and S. Salimian, Survey of rate constants in H/N/O systems, in W. C. Gardiner, Ed., *Combustion Chemistry,* 1984, 361.

4 PCGC-3

Scott C. Hill, Andrew M. Eaton, and L. Douglas Smoot

CONTENTS

0-8493-2000-X/00/$0.00+$.50

NOMENCLATURE

ARABIC SYMBOLS

a	absorptivity
A	pre-exponential factor
A	area
B	kg/m^2s^2, body force
B_D	m^2/s, diffusion coefficient
B_P	product constant
B_R	reactant constant
c	char bridge
C	$kmol/m^3$, molar concentration
C	turbulent model constant
d	diameter of particle
D	m^2/s, diffusion coefficient
E	wall velocity constant
E	kJ/kmolK, activation energy
E	emissive power
f	volatiles fraction
f	turbulent k-ε model parameter
f	mixture fraction
f_{O_2}	required oxygen ratio
F	force on particle
g	mixture fraction variance
g	gas
G	defined by 2.13
h	kJ/kg, enthalpy
I	radiative intensity
k	kJ, turbulent kinetic energy
k	m/s, reaction rate coefficient
k_m	mass transfer coefficient
l	m, length, mixing length
m	kg, mass
m_i	mass fraction
M_i	kg/kmol, molecular weight
n	m^{-3}, number density
N	total number of chemical species
q_r	kJ/s, radiative heat transfer rate
Q	m^2, radiative cross-section
Q	kJ/s, heat transfer rate
p,P	Pa, pressure
p_o	initial number in coal lattice
P	spherical harmonic moment
P_l	m, path (beam) length

r	kg/m²s, reaction rate
R	kJ/kmol K, universal gas constant
R	kg/m²s, reaction rate
s	conserved scalar
s	unit vector
S	kJ(kg)/m²s, source term
t	s, time
T	K, temperature
x	kmol/m³, molar concentrations
X	molar fraction
u_i	m/s, velocity component
u_i	m/s, gas velocity in Cartesian coordinates (tensor notation)
U	m/s, free stream velocity
U^+	dimensionless velocity
V	m/s, velocity
x	m, coordinate direction
x_i	Cartesian coordinates (tensor notation)
y	m, direction from wall
y^+	dimensionless position
Y	mass fraction
Y	fraction of volatiles

GREEK SYMBOLS

β	labile bridge population
δ_{ij}	Kroncker delta function
δ	side chain
ε	emissivity
ε	kinetic energy dissipation rate
ϕ	conserved scalar
ϕ_1	stoichiometric ratio, CO/CO_2
Φ	kJ/m²s, dissipation function
Φ	phase function
Γ	m²/s, turbulent energy diffusion coefficient
κ	absorption coefficient
μ	viscosity
μ_s	cosine of scattering angle
Π	product
ν	kinematic viscosity
ν	volatiles
ν_i'	product stoichiometric coefficient
ν_i''	reactant stoichiometric coefficient
τ	shear stress
ρ	kg/m³, density
σ	Schmidt number
σ	scattering coefficient
σ	Boltzman constant
$\sigma + 1$	lattice coordination number
ω	unit vector
Ω	phase angle

Subscripts

$_{-1,-2,-3}$	reverse reaction
$_{1,2,3}$	forward reaction, turbulent model constant
abs	absorption
b	black body
b	bridge
bridge	bridge
c, char	char
C	constant
Co	initial number of char bridge
CO	carbon monoxide
cross	crosslink
cluster	cluster
d	particle diameter
do	oxidizer diffusion
e	effective
e	eddy or turbulent
f	fraction (radiation)
f	mixture fraction
F	furnace
frag	fragment
frag	tar precursor fragment
g	mixture fraction variance
g, gas	gas
h	enthalpy
HCN	hydrogen cyanide
i	i^{th} species
i	location index
i	species index
j	location index
j	species index
k	location index
k	m^2/s^2, kinetic energy
l	species index
l,m	mixing length
m	mass
m	band index
NH_3	ammonia
NO	nitric oxide
o	initial number
o	oxidizer
op	oxidizer at particle surface
ox	oxidation
O_2	oxygen
p	primary
p	particle
p	product
r	radiation
reb	reburning
R	reactant

s	secondary
s	scattering
v	volatiles
w	wall
ε	dissipation
μ	viscosity
λ	stress
δ	side chain
τ	total moles

SUPERSCRIPTS

'	activated
'	fluctuating component
~	Favre-averaged mean value
—	mean value
a	apparent oxygen order
h	enthalpy
m	mass
n	reaction order
r	reactive
R	reaction
u	momentum
α	reaction order exponent
α	temperature exponent

ABBREVIATIONS

CFD	Computational Fluid Dynamics
CFX™	CFD code from AEA Technology
CPD	Chemical percolation-devolatilization
CPR	Controlled-Profile Reactor
DNS	Direct numerical solution
EB	Eddy breakup
FASTRAN™	Commercial software, CFD Research Corp.
FLASHCHAIN	Network devolatilization model
FLUENT™	Comprehensive combustion codes from Fluent, Inc.
FG-DVC	Functional group-depolymerization, vaporization, condensation
HP	Hewlett Packard
MF	Mixture fraction
MW_e	Megawatt, electric
NMR	Nuclear magnetic resonance
PDF	Probability density function
PCGC-3	Comprehensive combustion code, Pulverized Coal Gasification and Combustion, 3-dimensional
PISO	Numerical solution method for semi-implicit partial differential equations
RNG	Re-normalized group
RSM	Reynolds Stress Model
RTE	Radiative Transport Equation
SC	Species concentration

SIMPLE, SIMPLEC,
SIMPLER   Semi-implicit numerical solution methods for CFD codes
SSF   Stochastic separated flow
STP   Stochastic transport of particles
TG-FTIR   Thermogravimetric-Fourier transform infrared
WSGGM   Weighted sum of grey gases method

4.1   INTRODUCTION

Fossil fuels are the major source of energy in the world and will continue to be so for the foreseeable future.[1,2] Major efforts are therefore being invested in making the use of these fuels more efficient and environmentally acceptable. A significant portion of these efforts for the past quarter century has focused on the development of multi-dimensional, mathematical models of boilers, reactors, gasifiers, and pyrolyzers that can be used to design and analyze fossil-fuel combustion processes. These models serve as tools when characterizing reactive flow processes that complement and, in some cases, replace physical experiments with equivalent "numerical" experiments.

PCGC-2 (Pulverized Coal Gasification and Combustion: 2-Dimensions) was among the first of these multi-dimensional codes used to analyze fossil fuel combustion processes. Application of PCGC-2 was limited to laboratory-scale combustors which could be assumed to be two-dimensional in nature. Nevertheless, PCGC-2 was used for important development of the submodels and solution techniques that were eventually extended to various three-dimensional codes, including PCGC-3 (Pulverized Coal Gasification and Combustion: 3-Dimensions). PCGC-3 is a comprehensive combustion code developed at the Advanced Combustion Engineering Research Center (ACERC) at Brigham Young University. The focus of this chapter is to provide an in-depth discussion of the structure, submodels, and solution techniques used in PCGC-3, and provide several examples of the application of PCGC-3 to laboratory-scale and industrial-scale combustion systems.

Development of fossil-fuel combustion technology in the past was largely empirical in nature, being based primarily on years of accumulated experience in the operations of utility furnaces and on data obtained from sub-scale test facilities. Empirically based experience and data have limited applicability, however, when considering changes in process parameters, such as evaluating firing strategies for improving combustion efficiencies or mitigating pollutant formation. Combustion modeling technology bridges the gap between sub-scale testing that tends to be phenomenological in nature, and the operation of large-scale furnaces typically used for power generation by providing information about the combustion processes that experimental data alone cannot practically provide.

Combustion models exist in varying degrees of sophistication. At the highest level of sophistication is what is referred to as a *comprehensive combustion model*. PCGC-3 is a comprehensive combustion model, and is the focus of this chapter. The term "comprehensive" is used to signify that submodels for all pertinent physico-chemical mechanisms have been assembled into an integrated model with a solution approach that can adequately simulate the overall combustion process of interest. As an example, when modeling pulverized coal combustion, the framework for the solution approach is based on computational fluid dynamics (CFD) using numerical solutions of multi-dimensional, differential equations for conservation of mass, energy, and momentum. Other submodels are coupled within this framework to account for gaseous species mixing and chemical reactions, coal particle devolatilization and char oxidation, and radiant energy transport.[3] Information available from model predictions can include temperature distributions, gas composition, velocity, particle trajectories, extent of particle burnout, NOx formation and reduction, SO_x formation and capture, pressure distribution, particle size distributions, ash/slag accumulation, etc. PCGC-3 can provide spatial and temporal variations of such quantities.

Comprehensive combustion models offer many advantages in characterizing combustion processes, and they can effectively complement experimental programs. Code predictions are typically less expensive and take less time than experimental programs, and such models often provide additional information that cannot always be measured. Computational modeling thereby becomes a cost-effective, complementary alternative to exhaustive testing in the designing, retrofitting, analyzing, and optimizing the performance of fossil-fuel combustion and energy conversion systems.

While combustion modeling technology offers great potential, a major challenge in the use of such technology is obtaining confidence that a comprehensive modeling approach adequately characterizes, both qualitatively and quantitatively, the combustion process of interest. This is typically accomplished by making comparisons of code predictions with experimental data measured from flames in reactors that embody the pertinent aspects of the turbulent combustion of coal, oil, gas, or slurry fuels. Consequently, data from a range of different-sized facilities are necessary to validate the adequacy of code predictions and establish the degree of precision that the code can provide in simulating the behavior of industrial furnaces. Such detailed data also provide new insights into combustion processes and strategies. Data from both large- and small-scale systems are also necessary because the detailed measurements possible in the laboratory-scale facilities often complement the more coarse or sparse measurements of three-dimensional flow patterns and flame heat characteristics obtained in full-scale industrial and utility furnaces.

An overview of PCGC-3 as a comprehensive combustion model will be presented in two parts. The first part focuses on the basic features of PCGC-3. These features comprise the various submodels required to simulate the physical and chemical mechanisms of importance in the combustion process, and the strategies employed in coupling the submodels together into an integrated computational process model with an efficient solution procedure. Brief descriptions of the state-of-the-art of the submodels will be presented, and are limited to those required for characterizing gas and pulverized coal gasification and combustion processes.

The second part focuses on the application of PCGC-3 to simulate several pilot- and industrial-scale combustion facilities for comprehensive validation and evaluation. In particular, three sets of model simulations compared with validation data will be reported. One set with pulverized coal as the fuel, is for the pilot-scale combustor called the Controlled Profile Reactor (CPR). Another set is for a full-scale, corner-fired, 85-MW_e utility boiler. The third set is for an industrial glass-melting furnace.

4.2 REVIEW OF PCGC-3 MODEL FEATURES

PCGC-3 was designed for simulation of non-premixed, turbulent diffusion flames, with emphasis on pulverized coal combustion. For the non-premixed processes of interest, the oxidizer enters as a gas, while the fuel enters as a gas, as particles or droplets entrained in a gas, or as a combination of gas, entrained particles and/or entrained droplets. For simulations where entrained particles and/or droplets are present, PCGC-3 is only applicable to cases where the entrained phase can be classified as dilute (negligible volumetric fraction) and dispersed (no particle-particle interactions). The discussion that follows includes the features required for modeling pulverized coal gasification and combustion processes in furnaces and related combustors, but it should be emphasized that PCGC-3 is also applicable to gaseous combustion only. The basic submodels used in PCGC-3 are summarized in Table 4.1.

Turbulent, chemically reactive flow modeling for gaseous combustion processes has been extensively reviewed by others.[5-7] Multi-dimensional, entrained-flow coal combustion and gasification models were reviewed by Smoot and Smith[8,9] and by Niksa,[10] with a more recent update by Brewster et al.[3] Based on these reviews, the significant physical and chemical phenomena that are typically required as a minimum to classify a combustion model as *comprehensive* are illustrated in Figure 4.1, for a representative pulverized coal combustion process. Three of the illustrated phenomena are common to both gaseous flow and entrained flow combustion processes: (1) gaseous, turbulent fluid mechanics with heat transfer, (2) gaseous, turbulent combustion, and (3) radiative energy transport.

TABLE 4.1
Summary of PCGC-3 Combustion Modeling Features

Combustion Modeling Feature	PCGC-3 Implementation[a]
Gaseous turbulent flow	Finite volume, rectangular element mesh with staggered-gridding. SIMPLER solution algorithm. k-ε turbulence model, nonlinear k-ε turbulence model.
Gaseous turbulent combustion	Mixture-fraction with PDF. Two mixture fraction variables.
Radiative energy transport	Discrete-ordinates
Multiphase flow models for reacting flows	Lagrangian particle trajectories
Particle reactions	Devolatiliztion with single, two-step and CPD options. Char oxidation
NO pollutant formation	Thermal, fuel, reburn mechanisms

[a] Data from References 3, 8, 160.

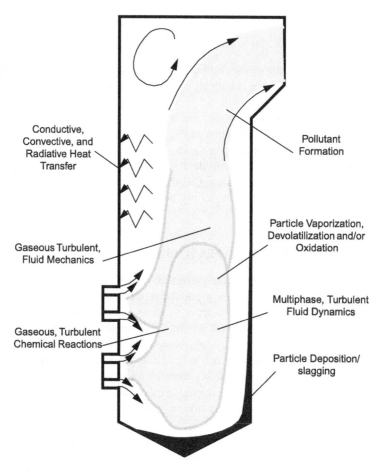

FIGURE 4.1 Major physical and chemical mechanisms characterized in a comprehensive combustion code for a pulverized-coal combustion reactor.

Additional phenomena and the related submodels that must be added for entrained particle or droplet combustion include: (4) multiphase, turbulent fluid mechanics, (5) liquid vaporization from the particles or droplets, (6) particle devolatilization, and (7) particle oxidation.

Finally, additional submodels for phenomena specific to the particular process of interest must be added. For example, with pulverized coal combustion, the following are significant: (8) soot formation, (9) pollutant formation and distribution, and (10) fouling/slagging behavior.

A discussion of the approach used in PCGC-3 for modeling these phenomena is presented in the same topical order as listed above. The discussion of the separate phenomenological models is then followed by a discussion of the solution strategy used in PCGC-3 to assemble the individual parts into an overall solution procedure. While different comprehensive combustion codes typically account for most or all of the phenomena indicated, the submodels used to characterize each phenomenon may differ both in the approaches used for characterizing the details of the physico-chemical mechanisms, and in the methodologies employed for coupling the submodels into an overall solution scheme.

4.2.1 Gaseous Turbulent Flow with Heat Transfer

4.2.1.1 Basic Equations

A fundamental assumption in the study of gaseous turbulent fluid flow is that the gas can be viewed as a continuum; that is, that the distance between molecules of gas and the mean free path of the molecules is very small compared with the physical dimensions of the geometric scale of interest. Given this assumption, for nonreacting gaseous flows, there are five basic variables that must be considered in modeling the fluid motion: three velocity components and two thermodynamic properties. Determining the values for these five variables as a function of space and time, which is the solution of the flow field, is the function of the mathematical models of fluid motion. Because five variables are of interest, the mathematical model must comprise at least five independent equations. Additional constitutive equations are added if additional variables are introduced. As an example, if a third thermodynamic property must be considered, an additional equation can be added in the form of an equation of state.

The derivation of the five equations that constitute a mathematical model of gaseous, turbulent fluid flow with heat transfer can be found in several references, such as Bird et al.,[11] Hoffman et al.,[12] or Crowe and Smoot.[13] In these derivations, three fundamental laws of physics are employed: (1) conservation of mass, (2) Newton's second law of motion, and (3) conservation of energy (first law of thermodynamics).

These laws can be used to derive integral relationships for control volumes, or differential relationships for local points in space. Differential relationships in the form of partial differential equations are the form most often employed in developing CFD numerical procedures and used in PCGC-3 as presented here.

If the flow is assumed to be a Newtonian fluid, the first of five partial differential equations, written in Cartesian tensor notation, required to model the flow is the continuity equation:

$$\frac{\partial \rho}{\partial t} + \frac{\partial(\rho u_i)}{\partial x_i} = S^m \tag{4.1}$$

where ρ is the fluid density, u_i are the velocity components in each of the x_i directions, and S^m represents a source of mass generated on a volumetric basis. The first term on the left side of this equation represents the time rate of change of mass per unit volume, and the second term represents the change of mass resulting from convective motion. The next three equations, summarized as one equation written in Cartesian tensor notation, are the momentum equations or the equations of motion, one for each of the coordinate directions.

$$\frac{\partial(\rho u_i)}{\partial t} + \frac{\partial(\rho u_i u_j)}{\partial x_j} = B_i - \frac{\partial p}{\partial x_i} + \frac{\partial}{\partial x_j}\left[\mu\left(\frac{\partial u_i}{\partial x_j} + \frac{\partial u_j}{\partial x_i} - \delta_{ij}\frac{2}{3}\left(\frac{\partial u_k}{\partial x_k}\right)\right)\right] + S_i^u \tag{4.2}$$

These equations are sometimes referred to as the Navier-Stokes equations. In these equations, p is the local pressure, μ is the dynamic viscosity of the fluid, S_i^u are the sources of momentum in each of the coordinate directions generated on a volumetric basis, and B_i are body forces acting on the fluid. The first term on the left side of this equation represents the time rate of change of momentum per unit volume, and the second term represents the change of momentum resulting from convective motion. On the right side of the equation, the terms combined with μ are viscous shear forces resulting from the motion of the fluid.

The fifth equation is the energy equation, in terms of enthalpy:

$$\frac{\partial(\rho h)}{\partial t} + \frac{\partial(\rho u_i h)}{\partial x_i} = \left(\frac{\partial}{\partial x_j}\right)\left(\Gamma_h \frac{\partial h}{\partial x_j}\right) + \frac{\partial p}{\partial t} + \frac{\partial u_i p}{\partial x_i} + \Phi + S^h \tag{4.3}$$

where h is the specific enthalpy, Γ_h is the ratio of effective viscosity and the Prandtl number, Φ is the dissipation function, and S^h is the source of enthalpy generated on a volumetric source basis. As in the previous equations, the first term on the left side of this equation represents the time rate of change of enthalpy per unit volume, and the second term represents the change of enthalpy resulting from convective motion. The first term on the right side represents the molecular diffusion of enthalpy based on Fourier's law of heat conduction. The terms on the right side of the equation with pressure represent reversible work done on the fluid, while Φ represents the irreversible work, or dissipation of energy resulting from the shear stresses, which is a heat source that raises the internal energy of the fluid. This term is given by the expression:

$$\Phi = \mu\left[\frac{\partial u_i}{\partial x_j} + \frac{\partial u_j}{\partial x_i} - \delta_{ij}\frac{2}{3}\left(\frac{\partial u_k}{\partial x_k}\right)\right]\frac{\partial u_i}{\partial x_j} \tag{4.4}$$

Various numerical techniques have been employed in the solution of the equations presented in the previous section. The solution strategies for the incompressible form of the equations of motion are different from the compressible form because the pressure variation is no longer coupled with the fluid density (which can be constant). This results because the dilatation of the fluid (the $\frac{\partial u_i}{\partial x_i}$ term in Equation (4.2)), which is a measure of the compressibility of fluid, can be shown using the continuity equation to be identically zero for an incompressible fluid. Even for a compressible fluid, or for flows with a changing density due to changes in temperature, the dilatation effect is minimal at low velocities (Mach number less than 0.25) and is often neglected.[11] Terms involving the dilatation of the fluid therefore drop out of the equations of motion and there is no longer a linkage between the momentum and continuity equations through the density that can be used to establish the value for pressure. Mathematical manipulations of the continuity and momentum equations are required to derive an additional relationship for pressure. Methods based on this approach are referred to as "segregated" solvers, because of the separation of the direct solution of the pressure from the velocity prediction. This segregation can be done in two different ways:

1. Vorticity-stream function formulation,[16] in which two variables, the fluid vorticity and the stream function, are substituted into the continuity equation and momentum equations. Two other equations in terms of the vorticity and stream function result that do not include the pressure. These equations can be solved to predict the velocity field, and the pressure is calculated from an additional equation referred to as the Poisson equation for pressure. This solution scheme loses much of its appeal when extended to three dimensions because of the lack of a simple stream function in three dimensions.
2. Primitive variable formulations, in which the equations of motion are solved directly for velocity and pressure (the "primitive" variables) with the introduction of an additional

equation for pressure, such as the Poisson equation, or those given in the family of approaches SIMPLE (Semi-Implicit Pressure Linked Equations) and SIMPLER,[15] SIMPLEC,[17] and the approach known as PISO.18,19

In PCGC-3, the incompressible form of the equations of motion using a finite-volume form of the discretication equations are solved with the SIMPLE-based approaches.

4.2.1.2 Modeling Turbulence with Finite-Volume Approximations

Because the distance between gridpoints in finite-volume approximations is generally not sufficient to resolve all the turbulent length scales, and because of the steady-state assumption often used in flow simulations, auxiliary relationships are required to account for the effects of turbulence on the transport processes. These relationships are developed by dividing the instantaneous properties in the conservation equations into mean and fluctuating components, as shown in the following equation:[11]

$$\phi = \bar{\phi} + \phi' \tag{4.5}$$

where ϕ is any conserved scalar variable. In addition, in combustion modeling, an alternative method of averaging the flow field variables known as Favre-averaging is defined using the following relation:

$$\tilde{\phi} = \frac{\overline{\rho\phi}}{\bar{\rho}} \tag{4.6}$$

where ρ is the density. As shown in this relation, Favre-averaging is denoted using $\tilde{\phi}$ instead of $\bar{\phi}$. An advantage of Favre-averaging, or density-weighted averaging, is the elimination of density-velocity cross-product terms in the momentum equations, which is an effective way to account for the effects of density fluctuations due to turbulence.8,20 Gorner and Zinser[21] elaborate more on the advantages and disadvantages of using conventional vs. Favre-averaging.

The substitution of these variable decompositions in Equation (4.5) for velocities in the equations of motion and continuity, and time-averaging the subsequent relations results in the following steady-state forms:

$$\frac{\partial \rho}{\partial t} + \bar{\rho}\frac{\partial u_i}{\partial x_i} = \tilde{S}^m \tag{4.7}$$

$$\frac{\partial(\bar{\rho}\tilde{u}_i)}{\partial t} + \bar{\rho}\frac{\partial(\tilde{u}_i\tilde{u}_j)}{\partial x_j} = B_i - \frac{\partial\bar{p}}{\partial x_i} + \frac{\partial}{\partial x_j}\left[\mu\left(\frac{\partial\tilde{u}_i}{\partial x_j} + \frac{\partial\tilde{u}_j}{\partial x_i}\right)\right] - \frac{\partial\overline{(\rho u_i'u_j')}}{\partial x_j} \tag{4.8}$$

The terms $\overline{\rho u'_i u'_j}$ are designated Reynolds stress terms. The modeling of these terms in the time-averaged conservation equations is known as turbulent closure, which is a major challenge of turbulence modeling in fluid dynamics solution schemes. Reviews of turbulence and its associated closure problems with respect to computational modeling have been given by Rodi[22] and more recently by Wilcox.[23] A review of turbulence models and their applications to swirling flows is given by Sloan et al.[24]

No final solution to the problem of turbulence modeling in CFD has yet been found, and this area continues to be an active field of research. For example, an on-going collaborative research effort continues to assess the state-of-the-art of turbulence models by maintaining a databank of test problems and providing information on the comparison of predictions with data.[25,26]

Most common turbulence models used in comprehensive CFD codes employ a form of the Boussinesq[11,27] hypothesis to model the Reynolds stress terms, in which these stresses are assumed to be analogous to viscous dissipation stresses. A modified version of the Boussinesq hypothesis for incompressible flow[22] is given by the following expression:

$$-\overline{\rho u_i' u_j'} = \mu_e \left(\frac{\partial \tilde{u}_i}{\partial x_j} + \frac{\partial \tilde{u}_j}{\partial x_i} \right) \qquad (4.9)$$

The constant of proportionality in this equation, μ_e, is known as the dynamic eddy viscosity, turbulent viscosity, or eddy diffusivity. The incentive for modeling the Reynolds stresses in this manner is that the molecular viscosity in the instantaneous equations can now be replaced with the eddy diffusivity that accounts for the effects of turbulent flow fluctuations, and the instantaneous conservation equations can be modeled with mean values. The disadvantage of using the Boussinesq hypothesis is the use of mean-flow quantities to represent instantaneous fluid behavior, and the inherent assumption of isotropic eddy diffusivity.

Given the Boussinesq hypothesis, a hierarchy of turbulence models has resulted from different methods for calculating the eddy diffusivity. These models are normally classified by the number of partial differential equations required for the calculation. Zero-equation models use an algebraic model to determine eddy viscosity. One of the most successful and well-known zero-equation approaches is the Prandtl mixing length model,[11] where the eddy viscosity is given by:

$$\nu_e = l^2 \left| \frac{\partial \overline{u}}{\partial x_j} \right| \qquad (4.10)$$

ν_e is the kinematic eddy viscosity (μ_e/ρ) and l is the mixing length. The evaluation of l generally varies as a function of the type of flow being analyzed, such as wall boundary layer, jet, or wake. Schlichting,[28] Kays and Crawford,[29] Patankar and Spalding,[30] and Wilcox[23] provide examples of the use of mixing length models. Well-known implementations include the Cebeci-Smith[31] model, which is applicable in situations where a well-defined boundary layer can be defined, and the Baldwin-Lomax model,[32] which is a more general formulation developed for handling separated flow phenomena.

The Prandtl mixing length model is provided as one of the turbulence models in PCGC-3. This model has very good stability characteristics, and is often used as the turbulence model at the beginning of a simulation when the flowfield is not well-defined. Once a reasonable flowfield has been established, a more sophisticated turbulence model can be used.

One obvious shortcoming of the zero-equation model is the dependence of the eddy viscosity on the gradient of the mean velocity, $\frac{\partial \tilde{u}}{\partial x_j}$. This suggests that turbulent effects disappear as this gradient approaches zero, which neither common sense nor experimental measurements justify. As a more physically based alternative, it has been assumed that ν_e is given by the expression:

$$\nu_e = C_k l k^{1/2} \qquad (4.11)$$

where $k = \frac{1}{2} \overline{(u_i' u_i')}$ is the kinetic energy (per unit mass) of turbulent velocity fluctuations, and C_k is a constant of proportionality. The introduction of k results in one-equation turbulence models. The one-equation is a partial differential equation for the transport of k, which can be derived by first formulating a total kinetic energy equation from the equations of motion, and then subtracting from this the mean flow kinetic energy equation.[14] This results in the following equation for the turbulent kinetic energy:

$$\frac{\partial k}{\partial t} + \rho \tilde{u}_i \frac{\partial k}{\partial x_i} = \frac{\partial}{\partial x_i} \left(\frac{\mu_e}{\sigma_k} \frac{\partial k}{\partial x_i} \right) + G - \varepsilon \tag{4.12}$$

In this equation, σ_k, referred to as the Schmidt number, represents the ratio of the eddy viscosity used in the equations of motion to an analogous eddy diffusivity in the turbulent kinetic energy transport equation. The variable ε is known as the kinetic energy dissipation. The variable G is given by the expression

$$G = \mu_e \left(\frac{\partial \bar{u}_i}{\partial x_j} + \frac{\partial \bar{u}_j}{\partial x_i} \right) \tag{4.13}$$

which represents the generation of turbulent kinetic energy. In one-equation models, the length scale l is still determined algebraically, and the accuracy of the predictions is typically based on the assumptions used in this determination. A review of one-equation models is provided by Launder and Spalding[33] and, more recently, by Wilcox.[23] A recent one-equation model that appears to be quite accurate for practical external turbulent-flow applications is that of Spalart and Allmaras.[34]

The next step in sophistication of eddy viscosity models is to eliminate the need to specify a length scale relationship. In doing so, another partial differential equation is introduced, which leads to two-equation turbulence models. While a variety of two-equation models, such as the k-ω, have been proposed, by far the most common and well-known is the k-ε model. The k-ε model was first suggested by Harlow and Nakayama.[36] In this approach, the turbulent kinetic energy, k, is related to its rate of dissipation, ε, by the Prandtl-Kolamagorov relationship

$$\nu_e = \frac{f_\mu C_\mu k^2}{\varepsilon} \tag{4.14}$$

where f_μ and C_μ are proportionality constants that are correlated with particular kinds of flow fields, such as jets or wakes. The transport equation for turbulent energy dissipation is given by the following expression[37]:

$$\frac{\partial \varepsilon}{\partial t} + \rho \tilde{u}_i \frac{\partial \varepsilon}{\partial x_i} = \frac{\partial}{\partial x_i} \left(\frac{\mu_e}{\sigma_\varepsilon} \frac{\partial \varepsilon}{\partial x_i} \right) + f_1 C_1 G \left(\frac{\varepsilon}{k} \right) - f_2 C_2 \rho \left(\frac{\varepsilon^2}{k} \right) \tag{4.15}$$

This equation is similar to the other equations of motion discussed previously. The second term on the right side of the equation represents a source for the generation of kinetic energy dissipation, while the third term represents a "dissipation of energy dissipation." As with the equation for the turbulent kinetic energy, a Schmidt number, σ_ε, is the ratio of the effective viscosity from the momentum equation and the effective turbulent diffusivity of ε. The application of this model has been described by Launder[33] and Spalding and Jones and Launder.[37]

The k-ε formulation has become the most widely used approach for the solution of practical fluid dynamics problems because of its general applicability. This is the main turbulence model used in PCGC-3. Several reviews of the k-ε model and its variations are available in the literature, such as those by Nallasamy,[38] Lakshminarayana,[39] Rodi,[22] and Wilcox.[23] One shortcoming of this turbulence model is that the turbulence is assumed to be isotropic, that is, all normal stresses are identical. With this limitation, secondary flowfield effects cannot be accurately predicted. This can be especially important in highly swirling flows.[24] This shortcoming of the k-ε model has led to the development of various approaches for introducing anisotropic effects. One example of such a model is the nonlinear k-ε model described by Speziale.[40] This model represents the Reynolds stress tensor as a nonlinear expansion of the Boussinesq hypothesis, yet keeps the mathematical complexity of the expression manageable. The inclusion of the nonlinear terms allows for more accurate prediction of normal Reynolds stress effects, and predicts secondary flows in noncircular ducts that have been observed experimentally.[40] This turbulence model is also available in PCGC-3.

A second major shortcoming of the k-ε model is that it does not relax to laminar flow conditions at low Reynolds numbers. The k-ε model assumes a high Reynolds number, fully turbulent flow regime and auxiliary methods are required to model the transitions between laminar flow behavior and turbulent flow behavior. An important example of this is the transition from the thin viscous sublayer flow region along a wall to the fully turbulent, free stream flow region. The accuracy of the near-wall representation of the flow field is crucial in the successful prediction of wall-bounded turbulent flows, although predictions of the flow away from the wall are generally not very sensitive to the near-wall turbulence model.[41]

4.2.1.3   Wall Functions

Two main approaches have evolved for modeling the transition from laminar-to-turbulent flow: wall functions, and modifications of k-ε for near-wall, low-Reynolds number situations. Both of these options are available in PCGC-3.

In the wall function approach, an analytical expression is used to model the velocity and temperature distribution from the wall to the nearest point in the computational grid. Thus, only one gridpoint is required to model the transition from the wall boundary to the fully turbulent flowfield. The analytical expression used for this wall function is the "universal" law-of-the-wall suggested by Prandtl,[11]

$$U^+ = \frac{1}{\kappa} \ln Ey^+ \tag{4.16}$$

where

$$U^+ = \frac{U}{u_\tau} \tag{4.17}$$

$$y^+ \equiv u_\tau \frac{y}{\nu} \tag{4.18}$$

and

$$u_\tau \equiv \sqrt{\frac{\tau_w}{\rho}} \tag{4.19}$$

In these equations, U is the free-stream velocity, E and κ are constants used in the correlation equation, and τ_w is the shear stress at the wall. Corresponding relationships are available for temperature and the wall heat flux. The adaptation of this technique by Launder and Spalding[33] has become the standard wall-function approach. Modifications to the standard approach have sought to improve its accuracy for a wider range of problems, such as dividing the near-wall region into layers[42-44] and "sensitizing" the law-of-the-wall to the pressure gradient.[45]

The wall function approach is an option available in PCGC-3, and has been found to be a robust, computationally efficient, and reasonably accurate method for establishing wall boundary conditions in turbulent fluid flows. It is a poor approach, however, in flow regions where the Reynolds number is low, where fluid properties are changing rapidly, where strong body forces or pressure gradients are present, and where boundary walls have blowing or suction.[26]

4.2.1.4 Low-Reynolds-Number Approaches

With low-Reynolds-number approaches, the turbulence model is modified to include terms that account for the transition from fully turbulent to laminar flow behavior.[37] In the case of the k-ε model, new terms have been included in the partial differential equation for k, Equation (4.12), or in both the equation for k and the equation for ε, Equation (4.15). These terms are in the form of "damping functions" that are a function of the distance from the wall, the molecular viscosity, and local values for k and/or ε. This relaminarization option is also available in PCGC-3. With the low-Reynolds-number approaches, an appropriate number of gridpoints (typically about 20 to 30) must be included in the near-wall region of the flow to properly capture the transition from fully turbulent flow to the laminar flow along the wall, compared with the single point needed for the wall-function approach.[46]

4.2.2 GASEOUS TURBULENT COMBUSTION

The proper simulation of non-premixed turbulent combustion processes requires an effective scheme for simultaneously modeling both the mixing and the reactions of relevant chemical species. The approach used in PCGC-3 is described below. As a starting point, in keeping with the approach used for the mass, momentum, and energy equations, a partial differential conservation equation can be written for each of the chemical species of interest, which has the following form:

$$\frac{\partial}{\partial t}\left(\rho m_i\right) + \frac{\partial}{\partial x_i}\left(\rho u_i m_i\right) = \frac{\partial}{\partial x_i}\left(\frac{\mu_e}{\sigma_m}\frac{\partial m_i}{\partial x_i}\right) + R_i + S_i \qquad (4.20)$$

where m_i is the mass fraction of the ith chemical species, σ_m is the ratio of effective diffusion coefficient for the ith species and the turbulent momentum diffusivity, R_i is the mass rate of creation or depletion by chemical reaction, and S_i represents other sources of species creation, such as addition from the dispersed phase. An equation of this form must be solved for $N - 1$ species, where N is the total number of chemical species present in the system. In writing conservation equations for species in turbulent, reacting systems, time-averaging is required, and Favre-averaging is recommended.[8] These equations can be discretized using the same methods discussed previously. The chemical production or depletion term, R_i, can be determined (on a mass/volume basis) by the law of mass action as follows:

$$R_i = \left(v_i'' - v_i'\right) M_i\, k \prod_l C_l^{v_l} \qquad (4.21)$$

where v_i'' and v_i' are the stochiometric coefficients for the ith species as a product and a reactant respectively, M_i is the molecular weight of the ith species, k is the specific reaction-rate constant, C_l is the molar concentration of the lth reactant species, and v_l is the stoichiometric coefficient of the lth reactant species. The reaction-rate constant k is, in general, expressed by the modified Arrhenius equation (with T^{α})

$$k = AT^{\alpha} \exp\left(\frac{-E}{RT}\right) \tag{4.22}$$

where α, A, and E are reaction rate parameters, R is the ideal gas constant, and T is the temperature. Combining these two relations results in:

$$R_i = \left(v_i'' - v_i'\right) M_i AT^{\alpha} \exp\left(\frac{-E}{RT}\right) \left(\prod_l C_l^{v_i}\right) \tag{4.23}$$

Time-averaging of this equation can be accomplished by substituting the sum of a mean and fluctuating value for each of the variables and then decomposing the instantaneous variables into their mean and fluctuating components, which gives rise to a time-averaged form of Equation (4.23), along with several additional terms involving fluctuating values of T and species concentrations, C_l. The magnitudes of the fluctuating values and associated correlation coefficients are not known, however. Modeling turbulent chemical reaction terms in a comprehensive combustion code therefore requires use of simplifying assumptions.

In employing simplifying assumptions to achieve feasible solutions of chemical reaction equations, coupled with the turbulent fluid mechanics, three modeling approaches have been used which can be classified according to the two hypothetical time scales associated with chemical reactions in turbulent flow: the reaction time scale and the turbulent mixing time scale. The reaction time scale is the typical time required for the species of interest to react completely to equilibrium. The turbulent mixing time scale is the typical time required for large-scale turbulent eddies to break up and reduce to the scale where molecular interactions can take place.

In the first modeling approach, which is a limiting case scenario, the reacting species are assumed to be premixed, or the turbulent mixing time scale is assumed to be very fast compared with the reaction time scale. In this case, turbulent mixing can be ignored, and the finite-rate chemistry and associated reaction rates can be based on mean flow properties. Bowman[47] and Bray[48] have reviewed literature for this approach. This limiting case is not applicable for non-premixed turbulent hydrocarbon diffusion flames however, which is the focus of this work, and will not be discussed further.

In the other two modeling approaches, the turbulent mixing time scale is either of the same order or much longer than the reaction time scales for the major species of interest. When the turbulent mixing and reaction time scales are of the same order, both must be considered in the modeling approach. When the turbulent mixing time scale can be assumed to be much longer than the reaction time scale, however, the reaction time scale can be ignored and the modeling focuses on turbulent mixing rates. Each of these modeling approaches is used in PCGC-3 for different flow configurations and will be discussed in turn.

4.2.2.1 Modeling Reaction and Mixing Time Scales Simultaneously

Attempts to simultaneously account for mixing and finite chemical reaction rates (turbulent mixing and reactions with similar time scales) in comprehensive combustion codes have been built mainly on variations of the method first presented by Magnussen and Hjertager,[49] which employs the eddy-breakup model suggested by Spalding[50] to account for the effects of turbulence on the chemical reaction rates. The eddy-breakup model relates the rate of reaction to the dissipation rate of turbulent eddies containing products and reactants. The dissipation rate of turbulent eddies is assumed to be proportional to the ratio of the turbulent kinetic dissipation and turbulent kinetic energy, ε/k.[50] One proportionality constant is used for eddies with reactants, and another for eddies with products.

Models of this type have been used for a wide range of applications, including laminar or turbulent reaction systems, and combustion systems including premixed or diffusion flames.

In this model, three different reaction rates are calculated and the smallest rate is assumed to be the governing rate. The first reaction rate uses the modified Arrhenius reaction rate presented above as Equation (4.23). The second reaction rate represents the rate of dissipation of turbulent reactant eddies and is given by the expression

$$R_i^R = -v_i' M_i \rho B_R \frac{\varepsilon}{k} \min(m_l)$$ (4.24)

where B_R is an empirical constant for the reactants, whose mass fraction for each lth reactant species is m_l. The minimum value of m_l determines R_i^R. The third reaction rate is the rate of dissipation of turbulent eddies with products and is given by the expression

$$R_i = v_i' M_i \rho B_P \frac{\varepsilon}{k} m_P$$ (4.25)

where B_P is an empirical constant for the products, whose total mass fraction is m_P. This approach can give good results when compared with data, once the proportionality constants B_R and B_P have been adjusted for the application of interest. These constants are established for each reactive flow application, however, and have not been generalized. This approach is available as an option in PCGC-3 for modeling premixed flames only because of its lack in generality for a wide range of diffusion flame applications.

4.2.2.2 Mixture Fraction Approach

In gas-phase diffusion flames, the fuel and oxidizer are initially contained in separate streams that must be intimately contacted on a molecular level before reaction can occur. When the mixing time scale is much longer than the reaction time scale, turbulent mixing is treated in detail, but local instantaneous (infinite rate) chemistry can be assumed and the reaction chemistry can be computed using an equilibrium algorithm. A convenient method with this assumption that has found extensive use in comprehensive combustion modeling applications is based on defining a conserved scalar variable called the mixture fraction. This is the approach used in PCGC-3 for simulating gaseous and particle-laden diffusion flames. The approach used in PCGC-3 for gaseous diffusion flames is discussed below.

For cases where there are two identifiable inlet streams or states that have uniform properties, it is convenient to define a conserved scalar f, the mixture fraction, that can be used to define the degree of "mixedness" of primary and secondary mass constituents:

$$f = \frac{m_p}{m_p + m_s}$$ (4.26)

where m_p and m_s are the masses that originate with the primary and secondary sources, respectively. This variable is equal to the mass fraction of the primary stream constituent. Typically, for a combustion application, the primary stream is the fuel and the secondary stream is the oxidizer. The advantage of the mixture fraction is that any other conserved scalar, s, that is a function of f, such as the fluid density, or the mass fraction of a given element, can be calculated from the local value of f:

$$s = f s_p + (1 - f) s_s$$ (4.27)

The mixture fraction approach greatly reduces the number of conserved scalars required to describe a combustion system, but the validity of this approach requires that the turbulent diffusivity of all gas-phase species be equal and that their boundary conditions be the same, which is not unreasonable. These conditions, taken together, constitute what is sometimes called Crocco Similarity.[8]

The equation for the Favre and time-averaged transport of the mixture fraction is given by:

$$\frac{\partial \tilde{f}}{\partial t} + \frac{\partial \left(\overline{\rho} \tilde{u}_j \tilde{f} \right)}{\partial x_j} = \frac{\partial}{\partial x_j} \left(\frac{\mu_e}{\sigma_f} \frac{\partial \tilde{f}}{\partial x_j} \right) + S_f \tag{4.28}$$

which is similar in form to the other partial differential transport equations presented previously. The solution of this equation, together with the fluid mechanics model, prescribes the mean fluid values for the flow and the mixing.

In a turbulent environment, the mixture fraction will fluctuate chaotically about its mean value at every point in the reaction chamber. To characterize the mixing process and the related chemical reactions, more information must be supplied regarding the statistics of the mixture fraction fluctuations.

As a starting point for modeling the fluctuations of the mixture fraction, a variance for the mixture fraction can be defined. If the mean of the distribution of the mixture fraction (first moment about the origin) is given by the equation above, the variance, g_f (second moment about the mean, or the mean square fluctuation), of the mixture fraction can be specified as

$$g_f = \overline{\left(f - \tilde{f} \right)^2} = \frac{1}{T} \int_0^T \left[f(t) - \tilde{f} \right]^2 dt \tag{4.29}$$

where T is large compared to the time scale of the local turbulence. As a means to model the variation of g_f in a flow field, Launder and Spalding[33] derived a transport equation for g_f with appropriate terms modeled in a manner analogous to, and consistent with, the equations in the k-ε turbulence model. While the variance of the mixture fraction has a physical basis, the transport of the variance is principally a modeling device. The Favre-averaged differential equation for the transport of the mixture fraction variance, g_f, is:

$$\frac{\partial \tilde{g}}{\partial t} + \frac{\partial}{\partial x_j} \left(\overline{\rho} \tilde{u}_j \tilde{g}_f \right) = \frac{\partial}{\partial x_j} \left(\frac{\mu_e}{\sigma_{gf}} \frac{\partial \tilde{g}_f}{\partial x_j} \right) + C_{g1} \frac{\mu_e}{\sigma_{gf}} \left(\frac{\partial \tilde{f}}{\partial x_j} \right)^2 - C_{g2} \frac{\overline{\rho} \varepsilon \tilde{g}_f}{k} \tag{4.30}$$

Appropriate constants for the combustion submodel typically used are $C_{g1} = 2.8$, $C_{g2} = 1.92$, $\sigma_f = 0.9$, and $\sigma_{gf} = 0.9$.

Besides $\tilde{f}$ and g_f, the functional form, or "shape," of the probability density function (PDF) associated with f is required to calculate time mean values. The PDF shape is often determined from experimentally observed fluctuations in coal flames. Smoot and Smith,[8] for example, discuss a uniform (top hat) distribution and clipped-Gaussian, which accounts for intermittency. Kent and Bilger[52] used a clipped-Gaussian distribution, and noted significant sensitivity of the predictions to the form of the PDF, particularly with respect to nitric oxide concentrations. Other references pertaining to PDF shape are found in Kuo[53] and Lockwood and Naguib.[54] In regions where g_f is small, the shape of the PDF makes little difference.

As an alternative to specifying the shape of the PDF, methods have recently been developed for reactive turbulent flows that use a joint PDF of velocity and any number of scalars, such as a mixture fraction.[55] A transport equation for the joint PDF is also derived from governing equations, and the evolution of the PDF is calculated along with the other conservation variables. The major advantage of this approach is that reaction effects can be included directly, and both premixed and non-premixed flames can be characterized. For computational efficiency when considering complex chemistry, the PDF solution is performed using a Monte Carlo approach. Incorporation of the Monte Carlo PDF calculations in a finite-volume flow solver has been described by Correa and Pope,[56] and a review of the these technologies has been given by Brewster et al.[4]

4.2.3 RADIATIVE ENERGY TRANSPORT

For many combustion processes, radiation is the dominant energy transport mechanism to surrounding surfaces, particularly when entrained particles are present. Typically, as the combustion length scale increases up to about one meter, the intensity of the radiation on a wall also increases.[57] Characterizing radiative energy transport is therefore a crucial element in modeling combustion systems, but it is also one of the most complex problems. For example, in a typical coal-fired furnace environment, radiation includes contributions from both particulates (coal/char, ash, and soot) and gases (mainly CO_2 and H_2O). The accuracy of the radiation calculation depends on a combination of the accuracy of the calculation method and the accuracy to which the properties of the radiating media and surrounding walls are known.

A large body of literature relevant to this subject exists and only a brief overview is provided here to summarize the basic principles and methods for modeling radiative heat transfer in combustion systems. The interested reader is referred to in-depth review articles by Viskanta and Menguc[57] and Menguc and Webb[58] for more information. Three fundamental topics related to radiative heat transfer are discussed here:

1. Governing equations for radiative heat transfer
2. Radiative properties of gases and entrained particles
3. Solutions of the radiative heat transfer equation

4.2.3.1 Governing Equations for Radiative Heat Transfer

Two complimentary approaches can be used to study the interaction of electromagnetic radiation with gases, solids, and/or liquids. In the first approach, which deals primarily with electromagnetic wave theory, the focus is on the interaction of radiation with a participating medium (gas, solid, or liquid) at the atomic or molecular level. This fundamental approach provides predictions of the macroscopic radiative properties of the medium. These properties are employed in the second approach used in the study of electromagnetic radiation, or heat transfer by radiation. In radiative heat transfer, the focus is on describing radiative interactions with a participating medium. This process is characterized by absorption, emission, and scattering of radiant energy.

The basis for the quantitative study of radiative heat transfer in a participating gray medium is the radiative transfer equation, which is given for a steady-state system by the following expression[58]:

$$\frac{dI(s,\omega)}{ds} = -(\kappa + \sigma)I(s,\omega) + \kappa I_b + \frac{\sigma}{4\pi}\int_{4\pi} I(s,\omega)\Phi d\omega \tag{4.31}$$

where I is the radiative intensity, (s, ω) are unit vectors in the directions of propagation, κ and σ are the local absorption and scattering coefficients, respectively, and Φ represents a phase function used to characterize the nature of the scattering media. The phase function (Φ) may be physically interpreted as the scattered radiative intensity in a given direction, divided by the scattered radiative intensity in the same direction if the scattering were isotropic. The term on the left side of the equation represents the gradient of intensity in the specified direction, ω. The three terms on the right side of the equation represent the changes in intensity due to absorption and out-scattering, emission, and in-scattering, respectively. The derivation of this equation, and several references describing various approaches for its derivation, are presented by Viskanta and Menguc.[57] The boundary conditions for the radiative transfer equation are defined in terms of wall absorption, emission, and reflection coefficients. Integration of Equation (4.31) results in an expression for the conservation of radiant energy, which provides the radiant energy source term for the energy equation.

The accuracy of the solution of the radiative transfer model is highly dependent on accurate knowledge of radiative properties of combustion product gases and entrained particles such as residual matter from fuel oil droplets, coal, ash, or soot. Usually, droplets evaporate quickly and do not enter significantly into the radiative heat transfer. The radiative intensity from the particles such as coal and soot, however, typically dominate because particles emit, absorb, and scatter radiation continuously over the entire wavelength spectrum, while combustion gases participate only in narrow bands centered around discrete radiative wavelengths.[58]

4.2.3.2 Radiative Properties of Gases

At the most fundamental level, radiation is absorbed and emitted in gases only at the discrete frequencies at which electrons become excited. Methods for calculating the absorption and emission properties of combustion gases have been developed using well-established mathematical techniques that depend only on temperature and partial pressures of the primary, participating gaseous species, which, in the case of most combustion processes, are H_2O, CO, and CO_2. While NOx and SO_x are also strong radiation absorbers and emitters, the concentrations of these species are typically small and can be neglected. Scattering effects in combustion gases can often be neglected.

The methods for predicting the radiative properties for gases can be classified according to the level of detail and sophistication required to determine the absorption at the discrete electron excitation frequencies, and fall into the following four categories:

1. *Line-by-line methods*, where quantum mechanics is used to calculate the absorption at each individual wavelength in the electromagnetic spectrum. This approach is very computationally intensive.
2. *Narrow-band methods*, in which several absorption "lines" are approximated by a narrow band. The lines are assumed to broaden (due to the effects of various physical mechanisms) into a shape that comprises several frequencies.
3. *Wide-band methods*, in which narrow-bands are approximated by a wide-band shape.
4. *Property correlations*, in which the radiative properties generated for each species of interest using methods 1, 2, or 3, are tabulated and fitted with polynomial expressions that yield the total property as a function of weighting factors for each of the species as functions of pressure and temperature.

In typical comprehensive combustion models used for engineering purposes, the first three approaches are not warranted, and satisfactory results (within the context of the assumptions of other models) can be obtained for typical combustion chambers using the property correlations. Several curve-fitted expressions are available in the literature for use in computer codes, some of which are given in terms of polynomials, and others expressed in terms of the weighted sum-of-gray gases (WSGGM) and variations of this model. The WSGGM approach is available as an option in PCGC-3 for predicting the radiative properties of the gas phase.

4.2.3.3 Radiative Properties of Entrained Particles

The radiative properties required for an entrained particle phase are the absorption coefficients and the scattering phase function, which depend on the particle concentration, size distribution, and effective complex refractive indices. However, optical properties of coal are not well-character-ized,[59,60] and considerable uncertainties exist, particularly regarding the size and concentration of soot and the refractive index of ash. The absorption and scattering efficiencies depend strongly on these properties, which limits the accuracy of the calculation of these properties. The approach used in PCGC-3 for calculating the radiative properties of entrained particles is outlined below.

Generally, as a starting point to arrive at a tractable method for calculating radiative properties, the particles are assumed to be spherical and homogeneous. Although it is known that typical reacting particles in combustion chambers are neither spherical nor homogeneous, the radiative characteristics of a cloud of irregularly shaped particles are not very sensitive to the geometrical shape of the particles, and the assumption of spherical particles seems adequate for combustion systems such as pulverized coal.[57] Given these assumptions, the absorption cross-sections can be calculated using Mie theory[61,62] based on a specified particle size, wavelength of the radiation, and the complex refractive index. Once the absorption and scattering efficiencies for individual particle sizes are known from Mie theory, the absorption and scattering coefficients of the char and ash particles can be evaluated by summing over the number of discrete particle sizes being used for a specific case. The absorption coefficient for these particles (κ_p) is then given as.

$$\kappa_p = \sum_j \kappa_j = \sum_j (\pi/4) Q_{aj} \bar{n}_j d_j^2 \tag{4.32}$$

and similiarly the scattering coefficient for these particles (σ_p) is given as:

$$\sigma_p = \sum_j \sigma_j = \sum_j (\pi/4) Q_{sj} \bar{n}_j d_j^2 \tag{4.33}$$

where Q_{aj} is the absorption efficiency of particle size j, Q_{sj} is the scattering efficiency of particle size j, n_j is the number density of particle size j, and d_j is the diameter of particle size j. Soot particles absorb highly, but scatter negligibly due to their microscopic size. Unfortunately, Mie theory is strictly applicable only to isolated particles interacting with plane waves. Because of this limitation and due to the highly forward-scattering properties of pulverized coal, scattering inten-sities are often approximated by phase functions. The scattering phase function is modeled with the Dirac-delta approximation outlined by Crosbie and Davidson.[63]

The overall absorption coefficient for the volume can then be obtained as

$$\kappa = \kappa_p + \kappa_g \tag{4.34}$$

and the total radiative source for the gas enthalpy equation (q_r) is:

$$q_r = \kappa_g \left[\int_{4\pi} I d\omega - 4E_b \right] \tag{4.35}$$

where E_b is the blackbody emission of the gas. For a gray analysis, the blackbody emissive power, E_b, for the gas is given by:

$$E_b = \sigma T_g^4 \tag{4.36}$$

where T_g is the local gas temperature.[64] Particle radiative properties and particle radiative emission are calculated using the source-in-cell technique[65] for the Eulerian radiation field calculations used in PCGC-3.

4.2.3.4   Radiative Heat Transfer Equation Solutions

The radiative heat transfer equation (RTE), Equation (4.31), is an integro-differential equation for which exact solutions are not available for practical engineering applications. Multidimensionality, nonhomogeneous media, and the spectral variation of radiative properties make the RTE solution quite difficult, but reasonably accurate numerical solutions of the RTE can be obtained by introducing certain approximations.[58] Because it is not possible to develop a single solution method that is equally applicable to a wide variety of different systems, several solution methods with varying degrees of approximation have been developed according to the nature of the physical system, characteristics of the medium, the degree of accuracy required, and the availability of computer facilities.[57] The major approaches can be summarized as follows: (1) statistical methods, (2) zonal method, (3) flux methods, including the discrete-ordinates approximation, (4) moment methods, (5) spherical harmonics approximation, and (6) hybrid methods.

Descriptions of the statistical methods, also known as the Monte Carlo methods, for radiative heat transfer calculations have been given by Howell[66] and Haji-Sheikh.[67] These are purely statistical approaches that can be used for any complex geometry, and spectral effects can be accounted for without much difficulty.

The zonal method, usually known as Hottel's zonal method, is one of the most widely employed models for calculating radiative transfer in combustion chambers. This approach has been described by Hottel and Sarofim[66] and Hottel and Cohen.[68] In this method, the surface and volume of the combustion chamber are divided into a numbers of zones, each of which has a uniform temperature and uniform radiative properties. An energy balance is written for each of the zones, which leads to a set of simultaneous equations for the unknown heat fluxes or temperatures. Because the approach is practical and powerful, it is attractive for many engineering calculations. However, it has not proven to be computationally efficient compared with other methods when coupled with finite-volume reactive fluid flow approaches used in comprehensive combustion models.[8,69]

Flux methods are based on separating the angular dependence of the radiant intensities that arise from the spatial dependence of the in-scattering source term (the integral part of Equation (4.31)). By employing the assumption that intensities are uniform over defined intervals of the solid angle, the integro-differential equation can be simplified into a series of coupled, linear, differential equations expressed in terms of average radiative intensities or fluxes. Examples of these techniques are given by Smoot and Smith[8] and Viskanta and Menguc.[57] These approaches have been particularly effective for simultaneous use with reacting flowfield solutions.

The discrete-ordinates approach is a particular case of a flux method that was originally developed by Chandrasekhar[70] for astrophysical applications and has also been used extensively in analyzing neutron transport, where detailed descriptions have been presented.[71,72] The discrete-ordinates approach is used in PCGC-3 and is discussed in more detail. In the discrete-ordinates model, a quadrature scheme is used to integrate the in-scattering term of Equation (4.31). The quadrature set consists of ordinates (radiant energy directions) and weights that are chosen by applying appropriate constraints, such as symmetry and moments matching, as well as conserving radiant energy within a control angle and the total control volume. The quadrature scheme used in PCGC-3 is the S_N approach, in which the entire solid angle is divided into N(N+2) angular subdomains, where N is determined by the order of the quadrature scheme. Quadrature orders of 2, 4, or 6 can be used in PCGC-3, with the computational time increasing with increasing quadrature order. A quadrature order of 2 has proven to be adequate for most combustion applications using PCGC-3. Applications of the S_N approach have been given by Fiveland,[69,73,74] where it was demonstrated to give both accurate and computationally efficient results compared with other approaches. Raithby and Chui[75,76] and, more recently, Chai et al.[77] have presented quadrature approaches based on a control volume scheme for the radiant heat transfer quadrature that is the same as used for the fluid mechanics and convective heat transfer calculations. Control volumes can be based on either structured or unstructured grid types. Because of the joint capabilities of common control volumes and grid structures for coupling the radiation and reactive fluid flow solutions, the discrete-ordinates method is becoming the method of choice in comprehensive combustion models, as well as in other fluid mechanics and heat transfer related processes where radiation is an important mode of energy transfer.

In other approximations of the radiative heat transfer equation, radiative intensity is expressed as a series of products of angular and spatial functions. This procedure can be used to eliminate the integral part of Equation (4.31) and yield a series of equations in terms of different orders of moments. A moment is defined as the integral of intensity multiplied by a power of a direction cosine over a predetermined solid angle division. If the angular dependence is expressed using a Taylor power series expansion, the method known as the moment method results; and if spherical harmonics are used to express the intensity, the spherical harmonics (P_N) approximation results. Application of the moment method has been presented by DeMarco and Lockwood[78] and Lockwood and Shah[79] for radiative transfer in three-dimensional rectangular enclosures. The spherical harmonics approximation in multi-dimensional geometries has been presented by many researchers (see, for example, Higenyi and Bayazitoglu[80] and Ratzel and Howell[81]), with general equations for the solution of P_1 and P_3 approximations in absorbing, emitting, and anisotropically scattering in cylindrical and three-dimensional rectangular enclosures developed by Menguc and Viskanta.[82,83]

Finally, combinations of the methods for solving the radiative transfer equation described above have been used to formulate hybrid methods, which attempt to compensate for the flaws of one approach with the strengths of another approach. Several of these hybrid approaches have been summarized by Viskanta and Menguc.[57] As an example of a hybrid approach, the "discrete transfer" model proposed by Lockwood and Shah[79] combines the virtues of the zonal, Monte Carlo, and discrete-ordinates methods. Although designed for computing radiation in absorbing, emitting, and scattering media, no results have been reported for scattering in media in multi-dimensional enclosures. This method has been applied in furnace simulations[84,85] and has been adapted for coupling with complex fluid mechanics and heat transfer grid topologies.[86]

4.2.4 PCGC-3 Multiphase Flow Models for Reacting Flows

Several commonly used combustion applications of PCGC-3 involve multiple phases of matter reacting and interacting simultaneously. The source of fuel in such processes is often a solid or liquid that interacts with a turbulent gas flow-field comprising the oxidizer and other reacting and nonreacting chemical species, and the products of reaction. PCGC-3 was originally designed specifically for multiple-phase combustion.

To reliably characterize the processes of interest, a comprehensive combustion model must be able to account for the mass, momentum, and energy transport for each of the phases of interest, as well as the exchange of these quantities between phases. The constitutive relationships that describe the exchange processes can be one of the most challenging parts of the model development, because the details of the physics and chemistry governing the inter-phase phenomena may not be completely understood and may be difficult to determine experimentally. The variety of multiphase flows, and the difficulty of determining the constitutive relationships, dictate that the comprehensive modeling approach be tailored to the problem of interest. For example, the model of a fixed or fluidized-bed combustion process, where the solid phase has a large volumetric fraction, will require a different computational methodology than a process such as pulverized coal combustion where the particles are entrained in the gas and have an almost insignificant volumetric fraction. PCGC-3 is limited to application of multiphase processes where the solid particles or liquid droplets entrained in a gas flow are dilute (negligible volume fraction) and dispersed (particle-particle interactions can be neglected).

4.2.4.1   Modeling Dilute, Dispersed Multiphase Flow

Extensive research has been done to characterize a variety of dilute, dispersed flow fields such as particle-laden gas and liquid jets,[87,88] evaporating, nonevaporating, and combusting sprays,[89-91] noncondensing and condensing bubbles in jets,[92-94] free shear flows,[95] and particle dispersion in a uniform, grid-generated turbulent flow.[96] Reviews by Crowe et al.,[97] Stock, Faeth,[99,100] Lumley,[101] Crowe,[102] Kuo,[53] Sirignano,[103] and Shirolkar et al.[104] provide good summaries of current research results and directions.

In the analysis of dilute, dispersed flows, methods can be divided into two general categories based on the approach used to treat interphase transport rates (the constitutive relationships discussed earlier). The first category has been referred to by Faeth[100] as locally homogeneous flow (LHF), where interphase exchange rates for mass, momentum, and energy are assumed to be infinitely fast. This implies that the phases are constantly in dynamic and thermodynamic equilibrium at each point in the fluid. In the special case where the density of the particle is similar to that of the fluid, particle inertial effects can be neglected and the particles follow the fluid motion like a point in the fluid. Because of this equilibrium constraint, LHF analysis only accurately represents dispersed flows whose dispersed-phase elements are very small. When the values of parameters required for a more complete analysis are unknown, however, or have a large degree of uncertainty, LHF can be a viable alternative. Significant advantages of LHF are the relatively little amount of information required to specify initial conditions, a simplified computation formulation because the model is equivalent to a single-phase flow analysis, and the relative ease of handling such complex flow phenomena as the creation and disappearance of a variety of material phases.

The other category of dilute, dispersed phase analytical methods is separated flow (SF) methods, where finite interphase transport rates are considered. In attempting to best characterize the constitutive relationships describing the transport rates, three main SF formulations have evolved[53,103]: (1) Eulerian-Eulerian, (2) Eulerian-Lagrangian, and (3) probabilistic. The Eulerian-Lagrangian approach is used in PCGC-3.

4.2.4.2   Eulerian-Eulerian Separated Flow Model

In this approach, both the gas-phase and the entrained particle-phase are considered to be continuous in space and time, and a similar set of discretization equations is solved for each phase. If the volume of the entrained phase cannot be neglected, an additional relationship describing the local volumetric fraction of each phase is required to close the set of conservation equations. As applied to the case of dilute, dispersed multiphase systems, an important assumption in using this approach is that the presence of the particles is well-represented by averaged property values in a control volume, and that the local volumetric fraction of the particle is negligible. This approach for dilute particle loadings has been described by Sirignano[103] and Elghobashi et al.[105] In particular, a two-equation turbulence model for a two-phase system was developed and applied to the prediction of a particle-laden gas jet. This method suffers, however, from a severe problem of numerical diffusion for the dilute phase.

4.2.4.3   Lagrangian-Eulerian Separated Flow Model

This is the model used in PCGC-3 for modeling the particle phase in combustion modeling and is discussed in more detail. In this method, the gas phase is modeled as a continuum using the discretization approaches discussed earlier. The entrained particle phase is modeled using the following set of ordinary differential equations for mass, momentum, and energy that characterize the change in particle properties along particle trajectory as it moves through the gas continuum:

$$\frac{dm_p}{dt} = -R_p \tag{4.37}$$

$$m_p \frac{du_p}{dt} = \sum F_p \qquad (4.38)$$

$$\frac{d(mh)_p}{dt} = \sum Q_p \qquad (4.39)$$

The variable m_p is the mass of the particle and R_p is rate of change of mass due to phenomena such as droplet vaporization, particle devolatilization, or char oxidation; u_p is the velocity of the particle and F_p is the sum of the various forces acting on the particle such as drag and gravity; $(mh)_p$ is the enthalpy of the particle and Q_p is the sum of the various sources of energy, such as conduction or convection from the gaseous flow, radiative energy transport, or heat-of-reaction. An example of such an approach that is widely used is the PSI-Cell (Particle-Source In Cell) method.[106] At each point in the trajectory, the local gas properties are used to determine the boundary conditions for the particle, and constitutive relationships for drag and heat transfer coefficients are used to calculate momentum and heat transfer exchange rates. Additional relationships are used to calculate particle mass loss rates, such as by vaporization, devolatilization, or char oxidation, which are used to determine the mass exchange rates. The amount of mass, energy, and momentum exchange is balanced by the use of source terms in the ordinary differential equations for the particles, and the discretization equations for the gas contiuum.

One complication in the use of the Eulerian-Lagrangian formulation is the characterization of dispersion of the particles resulting from the turbulent motion of the gas. This is often done by adding a fluctuating component or turbulent diffusive velocity to the mean velocity. The turbulent diffusive velocity has been determined by a variety of methods, including random-walks,[107] proportional relations between the particle diffusion velocity and the gas fluctuating velocity,[108] and with a diffusion velocity based on particle number density gradients similar to the Fickian diffusion phenomenon.[8] One of the most rigorous techniques in modeling turbulent dispersion of particles is the stochastic, separated flow (SSF) model, which is based on a Monte Carlo simulation of the particle flowfield.[109,110] This is a computationally intensive method that characterizes the overall particle phase behavior using thousands of particle trajectories. The particle diffusion velocity approach and the SSF model are the two options available in PCGC-3 for modeling particle diffusion.

4.2.5 PARTICLE REACTION SUBMODELS

An additional constitutive relationship required for modeling combusting particles such as coal, char, or liquid fuels is the rate of change in mass of the droplet or particle due to processes such as vaporization, devolatilization, and char oxidation. Most often, this rate is not known, nor is the assumption of a constant rate valid, in which case the rate must be calculated as a function of the particle size, composition and temperature, and the local gas environment.

In the case of droplets, and for the portion of a particle that may be liquid, the vaporization rate can often be assumed to be directly proportional to the droplet heating rate, once the droplet reaches the vaporization temperature. The proportionality constant is simply the heat of vaporization. Additional details describing the droplet heatup may be required, however, when it cannot be assumed that the droplet temperature and composition are homogeneous. In PCGC-3, both coal devolatilization and char oxidation processes are modeled.

4.2.5.1 Particle Devolatilization

PCGC-3 was developed for coal combustion applications, although the methodologies for coal can be applied to the pyrolysis of other organic substances, such as wood or various waste materials. Many coal devolatilization models have been developed and several reviews of these models have been published.[111-114] More simple approaches that have been used in comprehensive combustion models use correlations of volatile yield with particle temperature,[115] or define devolatilization rates with single- or two-step Arrhenius reaction schemes,[116-118] such as

$$r_v = f_v m_p A_v \exp\left(-\frac{E_v}{RT_p}\right) \qquad (4.40)$$

where r_v is the volatile production rate, f_v is the fraction of volatiles in the mass of the coal particle, m_p is the particle mass, and A_v and E_v are the rate constants, R is the universal gas constant, and T_p is the temperature of the particle. The two-equation method[118] for coal devolatilization is used in PCGC-3:

$$coal \xrightarrow{k_{1v}} \left(1 - Y_1\right)_{char} + Y_{1volatile} \qquad (4.41)$$

$$coal \xrightarrow{k_{2v}} \left(1 - Y_2\right)_{char} + Y_{2volatile} \qquad (4.42)$$

where Y_1 and Y_2 are stoichiometric coefficients representing the partitioning of coal into char and volatiles, and k_1 and k_2 have the form of Equation (4.40). Typically, the activation energy E_{v2} is much greater than E_{v1}, and Y_1 is near the proximate (daf) volatile fraction, while Y_2 is higher and closer to unity. With this method, the volatiles percentage released from the coal is temperature dependent, which matches experimental observations.[117]

More generalized approaches, which become correspondingly more complex, consider the evolution of different volatile species based on network models of coal devolatilization. While this mechanistic approach for modeling devolatilization is more complicated than the standard empirical rate models, greater flexibility, reliability, and more general applicability can be obtained. These types of models include the Functional Group-Depolymerization, Vaporization and Cross-linking (FG-DVC) model,[119,120] the FLASHCHAIN model,[121-123] and the Chemical Percolation Devolatilization (CPD) model.[124-126] Smith et al.[114] provides a detailed comparison of these three coal-structure-based submodels, and a brief summary of that comparison follows. The CPD model is provided as an optional devolatilization model in PCGC-3, and is discussed below.

The CPD model incorporates detailed information about the coal composition and structure determined from NMR (nuclear magnetic resonance) analysis, ultimate analyses, swelling data, and experimental tar yields. It defines a simplified coal structure based on this information, which is used in describing the coal devolatilization process.

According to the CPD model for coal pyrolysis,[124] two general sets of fragments are devolatilized from the coal. One set of fragments has low molecular weight (and correspondingly high vapor pressure) that escape from the coal particle as light gases. The other set of fragments are tar gas precursors that have a relatively high molecular weight (and correspondingly low vapor pressure), and tend to remain for longer periods of time in the coal during typical devolatilization conditions. During this time, reattachment with the coal lattice can occur, which is referred to as cross-linking. The high molecular weight compounds plus the residual lattice are referred to as metaplast. The softening behavior of a coal particle is determined by the quantity and nature of the metaplast generated during devolatilization. The portion of the lattice structure that remains after devolatilization comprises char and mineral matter.

The CPD model characterizes the chemical and physical processes by considering the coal structure as a simplified lattice or network of chemical bonds or bridges that link the aromatic clusters. The description of the breaking of the bonds and the generation of light gases, char, and tar precursors is then considered to be analogous to the following chemical reaction scheme[114]:

$$(4.43)$$

The variable β represents the original population of labile bridges in the coal lattice. Upon heating, these bridges become the set of reactive bridges, β'. For the reactive bridges, two competing paths are available. In one path, the bridges react to form side chains, δ. The side chains may detach from the aromatic clusters to form gases, g_1. As bridges between neighboring aromatic clusters are cleaved, a certain fraction of the coal becomes detached from the coal lattice. These detached aromatic clusters are the heavier molecular weight tar precursors that form the metaplast. The metaplast vaporizes as coal tar. The metaplast can also reattach to the coal lattice matrix (cross-linking). In the second path, the bridges react and become a char bridge, c, with the release of an associated light gas product, g_2. The following set of reaction rate expressions describes this reaction scheme assuming that the reactive bridges are destroyed at the same rate that they are created[114,124] (i.e., $(\partial\beta')/(\partial t) = 0$):

$$\frac{d\beta}{dt} = -k_b\beta \tag{4.44}$$

$$\frac{dc}{dt} = -k_b\frac{\beta}{\left(k_c+1\right)} \tag{4.45}$$

$$\frac{d\delta}{dt} = \left[2k_ck_b\frac{\beta}{\left(k_c+1\right)}\right] - k_g\delta \tag{4.46}$$

$$\frac{dg_1}{dt} = k_g\delta \tag{4.47}$$

$$\frac{dg_2}{dt} = 2\frac{dc}{dt} \tag{4.48}$$

The variables k_b, k_c, k_g are rate constants with Arrhenius form

$$k = A\exp\left(-\frac{E}{RT}\right) \tag{4.49}$$

Given the above set of reaction equations for the coal structure parameters, it is necessary to relate these quantities to changes in coal mass and the related release of volatile products. The fractional change in the coal mass as a function of time is divided into three parts: light gas, f_{gas}, tar precursor fragments, f_{frag}, and char, f_{char}, and algebraic expressions are obtained for each using percolation lattice statistics.

In accounting for mass in the metaplast (tar precursor fragments), the part that vaporizes is treated in a manner similar to flash vaporization, where it is assumed that the finite fragments undergo vapor/liquid phase equilibration on a time scale that is rapid with respect to the bridge reactions. As an estimate of the vapor/liquid that is present at any time, a vapor pressure correlation based on a simple form of Raoult's law is used. For the part of the metaplast that reattaches to the coal lattice, a cross-linking rate expression is used[114]:

$$\frac{dm_{cross}}{dt} = k_{cross} m_{frag} = m_{frag} A_{cross} \exp\left(-\frac{E_{cross}}{RT}\right)$$ (4.50)

where m_{cross} is the amount of mass reattaching to the matrix, m_{frag} is the amount of mass in the tar precursor fragments (metaplast), and A_{cross} and E_{cross} are rate parameters.

For this set of equations and corresponding rate constants introduced for the CPD model,[125,126] five parameters are coal-specific[114]:

1. Lattice coordination number, $\sigma + 1$
2. Initial number of bridges in the coal lattice, p_o
3. Mass of the coal in the clusters, $n_{cluster}$
4. Mass of coal in the bridges, n_{bridge}
5. Initial number of char bridges, c_o

The first four of these are coal structure quantities that can be obtained from solid-state NMR experimental data.[114] The last quantity is estimated from the initial char content of the coal. Table 72 of Smith et al.[114] tabulates values for these parameters for 20 coals of various rank, while other parameters in the CPD submodel that do not vary with the coal are shown in their Table 71. Much more detail on this network-based devolatilization is given by Smith et al.[114]

Network models for devolatilization such as CPD add much to the complexity of the calculation. The trade-off for the added complexity is a more mechanistic, generalized approach that can be more applicable to a variety of coals. Recent research in refining network-based devolatilization submodels has included work on selective fuel-nitrogen species release during primary and secondary coal pyrolysis.[128]

4.2.5.2   Char Oxidation

Char oxidation in coal combustion is a complex process that involves balancing the rate of mass diffusion of the oxidizing chemical species to the surface of a coal particle with the surface reaction of these species with the char. Both the diffusion of oxygen to the surface and the rate of surface reaction are considered simultaneously in a quasi-steady-state reaction scheme. The approach used in PCGC-3 is briefly outlined here, with further details available in Smoot and Smith,[8] Mitchell,[129] and Hurt and Mitchell.[130]

The rate of surface oxidation is given by the kinetic relationship presented in Mitchell[129]:

$$r_{ox} = A_{ox} \exp\left(-\frac{E_{ox}}{RT}\right) P_{op}^n$$ (4.51)

where A_{ox}, E_{ox}, and n are rate parameters and P_{op} is the partial pressure of oxygen at the surface of the particle, which is not known a priori. A value of $n = 0.5$ best correlated their data.[130] The net rate of diffusion of the oxidizer, r_{do}, accounts for the rate of diffusion of oxidizer to the surface and the rate of bulk mass transport of oxidizer away from the surface. A simple expression[8,114] for this relationship is:

$$r_{do} = k_m M_o A_p \left(C_{og} - C_{op}\right) - r_{do}\left(\frac{M_o C_{op}}{M_{cg} C_{cg}}\right)$$ (4.52)

The first term on the right side represents the transport of the oxidizer to the surface of the particle, where k_m is the mass transfer coefficient, M_o is the molecular weight of the oxidizer, A_p is the surface area of the particle, and C_{og} and C_{op} are the molar concentrations of the oxidizer in the bulk flow and at the surface of the particle, respectively. The second term on the right side represents the bulk transport of the oxidizer away from the surface of the particle, where r_d is the total mass transfer rate away from the surface of the particle, and M_{cg} and C_{cg} are the molecular weight and molar concentration, respectively, of the gases leaving the particle and entering the bulk gas flow. If it is assumed that the char oxidation rate is equal to the oxidizer diffusion rate (i.e., quasi-steady state), then:

$$r_{ox} = r_{do} \phi_l \frac{M_c}{M_o} \tag{4.53}$$

ϕ_l is the stochiometric ratio of moles of char per mole of oxidizer. Then Equation (4.52) can be rearranged into the following form:

$$r_{ox} = \frac{A_p M_c \rho_l k_m \left(C_{og} - C_{op} \right)}{1 + \left(-\dfrac{dM_p}{dt} \left(\dfrac{\phi_l M_c C_{op}}{r_{ox} M_{cg} C_{cg}} \right) \right)} \tag{4.54}$$

The value of ϕ_l depends on whether the gaseous product of carbon oxidation at the surface is CO or CO_2.[129] Hurt and Mitchell[130] provide the following relationship to evaluate ϕ_l from a correlation of pulverized coal char data for 19 chars:

$$\frac{\text{moles} - CO}{\text{moles} - CO_2} = A_{co} \exp\left(-\frac{E_{co}}{RT} \right) \tag{4.55}$$

where A_{co} and E_{co} are the Ahrrenius constants and differ in values among the various coal char ranks. In general, CO is the main product at higher char surface temperatures, while some CO_2 is formed at lower temperatures.[114]

Assuming quasi-steady-state conditions, the rate of oxidation due to the heterogeneous reaction (Equation (4.51)) at the surface is equated with the net rate at which oxygen can reach the surface of the particle by diffusion (Equation (4.52)), thereby providing values for the char oxidation rate and C_{op}, which are typically determined in an iterative solution procedure, in conjunction with an energy balance for the char particle to provide T_p. Hurt and Mitchell[130] provide a correlation for the pre-exponential factor, A_{ox}, and activation energy, E_{ox}, with varying carbon content for the family of 19 coal chars. This correlation provides a basis for estimating A_{ox} and E_{ox} for coal chars where no data are available.

4.2.6 NOx Pollutant Formation Models

Oxides of nitrogen (NOx) are atmospheric pollutants (i.e., NO, NO_2, N_2O) that contribute to the formation of acid rain and photochemical smog, and to the depletion of ozone.[131] Stringent standards and guidelines have been established to limit the amount of NOx that can be generated in combustion processes, and the prediction of NOx has been a major driving force in comprehensive combustion modeling.[131] The development of an effective model to predict formation of this pollutant in combustion processes requires an adequate characterization of homogeneous and heterogeneous reactions among the nitrogen, fuel, and oxidizer species, with additional descriptions, when necessary, of the nitrogen conversion from fuel and soot to volatile gaseous nitrogen species. The kinetic mechanisms must be limited to sufficiently few homogeneous reactions to allow for effective coupling with the turbulent mixing process. The approach used for calculating each reaction rate is therefore dependent on both the relative time scales of reaction and turbulent fluctuations in the gas temperature and species concentrations.

Typically, the NOx emitted to the atmosphere from combusting fuels consists mostly of nitric oxide (NO), with much lower contributions from nitrogen dioxide (NO_2) and nitrous oxide (N_2O), although production of N_2O becomes more important in lower temperature, coal-fired fluidized beds and in lean, premixed gaseous combustion.[4] Typically, NOx models focus only on predicting the concentration of NO in the reactive flowfield of interest.[8,131] NO pollutant modeling has been reviewed by Boardman and Smoot,[131] and more recently by Hill and Smoot.[132] These reviews provide an overview of NOx mechanisms and modeling, while Hill and Smoot[132] compare published NOx submodels from several investigators. Possibly, the first NOx submodels to be developed and incorporated into a comprehensive code, and representative of the approach used in PCGC-3, is that of Smith et al.,[133,134] Hill et al.,[135] Boardman and Smoot,[136] and Smoot et al.[137]

In PCGC-3, the amount of NO produced in the combustion process is characterized using the following steady-state transport equation for the mass flow of NO:

$$\frac{\partial \rho u_i Y_{NO}}{\partial x_i} = \frac{\partial}{\partial x_i}\left(\rho D \frac{\partial Y_{NO}}{\partial x_i}\right) + S_{NO} \tag{4.56}$$

where Y_{NO} is the mole fraction of NO. This equation, which is similar to the other conservation relationships discussed previously, is general in accounting for the convection, diffusion, production, and consumption of NO in the process being modeled. Because the concentration of NO is typically very small compared with the concentrations of other species of interest in the combustion process, the NO transport equation can be solved for a given combustion flowfield solution. In PCGC-3, the NO model is therefore employed in a "post-processing" fashion, where a converged combustion flow field solution is first obtained before performing the NO prediction. It also follows that the quality of the NO prediction is dependent on the quality of the flowfield prediction.

The source term, S_{NO} in Equation (4.56), which represents the production and consumption of NO, is predicted from contributions from the following four predominant mechanisms[131]: (1) thermal, which is the fixation of molecular nitrogen by oxygen atoms at high temperatures; (2) fuel, which is the oxidation of nitrogen contained in the fuel during the combustion process, and includes both homogeneous and heterogeneous processes; (3) prompt, which is the attack of a hydrocarbon free radical on molecular nitrogen producing NO precursors; and (4) reburning, which is the reduction of NO by reaction with hydrocarbons.

The major effort in the development of the NO model has been proper characterization of these mechanisms, which has been the subject of substantial reviews.[131,138-140] A brief review of the NOx formation and distribution processes modeled in PCGC-3 follows.

4.2.6.1 Thermal NO

The formation of thermal NO is determined by the following set of three highly temperature-dependent chemical reactions know as the extended Zel'dovich mechanism[141,142]:

$$O + N_2 \xrightleftharpoons{k_1} N + NO \tag{4.57}$$

$$N + O_2 \xrightleftharpoons{k_2} N + NO \tag{4.58}$$

$$OH + N \xrightleftharpoons{k_3} H + NO \tag{4.59}$$

Using these three relationships, the net rate of change in the concentration of NO is given by:

$$\frac{d[NO]}{dt} = k_1[O][N_2] + k_2[N][O_2] + k_3[N][OH]$$

$$- k_{-1}[NO][N] - k_{-2}[NO][O] - k_{-3}[NO][H]$$

(4.60)

The first three terms on the right side of this equation represent the forward reactions where NO is produced, while the last three terms represent the corresponding reverse reactions where NO is consumed. The rate constants have been reported in many investigations,[140,143-145] and these studies have been critically evaluated by Hanson and Salimian.[146] As an example, the following set was given by Miller et al.[145] and Miller and Bowman[140]:

$$k_1 = 1.36 \times 10^8 \exp\left(-\frac{315,900}{RT}\right), \quad \left(m^3 mol^{-1} s^{-1}\right)$$

(4.61)

$$k_{-1} = 3.27 \times 10^6 T^{0.3}, \quad \left(m^3 mol^{-1} s^{-1}\right)$$

(4.62)

$$k_2 = 6.40 \times 10^3 T \exp\left(-\frac{26,300}{RT}\right), \quad \left(m^3 mol^{-1} s^{-1}\right)$$

(4.63)

$$k_{-2} = 1.50 \times 10^3 T \exp\left(-\frac{162,100}{RT}\right), \quad \left(m^3 mol^{-1} s^{-1}\right)$$

(4.64)

$$k_3 = 3.80 \times 10^7, \quad \left(m^3 mol^{-1} s^{-1}\right)$$

(4.65)

$$k_{-3} = 2.00 \times 10^8 \exp\left(-\frac{196,600}{RT}\right), \quad \left(m^3 mol^{-1} s^{-1}\right)$$

(4.66)

Using the quasi-steady assumption (rates of free nitrogen radical formation and consumption are set equal), a general expression for the rate of formation of NO was derived by Westenberg[147]:

$$\frac{d[NO]}{dt} = 2[O] \left\{ \frac{k_1[N_2] - \dfrac{k_{-1}k_{-2}[NO]^2}{k_2[O_2]}}{1 + \dfrac{k_{-1}[NO]}{k_2[O_2] + k_3[OH]}} \right\}$$

(4.67)

If the reverse reactions and the third Zeldovich mechanism involving OH are neglected, the following simplified expression results:

$$\frac{d[NO]}{dt} = 2k_1[O][N_2]$$

(4.68)

The [O] concentration is typically estimated from equilibrium with O_2. However, use of this simple expression is not advised for the general treatment of practical cases.

4.2.6.2 Prompt NO

Prompt NO is a mechanism resulting from the reaction of hydrocarbons with molecular nitrogen. This mechanism was first identified by Fenimore[148] and can become significant in fuel-rich conditions where temperatures are relatively low and residence times are short, such as surface burners, staged-combustion systems, and gas turbines.[149] In most other combustors, particularly coal-fired systems, the contribution from prompt NO is small and, currently, prompt NO formation is not modeled in PCGC-3. From a large list of possible reactions involving hydrocarbon fragments, two are believed to be particularly significant[140]:

$$CH + N_2 \rightarrow HCN + N \tag{4.69}$$

$$C + N_2 \rightarrow CN + N \tag{4.70}$$

The cyanide species is further oxidized to NO. DeSoete[150] has derived a global kinetic parameter for the prompt NO formation rate based on measured O_2, N_2, and fuel concentrations, with further modifications by Backmier et al.[151]

4.2.6.3 Fuel NO

The nitrogen present in fossil fuels such as coal and fuel oil is typically the most significant source of NO formed during combustion. Referred to as fuel NO, it typically accounts for 75 to 95% of the total NO in coal combustors and more than 50% of the total NO in fuel-oil combustors.[131] Experiments by Pershing and Wendt[152] demonstrated that thermal NO contributions do not become significant until temperatures in coal flames are greater than 1650K. Fuel NO often dominates because of the moderately low temperatures and locally fuel-rich nature of most flames, and the weaker nitrogen bonds common in fuel-bound nitrogen compared with molecular nitrogen.[131]

The fuel N primarily evolves as HCN from the products and, to a lesser extent, as NH_3 from the amines present in the coal.[131] The following steady-state transport equations, in addition to the NO transport equation, are therefore required to track these species:

$$\rho \frac{\partial Y_{HCN}}{\partial t} + \rho u_i \frac{\partial Y_{HCN}}{\partial x_i} = \frac{\partial}{\partial x_i} \left(\rho D \frac{\partial Y_{HCN}}{\partial x_i} \right) + S_{HCN} \tag{4.71}$$

$$\rho \frac{\partial Y_{NH_3}}{\partial t} + \rho u_i \frac{\partial Y_{NH_3}}{\partial x_i} = \frac{\partial}{\partial x_i} \left(\rho D \frac{\partial Y_{NH_3}}{\partial x_i} \right) + S_{NH_3} \tag{4.72}$$

The procedure for calculating the source terms, S_{HCN} and S_{NH_3}, is based on the devolatilization and char oxidation rates for production of fuel-N from coal and the use of global kinetic parameters for fuel-N consumption in the gaseous state.[131] While comprehensive kinetics modeling has been done that describes the detailed C/H/N/O mechanisms to ultimately determine the formation and destruction of NO starting with HCN and NH_3,[47] this approach is currently prohibitive in comprehensive combustion codes for practical turbulent combustors. The major focus in NO modeling has therefore been to identify global kinetic parameters that can be used with the essential species HCN, NH_3, N_2, O_2, and NO in the following four global reaction expressions:

$$HCN + O_2 \rightarrow NO + \ldots \tag{4.73}$$

$$HCN + NO \rightarrow N_2 + \ldots \tag{4.74}$$

$$NH_3 + O_2 \rightarrow NO + \ldots \tag{4.75}$$

$$NH_3 + NO \rightarrow N_2 + \ldots \tag{4.76}$$

Global rate expressions for these reaction rates have been reported by DeSoete,[150] Mitchell and Tarbell,[153] and Bose et al.[154] The fuel NO reaction mechanisms from these three investigators are available in PCGC-3. As an example, the following rates were given by DeSoete for Eqs. (4.73) to (4.76):

$$R_{4.73} = 3.5 \times 10^{10} X_{HCN} X_{O_2}^a \exp\left(-\frac{280,300}{RT}\right) \tag{4.77}$$

$$R_{4.74} = 3.0 \times 10^{12} X_{HCN} X_{NO} \exp\left(-\frac{251,000}{RT}\right) \tag{4.78}$$

$$R_{4.75} = 4.00 \times 10^6 X_{NH_3} X_{O_2}^a \exp\left(-\frac{133,900}{RT}\right) \tag{4.79}$$

$$R_{4.76} = 1.80 \times 10^8 X_{NH_3} X_{NO} \exp\left(-\frac{113,000}{RT}\right) \tag{4.80}$$

The oxygen reaction order a in Eqs. (4.77) and (4.79) depends on the flame conditions and is given by DeSoete[150] based on the oxygen mole fraction in the flame.

4.2.6.4 Reburning

The approach used for modeling reburning in PCGC-3 is discussed below. Smoot et al.[155] provide a recent review of the fundamentals and practical applications of reburning technology. In this technology, a hydrocarbon fuel such as natural gas is introduced into the combustor downstream of the combustion zone in order to react with and reduce NO.

This reburning NO pathway is:

$$CH_i + NO \rightarrow HCN + Products \tag{4.81}$$

Chen and co-workers[156,157] have shown, by sensitivity analysis and simulations of premixed flames, that the reburning pathway is significant in fuel-rich flames. The rate expression of the global reburning reaction can be written as:

$$r_{reb} = k_{reb} X_{HC} X_{NO} \tag{4.82}$$

where X_{HC} is the sum of hydrocarbon molar concentrations, and X_{NO} is the nitric oxide molar concentration. By the method described by DeSoete,[150] the rate constant can be calculated as:

$$k_{reb} = \frac{k_2 X_{HCN} X_{O_2}^a - \dfrac{dX_{NO}}{dt} - k_e X_{NO} X_{HCN}}{X_{NO} X_{HC}} \tag{4.82}$$

where the first and third terms in the numerator on the right side of the expression are the same as the HCN depletion rates by DeSoete discussed in the fuel NO section above. Using this relationship, a global reburning-NO rate expression was deduced from a combination of elemental reactions by correlating predicted species profiles from simple hydrocarbon laminar flames. This global expression with its rate constants can be expressed as[156]:

$$k_{reb} = 2.7 \times 10^6 \exp\left(\frac{-18,800}{RT}\right) \tag{4.84}$$

The following species were used to determine X_{HC} and should be included with other species in the combustion-chemistry calculation: CH_4, C_2H_2, C_2H_4. The source term for the reburning mechanism in the NO transport equation can then be calculated as:

$$S_{NO,reb} = -r_{reb} M_{NO} \frac{P}{RT} \tag{4.85}$$

4.2.6.5   Advanced Reburning

Advanced reburning is another technique used to reduce NO emissions. In this approach, ammonia, urea, or other similar gas is injected downstream of the main reburning zone in a slightly fuel-rich zone, followed by additional combustion air to further reduce NO concentrations. Smoot et al.[155] also review advanced reburning technologies and provide additional details about this technique. Work by Xu et al.[158,159] identifies the key overall advanced reburning reactions and provides details about incorporating this mechanism into PCGC-3. The approach used for modeling advanced reburning in PCGC-3 is discussed below.

A four-step, eight-species reduced mechanism was developed from a 312-step, 50-species full mechanism using a systematic reduction method.[171] The four-step reduced mechanism showed good agreement with the full mechanism for most laminar flow cases, and agrees qualitatively with three sets of experimental data.[171] The reactions used in the advanced reburning NO mechanism in PCGC-3 are shown below:

$$O_2 + CO + H_2O \rightarrow 2OH + CO_2 \tag{4.86}$$

$$O_2 + 2CO \rightarrow 2CO_2 \tag{4.87}$$

$$O_2 + N_2 \rightarrow 2NO \tag{4.88}$$

$$OH + 3NO + NH_3 \rightarrow O_2 + H_2O + 2N_2 \tag{4.89}$$

Transport equations are solved for the species: NO, NH_3, OH, O_2, and CO. The major species — H_2O, CO_2, N_2, and CO — are assumed to be in local chemical equilibrium. The advanced reburning model is also decoupled from the main combustion model and is solved as a post-processor following solution of the flowfield and flame structure have been predicted. The thermal-NO and fuel-NO submodels are not used in the regions where the advanced reburning submodel is used, because these reaction processes are usually not important in the post-combustion zone. The reburning model is used in combination with the advanced reburning model. Using this approach, the advanced reburning model increased the computational time of the NO model by about 15%.[171]

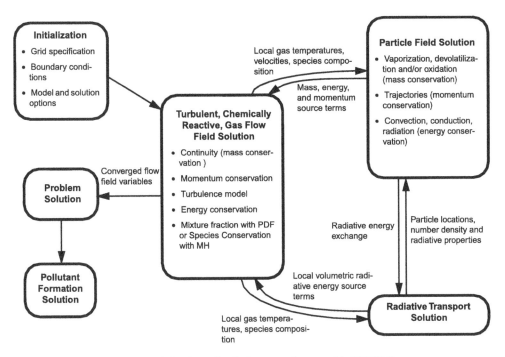

FIGURE 4.2   Overall solution procedure used in PCGC-3.

4.2.7   PCGC-3 SOLUTION STRATEGY

To effectively model a combustion process, the assortment of submodels described in the previous sections must be coupled with the basic turbulent, reacting flow equations outlined above into an effective solution strategy. Typically, the various submodels provide the numerical values for the source terms in the CFD equations. In PCGC-3, for pulverized coal combustion and gasification processes, this strategy is to separate the overall solution procedure into four main parts, as summarized in Figure 4.2: (1) initialization, (2) turbulent, chemically reactive, gaseous flow field solution, (3) particle field solution, and (4) radiative transport solution.

Figure 4.2 also illustrates the pertinent information that must be exchanged in PCGC-3 to couple the parts into an integrated solution strategy. The reason for the division is that each of these parts requires a different numerical solution approach. The radiation part is the solution of the radiative transport equation, Equation (4.31), using the technique described in Section 4.2.3.1. The particle or entrained phase part is the solution of Eqs. (4.37) through (4.39) discussed in Section 4.2.4.3, as well as the particle-specific phenomenological submodels such as vaporization and devolatilization. In PCGC-3, the turbulent gas flowfield solution comprises the following phenomena: (1) continuity (usually incorporated with solution of pressure in the equation of motion); (2) momentum; (3) energy (enthalpy); (4) turbulence; and (5) gaseous chemical reactions, in the form of mixture-fraction equation with a variance based on a specified probability density function (MF/PDF).

A review of the conservation equations presented for each of these gaseous flowfield phenomena shows that all can be written as a partial differential equation for steady-state in the following form:

$$\frac{\partial\left(\rho u_j \phi\right)}{\partial x_j} = \frac{\partial}{\partial x_j}\left(D_\phi \frac{\partial \phi}{\partial x_j}\right) + S^\phi \tag{4.90}$$

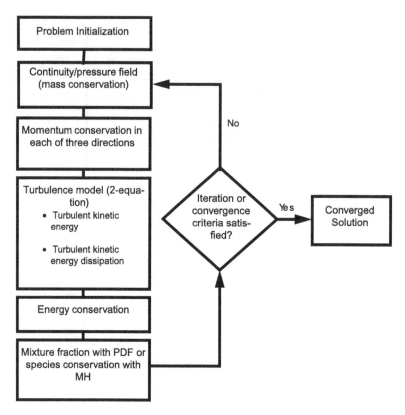

FIGURE 4.3   Phenomenological model solution sequence used in PCGC-3 for a turbulent reactive flow.

where ϕ is the conserved variable of interest (e.g., momentum, enthalpy, species concentration). The left-hand side of the equation is the transport of the conserved scalar by convection, the partial derivative term on the right-hand side is the transport by diffusive processes (both molecular and effective turbulent mechanisms, represented by the coefficient, D_ϕ), and the source term, S^ϕ, combines all of the other gains or losses of ϕ, such as by reaction, body forces, exchanges with the particles, or heat transfer. The advantage of having the same form is that the same numerical discretization solution algorithm can be used for each equation. In PCGC-3, this set of equations is solved sequentially as shown in Figure 4.3. These equations are elliptic in form and inherently nonlinear, and must be solved in an iterative fashion. An initial guess is made for the velocity flowfield to start the process, and then each equation is solved in sequence using the most current values for the variables required from the other equation solutions until a satisfactory level is reached where the values for the variables do not change significantly from one iteration to the next.

4.3   EVALUATION OF PCGC-3 PREDICTIONS

4.3.1   DATA REQUIREMENTS

The evaluation of the quality of predictions available from any comprehensive combustion models requires a level of detail and accuracy from experimental measurements that corresponds to the level of detail of information available from the model predictions. Germane et al.[161] have provided a review of process data and measurement strategies. Ideally, data for validating pulverized coal combustion predictions will include precise information for the following: (1) details of combustor geometry; (2) inlet gas stream conditions, including mass flow rate(s), velocity profiles, species composition, temperature, and initial turbulence intensities; (3) inlet particle stream conditions, including mass flow rate(s), velocity profile, temperature, heating value, ultimate and proximate analysis of the coal, coal particle size distribution, and coal structural properties (in some cases); and (4) wall boundary conditions, including local values of temperature and/or heat flux, and radiative properties such as absorptivity, emissivity, and reflectivity.

Data measured in the combustion chamber for validating predictions should include: (1) locally measured values at several locations, including the flame zone and near-burner regions, of the gaseous flowfield velocity (both magnitude and direction), temperature, and species composition; (2) spatially and temporally resolved entrained particle burnout histories, including number density, velocity (both magnitude and direction), temperature, and composition; and (3) local values at several locations of wall temperatures and/or heat flux. Complete evaluation of code predictions should also include comparisons with measurements from a wide variety of combustors and furnaces that range in scale from very small laboratory combustors (0.01 to 0.5 MW), to industrial furnaces (1 to 10 MW), to large utility boilers (up to 2200 MW).

The detail and range of possible measurements are generally inversely related to the scale of the furnace facility. Detailed, spatially resolved and temporally resolved data are more readily and inexpensively obtained in laboratory-scale facilities, while measurements obtained from large commercial furnaces are more coarse in nature, but represent more practical systems.

4.3.1.1 Laboratory-scale Data

Laboratory-scale data have become very important to comprehensive model validation for several reasons. First, in most cases, the instrumentation needed to characterize the transient behavior in a turbulent flame, such as laser-based instruments necessary for nonintrusive measurements, cannot be as readily used in industrial-scale or utility-scale furnaces, but can be used more conveniently in smaller laboratory reactors. Second, a wider range of test conditions and better control of the operating conditions in the reactor are possible for a smaller facility that is under the full control of the group conducting the test program. Third, not only can the smaller reactor be carefully controlled, but because of its size and optical access provided, the entire reactor volume can be traversed with small incremental steps, so that much more detailed data sets can be obtained. This allows a much higher resolution in the variation of the combustion properties across the flame zone of the combustor than would be possible in large-scale furnaces. Other advantages of small laboratory facilities are relatively low operating cost, flexibility and accessibility, and ability to carefully control and define the boundary and inlet conditions. They can be large enough to give sufficient spatial resolution and to create a near-burner furnace environment, and small enough to utilize the advanced measurement techniques that are essential to providing accurate and complete data for model evaluation and for detailed understanding of combustion processes.

4.3.1.2 Large-scale Data

In large-scale facilities, detailed measurements are difficult to make, inlet conditions are often not well-defined, operating costs for obtaining data can be high, and the facility may have instrumentation limitations. For example, detailed model evaluation measurements of species and temperature within or across large flames cannot easily be made, although such spatially resolved profile data are often best for comparing flame characteristics, near-field burner performance, and jet mixing behavior with predictions. Such detailed measurements provide important information concerning flame response to parametric variation. For many purposes, however, effluent measurements (measurements at the outlet of the system or process), which are a subset of complete profile data, can be useful and may be sufficient to provide the required information for many types of process analyses. This is particularly so when effluent data measurements show the effects of a key system variable, such as excess air percentage, firing rate, or burner tilt angle. In some facilities, effluent data are the only kind available because of furnace size or access constraints, or because that was the only objective for making the measurements. Effluent data are generally sufficient for evaluating zero-dimensional or one-dimensional model predictions, which may be a small part of an overall plant operations model. Such combustion codes may treat the entire furnace with overall thermodynamic parameters or correlations with operations parameters, so that detailed space-resolved measurements are not necessary for code evaluation. Effluent data are most valuable and useful when key process parameters have been varied systematically and correlated with overall process performance. While simple effluent measurements can be useful for comparing overall model predictions with respect to furnace operation, process efficiency, pollutant production, and emissions compliance, detailed, space-resolved, and time-resolved data are best for comprehensive evaluations of multi-dimensional combustion computational codes.

4.4   PCGC-3 MODEL PREDICTIONS AND COMPARISONS WITH DATA

4.4.1   CPR GAS PREDICTIONS

PCGC-3 predictions with data from a laboratory-scale combustor used measurements from ACERC's Controlled Profile Reactor (CPR).[161] One of the primary considerations in the design of this facility was to obtain data that could be used for computer model validation as well as for the study of combustion phenomena. The CPR is a 0.5-MW$_t$, cylindrical, down-fired reactor with an internal diameter of 80 cm and a length of 240 cm (Figure 4.4). Swirl is imparted to the secondary air stream using a movable-block swirl generator. The reactor is highly instrumented so that accurate combustion measurements as well as flow and thermal boundary control and measurement are possible. The reactor body is sectional, with an interchangeable top section so that different burner types can be used. The reactor provides ample optical accessibility for the use of nonintrusive laser

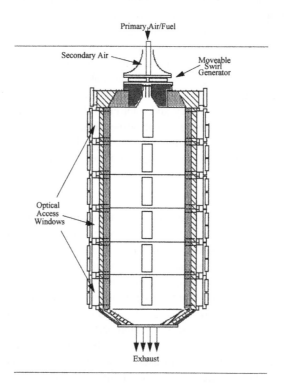

FIGURE 4.4   ACERC Controlled Profile Reactor geometry.[161,162]

diagnostic systems, visual observations, radiometer access, and photography. Accurate combustion zone measurements of radiative heat flux, species concentrations, temperatures, and velocities can be made. Each section of the CPR has four vertical windows every quarter circle around the periphery of the section. These windows extend nearly the full height of the section to allow for fine axial resolution of the flame. These four windows allow for flame measurements to check flame symmetry as well as to make optical measurements in two separate vertical planes. These access ports are filled with a refractory plug simulating the remainder of the reactor insulation when not in use, in order to minimize flow disturbances. Wall heat flux and temperature are controlled by means of electrical heaters imbedded in the reactor walls.[164] Both inlet and outlet flow rates are measured for use as code input as well as calculating mass balances.[162]

For the comparisons in this chapter, one set of natural gas combustion data (reported by Eatough[162] and Boardman et al.[165]) and one set of pulverized coal combustion data (reported by Sanderson[166] and Butler[167] taken in the CPR) have been selected for use in the model evaluation described herein. Parameters and data required for input to comprehensive models are summarized in Tables 4.2 and 4.3 for these two laboratory-scale cases.

To manage the large number of signals requiring measurement for the purpose of data collection, system control, material and thermal balances, and operating condition specification, a computerized data acquisition and process control system was developed and installed on the CPR. A suction pyrometer was used to measure gas temperatures and a five-hole pitot tube was used for measuring gas velocities. A water-quenched sample probe was utilized to collect gas samples for species determination. Combustion gas samples were removed from the combustion zone and water-quenched with a stainless steel, water-cooled probe. The combustion sample was subsequently dried using an ice bath. The resulting dry gas sample was analyzed for major species with a gas chromatograph as well as infrared online instrumentation. NOx was measured with an online chemiluminescent analyzer.

For the CPR gas predictions, an axisymmetric grid with approximately 10,000 gridpoints was used for the PCGC-3 predictions. The grid incorporated a primary inlet duct for the injection of the natural gas, and a secondary inlet duct for the injection of the swirling secondary air stream with a swirl number of 1.50. The turbulent intensity of the inlet gas streams was assumed to be 10% of the inlet velocities. The furnace wall temperature was assumed to have a constant value of 1000K. A converged solution was obtained in about 650 iterations, which required 152 minutes on an HP 9000, Model 735 UNIX workstation with PCGC-3.

One of the key characteristics of a combustor burner with high swirl level can be the creation of the internal, symmetric recirculation zone along the centerline of the reactor downstream from the burner outlet. Modeling this behavior demonstrates the complex nature of turbulent, swirling

---

TABLE 4.2
Input Data for CPR Gaseous Combustion Case[a]

Natural gas flow rate (kg/hr)	10.1
Natural gas inlet temperature (K)	298
Air flow rate (kg/hr)	165
Air inlet temperature (K)	298
Natural gas composition (% weight)	CH_4 – 80
	C_2H_2 – 12
	C_2H_6 – 5
	CO_2 – 2
	C_3H_8 – 1

[a] Additional geometric details are provided in Reference 162.

TABLE 4.3
Input data for CPR Coal Combustion Case[a]

Coal flow rate (kg/hr)	11.4
Coal inlet temperature (K)	289
Primary air flow rate (kg/hr)	15
Primary air inlet temperature (K)	289
Secondary air flow rate (kg/hr)	127
Secondary air inlet temperature (kg/hr)	533
Secondary swirl number	1.4
Coal proximate analysis (% mass)	Moisture – 8
	Ash – 13
	Volatile – 39
	Fixed C – 40
Coal ultimate analysis (% weight, dry, ash-free)	Carbon – 81
	Hydrogen – 6
	Nitrogen – 2
	Oxygen – 11
Coal High Heating Value (kJ/kg)	26,590

[a] Additional geometric details are provided in References 163, 166.

flow combustion calculations. The size and velocity magnitude of the recirculation zone are functions of the turbulent gas dynamics, fuel and oxidizer mixing rates, and chemical reaction rates. Turbulent mixing dominates the overall rate of combustion, which in turn affects the local gas temperature, density, and species concentrations.

PCGC-3 predicted the length and strength of the recirculation zone within the limitations of the data, as shown by the centerline axial velocity comparisons with measured data shown in Figure 4.5. Differences between the predictions and measurements shown in the comparisons of radial profiles for axial and tangential velocity (Figures 4.6 and 4.7), and temperature (Figure 4.8)

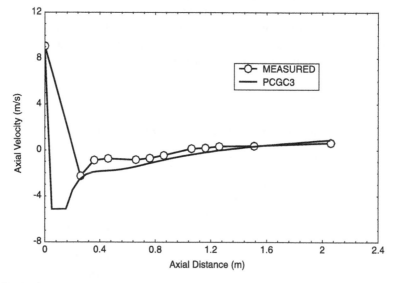

FIGURE 4.5 Predicted and experimentally measured axial velocities along the reactor centerline for the CPR gaseous combustion case.

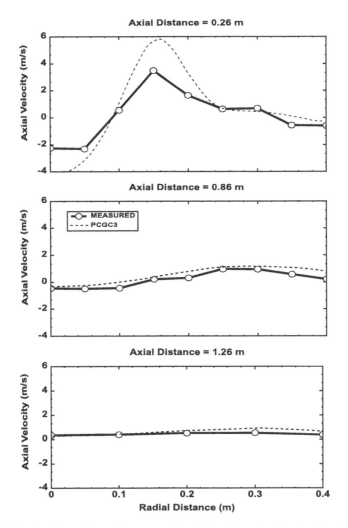

FIGURE 4.6 Predicted and experimentally measured radial profiles of axial velocities at distances of 0.26, 0.86, and 1.26 m from the burner for the CPR gaseous combustion case.

indicate that the mixing and chemical reactions are occurring closer to the inlet than predicted. Because the predicted and measured O_2 and CO_2 concentrations are essentially the same as equilibrium values at the points where measurements were taken, plots for these equilibrium quantities are not shown.

Possible reasons for the observed differences between predictions and measurements could be attributed to: (1) inadequacy of the two-equation turbulence model, which assumes that the turbulence is isotropic, which is not the case in a highly swirling flow; (2) difficulty in predicting the precise location for onset of ignition, which changes the flow structure substantially; and (3) alteration of the flow caused by the probe intrusion while the measurements were taken.

4.4.2 CPR COAL PREDICTIONS

The CPR coal combustion case not only has the complexities of a gaseous combustion, but adds the complexities of reacting coal particles. The development of the flowfield depends on coal reactions leading to the release of volatiles and char-oxidation products in a gaseous form, and on subsequent mixing of these gaseous products with other process gases.

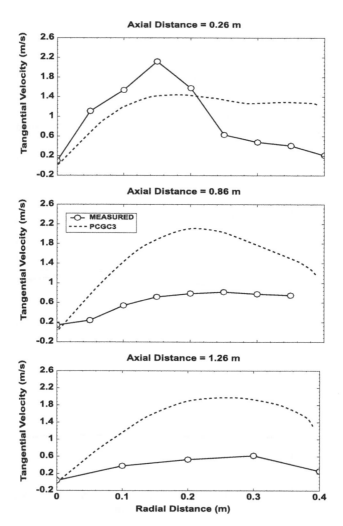

FIGURE 4.7 Predicted and experimentally measured radial profiles of tangential velocities at distances of 0.26, 0.86, and 1.26 m from the burner outlet for the CPR gaseous combustion case.

For these predictions, an axisymmetric grid with approximately 10,000 grid points, similar to the CPR gaseous combustion case, was used. The grid incorporated two different inlet streams: a primary inlet stream where the coal particles are injected with the primary air stream, and a secondary inlet stream where the secondary air is injected with the appropriate swirl velocity (swirl number of 1.5). The turbulent intensity of the inlet gas stream was assumed to be 10% of the magnitude of the inlet velocity. The furnace wall temperature was assumed to have a constant value of 1000K. In addition to the gaseous flowfield calculations, the pulverized coal flowfield was modeled in Lagrangian fashion. Five different particle sizes with 18 starting locations spread across primary inlet area (a total of 90 separate particle trajectories) were used to characterize the coal dust stream. In this solution scheme, gaseous flowfield calculations performed for a specified number of iterations (150 to 200) alternated with the particle trajectory computations. For the CPD model, the five input parameters corresponding to the coal types are $\sigma + 1 = 5.1$, $p_o = 0.49$, $m_c = 366$, $m_b = 36$, and $C_o = 0$. About 15 sets of these alternating computations were required to achieve a solution with satisfactory convergence. The solutions required a run-time of about 1200 minutes on an HP 9000, Model 735 UNIX workstation.

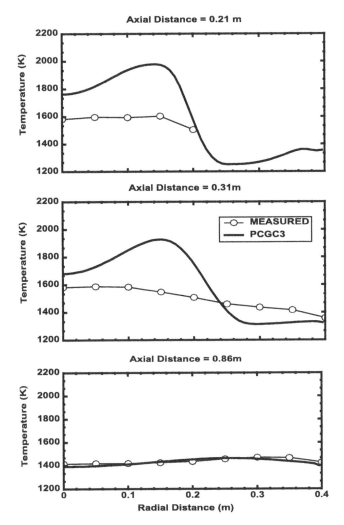

FIGURE 4.8   Predicted and experimentally measured radial profiles of gas temperatures at distances of 0.21, 0.31, and 0.86 m from the burner outlet for the CPR gaseous combustion case.

The significant features of the gaseous flow field are illustrated with the gaseous flow streamlines and temperature field predictions from PCGC-3 for the region near the quarl in Figure 4.9.

As shown by the gaseous flow traces in Figure 4.9, there is no centerline internal recirculation zone predicted for this case, although a large recirculation zone develops away from the centerline. Although no measured velocity data were available for confirmation, this prediction seems reasonable because of the relatively high primary jet velocity of 29 m/s, compared with a more typical primary inlet velocity (such as the CPR gaseous combustion case discussed in the previous section) of around 9 m/s. The secondary axial velocity was about 9 m/s.

Radial profile comparisons of data with predictions from PCGC-3 for gas temperature, O_2 concentrations, and CO_2 concentrations at selected axial locations are presented in Figures 4.10, 4.11, and 4.12, respectively. The solutions for the particle combustion case generally gave improved agreement with the data, compared with the gaseous case, although agreement near the burner was often unsatisfactory. The most likely reason for this is the non-prediction of a centerline recirculation zone, which results in a somewhat less-complex flowfield, especially in the flame region, than the gaseous combustion case. This would also result in less disruption of the flowfield by any intrusive

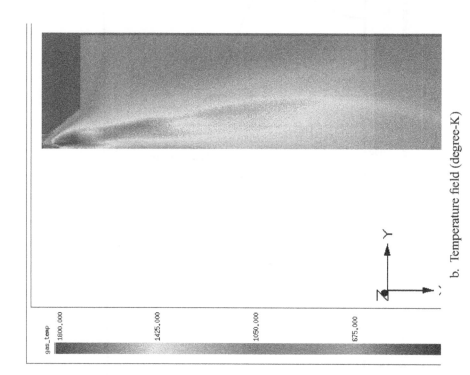

gas_temp

1800.000

1425.000

1050.000

675.000

b. Temperature field (degree-K)

a. Velocity traces

FIGURE 4.9   Gas velocity traces and temperature field predicted by PCGC-3 for the CPR coal combustion case.

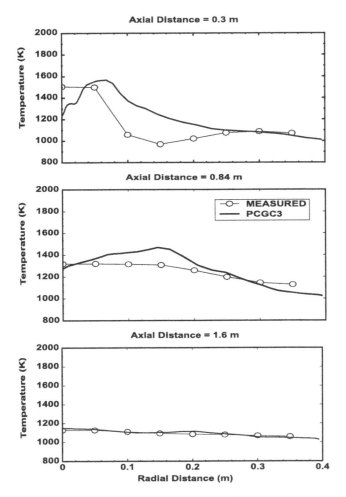

FIGURE 4.10   Predicted and experimentally measured radial profiles of gas temperatures at 0.30, 0.84, and 1.60 m from the burner outlet for the CPR coal combustion case.

probes used to make measurements. Use of the more fundamental, networkbased CPD devolatilization model did not always improve the comparison over the more empirical two-step devolatilization model for these sets of profiles.

A major motivation of many combustion calculations is the prediction of emissions of pollutants such as CO and NO. This is especially true for NO in coal combustion systems, due to the high fuel-nitrogen content relative to gaseous fuels. Accurate prediction of these pollutant quantities, which are predicted in a post-processing fashion in both cases once a fully converged flow solution is obtained, are highly dependent on the quality of the flowfield solution. With pulverized coal combustion, NO results principally from the nitrogen in the coal particles, and thermal NO production is often negligible. The comparison of measured data with the fuel NO predictions is shown in Figure 4.13. Measured values peak at about 600 ppm in the hot combustion zone off the centerline and decay radically and axially to near 100 ppm. The computed maximum was 400 ppm, decaying also to about 100 ppm.

4.4.3   Utility Furnace Predictions

For utility furnace comparisons with PCGC-3 predictions, data from testing performed at the Goudey Station boiler #13 in Johnson City, New York, were selected.[168] The primary purpose of this testing

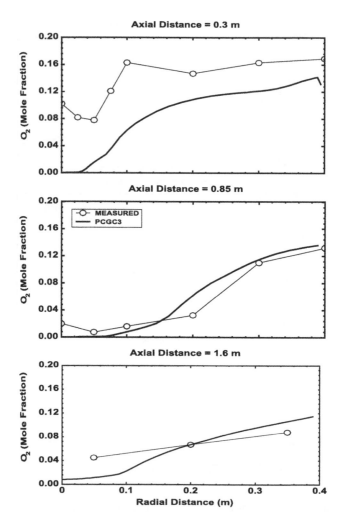

FIGURE 4.11   Predicted and experimentally measured radial profiles of O_2 concentrations at 0.30, 0.85, and 1.60 m from the burner inlet for the CPR coal combustion case.

was to obtain a well-defined set of large-scale furnace data for use in validating the reliability of comprehensive combustion model predictions. These data form a complete set of *in situ*, steady, spatially resolved, contiguous large-scale combustion measurements from the archival literature.

This boiler is a corner-fired unit, constructed in 1952 with a nominal 85-MW_e generation rate. The general configuration of the furnace geometry is illustrated in Figure 4.14. The Goudey station had been used as a test bed in the past and has a significant number of preexisting access ports, thus facilitating spatially resolved data measurements.[168] The unit has about a 7.6 m × 7.6 m cross-section, which allows the use of water-cooled probes of reasonable length to reach half-way across the boiler interior. A full traverse across the boiler can be made where opposing ports exist. Eight tests were made with variation in coal type, particle size distribution, furnace load, burner tilt, and percent excess air. The data selected for comparisons herein corresponded to Test Four and input parameters are summarized in Table 4.4.

Quantities measured included spatially resolved gas velocities, temperatures, and species concentrations, particle size distribution, particle velocities, particle number densities, and radiative heat fluxes. Gas and particle sampling were performed with a water-cooled suction probe equipped with a particle filter system. The sampling probes were 3 m long, with the maximum probe insertion

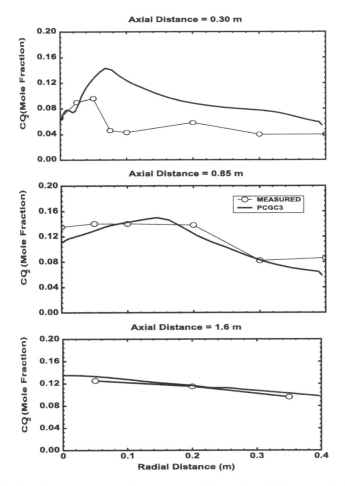

FIGURE 4.12 Predicted and experimentally measured radial profiles of CO_2 concentrations at 0.30, 0.85, and 1.60 m from the burner inlet for the CPR coal combustion case.

depth of about 2.5 m from the inner boiler wall. The probe was used to simultaneously draw combustion gases for online analysis and solids for subsequent chemical characterization. The gas samples were analyzed using electrochemical and infrared online instrumentation. Gas temperatures were measured using a triply shielded, water-cooled suction pyrometer.[163] This probe was 4 m long and could be inserted 3.2 m into the boiler, as measured from the inside wall. Use of the data for evaluation of comprehensive combustion codes follows.

The utility furnace case has the additional complexities of large-scale and asymmetry compared with the CPR coal combustion case. These simulations used the two-step coal devolatilization model (reaction constants given by Ubhayakar et al.[118]). The grid incorporated two different inlet streams: a primary inlet where the coal particles are injected with the primary air stream, and a secondary air stream. These inlets are located on the corners of the furnace at the four elevations. The orientation of the inlets results in a swirling core of combustion gases in the center of the furnace. Five different particle sizes with 80 starting locations spread across the primary inlet area (a total of 400 separate particle trajectories) were used to characterize the coal stream. In this solution scheme, gaseous flowfield calculations are performed for a specified number of iterations (150 to 200), and then alternated with the particle trajectory computations. About 15 sets of these alternating computations were required to achieve satisfactory convergence. The solution of this case required a run-time of about 280 hours on an HP-9000, Model 735 workstation.

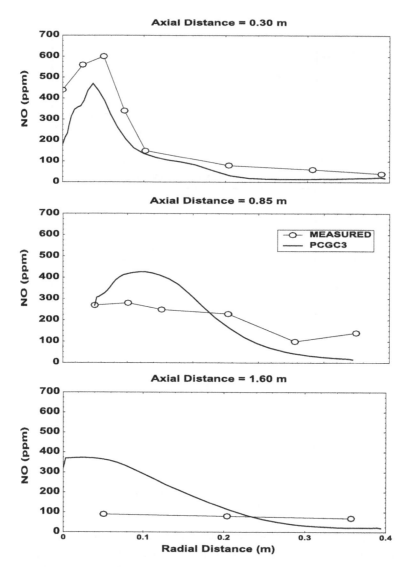

FIGURE 4.13 Predicted and experimentally measured radial profiles of NO concentrations at 0.35, 0.85, and 1.60 m from the burner outlet for the CPR coal combustion case.

Comparisons of PCGC-3 predictions with the measured data for utility furnace for gas temperatures, O_2 and CO_2 concentrations are shown in Figures 4.15 through 4.17, respectively. Temperature predictions seem best at port 5 just above the upper burner sets. The predicted values show the same shape as the data, but are about 200 to 300K hotter in the upper zones. For CO_2 and O_2 compositions, differences from 1 to 6 mol% are observed between the measured and predicted values. Comparisions of predictions with measured NO concentrations from the utitlity furnace are shown in Figure 4.18. Predicted NO concentrations were within about 60 to 80 ppm of the observed value of 500 ppm at port 5a. Prediction of utility furnace *in situ* behavior is a very stringent test for a comprehensive combustion model, as these comparisons show. PCGC-3 is representative of the state-of-the-art in combustion modeling, and can provide insight into furnace behavior, but much work remains. With more detailed measurements, it may be possible to identify causes of differences between the predictions and the measured data, and make improvements in the relative submodels.

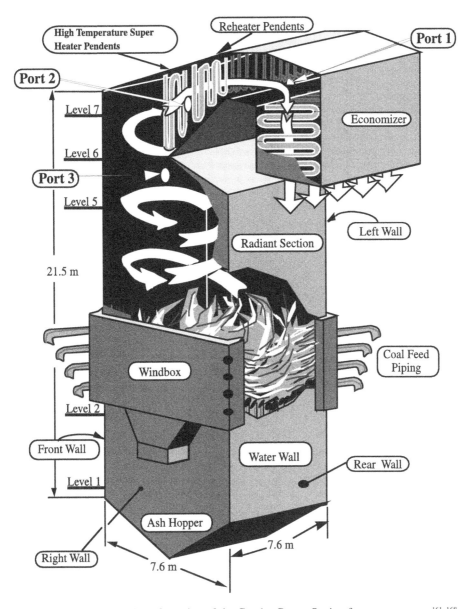

FIGURE 4.14 General configuration of the Goudey Power Station furnace geometry.[161,.168]

4.4.4 INDUSTRIAL GAS FURNACE

PCGC-3 has also been used to model an industrial glass furnace. A schematic of the furnace is shown Figure 4.19. The furnace is approximately 2.6 m high from the glass line to the bottom of the crown, 11 m wide, and 21.5 m long. A regenerator is located on the north and south sides of the furnace. Between each regenerator and the furnace is a set of six portnecks, each approximately 3.35 m long. Experimental measurements were made along the crown of the furnace at the centerline of portneck 3 near the end of the furnace's useful life, and just prior to the rebuild of the furnace. These measurements were made at 1.1 m, 2.3 m, 3.4 m, 4.4 m, 5.3 m, and 7.6 m into the furnace from the north furnace wall. Average velocities were measured with a water-cooled Pitot tube. Gas species concentrations were measured with a water-cooled quench probe and online chemical analyzers. Gas temperature data were acquired using suction pyrometry with triple radiation shields.

TABLE 4.4
Input Data for Goudey Coal Combustion Case[a]

Coal flow rate (kg/hr)	29,100
Coal inlet temperature (K)	360
Primary air flow rate (kg/hr)	76,300
Primary air inlet temperature (K)	360
Secondary air flow rate (kg/hr)	411,100
Secondary air inlet temperature (K)	540
Coal proximate analysis (% weight)	Moisture – 5
	Ash – 7
	Volatile – 36
	Fixed C – 52
Coal ultimate analysis (% weight, dry, ash-free)	Carbon – 88
	Hydrogen – 4
	Nitrogen – 1
	Oxygen – 5
	Sulfur – 2
Coal High Heating Value (kJ/kg)	31,250

[a] From References 161, 164, 167, 168. Additional metric details are provided in these references.

Crown incident radiant flux was characterized using an ellipsoidal, hemispherical radiometer. Additional information about the measurement devices, their calibration, and measurement procedures is reported elsewhere.[169] Furnace operating conditions are listed in Table 4.5. These conditions were maintained at or near these values for the duration of the test period. The side-wall temperatures in Table 4.5 are based on optical temperature measurements made through an access hole in the side of the regenerator coinciding with portneck 4. These measurements were made during the approximately thirty-30 interval between flame reversals when no flame was present. The overall furnace equivalence ratio reported in Table 4.5 was calculated from the total fuel and air-flow rates metered by the plant.

Typical industrial glass furnaces exhibit strongly three-dimensional fluid flow and heat transfer characteristics. Rigorously, the thermal and chemical phenomena in glass melting furnaces exhibit intimate coupling between the combustion space, the molten glass tank, and the batch blanket (raw materials). The transport is thus a complex chemical and thermal interaction between these three physical domains, with matching of boundary conditions at the shared interfaces. The approach taken here is the prediction of turbulent mixing, chemical reaction, and heat transfer in the combustion space only. Therefore, assumptions must be made in the imposition of thermal boundary conditions at the combustion space/melt tank interface. Additionally, industrial float-glass furnaces have multiple combustion ports, and to simulate the entire furnace would require both a very large computational grid and considerable computer resources. By exploiting the port-by-port modular nature of the furnace, a more manageable computational grid can be constructed requiring considerably less computer time. For the simulations reported here, a single portneck was modeled. By assuming symmetry along the centerplane of each port, the computational domain begins at the centerplane of a portneck and ends halfway between the next adjacent portneck. Figure 4.20 represents this schematically, showing various characteristics of the model including the measurement locations along the crown of the furnace. This model assumes that there is no influence of adjacent portnecks. The calculation domain extends from the measurement location in the inlet portneck to the measurement location in the exhaust portneck. Simulations were performed with PCGC-3 using three different grids to determine grid independence. The computational grids used to represent a portneck module were 34,500 (100 × 15 × 23); 62,100 (100 × 23 × 27); and 96,228

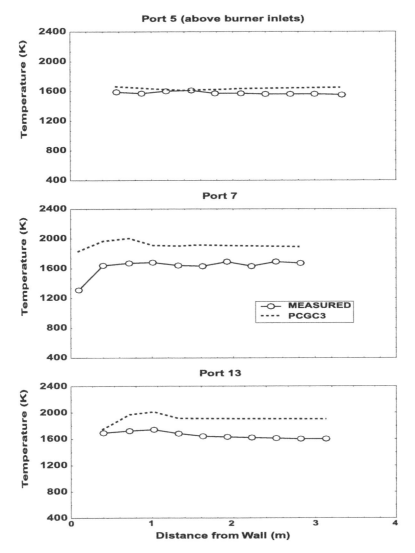

FIGURE 4.15   Predicted and experimentally measured profiles of gas temperatures at ports 5a, 7, and 13a for the utility furnace.

(108 × 27 × 33) cells. Significant differences were observed between the predicted results using the 34,500 and the 62,100 grids, but the finest grid (96,228 cells) showed grid independence when compared to the 62,100-cell grid, and was used for all of the simulations reported here.

Boundary and inlet conditions used for the PCGC-3 simulations with are listed in Table 4.6. The natural gas composition, temperature, and flow rate per port, and the emissivity and temperature of the walls and glass surface were supplied by Ford plant personnel, or were determined from measurements reported by Newbold et al.[169] Although the real furnace environment exhibits thermal coupling between the combustion space, melt tank, and batch blanket, in these predictions the glass surface temperature was assumed. The imposed temperature boundary condition decreased linearly from 1810K at the furnace axial centerline to 1760K at both furnace walls, and was assumed invariant in the direction normal to the primary portneck flow. Because the portneck walls were insulated refractory, they were assumed adiabatic. The characteristic length was assumed to be the hydraulic radius of each inlet. The turbulence intensity for each inlet stream was arbitrarily set to 10%.

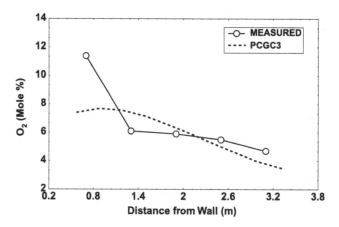

FIGURE 4.16   Predicted and experimentally measured profiles of O_2 concentrations at port 5a for the utility furnace.

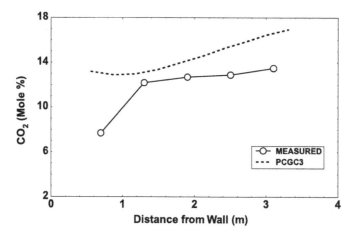

FIGURE 4.17   Predicted and experimentally measured profiles of CO_2 concentrations at port 5a for the utility furnace.

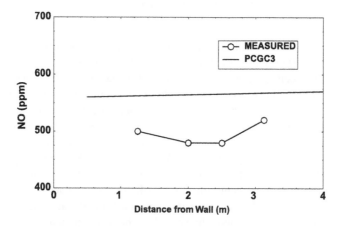

FIGURE 4.18   Predicted and experimentally measured profiles of NO concentration at port 5a for the utility furnace.

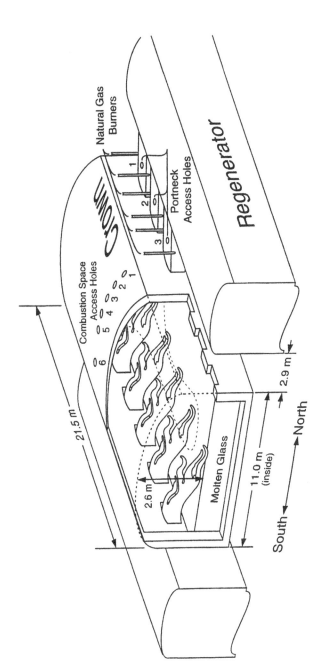

FIGURE 4.19 Schematic representation of an industrial glass furnace.

TABLE 4.5
Input Data for Industrial Glass Furnace Case

Wall temperatures (K)	
Optical Measurements	
North	1910
South	1900
Fuel Composition (vol%)	
Methane	87.49
Nitrogen	6.79
Ethane	3.66
Propane	0.77
Carbon Dioxide	0.58
Oxygen	0.46
Hexane	0.10
n-Butane	0.08
i-Butane	0.05
n-Pentane	0.02
Stoichiometry (F*)	
Overall (plant value)	0.61
Portneck 1	0.49
Portneck 2	0.54
Portneck 3	0.66
Portneck 4	0.72
Portneck 5	0.65
Portneck 6	0.58
Flow Rates (kg/hr)	
Fuel	2763
Air	66,557
Species generation from melt reaction (kg/hr)	
CO_2	2500
H_2O	600

Simulations were performed on HP 9000/735 workstations, and required approximately 30 hours of computer time to achieve a converged solution. Comparisons of representative predicted and measured results of local gas velocity, gas temperature, species concentration (O_2, CO, and CO_2), and incident radiative heat flux on the crown are discussed below. Figure 4.21 shows comparisons of measured and PCGC-3 predicted u-velocities in the combustion space at holes 2, 4, and 6. Generally, the flow is characterized by a dominant flame jet spreading over the glass surface, and a large recirculation zone in the upper portion of the furnace near the crown. Comparison of the predictions with experimental data shows reasonably good agreement, and indicates that PCGC-3 predictions give a reasonable representation of the mean flow conditions in the combustion chamber of the furnace. Figure 4.22 shows comparisons of measured and predicted gas temperatures in holes 1 and 2 for the North-to-South firing direction. These holes are near the entrance to the furnace with this firing configuration. Measurements at other holes were not available due to experimental difficulties. PCGC-3 had difficulty predicting the detailed local gas temperature, and errors as high as 400 to 600K are noted locally. Possible reasons for discrepancies between measured and predicted values are discussed at the end of this section. Figure 4.23 compares measured and predicted oxygen concentrations in holes 2, 4, and 6. While the predictions generally describe the oxygen depletion in the core of the flame jet and the higher O_2 concentrations in the recirculation

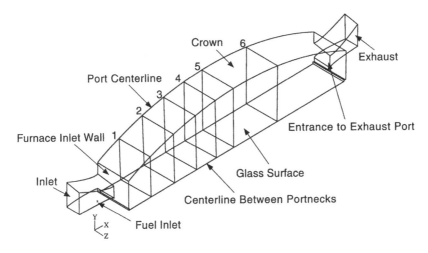

FIGURE 4.20 Schematic representation of a single portneck in an industrial glass furnace modeled using PCGC-3.

TABLE 4.6
Boundary and Inlet Conditions for Glass Furnace Case

Portneck walls	
Emissivity	0.6
Heat Flux (W m-2)	0.0
Crown and furnace walls	
Emissivity	0.6
Temperature (K)	1910
Glass surface	
Emissivity	0.9
Temperature (K)	Piecewise linear varying from 1810 in the middle to 1760 at the furnace walls
Air inlet	
Temperature (K)	1391
Turbulence intensity (%)	10.0
Characteristic length (m)	0.29
u-Velocity (m/s)	Velocity data measured in portneck 3 for N-S burn2
Fuel Inlet	
Fuel	Same as in Table 4.5
Temperature (K)	283
Turbulence intensity (%)	10.0
Characteristic length (m)	0.022
u-Velocity (m/s)	61.42
w-Velocity (m/s)	−61.42

zone above, the magnitude of predicted concentrations differs substantially from the measured values. Figure 4.24 compares measured and predicted CO concentrations in holes 2, 4, and 6. The flame core is evident near the glass surface by the high CO concentrations found there. In the flame region, the predicted values of CO concentration are considerably higher than the measured values; however, the measurements in this region exceeded the 40,000 ppm limit of the electrochemical

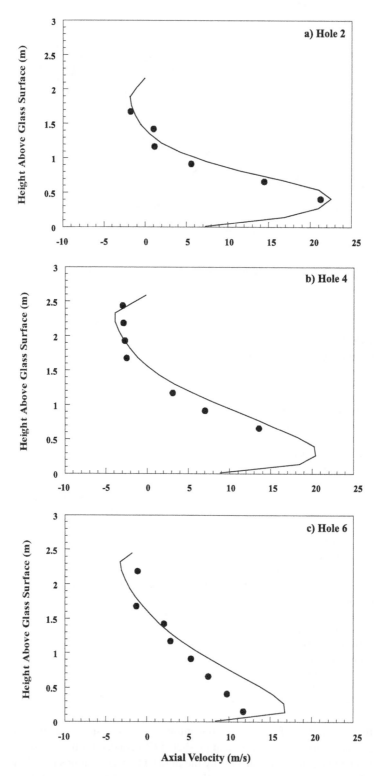

FIGURE 4.21 Comparisons of predicted (—) and measured (•) u-velocities in the combustion space of a glass furnace at holes 2, 4, and 6.

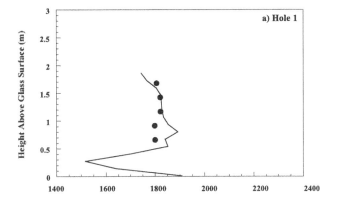

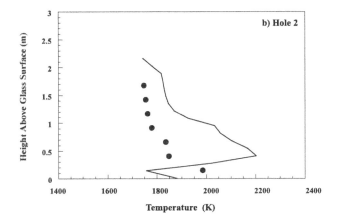

FIGURE 4.22   Comparisons of predicted (—) and measured (•) gas temperatures in the combustion space of a glass furnace at holes 1 and 2.

cell in the instrument used to measure CO.[169] Thus, the actual maximum CO concentration in the flame region is not known.

The average predicted exhaust CO concentration weighted by the mass flow rate is 1680 ppm, while the average measured concentration is 8125 ppm. Higher exit CO concentrations result from incomplete combustion, which in this case could result from several factors, including (1) incomplete mixing of fuel and oxidizer, (2) insufficient time for complete combustion, and (3) fuel-rich combustion. Higher exit CO concentrations also suggest that a longer flame occurs in the actual operation than is predicted by PCGC-3. The discrepancies between measured and predicted CO values may be due in part to the equilibrium assumption used in PCGC-3 that assumes instantaneous kinetics. This would explain why a more rapid decay of CO is predicted, but in fact the measured CO values would indicate that the decay process may be kinetically controlled, and there is insufficient residence time in the furnace for complete combustion.

Comparisons of predicted and measured CO_2 concentrations for holes 2, 4, and 6 are shown in Figure 4.25. Predicted profiles and measured data show low CO_2 concentrations near the glass surface with higher values near the crown. The predicted values of CO_2 concentration are generally lower at all locations. The under-prediction of CO_2 concentration relative to the measured data may be explained primarily by the fact that nearly half of the CO_2 produced in a glass furnace may originate from the reactions in the glass itself.[169] Most of the CO_2 generated in the melt occurs near the batch feeder early in the furnace (e.g., portnecks 1 to 3). Because neither code accounts

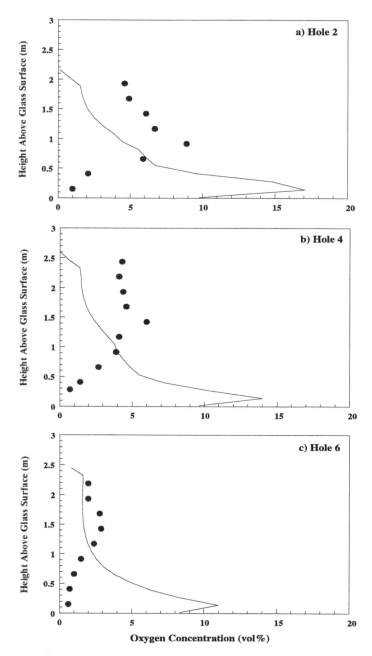

FIGURE 4.23 Comparisons of predicted (—) and measured (•) oxygen concentrations in the combustion space of a glass furnace at holes 2, 4, and 6.

for this source of CO_2, it would partially explain the lower predicted concentrations compared to the measured data, and suggest that the CO_2 evolving from the glass melt has a larger impact moving toward the center of the furnace. Chemical coupling between the melt and the combustion space, with the associated evolution of gases in the molten glass firing process and their introduction into the combustion space, would account for the higher CO_2 concentrations in the furnace.

A comparison of the predicted and measured incident flux on the crown of the furnace is shown in Figure 4.26 as a function of distance from the furnace inlet wall. The figure shows that the

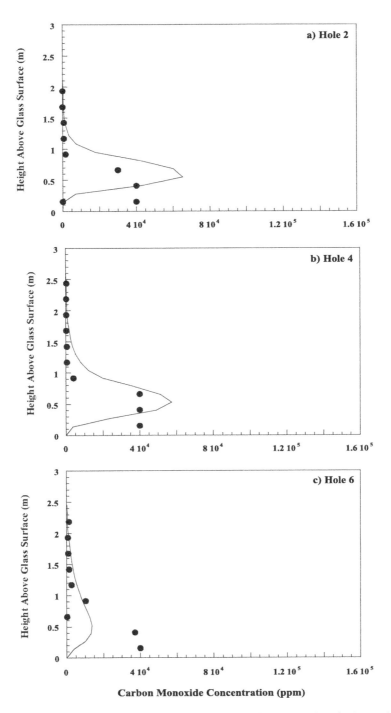

FIGURE 4.24 Comparisons of predicted (—) and measured (•) CO concentrations in the combustion space of a glass furnace at holes 2, 4, and 6.

incident flux is relatively constant across the crown and is predicted reasonably well. Given the discrepancies in predicted temperature and species concentrations, the good agreement with radiant flux is fortuitous, and results because 60 to 90% of the radiant heat transfer incident on the crown originates from the glass surface.[170] If the glass surface temperature and emissivity are well approximated, the predicted crown flux will exhibit good agreement with the measured data.

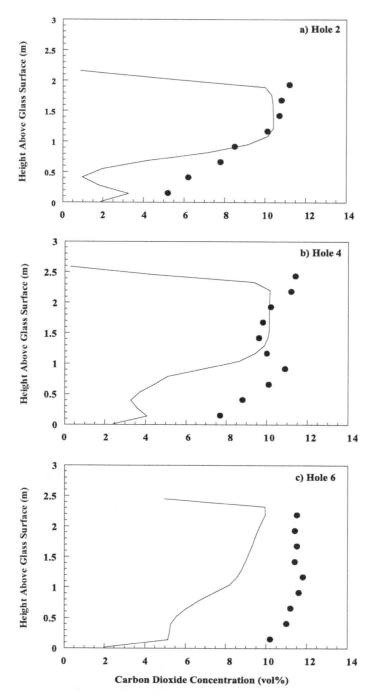

FIGURE 4.25 Comparisons of predicted (——) and measured (•) CO_2 concentrations in the combustion space of a glass furnace at holes 2, 4, and 6.

In summary, the values of gas velocities and incident heat flux predicted using geometrical and boundary condition information based on furnace blueprint and plant operating values for fuel and airflow rates compared well with measured values. However, predicted values of gas species concentrations and temperatures did not compare well with experimental quantities. Analysis of

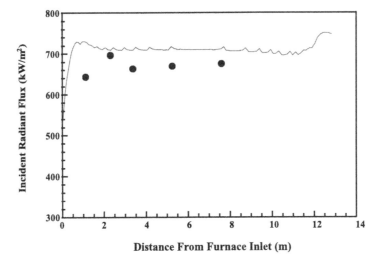

FIGURE 4.26 Comparisons of predicted (—) and measured (•) incident heat flux on the crown of the glass furnace.

the exhaust gas compositions suggested that there was considerable uncertainty in the plant-measured air flow rate. Without the luxury of furnace effluent measurements, the modeler must rely on furnace operation data provided by the plant operators, which can result in considerable errors in the input and boundary conditions. The discrepancies between prediction and measurement may be explained by

1. Uncertainties in the distributions of mean velocity and turbulence in the portneck
2. Uncertainties in the port-by-port stoichiometry
3. Grid-based approximations to the furnace geometry inherent with computer simulations
4. The assumption of infinitely fast chemistry made in the chemical reaction model
5. Simplifying assumptions made in the simulations regarding the complex coupling between the combustion space, batch blanket, and melt tank.

These comparisons illustrate the critical need for accurate boundary conditions (inlet air and fuel flow distributions, boundary surface temperatures, etc.), and the importance of representative furnace geometry in simulating these complex industrial combustion systems.

4.5 SUMMARY

PCGC-3, as a comprehensive combustion model, incorporates features that seek to account for the major physico-chemical phenomena of importance in gaseous and dilute-dispersed solid particle and droplet fossil-fuel combustion processes such as pulverized coal gasification and combustion. These submodels, however, do not necessarily have the level of sophistication and complexity that stand-alone submodels for the same phenomena would have. A comprehensive model must balance submodel sophistication with computational practicality. In other words, a comprehensive model must incorporate submodels that not only simulate the phenomena of interest, but do so in a way that can be coupled with the other submodels and solved in a reasonable amount of time.

Comprehensive combustion codes still have limitations in their capabilities to predict complex flow fields. Consequently, validation and demonstration of capabilities are very important in evaluating how well the code predictions match the observations, both qualitatively and quantitatively.

Validation and application of PCGC-3 has demonstrated that it can give practical results that provide useful qualitative and quantitative information that describes the experimental measurements.

The results of the comparisons presented in this chapter provide an indication of the kinds of results that might be expected from comprehensive combustion model predictions of practical combustors and furnaces. Published comparisons of comprehensive combustion models with test data often show superior agreement to that reported herein. The laboratory-scale CPR data are for a highly swirling flow, while the utility-scale 3-D data are for a complicated configuration. These computations provide an indication of the quality of agreement that might be anticipated for such cases.

Given the very large number of parameters and variables in comprehensive combustion models, it is difficult to explicitly identify the reasons for observed lack of agreement when it exists. However, after more than a decade of repeated applications of PCGC-3 to test results, the elements of comprehensive combustion models that are thought to require the most improvement include the following:

1. The turbulence model, with the commonly used k-ε method being inadequate, particularly for strongly swirling flows
2. The prediction of the onset of gaseous ignition and its inter-relationships to coal devolatilization rates and particle dispersion behavior, because ignition onset with strong heat release greatly impacts the combusting gas structure
3. Questions of initial and boundary conditions, measurement accuracy, and grid reduction.

Even when some uncertainty exists about the exact values predicted by the models for problem variables, useful parametric studies can be performed to evaluate the effects that changing parameters will have on the overall processes of interest.

ACKNOWLEDGMENTS

Support for development of PCGC-3 was provided by the Advanced Combustion Engineering Research Center (ACERC) at Brigham Young University. ACERC funding sources included the National Science Foundation's Engineering Education and Centers Division, the State of Utah, Brigham Young University, and over 40 other industrial and governmental participants. Appreciation is also expressed to a large and productive group of colleagues and students far too numerous to mention by name who have, over the past decade and more, contributed to the development, application, and evaluation of PCGC-3.

REFERENCES

1. Smoot, L.D., *Energy & Fuels*, 7, 659, 1993.
2. Annual Energy Review 1997, DOE/EIA-0384(97), U.S. Department of Energy, Energy Information Agency, Washington D.C., 1998.
3. Brewster, B.S., Hill, S.C., Radulovic, P.T., and Smoot, L.D., *Fundamentals of Coal Combustion for Clean and Efficient Use*, Smoot, L.D., Ed., Elsevier, Amsterdam, 1993, 567.
4. Brewster, B.S., Meng, F., and Cannon, S., *Prog. Energy Combust. Sci.*, in press, 1999.
5. Chung, T.J., *Numerical Modeling in Combustion*, Taylor & Francis, Washington, D.C., 1993.
6. Ramos, J.L., *Internal Combustion Engine Modeling*, Hemisphere, New York, 1989.
7. Jones, W.P. and Whitelaw, J.H., *Combust. & Flame*, 48, 1, 1982.
8. Smoot, L.D. and Smith, P.J., *Coal Combustion and Gasification*, Plenum Press, New York, 1985.
9. Smoot, L.D., *Prog. Energy Combust. Sci.*, 10, 229, 1984.
10. Niksa, S., IEA PER/31, IEA Coal Research, London, UK, 1996.
11. Bird, R.B., Stewart, W.E., and Lightfoot, E.N., *Transport Phenomena*, Wiley, New York, 1960.

12. Hoffman, K.A., Chiang, S.T., Siddiqui, S., and Papadakis, M., *Fundamental Equations of Fluid Mechanics,* Engineering Education System, Witchita, Kansas, 1996.

13. Crowe, C.T. and Smoot, L.D., Chapter 11 in *Pulverized Coal Combustion and Gasification,* Smoot, L.D. and Pratt, D.T., Eds., Plenum, New York, 1979, 15.

14. Tennekes, H. and Lumley, J.L., *A First Course in Turbulence,* MIT Press, Cambridge, MA, 1972.

15. Patankar, S.V., *Numerical Heat Transfer and Fluid Flow,* Hemisphere, Washington, 1980.

16. Aregbesola, Y.A.S. and Burley, D.M., *J. Comput. Phys.*, 24, 398, 1977.

17. Van Doormal, J.P. and Raithby, G.D., *Numerical Heat Transfer,* 7, 147, 1984.

18. Issa, R.I., *J. Comput. Phys.*, 62, 40, 1985.

19. Issa, R.I., *J. Comput. Phys.*, 62, 66, 1985.

20. Bilger, R.W., *Prog. Energy Combust. Sci.,* 1, 87, 1976.

21. Gorner, K. and Zinser, W., Predictions of Three-Dimensional Flows in Utility Boiler Furnaces and Comparison with Experiments. *The Annual ASME conference,* San Diego, CA, 1987.

22. Rodi, W., *Turbulence Models and Their Application in Hydraulics*, Institute for Hydromechanics, University of Karlsruhe, Karlsruhe, Federal Republic of Germany, 1984.

23. Wilcox, D.C., *Turbulence Modeling for CFD,* DCW Industries, Inc., La Canada, California, 1998.

24. Sloan, D.G., Smith, P.J., and Smoot, L.D., *Prog. Energy Combust. Sci.*, 12, 163, 1986.

25. Kline, S.J., Cantwell, B.J., and Lilley, G.M., Eds., *Complex Turbulent Shear Flows: Comparison of Computation and Experiment*, Thermosciences Division, Mech. Eng. Dept., Stanford University, Palo Alto, CA, 1982.

26. Bradshaw, P., Launder, B.E., and Lumley, J.L., *J. Fluids Eng.,* 118, 242, 1996.

27. Bousinessq, T.V., *Mem. Pres. Acad. Sci.*, 3rd ed., Paris, 23, 46, 1877.

28. Schlichting, H., *Boundary Layer Theory,* McGraw-Hill, New York, 1979.

29. Kays, W.M. and Crawford, M.E., *Convective Heat and Mass Transfer,* McGraw-Hill, New York, 1980.

30. Patankar, S.V. and Spalding, D.B., *Heat and Mass Transfer in Boundary Layers,* Intertext, London, 1970.

31. Smith, A.M.O. and Cebeci, T., Douglas Aircraft Division Report DAC 33735, 1967.

32. Baldwin, B.S. and Lomax, H., Thin-layer Approximation and Algebraic Model for Separated Turbulent Flows, AIAA Paper 78-257, 1978.

33. Launder, B.E. and Spalding, D.B., *Lectures in Mathematical Models of Turbulence,* Academic Press, 1972.

34. Spalart, P.R. and Allmaras, S.R., A One-Equation Turbulence Model for Aerodynamic Flows, AIAA Paper 92-439, 1992.

35. Anderson, D.A., Tannehill, J.C., and Pletcher, R.H., *Computational Fluid Mechanics and Heat Transfer,* McGraw-Hill, New York, 1984.

36. Harlow, F.H. and Nakayama, P.I., Transport of Turbulence Energy Decay Rate, Los Alamos National Laboratory Report No. LA-3854, Los Alamos, NM, 1968.

37. Jones, W.P. and Launder, B.E., *Int. J. Heat Mass Transfer*, 15, 301, 1972.

38. Nallasamy, M., *Computers and Fluids*, 15, 151, 1987.

39. Lakshminarayana, B., *AIAA J.*, 24, 1900, 1986.

40. Speziale, C.G., On nonlinear K-1 and K-ε models of turbulence, *J. Fluid Mech.*, 178, 459, 1987.

41. Djilali, N., Gartshore, I., and Salcudean, M., *Numerical Heat Transfer, Part A.,* 16, 189, 1989.

42. Launder, B.E., *Int. J. Heat Fluid Flow,* 10, 282, 1989.

43. Chiang, C.C. and Launder, B.E., *Numer. Heat Transfer,* 3, 189, 1980.

44. Amano, R.S., *Numer. Heat Transfer,* 7, 59, 1984.

45. Huang, P.G. and Bradshaw, P., *AIAA J.*, 33, 624, 1994.

46. Norris, L.H. and Reynolds, W.C., *Annu. Rev. Fluid Mech.,* 8, 183, 1976.

47. Bowman, C.T., *Prog. Energy Combust. Sci.,* 1, 1, 1975.

48. Bray, K.N.C., Turbulent flows with premixed reactants, in *Turbulent Reacting Flows*, Libby, P.A. and Williams, F.A., Eds., Topics in Applied Physics, Springer-Verlag, New York, chap. 4, 115, 1980.

49. Magnussen, B.F. and Hjertager, B.H., *16th Symp. (Int.) on Combustion*, The Combustion Institute, 1976, 719.

50. Spalding, D.B., *13th Symp. (Int.) on Combustion,* The Combustion Institute, 1970, 649.

51. Weber, R., Peters, A.A.F., and Breithaupt, P.P., *Combustion Modeling, Confiring and NO$_x$ Control,* Gupta, A.K., Mehta, A., Moussa, N.A., Presser, C., Rini, M.J., Saltiel, C., Warchol, J., and Whaley, H., Eds., FACT Vol. 17, ASME Book No. H000827, 1993.

52. Kent, J.H. and Bilger, R.W., *16th Symp. (Int.) on Combustion,* The Combustion Institute, 1977, 1643.

53. Kuo, K.K.Y., *Principles of Combustion,* John Wiley & Sons, New York, 1986.

54. Lockwood, F.C. and Naguib, A.S., *Combust. Flame,* 24, 109, 1975.

55. Pope, S.B., *Prog. Energy Combust. Sci.,* 11, 119, 1985.

56. Correa, S.M., and Pope, S.B., *24th Symp. (Int.) on Combustion,* The Combustion Institute, 1992, 279.

57. Viskanta, R. and Menguc, M.P., *Prog. Energy Combust. Sci.,* 10, 97, 1987.

58. Menguc, P.P., and Webb, B.W., Radiative Heat Transfer, Ch. 5., *Fundamentals of Coal Combustion for Clean and Efficient Use,* Smoot, L.D., Ed., Elsevier, Amsterdam, 1993, 375.

59. Brewster, M.Q. and Kunimoto, T., *ASME J. Heat Transfer,* 106, 678, 1984.

60. Baxter, L.L., Fletcher, T.H., and Otteson, D.K., *Energy & Fuels,* 2, 423, 1988.

61. Van De Hulst, H.C., *Light Scattering by Small Particles,* Wiley, New York, 1957; also Dover, New York (1981).

62. Kerker, M., *The Scattering of Light,* Academic Press, New York, 1969.

63. Crosbie, A.L. and Davidson, B.W., *J. Quant. Spect. Radiat. Transfer,* 33, 391, 1985.

64. Siegel, R. and Howell, J.R., *Thermal Radiation Heat Transfer,* 2nd ed., Hemisphere, Washington D.C., 1981.

65. Crowe, C.T., Gas-Particle Flow, Ch.6, *Pulverized Coal Combustion and Gasification,* Smoot, L.D. and Pratt, D.T., Eds., Plenum, New York, 1979.

66. Hottel, H.C. and Sarofim, A.F., *Radiative Transfer,* McGraw-Hill, New York, 1959.

67. Haji-Sheikh, *Handbook of Numerical Heat Transfer,* Minkowycz, W.J., Sparrow, E.M., Pletcher, R.H., and Schneider, G.E., Eds., Wiley, New York, 1988.

68. Hottel, H.C. and Cohen, E.S., *AIChE J.,* 4, 3, 1958.

69. Fiveland, W.A., *J. of Heat Transfer,* 106, 699, 1984.

70. Chandrasekhar, S., *Radiative Transfer,* Oxford University Press, London, 1950; also Dover Publications, New York (1960).

71. Carlson, B.G. and Lathrop, K.D. *Transport Theory — The Method of Discrete-Ordinates in Computing Methods in Reactor Physics,* Greenspan, H., Kelber, C., and Okrent, D., Eds., Gordon and Breach, New York, 1968.

72. Lewis, E.E. and Miller, W.F., Jr., *Computational Methods of Neutron Transport,* Wiley, New York, 1984.

73. Fiveland, W.A., ASME Paper No. 82-HT-20 (1982).

74. Fiveland, W.A., *J. Thermophysics and Heat Transfer,* 2, 309, 1988.

75. Raithby, B.D and Chui, E.H., *J. Heat Transfer,* 112, 415, 1990.

76. Chui, E.H. and Raithby, G.D., *Numerical Heat Transfer,* 23 (Pt. B), 269, 1993.

77. Chai, J.C., Lee, H.S., and Patankar, S.V., *J. Thermophysics and Heat Transfer,* 8, 421, 1994.

78. DeMarco, A.G. and Lockwood, F.C., *La Rivista dei Combustible,* 22, 184, 1975.

79. Lockwood, F.C. and Shah, N.G., *Heat Transfer — 1978,* Hemisphere, Washington, D.C., 2, 33, 1978.

80. Higenyi, J. and Bayazitoglu, Y., *J. Heat Transfer,* 102, 719, 1980.

81. Ratzel, A.C., III and Howell, J.R., *J. Heat Transfer,* 105, 333, 1983.

82. Menguc, M.P. and Viskanta, R., *J. Quant. Spectrosc. Radiat. Transfer,* 33, 533, 1985.

83. Menguc, M.P. and Viskanta, R., *J. Heat Transfer,* 108, 271, 1986.

84. Fiveland, W.A. and Wessel, R.A., ASME Paper No. 86-HT-35, ASME, New York, 1986.

85. Gosman, A.D., Lockwood, F.C., Megahed, I.E.A., and Shah, N.G., *J. Energy,* 6, 353, 1982.

86. Murthy, J.Y. and Choudhury, D., *HTD-Vol. 203, Developments in Radiative Heat Transfer,* 153, ASME, 1992.

87. Shuen, J.S., Chen, L.D., and Faeth, G.M., *AIAA J.,* 21, 1483, 1983.

88. Parthasarathy, R.N., and Faeth, G.M., *Int. J. Multiphase Flow,* 13, 699, 1987.

89. Solomon, A.S.P.,et al., *AIAA J.,* 23, 1724, 1985.

90. Solomon, A.S.P., et al., *J. Heat Transfer,* 107, 679, 1985.

91. Shuen, J.S., Solomon, A.S.P., and Faeth, G.M., *AIAA J.,* 24, 101, 1986.

92. Sun, T.Y., and Faeth, G.M., *Int. J. Multiphase Flow,* 12, 99, 1986.

93. Sun, T.Y., and Faeth, G.M., *Int. J. Multiphase Flow,* 12, 115, 1986.

94. Sun, T.Y., Parthasarathy, R.N. and Faeth, G.M., *J. Heat Transfer,* 108, 951, 1986.

95. Crowe, C.T., Chung, J.N., and Troutt, T.R., *Prog. Energy Combust. Sci.,* 14, 171, 1988.

96. Snyder, W.H. and Lumley, J.L., *J. Fluid Mech.,* 48, 41, 1971.

97. Crowe, C.T., Troutt, T.R., and Chung, J., *Annu. Rev. Fluid Mechanics,* 28, 11, 1996.

98. Stock, D.E., *J. Fluids Eng.*, 118, 4, 1996.
99. Faeth, G.M., *Prog. Energy Combust. Sci.,* 9, 1, 1983.
100. Faeth, G.M., *Prog. Energy Combust. Sci.,* 13, 293, 1987.
101. Lumley, J.L., *Turbulence,* Bradshaw, P., Ed., Springer-Verlag, 2nd ed., 1978, 289.
102. Crowe, C.T., *J. Fluids Eng.,* 104, 297, 1982.
103. Sirignano, W.A., *Numerical Modeling in Combustion*, Chung, T.J., Ed., 1993, 457.
104. Shirolkar, J.S., Coimbra, C.F.M., and McQuay, M.Q., *Prog. Energy Combust. Sci.,* 22, 363, 1996.
105. Elghobashi, S., Abou-Arab, T., Rizk, M., and Mostafa, A., *Int. J. Multiphase Flow,* 10, 697, 1984.
106. Crowe, C.T., Sharma, M.P., and Stock, D.E., *J. Fluids Engineering*, 99, 325, 1977.
107. Boysan, F., Ayers, W.H., and Swithenbank, J., *Trans. Inst. Chem. Eng.,* 60, 222, 1982.
108. Melville, W.K. and Bray, K.N.C., *Int. J. Heat Transfer,* 22, 647, 1979.
109. Shuen, J.S., Chen, L.D., and Faeth, G.M., *AIChE J.,* 29, 167, 1983.
110. Gosman, A.D. and Ioannides, E., *AIAA 19th Aerospace Science Mtng.,* 81-0323, 1981.
111. Gavalas, G.R., *Coal Pyrolysis*, Elsevier, New York, 1982.
112. Solomon, P.R., *Chemistry of Coal Conversion*, Schlosberg, R.H., Ed., Plenum, New York, 1985, 121.
113. Grant, D.M., Pugmire, R.J., Fletcher, T.H., and Kerstein, A.R., *Energy and Fuels*, 3, 175, 1989.
114. Smith, K.L., Smoot, L.D., Fletcher, T.H., and Pugmire, R.J., *The Structure and Reactions of Coal,* Plenum, NY, 1994.
115. Solomon, P.R., Hamblen, D.G., Carangelo, R.M., Serio, M.A. and Deshpande, G.V., *Fuel*, 65, 182, 1986.
116. Truelove, J.S. and Jamaluddin, A.S., *Combust. Flame*, 64, 369, 1986.
117. Kobayashi, H., Howard, J.B., and Sarofim, A.F., *16th Symp. (Int.) on Combustion*, The Combustion Institute, 1976, 411.
118. Ubhayakar, S.K., Stickler, D.B., von Rosenberg, C.W. and Gannon, R.E., *16th Symp. (Int.) on Combustion*, The Combustion Institute, 1976, 427.
119. Solomon, P.R., Serio, M.A., Hamblen, D.G., Yu, Z.Z., and Charpenay, S., preprint, *Am. Chem Soc., Fuel Chem.,* 35, 479, 1990.
120. Solomon, P.R., Serio, M.A., and Suuberg, E.M., *Prog. Energy Combust. Sci.,* 18, 133, 1992.
121. Niksa, S. and Kerstein, A.R., *Energy & Fuel*, 5, 647, 1991.
122. Niksa, S., *Energy & Fuels,* 5, 665, 1991.
123. Niksa, S., *Energy & Fuels,* 5, 673, 1991.
124. Grant, D.M., Pugmire, R.J., Fletcher, T.H., and Kerstein, A.K., *Energy & Fuels,* 3, 175, 1989.
125. T.H. Fletcher, Kerstein, A.R., Pugmire, R.J., and Grant, D.M., *Energy & Fuels,* 4, 54, 1990.
126. Fletcher, T.H., Kerstein, A.R., Pugmire, R.J., and Grant, D.M., *Energy & Fuels,* 6, 414, 1992.
127. Fletcher, T.H., Solum, M.S., Grant, D.M., and Pugmire, R.J., *Energy & Fuels,* 6, 643, 1992.
128. Fletcher T.H., personal communication, work in process, Brigham Young University, Provo, Utah, 1999.
129. Mitchell, R.E., *22nd Symp. (Int.) on Combustion*, The Combustion Institute, 1988, 69.
130. Hurt, R.H. and Mitchell, R.E., *24th Symp. (Int.) on Combustion*, The Combustion Institute, 1992, 1243.
131. Boardman, R.D. and Smoot, L.D., Pollutant Formation and Control, in *Fundamentals of Coal Combustion for Clean and Efficient Use,* Smoot, L.D., Ed., Elsevier, 1993, 433.
132. Hill, S.C. and Smoot, L.D., *Prog. Energy Combust. Sci.,* in review, 1999.
133. Smith, P.J., Hill, S.C., and Smoot, L.D., *19th Symp. (Int.) on Combustion,* The Combustion Institute, 1982, 1263.
134. Smith, P.J., Smoot, L.D., and Hill, S.C., *AIChE J.,* 32, 1917, 1986.
135. Hill, S.C., Smoot, L.D., and Smith P.J., *20th Symp. (Int.) on Combustion,* The Combustion Institute, 1984, 1391.
136. Boardman, R.D. and Smoot, L.D., *AIChE J.,* 34, 1573, 1988.
137. Smoot, L.D., Boardman, R.D., Brewster, B.S., Hill, S.C., and Foli, A.K., *Energy and Fuels*, 7, 786, 1993.
138. Hayhurst, A.N. and Lawrence, A.D., *Prog. Energy Combust. Sci.,* 18, 529, 1992.
139. Kramlich, J.C. and Linak, W.P., *Prog. Energy Combust. Sci.,* 20, 149, 1994.
140. Miller, J.A. and Bowman, C.T., *Prog. Energy Combust. Sci,* 15, 287, 1989.
141. Zel'dovich, Y.B., Sadovnikov, P.Y., and Frank-Kamentskii, D.A., (translated by Shelef), Oxidation of Nitrogen in Combustion, Academy of Sciences of USSR, 1947.
142. Lavoie, G.A., Heywood, J.B., and Keck, J.C., *Combust. Sci. Tech.,* 1, 313, 1970.
143. Flower, W.L., Hanson, R.K., and Kruger, C.H., *15th Symp. (Int.) on Combustion*, The Combustion Institute, 1975, 823.

144. Monat, J.P., Hanson, R.K., and Kruger, C.H., *17th Symp. (Int.) on Combustion,* The Combustion Institute, 1979, 543.
145. Miller, J.A., Branch, M.C., McLean, W.J., Chandler, D.W., and Smooke, M.D., *20th Symp. (Int.) on Combustion,* The Combustion Institute, 1984, 673.
146. Hanson, R.K., and Saliman, S., *Combustion Chemistry,* W.C. Gardiner, Ed., 361, Springer-Verlag, New York, 1984.
147. Westenberg, A.A., *Combust. Sci. Tech.,* 9, 59, 1971.
148. Fenimore, C.P., *13th Symp. (Int.) on Combustion,* The Combustion Institute, 1971, 373.
149. Barnes, F.J., Bromly, J.H., Edwards, T.J., and Madngezewsky, R., *J. Institute of Energy,* 155, 184, 1988.
150. DeSoete, G.C., *15th Symp. (Int.) on Combustion,* The Combustion Institute, 1975, 1093.
151. Backmeir, F., Eberius, K.H., and Just, T., *Combustion Sci. Tech.,* 7, 77, 1973.
152. Pershing, D.W. and Wendt, J.O.L., *16th Symp. (Int.) on Combustion,* The Combustion Institute, 1977, 389.
153. Mitchell, J.W. and Tarbell, J.M, *AIChE J.,* 28, 302, 1982.
154. Bose, A.C., Dannecker, K.M., and Wendt, J.O.L., *Energy & Fuels,* 2, 301, 1988.
155. Smoot, L.D., Xu, H., and Hill, S.C., *Prog. Energy Combust. Sci.,* 24, 385, 1998.
156. Chen, W., Smoot, L.D., Fletcher, T.H., and Boardman, R.D., *Energy & Fuels,* 10, 1036, 1996.
157. Chen, W., Smoot, L.D., Hill, S.C., and Fletcher, T.H., *Energy & Fuels,* 10, 1046, 1996.
158. Xu, H., Smoot, L.D., and Hill, S.C., *Energy & Fuels,* 12, 1278, 1998.
159. Xu, H., Smoot, L.D., and Hill, S.C., *Energy & Fuels,* 13, 411, 1999.
160. Hill, S.C. and Smoot, L.D., *Energy & Fuels,* 7, 874, 1993.
161. Germane, G.J., Eatough, C.N., and Cannon, J.N., Process Data and Strategies, Ch. 2, *Fundamentals of Coal Combustion,* Smoot, L.D., Ed., Elsevier, Amsterdam, 1992.
162. Eatough, C.N., Ph.D. Dissertation, Brigham Young University, 1991.
163. Tree, D.R., Black, D.L., Rigby, J.R., McQuay, M.Q., and Webb, B.W., *Prog. Energy Combust. Sci.,* 24, 355, 1998.
164. Butler, B.W., M.S. thesis, Brigham Young University, 1988.
165. Boardman, R.D., Eatough, C.N., Germane, G.J., and Smoot, L.D., *Combustion Sci. Tech.,* 93, 193, 1993.
166. Sanderson, D.K., M.S. thesis, Brigham Young University, 1993.
167. Butler, B.W., Ph.D. Dissertation, Brigham Young University, 1991.
168. Cannon, J.N., Webb, B.W., and Queiroz, M., Final Report, ESEERCO Project EP 89-09, Brigham Young University, 1995.
169. Newbold, J., Webb, B. W., and McQuay, M.Q., Combustion measurements in an industrial gas-fired flat-glass furnace, *J. Inst. Energy,* 70, 71-81, 1997.
170. Hayes, R.R., Brewster, B.S., Webb, B.W., and McQuay, M.Q., Crown Incident Radiant Heat Flux Measurements in an Industrial, Regenerative, Gas-Fired, Flat-Glass Furnace, Experimental Thermal and Fluid Science (in review).
171. Xu, H., Smoot, L.D., and Hill, S.C., Computational Model for NOx Reduction by Advanced Reburning, *Energy & Fuels,* 13, 411-420, 1999.

5 Local Integral Moment (LIM) Simulations

Werner J.A. Dahm, Grétar Tryggvason,
Richard D. Frederiksen, and Mark J. Stock

CONTENTS

5.1   INTRODUCTION

Most computational fluid dynamics codes for industrial application were developed as general-purpose software packages applicable to a very wide range of problems. Such codes are typically written to address both laminar and turbulent flows, often in both the compressible and incompressible flow regimes, and usually include submodels for a variety of other physical processes such as mixing, chemical kinetics, heat transfer, and phase changes. This generality is needed for such codes to address a sufficiently large market of applications to sustain a commercial entity that develops and maintains the code and provides technical support for its user base.

However, this same generality also dictates that the solution methodology in such codes must be sufficiently general to apply for the entire range of flow regimes and physical processes to which they can be applied. The methodology cannot be specialized to take advantage of the characteristics of any one flow regime or any narrower set of applications. Limitations in the accuracy of results from such general-purpose codes can therefore be as much a consequence of the need to maintain their generality as they are due to any incomplete understanding of the underlying flow regimes or physical processes themselves or a lack of better submodels to treat them.

This suggests that certain industrial applications could benefit from a more specialized code specifically designed for a much narrower range of flow regimes or processes. Such a code would be restricted to a more limited set of industrial applications, but could address these with a solution methodology tailored to the characteristics of the more specialized range of flow regimes or processes for which the code is intended.

The Local Integral Moment (LIM) code was conceived and developed as such a specialized computational model for turbulent flow, mixing, and combustion processes. The solution methodology developed for LIM is fundamentally different from all traditional CFD models. Unlike traditional time-averaged codes, LIM is an unsteady simulation code that computes the time evolution of the large scales in the underlying turbulent flow, which are unique to each problem and determine the resulting local entrainment and mixing rates. LIM also makes use of the self-similar structure at the diffusive scales of turbulent flows through a local parabolization of the governing transport equations around a time-evolving computational surface on which velocity gradients and scalar gradients are naturally concentrated in turbulent flows. This transforms the original partial differential equations that must be solved throughout the three-dimensional volume to a set of ordinary differential equations that can be solved on a time-evolving surface. Information is stored and updated only on this time-evolving surface, rather than on a fixed grid throughout the domain. The equations that are solved on this surface describe the evolution of local gradient profiles around the surface in terms of their local integral moments. The solution throughout the computational domain can then be reconstructed at any time from the integral moments on this surface.

LIM was originally formulated as a two-dimensional model for problems based on planar, axisymmetric, and axisymmetric swirling flows. It has more recently been successfully extended to fully three-dimensional problems. Although originally developed as a research tool, the LIM code has been successfully applied in practice to a number of large-scale industrial problems that are suited to the assumptions inherent in the underlying solution methodology.

This chapter describes the LIM approach and how it can be applied to industrial applications. Section 5.2 gives an overview of the characteristics of turbulent combustion that provide the basis for the LIM methodology. Following this, Section 5.3 outlines how these fundamental characteristics can be used to simplify the fully general problem of turbulent combustion, and Section 5.4 describes how these simplifications lead to a specialized computational method applicable to problems for which these simplifying approximations are valid. Section 5.5 presents demonstrations of LIM modeling applied to some relatively simple configurations, and Section 5.6 gives some examples of how LIM has been applied to assist in the development of reburn systems for large utility boilers. Section 5.7 describes the relatively recent extension of the LIM methodology to fully three-dimensional geometries, and Section 5.8 provides concluding remarks as to the applicability of such specialized computational approaches and the role they can play in industrial combustion.

5.2 FOUNDATIONS FOR A NEW TURBULENT COMBUSTION MODEL

In all traditional CFD codes, the time-averaged turbulent transport of mass, momentum, energy, and scalars are modeled with an assumption first introduced in 1877 called the "gradient diffusion hypothesis." This presumes that transport in turbulent flows is by small-scale, stochastic eddies, analogous to transport by molecules in kinetic theory, and hence proportional to the average gradient via a turbulent diffusivity that depends on the local turbulence properties. The numerous turbulence models used in various codes are *all* inherently based on this gradient diffusion assumption, and differ only in the comparatively minor issue of how they relate the fictitious turbulent diffusivities back to the averaged quantities.

However, basic research over the past 25 years has revealed that this gradient diffusion assumption, while conveniently general in that it models turbulent flows like laminar flows with an enhanced diffusivity, is fundamentally incorrect. These studies have also provided a number of specific insights that allow formulation of an entirely different approach to turbulent combustion modeling.

5.2.1 LARGE-SCALE STRUCTURES

The gradient diffusion assumption would be valid if transport in turbulent flows were indeed dominated by small-scale, stochastic eddies. However beginning with Brown and Roshko,[1] basic research studies have established that transport in turbulent shear flows is instead dominated by large-scale, quasi-periodic, organized motions. Examples of these motions in two-dimensional turbulent mixing layers are shown in Figure 5.1. Often termed "large-scale coherent structures," such organized motions have been found to dominate turbulent transport in all turbulent shear flows. The presence of such large-scale structures inherently invalidates the assumptions of the gradient diffusion hypothesis, irrespective of how the fictitious turbulent diffusivities are modeled. Moreover, the large-scale structures in any given problem correspond to the most unstable mode of the local mean flow, and thus are unique to each problem. This implies that, not only is transport in turbulent flows manifestly not of the gradient diffusion type, but it inherently cannot be treated with any universal turbulence model. Collectively, these insights suggest that improved modeling of turbulent flow, mixing, and combustion requires a code that departs from the traditional time-averaged approach and solves directly for the time-varying large-scale structures in each problem.

5.2.2 FINE-SCALE STRUCTURES

Other basic research has investigated the fine-scale structure associated with velocity gradients and scalar gradients in turbulent flows. Of particular importance are local solutions to the governing equations over regions that are sufficiently small for the strain rate tensor to be spatially uniform. These[2,3] show that there are just two classes of fine-scale structures possible. Where the space- and time-varying strain rate tensor has two compressional and one extensional principal values, it forms locally "line-like" structures in the vorticity and scalar gradient fields, and where it has one compressional and two extensional principal strain rates, it forms locally "sheet-like" structures. The stretching term in the vorticity dynamics allows both types of structures to be maintained in the vorticity, but in the scalar gradient dynamics only the sheet-like structures can survive. The fine-scale structure of scalar mixing in turbulent flows is therefore remarkably simple, and consists entirely of such locally convoluted, sheet-like structures. These fine-scale structures are readily apparent in highly resolved measurements of scalar gradient fields in turbulent shear flows, as shown in Figures 5.2 and 5.3. The internal structure within these fine-scale structures is self-similar, with their thickness set by the competing effects of the local strain rate and the molecular diffusivity. Consequently, these fine-scale structures can be reconstructed from just a few low-order moments of their internal profiles in a computational method that is specifically designed to track these moments.

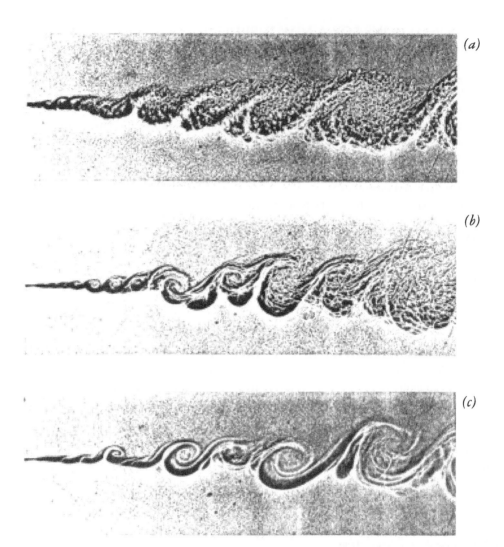

FIGURE 5.1 Results from flow visualization of a mixing layer at (a) high, (b) medium, and (c) low Reynolds numbers. Note the distinctive large scale structure of the flow, which is Reynolds number invariant. LIM simulations improve upon conventional computational fluid dynamics techniques by directly simulating these large scale structures and avoiding the gradient transport hypotheses of traditional computational fluid dynamics simulations. (From Brown, G.L. and Roshko, A., *J. Fluid Mechanics,* 64, 775-816, 1974. With permission.)

5.2.3 MIXING-CHEMISTRY COUPLING

The introduction of conserved scalar mixture fraction variables based on the mass fractions of elements involved in the chemical system greatly simplifies the understanding and modeling of turbulence-chemistry coupling in initially non-premixed or partially premixed combustion systems.[4] Note that any such elemental mixture fraction ζ_i may be written as a sum over the chemical species fields $Y_j(\mathbf{x},t)$ as

$$\zeta_i(\mathbf{x},t) = \sum_{j=1}^{N} a_{i,j} Y_j(\mathbf{x},t) \qquad i = 1,2,\ldots,m \tag{5.1}$$

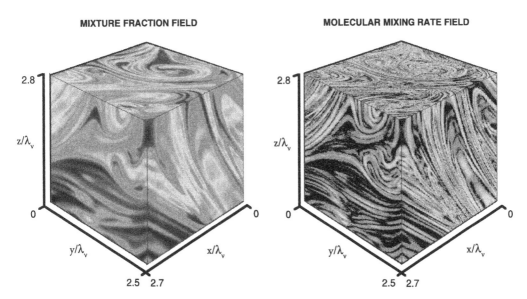

FIGURE 5.2 Typical experimental results from instantaneous, three-dimensional (256^3), laser-induced fluorescence measurements of the fine scale structure of molecular mixing between two different fluids in a turbulent flow. Shown at left is the instantaneous mixture fraction field $\zeta(\mathbf{x},t)$, giving the local extent of mixing at nearly 17 million spatial points throughout a cube spanning slightly less than 3 inner (Kolmogorov) scales λ_ν on each side. The corresponding instantaneous scalar energy dissipation rate field $\nabla\zeta\cdot\nabla\zeta(\mathbf{x},t)$ at right gives the local molecular mixing rate. The cube on the right shows that the molecular mixing process is entirely concentrated on wrinkled sheet-like structures that are stretched and folded by the underlying turbulent flow field. (Data from Dahm, W.J.A. and Southerland, K.B., Flow Visualization: Techniques and Examples, Smits, A.J., Lim, T.T., Eds., Cambridge University Press, in press, 1999. With permission.)

where N is the number of chemical species and m the number of elements involved in the chemical system. Because mixture fractions are conserved scalar quantities, in the absence of significant differential diffusion, they suffice to quantify the local elemental composition in the underlying turbulent flow. In the limit of adiabatic equilibrium chemistry, this fully determines the species concentrations and temperature, and by using the mixture fraction $\zeta(\mathbf{x},t)$ and the local mixing rate $\nabla\zeta\cdot\nabla\zeta(\mathbf{x},t)$ various nonequilibrium chemistry models[5] have been successful in predicting chemical species in turbulent flames. Such conserved scalar approaches permit detailed chemical kinetics libraries to be coupled to turbulent flow and mixing simulations without the need for direct calculations of the complex chemistry.

5.2.4 HEAT RELEASE EFFECTS

The coupling from the chemical reactions back to the flow and mixing processes is a consequence of various heat release effects. These include changes in the fluid viscosity and diffusivities with temperature; however, these molecular transport properties only lead to an increase in the small scales and play no direct role in setting the entrainment and mixing rates of the turbulent flow. The velocities induced by the volume source field due to the Lagrangian rate of change in density due to heat release are small in comparison with those induced by the underlying vorticity field. Heat release also generates baroclinic vorticity from interactions of density gradients with both hydrodynamic and hydrostatic pressure gradients. The latter are buoyancy effects, which can be quite pronounced in certain applications. However, when buoyancy is negligible, the dominant effect of heat release on the entrainment and mixing rates, as well as other properties of turbulent shear flows, can be

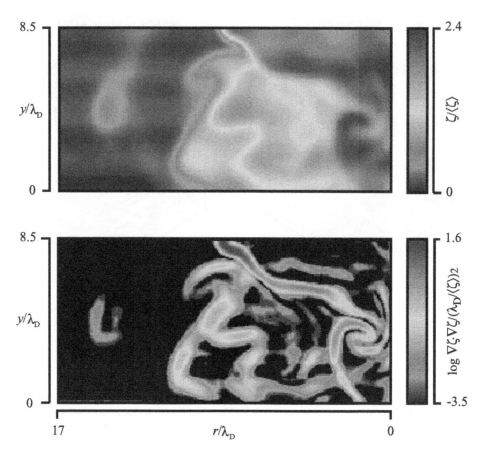

FIGURE 5.3 Instantaneous, two-dimensional laser Rayleigh scattering measurements of the fine-scale structure of molecular mixing of two gases in a turbulent flow, in this case propane gas injected into a coflowing air stream. Shown at top is the instantaneous mixture fraction field, $\zeta(\mathbf{x},t)$, giving the local extent of mixing, and at bottom the corresponding instantaneous scalar energy dissipation rate fields, $\nabla\zeta\cdot\nabla\zeta(\mathbf{x},t)$, giving the local molecular mixing rate. Each such data plane spans from the jet centerline, at the right edge, to pure air in the coflowing airstream, at the left edge. Colors identify the instantaneous propane concentrations and molecular mixing rates. As in Figure 5.2, all the molecular mixing is concentrated on locally one-dimensional wrinkled sheet-like structures, which have a self-similar internal structure and are continually stretched and folded by the underlying turbulent flow. These structures form the basis of the LIM model, which incorporates them directly into a new solution method for the partial differential equations (PDEs) governing the flow and mixing processes. This reduces the PDEs to a much simpler set of ordinary differential equations (ODEs) that can be solved on the time-evolving surface on which the structures are centered. (From Buch, K.A. and Dahm, W.J.A., *Journal of Fluid Mechanics*, 364, 1-29, 1998. With permission.)

accounted for by a simple rescaling of the ambient fluid density.[6] Under these conditions, the density may be treated as constant and subsequently rescaled to account for heat release.

5.3 FUNDAMENTAL CONSIDERATIONS AND APPROXIMATIONS

The physical insights in Section 5.2, which apply only to turbulent flows, can be used to simplify the fully general problem of reacting fluid dynamics to a more tractable form, which provides the basis for the numerical method in Section 5.4.

5.3.1 Flowfield u(x,t)

The Helmholtz decomposition for any vector field, when applied to the fluid velocity field $\mathbf{u}(\mathbf{x},t)$, shows that

$$\mathbf{u}(\mathbf{x},t) = \underbrace{\frac{1}{4\pi}\iiint_{\mathbf{x}'}\omega(\mathbf{x}',t) \times \mathbf{K}(\mathbf{x},\mathbf{x}')d^3\mathbf{x}'}_{\mathbf{u}_\omega(\mathbf{x},t)} + \underbrace{\frac{1}{4\pi}\iiint_{\mathbf{x}'}q(\mathbf{x}',t)\mathbf{K}(\mathbf{x},\mathbf{x}')d^3\mathbf{x}'}_{\mathbf{u}_q(\mathbf{x},t)} \tag{5.2}$$

where the kernel is

$$\mathbf{K}(\mathbf{x},\mathbf{x}') \equiv \frac{\mathbf{x}-\mathbf{x}'}{|\mathbf{x}-\mathbf{x}'|^3} \tag{5.3}$$

In Equation (5.2), $\mathbf{u}_\omega(\mathbf{x},t)$ is the divergence-free part of the velocity, with $\omega(\mathbf{x},t) \equiv \nabla \times \mathbf{u}(\mathbf{x},t)$ being the vorticity field. The integral that relates $\omega(\mathbf{x},t)$ and $\mathbf{u}_\omega(\mathbf{x},t)$ in Equation (5.2) is the Biot-Savart law. As noted in Section 5.2.2 and in Figures 5.2 and 5.3, with increasing Reynolds number, the competing effects of strain and diffusion concentrate gradient quantities such as the vorticity onto an increasingly smaller and highly intertwined subset of the entire flow domain. Under such conditions, a numerical method based on tracking these concentrated regions of vorticity can provide an extremely efficient approach for simulating the flowfield.

The second term in Equation (5.2), namely $\mathbf{u}_q(\mathbf{x},t)$, is the curl-free part of the velocity, where $q(\mathbf{x},t) \equiv \nabla \cdot \mathbf{u}(\mathbf{x},t)$ is the volumetric dilatation field. Consistent with the intended applications, density changes due to pressure variations are taken to the negligible; this is the zero Mach number approximation. From this together with the equation for mass conservation and the assumption of constant pressure in the ideal gas law, it follows that

$$q(\mathbf{x},t) = \frac{1}{T}\frac{DT}{Dt} \tag{5.4}$$

where $T(\mathbf{x},t)$ is the fluid temperature field. The energy equation then provides

$$q(\mathbf{x},t) = \frac{1}{T}\nabla \cdot (k\nabla T) + \frac{1}{T}\dot{w}_T(\mathbf{x},t) \tag{5.5}$$

with k the thermal conductivity and $w_T(\mathbf{x},t)$ the Lagrangian time rate of change of temperature due to chemical reaction. In Equation (5.5) the dilatation due to reaction heat release dominates that due to heating or cooling by conduction and, consequently, the first term on the right may be neglected. For typical hydrocarbon combustion in turbulent flows, the velocity field $\mathbf{u}_q(\mathbf{x},t)$ induced by the second term on the right in Equation (5.5) is negligible in comparison with $\mathbf{u}_\omega(\mathbf{x},t)$ in Equation (5.2).

Under these conditions, the curl-free part of the velocity field in Equation (5.2) may be neglected, and it is only necessary to compute the divergence-free part, namely the vorticity $\omega(\mathbf{x},t)$. As noted in Section 5.2.4, baroclinic generation of vorticity generally becomes dominant only in flows with strong buoyancy. Excluding these, the vorticity satisfies

$$\frac{\partial\omega}{\partial t} + \mathbf{u}\cdot\nabla\omega - \nabla\cdot\nu\nabla\omega = \omega\cdot\nabla\mathbf{u} \tag{5.6}$$

where consistent with Section 5.2.4, the variations in the viscosity can be neglected. If one specializes further to two-dimensional flows, then in planar simulations the stretching term on the right-hand side vanishes, while in axisymmetric and fully three-dimensional flows, it is retained.

5.3.2 CHEMICAL SPECIES $Y_j(x,t)$ AND TEMPERATURE $T(x,t)$

Because the locally sheet-like scalar gradient layers in Figures 5.2 and 5.3 result entirely for hydrodynamical reasons, all conserved scalar quantities must have their gradients concentrated in such locally one-dimensional layer-like structures. This includes mixture fraction variables formed as in Equation (5.1) from the concentration of chemical species evolving in a reacting flow. Moreover, because this is a consequence solely of the strain-diffusion balance of conserved scalar mixing, this must remain true *irrespective* of the degree of chemical nonequilibrium to which the chemical species fields are subjected.

Because the mixture fractions in Equation (5.1) are conserved scalars, they satisfy

$$\frac{\partial \zeta_i}{\partial t} + \mathbf{u} \cdot \nabla \zeta_i - \frac{1}{\rho} \nabla \cdot \left(\rho D_{\zeta_i} \nabla \zeta_i \right) = 0 \tag{5.7}$$

The locally one-dimensional scalar gradients within the layers then allows local parabolization of Equation (5.7) as

$$\frac{\partial \zeta_i}{\partial t} - \sigma(t) n \frac{\partial \zeta_i}{\partial n} - \frac{1}{\rho} \frac{\partial}{\partial n} \cdot \left(\rho D_{\zeta_i} \frac{\partial \zeta_i}{\partial n} \right) = 0 \tag{5.8}$$

where n is the Howarth-transformed coordinate and $\sigma(t)$ is the local time-varying strain rate along the layer-normal direction. Replacing ζ_i in Equation (5.8) with Equation (5.1) together with the requirement that, since the ζ_i are conserved, the weighted species reaction rate terms $w_j(\mathbf{x},t)$ must sum to zero, gives

$$\frac{\partial Y_j}{\partial t} - \sigma(t) n \frac{\partial Y_j}{\partial n} - \frac{1}{\rho} \frac{\partial}{\partial n} \cdot \left(\rho D_{\zeta_i} \frac{\partial Y_j}{\partial n} \right) = \dot{w}_j / \rho \tag{5.9}$$

The local one-dimensionality of the conserved scalar field within each scalar gradient layer thus implies a locally one-dimensional structure for the underlying chemical species fields within the layer. This is precisely the opposite point of view from that usually taken in deriving the classical "flamelet" model, where conditions are assumed for which the flamelet is thin and the species transport equations then formally reduce to locally one-dimensional equations. Here, the observation that the mixture fraction fields must be locally one-dimensional implies that the constituent chemical species fields $Y_j(\mathbf{x},t)$ are also locally one-dimensional as in Equation (5.9).

Solutions of Equation (5.9) are libraries parametrized by the local strain rate σ and the chemical species boundary values $Y_j^\pm$. For pure fuel and air, these become equivalent to classical flamelet libraries. The mapping from $(\zeta, \chi, \zeta^+, \zeta^-)$ to the strain rate ε depends strongly on $\zeta^\pm$, and thus local departures from chemical equilibrium in each of the species concentrations will be determined principally by the local scalar value ζ and the local strain rate $\sigma(\zeta, \chi, \zeta^+, \zeta^-)$.

5.4 NUMERICAL IMPLEMENTATION

This section outlines how the simplifications in Sections 5.2 and 5.3 lead to the numerical methodology on which LIM simulations are based.

5.4.1 LAYER-NORMAL GRADIENT PROFILES

Consistent with the arguments already noted for treating the density and all molecular transport properties as constant, each of the mixture fraction variables in Equation (5.7) follows an advection-diffusion equation of the form

$$\frac{\partial \zeta}{\partial t} + \mathbf{u} \cdot \nabla \zeta - D\nabla^2 \zeta = 0 \tag{5.10}$$

At large Reynolds numbers, the diffusive term in Equation (5.10) involves a singular limit that leads to the formation of internal "scalar gradient boundary layers" of the type noted in Section 5.2.2. These scalar gradient layers result solely from the hydrodynamics of scalar mixing in turbulent flows, and imaging measurements as well as direct numerical simulations indicate that they remain present even in flows undergoing exothermic combustion reactions. The same advection-diffusion balance in the vorticity transport equation leads to a similar localization of velocity gradients as well.

Owing to the locally one-dimensional internal structure, spatial derivatives of the scalar in the plane of the layer are small in comparison with the derivative along the local layer-normal direction. Discarding spatial derivatives in all but the layer-normal direction n gives the locally parabolized form of Equation (5.10) as

$$\frac{\partial \zeta}{\partial t} - \sigma(t)n\frac{\partial \zeta}{\partial n} - D\frac{\partial^2 \zeta}{\partial n^2} = 0 \tag{5.11}$$

where $\sigma(t)$ is the local strain rate along the local layer-normal direction. In both cases, the one-dimensional self-similar structure within these gradient layers allows development of a computational method that exploits this simplifying feature by incorporating it directly into the solution method.

In particular, because the gradients within layers of the type in Figures 5.2 and 5.3 are self-similar,[2,3] they can be represented by a family of profile shapes having just a few degrees of freedom. Solutions to Equations (5.6) and (5.11) under time-varying but spatially uniform and locally planar or axisymmetric strain rate fields lead to gaussian internal profiles within these fine-scale structures, as indicated in Figure 5.4. The precise shape is less important than the fact it can be determined from a limited number of low-order integral moments of the local profile.

5.4.2 THE LIM EQUATIONS

From the self-similarity of the scalar profiles within these layers, local integral moments G_j of all orders j for the scalar gradient profile across each layer are defined as

$$G_j \equiv \int_{-\infty}^{+\infty} n^j \frac{\partial \zeta}{\partial n} dn, \quad j = 0,1,2,... \tag{5.12}$$

The exact set of transport equations for these scalar gradient moments can be readily derived from Equation (5.11), giving the ordinary differential equations (ODEs) for the time-evolution of the local moment as

$$\frac{dG_0(s)}{dt} = 0 \tag{5.13a}$$

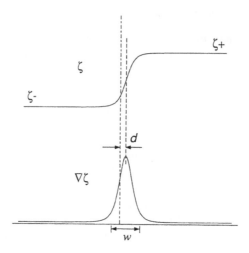

FIGURE 5.4 The family of self-similar layer profile shapes used to model the internal structure within the scalar diffusion layers. Shown (top) is the profile for the scalar (ζ), and (bottom) the corresponding scalar gradient ($\nabla\zeta$) profile. (From Suresh, N.C., A Local Integral Moment Method to Simulate Flow, Mixing and Chemistry in Complex Flows, Ph.D. thesis, The University of Michigan, Ann Arbor, 1997. With permission.)

$$\frac{dG_1(s)}{dt} = -\sigma(s)G_1(s) \tag{5.13b}$$

$$\frac{dG_2(s)}{dt} = -2\sigma(s)G_2(s) - 2DG_0(s) \tag{5.13c}$$

$$\vdots$$

where the coordinate s identifies the location on the computational surface on which the gradients are concentrated. These local integral moment equations are exact. The set of ODEs for the integral moments G_j for $j = \{1, \ldots, \infty\}$ is equivalent to the partial differential equation in Equation (5.11). Owing to the linearity of Equation (5.11) in ζ, the local integral moment equations are closed at any level of truncation. Thus, the equations for the integral moments G_j for $j = \{1, \ldots, k\}$ involve only the moments of order k and lower. As a result, the evolution of the local moments up to any desired order can be determined from the local integral equations without any closure approximation.

Vorticity concentrations that result from the advection-diffusion balance in the vorticity transport equation allow a similar local integral treatment in the flowfield that underlies the solution of Equation (5.6). This leads to a set of local integral moment equations for the moments Γ_j for $j = \{1, \ldots, \infty\}$ of the vorticity. For simplicity, only the zeroth integral moment is tracked. In planar flows and in axisymmetric flows without swirl, this satisfies a similar equation as in Equation (5.13a), namely

$$\frac{d\Gamma_0(s)}{dt} = 0 \tag{5.14}$$

where $\Gamma_0(s)$ is the zeroth moment of the local profile of the vorticity normal to the computational surface. In axisymmetric flows with swirl, the axial and azimuthal components of circulation are coupled. With the local axial (swirling) component of the circulation $\Gamma_{ax}(r) \equiv 2\pi r u_\theta(r)$ and $u_\theta(r)$ the tangential component of velocity, then from Tryggvason (1984) the azimuthal component of vorticity satisfies a moment equation for $\Gamma_0(s)$ as in Equation (5.14) but with a non-zero right-hand side determined by the local inclination of the surface and the change in u_θ across the surface.[7]

The LIM model determines the integral moments everywhere on the time-evolving material surface on which the scalar gradients were concentrated by the initial and boundary conditions. The scalar field evolution is coupled to the velocity field through the local strain rate $\sigma(s)$, which is determined from the instantaneous deformation rate at each point on the material surface. The material surfaces on which the integral moments $G_j(s)$ of the scalar gradient profiles and the integral moments $\Gamma_0(s)$ of the vorticity profiles are located evolves from the Biot-Savart induced velocities resulting from the vorticity in the flow.

5.4.3   LIM SURFACE EVOLUTION

The computational surfaces containing the local integral moment values of the vorticity and scalar gradient profiles are defined by an array of numbered points x_i connected to numbered surface elements s_i. Each surface element carries with it the identity of its endpoints. The endpoints carry their coordinate information, and their connectivity is determined only by the elements. The vorticity and scalar gradients associated with these elements are assumed to be concentrated at the element centers. Linked lists and pointers are used to keep track of the endpoints and the elements. An efficient data structure is used that accommodates both open and closed surfaces.

At each time step, the local strain rate $\sigma(s,t)$ acting on each element is determined by $d(\log l)/dt$, where $l(s,t)$ is the element length as determined by its endpoints. Advecting the endpoints that define the elements on which the vorticity moments and scalar gradient moments are carried involves evaluating the Biot-Savart integral in Equation (5.2) to find the velocity field $\mathbf{u}(\mathbf{x},t)$ at each endpoint from the vorticity field $\omega(\mathbf{x},t)$ over the entire computational surface. A "direct summation" approach such as would result from simply discretizing the Biot-Savart integral leads to an $O(N^2)$ scheme for the vorticity field, which would become computationally prohibitive as the number of elements defining the surface becomes large. Instead, an $O(N \log N)$ evaluation is possible by using any of a number of more efficient means for computing the integral. Here, one adopts a method known as "vortex-in-cell" (VIC). The singular points at which vorticity is concentrated at the surface-element centers are regularized by distributing their circulations onto a temporary VIC grid $\mathbf{X}_i$, as indicated in Figure 5.5. This is done in a conservative manner using an area-weighting scheme appropriate for two-dimensional or axisymmetric flow. The circulation from each of the points defining the vorticity surface is distributed and summed. The streamfunction $\psi(\mathbf{X}_i, t)$ is then obtained from the Poisson equation

$$\nabla^2\psi = -\omega\left(\mathbf{X}_i,t\right) \tag{5.15}$$

From the streamfunction, the resulting velocity components $u = \partial\psi/\partial y$ and $v = -\partial\psi/\partial x$ are then interpolated back onto the points comprising the surface using the same area weighting scheme. The points defining the computational surface are then advected to the next time step. At every time step, the local integral moment values carried by each element of the computational surfaces are updated via Equations (5.13a-c) from the resulting strain rate $\sigma(s,t)$ for the element.

Because the surfaces of all objects in the flow are streamlines, all points on connected boundaries will have the same streamfunction value. In effect, fixed boundaries are treated in the same manner as is the LIM vorticity surface, but the corresponding surface is fixed. The boundary is thus discretized into vorticity surface elements as well, with their strength iteratively adjusted until the appropriate streamfunction value is obtained.

Vorticity shed from all convex corners in the computational domain is computed in the same manner as the boundary vorticity. The surface elements containing the shed vorticity are initially attached to the specified shedding points on the boundary, and thus have the same streamfunction

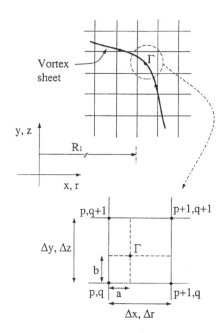

FIGURE 5.5 An illustration of the circulation distribution scheme used in the LIM method. The circulation Γ associated with each element of the LIM vorticity surface is distributed onto the four nearest points of the temporary grid using an area weighting scheme. The schematic is valid for both the two-dimensional and axisymmetric formulations, as indicated. (From Suresh, N.C., A Local Integral Moment Method to Simulate Flow, Mixing and Chemistry in Complex Flows, Ph.D. thesis, The University of Michigan, Ann Arbor, 1997. With permission.)

value as the boundary to which they are attached. This automatically enforces the Kutta condition and ensures that the shedding from the boundary occurs smoothly. Once the shed surface enters the flow, it is treated like all other parts of the computational surface.

5.4.4 LIM Surface Restructuring

As the computational surface evolves and the separation between adjacent points on the surface grows, additional surface elements are introduced to preserve the accuracy of the scheme and the definition of the surface. When an element is stretched beyond a preset threshold length, a point is added within the element and a new surface element is created, with integral moment properties carried by the element being adjusted to satisfy conservation laws. Similarly, when element lengths fall below a preset threshold value, the element is removed from the surface by deleting one of its endpoints and distributing the local integral moment information it carries in a conservative manner onto the adjacent surface elements.

Additionally, as the vorticity and scalar gradient surfaces are stretched and folded by the velocity field, surface elements come in sufficiently close proximity that their vorticity or scalar gradients overlap. The individual surface elements in the overlapped region can be replaced by a single element that preserves the local integral moment properties in a process termed "surface restructuring." As regards the endpoints, the same operations as in element deletion are performed. The orientation of the elements is taken into account to determine the merging rules. For the vorticity surface elements, simple redistribution of individual circulations is sufficient to retain the dominant physics while reducing the complexity of the surface. It may be shown that this applies to both swirling and nonswirling vorticity elements. For the scalar gradient elements, the moments are distributed in a manner that conserves the total amount of scalar carried by the individual elements, and at the same

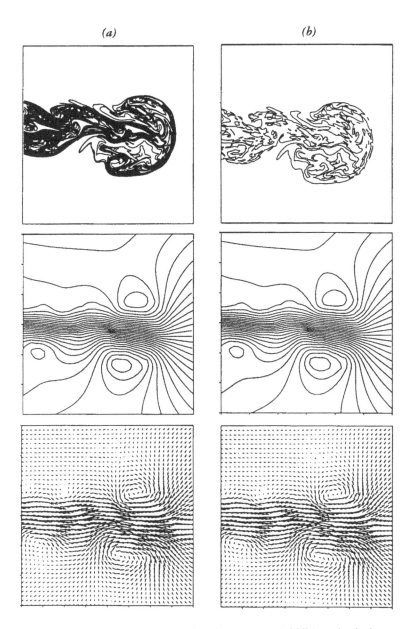

FIGURE 5.6   LIM vorticity surface (top), streamfunction contours (middle), and velocity vectors (bottom) for (a) an unreconstructed surface (40,000 elements) and (b) a restructured surface (2000 elements). Although restructuring greatly reduces the number of elements in the vorticity surface, it does not significantly alter the resulting flow fields. (From Suresh, N.C., A Local Integral Moment Method to Simulate Flow, Mixing and Chemistry in Complex Flows, Ph.D. thesis, The University of Michigan, Ann Arbor, 1997. With permission.)

time preserves the peak scalar value. Details of the conservation rules involved in the restructuring scheme for the computational surfaces are rather complex and can be found in Suresh.[7]

Figure 5.6 shows an example of the benefits and effects of the surface restructuring. The top panel in the left-hand column shows the computational surface for the vorticity at a relatively early stage in the simulation of a starting jet. In this case, surface restructuring was intentionally turned off, and the resulting vorticity surface consists of over 40,000 elements at the time shown. The top panel in the right-hand column shows the same surface after the restructuring algorithm has been applied. The lower panels compare the streamfunction contours and the velocity vectors before and after restructuring, where it can be seen that the conservation rules produce virtually identical velocity fields despite the fact that the surface has been reduced to only about 2000 elements.

5.4.5 SCALAR FIELD RECONSTRUCTION

The distribution of local integral moments $G_j(s)$ of the scalar gradient profile over the computational surface is used to reconstruct the scalar field $\zeta(\mathbf{x},t)$ at any desired set of points throughout the domain. Any family of self-similar profile shapes with the same number of degrees of freedom as the number of local integral moments being tracked in the simulation can be used to construct the local scalar gradient field. Used here is an axisymmetric Peskin function and summation over all the elements that comprise the scalar gradient surface as

$$\nabla\zeta(\mathbf{x},t) = \sum \left[\frac{1}{4w_i^2} \left(1 + \cos\frac{\pi(r_x)_i}{w_i} \right) \left(1 + \cos\frac{\pi(r_y)_i}{w_i} \right) \mathbf{n}_i \right] \tag{5.16}$$

where $\mathbf{n}_i$ is the local unit normal to the i-th element of the scalar gradient surface, $(r_x)_i$ and $(r_y)_i$ are the components of the vector from the element center to the reconstruction point, and w_i is the local gradient profile width. The gradient field $\nabla\zeta(\mathbf{x},t)$ is then differentiated to evaluate the divergence $\nabla\cdot\nabla\zeta(\mathbf{x},t)$, which forms a Poisson equation

$$\nabla^2\zeta = \nabla\cdot\nabla\zeta(\mathbf{x},t) \tag{5.17}$$

that is then solved numerically to give the scalar field $\zeta(\mathbf{x},t)$.

5.4.6 CHEMISTRY

While the LIM methodology can be used for direct calculations of the space- and time-evolving chemical species and temperature fields in a turbulent reacting flow with simple chemistry (e.g., Section 5.5), such calculations are typically too time-consuming for most practical industrial applications. For this reason, like many codes for modeling turbulent reacting flows, the chemical reaction processes are generally separated from the underlying flow and mixing processes via a conserved scalar formulation. The coupling from the space- and time-varying mixing to the resulting chemistry can be accomplished via any one of a variety of standard methods that map the extent of mixing, quantified by the conserved scalar value $\zeta(\mathbf{x},t)$, and the molecular mixing rate quantified by the scalar energy dissipation rate $\nabla\zeta\cdot\nabla\zeta(\mathbf{x},t)$, to a library of chemical species concentrations and temperature from Equation (5.9). The appropriate mapping, including equilibrium chemistry and various flamelet methods, depends on the degree of chemical nonequilibrium for the problem at hand. The particular choice for any given problem is not germane to the LIM solution methodology, which at its most basic level is principally a numerical method for simulating the flow and mixing processes.

5.5 LIM MODEL DEMONSTRATIONS

Figures 5.7 through 5.16 show sample results obtained when the LIM model described in the previous section is applied to a series of demonstration problems of successively increasing complexity.

The simplest of these problems is in Figure 5.7, which demonstrates direct calculations of reaction chemistry without a conserved scalar mapping approach. Details are given in Chang et al.[8] A two-dimensional vortex sheet between parallel fuel and oxidizer streams moving at different speeds undergoes Kelvin-Helmholtz instability and rolls up to form concentrated vortical structures. A single-step finite-rate chemical reaction proceeds between the reactants. The temperature dependence of the Arrhenius reaction rate produces nonequilibrium chemistry effects that are coupled directly to the mixing rate produced by the underlying flow. For this simple problem, direct numerical simulations allow detailed comparisons with the LIM results. These are shown for two cases, one corresponding to low flow speed (or fast chemistry) conditions giving a global Damköhler number $Da = 300$, and the other to high-speed (or slow chemistry) conditions giving $Da = 30$. In both cases, the agreement between the LIM model results and the direct simulations in Figures 5.7a-d is good. Under the low-speed conditions ($Da = 30$), the reaction proceeds in the braids between the vortices but has essentially stopped in the vortex cores due to consumption of the fuel. At $Da = 300$, reaction in the braids between the vortices is strained out but continues in the cores where the strain rates are lower. Quantitative comparisons between the LIM model results and the direct simulations are shown in Figure 5.8.

Figure 5.9 shows results for a simple methane-air opposed flow diffusion flame simulation with the LIM model, where a flamelet mapping approach has been used for the chemistry. Details are given in Suresh.[7] Figure 5.10 shows a more complex flow without chemistry, where LIM simulation results for the flow in the interior of a wind instrument are compared with flow visualization results. Note in particular that the development of the large-scale vortical structures and their evolution in the LIM simulations very accurately match those seen in the experiments.

Figure 5.11 shows results from LIM simulations of the effects of confinement on two-dimensional and axisymmetric bluff-body recirculation flows. The simulations indicate a larger effect of confinement for planar bluff-body flows and a smaller effect for axisymmetric bluff bodies. This can be seen by comparing the relatively small difference in recirculation zone size and structure for the axisymmetric flow in Figures 5.11b,d with the much larger differences apparent in the corresponding planar results in Figures 5.11a,c. These differences agree with experiments on confinement effects in bluff-body flows.

Figures 5.12 and 5.13 show LIM simulation results for the flow, mixing, and chemical reaction in a planar turbulent jet. A typical instantaneous view of the time-evolving computational surface on which the vorticity moments are computed, together with the resulting instantaneous velocity vector field, are shown in Figure 5.12. The velocity decay rate and lateral growth rate of the jet agree with experimental results from planar turbulent jets (see Suresh[7]). The conserved scalar field $\zeta(\mathbf{x},t)$ and scalar energy dissipation rate field $\nabla\zeta\cdot\nabla\zeta(\mathbf{x},t)$ at the same instant of time are shown in Figures 5.13a,b. From these fields, chemical species fields were obtained via a flamelet mapping, with sample results for the temperature and OH mass fraction fields shown in Figures 5.13c,d.

Analogous results are shown in Figure 5.14 for the flow and mixing processes in a somewhat more complex burner. Details are given in Suresh.[7] The resulting integral moments on the computational surface yield the $\zeta(\mathbf{x},t)$ and $\nabla\zeta\cdot\nabla\zeta(\mathbf{x},t)$ field shown. For various levels of swirl, time-averaged results obtained from LIM simulations are given in Figure 5.15. It can be seen that the small bluff-body recirculation zone present under zero-swirl conditions gradually forms into the large and rather different recirculation zone that is typical of swirl-stabilized burners. Comparisons of LIM simulation results for the average streamwise velocity field obtained[9] in experiments on this burner at four different scales ranging from 300 kW to 12 MW are shown in Figure 5.16 (see also Suresh[7]). As in the other demonstration cases, the LIM simulation methodology appears to capture the range of widely differing phenomena that are seen to dominate the various problems considered.

5.6   LIM SIMULATIONS OF GAS REBURN PROCESSES

LIM was originally developed as a research tool, and has subsequently been applied to model gas reburn systems in the upper furnace of large, coal-fired utility boilers. Because these typically involve rows of gas injectors spanning across the entire furnace, such systems can be reasonably approximated by the two-dimensional formulation of the LIM code. The upper furnace geometries also typically involve division walls that span across much of the furnace, thereby further accentuating the two-dimensional nature of the flow and mixing processes.

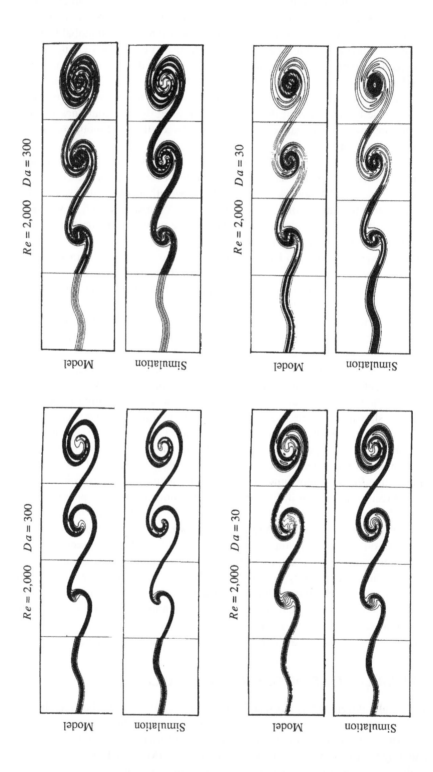

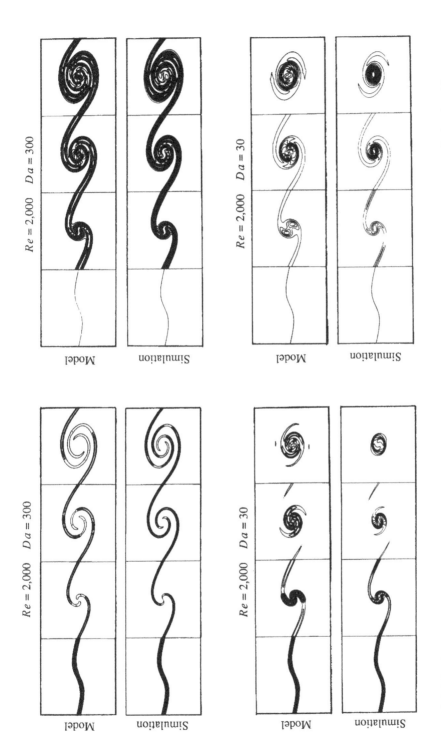

FIGURE 5.7 Comparison of the LIM model results with direct simulations for mixing and chemical reaction between fuel and oxidizer streams in a two-dimensional shear layer under conditions of high-speed flow (slow chemistry, $Da = 30$) and low-speed flow (fast chemistry, $Da = 300$). Note that all results from the LIM model are in good agreement with direct simulations under all flow conditions. (From Chang, C.H.H., Dahm, W.J.A., and Tryggvason, G., *Physics of Fluids*, A3, 1300-1311, 1991. With permission.)

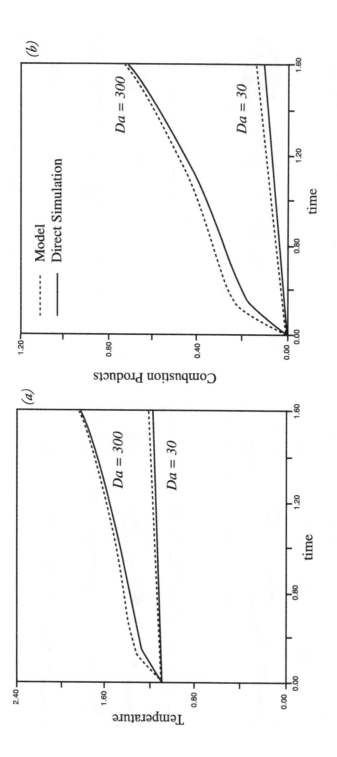

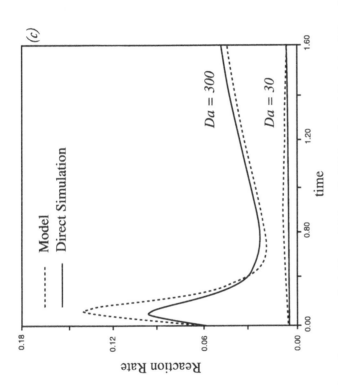

FIGURE 5.8 Quantitative comparisons between results in Figure 5.7 from the LIM model and direct simulations for shear layer mixing and reaction under both slow (Da = 30) and fast (Da = 300) chemistry. Shown are time variations in (a) the temperature $T(\mathbf{x},t)$, (b) the total chemical product concentration $Y_p(\mathbf{x},t)$, and (c) the reaction rate $w_p(\mathbf{x},t)$. Note that the LIM model results correctly show all essential features of the mixing and chemical reaction processes, including the approach to steady burning in the fast chemistry case, and the local extinction of reactions due to strain-out in the braids between the vortices that form in the shear layer. (From Chang, C.H.H., Dahm, W.J.A., and Tryggvason, G., *Physics of Fluids*, A3, 1300–1311, 1991. With permission.)

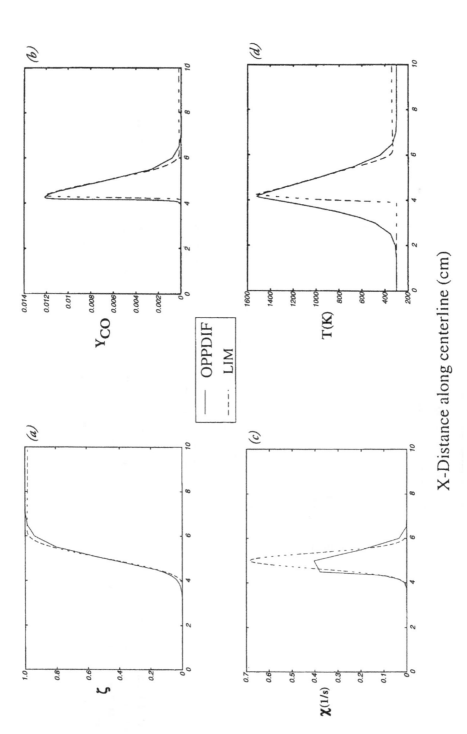

FIGURE 5.9 Comparison of the (a) ζ, (b) Y_{CO}, (c) χ, and (d) temperature profiles across a single layer calculated by the OPPDIF and LIM codes. (From Suresh, N.C., A Local Integral Moment Method to Simulate Flow, Mixing and Chemistry in Complex Flows, Ph.D. thesis, The University of Michigan, Ann Arbor, 1997. With permission.)

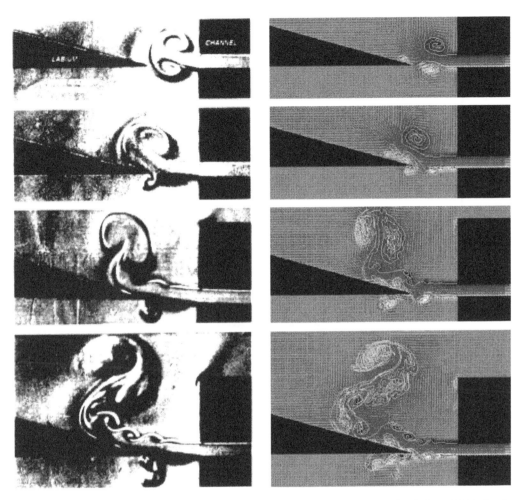

FIGURE 5.10 Comparison of experimental flow visualizations (left) with LIM simulations (right) for the flow in a wind instrument. Note good agreement of the time-dependent development of vortical structures in the flowfield and their interaction with the solid surface. (Experimental visualizations by A.P.J. Wijnands, W. Mahu and A. Hirschberg (Fluid Dynamics Laboratory, Faculty of Applied Physics, Eindhoven University of Technology); see Verge, M.P., *J. Acoustical Soc. Am.,* 95, 1119-1132, 1994. With permission.)

5.6.1 SIMULATIONS OF FLGR™ IN A ROOF-FIRED BOILER

5.6.1.1 Fuel-Lean Gas Reburning

A technology called "fuel-lean gas reburning" (FLGR™) has recently been proposed for achieving moderate NOx reductions at much lower gas input levels than are typical of conventional reburning, and without the need for an overfire air system to achieve CO burnout. Natural gas is injected into the reburn zone at sufficiently low levels to maintain overall fuel-lean conditions in the furnace. The NOx reburning reactions proceed within locally fuel-rich regions formed by the injection and mixing process. CO burnout is achieved by the excess O_2 available in the overall fuel-lean furnace gas. This fuel-lean approach to gas reburning offers the potential to meet the NOx emissions targets applicable to a large number of utility boilers at much lower initial capital costs and lower operating costs than are required for conventional gas reburning. LIM modeling has been used to help adapt the FLGR process to various utility boiler furnace designs.

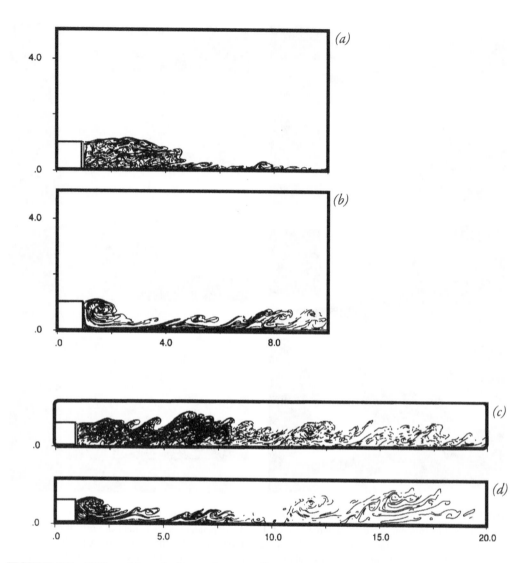

FIGURE 5.11 LIM model results for confinement effects in two-dimensional and axisymmetric bluff body flows, showing the more pronounced effects of confinement in two-dimensional flows, and the differences in recirculation zone structures. Shown are instantaneous computational surfaces for: (a) two-dimensional bluff body with low confinement, (b) axisymmetric bluff body with low confinement, (c) two-dimensional bluff body with high confinement, and (d) axisymmetric bluff body with high confinement. Note that the primary recirculation zone length changes significantly with confinement in the two-dimensional case, but not in the axisymmetric case, in agreement with experimental results. Note also the fundamentally different vortical structure of the recirculation zones in the two-dimensional cases and the axisymmetric cases, which is also in agreement with experimental results. (From Suresh, N.C., A Local Integral Moment Method to Simulate Flow, Mixing and Chemistry in Complex Flows, Ph.D. thesis, The University of Michigan, Ann Arbor, 1997. With permission.)

5.6.1.2 Elrama Unit 2

Elrama Unit 2, a 112-MWe, roof-fired, coal-burning utility boiler shown in Figure 5.17, was the first field demonstration of the FLGR technology on a full-scale utility boiler. Roof-fired boilers such as this are a relatively unusual design. Pulverized coal is introduced through eight intertube burner blocks, each comprised of a three-by-four matrix of coal nozzles that inject coal and primary air downward from the roof into the furnace. Secondary air is introduced around the burner nozzles

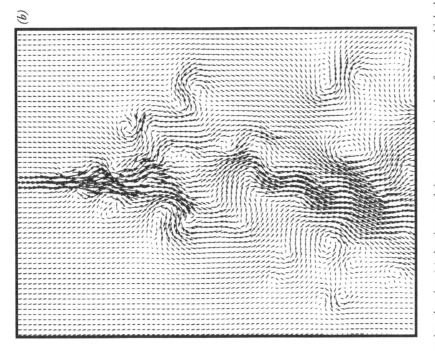

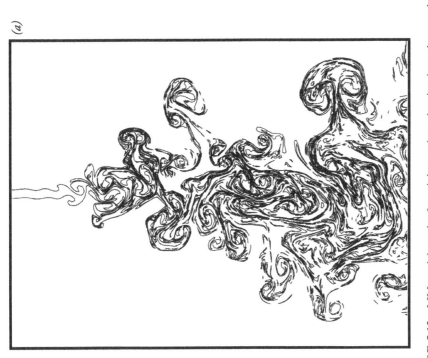

FIGURE 5.12 LIM model results for mixing and combustion in a planar turbulent jet, showing (a) the time-evolving computational surface on which the local integral moment equations are solved, and (b) the velocity vector field as obtained from the time-evolving vorticity field. The moments at every point on this surface are computed from the deformation of the surface by the vorticity field. These moments produce the conserved scalar field $\zeta(\mathbf{x},t)$ and the associated scalar energy dissipation rate field $\nabla\zeta \cdot \nabla\zeta(\mathbf{x},t)$ shown in Figure 5.13a,b, from which chemical species and reaction rate fields such as those in Figure 5.13c,d are obtained using the strain dissipation and reaction layer mapping approach. (From Suresh, N.C., A Local Integral Moment Method to Simulate Flow, Mixing and Chemistry in Complex Flows, Ph.D. thesis, The University of Michigan, Ann Arbor, 1997. With permission.)

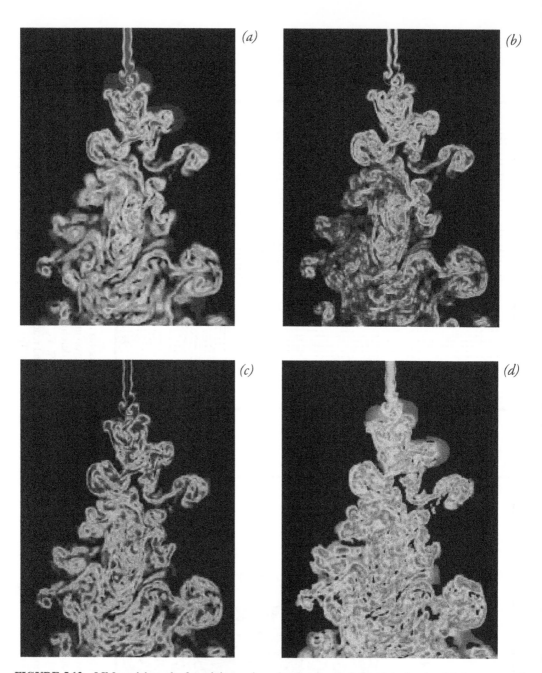

FIGURE 5.13 LIM model results for mixing and combustion in a turbulent jet, showing (a) the conserved scalar field $\zeta(\mathbf{x},t)$ and (b) the associated scalar energy dissipation rate field $\nabla\zeta \cdot \nabla\zeta(\mathbf{x},t)$. These fields are obtained from the moment values at every point on the computational surface, shown at the same instant of time in Figure 5.12. Also shown are the (c) OH species mass fraction and (d) reaction rate fields obtained via the strained dissipation and reaction layer mapping approach, which maps the local extent and rate of molecular mixing to the corresponding state of nonequilibrium chemistry. This physically based mapping approach allows arbitrarily large chemical kinetics sets to be readily incorporated in the model even for mixing and combustion in relatively complex geometries. (From Suresh, N.C., A Local Integral Moment Method to Simulate Flow, Mixing and Chemistry in Complex Flows, Ph.D. thesis, The University of Michigan, Ann Arbor, 1997. With permission.)

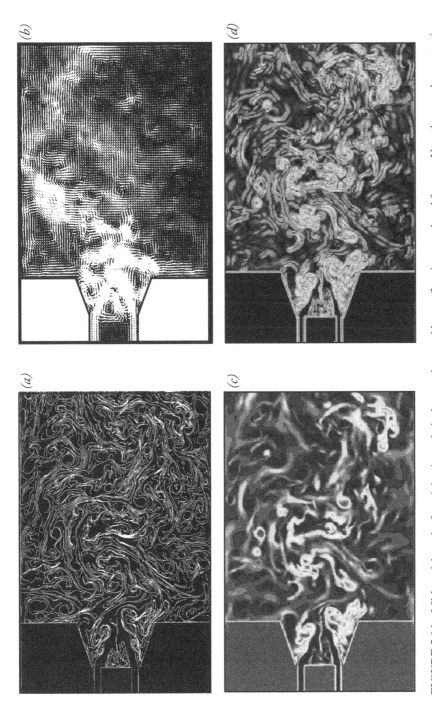

FIGURE 5.14 LIM model results for mixing in a relatively complex quarl burner flow in an enclosed furnace. Note the complex separation and resulting recirculation zones that form in the quarl and behind the fuel gun, in which much of the mixing occurs. Shown are (a) an example of the typical instantaneous shape of the time-evolving computational surface on which the LIM equations for the local scalar gradient moments are solved, (b) the corresponding velocity vector field at the same instant, (c) the mixture fraction field giving the instantaneous pattern of fuel-air mixing, and (d) the scalar energy dissipation rate field giving the instantaneous rate of molecular mixing at every point in the flow. (From Suresh, N.C., A Local Integral Moment Method to Simulate Flow, Mixing and Chemistry in Complex Flows, Ph.D. thesis, The University of Michigan, Ann Arbor, 1997. With permission.)

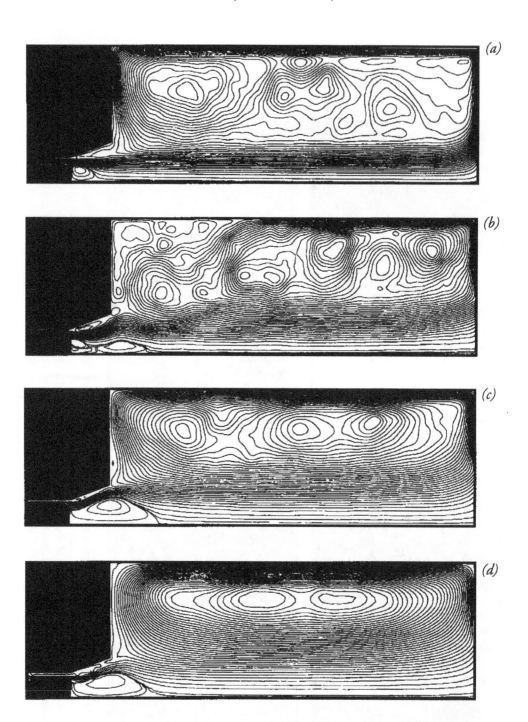

FIGURE 5.15   LIM model results for flow patterns produced by an axisymmetric swirling quarl burner in an enclosed furnace under varying levels of swirl. Shown are mean flow streamlines resulting under conditions of: (a) zero swirl, showing the relatively small bluff-body recirculation zone formed from separation occurring on the fuel gun; (b) low swirl, showing small remaining bluff-body recirculation zone and formation of a comparably sized classical swirl-induced recirculation zone; and (c) high swirl, showing disappearance of the bluff-body recirculation zone and formation of a large swirl-induced recirculation zone. The bottom panel (d) shows the long-time average streamline pattern under conditions of high swirl. (From Suresh, N.C., A Local Integral Moment Method to Simulate Flow, Mixing and Chemistry in Complex Flows, Ph.D. thesis, The University of Michigan, Ann Arbor, 1997. With permission.)

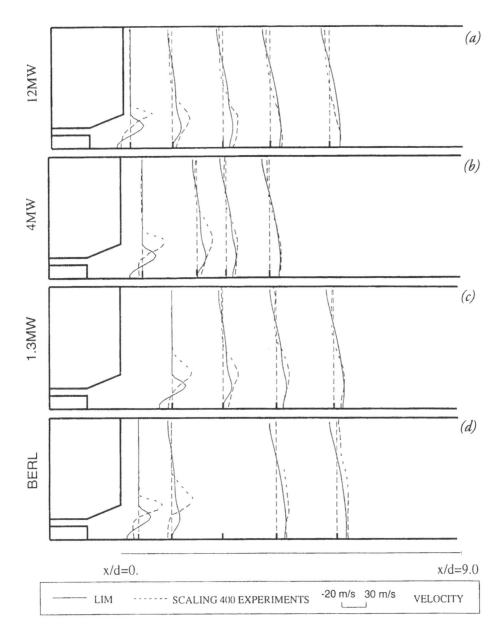

FIGURE 5.16 Velocity comparisons between the LIM model results and SCALING 400 experimental measurements for swirl stabilized quarl burners. Comparisons are shown for the (a) 12 MW, (b) 4 MW, (c) 1.3 MW, and (d) BERL burners. Note that the LIM model reproduces the bluff-body recirculation zone on the burner centerline. (From Suresh, N.C., A Local Integral Moment Method to Simulate Flow, Mixing and Chemistry in Complex Flows, Ph.D. thesis, The University of Michigan, Ann Arbor, 1997. With permission.)

through a duct arrangement shown in Figure 5.17c. The Unit 2 furnace had been modified for lower emissions by the addition of a parallel separated overfire air (SOFA) system that provides air staging capability by diverting secondary air from the coal nozzles. This SOFA system, also shown in Figure 5.17c, consists of two independent SOFA ducts that allow biasing between the front and rear SOFA injection nozzles to control the staging. The separated overfire air from the SOFA

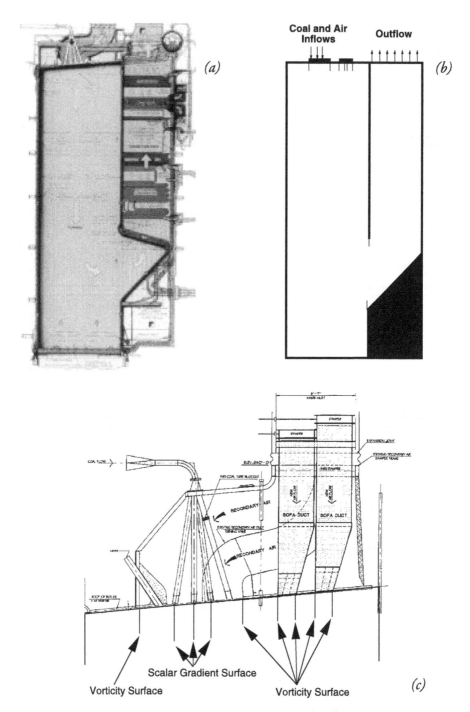

FIGURE 5.17   Diagrams indicating (a) the layout of the Elrama Unit 2 roof-fired utility boiler and (b,c) the initial conditions used in the LIM modeling of the boiler. (a) shows the primary flow in the furnace which is downward in the main furnace and upward through the convective pass, while (b,c) show the initial conditions used in the LIM simulations with vorticity surfaces introduced at all inflows and at any internal location where separation may occur. Scalar gradient surfaces were introduced at all coal injection locations. (From Frederiksen, R.D., Dahm, W.J.A., Pratapas, J.M., Serauskas, R.V., and Bartok, W., *Proc. 1997 Int. Symp. American Flame Res. Comm.*, Chicago, IL, 1997. With permission.)

nozzles is injected downward and nearly parallel to the coal and secondary air inflows to provide staging control over the boiler load range.

Prior to these LIM simulations, an injection system for fuel-lean gas reburning had been installed on the boiler. The system injected gas at the 5th-floor elevation through 12 wall injectors — 6 mounted in the furnace front wall and 3 on each sidewall. Interchangeable gas nozzles allowed injection perpendicular to the wall as well as 45° downward injection. Performance testing of this system over a wide range of conditions showed that NOx reductions improved when the furnace was operated at increasingly higher SOFA settings. Injecting more gas often did not yield higher levels of NOx reduction. Injecting gas downward through the 45° injectors typically yielded a 10 to 15% incremental improvement in the level of NOx reduction achieved. The use of side-wall injectors produced less effective reburn performance than did front-wall injectors alone. These aspects of the performance were unexpected and represented limitations on the level of NOx reduction achievable. Accordingly, a study based on LIM simulations of the flow and mixing was conducted to identify the origins of these performance limitations, and indicate modifications to the gas reburn system to allow higher levels of NOx reduction.

5.6.1.3. LIM Simulations

The burner blocks and SOFA systems, as well as the wall injectors in the FLGR system, could be reasonably modeled as a two-dimensional planar geometry. The resulting LIM model geometry for the furnace is shown in Figures 5.17b,c. All surfaces with the exception of inflows and outflows are represented by zero flow-through boundary conditions. Viscous effects at the walls are incorporated by vorticity surfaces shedding from all corners in the furnace interior. The coal and secondary air inflows through the burner are modeled as having the rear-most row of coal nozzles blocked on all burner blocks. The inflow rate through this section matches the actual coal and primary/secondary air inflows to the boiler. Because primary interest is in the mixing properties within the furnace interior and with the injected gas from the reburn system, detailed coal volatilization processes are not modeled; the coal inflows represent coal gas. The convective pass is modeled as a simple duct with a uniform mass outflow at the exit that matches the inflows to the boiler.

Unlike time-averaged models, which only require boundary conditions, LIM simulations are time dependent and thus begin from a set of initial conditions. The initial conditions correspond to the onset of fuel and air inflows into the initially quiescent furnace. These are specified via vorticity surfaces that shed from all inflow edges, as well as from all corners in the furnace interior, and scalar gradient surfaces that shed from the coal nozzles (see Figures 5.17b,c). These vorticity and scalar gradient surfaces, on which the LIM equations are solved, are completely free to move with the time-varying flow that results in the furnace interior, and thus are rapidly deformed by the underlying flow field, as is evident in Figures 5.18a,b.

Figure 5.19 shows typical results for instantaneous coal gas mixture fraction fields. Colors give the coal gas mixture fractions, with black denoting zero and red through yellow denoting increasingly higher mixture fractions. Note the flows issuing from the coal pipes and the SOFA ducts, and the resulting pattern of coal gas mixture fractions within the furnace. Figure 5.20 shows the mixing rate fields $\nabla\zeta\cdot\nabla\zeta(\mathbf{x},t)$ at the same three times. Of relevance to the reburning performance, the SOFA system is seen to create a region of high mixture fraction values near the front wall, extending downward below the 5th-floor location of the reburn gas injectors. A region of high O_2 concentrations forms behind this region due to the SOFA system and extends to the back wall. At the 5th-floor elevation, the oxygen-rich zone begins about 8 feet from the front wall, in good agreement with furnace probing measurements.

When gas is injected through the front wall, the resulting jets can penetrate into this oxygen-rich region, as seen in Figures 5.21 and 5.22. Under such conditions, the reburn gas makes little contribution to NOx removal. Better NOx reduction would result if the gas jets penetrate only up to the oxygen mixing layer, where oxygen concentrations are low and NOx concentrations are high,

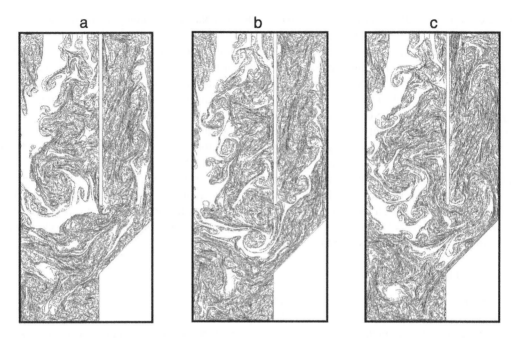

FIGURE 5.18a   The LIM computational surrface on which vorticity field moments are tracked, shown at three typical isntants of time. At each time step, the vorticity field $\omega(\mathbf{x},t)$ is constructed from the local integral moments at each point on this surface. The resulting velocity field $\mathbf{u}(\mathbf{x},t)$ then advances this surface to the next time step, and also advances the conserved scalar surface shown below to the next time step. In this manner, the flow and mixing fields evolve from the simple initial conditions. (From Frederiksen et al. 1997. With permission.)

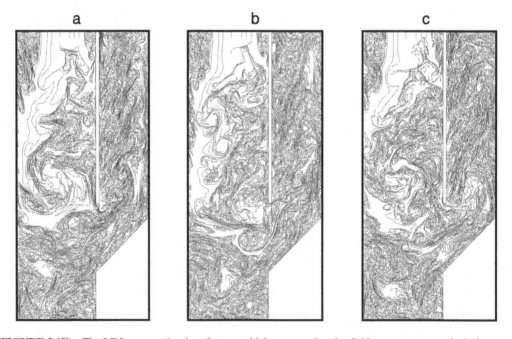

FIGURE 5.18b   The LIM computational surface on which conserved scalar field moments are tracked, shown at the same three instants of time as the vorticity surfaces above. At each time step, the local deformation of this surface by the instantaneous velocity field determines the evolution of the local integral moments. The instantaneous scalar field $\xi(\mathbf{x},t)$ and molecular mixing rate field $\nabla\xi\cdot\nabla\xi(\mathbf{x},t)$ are then determined from the instantaneous local integral moment values. (From Frederiksen et al. 1997. With permission.)

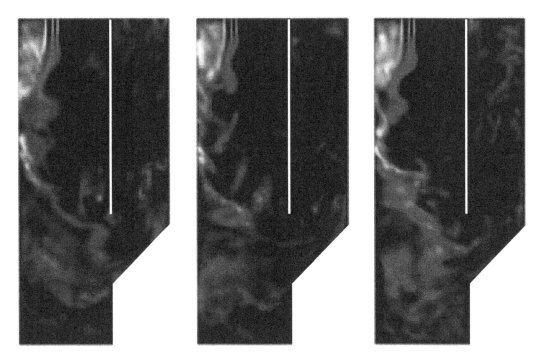

FIGURE 5.19   Typical space- and time-varying coal gas mixture fraction fields $\zeta(\mathbf{x},t)$ obtained from the LIM simulations of Elrama Unit 2, shown at three typical instants of time. Black denotes zero mixture fraction, and colors from red through yellow denote increasing mixture fraction values. (From Frederiksen, R.D., Dahm, W.J.A., Pratapas, J.M., Serauskas, R.V., and Bartok, W., *Proc. 1997 Int. Symp. American Flame Res. Comm.,* Chicago, IL, 1997. With permission.)

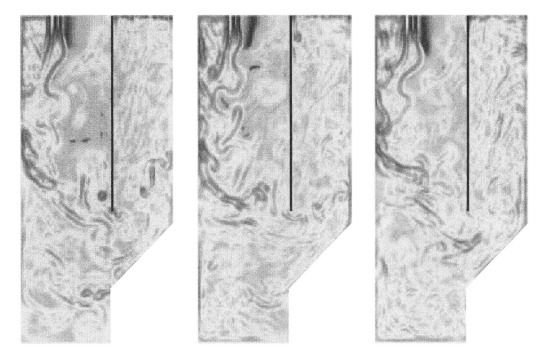

FIGURE 5.20   The molecular mixing rate field $\nabla\zeta\cdot\nabla\zeta(\mathbf{x},t)$ shown at the same three instants of time as in Figure 5.19, with colors from blue through red denoting logarithmically increasing mixing rates. The $\zeta(\mathbf{x},t)$ and $\nabla\zeta\cdot\nabla\zeta(\mathbf{x},t)$ fields permit detailed nonequilibrium chemistry to be incorporated into the LIM simulation. (From Frederiksen, R.D., Dahm, W.J.A., Pratapas, J.M., Serauskas, R.V., and Bartok, W., *Proc. 1997 Int. Symp. American Flame Res. Comm.,* Chicago, IL, 1997. With permission.)

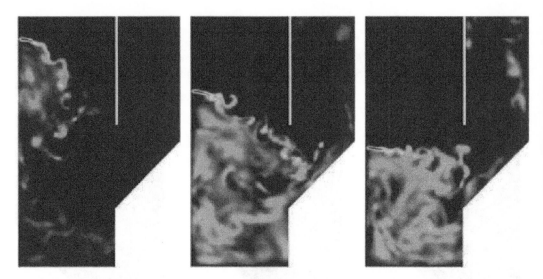

FIGURE 5.21   The injected reburn gas mixture fraction field $\zeta(\mathbf{x},t)$ obtained from the LIM simulations of Elrama Unit 2 at a typical instant of time, shown for gas injection at the (left) 6th floor, (middle) 5th floor, and (right) 4th floor elevations. Colors from blue through red denote increasing mixture fraction values. (From Frederiksen, R.D., Dahm, W.J.A., Pratapas, J.M., Serauskas, R.V., and Bartok, W., *Proc. 1997 Int. Symp. American Flame Res. Comm.,* Chicago, IL, 1997. With permission.)

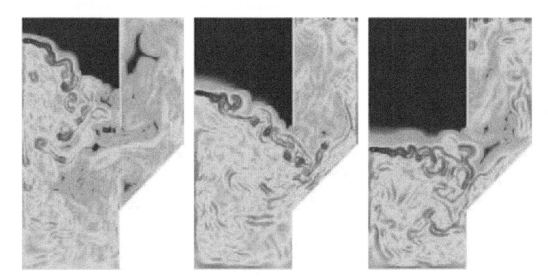

FIGURE 5.22   The molecular mixing rate field $\nabla\zeta\cdot\nabla\zeta(\mathbf{x},t)$ corresponding to the mixture fraction fields shown above, with colors from blue through red denoting logarithmically increasing mixing rates. These $\zeta(\mathbf{x},t)$ and $\nabla\zeta\cdot\nabla\zeta(\mathbf{x},t)$ fields allow detailed reburn chemistry to be incorporated into the LIM simulations. (From Frederiksen, R.D., Dahm, W.J.A., Pratapas, J.M., Serauskas, R.V., and Bartok, W., *Proc. 1997 Int. Symp. American Flame Res. Comm.,* Chicago, IL, 1997. With permission.)

and thus significant reburning can occur before the CH radicals are depleted. The simulation results show that this is why the downward-pointing nozzles gave better reburn performance than did the straight injectors, and why injecting more gas (resulting in higher penetrations) did not lead to higher NOx removal effectiveness. This is also why gas injection through the side-wall injectors produced less effective reburn performance than did the front-wall injectors alone, since the two rear-most pairs of side-wall injectors place gas directly into the high oxygen-containing back portion

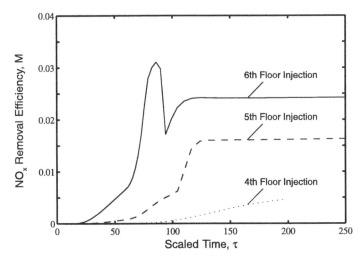

FIGURE 5.23 Lagrangian histories for gas injection at the 4th, 5th, and 6th floor elevations shown as NOx removal efficiency in the proper mixing-scaled time variable. In these variables, the effect of various injection parameters on the NOx reburn chemistry becomes more readily apparent. (From Frederiksen, R.D., Dahm, W.J.A., Pratapas, J.M., Serauskas, R.V., and Bartok, W., *Proc. 1997 Int. Symp. American Flame Res. Comm.,* Chicago, IL, 1997. With permission.)

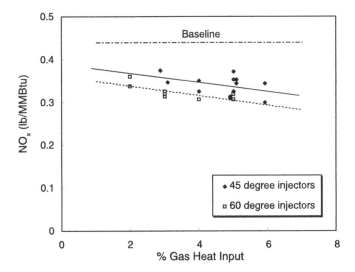

FIGURE 5.24 Comparison of NOx emissions at Elrama Unit 2 obtained using the original 45° gas injectors to those obtained with the recommended 60° gas injectors. Implementing this recommendation led to an extra 5% decrease in the NOx emissions and a 20% relative improvement in the NOx reduction. (Data from Energy Systems Associates. With permission.)

of the furnace. The effect of gas injection at different furnace elevations but at otherwise comparable conditions is also shown in Figures 5.21 and 5.22. Examples of the accompanying reburn chemistry are shown in Figures 5.23 and 5.24.

Based on these insights from the LIM simulations, it was recommended that gas should be injected through the six front-wall injectors, as well as through the two foremost side-wall injectors, and should be introduced downward with 60° injector inclinations with holes enlarged to 5/8-in. to avoid over-penetration past the oxygen mixing layer. Figure 5.24 shows the improvement in NOx reduction that was obtained when these recommendations were implemented.

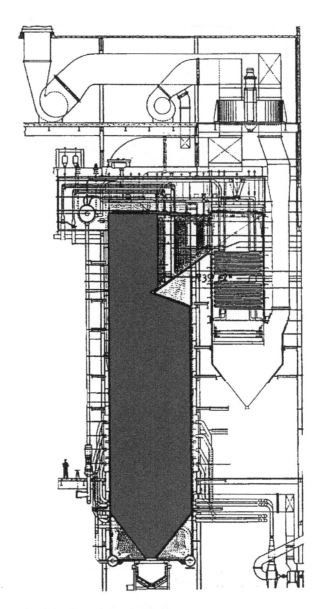

FIGURE 5.25   Schematic of the Riverbend Unit 7 boiler, a tangentially fired coal-burning unit for which a fuel-lean gas reburn (FLGR™) system was developed. (From Frederiksen, R.D., Dahm, W.J.A., Tryggvason, G., Breen, B.P., Glickert, R., Urich, J.A., Pratapas, J.M., and Serauskas, R.V., *Proc. 1998 Joint American/Japanese Flame Res. Comm. Int. Symp.,* Maui, Hawaii, 1998. With permission.)

5.6.2   Simulations of FLGR™ in a Tangentially Fired Boiler

Riverbend Unit 7, a 140-MWe, coal-burning utility boiler shown in Figure 5.25, was the first application of FLGR to a tangentially fired furnace. The utility estimated that adaptation of the FLGR technology to meet the emissions requirements for the 29 such tangentially fired boilers in its generating system could reduce its system-wide cost of compliance by $70 million relative to other compliance options. Duke Power therefore initiated a study consisting of physical testing, conventional modeling, and LIM modeling to develop an FLGR system suitable for this boiler.

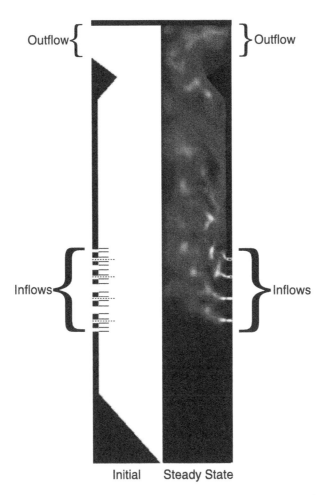

FIGURE 5.26 Initial and steady-state conditions for the LIM simulations of the Riverbend Unit 7 boiler, showing initial vortex sheets (solid lines) and scalar gradient sheets (dotted lines). (From Frederiksen, R.D., Dahm, W.J.A., Tryggvason, G., Breen, B.P., Glickert, R., Urich, J.A., Pratapas, J.M., and Serauskas, R.V., *Proc. 1998 Joint American/Japanese Flame Res. Comm. Int. Symp.,* Maui, Hawaii, 1998. With permission.)

5.6.2.1   Riverbend Unit 7 Model

The axisymmetric model shown in Figure 5.26 was developed to simulate the actual furnace geometry. The lower part of the furnace model matches the actual cross-sectional area, while in the upper part, the nose was represented in a manner that produces the same equivalent cross-sectional area. The entry to the convective section was represented by an axisymmetric duct that matches the cross-sectional area of the actual furnace.

The boiler operates with a Low NOx Concentric Firing System with Close Coupled Overfire Air, shown in Figure 5.27. This consists of four coal nozzles separated by combination air registers, with an additional CCOFA register at the top. Four identical burners of this type — one located in each corner of the 40-ft × 26-ft rectangular furnace cross-section — comprise the only inflows to the boiler. The firing arrangements is based on a 6.5° main tangential firing circle, with the CFS air registers in Figure 5.27 introducing air tangentially at 22° off the main firing circle. The burners were modeled as in Figure 5.28, with the coal nozzles and air registers represented by alternating coal and air inflows of various sizes separated by gaps chosen to best simulate the actual burners.

All boundaries except of the inflows at the coal nozzles and air registers and the outflow at the convective section are represented by zero flow-through boundary conditions. Viscous effects at

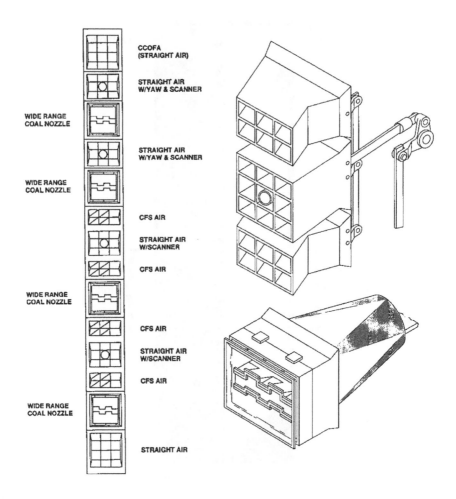

FIGURE 5.27   Coal nozzles and air registers in the ABB/CE low-NOx concentric firing system (LNCFS-1) with close-coupled overfire air (CCOFA) used at Riverbend Unit 7. CFS and straight air nozzles are shown at top right, and a coal nozzle is shown at bottom right. (From Frederiksen, R.D., Dahm, W.J.A., Tryggvason, G., Breen, B.P., Glickert, R., Urich, J.A., Pratapas, J.M., and Serauskas, R.V., *Proc. 1998 Joint American/Japanese Flame Res. Comm. Int. Symp.*, Maui, Hawaii, 1998. With permission.)

the walls are incorporated by vorticity surfaces shedding from all corners in the furnace interior. The coal and air inflows in the LIM simulations match the actual coal and air inflow rates to the boiler. As in Section 5.6.1.3, the coal inflows represent fully volatilized coal gas. A uniform mass outflow matching the inflow rate is specified at the entrance to the convective section. Initial conditions are specified via vorticity surfaces that shed from all inflow edges, as shown by solid lines at the edges of the coal and air inflows in Figure 5.26, and from all corners in the furnace interior. The moments on these surfaces carry information about both the azimuthal vorticity as set by the radial inflows to the boiler and the axial vorticity set by the tangential swirl imparted on the inflows by details of the burner design. For the scalar field, initial conditions are scalar gradient surfaces shed from the coal nozzles, shown by the dotted lines in Figure 5.26.

5.6.2.2   LIM Simulations

Typical instantaneous computational surfaces, coal gas mixture fraction fields, and mixing rate fields from these LIM simulations are shown in Figure 5.29. As in any axisymmetric simulation,

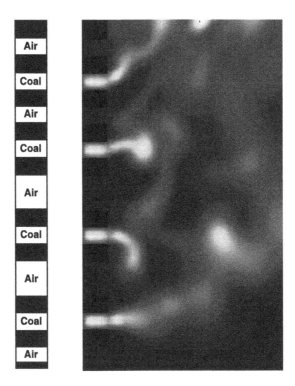

FIGURE 5.28 Representation of the coal and air inflows in the LIM simulations of Riverbend Unit 7, showing a close-up view of the resulting instantaneous coal gas mixture fraction field $\zeta(\mathbf{x},t)$ in the vicinity of the burners. (From Frederiksen, R.D., Dahm, W.J.A., Tryggvason, G., Breen, B.P., Glickert, R., Urich, J.A., Pratapas, J.M., and Serauskas, R.V., *Proc. 1998 Joint American/Japanese Flame Res. Comm. Int. Symp.*, Maui, Hawaii, 1998. With permission.)

there is no flux across the centerline of the furnace. When the mixture fraction fields are averaged over several boiler residence times, they show relatively high O_2 concentrations along the boiler walls, in good agreement with probing measurements conducted on the furnace. This indicates that the gas injection system should be designed to avoid placing gas near the boiler walls. Simulations were conducted to evaluate the suitability of the 704-ft, 714-ft, and 722-ft boiler elevations for gas injection, with results for the 704-ft elevation shown in Figure 5.30. These indicate that the 714-ft elevation represents the best compromise between high NOx removal and low CO emissions.

Further simulations of the type in Figure 5.31 were conducted to determine the optimal layout for the gas injectors at this elevation. The figure shows the gas penetration and mixing in a two-dimensional cross-section through the boiler at a fixed elevation. Of key interest is the interaction of the injected gas with the tangential swirl flow in the boiler, and the effect that this has on gas placement and mixing with regard to reburn.

Collectively, LIM simulation results such as these contributed significantly to the development of an effective gas injection system for fuel-lean reburning in tangentially fired boilers of this type.

5.6.3 SIMULATIONS OF AEFLGR™ IN A WALL-FIRED BOILER

Amine-Enhanced Fuel-Lean Gas Reburn (AEFLGR) is an extension of the FLGR NOx reduction technology and is based on synergistic injection of natural gas with an aqueous urea spray to achieve relatively high levels of NOx reduction. The first full-scale demonstration of AEFLGR was conducted in PSE&G's Mercer Station Furnace 22, the reheat furnace of a 324-MWe, wall-fired, coal-burning utility boiler. The furnace is fired by 12 front-wall-mounted burners, arranged in 3 levels

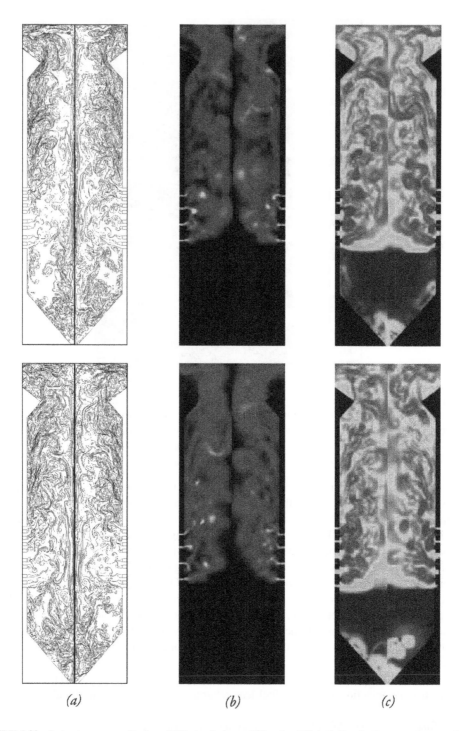

(a)   (b)   (c)

FIGURE 5.29 Instantaneous results from LIM simulations of Riverbend Unit 7. Results from two instants in time are shown (top, bottom) for (a) the LIM computational vorticity surface, (b) the coal gas mixture fraction field, and (c) the instantaneous molecular mixing rate field. In (b), colors ranging form dark red to bright yellow denote linearly increasing values of the local, instantaneous coal gas mixture fraction, while in (c) colors ranging from dark blue to red indicate logarithmically increasing mixing rates. Note the highly time-dependent nature of the results. (From Frederiksen, R.D., Dahm, W.J.A., Tryggvason, G., Breen, B.P., Glickert, R., Urich, J.A., Pratapas, J.M., and Serauskas, R.V., *Proc. 1998 Joint American/Japanese Flame Res. Comm. Int. Symp.*, Maui, Hawaii, 1998. With permission.)

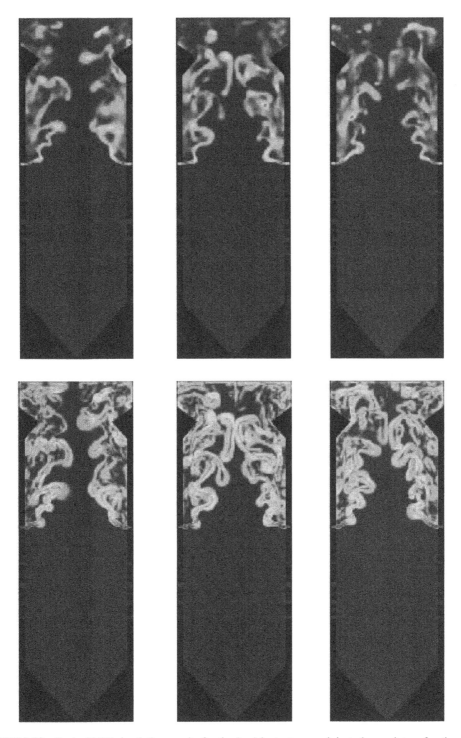

FIGURE 5.30 Typical LIM simulation results for the (top) instantaneous injected gas mixture fraction fields $\zeta(\mathbf{x},t)$ and (bottom) mixing rate fields $\nabla\zeta\cdot\nabla\zeta(\mathbf{x},t)$ for gas injection at Riverbend Unit 7 at the 704-ft elevation. Color scales are the same as in Figure 5.29. (From Frederiksen, R.D., Dahm, W.J.A., Tryggvason, G., Breen, B.P., Glickert, R., Urich, J.A., Pratapas, J.M., and Serauskas, R.V., *Proc. 1998 Joint American/Japanese Flame Res. Comm. Int. Symp.,* Maui, Hawaii, 1998. With permission.)

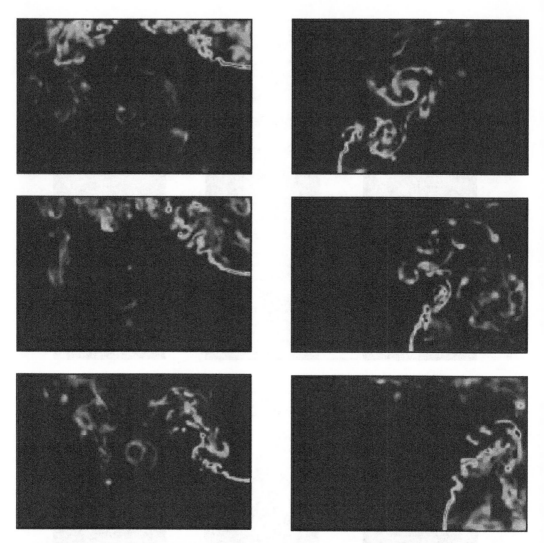

FIGURE 5.31 Typical LIM simulation results, as seen from the top of Riverbend Unit 7 looking down, showing the injected gas mixture fraction fields $\zeta(\mathbf{x},t)$ for various gas injection locations at the 714-ft elevation. The color scale is the same as for $\zeta(\mathbf{x},t)$ in Figures 5.29 and 5.30. (From Frederiksen, R.D., Dahm, W.J.A., Tryggvason, G., Breen, B.P., Glickert, R., Urich, J.A., Pratapas, J.M., and Serauskas, R.V., *Proc. 1998 Joint American/Japanese Flame Res. Comm. Int. Symp.*, Maui, Hawaii, 1998. With permission.)

of 4 burners each. Three division walls centered between each burner column divide the main furnace and accentuate the two-dimensionality of the flow and mixing processes.

Two submodels were developed for LIM to simulate the AEFLGR process. The first involved a change in the chemical kinetics used to simulate the combined natural gas and urea chemistry. The second involved adding a droplet submodel to LIM to track the evaporating urea droplets in the flow. Individual droplets were modeled as spheres of aqueous urea solution, with the droplet mass varied based on a convective correction to the steady-state droplet vaporization rate. The droplets were tracked taking into account their momentum, drag, and evaporation. Due to the large molecular weight differences between the water and urea in the droplets, differential vaporization of the two components leads to a two-step process in which the water evaporates first, followed by the urea.

Figure 5.32 shows typical velocity fields, coal gas mixture fraction fields, and mixing rate fields in the furnace. These clearly showed a coal-rich region that extends up the rear wall of the furnace

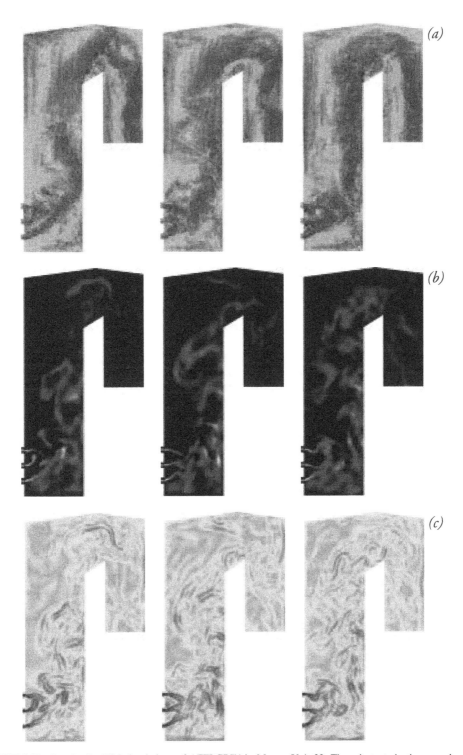

FIGURE 5.32 Results for LIM simulations of AEFLGR™ in Mercer Unit 22. Three instants in time are shown for (a) the instantaneous velocity vectors superimposed over the instantaneous vorticity field, (b) instantaneous coal gas mixture fraction field, $\zeta(\mathbf{x},t)$, and (c) the instantaneous coal gas molecular mixing rate field, $\nabla\zeta\cdot\nabla\zeta(\mathbf{x},t)$. These fields clearly show the presence of a high-temperature coal-rich core extending up the rear wall of the furnace. (From Frederiksen, R.D., Dahm, W.J.A., Tryggvason, G.T., Pratapas, J.M., and Serauskas, R.V., *Proc. 1999 EPRI/DoE/EPA Combined Utility Air Pollutant Control Symp.*, Atlanta, GA, 1999. With permission.)

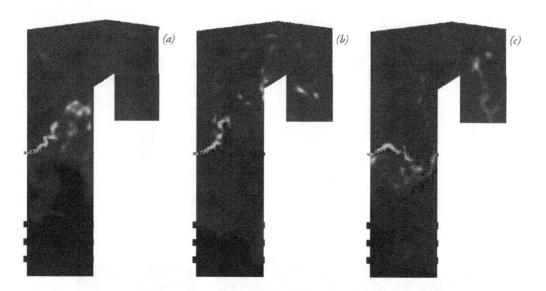

FIGURE 5.33   LIM simulation results for combined gas and urea injection at the Zone 5 elevation in Mercer Furnace 22 for three different drop diameters: (a) 200 μm, (b) 300 μm, and (c) 400 μm. These results show an increased tendency for the urea droplets to separate from the gas with increasing droplet diameter. This can lead to poor NOx reduction performance due to the loss of any synergistic effects and the occurrence of selective noncatalytic reactions at elevated temperatures. (From Frederiksen, R.D., Dahm, W.J.A., Tryggvason, G.T., Pratapas, J.M., and Serauskas, R.V., *Proc. 1999 EPRI/DoE/EPA Combined Utility Air Pollutant Control Symp.,* Atlanta, GA, 1999. With permission.)

opposite the wall burners. Injecting gas into this low-O_2 region may produce effective NOx reduction, but may also result in high CO emissions due to interactions between the gas and the coal particle burnout process. The results also showed a large but relatively weak circulation zone along the upper front wall of the furnace, and a strong upflow along the rear wall.

LIM simulations were conducted for combined gas and aqueous urea droplet injection at three injector elevations and for various injector inclinations, with initial droplet sizes ranging from 100 μm to 400 μm. Typical results for straight injection at the Zone 5 elevation are in Figure 5.33, showing the injected gas mixture fraction fields and superimposed droplets, and giving the relative drop sizes and locations for three different initial droplet sizes. The difficulty in keeping the urea droplets together with the natural gas is apparent in these results. As the initial droplet sizes increase, their momentum typically causes them to separate from the injected gas and enter the relatively high O_2 furnace gases. This is quantified in Figure 5.34, which shows the scalar value versus droplet size from tracking of a relatively large number of such typical droplets in LIM simulations. Of key importance to the process effectiveness is the scalar value at which the drop size reaches zero, because this gives the environment in which the urea is released. The 100-μm droplets follow the gas well and release urea in relatively oxygen-depleted surroundings (high scalar values). Many of the 300-μm and 400-μm droplets, however, release urea at much higher O_2 concentrations, for which the selectivity of the NOx-reducing amine reactions is decreased. Results such as these allow the optimal droplet sizes for each injector location to be determined.

Figure 5.35 shows detailed information from LIM simulations for the urea release sites from droplets of various sizes injected at various locations. The results show that large urea droplets can impact on the furnace walls, significantly decreasing AEFLGR performance. This is most apparent in the 300- and 400-μm droplets, for which a considerable fraction evaporate after accumulating on the rear wall of the furnace. The results in Figure 5.35 also demonstrate that the concept of targeting specific regions of the furnace with specific injectors only works with droplet sizes of

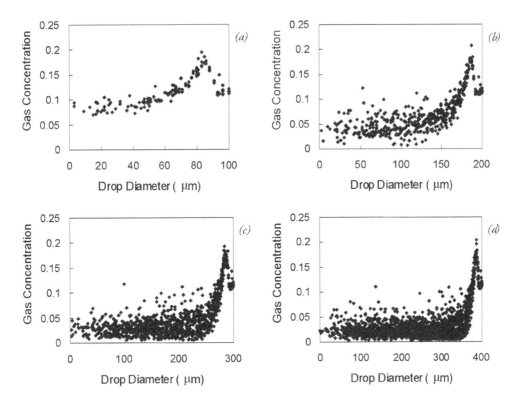

FIGURE 5.34 LIM simulation results showing the gas concentration and droplet size histories for gas and urea injection at Zone 5 of Mercer Furnace 22 for four different drop diameters: (a) 100 μm, (b) 200 μm, (c) 300 μm, and (d) 400 μm. These results show the local scalar value experienced by individual droplets as they move about the furnace and evaporate. From these results it is apparent that large droplets quickly separate from the natural gas and cannot produce any synergistic NOx reduction. (From Frederiksen, R.D., Dahm, W.J.A., Tryggvason, G.T., Pratapas, J.M., and Serauskas, R.V., *Proc. 1999 EPRI/DoE/EPA Combined Utility Air Pollutant Control Symp.*, Atlanta, GA, 1999. With permission.)

approximately 200 mm or less. Based on these simulation results, Zones 6 and 7 were determined to provide the best coverage of the hot furnace gases near the rear wall of the furnace.

These examples demonstrate how LIM simulations have been used in relatively complex industrial applications to provide crucial insight into process adaptation and optimization that would otherwise have been difficult to obtain.

5.7 THREE-DIMENSIONAL LIM MODELING (LIM3D)

Although two-dimensional simulations of planar, axisymmetric, and swirling flows can be used by the specialist to examine certain aspects of three-dimensional problems, in many cases it is necessary to conduct fully three-dimensional simulations. Such simulations involve greatly increased computational times, irrespective of the modeling approach being used. However, the efficiencies gained from the LIM methodology in two-dimensional simulations provide even greater reductions in three-dimensional simulations. For this reason, LIM is inherently suited for fully three-dimensional simulations of practical problems involving flow, mixing, and combustion processes under conditions for which the simplifying assumptions of this specialized solution methodology apply.

Fundamentally, the LIM methodology is largely unchanged in going from two dimensions to three dimensions; however, there are two practical challenges that must be met. The first is how

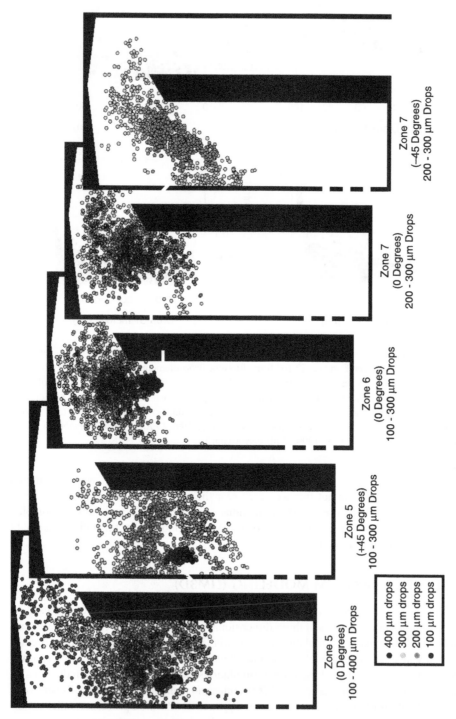

FIGURE 5.35 LIM simulation results showing the urea release sites of droplets of various sites injected at various zones in the furnace. Injection at Zone 6 and Zone 7 delivers the best coverage of the hot furnace gases near the rear wall of the furnace. Large droplets (>300 μm) injected in Zone 5 show an increased tendency to impact the furnace walls and in some cases may drop down into the main burner zone, dramatically decreasing the NOx reduction performance of the AEFLGR™ system. (From Frederiksen, R.D., Dahm, W.J.A., Tryggvason, G.T., Pratapas, J.M., and Serauskas, R.V., *Proc. 1999 EPRI/DoE/EPA Combined Utility Air Pollutant Control Symp.*, Atlanta, GA, 1999. With permission.)

to represent three-dimensional surfaces and the vorticity and scalar gradients carried with them in an efficient way that accounts for the rapid deformation of these surfaces. The second is the development of efficient algorithms to reduce the exponential growth of this computational surface by restructuring in three dimensions, analogous to those described for two-dimensional simulations in Section 5.4.4.

In three dimensions, the same simplifying approximations apply as in the two-dimensional formulation; however, the circulation $\Gamma(\mathbf{s})$ is how determined by the vector change in velocity $\mathbf{u}(\mathbf{x},t)$ across the surface on which the vorticity is concentrated. The computational surfaces on which the vorticity moments and scalar gradient moments are tracked are now composed of triangular surface elements, which connect points that are moved with the flow. The vorticity is represented by vortex lines of constant circulation that coincide with the edges of the triangular elements. The vorticity still satisfies Equation (5.6), and following the same reasoning as in the two-dimensional formulation, only its zeroth moment is tracked. The velocities are again found from the vorticity moments over the computational surface by the Biot-Savart law in Equation (5.2), in this case using a three-dimensional vortex-in-cell approach. The scalar gradient surfaces are discretized in the same manner. The scalar field still satisfies Equation (5.10), and thus the scalar gradient moments on these surfaces still satisfy Equation (5.13a-c). The local strain rate $\sigma(\mathbf{s})$ in three-dimensions now comes from the change in area of the triangular surface elements as $d \log A(\mathbf{s})/dt$, where $A(\mathbf{s})$ is the area of the element located at any point $\mathbf{s}$ on the surface.

As the simulation time progresses and the vorticity and scalar gradient surfaces deform, new elements are inserted where appropriate by splitting the longest edge of the triangular surface element and distributing its circulation among the new elements to conserve the total circulation and its vector direction. Small elements are merged in a similar manner to form larger elements. Similarly, a surface restructuring algorithm was developed that identifies elements from different parts of the surface that are sufficiently close and nearly parallel, and merges these conservatively into a single element. Extensive tests analogous to those in Figure 5.6 have shown that, when properly done, the flow is essentially unaffected by this restructuring algorithm while it greatly reduces the amount of computational surface that must be tracked.

A simple example of a flow computed by this three-dimensional extension of the LIM methodology is shown in Figure 5.36, corresponding to the early stages of a jet issuing at high Reynolds number from a circular nozzle. The jet is represented by the initially cylindrical surface on which vorticity and scalar gradients are concentrated at the jet exit. This computational surface, on which the local integral moments are being tracked, is shown at two times by rendering it behind the centerplane of the jet, thus revealing the internal structure of the resulting flow. Beyond the nozzle exit, the rapid deformation of this surface and its roll-up to form concentrated toroidal vortex structures in the jet near-field is readily apparent. The development of azimuthal instabilities along these torroidal shear layer structures and the subsequent three-dimensionality is also clear evident. All of these features are seen in physical experiments as well. Test cases such as this with the LIM3D code show that it represents an inherently efficient approach to simulations of fully three-dimensional problems involving flow, mixing, and combustion processes under conditions for which the simplifying assumptions of this specialized solution methodology apply.

5.8 CONCLUDING REMARKS

Unlike traditional general-purpose codes, the LIM simulation approach was conceived and developed specifically as a computational model for turbulent flow, mixing, and combustion processes. By remaining confined to this narrower field of applications, it is able to make use of an efficient solution methodology that is tailored to the characteristics of the specialized range of flow regimes and physical processes for which it is intended. The approach makes direct use of several key insights from basic research over the past 25 years into turbulent flow, mixing, and combustion processes. In doing so, it is able to completely circumvent the "gradient diffusion assumption" that

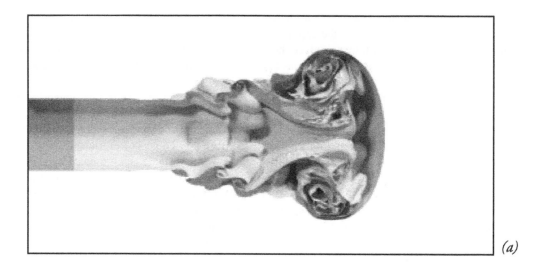

(a)

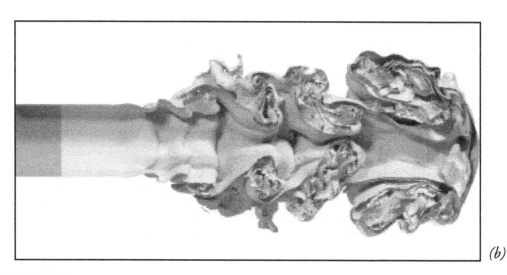

(b)

FIGURE 5.36 Instantaneous LIM3D vorticity surface for a confined, coflowing three-dimensional round jet at two different time steps: (a) At t = 2.3 with 214,000 triangular elements, and (b) at t = 3.0 with 475,000 elements. Each surface is shown with its front half removed to expose the internal structure of the jet. The dark gray segment at the left of each image is the nozzle.

underlies all other computational models for turbulent reacting flow and thereby avoid the main impediment that has limited the accuracy of such traditional models.

However, as with any fundamentally new numerical method, acceptance of the LIM model is slowed by the same specialized nature of the underlying numerical method that is also the main strength of the code. Because the numerical method on which LIM is based is completely different from methods traditionally used in conventional CFD models, there is a large and rather steep "learning curve" required to use LIM effectively. With the two-dimensional implementation in particular, considerable practical experience with the code is needed to identify the types of problems that can be accurately modeled and how these are best set up. While such experience is essential to productively use any simulation code, this is especially true for LIM because the underlying flowfields are manifestly two-dimensional, while traditional time-averaged models impose two-dimensionality only on average fields. This two-dimensionality leads to a variety of phenomena not encountered in traditional models. For this reason, the experience base needed to accurately use LIM is very different from that needed for traditional models, and experience with conventional time-averaged models does not carry over well to this new methodology. The three-dimensional code, however, inherently avoids most of these difficulties in the two-dimensional formulation, although it does so at the expense of a considerable increase in computational requirements. For these reasons, it is likely to be the three-dimensional version of the LIM code that is best-suited for use by nonspecialists in industrial practice.

Additional information about the LIM methodology and its application to modeling reacting flows can be found in References 7, 8, and 10 through 15.

ACKNOWLEDGMENTS

Development of the LIM methodology was supported by the Gas Research Institute (GRI) under GRI Contract Nos. 5088-260-1692 and 5093-260-2723, with James A. Kezerle, Robert V. Serauskas, and John M. Pratapas as contract monitors. Numerical implementation of this simulation methodology was developed at The University of Michigan with graduate students Chester H.H. Chang and Nallan C. Suresh. Discussions with Quang-Viet Nguyen, Robert W. Dibble, and William Bartok contributed significantly to our approach to gas reburn modeling.

REFERENCES

1. Brown, G.L. and Roshko, A. (1974), On density effects and large structure in turbulent mixing layers. *Journal of Fluid Mechanics,* 64, 775-816, 1974.
2. Buch, K.A. and Dahm, W.J.A. (1996), Experimental study of the fine-scale structure of conserved scalar mixing in turbulent flows. Part I. Sc » 1. *Journal of Fluid Mechanics,* 317, 21-71.
3. Buch, K.A. and Dahm, W.J.A. (1998), Experimental study of the fine-scale structure of conserved scalar mixing in turbulent flows. Part II. Sc ≈ 1. *Journal of Fluid Mechanics,* 364, 1-29.
4. Bilger, R.W. (1980), Turbulent flows with nonpremixed reactants, in *Turbulent Reacting Flows* (P.A. Libby and F.A. Williams, Eds.), 65-113, Springer Verlag, Berlin.
5. Bish, E.S. and Dahm, W.J.A. (1995), Strained dissipation and reaction layer analyses of nonequilibrium chemistry in turbulent reacting flows. *Combustion and Flame,* 100, 457-464.
6. Tacina, K.M. and Dahm, W.J.A. (1999), Effects of heat release on turbulent shear flows. 1. A general equivalence principle for nonbuoyant flows and its application to turbulent jet flames. *Journal of Fluid Mechanics,* 415, 23-34.
7. Suresh, N.C. (1997), A Local Integral Moment Method to Simulate Flow, Mixing and Chemistry in Complex Flows, *Ph.D. thesis,* The University of Michigan, Ann Arbor.
8. Chang, C.H.H., Dahm, W.J.A., and Tryggvason, G. (1991), Lagrangian model simulations of molecular mixing, including finite rate chemical reactions, in a temporally developing shear layer. *Physics of Fluids A* 3, 1300-1311.
9. Hsieh, A., Dahm, W.J.A., and Driscoll, J.F. (1996), Scaling laws for NOx emissions performance of burners and furnaces from 30 kW to 12 MW. *Combustion and Flame,* 114, 54-80.
10. Tryggvason, G., Dahm, W.J.A., and Sbeih, K. (1990), Fine structure of vortex sheet rollup by viscous and inviscid simulation. *ASME of Journal of Fluids Engineering,* 113, 31-36.
11. Tryggvason, G. and Dahm, W.J.A. (1990), An integral method for mixing, chemical reactions, and extinction in unsteady strained diffusion layers. *Combustion and Flame,* 83, 207-220.
12. Dahm, W.J.A., Tryggvason, G., and Zhuang, M. (1996), Integral method solution of time-dependent strained diffusion-reaction equations with multi-step kinetics. *SIAM Journal of Applied Mathematics,* 56, 1039-1059.
13. Frederiksen, R.D., Dahm, W.J.A., Pratapas, J.M., Serauskas, R.V., and Bartok, W. (1997), LIM simulations of gas injection and reburning in two coal-fired utility boilers, in *Proceedings of the 1997 International Symposium of the American Flame Research Committee,* Chicago, IL.

14. Frederiksen, R.D., Dahm, W.J.A., Tryggvason, G., Breen, B.P., Glickert, R., Urich, J.A., Pratapas, J.M., and Serauskas, R.V. (1998), Fuel-lean gas reburn (flgr™) technology for achieving nox emissions compliance: application to a tangentially-fired boiler, *Proceedings of the 1998 Joint American/Japanese Flame Research Committee International Symposium,* Maui, Hawaii.

15. Frederiksen, R.D., Dahm, W.J.A., Tryggvason, G.T., Pratapas, J.M., and Serauskas, R.V. (1999), Advanced simulation of amine-enhanced fuel-lean gas reburn (AEFLGR™) systems for utility boiler NOx reduction, *Proceedings of the 1999 EPRI/DoE/EPA Combined Utility Air Pollutant Control Symposium,* Atlanta, GA.

ADDITIONAL REFERENCES

Tryggvason, G. (1984), Numerical Studies of Flows with Sharp Interfaces, *Ph.D. thesis,* Brown University, Providence.

Tryggvason, G. (1989), Simulation of vortex sheet roll-up by vortex methods, *Journal of Computational Physics,* 80, 1-16.

Verge, M.P. (1994), *Journal of the Acoustical Society of America,* 95, 1119-1132.

6 Numerical Modeling of Multidimensional Laminar Flames

Mitchell D. Smooke and Beth Anne V. Bennett

CONTENTS

6.1 INTRODUCTION

In the last few years, the focus of combustion research has shifted. The push for higher combustion efficiency in propulsion applications that dominated much of research in the past few decades is gradually being replaced by a drive toward *cleaner* combustion. A direct result of environmental consciousness, this shift has been translated into stricter air quality legislation. Although it originated as a reaction to regulatory pressure, research in the area of pollutant formation and control will become economically indispensable to the export of combustion-related technologies and products worldwide. In particular, as emissions legislation becomes more restrictive, a detailed understanding of the processes by which NOx, soot, and unburned hydrocarbons are formed in flames will be critical for the design of pollutant abatement strategies and for the preservation of the competitiveness of combustion-related industries.

    In many practical combustion devices, such as gas turbines and commercial burners, the flame is often non-premixed (diffusion) or partially premixed. Such flames encompass complex interactions of heat and mass transfer with chemical reactions. The ability to predict the coupled effects of complex transport phenomena with detailed chemical kinetics in these systems is critical for modeling turbulent reacting flows, for improving engine efficiency, and for understanding the processes by which pollutants are formed. Many commercial power-generating units and home heating devices also employ diffusion flames or partially premixed flames as their primary flame type.

During combustion, most of the oxides of nitrogen are formed when part of the oxygen combines with atmospheric nitrogen (rather than with the fuel), so the widespread burning of hydrocarbon fuels can produce large quantities of nitrogen dioxide and nitric oxide, collectively referred to as NOx. Both compounds are considered toxic, and nitric oxide is related to the formation of photo-chemical smog.[1] In addition, combustion-generated soot particles from these systems pose a significant health risk and are the subject of stringent new EPA regulations. Besides regulatory issues, soot contributes to thermal radiation loads on combustor liners and turbine blades. Other undesirable effects of soot emissions include the enhancement of contrail formation and tactical visibility of military aircraft. Furthermore, impaction of soot on low observable surfaces can compromise the radar signature of aircraft. Quantitative understanding of the soot growth and oxidation mechanisms and the ability to model these processes accurately may be critical to the development of strategies to control emissions.

Modeling laminar non-premixed and partially premixed flames requires the solution of the coupled conservation equations of mass, momentum, individual species mass, and energy. Such a model must include submodels for thermodynamic and transport relations along with finite-rate chemistry. These equations are solved for density, velocities, species mass or mole fractions, temperature, and various soot size classes. The interaction of heat and mass transfer with chemical reactions in practical combustion systems requires a multidimensional study. While three-dimensional models combining both fluid-dynamical effects with finite-rate chemistry are not yet computationally feasible, they will become a reality in the next 3 to 5 years. The modeling of two-dimensional (axisymmetric or rectangular) systems is well established and is the subject of this chapter, in which a variety of related computational issues are discussed. More specifically, this chapter examines various fluid-dynamic-thermochemistry models, along with algorithmic issues involving the solution of the resulting non-linear equations. The chapter concludes with several detailed examples.

6.2   PROBLEM FORMULATION

Conclusions derived from studies of laminar flames are important in characterizing the combustion processes occurring in turbulent flames, in improving engine efficiency, and in understanding the formation of combustion-based pollutants. Thus far, applications of the laminar flamelet model (see, e.g., Reference 2) to turbulent non-premixed combustion have been in relatively simple geometric configurations. The ability to predict reliably the instantaneous local extinction of a turbulent diffusion flame may require application of the flamelet model to more complex multidimensional systems. In addition, an understanding of the factors affecting flame extinction is critical for improving engine efficiency. Combustor geometry and flow field patterns may produce flames with sharp edges; extinction along these edges results in incomplete combustion overall, hence decreasing fuel efficiency. By studying laminar diffusion flames, one can identify the important reactions controlling extinction, along with the important species involved in pollutant formation, while providing information on the fluid mechanics of the flames.

One of the simplest two-dimensional flame configurations of practical importance is the axisymmetric coflow configuration. Although axisymmetric flames are important in combustion applications, it is only recently that they have received substantial attention in theoretical flame studies. Part of this earlier neglect was due to the two-dimensional nature of the problem, coupled with the complexities associated with the combined effects of transport phenomena and chemical processes. This chapter examines axisymmetric non-premixed (or partially premixed) flames in which an inner tube discharges fuel (or fuel/oxidizer) and an outer annular tube discharges air. The inner and outer

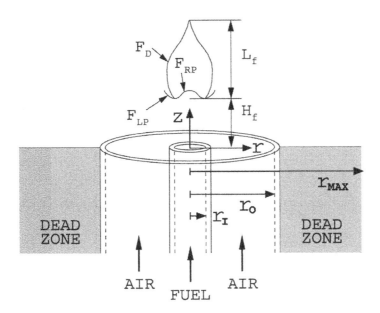

FIGURE 6.1 Schematic of an axisymmetric coflow burner. In the configuration shown, with fuel in the central tube and oxidizer (air) in the outer coflow, a diffusion flame is generated. If, instead, a fuel and oxidizer mixture flows from the central tube, then a partially premixed flame results.

tubes are concentric and have radii R_I and R_O, respectively. The two gas streams make contact at the outlet of the inner tube, and the resulting flames resembles that of a candle (see Figure 6.1).

While the thermochemistry portion of the problem is often solved in terms of species equations for the mass or mole fractions and an energy equation for the temperature or the enthalpy, several options exist for the fluid-dynamic modeling of the problem. This section examines three such possibilities: primitive variables; streamfunction-vorticity; and vorticity-velocity. The advantages and disadvantages of each model are discussed and compared below. Next, the soot modeling is examined in detail, and finally, some starting estimates for full-chemistry flame problems are discussed.

6.2.1 PRIMITIVE VARIABLE FORMULATION

Consider the full set of two-dimensional governing equations in an axisymmetric laminar flame model, where r and z denote the radial and axial coordinates, respectively. In primitive variables, the governing equations can be written in the following form.

Continuity:

$$\frac{1}{r}\frac{\partial}{\partial r}\left(r\rho v_r\right)+\frac{\partial}{\partial z}\left(\rho v_z\right)=0 \tag{6.1}$$

Radial momentum:

$$\left[\rho v_r\frac{\partial v_r}{\partial r}+\rho v_z\frac{\partial v_r}{\partial z}\right]-2\frac{\partial}{\partial r}\left(r\mu\frac{\partial v_r}{\partial r}\right)-\frac{\partial}{\partial z}\left(r\mu\frac{\partial v_r}{\partial z}\right)+\frac{2}{3}\frac{\partial}{\partial r}\left(\mu\frac{\partial}{\partial r}\left(rv_r\right)\right)$$

$$+\frac{2}{3}\frac{\partial}{\partial r}\left(r\mu\frac{\partial v_z}{\partial z}\right)-\frac{\partial}{\partial z}\left(r\mu\frac{\partial v_z}{\partial r}\right)+2\mu\frac{v_r}{r}-\frac{2}{3}\frac{\mu}{r}\frac{\partial}{\partial r}\left(rv_r\right)-\frac{2}{3}\mu\frac{\partial v_z}{\partial z}+r\frac{\partial p}{\partial r}=0 \tag{6.2}$$

Axial momentum:

$$\left[r\rho v_r \frac{\partial v_z}{\partial r} + r\rho v_z \frac{\partial v_z}{\partial z} \right] - \frac{\partial}{\partial r}\left(r\mu \frac{\partial v_z}{\partial r} \right) - 2\frac{\partial}{\partial z}\left(r\mu \frac{\partial v_z}{\partial z} \right) + \frac{2}{3}\frac{\partial}{\partial z}\left(\mu \frac{\partial}{\partial r}(rv_r) \right)$$

$$+ \frac{2}{3}\frac{\partial}{\partial z}\left(r\mu \frac{\partial v_z}{\partial z} \right) - \frac{\partial}{\partial r}\left(r\mu \frac{\partial v_r}{\partial z} \right) + r\frac{\partial p}{\partial z} - r\rho g = 0 \tag{6.3}$$

Species:

$$\left[r\rho v_r \frac{\partial Y_k}{\partial r} + r\rho v_z \frac{\partial Y_k}{\partial z} \right] + \frac{\partial}{\partial r}\left(r\rho Y_k V_{k_r} \right) + \frac{\partial}{\partial z}\left(r\rho Y_k V_{k_z} \right) - rW_k \dot{w}_k = 0, \quad k = 1,2,\ldots,K \tag{6.4}$$

Energy:

$$c_p\left[r\rho v_r \frac{\partial T}{\partial r} + r\rho v_z \frac{\partial T}{\partial z} \right] - \frac{\partial}{\partial r}\left(r\lambda \frac{\partial T}{\partial r} \right) - \frac{\partial}{\partial z}\left(r\lambda \frac{\partial T}{\partial z} \right)$$

$$+ r\sum_{k=1}^{K}\left\{ \rho c_{pk} Y_k \left(V_{k_r} \frac{\partial T}{\partial r} + V_{k_z} \frac{\partial T}{\partial z} \right) \right\} + r\sum_{k=1}^{K} h_k W_k \dot{w}_k - \nabla \cdot q_r = 0 \tag{6.5}$$

Equation of state:

$$\rho = \frac{p\overline{W}}{RT} \tag{6.6}$$

The system is closed with appropriate boundary conditions on each side of the computational domain, as follows.

Axis of symmetry (r = 0):

$$\frac{\partial p}{\partial r} = v_r = \frac{\partial v_z}{\partial r} = \frac{\partial Y_k}{\partial r} = \frac{\partial T}{\partial r} = 0, \quad k = 1, 2, \ldots, K \tag{6.7}$$

Exit (z → ∞):

$$p - p_{atm} = \frac{\partial v_r}{\partial z} = \frac{\partial v_z}{\partial z} = \frac{\partial Y_k}{\partial z} = \frac{\partial T}{\partial z} = 0, \quad k = 1, 2, \ldots, K \tag{6.8}$$

Inlet (z = 0):

$r \leq R_I$

$$v_r = v_z - v_I = Y_k - Y_{k_I} = T - T_I = 0, \quad k = 1, 2, \ldots, K \tag{6.9}$$

$R_I < r \leq R_O$

$$v_r = v_z - v_O = Y_k - Y_{k_O} = T - T_O = 0, \quad k = 1, 2, \ldots, K \tag{6.10}$$

Outer zone ($r = R_O$):

$$\frac{\partial v_r}{\partial r} = \frac{\partial v_z}{\partial r} = \frac{\partial Y_k}{\partial r} = \frac{\partial T}{\partial r} = 0, \quad k = 1, 2, ..., K \tag{6.11}$$

The subscripts *I* and *O* refer to the inner and outer jets, respectively, and v_I, v_O, Y_{k_I}, Y_{k_O}, T_I, T_O, and p_{atm} are specified quantities.

In addition to the variables already defined, T denotes the temperature; Y_k, the mass fraction of the *k*th species; p, the pressure; v_r and v_z, the velocities of the fluid mixture in the radial and axial directions, respectively; ρ, the mass density; W_k, the molecular weight of the *k*th species; $\overline{W}$, the mean molecular weight of the mixture; R, the universal gas constant; λ, the thermal conductivity of the mixture; c_p, the constant pressure heat capacity of the mixture; c_{p_k}, the constant pressure heat capacity of the *k*th species; $\dot{w}_k$, the molar rate of production of the *k*th species per unit volume; h_k, the specific enthalpy of the *k*th species; g, the gravitational constant; μ, the viscosity of the mixture; V_{k_r} and V_{k_z}, the diffusion velocities of the *k*th species in the radial and axial directions, respectively; and $\nabla \cdot q_r$, the divergence of the net radiative flux in the optically-thin limit.

Formulation of the fluid-dynamic portion of the problem in terms of primitive variables enables the computation of the velocities and the pressure directly from the governing equations. Boundary conditions can be specified in terms of velocities and pressures. In addition, the formulation can be generalized for the solution of three-dimensional as well as time-dependent systems. The difficulty with the primitive variable formulation lies in the treatment of the pressure terms in the momentum equations and the velocity components in the continuity equation. If the first-order derivatives of the pressure in the momentum equations are discretized using a central difference operator on a regular grid, a zigzag pressure field will be produced.[3] A similar type of difficulty arises in discretizing the continuity equation. These problems can be avoided by employing a staggered grid technique[4] or by using a one-sided difference operator on a single grid.[5] In the latter case, the discretization is only accurate to first order, but a nonuniform grid with finer spacing in critical regions is employed to increase overall solution accuracy.

6.2.2 Streamfunction-Vorticity Formulation

A first alternative to primitive variable modeling of the fluid-dynamic portion of the problem is the streamfunction-vorticity formulation, in which the number of governing equations to be solved is reduced by introducing the streamfunction ψ and the vorticity ω.[6] The streamfunction is used to replace the radial and axial components of the velocity vector by a single function. It is defined as follows, such that the continuity equation is satisfied identically.

$$\rho r v_r = -\frac{\partial \psi}{\partial z} \tag{6.12}$$

$$\rho r v_z = \frac{\partial \psi}{\partial r} \tag{6.13}$$

The vorticity is a measure of the counterclockwise rotation in the flow, and formulation of the vorticity transport equation serves to eliminate the pressure as one of the dependent variables. The vorticity is defined as:

$$\omega = \frac{\partial v_r}{\partial z} - \frac{\partial v_z}{\partial r} \tag{6.14}$$

With the definitions in Equations (6.12) to (6.14), the governing equations can be written as follows.

Streamfunction:

$$\frac{\partial}{\partial z}\left(\frac{1}{r\rho}\frac{\partial\psi}{\partial z}\right) + \frac{\partial}{\partial r}\left(\frac{1}{r\rho}\frac{\partial\psi}{\partial r}\right) + \omega = 0 \tag{6.15}$$

Vorticity:

$$r^2\left[\frac{\partial}{\partial z}\left(\frac{\omega}{r}\frac{\partial\psi}{\partial r}\right) - \frac{\partial}{\partial r}\left(\frac{\omega}{r}\frac{\partial\psi}{\partial z}\right)\right] - \frac{\partial}{\partial z}\left(r^3\frac{\partial}{\partial z}\left(\frac{\mu}{r}\omega\right)\right) - \frac{\partial}{\partial r}\left(r^3\frac{\partial}{\partial r}\left(\frac{\mu}{r}\omega\right)\right)$$

$$+ r^2 g\frac{\partial\rho}{\partial r} + r^2\nabla\left(\frac{v_r^2 + v_z^2}{2}\right)\cdot\text{iso}\left(\rho\right) = 0 \tag{6.16}$$

Species:

$$\left[\frac{\partial}{\partial z}\left(Y_k\frac{\partial\psi}{\partial r}\right) - \frac{\partial}{\partial r}\left(Y_k\frac{\partial\psi}{\partial z}\right)\right] + \frac{\partial}{\partial r}\left(r\rho Y_k V_{k_r}\right) + \frac{\partial}{\partial z}\left(r\rho Y_k V_{k_z}\right) - rW_k\dot{w}_k = 0, \quad k = 1, 2, ..., K \tag{6.17}$$

Energy:

$$c_p\left[\frac{\partial}{\partial z}\left(T\frac{\partial\psi}{\partial r}\right) - \frac{\partial}{\partial r}\left(T\frac{\partial\psi}{\partial z}\right)\right] - \frac{\partial}{\partial r}\left(r\lambda\frac{\partial T}{\partial r}\right) - \frac{\partial}{\partial z}\left(r\lambda\frac{\partial T}{\partial z}\right)$$

$$+ r\sum_{k=1}^{K}\left\{\rho c_{p_k}Y_k\left(V_{k_r}\frac{\partial T}{\partial r} + V_{k_z}\frac{\partial T}{\partial z}\right)\right\} + r\sum_{k=1}^{K}h_k W_k\dot{w}_k - \nabla\cdot q_r = 0 \tag{6.18}$$

In these equations, the components of the iso operator are given by $(\partial(\circ)/\partial z, -\partial(\circ)/\partial r)$. The boundary conditions in the streamfunction-vorticity formulation are written as follows.

Axis of symmetry ($r = 0$):

$$\psi = \omega = \frac{\partial Y_k}{\partial r} = \frac{\partial T}{\partial r} = 0, \quad k = 1, 2, ..., K \tag{6.19}$$

Exit ($z \rightarrow \infty$):

$$\frac{\partial\psi}{\partial z} = \frac{\partial\omega}{\partial z} = \frac{\partial Y_k}{\partial z} = \frac{\partial T}{\partial z} = 0, \quad k = 1, 2, ..., K \tag{6.20}$$

Inlet ($z = 0$):

$r \leq R_I$

$$\psi = \frac{1}{2}\rho_I v_I R_I^2, \quad \omega = -\frac{\partial}{\partial z}\left(\frac{1}{r\rho}\frac{\partial\psi}{\partial z}\right) - \frac{\partial}{\partial r}\left(\frac{1}{r\rho}\frac{\partial\psi}{\partial r}\right),$$

$$Y_k - Y_{k_I} = T - T_I = 0, \quad k = 1, 2, ..., K \tag{6.21}$$

$R_I < r \le R_O$

$$\psi = \frac{1}{2}\rho_I v_I R_I^2 + \frac{1}{2}\rho_0 v_0 \left(r^2 - R_I^2\right), \quad \omega = -\frac{\partial}{\partial z}\left(\frac{1}{r\rho}\frac{\partial \psi}{\partial z}\right) - \frac{\partial}{\partial r}\left(\frac{1}{r\rho}\frac{\partial \psi}{\partial r}\right),$$

(6.22)

$$Y_k - Y_{k_O} = T - T_O = 0, \quad k = 1, 2, \ldots, K$$

Outer zone ($r = R_O$):

$$\psi = \psi_{max} = \frac{1}{2}\rho_I v_I R_I^2 + \frac{1}{2}\rho_0 v_0 \left(R_O^2 - R_I^2\right),$$

(6.23)

$$\frac{\partial \omega}{\partial r} = \frac{\partial Y_k}{\partial r} = \frac{\partial T}{\partial r} = 0, \quad k = 1, 2, \ldots, K$$

Although the use of the streamfunction-vorticity approach eliminates the pressure as a dependent variable, this reduction brings with it some side effects. To incorporate nonzero vorticity boundary conditions at the base of the flame and to solve the resulting nonlinear system is an extremely difficult task. Because the vorticity is defined as the curl of the velocity field, and because the velocities are, in turn, related to the derivatives of the streamfunction, the vorticity boundary condition at the inlet (the burner surface) is written in terms of streamfunction second derivatives. Therefore, when Newton's method is used to solve the governing equation system, the corresponding off-diagonal terms of the method's Jacobian matrix scale as $1/(\Delta z)^2$, where Δz is the axial grid spacing. The resulting linear system is non-diagonally-dominant, requiring an exceptionally small time step to relax to steady state when employing pseudo-transient techniques. Consequently, the vorticity boundary conditions specified at the burner surface are often simplified (e.g., $\omega = 0$), producing an inaccurate flame structure near the fuel jet as well as incorrect flame liftoff heights.[7] Moreover, the streamfunction-vorticity approach intrinsically limits the modeling of reacting flows to two-dimensional problems. The three-dimensional equivalent of the streamfunction, namely the vector potential, can be used with the vorticity to model three-dimensional flows, but such a formulation introduces additional dependent variables,[7] as compared to a three-dimensional primitive variable approach.

6.2.3 VORTICITY-VELOCITY FORMULATION

A second alternative to primitive variable modeling of the fluid dynamics is the vorticity-velocity formulation. This formulation is relatively new to the solution of chemically reacting flows,[8,9] although it has been used for incompressible fluid flow computations for some time.[10] The key to the vorticity-velocity approach is to eliminate the pressure from the governing equations while replacing the first-order continuity equation with additional second-order equations. Unlike the streamfunction-vorticity formulation, the vorticity-velocity approach is easily extended to three dimensions,[8] and it allows accurate formulation of boundary conditions in a numerically compact way. By taking the curl of the momentum equations and introducing the vorticity, one can generate a vorticity transport equation much as was done to produce Equation (6.16). A nonlinear Poisson equation is obtained for each velocity component by combining the gradient of the vorticity definition with the continuity equation. The system of governing equations is then written as follows.

Radial velocity:

$$\frac{\partial^2 v_r}{\partial r^2} + \frac{\partial^2 v_r}{\partial z^2} = \frac{\partial \omega}{\partial z} - \frac{\partial}{\partial r}\left(\frac{v_z}{\rho}\frac{\partial \rho}{\partial z}\right) - \frac{\partial}{\partial r}\left(\frac{v_r}{r}\right) - \frac{\partial}{\partial r}\left(\frac{v_r}{\rho}\frac{\partial \rho}{\partial r}\right)$$

(6.24)

Axial velocity:

$$\frac{\partial^2 v_z}{\partial r^2} + \frac{\partial^2 v_z}{\partial z^2} = -\frac{\partial \omega}{\partial z} - \frac{\partial}{\partial z}\left(\frac{v_z}{\rho}\frac{\partial \rho}{\partial z}\right) - \frac{\partial}{\partial z}\left(\frac{v_r}{r}\right) - \frac{\partial}{\partial z}\left(\frac{v_r}{\rho}\frac{\partial \rho}{\partial r}\right) \tag{6.25}$$

Vorticity transport:

$$\frac{\partial^2}{\partial r^2}(\mu\omega) + \frac{\partial^2}{\partial z^2}(\mu\omega) + \frac{\partial}{\partial r}\left(\frac{\mu\omega}{r}\right) = \left[\rho v_r \frac{\partial \omega}{\partial r} + \rho v_z \frac{\partial \omega}{\partial z}\right] - \frac{\rho v_r \omega}{r} + \text{iso}(\rho) \cdot \nabla\left(\frac{v_r^2 + v_z^2}{2}\right)$$

$$+ 2\left[\text{iso}\left(\frac{1}{r}\frac{\partial(r v_r)}{\partial r} + \frac{\partial v_z}{\partial z}\right) \cdot \nabla\mu - \nabla v_r \cdot \text{iso}\left(\frac{\partial \mu}{\partial r}\right) - \nabla v_z \cdot \text{iso}\left(\frac{\partial \mu}{\partial z}\right)\right] + g\frac{\partial \rho}{\partial r} \tag{6.26}$$

Species:

$$\left[r\rho v_r \frac{\partial Y_k}{\partial r} + r\rho v_z \frac{\partial Y_k}{\partial z}\right] + \frac{\partial}{\partial r}\left(r\rho Y_k V_{k_r}\right) + \frac{\partial}{\partial z}\left(r\rho Y_k V_{k_z}\right) - rW_k \dot{w}_k = 0, \quad k = 1, 2, ..., K \tag{6.27}$$

Energy:

$$c_p\left[r\rho v_r \frac{\partial T}{\partial r} + r\rho v_z \frac{\partial T}{\partial z}\right] = \frac{\partial}{\partial r}\left(r\lambda \frac{\partial T}{\partial r}\right) - \frac{\partial}{\partial z}\left(r\lambda \frac{\partial T}{\partial z}\right)$$

$$+ r\sum_{k=1}^{K}\left\{\rho c_{p_k} Y_k\left(V_{k_r}\frac{\partial T}{\partial r} + V_{k_z}\frac{\partial T}{\partial z}\right)\right\} + r\sum_{k=1}^{K} h_k W_k \dot{w}_k - \nabla \cdot q_r = 0 \tag{6.28}$$

The system is closed with appropriate boundary conditions on each side of the computational domain, as follows:

Axis of symmetry ($r = 0$):

$$v_r = \frac{\partial v_z}{\partial r} = \omega = \frac{\partial Y_k}{\partial r} = \frac{\partial T}{\partial r} = 0, \quad k = 1, 2, ..., K \tag{6.29}$$

Exit ($z \to \infty$):

$$\frac{\partial v_r}{\partial z} = \frac{\partial v_z}{\partial z} = \frac{\partial \omega}{\partial z} = \frac{\partial Y_k}{\partial z} = \frac{\partial T}{\partial z} = 0, \quad k = 1, 2, ..., K \tag{6.30}$$

Inlet ($z = 0$):

$r \le R_I$

$$v_r = v_z - v_I = \omega - \frac{\partial v_r}{\partial z} + \frac{\partial v_z}{\partial r} = Y_k - Y_{k_I} = T - T_I = 0, \quad k = 1, 2, ..., K \tag{6.31}$$

$R_I < r \leq R_O$

$$v_r = v_z - v_O = \omega - \frac{\partial v_r}{\partial z} + \frac{\partial v_z}{\partial r} = Y_k - Y_{k_O} = T - T_O = 0, \quad k = 1, 2, ..., K \quad (6.32)$$

Outer zone $(r = R_O)$:

$$\frac{\partial v_r}{\partial r} = \frac{\partial v_z}{\partial r} = \omega = \frac{\partial Y_k}{\partial r} = \frac{\partial T}{\partial r} = 0, \quad k = 1, 2, ..., K \quad (6.33)$$

The vorticity-velocity formulation is particularly effective in the solution of steady-state reacting flow problems. The elimination of the pressure as a dependent variable and the ability to specify the vorticity boundary conditions directly in terms of the velocities can significantly reduce the CPU time required to solve a steady-state flame problem, as compared to the primitive variable and the streamfunction-vorticity approaches. The vorticity-velocity formulation, however, is not as easily implemented for time-dependent problems, because some of the unsteady terms are mixed derivatives with respect to time and space, requiring careful discretization.

6.2.4 SOOT MODELING

Modeling soot formation in practical combustion systems is an extremely challenging problem. Reacting flow problems for simple hydrocarbon fuels such as methane can require the solution of more than 50 chemical species in addition to the temperature *and* the fluid-dynamic variables. The inclusion of soot inception, growth, and oxidation processes in these models greatly increases the complexity of the problem. Consequently, the investigation of soot formation with detailed chemistry in a generic, possibly turbulent, multidimensional configuration is still beyond current computational capabilities. The laminar coflow flame, however, provides a multidimensional environment — one that is currently computationally feasible — in which to investigate the interaction of soot formation with detailed gas-phase chemistry.

Most studies using detailed chemical kinetics and coupled models of soot production and oxidation have focused on one-dimensional geometries. A few recent studies[11,12] have focused on jet diffusion flames, using simplified, monodisperse, soot formation models with skeletal kinetic mechanisms. This chapter considers modifications of the sectional soot formation model developed in Colket and Hall[13] and Hall[14] for incorporation into a laminar, axisymmetric, diffusion flame in which a cylindrical fuel steam is surrounded by a coflowing oxidizer jet. To describe the different soot particle size classes, appropriate sectional equations are included, described in detail below, each of which incorporates a convective, diffusive, and chemical production balance. These equations, along with surface growth, oxidation, and radiation, are fully integrated with the governing equations.

Soot kinetics are modeled as coalescing, solid carbon spheroids undergoing surface growth in the free-molecule limit. The particle mass range of interest is divided into sections,[15] and a dynamical equation including coalescence, surface growth, and oxidation is written for each section. Sectional analysis makes it possible to obtain the particle size distribution without *a priori* assumptions about the form of the distribution. For the first sectional bin, an inception source term is included. The incorporation of these dynamical equations into a transport/conservation equation for each section includes both thermophoresis and an effective bin diffusion rate. In the gas-phase species conservation equations, additional source terms are included to account for scrubbing or generation of gaseous species arising from the particle growth and oxidation processes. The gas and the soot are additionally coupled through nonadiabatic radiative loss from both the gas and the soot in the optically thin approximation.

For one-atmosphere calculations, the sectional coefficients for transport, coalescence, and surface processes are calculated in the free-molecule limit,[15] with surface growth and oxidation proportional to particle surface area. In the free-molecule limit, the particle mass and medium dependencies of the sectional coefficients are factored such that they only have to be evaluated once. Oxidation of soot by O_2 and OH is treated as described in Reference 14. The inception model employed here is based on an estimate of the formation rate of two- and three-ringed aromatic species, and it is a function of local acetylene, benzene, phenyl, and molecular hydrogen concentrations. Assuming steady-state values of intermediates, and also that $[H_2] \gg [C_2H_2]$ (where square brackets denote species concentrations), the rate of production of the polyaromatic species can be estimated to be given by the following expressions[14]:

$$\frac{d[C_{10}H_7]}{dt} = 10^{11.88} e^{(-4378/T)} \left[\frac{C_2H_2}{H_2}\right]^2 [C_6H_5] \text{ cc/mole/s} \tag{6.34}$$

and

$$\frac{d[C_{14}H_{10}]}{dt} = 10^{12.50} e^{(-6390/T)} \left[\frac{C_2H_2}{H_2}\right] [C_6H_6][C_6H_5] \text{ cc/mole/s} \tag{6.35}$$

where the gas-phase concentrations and temperatures are evaluated at local conditions. With the further assumptions that inception is limited by the formation of polyaromatics and that oxidation/decomposition of such species can be neglected, the inception rate S_i, in grams/cc/s, was initially assigned to

$$S_i = 127 \times \frac{d[C_{10}H_7]}{dt} + 178 \times \frac{d[C_{14}H_{10}]}{dt} \tag{6.36}$$

where the constants (molecular weights) are provided to convert from molar to mass units. The second term in Equation (6.36) plays a negligible role in methane/air diffusion flames but has been included here for generality. The contributions from both inception processes are incorporated in the first sectional bin, whose lower mass boundary is set equal to the mass of the smallest inception species. In the sectional representation, the sectional mass boundaries vary linearly on a logarithmic scale.

The surface growth rate is that proposed by Colket and Hall, denoted by MODFW in Reference 13. The mechanism treats both acetylene addition and elimination, and hence simulates the rapid falloff of surface growth rates at elevated temperatures. Use of an expression with only a simple Arrhenius form can result in a gross overprediction of soot in hydrocarbon systems.

The inception and surface growth models used here agree well with available ethylene opposed jet data. However, the authors do not suggest that they represent a universal solution to the problem of soot formation, because comparisons with soot data in other fuels have not been performed. The possibility of surface growth rate aging has not been taken into consideration, nor are the results with finite coalescence appropriate for aggregate formation. These effects should be included in future modeling efforts. Fundamental questions relating to inception, temperature dependence of surface growth, and precise dependence of surface growth on gas-phase species concentrations remain. Whether soot growth rates derived from premixed and opposed jet studies are self-consistent is still an open issue as well.

A primary parameter of interest is the soot volume fraction, for which numerical convergence is generally obtained with a smaller number of sections than is the case for particle size and number density.[14] The number of sections required for convergence must be examined in each case, although evidence points to the relative magnitudes of surface growth and inception as the important parameter. When the contribution of inception to total volume fraction is significant, although not necessarily dominant, the number of sections required can be very small,[14] with larger numbers of sections necessary when the contribution of surface growth to volume fraction becomes very much larger than that from inception.

To incorporate the sectional soot model into the gas-phase system, one must modify the species and energy equations to account for gas-phase scrubbing and soot radiation. In addition, one needs to include a section equation for each of the soot size classes. These equations take the following form.

Species:

$$\left[r\rho v_r \frac{\partial Y_k}{\partial r} + r\rho v_z \frac{\partial Y_k}{\partial z} \right] + \frac{\partial}{\partial r}\left(r\rho Y_k V_{k_r} \right) + \frac{\partial}{\partial z}\left(r\rho Y_k V_{k_z} \right) - rW_k\left(\dot{w}_k + \dot{w}_k^s \right) = 0, \quad k = 1, 2, \ldots, K \quad (6.37)$$

Energy:

$$c_p \left[r\rho v_r \frac{\partial T}{\partial r} + r\rho v_z \frac{\partial T}{\partial z} \right] - \frac{\partial}{\partial r}\left(r\lambda \frac{\partial T}{\partial r} \right) - \frac{\partial}{\partial z}\left(r\lambda \frac{\partial T}{\partial z} \right)$$

$$+ r \sum_{k=1}^{K} \left\{ \rho c_{pk} Y_k \left(V_{k_r} \frac{\partial T}{\partial r} + V_{k_z} \frac{\partial T}{\partial z} \right) \right\} + r \sum_{k=1}^{K} h_k W_k \left(\dot{w}_k + \dot{w}_k^s \right) - \nabla \cdot q_r = 0 \quad (6.38)$$

Sections:

$$\left[r\rho v_r \frac{\partial Y_k}{\partial r} + r\rho v_z \frac{\partial Y_k}{\partial z} \right] + \frac{\partial}{\partial r}\left(r\rho Y_k \left(V_{k_r}^s + V_{T_r} \right) \right) + \frac{\partial}{\partial z}\left(r\rho Y_k \left(V_{k_z}^s + V_{T_z} \right) \right) - \dot{q}_k = 0,$$

$$k = K+1, K+2, \ldots, K+M \quad (6.39)$$

In addition to the variables already defined, M represents the number of soot (size) sections; Y_k, the $(k - K)th$ soot size class ($k > K$): $\dot{w}_k^s$, the molar rate of production of the kth species per unit volume due to scrubbing/replenishment by the soot growth/oxidation processes; $\dot{q}_k$, the rate of change of section k due to inception, surface growth, oxidation, and coalescence; $V_{T_r}, V_{T_z}, V_{k_r}^s, V_{k_z}^s$, the sectional thermophoretic and diffusion velocities of the kth soot size class; and $\nabla \cdot q_r$, the divergence of the net radiative flux for gas bands and soot in the optically thin limit.

One can write the diffusion velocities in the r and z directions in the form

$$V_{k_r} = -\left(1/X_k \right) D_k \frac{\partial X_k}{\partial r}, \quad k = 1, 2, \ldots, K \quad (6.40)$$

$$V_{k_z} = -\left(1/X_k \right) D_k \frac{\partial X_k}{\partial z}, \quad k = 1, 2, \ldots, K \quad (6.41)$$

where X_k is the mole fraction of the kth species, and D_k is related to the binary diffusion coefficient through the expression (see, e.g., Reference 16)

$$D_k = \frac{\left(1 - Y_k\right)}{\sum_{j \neq k}^{K} X_j / D_{jk}} \tag{6.42}$$

The binary diffusion coefficients, the viscosity, the thermal conductivity of the mixture, the chemical production rates, and the thermodynamic quantities are evaluated using vectorized and highly optimized transport and chemistry libraries.[17]

The sectional thermophoretic velocities in the free-molecule regime are given by[14]:

$$V_{T_r} = -0.55 \frac{\mu}{\rho} \frac{1}{T} \frac{\partial T}{\partial r} \tag{6.43}$$

$$V_{T_z} = -0.55 \frac{\mu}{\rho} \frac{1}{T} \frac{\partial T}{\partial z} \tag{6.44}$$

The sectional diffusion velocities are written as in Equations (6.40) and (6.41) with a mass-weighted mean diffusion coefficient for each bin.[14]

Anticipating that radiative losses could significantly influence soot levels, an optically thin radiation model is included in the calculations.[18,19] For methane/air mixtures, it is assumed that the only significant radiating species are H_2O, CO, and CO_2. By utilizing an optically thin limit in which self-absorption of radiation is neglected, the divergence of the net radiative flux can be written as:

$$\nabla \cdot q_R = C f_v T^5 + 4\pi \sum_{ik} \alpha_{ik} \rho_k I_{b_{ik}}, \tag{6.45}$$

where f_v is the soot volume fraction and $I_{b_{ik}}$ is the Planck function evaluated at the gas-band centers of the contributing vibrational-rotational or pure rotational bands, whose integrated intensities are given by α_{ik}.

6.2.5 STARTING ESTIMATES

The governing equations ((Equations 6.1–6.5), (6.15–6.18), or (6.24–6.28)) are highly nonlinear and require a starting estimate for the discrete solution method discussed in the next section. The determination of a sufficiently "good" initial solution estimate in two-dimensional problems can be a difficult task, mainly because of the exponential dependence of the chemistry terms on the temperature and the multidimensional nonlinear coupling of the fluid-dynamic and thermochemistry solution fields.

The burning rate in a diffusion flame is controlled by the rate at which the fuel and the oxidizer are brought together in the proper proportions; by contrast, the burning rate in a premixed flame is controlled by chemical reactions. In diffusion flames of practical interest, the oxidation of the fuel to form intermediates and products proceeds through a detailed kinetics mechanism. In these problems, combustion takes place at a finite rate, and some fuel and some oxidizer co-exist on either side of the reaction zone. Nevertheless, the use of a thin, infinitely fast, global reaction model (commonly called a flame-sheet model) is a natural starting point for the determination of a "good" initial solution estimate for this finite-rate axisymmetric diffusion flame model (see also References 20 to 23). For premixed or partially premixed flames, one can utilize a converged non-premixed solution as the starting estimate and then gradually mix oxidizer into the fuel jet until the appropriate equivalence ratio is reached. For this reason, the discussion below focuses on starting estimates for diffusion flames only.

For a diffusion flame in the limit of infinitely fast kinetics, the fuel and the oxidizer are separated by a thin exothermic reaction zone, referred to as a flame sheet. In this zone, the fuel and the oxidizer are in stoichiometric proportion, and the temperature and products of combustion are maximized. In such an ideal situation, no oxidizer is present on the fuel side of the flame, and no fuel is present on the oxidizer side. The fuel and the oxidizer diffuse toward the reaction zone as a result of concentration gradients in the flow.

It is assumed that the fuel and the oxidizer obey a single overall irreversible reaction of the type

$$\text{Fuel } (F) + \text{Oxidizer } (X) \rightarrow \text{Products } (P) \tag{6.46}$$

in the presence of an inert gas (N). More specifically, one can write

$$v_F F + v_X X \rightarrow v_P P \tag{6.47}$$

where v_F, v_X, and v_P are the stoichiometric coefficients of the fuel, the oxidizer, and the product, respectively. In addition, thermal diffusion and radiation are neglected, c_p and c_{p_k} are taken to be constant, and also the ordinary mass diffusion velocities are written in terms of Fick's law. With these approximations, one can write the governing equations (in the vorticity-velocity formulation) as follows.

Radial velocity:

$$\frac{\partial^2 v_r}{\partial r^2} + \frac{\partial^2 v_r}{\partial z^2} = \frac{\partial \omega}{\partial z} - \frac{\partial}{\partial r}\left(\frac{v_z}{\rho}\frac{\partial \rho}{\partial z}\right) - \frac{\partial}{\partial r}\left(\frac{v_r}{r}\right) - \frac{\partial}{\partial r}\left(\frac{v_r}{\rho}\frac{\partial \rho}{\partial r}\right) \tag{6.48}$$

Axial velocity:

$$\frac{\partial^2 v_z}{\partial r^2} + \frac{\partial^2 v_z}{\partial z^2} = -\frac{\partial \omega}{\partial z} - \frac{\partial}{\partial z}\left(\frac{v_z}{\rho}\frac{\partial \rho}{\partial z}\right) - \frac{\partial}{\partial z}\left(\frac{v_r}{r}\right) - \frac{\partial}{\partial z}\left(\frac{v_r}{\rho}\frac{\partial \rho}{\partial r}\right) \tag{6.49}$$

Vorticity transport:

$$\frac{\partial^2}{\partial r^2}(\mu\omega) + \frac{\partial^2}{\partial z^2}(\mu\omega) + \frac{\partial}{\partial r}\left(\frac{\mu\omega}{r}\right) = \left[\rho v_r \frac{\partial \omega}{\partial r} + \rho v_z \frac{\partial \omega}{\partial z}\right] - \frac{\rho v_r \omega}{r} + \text{iso}(\rho)\cdot\nabla\left(\frac{v_r^2 + v_z^2}{2}\right)$$

$$+ 2\left[\text{iso}\left(\frac{1}{r}\frac{\partial(rv_r)}{\partial r} + \frac{\partial v_z}{\partial z}\right)\cdot\nabla\mu - \nabla v_r \cdot \text{iso}\left(\frac{\partial\mu}{\partial r}\right) - \nabla v_z \cdot \text{iso}\left(\frac{\partial\mu}{\partial z}\right)\right] + g\frac{\partial\rho}{\partial r} \tag{6.50}$$

Species:

$$\left[r\rho v_r \frac{\partial Y_F}{\partial r} + r\rho v_z \frac{\partial Y_F}{\partial z}\right] - \frac{\partial}{\partial r}\left(r\rho D_F \frac{\partial Y_F}{\partial r}\right) - \frac{\partial}{\partial z}\left(r\rho D_F \frac{\partial Y_F}{\partial z}\right) + rW_F v_F \dot{w} = 0 \tag{6.51}$$

$$\left[r\rho v_r \frac{\partial Y_X}{\partial r} + r\rho v_z \frac{\partial Y_X}{\partial z}\right] - \frac{\partial}{\partial r}\left(r\rho D_X \frac{\partial Y_X}{\partial r}\right) - \frac{\partial}{\partial z}\left(r\rho D_X \frac{\partial Y_X}{\partial z}\right) + rW_X v_X \dot{w} = 0 \tag{6.52}$$

$$\left[r\rho v_r \frac{\partial Y_P}{\partial r} + r\rho v_z \frac{\partial Y_P}{\partial z}\right] - \frac{\partial}{\partial r}\left(r\rho D_P \frac{\partial Y_P}{\partial r}\right) - \frac{\partial}{\partial z}\left(r\rho D_P \frac{\partial Y_P}{\partial z}\right) - rW_P v_P \dot{w} = 0 \qquad (6.53)$$

$$\left[r\rho v_r \frac{\partial Y_N}{\partial r} + r\rho v_z \frac{\partial Y_N}{\partial z}\right] - \frac{\partial}{\partial r}\left(r\rho D_N \frac{\partial Y_N}{\partial r}\right) - \frac{\partial}{\partial z}\left(r\rho D_N \frac{\partial Y_N}{\partial z}\right) = 0 \qquad (6.54)$$

Energy:

$$c_p\left[r\rho v_r \frac{\partial T}{\partial r} + r\rho v_z \frac{\partial T}{\partial z}\right] - \frac{\partial}{\partial r}\left(r\lambda \frac{\partial T}{\partial r}\right) - \frac{\partial}{\partial z}\left(r\lambda \frac{\partial T}{\partial z}\right) - r\left(W_F v_F h_F + W_X v_X h_X - W_P v_P h_P\right)\dot{w} = 0 \quad (6.55)$$

In the above equations,

$$\dot{w} = -\frac{\dot{w}_F}{v_F} = -\frac{\dot{w}_X}{v_X} = \frac{\dot{w}_P}{v_P} \qquad (6.56)$$

is the rate of progress of the reaction.

Introducing the heat release per unit mass of the fuel, denoted by Q, where

$$Q = h_F + \frac{W_X v_X}{W_F v_F} h_X - \frac{W_P v_P}{W_F v_F} h_P \qquad (6.57)$$

and assuming that each of the Lewis numbers

$$\mathrm{Le}_F = \frac{\lambda}{\rho D_F c_p}, \quad \mathrm{Le}_X = \frac{\lambda}{\rho D_X c_p}, \quad \mathrm{Le}_P = \frac{\lambda}{\rho D_P c_p}, \quad \mathrm{Le}_N = \frac{\lambda}{\rho D_N c_p} \qquad (6.58)$$

is equal to one, then each of the Shvab-Zeldovich variables

$$Z_F = Y_F - Y_{F_o} + \frac{c_p}{Q}\left(T - T_o\right) \qquad (6.59)$$

$$Z_X = Y_X - Y_{X_o} + \frac{c_p}{Q}\frac{W_X v_X}{W_F v_F}\left(T - T_o\right) \qquad (6.60)$$

$$Z_P = Y_P - Y_{P_o} - \frac{c_p}{Q}\frac{W_P v_P}{W_F v_F}\left(T - T_o\right) \qquad (6.61)$$

$$Z_N = Y_N - Y_{N_o} \qquad (6.62)$$

satisfies the differential equation

$$\left[r\rho v_r \frac{\partial Z_k}{\partial r} + r\rho v_z \frac{\partial Z_k}{\partial z}\right] - \frac{\partial}{\partial r}\left(r\rho D_k \frac{\partial Z_k}{\partial r}\right) - \frac{\partial}{\partial z}\left(r\rho D_k \frac{\partial Z_k}{\partial z}\right) = 0, \quad k = F, X, P, N \qquad (6.63)$$

One can show that all of the Z_k are proportional to each other and to a conserved scalar S which satisfies an equation similar in form to Equation (6.63).

To complete the specification of the starting estimate, one must be able to recover the temperature and the major species profiles from the conserved scalar. Of critical importance to this procedure is an estimate of the location of the flame front. In the Shvab-Zeldovich formulation, the fuel and the oxidizer cannot co-exist. Hence, on the fuel side of the flame, $Y_X = 0$; and on the oxidizer side, $Y_F = 0$. If one denotes variables at the flame front with the subscript f, then it can be shown (see also Reference 20) that, for a fixed value of the axial coordinate z, the location of the flame front is defined such that

$$S\left(r_f\right)\bigg|_{\text{fixed } z} = \frac{Y_{X_O}}{Y_{X_O} + \frac{W_X \nu_X}{W_F \nu_F} Y_{F_I}} \equiv S_f \tag{6.64}$$

The location of the flame front can be obtained by solving Equation (6.64) at each axial coordinate level.

Utilizing the proportionality of the Z_k's to S along with the expressions in Equations (6.59) to (6.62), one can derive expressions for the temperature and species on the fuel and oxidizer sides of the flame. On the fuel side, one obtains

$$T = T_I S + \left(T_O + Y_{X_O} \frac{Q}{c_p} \frac{W_F \nu_F}{W_X \nu_X} \right)(1 - S) \tag{6.65}$$

$$Y_F = Y_{F_I} S - Y_{X_O} \frac{W_F \nu_F}{W_X \nu_X}(1 - S) \tag{6.66}$$

$$Y_X = 0, \tag{6.67}$$

$$Y_P = Y_{X_O} \frac{W_P \nu_P}{W_X \nu_X}(1 - S) \tag{6.68}$$

and

$$Y_N = Y_{N_I} S + Y_{N_O}(1 - S) \tag{6.69}$$

On the oxidizer side, one obtains

$$T = T_O(1 - S) + \left(T_I + Y_{F_I} \frac{Q}{c_p} \right) S \tag{6.70}$$

$$Y_F = 0, \tag{6.71}$$

$$Y_X = Y_{X_O}(1 - S) - Y_{F_I} \frac{W_X \nu_X}{W_F \nu_F} S \tag{6.72}$$

$$Y_P = Y_{F_I} \frac{W_P \nu_P}{W_F \nu_F} S \tag{6.73}$$

and

$$Y_N = Y_{N_O}(1 - S) + Y_{N_I}S \tag{6.74}$$

Equations (6.48) to (6.50) and (6.63) are solved for the two velocities, the vorticity, and the conserved scalar. For a given profile of the conserved scalar, one solves Equation (6.64) for the location of the flame front at each axial level. One then utilizes the relations in Equations (6.65) to (6.74) to obtain expressions for T, Y_F, Y_X, Y_P, and Y_N. The recovered temperature profile is used in the ideal gas law to evaluate the density. The temperature is also required for forming the viscosity and the diffusion coefficient (D_k is replaced by D). If one introduces the Prandtl number.

$$Pr = \frac{\mu c_p}{\lambda} \tag{6.75}$$

and recalls that each of the Lewis numbers is equal to one, then one can write

$$\rho D = \frac{\lambda}{c_p} = \frac{\mu}{Pr_{ref}}, \tag{6.76}$$

where Pr_{ref} is a reference Prandtl number. Specifically, one can use a value approximately that of air, $Pr_{ref} = 0.75$. Hence, determination of ρD is reduced to the specification of a transport relation for the viscosity. The authors use the simple power law

$$\mu = \mu_0 \left(\frac{T}{T_0}\right)^r \tag{6.77}$$

where $r = 0.7$, $T_0 = 298$ K, and $\mu_0 = 1.85 \times 10^{-4}$ gm/cm-s; the latter is again a reference value for air.[24] The temperature exponent has been determined by fitting Equation (6.77) to the mixture viscosity and temperature data of a representative one-dimensional finite-rate chemistry calculation. The scaled heat release parameter Q/c_p can be determined from an estimate of the peak temperature (e.g., from an experiment) or from the heat of combustion of the system under consideration and a representative heat capacity.

6.3 METHOD OF SOLUTION

The goal is to obtain a discrete solution of the governing equations in two dimensions on a grid (or a mesh) which is denoted by $\mathcal{G}_2$. The *initial* points of this grid are defined by the intersection of the lines of the one-dimensional grid $\mathcal{G}_r$

$$\mathcal{G}_r = \left\{0 = r_0 < r_1 < \ldots < r_i < \ldots < r_{G_r} = R_{max}\right\} \tag{6.78}$$

and the one-dimensional grid $\mathcal{G}_z$

$$\mathcal{G}_z = \left\{0 = z_0 < z_1 < \ldots < z_j < \ldots < z_{G_z} = Z_{max}\right\} \tag{6.79}$$

Computationally, a steady-state and a time-dependent solution method are combined. A time-dependent approach is employed to obtain a converged numerical solution on an initial coarse grid using the flame-sheet starting estimate. Gridpoints are then inserted adaptively, and the steady-state solution procedure is used to complete the problem. The components of this solution procedure are discussed in detail below.

6.3.1 NONLINEAR SOLVER

One can approximate the spatial operators in the governing partial differential equations using finite-difference expressions. Diffusion terms are discretized with centered differences and convective terms with upwind approximations. The problem of finding an analytical solution to the equations is thereby converted into one of finding an approximate solution at each point (r_i, z_j) of the grid in two dimensions. With the difference equations written in residual form, one wishes to obtain the solution of the system of nonlinear equations

$$F(U) = 0 \tag{6.80}$$

For an initial solution estimate U^0 sufficiently close to U^*, the system of nonlinear equations in Equation (6.80) can be solved using Newton's method, which leads to the iteration

$$J(U^n)(U^{n+1} - U^n) = -\lambda^n F(U^n), \quad n = 0, 1, 2, \dots \tag{6.81}$$

$J(U^n) = \partial F(U^n)/\partial U$ is the Jacobian matrix, and λ^n ($0 < \lambda \le 1$) is the nth damping parameter.[25]
Note that with the spatial discretizations used in forming Equation (6.80), the Jacobian matrix can be written in block-nine-diagonal form. For problems involving detailed transport and complex chemistry, it is often more efficient to evaluate the Jacobian matrix numerically instead of analytically. The numerical procedure implemented here extends the ideas outlined by Curtis, Powell, and Reid.[26] Several columns of the Jacobian are formed simultaneously using vector function evaluations and taking advantage of the Jacobian's given sparsity structure. If, with each column of the Jacobian, one associates the i and j values of the node corresponding to the column's diagonal block, then all columns of the Jacobian having the same value of the parameter

$$\alpha = (i + 3j) \bmod 9 \tag{6.82}$$

can be evaluated simultaneously. Ideas along these lines have also been explored by Newsam and Ramsdell[27] and Coleman and More.[28]
The formation of the Jacobian accounts for a substantial part of the cost of the flame calculation. Therefore, the use of a modified Newton method is recommended, in which the Jacobian is reevaluated only periodically. The immediate implication of applying the modified Newton method is that the partial factorization of the Jacobian can be stored, and each modified Newton iteration can be obtained by performing relatively inexpensive block-line iterative back substitutions. One difficulty when applying the modified method lies in determining whether the convergence rate is fast enough. If the rate is too slow, one will want to revert back to a full Newton method and make use of new Jacobian information. If the rate is acceptable, one will want to continue performing modified Newton iterations. The Newton iterations continue until the size of $\|U^{n+1} - U^n\|_2$ is reduced below a preset tolerance.
When points are added, as discussed in the next section, and the size of the grid spacing thus decreases, it is anticipated that the solution, interpolated from one grid to the next, should become a better starting estimate for Newton's method on the next finer grid. For a class of nonlinear

boundary value problems, Smooke and Mattheij[29] have shown that there exists a critical mesh spacing such that the interpolated solution lies in the domain of convergence of Newton's method on the next grid. As a result, the hypotheses of the Kantorovich theorem[30] are met, and the sequence of successive modified Newton iterates can be shown to satisfy a recurrence relation scaled by the first Newton step[31]. As a result, if in the course of a calculation, one determines that the size of the $n + 1^{st}$ modified Newton step is larger than the value predicted by the theorem, then one forms a new Jacobian and restarts the iteration count.

The use of a direct solver such as Gaussian elimination in the solution of the Newton equations (Equation (6.81)) is not feasible due to the size of the system and the Jacobian's given sparsity structure; for most problems of practical interest, the cost would be prohibitive. Consequently, iterative methods are generally used in the solution of the block-nine-diagonal linear equations in Equation (6.81). Although a variety of iterative techniques ranging from point Jacobi to block-line SOR methods have been used in solving the Newton equations, conjugate-gradient-based methods such as GMRES[32] and Bi-CGSTAB[33] are now the most widely implemented methods to solve these nonsymmetric linear systems.

To improve the convergence properties of such solvers, a preconditioning technique may be used, in which the linear system is effectively multiplied through by an approximate (and very computationally cheap) inverse of the Jacobian, thereby transforming the linear system into one with a smaller condition number and better convergence properties.[34] Both the GMRES and Bi-CGSTAB solvers are preconditioned with a Gauss-Seidel (GS) left preconditioner. The effectiveness of this preconditioner has been illustrated on a flame-sheet problem by Ern et al.[34]. Moreover, in the present flame problems with one predominant flow direction, the GS preconditioner fully retains the upstream convective coupling in the Jacobian matrix. The preconditioned matrix-vector multiply consists of a lower triangular system solve combined with a block tridiagonal solver. The convergence of the GMRES and Bi-CGSTAB algorithms is based on the weighted norm of the left preconditioned linear residual using an absolute tolerance equal to one-tenth of the Newton tolerance.

6.3.2   Pseudo-Time Stepping

There are two fundamental mathematical approaches for solving flame problems: one type employing a transient method, and the other solving the steady-state boundary value problem directly. Generally speaking, transient methods are robust but computationally inefficient compared to boundary value methods, which are efficient but have less desirable convergence properties. For the two-dimensional flame problems, use of the flame-sheet starting estimate eliminates many of the convergence difficulties associated with solving the governing equations directly. Nevertheless, to obtain on the initial grid a starting estimate that lies in the convergence domain of Newton's method, one can apply a time-dependent iteration to the flame-sheet solution.

In a time-dependent method, the original nonlinear two-point boundary value problem is converted into a nonlinear parabolic mixed initial-boundary value problem. This conversion is accomplished by appending the term $\partial(\bigcirc)/\partial t$ to the left-hand side of each conservation equations, where $\bigcirc$ represents the quantity being conserved. This same procedure is used in the two-dimensional calculations, obtaining

$$\frac{\partial U}{\partial t} = F(U) \tag{6.83}$$

with appropriate initial conditions. If the time derivative is replaced, for example, by a backward Euler approximation, the governing equations can be written in the form

$$\Im(U^{n+1}) = F(U^{n+1}) - \frac{(U^{n+1} - U^n)}{\tau^{n+1}} = 0 \tag{6.84}$$

where, for a function $g(t)$, the shorthand notation $g(t^n) \equiv g^n$ is used and where the time step τ^{n+1} is defined as $t^{n+1} - t^n$.

At each time level, one must apply Newton's method to solve this system of nonlinear equations in Equation (6.84) that look very similar to the nonlinear equations in Equation (6.80). The important difference between the systems in Equation (6.80) and in Equation (6.84) is that the diagonal of the time-dependent Jacobian is approximately that of the steady-state Jacobian, but with an additive weighting of $1/\tau^{n+1}$. This increase in the size of the diagonal produces a better conditioned system, and the solution from the nth time step usually provides an excellent starting guess to the solution at the $n + 1$st time level. The work per time step is similar to that for the modified Newton iteration, but the time-like continuation of the numerical solution produces an iteration strategy that will, in general, be less sensitive to the initial starting estimate than if Newton's method were to be applied to Equation (6.80) directly. As a result, when one ultimately implements Newton's method on the steady-state equations, using as an initial guess the output of the time-stepping iterations from Equation (6.84), one obtains a converged numerical solution with only a few additional iterations. This time-dependent starting procedure can also be used on grids other than the initial one. The size of each time step is chosen by monitoring the local truncation error of the time discretization process (see also Reference 35).

6.3.3 ADAPTIVE GRID REFINEMENT

The ability to resolve flame fronts oriented in any direction is critical to the accuracy of a calculation, and the number of points utilized impacts the overall efficiency of the method. The solution of the governing equations in the axisymmetric problem contains regions in which the dependent variables have steep fronts and sharp peaks, and an efficient solution technique requires that these regions be resolved adaptively. Adaptive grid refinement in two dimensions can proceed along several different routes.

The simplest procedure involves determining the gridpoints of $\mathcal{G}_z$ by equidistributing positive weight functions over grid intervals in each of the r and z coordinate directions. Specifically, one attempts to equidistribute the grid $\mathcal{G}_r$ with respect to the nonnegative function $\mathcal{W}_r$ and constant C_r for each of the $G_z + 1$ horizontal grid lines; for each of $j = 0, 1, \ldots, G_z$, one writes

$$\int_{r_i}^{r_{i+1}} \mathcal{W}_r \, dr < C_r, \quad i = 0, 1, \ldots, G_r - 1 \tag{6.85}$$

Similarly, one attempts to equidistribute the grid $\mathcal{G}_z$ with respect to the nonnegative function W_z and constant C_z for each of the $G_r + 1$ vertical grid lines; for each of $i = 0, 1, \ldots, G_r$, one writes

$$\int_{z_j}^{z_{j+1}} \mathcal{W}_z \, dz < C_z, \quad j = 0, 1, \ldots, G_z - 1. \tag{6.86}$$

The resulting grid $\mathcal{G}_r \times \mathcal{G}_z$ has a tensor product structure, in which each grid line starts at one domain boundary and continues to the opposite boundary. Unfortunately, each time a grid line is added with the intent of reducing the error in a particular region, points are unnecessarily introduced at each intersection of this new line with all perpendicular ones.

A more complicated but less frequently used type of adaptive gridding is that of unstructured "locally-refined" rectangular gridding. Such methods usually begin with a coarse, nonuniform, rectangular tensor product grid. According to a solution-dependent criterion, new points are added (or removed), subdividing a given cell into four smaller cells. This process refines the grid without the global introduction of many unnecessary new points. One disadvantage, however, is that the loss of the tensor product structure requires indexing arrays to find neighboring points, and another challenge lies in properly treating points at the interfaces between different refinement levels of the grid. Methods of this type have been applied to full-chemistry combustion problems by several researchers (see References 36 to 39), although in combination with non-Newton iterative solution techniques.

For solving the diffusion or partially premixed flames described previously, the authors employ the Local Rectangular Refinement (LRR) method, discussed in detail in References 40 to 42. This adaptive gridding technique is used in conjunction with an iterative Newton solver. The LRR method begins with a coarse initial grid (typically a nonuniform tensor product grid such as that given above by $\mathcal{G}_r \times \mathcal{G}_z$), referred to as the base grid. Once the converged full-chemistry solution has been determined on the base grid using the numerical methods described in the previous sections, the adaption process begins. Points can be either added or removed; because the point removal process is the reverse of the point addition process, only the latter is described below.

During the adaption process, individual boxes of the grid are flagged for subsequent refinement via subequidistribution of positive weight functions $\mathcal{W}$, one for each of the dependent variables (temperature, velocity, etc.), represented below by U. The LRR weight functions are evaluated at the center of each box b and incorporate solution derivatives, which approximate truncation error trends.

$$\mathcal{W}_b = 1 + \left[\alpha\|\nabla U\|_b\right]\left[\max_b\|\nabla U\|_b\right]^{-1} \tag{6.87}$$

The α are user-specified coefficients. To remove any roughness or oscillation, the unsmoothed $\mathcal{W}_b$ are smoothed using a procedure involving a modified Laplace filter.

$$\frac{\partial^2 \mathcal{W}_b}{\partial r^2} + \frac{\partial^2 \mathcal{W}_b}{\partial z^2} = 0 \tag{6.88}$$

The number of smoothing passes, N_{smth}, must be specified and is typically 8 to 10. By requiring that each smoothed weight function be subequidistributed, satisfying

$$\iint_b \mathcal{W}_b \, dr \, dz \leq C, \tag{6.89}$$

various cells may be automatically flagged for refinement (or for coarsening). Each constant C, one for each dependent variable, can be found by numerically integrating Equation (6.89) over the entire domain. Note that the overall adaption process terminates when the smoothed weight functions are subequidistributed to within 5%.

Refining a single cell entails placing a nine-point stencil within it, so that the existing corners of the box become points SW, SE, NW, and NE of the new stencil. Point P is added at the box's center, and, if necessary, some or all of points W, E, S, and N are also added, depending on the configuration of the surrounding grid. An example appears in Figure 6.2a. Along interfaces between different levels of grid refinement, some gridpoints (referred to as internal boundary points) do not have full nine-point stencils. To limit the rapidity with which grid spacing changes and, in turn, control the truncation error, LRR grids cannot have two or more adjacent internal boundary points. In addition, the boxes meeting at any single point are not allowed to differ by more than one grid level. Such grid imperfections can be remedied by additional selective refinements. The least upper bound (LUB) on the number of possible refinements caused by creation of a single level $\mathcal{M}$ stencil is

$$\text{LUB} = 3 \max\left(\mathcal{M} - 1, 0\right) + 5 \max\left(\mathcal{M} - 2, 0\right) \tag{6.90}$$

Application of these grid-structure constraints results in a limited number of allowable grid configurations.[40]

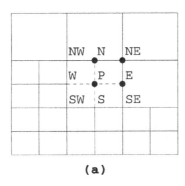

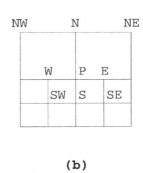

(a)   (b)

FIGURE 6.2 Examples of (a) the refinement of a single box within an existing LRR grid, in which the added points are represented by black circles, and (b) an LRR multiple-scale computational stencil.

If too small an area is refined, the front of interest may try to equilibrate at a position beyond the refined area, becoming falsely trapped at the coarse-fine interface. In addition, discretization errors at internal boundary points could be reduced if grid-level interfaces were made to occur further from the region of high gradients. Therefore, the region of refinement is extended through a layering process. Any cells that are within N_{layer} boxes of recently refined boxes are refined as well, thus surrounding the originally refined area with N_{layer} layers of refined cells. The layering parameter is calculated based on the user-specified number of layers of refinement $N_{layer,1}$ desired during the first adaption and on the adaption number itself.[40–42] In practice, typical $N_{layer,1}$ values do not exceed 5, and a value of 2 or 3 suffices for most problems.

The LRR method employs multiple-scale nine-point stencils, which capture the natural coupling between information of differing length scales by using dependent variable values at points belonging to different grid levels. At a given point P, the multiple-scale stencil's W (E, S, N) is the first point encountered when proceeding westward (eastward, southward, northward) from P. Point SW is the lower left corner of the box to the lower left of P, etc. Thus, the points closest to P comprise its multiple-scale stencil, an example of which is shown in Figure 6.2b. First and second derivative discretizations, as well as grouped discretizations (for derivatives such as $\partial/\partial x(f \partial g/\partial y)$) exist for each multiple-scale stencil.[40] At the internal boundary points, instead of employing commonly used spatial interpolants, one can discretize the governing equations using "pseudo-nine-point" stencils developed in References 40 to 42. Convective derivatives are upwinded throughout the grid.[40]

Interpolation is used to form an initial guess for the dependent variables on the new grid, on which the governing equations will be re-discretized and iteratively re-solved; interpolation is never used to produce a final solution. Because of the unstructured nature of the LRR grids, the Jacobian no longer retains the traditional block-nine-diagonal sparsity structure. However, the Jacobian's bandwidth is dramatically reduced by using the scheme incorporated into the LRR method, which numbers the points beginning in the grid's lower left corner and ending in the upper right. This ordering produces a matrix of small bandwidth and one fairly close in sparsity structure to the traditional block-nine-diagonal form, despite the grid's unstructured nature. The standard linear algebra solvers have been modified to accommodate the placement of these block "diagonals," because points frequently participate in more than nine stencils.

6.3.4. PARALLEL IMPLEMENTATION

The parallelization of a complex chemistry combustion code requiring large amounts of memory is an extremely difficult process. The approach the authors have used in solving the elliptic governing partial differential equations employs a combined time-dependent/Newton solution algorithm. This algorithm can be divided into two computationally intensive portions — the formation of the Jacobian matrix and the subsequent iterative solution of the Newton equations. The cost of adaptively refining the grid is negligible compared to these other two activities. In the parallelization strategy,[43] one exploits the block-sparse structure of the Jacobian, which has $9N$ nonzeros per row on average, where $N = 4 + K + M$ is the number of dependent variables. Hence, the Jacobian uses a considerable amount of memory compared to the solution vector.

The general computing methodology employed assumes that there is one processor that directs all of the processors to do various computing tasks. These tasks include having processors do their part of the parallel iterative solution, compute a Jacobian, evaluate a nonlinear residual, and form a matrix-vector multiply or inner product. A sparse-matrix domain decomposition method is used. The Jacobian matrix is considered a two-dimensional domain and is decomposed block-row by block-row, which corresponds to a strip domain decomposition method. In this way, only the unknowns associated with end blocks must be transferred between processors. The preconditioner is local to each processor. Therefore, the total number of iterations is not always the same as the serial computer equivalents.

Each block of the Jacobian is composed of similar subblocks, each of which is dense and of size $N \times N$. In two dimensions, formation of the Jacobian requires $9N + 1$ function evaluations. Given a solution estimate U, each function evaluation can be performed independent of the others. The $9N + 1$ function evaluations are partitioned across the $\mathcal{P}$ processors of the machine. In this way, each processor is able to compute a portion of the Jacobian matrix in parallel with the other processors. One advantage of the Jacobian having so many nonzeros is that one can afford to store the complete solution vector on a single processor without using much extra memory. If a particular operation is quick and does not parallelize well, one can actually gather all of the data, do the operation on a single node, and then scatter data back to the remaining processors.

Once the Jacobian is formed, the linear Newton equations are solved iteratively. As an example, consider a preconditioned Bi-CGSTAB procedure on the initial grid, which requires the formation of $\mathcal{P}$ preconditioners followed by the solution of block-line linear systems of equations of size $G_z/\mathcal{P}$. Each block-line system corresponds to a given axial ($j = 0, 1, \ldots, G_z$) grid line. After the system for each axial line is solved, the resulting solution information can be utilized in the solution of the system at the next axial level. Each processor works with $G_z/\mathcal{P}$ consecutive grid lines simultaneously. Solution information is passed from one line to another within the $G_z/\mathcal{P}$ grid line domains. After the systems within each domain are solved, the solution information is shared among the $\mathcal{P}$ processors and the next iteration begins.

6.4   NUMERICAL RESULTS

This section presents several numerical results obtained by applying the numerical procedures discussed in Section 6.3 to: (1) an unconfined, axisymmetric, methane/air diffusion flame; (2) an unconfined, axisymmetric, methane/air partially premixed flame; and (3) an unconfined, axisymmetric, sooting, ethylene/air diffusion flame. Complex transport and finite-rate chemistry models are used along with a vorticity-velocity formulation of the Navier-Stokes equations.

6.4.1   METHANE/AIR DIFFUSION FLAME

The first flame considered is an unconfined, methane/air diffusion flame. The experimental config-uration is such that the radius of the inner fuel jet R_I is 0.2 cm, and the radius of the outer oxidizer jet R_O is 2.54 cm. Fuel and nitrogen are introduced through the inner tube and air through the outer coflow. The boundary conditions at the inlet are as follows:

$r \leq R_I$

$$v_r = 0.0 \text{ cm/s}, \quad v_z = 35.0 \text{ cm/s}$$

$$Y_{CH_4} = 0.5149, \quad Y_{N_2} = 0.4851, \quad Y_k = 0.0000, \quad k \neq CH_4, N_2 \tag{6.91}$$

$$T = 298K$$

$R_I < r \le R_O$

$$v_r = 0.0 \text{ cm/s}, \quad v_z = 35.0 \text{ cm/s}$$

$$Y_{O_2} = 0.2320, \quad Y_{N_2} = 0.7680, \quad Y_k = 0.0000, \quad k \ne O_2, N_2 \qquad (6.92)$$

$$T = 298\text{K}$$

The boundary condition for the vorticity is based on its definition in terms of the velocity gradients. The flame-sheet model provides initial solution profiles for the two velocities, the vorticity, the temperature, and the major species (i.e., CH_4, O_2, N_2, CO_2, and H_2O). The flame-sheet starting estimate requires approximately 100 adaptive time steps and 5 Newton iterations to reduce the norm of the steady-state residuals below 1.0×10^{-3} on an initial 62×75 nonuniform, structured, rectangular tensor product grid covering a region of $0 \le r \le 7.50$ cm by $0 \le z \le 20.00$ cm. Because the flame will ultimately sit slightly above the fuel and oxidizer tubes, the imposed grid is spaced more finely in these regions.

Once the flame-sheet estimate is calculated, the full set of governing equations is solved in a two-step procedure. First, a solution to the governing Equations (6.24–6.27) is determined based on the flame-sheet temperature profile. The starting estimates for the minor species in the full-chemistry solution are approximated by Gaussian profiles centered at the location of the flame sheet on each axial level, with peak heights of at most a few percent. To conserve mass in the starting estimate, the N_2 mass fraction is reduced accordingly. This flame-sheet fixed-temperature solution is then used as input to the full fluid-dynamic-thermochemistry model Equations (6.24)–(6.28), in which the energy equation is included. This procedure reduces convergence difficulties as well as total CPU time, and it is similar to the two-pass solution method used in the solution of adiabatic premixed laminar flames.[44] After a converged full-chemistry solution is obtained on the initial grid, to Newton tolerance of 1.0×10^{-2}, the grid is refined using the LRR algorithm discussed in Section 6.3.3; the full-chemistry problem is then re-solved on the successive LRR grids.

For the LRR method, the number of weight-function smoothing passes N_{smth} is set to 10, and nonzero coefficients α in the weight function are set to 1.0×10^6. (A weight-function normalization procedure incorporated in the equidistribution part of the algorithm renders the adaptive grids largely insensitive to the specific values for the chosen nonzero α's, as long as each nonzero $\alpha \gg 1$.) Although the diffusion flame liftoff height H_f indeed changes as the grid undergoes refinement, $N_{layer,1} = 3$ is sufficient to allow the solution to restabilize on each new grid without interference between the flame front and the grid-level interfaces. This reasoning is based on the small role played by convection in the diffusion flame and the ensuing decreased effect of artificial diffusion.

On the Adaption 1 grid, the Newton tolerance is 2.5×10^{-3}, or one-fourth that of the initial grid. Adaption 2 is solved to a tolerance of 6.25×10^{-4}, or one-fourth of the previous grid's tolerance, etc. This procedure is justified on the basis that each grid is better able than the previous one to represent an accurate solution. As the computational grid is refined, Newton's method typically converges with a smaller number of time steps than on the coarser grids. Typically, two levels of refinement are needed to obtain a grid containing 15,000 points. The final grid spacing was such that over 12 million equispaced points would have been needed to obtain comparable accuracy. All computations were performed on an IBM RS/6000 Model 590 workstation with GRI-Mech version 2.11[45] without the NOx reaction set, resulting in a total of 35 dependent variables being solved at each gridpoint.

Figure 6.3 compares computed and experimentally determined temperature and major species profiles for the diluted methane/air diffusion flame. The experimental measurements were made using a difference spontaneous Raman scattering technique along with linear laser-induced fluorescence. Details of the experimental procedures can be found in Reference 46. The displayed maximum values for each experimental plot have been clipped to match those of its computational

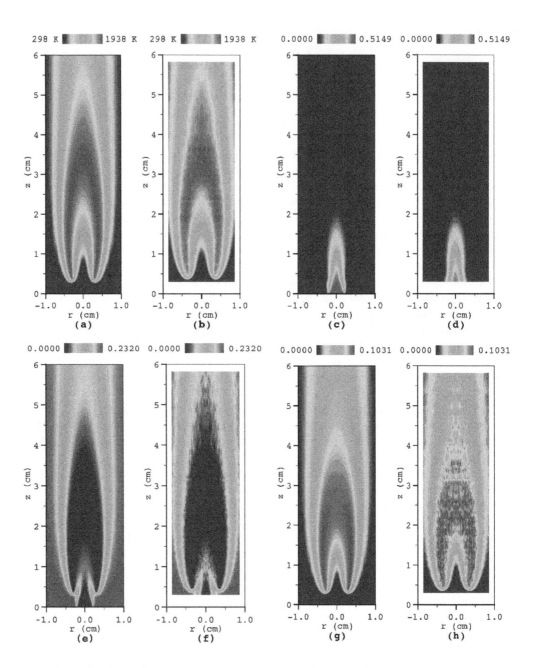

FIGURE 6.3 In the axisymmetric laminar diffusion flame, (a) computed and (b) experimental isotherms; (c) computed and (d) experimental isopleths of Y_{CH_4}; (e) computed and (f) experimental isopleths of Y_{O_2}; (g) computed and (h) experimental isopleths of Y_{H_2O}; (i) computed and (j) experimental isopleths of Y_{CO_2}; (k) computed and (l) experimental isopleths of Y_{H_2}; (m) computed and (n) experimental isopleths of Y_{CO}; and (o) computed and (p) experimental isopleths of Y_{N_2} are shown in a portion of the computational domain. Numerical results have been computed using GRI-Mech without NOx chemistry, on an LRR adapted grid formed with Y_{OH} as the adaption variable. Experimental data have been clipped so that their maxima match those of the computed data. For each plot, the displayed color scale is linear. (From Bennett, B.A.V. and Smooke, M.D., *Combust. Theory Model.*, 2, 221, 1998. With permission.)

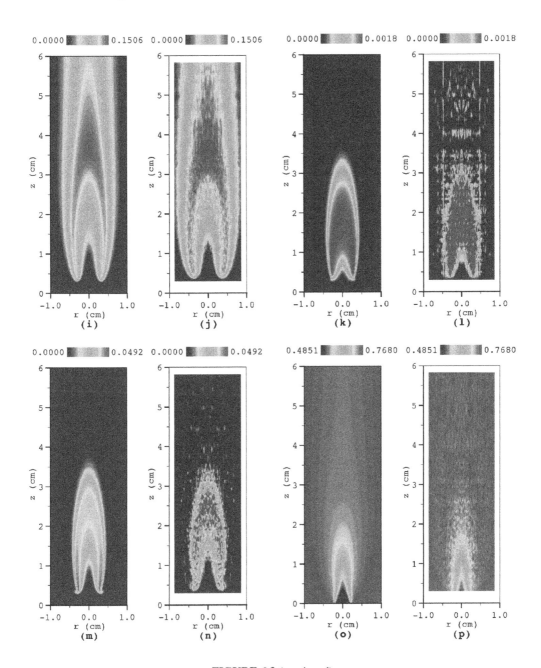

FIGURE 6.3 (continued)

counterpart. In addition, the experimental imaging technique did not allow data acquisition within 3 mm of the burner surface, which resulted in a lack of data in the lower region of each experimental figure. The presented computational data are those calculated on the final LRR grid — an Adaption 2 grid generated using Y_{OH} as the adaption variable.

Figures 6.3a,b illustrate the computational and experimental isotherms as functions of r and z. Note the high-temperature region beginning several millimeters above the burner surface, clearly illustrating the lifted diffusion flame structure and the presence of a low-temperature core above the burner along the axis of symmetry. These figures also indicate the very high temperature gradients directly above the burner surface, where the temperature rises from 298K to nearly 2000K over approximately 0.8 mm. It is in this region that the fuel and the oxidizer first meet in stoichiometric proportion, and combustion occurs only in a thin region above the inlet.

An important aspect of the lifted diffusion flame is the triple flame present at its base. Methane diffuses rapidly toward the flame front, where it is almost completely consumed, but a small amount diffuses and convects outward from the leading edge of the flame. At the same time, a substantial amount of oxygen penetrates into the fuel region. These occurrences yield a triple-flame structure consisting of a fuel-lean premixed flame F_{LP}, a fuel-rich premixed flame F_{RP}, and a diffusion flame F_D. The three flames originate from a common junction point and the diffusion flame trails the two premixed flames, as illustrated in Figure 6.1. The penetration of fuel into the oxidizer zone and that of oxidizer into the fuel zone is illustrated in Figures 6.3c,d and 6.3e,f, respectively, where the mass fraction isopleths for methane (CH_4) and oxygen (O_2) mass fractions are presented. Along the centerline, the CH_4 mass fraction displays a monotonic decrease with height; and at heights greater than 3 cm, the methane has been completely burnt out. On the other hand, a substantial amount of oxygen penetrates inside the fuel region. Along the centerline, the O_2 mass fraction peaks at a value of $\approx 6.0 \times 10^{-2}$ for $z = 1.2$ cm and then decreases up to $z = 3.3$ cm. From $z = 2.4$ cm until $z = 3.8$ cm, no oxygen is present in any amount larger than a few parts per thousand. Further downstream, the O_2 mass fraction exhibits a monotonic increase due to diffusion from the surrounding air — an increase that starts slightly before the flame closes and extends up to the outflow boundary.

Figures 6.3g,h; 6.3i,j; 6.3k,l; and 6.3m,n illustrate computational and experimental isopleths for water (H_2O), carbon dioxide (CO_2), hydrogen (H_2), and carbon monoxide (CO), respectively. These figures reveal that large quantities of H_2O, CO, and H_2 are produced soon after the methane has been consumed. In this region, the methane is attacked by O, H, and OH radicals to form methyl (CH_3); only small amounts of OH, H, and O exist here due to the high affinity of methane for these radicals. The oxidation of formaldehyde (CH_2O) to HCO and the subsequent formation of CO occur in regions of high methyl and formaldehyde concentrations.

The oxidation of CO to CO_2 proceeds primarily via the reaction

$$CO + OH \rightarrow CO_2 + H \tag{6.93}$$

Hence, the rate of CO oxidation depends on the availability of OH radicals. However, as Westbrook and Dryer point out,[47] the presence of most hydrocarbon species inhibits the oxidation of CO. This behavior can be attributed to the fact that the rate of the reaction

$$H + O_2 \rightarrow OH + O \tag{6.94}$$

is considerably smaller than the reaction rates of H atoms with hydrocarbon species, and the rate of the CO oxidation reaction is also smaller than the reaction rates of hydrocarbon species with OH. As a result, small quantities of hydrocarbons can effectively restrict the oxidation of CO to CO_2. Although CO and H_2 are formed during the oxidation of the hydrocarbon species, it is not until after the hydrocarbons and their fragments have been consumed that the OH level rises and CO_2 is formed. Observe in Figure 6.3i,j that CO_2 forms downstream of the regions of high CO concentration.

6.4.2 METHANE/AIR PARTIALLY PREMIXED FLAME

The second problem examined is that of a methane/air partially premixed flame, formed when a rich mixture of methane and primary air (34% methane and 66% air, by volume) flows from the central burner tube, and so-called secondary air flows from the annular region between the central and outer tubes. The primary air flowrate required for complete combustion, divided by the *actual* primary air flowrate (referred to as the primary equivalence ratio Φ), is 4.1. The flame is confined, meaning that a cylindrical shield is placed around the flame at a radius equal to the outer diameter of the outer tube. No-slip boundary conditions are employed at the shield's surface. At the burner surface (the inlet), the inner jet has a plug flow velocity profile of 16.5 cm/s, and the coflow velocity is also flat, fixed at 10.5 cm/s.

The inlet boundary conditions are the same as those specified earlier, except for the species boundary condition, given by a conservation of species mass, as follows:

$$\rho v_z \left(Y_k - Y_{k,B} \right) = \frac{1}{\mathrm{Le}_k} \left(\frac{\lambda}{c_p} \right) \frac{\partial Y_k}{\partial z} \tag{6.95}$$

where v_z is either 16.5 cm/s or 10.5 cm/s, $Y_{k,B}$ is the mass fraction of the kth species at the burner surface, and Le_k is the Lewis number of the kth species. Temperature and species concentrations in this flame have also been measured experimentally along the axis of symmetry, using both probe (thermocouple and gas-sampling techniques) and optical diagnostic methods (laser-induced fluorescence, LIF).[48]

All computations were performed on an IBM RS/6000 Model 590 workstation and the chemical kinetics were modeled via GRI-Mech version 2.11[45] without the NOx reaction set, resulting in a total of 35 dependent variables being solved at each gridpoint. The initial grid is a nonequispaced tensor product grid of size 66 × 84, with finer spacing in the region immediately above the burner surface than in, for example, parts of the domain very far removed from the inlet. Specifically, this initial grid has $\Delta r = 0.02$ cm for $0 \leq r \leq 0.90$ cm, with increasingly larger spacing for $0.90 \leq r \leq 5.1$ cm; in the axial direction, $\Delta z = 0.03$ cm from $z = 0$ to $z = 0.90$ cm, with increasingly larger spacing for 0.90 cm $\leq z \leq 20.0$ cm. Developing an acceptable starting estimate for the steady-state Newton's method on this grid is difficult because of the strong nonlinearities in the governing equations, so a flame-sheet starting estimate is used, in the same way was described in Section 6.4.1, to aid in developing a converged full-chemistry solution on the initial grid.

Once this initial-grid solution is obtained, the LRR method described in Section 6.3.3 is applied to refine the grid automatically, based on gradients of Y_{CH_4} and $Y_{C_2H_6}$; the majority (roughly 95%) of the adaption occurs because of gradients in the former. These adaption variables are chosen so that the refinement primarily occurs in the fuel-consumption region. The governing equations are re-discretized on the adapted grid and re-solved, etc. The Adaption 0, Adaption 1, and Adaption 2 grids are displayed in the left halves of Figures 6.4a,b,c, respectively, with the corresponding CH_4 mass fraction isopleths displayed in the right halves. It is apparent that refinement occurs in the regions of high solution activity, and not as visible is the fact that the flame length increases slightly as the grid undergoes refinement. This latter behavior results from the fact that as the grid spacing decreases, the upwinded discretizations for the convective terms become more accurate (less artificial diffusion). The *rate* of flame-length increase decreases, indicating that the value is approaching an asymptote. Much more acute displays of this behavior are observed in References 41 and 49 for flames with much smaller primary equivalence ratios; the authors conclude that the most accurate flame length is that represented on Adaption 2 grid.

Figure 6.5 presents computed isopleths of heat release and temperature, as well as those of some major and minor species concentrations, in a portion of the computational domain, for the methane/air partially premixed flame. In Figure 6.5a, the maximum heat release occurs in a small annular region just above the burner surface, where the heat release rate is an order of magnitude larger than anywhere else in the domain. Therefore, the chosen color scale runs from the minimum value to one-tenth of the maximum, in order to make the structure of the other regions more visible. Heat is released in two bands that approach the centerline near $z = 4.50$ cm; this region of heat

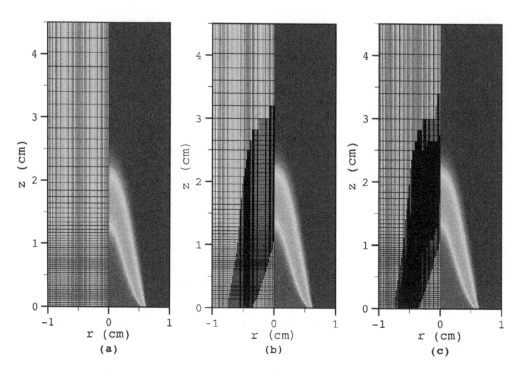

FIGURE 6.4 The left half of each figure displays a portion of the grid generated for (a) Adaption 0, (b) Adaption 1, and (c) Adaption 2, and the right half of each figure shows isopleths of Y_{CH_4}, ranging from computed minimum (0.0000) to computed maximum (0.2241).

release is referred to as the outer flame front, which is a nonpremixed flame with negligible heat release on the centerline. An inner (premixed) flame front is also present, intersecting the centerline at $z = 2.30$ cm — approximately half the height of the outer flame front. Unlike the outer non-premixed flame, the inner flame front's highest heat release rate occurs near the centerline.

From the isotherms of Figure 6.5b, one can see that the high-temperature region has a "wishbone" structure. The temperature gradient is steep at the conical inner flame front, and one notes a corre-spondence in centerline axial location between the peak centerline temperature and the outer flame front. Figures 6.5c and 6.5d show isopleths of CH_4 and O_2 mass fractions, respectively. These species co-exist in a conical region between the burner surface and the inner (premixed) flame front. Methane disappears rapidly over a narrow region near the inner flame front. Isopleths of OH mass fraction appear in Figure 6.5f. Because CH_4, H_2, C_2H_2, and other hydrocarbons react quickly with OH, O, and H radicals,[47] negligible OH is present in the fuel-rich region due to the high OH destruction rate. It is only in the vicinity of the non-premixed flame front that the OH production finally exceeds its destruction. The HCO isopleths presented in Figure 6.5g are shown on a color scale that makes their spatial structure more visible, running from the minimum value to one-tenth of the maximum, as has been already done for the heat release of Figure 6.5a. One can see that HCO is a good flame front indicator[50] for both the outer non-premixed flame and the inner premixed flame.

In Figure 6.6, the computations can be compared with the experimental data along the axis of symmetry of the configuration. Figure 6.6a displays the axial temperature profile along the center-line. The peak temperatures differ by about 100 K, which is consistent with previously observed behavior, and the shape of the profile (a sharp rise with a "shoulder" occurring near the inner flame front, a second peak near the outer flame front, and a subsequent falloff) is represented in both computation and experiment. In Figure 6.6b, CH_4 mole fractions are shown; computational mass

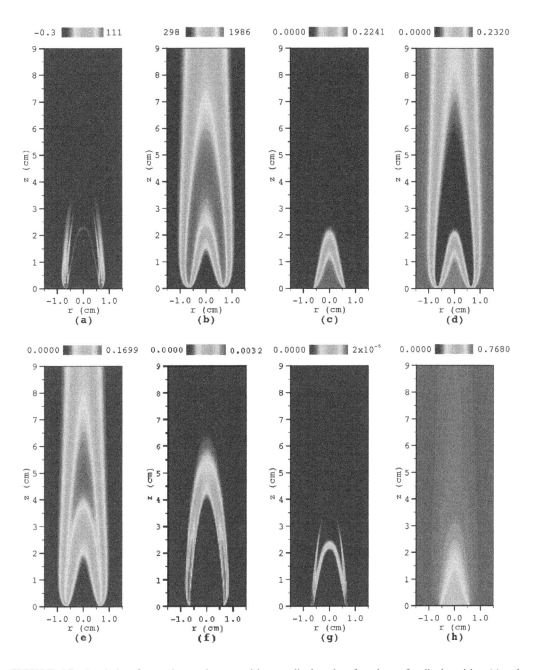

FIGURE 6.5 Isopleths of some interesting quantities are displayed as functions of radical position (r) and axial position (z) in a portion of the computational domain. Shown are (a) isopleths of heat release in W/cm^3; (b) isotherms in K; (c) isopleths of CH_4 mass fraction (Y_{CH_4}); (d) isopleths of O_2 mass fraction (Y_{O_2}); (e) isopleths of CO_2 mass fraction (Y_{CO_2}); (f) isopleths of OH mass fraction (Y_{OH}); (g) isopleths of HCO mass fraction (Y_{HCO}); and (h) isopleths of N_2 mass fraction (Y_{N_2}). In each case except (a) and (g), the plotted isopleths range from the computed minimum to the computed maximum. Note that for each of (a) and (g), the plotted isopleths range from the computed minimum to *one-tenth* of the computed maximum.

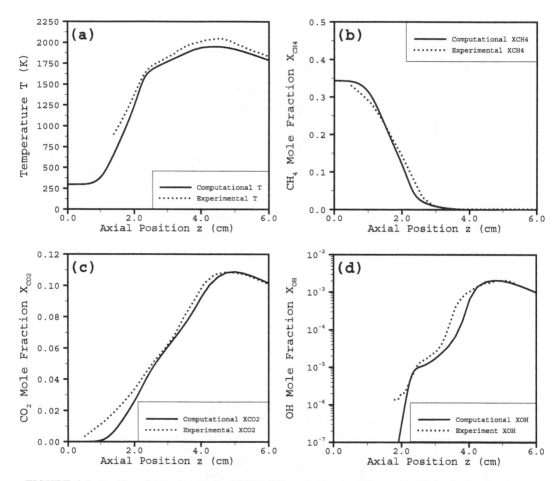

FIGURE 6.6 Profiles of (a) temperature (T); (b) CH_4 mole fraction (X_{CH_4}); (c) CO_2 mole fraction (X_{CO_2}); and (d) OH mole fraction (X_{OH}) are shown along the flame centerline, as functions of the axial position z above the burner surface. Computations are denoted by solid curves and experiments by dotted curves.

fractions have been converted to mole fractions for easier comparison with experimental data. Very good agreement can be observed, with the CH_4 vanishing near the inner flame front in both profiles. Figure 6.6c illustrates the CO_2 mole fractions; again, the shapes of the profiles agree quite well, although the computational CO_2 peaks further downstream than the experimental CO_2. Finally, the OH mole fraction profiles in Figure 6.6d are very similar in shape, with the computational profile shifted slightly downstream of the experimental profile. The computation clearly captures the profile's shoulder near the inner flame front, as well as the magnitude of the peak at the outer flame front.

6.4.3 ETHYLENE/AIR DIFFUSION FLAME

The final problem considered is a set of sooting, ethylene diffusion flames generated with a burner in which the fuel flows from an uncooled, 4.0-mm inner diameter, vertical brass tube (wall thickness 0.038 mm) and the oxidizer flows from the annular region between this tube and a 50-mm diameter concentric tube. The oxidizer is air, while the fuel is a mixture containing ethylene (C_2H_4) and nitrogen. Both probe (thermocouple and gas-sampling techniques) and optical diagnostic methods (Rayleigh scattering and laser-induced incandescence, LII) have been used to measure the temperature, gas species, and soot volume fractions.[51] The chemical kinetic mechanism for ethylene

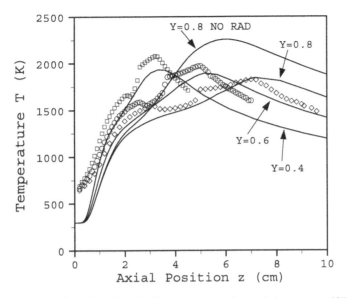

FIGURE 6.7 Comparison of predicted (solid lines) and experimental (squares = 40%, circles = 60%, diamonds = 80%) centerline temperatures. The radiation-free 80% case is also included.

combustion has 52 species and 255 reactions and is an expanded version of that used in Reference 51. It is based on comparisons to experimental ethylene data from perfectly stirred reactors, flow reactors, and ignition delay data. The mechanism includes reactions describing the formation and oxidation of benzene (C_6H_6) and related species.

Ethylene and nitrogen are introduced through the center tube, utilizing a parabolic velocity profile whose average velocity is 35 cm/s, and air is introduced through the outer coflow with a plug flow (flat) velocity profile at 35 cm/s. These profiles are also employed in the experiments. The ethylene mole fractions at the burner exit are 0.4, 0.6, and 0.8 for each of the three flames, with nitrogen as the remaining species. Reactant temperatures are assumed to be 298K, and all radial velocities are assumed to vanish at the flame base. Calculations were performed on an IBM RS/6000 Model 590 computer. In the computations presented, 20 soot size classes are included in the model, with approximately 10,000 adaptively placed gridpoints. Starting from a converged solution for an ethylene/air flame without the sectional equations, one typically obtains converged solutions for the complete gas-soot problem in several hours of computer time.

Figure 6.7 illustrates temperatures as functions of height along the centerline for these three flames. The high experimental (thermocouple) temperatures very close to the burner are likely a result of limitations in the experimental methods at these locations. Also shown in the figure is the centerline profile for the 80% ethylene case in which radiation is ignored. Nonadiabatic radiative loss (primarily from soot, in this case) drastically modifies the temperature profile and indicates that radiative loss lengthens the flame by about 15%. A comparison of isotherms for the 60% flame is provided in Figure 6.8.

Acetylene (C_2H_2) is a principal surface-growth species and also contributes to inception. Accurate simulation of acetylene is critical to predictions of soot. Comparisons of acetylene along the centerline are shown in Figure 6.9. They are well simulated, and the changes with decreasing dilution are also reproduced fairly closely. The comparison between the computed and measured benzene concentrations is acceptable for the 40% flame; but as the level of dilution decreases, the disagreement between the experimental measurements and the computations increases. It is most likely the case that interactions with larger aromatic species or soot contribute to the disagreement in the benzene concentrations. The inability of the model to predict this feature may contribute to an underprediction of soot along the centerline of the flame.

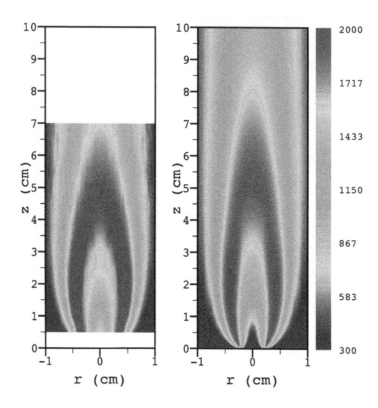

FIGURE 6.8   Experimental (left) and computational (right) isotherms for the 60% ethylene flame.

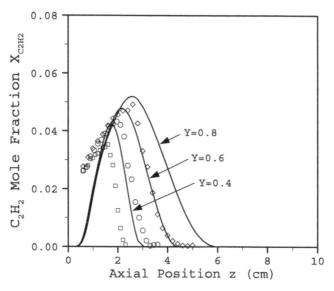

FIGURE 6.9   Predicted (solid lines) and experimental (squares = 40%, circles = 60%, diamonds = 80%) centerline acetylene mole fractions.

Finally, Figure 6.10 compares the experimental and theoretical soot volume fractions for the 40%, 60%, and 80% flames. Experimentally, a dramatic shift in the location of maximum soot away from the centerline to the wings occurs as the ethylene is increased. The model, although it

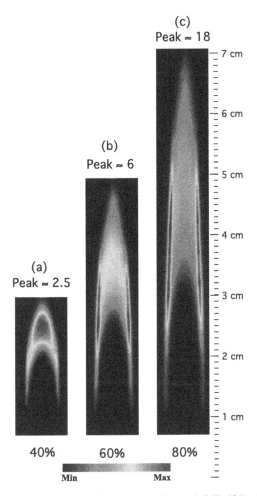

FIGURE 6.10 Soot volume fraction contours from (a) experimental (LII) 40%; (b) experimental (LII) 60%; (c) experimental (LII) 80%; (d) computed 40%; (e) computed 60%; and (f) computed 80%.

underpredicts the relative amount of soot on the centerline, agrees qualitatively with this trend. Soot formed in the wings of the flame remains inside the flame front due to thermophoretic forces and continues to grow along the full length of the flame. Thus, as the flame lengthens due to decreased levels of dilution, soot levels in the wings increase substantially because of both the increased residence time and the increased levels of benzene. Uncertainties in the prediction of benzene and polyaromatic hydrocarbons are major contributors to the uncertainty in the prediction of soot along the centerline and to the prediction of the peak soot volume fraction in the wings. Additional research is needed to fully understand this problem.

6.5 CONCLUSIONS

Recent advances in the development of computational algorithms, reaction kinetics, and mainframe supercomputers have enabled the combustion scientist to investigate chemically reacting systems that were computationally infeasible only a few years ago. In particular, the coupling of local adaptive numerical methods with large-memory parallel computers has produced an extremely powerful tool with which to probe flame structure. The difficulties in solving high heat release combustion problems, such as flames, center on the large number of equations that must be solved for the elementary chemical species, the exponential nonlinearities that occur in the governing

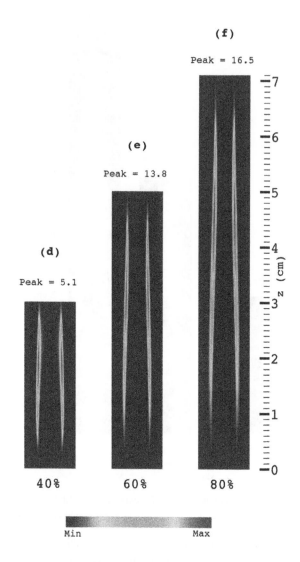

FIGURE 6.10 (continued)

partial differential equations, and the disparate length scales that must be resolved in the computed solution. As a result, the modeling of chemically reacting flows generally has proceeded along two independent paths. In one case, chemistry is given priority over fluid mechanical effects and these models are used to assess the important elementary reaction paths in, for example, hydrocarbon fuels. In the other case, multidimensional fluid-dynamical effects are emphasized, with chemistry receiving little priority. While the ultimate goal in combustion modeling is the solution of three-dimensional turbulent reacting flows with finite-rate chemistry, there are still important, less complex systems that can be analyzed in detail with current computational resources. Axisymmetric laminar non-premixed and partially premixed flames are examples of such systems. They are the flame type of most practical combustion devices.

    This chapter has discussed the numerical solution of these systems with detailed transport and finite-rate chemistry. It has focused on the formulation of the problem in terms of the fluid-dynamic and thermochemistry governing equations and indicated the various transport, chemistry, and thermodynamic submodels needed in the complete specification of the problem. As these models are highly nonlinear, starting estimates play an important role in the efficient solution of the equations. The authors have employed a flame-sheet model to generate an initial starting estimate for the velocities, vorticity, temperature, and the major species, and have indicated how this estimate is utilized in the solution of a finite-rate chemistry problem. Computationally, the authors have discretized the governing equations initially on a structured tensor product grid, and employed Newton's method to solve the resulting system of nonlinear algebraic equations. Theoretical estimates along with various Jacobian approximations and preconditioned iterative linear equation solvers were used to increase the efficiency with which the Newton equations are solved. Once a solution has been obtained on a specified initial grid, the authors utilized local grid refinement techniques to increase the resolution and accuracy of the solutions. The authors also examined both non-premixed and partially premixed flames and have compared the computations with experimental data.

The model and solution algorithm discussed in this chapter can be easily adapted to more complex fuels whose kinetic models contain a higher number of chemical species and elementary reactions. Finally, the authors anticipate that within the next 3 to 5 years, fully three-dimensional models of flame structure with detailed transport and chemistry will become a reality.

ACKNOWLEDGMENTS

This work was supported in part by the United States Department of Energy, Office of Basic Energy Sciences, NASA, and the Air Force Office of Scientific Research. The authors are indebted to colleagues Dr. M.B. Colket, Dr. A. Ern, Dr. V. Giovangigli, Dr. R.J. Hall, Dr. M.B. Long, Dr. D.F. Marran, Dr. C.S. McEnally, and Dr. L.D. Pfefferle for their numerous helpful discussions and suggestions.

REFERENCES

1. Sax, N.R., *Dangerous Properties of Industrial Materials,* Reinhold, New York, 1968.
2. Rogg, B., Behrendt, F., and Warnatz, J., Turbulent non-premixed combustion in partially premixed diffusion flamelets with detailed chemistry, *Twenty-First Symposium (International) on Combustion,* Reinhold, New York, 1986, 1533.
3. Patankar, S.V., *Numerical Heat Transfer and Fluid Flow,* McGraw-Hill, New York, 1980.
4. Xu, Y. and Smooke, M.D., Application of a primitive variable Newton's method for the calculation of axisymmetric laminar diffusion flames, *J. Comput. Phys.,* 104, 99, 1993.
5. Mohammed, R.K., Tanoff, M.A., Smooke, M.D., Schaffer, A.M., and Long, M.B., Computational and experimental study of a forced, time-varying, axisymmetric, laminar diffusion flame, *Twenty-Seventh Symposium (International) on Combustion,* The Combustion Institute, Pittsburgh, 1998, 693.
6. White, F.M., *Fluid Mechanics,* McGraw-Hill, New York, 1986, 222.
7. Xu, Y., Numerical Calculations of an Axisymmetric Laminar Diffusion Flame with Detailed and Reduced Kinetics, Ph.D. thesis, Yale University, 1991.
8. Ern, A. and Smooke, M.D., Vorticity-velocity formulation for three-dimensional steady compressible flows, *J. Comput. Phys.,* 105, 58, 1993.
9. Ern, A., Vorticity-Velocity Modeling of Chemically Reacting Flows, Ph.D. thesis, Yale University, 1994.
10. Gatski, T.B., Review of incompressible fluid flow computations using the vorticity-velocity formulation, *Appl. Numer. Math,* 7, 227, 1991.
11. Kaplan, C.R., Shaddix, C.R., and Smyth, K.C., Computations of enhanced soot production in time-varying CH_4/air diffusion flames, *Combust. Flame,* 106, 392, 1996.
12. Kennedy, I.M., Rapp, D.R., Santoro, R.J., and Yam, C., Modeling and measurements of soot and species in a laminar diffusion flame, *Combust. Flame,* 107, 386, 1996.

13. Colket, M.B. and Hall, R.J., Successes and uncertainties in modeling soot formation in laminar, premixed flames, in *Soot Formation in Combustion, Mechanisms and Models,* H. Bockhorn, Ed., Springer Series in Chemical Physics, 59, Springer-Verlag, 1994, 442.

14. Hall, R.J., Smooke, M.D., and Colket, M.B., Predictions of soot dynamics in opposed jet diffusion flames, in *Physical and Chemical Aspects of Combustion: A Tribute to Irvin Glassman,* R.F. Sawyer and F.L. Dryer, Eds., Combustion Science and Technology Book Series, Gordon and Breach, 1997.

15. Gelbard, F. and Seinfeld, J.H., Simulation of multicomponent aerosol dynamics, *J. Coll. Int. Sci.,* 78, 485, 1980.

16. Curtiss, C.F. and Hirschfelder, J.O., Transport properties of multicomponent gas mixtures, *J. Chem. Phys.,* 17, 550, 1949.

17. Giovangigli, V. and Darabiha, N., Vector computers and complex chemistry combustion, in *Proc. Conf. Mathematical Modeling in Combustion,* Lyon, France, NATO ASI Series, 1987.

18. Hall, R.J., The radiative source term for plane-parallel layers of reacting combustion gases, *J. Quant. Spec. Rad. Tran.,* 49, 517, 1993.

19. Hall, R.J., Radiative dissipation in planar gas-soot mixtures, *J. Quant. Spec. Rad. Tran.,* 51, 635, 1994.

20. Keyes, D.E. and Smooke, M.D., Flame sheet starting estimates for counterflow diffusion flame problems, *J. Comput. Phys.,* 73, 267, 1987.

21. Burke, S.P. and Schumann, T.E.W., Diffusion flames, *Industrial Engineering Chemistry,* 29, 998, 1928.

22. Mitchell, R.E., Sarofim, A.F., and Clomburg, L.A., Experimental and numerical investigation of confined laminar diffusion flames, *Combust. Flame,* 37, 227, 1980.

23. Smooke, M.D., Mitchell, R.E., and Grcar, J.F., Numerical solution of a confined laminar diffusion flame, in *Elliptic Problem Solvers II,* G. Birkhoff and A. Schoenstadt, Eds., Academic Press, New York, 1984, 557.

24. Kanury, A.M., *Combustion Phenomena,* Gordon and Breach, New York, 1982.

25. Deuflhard, P.A., A modified Newton method for the solution of ill-conditioned systems of nonlinear equations with application to multiple shooting, *Numer. Math.,* 22, 289, 1974.

26. Curtis, A.R., Powell, M.J., and Reid, J.K., On the estimation of sparse Jacobian matrices, *J. Inst. Math. Appl.,* 13, 117, 1974.

27. Newsam, G.N. and Ramsdell, J.D., Estimation of sparse Jacobian matrices, *Harvard Univ. Rep.,* TR-17-81, 1981.

28. Coleman, T.F. and More, J.J., Estimation of sparse Jacobian matrices and graph coloring problems, *Argonne National Laboratory Report* ANL-81-39, 1981.

29. Smooke, M.D. and Mattheij, R.M.M., On the solution of nonlinear two-point boundary value problems on successively refined grids, *Appl. Num. Math.,* 1, 463, 1985.

30. Kantorovich, L.V. and Akilov, G.P., *Functional Analysis in Normed Spaces,* Pergamon Press, New York, 1964.

31. Smooke, M.D., An error estimate for the modified Newton method with applications to the solution of nonlinear two-point boundary value problems, *J. Opt. Theory and Appl.,* 39, 489, 1983.

32. Saad, Y. and Schultz, M.H., GMRES: a generalized minimum residual algorithm for solving nonsymmetric linear systems, *SIAM J. Sci. Stat. Comput.,* 7, 856, 1986.

33. van der Vorst, H.A., Bi-CGSTAB: a fast and smoothly converging variant of Bi-CG for the solution of nonsymmetric linear systems, *SIAM J. Sci. Stat. Comput.,* 13, 631, 1992.

34. Ern, A., Giovangigli, V., Keyes, D.E., and Smooke, M.D., Towards polyalgorithmic linear system solvers for nonlinear elliptic problems, *SIAM J. Sci. Comput.,* 15, 681, 1994.

35. Smooke, M.D., Miller, J.A., and Kee, R.J., Solution of premixed and counterflow diffusion flame problems by adaptive boundary value methods, in *Numerical Boundary Values ODEs,* U.M. Ascher and R.D. Russell, Eds., Birkhäuser, Boston, 1985, 303.

36. Coelho, P.J. and Pereira, J.C.F., Calculation of a confined axisymmetric laminar diffusion flame using a local grid refinement technique, *Combust. Sci. Tech.,* 92, 243, 1993.

37. Mallens, R.M.M., de Lange, H.C., van de Ven, C.H.J., and de Goey, L.P.H., Modeling of confined and unconfined laminar premixed flames on slit and tube burners, *Combust. Sci. Tech.,* 107, 387, 1995.

38. Pember, R.B., Howell, L.H., Bell, J.B., Colella, P., Crutchfield, W.Y., Fiveland, W.A., and Jessee, J.P., An adaptive projection method for unsteady, low-Mach number combustion, *Combust. Sci. Tech.,* 140, 123, 1998.

39. Day, M.S. and Bell, J.B., Numerical simulation of laminar reacting flows with complex chemistry, *Combust. Theory Model.,* submitted, 2000.

40. Valdati, B.A., Solution-Adaptive Gridding Methods with Application to Combustion Problems, Ph.D. thesis, Yale University, 1997.

41. Bennett, B.A.V. and Smooke, M.D., Local rectangular refinement with application to axisymmetric laminar flames, *Combust. Theory Model.,* 2, 221, 1998.

42. Bennett, B.A.V. and Smooke, M.D., Local rectangular refinement with application to nonreacting and reacting fluid flow problems, *J. Comput. Phys.,* 151, 684, 1999.

43. Ern, A., Douglas, C.C., and Smooke, M.D., Detailed chemistry modeling of laminar diffusion flames on parallel computers, *Int. J. Supercomp. Appl.,* 9, 167, 1995.

44. Smooke, M.D., Miller, J.A., and Kee, R.J., Determination of adiabatic flame speeds by boundary value methods, *Combust. Sci. Tech.,* 34, 79, 1983.

45. Bowman, C.T., Hanson, R.K., Davidson, D.F., Gardiner, Jr., W.C., Lissianski, V., Smith, G.P., Golden, D.M., Frenklach, M., Wang, H., and Goldenberg, M., *GRI-Mech* version 2.11, http://www.gri.org, 1995.

46. Marran, D.F., Quantitative Two-dimensional Laser Diagnostics in Idealized and Practical Combustion Systems, Ph.D. thesis, Yale University, 1997.

47. Westbrook, C.K. and Dryer, F.L., Chemical kinetic modeling of hydrocarbon combustion, *Prog. Energy Combust. Sci.,* 10, 1, 1984.

48. McEnally, C.S. and Pfefferle, L.D., Experimental study of nonfuel hydrocarbon concentrations in coflowing partially premixed methane/air flames, *Combust. Flame,* 118, 619, 1999.

49. Bennett, B.A.V. and Smooke, M.D., A comparison of the structures of lean and rich axisymmetric laminar Bunsen flames: application of local rectangular refinement solution-adaptive gridding, *Combust. Theory Model,* 3, 657, 1999.

50. Najm, H.N., Paul, P.H., Mueller, C.J., and Wyckoff, P.S., On the adequacy of certain experimental observables as measurements of flame burning rate, *Combust. Flame,* 113, 312, 1998.

51. McEnally, C.S., Schaffer, A.M., Long, M.B., Pfefferle, L.D., Smooke, M.D., Colket, M.B., and Hall, R.J., Computational and experimental study of soot formation in a coflow, laminar ethylene diffusion flame, *Twenty-Seventh Symposium (International) Combustion,* The Combustion Institute, Pittsburgh, 1998, 1497.

Section II

Industrial Applications

7 CFD in Burner Development

Vladimir Y. Gershtein and Charles E. Baukal, Jr.

CONTENTS

7.1   INTRODUCTION

7.1.1   BURNER DEVELOPMENT

A combustion system consists of the combustor, the heat load, the heat generator, and in some cases a heat recuperation system. In industrial combustion systems, the heat generator consists of one or more burners where a fuel is combusted with an oxidizer ranging from air to pure oxygen. The effects of the oxidizer composition are considered in this chapter. There are many types of burners, which are tailored to meet the specific needs of the heat load and the combustor. With the increasing emphasis on reducing pollutant emissions and increasing thermal efficiency, which are often at odds with each other, there has been a need for continuous improvement in the design of burners. The various methods used to design burners are briefly considered in this chapter. The increasing demands on burner performance have caused an increase in the use of advanced computational techniques to design and optimize burners in a wide range of industrial applications.

7.1.2 MODELING BURNERS

Up until fairly recently, there has not been computer power nor enough models developed to simulate full-scale industrial combustion problems. These simulations are among the most difficult to make as they combine complicated turbulent fluid dynamics, both fine and coarse length scales, nonlinear spectral radiant heat transfer, numerous chemical reactions with many species, possibly multiple phases, and complex geometries that may include porous media. The dramatic improvements in both computer hardware and computational fluid dynamics models have led to numerous papers on modeling industrial burners. A sampling of references are given for modeling industrial burners:

- Radiant tube burners[1-3]
- Swirl burners[4]
- Pulse combustion burner[5]
- Porous radiant burners[6-11]
- An industrial hydrogen sulfide burner[12]

Butler et al. (1986) gave a general discussion of modeling burners using the finite-volume technique.[13] Schmücker and Leyens (1998) described the use of CFD to design a new nozzle-mix burner referred to as the Delta burner.[14] Schmidt et al. (1998) described the use of CFD to redesign a burner, originally firing on coal, to fire on natural gas for use in a rotary kiln.[15]

Modeling radiant burners poses the additional challenge of simulating a porous medium.[16] Perrin et al. (1986) discussed the use of a numerical model for the design of a single-ended radiant tube for immersion in and heating of a bath of molten zinc.[17] Hackert et al. (1998) simulated the combustion and heat transfer in two-dimensional porous burners.[18] Fu et al. (1998) used a one-dimensional model to simulate the performance of a porous radiant burner.[19]

Weber et al. (1993) classified models for designing industrial burners into three categories.[20] First-order methods give rough qualitative estimates of heat fluxes and flame shapes. Second-order methods give higher accuracy results than first-order methods for temperature, oxygen concentration, and heat flux. Third-order methods further improve accuracy over second-order methods and give detailed species predictions in the flame that are useful for pollutant formation rates. The order used will in large part depend on the information and accuracy that are needed.

7.2 OVERVIEW OF BURNERS

There are many different types of burners and there are also many ways of classifying them. Some methods to classify burners include by:

- Design (porous radiant, swirl, etc.)
- Mixing method (premixed, partial premixed, or diffusion)
- Burner location (hearth, wall, roof, etc.)
- Oxidizer supply method (forced draft or natural draft)

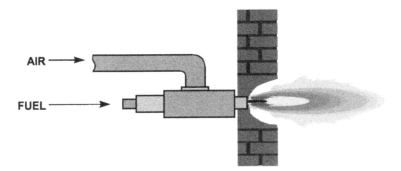

FIGURE 7.1 Schematic of an air/fuel, premix burner. (Courtesy of CRC Press LLC, Boca Raton, FL.)

- Flame shape (round, flat, etc.)
- Fuel (gaseous, liquid, solid, mixed)
- Emissions performance (low NOx, ultra-low NOx, etc.)
- Method of heat recuperation (none, furnace gas recirculation, external flue gas recirculation)
- Oxidizer temperature (non-preheated or preheated)

Here, they are classified by the type of oxidizer that is used, as the burner design examples given later relate to the oxidizer composition.

7.2.1 AIR/FUEL BURNERS

Figure 7.1 shows a cartoon of a typical air/fuel burner. Air/fuel burners are the workhorses of industry and are used in the vast majority of applications. Forced-draft burners, where the combustion air is supplied by a fan or blower, are most commonly used. The cost of the air is minimal and involves only the cost of the electricity for the blower and the associated maintenance costs for the equipment. In the petrochemical and hydrocarbon industries, natural draft burners (see Figures 1.8 and 1.9) are commonly used wherein the air is drawn into the burner by the suction created by the buoyancy of the hot gases rising through the heaters. For natural draft burners, there is no cost for the air, but there is also often less control over the air supply, which can be substantially affected by things like the ambient wind conditions. Air/fuel burners may be fully premixed as shown in Figure 1.4a (air and fuel mix prior to exiting the burner), partially premixed as shown in Figure 1.4c (some of the air and fuel mix prior to exiting the burner), or fully diffusion mixed as shown in Figure 1.4b (the air and fuel mix after leaving the burner). In the higher temperature applications like metals production and glass manufacturing, preheated air is often used to improve the thermal efficiency and increase the flame temperature for higher heating rates.

7.2.2 AIR-OXY/FUEL BURNERS

Most industrial heating processes require substantial amounts of energy, which is commonly generated by combusting hydrocarbon fuels such as natural gas or oil. Most combustion processes use air as the oxidant. In many cases, these processes can be enhanced by using an oxidant that contains a higher proportion of O_2 than that in air. This is known as *oxygen-enhanced combustion* or OEC.[21] Air consists of approximately 21% O_2 and 79% N_2, by volume. One example of OEC is using an oxidant consisting of air blended with pure O_2. Another example is using high-purity O_2 as the oxidant, instead of air. This is usually referred to as *oxy/fuel* combustion (see Section 7.2.3).

Figure 7.2 shows schematically an air-oxy/fuel burner. In some cases, an existing air/fuel burner may be easily retrofitted by inserting an oxy/fuel burner through it.[22] In other cases, a specially designed burner can be used.[23] The operating costs are less than for oxy/fuel, which uses very high

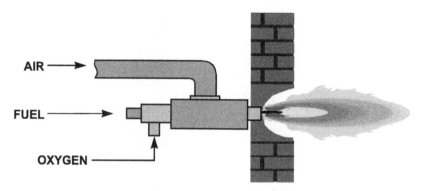

FIGURE 7.2   Schematic of an air-oxy/fuel burner. (Courtesy of CRC Press LLC, Boca Raton, FL.)

levels of O_2. The flame shape and heat release pattern may be adjusted by controlling the amount of O_2 used in the process. With this method, the oxidizer composition may be specified in an alternate way. Instead of giving the overall O_2 concentration in the oxidizer, the oxidizer can be given as the fraction of the total oxidizer that is air and the fraction of the total oxidizer that is pure O_2. The equivalent overall O_2 in the oxidizer can be calculated as follows:

$$\Omega = \frac{20.9}{0.209\left(vol.\%\,O_2\right)+\left(vol.\%\ air\right)} \times 100\% \tag{7.1}$$

For example, the oxidizer may be specified as a blend of 60% O_2 and 40% air. That ratio of O_2 to air produces an equivalent of 39.8% overall O_2 in the oxidizer. The total oxygen enrichment or TOE (Ω) is expressed as a percentage.

7.2.3   OXY/FUEL BURNERS

Figure 7.3 shows a technique commonly referred to as *oxy/fuel combustion*. In nearly all cases, the fuel and the oxygen remain separated inside the burner. This type of mixing is commonly referred to as a nozzle-mix burner that produces a diffusion flame. There is no premixing of the gases for safety reasons. Because of the extremely high reactivity of pure O_2, there is the potential for an explosion if the gases are premixed. In this method, high-purity oxygen (>90% O_2 by volume) is used to combust the fuel. In an oxy/fuel system, the actual purity of the oxidizer will depend on how the O_2 was generated. Oxy/fuel combustion has the greatest potential for improving process efficiency, but it also may have the highest operating cost.[24]

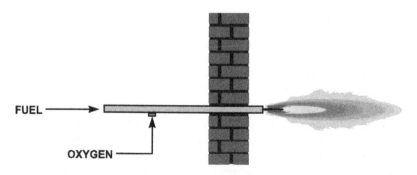

FIGURE 7.3   Schematic of an oxy/fuel burner. (Courtesy of CRC Press LLC, Boca Raton, FL.)

7.2.4 APPLICATIONS OF OEC

Many industrial heating processes may be enhanced by replacing some or all of the air with high-purity oxygen.[25,26] Typical applications include metal heating and melting, glass melting, and calcining. In a report done for the Gas Research Institute,[27] the following applications were identified as possible candidates for OEC:

- Processes that have high flue gas temperatures, typically in excess of 2000°F (1400K)
- Processes that have low thermal efficiencies, typically due to heat transfer limitations
- Processes that have throughput limitations which could benefit from additional heat transfer without adversely affecting product quality
- Processes that have dirty flue gases, high NOx emissions, or flue gas volume limitations

When air is used as the oxidizer, only the O_2 is needed in the combustion process. By eliminating N_2 from the oxidizer, many benefits can be realized.

7.3 CONVENTIONAL BURNER DEVELOPMENT PROCESS

7.3.1 OBJECTIVES

There are usually two primary objectives in most burner development projects: original design and scaling. The tools used in both cases are similar. Both development objectives are briefly considered next.

7.3.1.1 Original Design

At the beginning of a new burner development project, normally a single size (i.e., firing rate) is selected based on market needs and commercial goals. After that size has been developed, the next phase involves scaling that design for other burner sizes. The scaling aspect is discussed in Section 7.3.1.2. There are often many specific objectives for a new burner development. These may include emission performance (NOx, CO, SOx, etc.), noise, flame shape (width, length, geometry, etc.), flame luminosity, heat flux distribution, stability, firing rate range (turndown), fuel composition, flame intensity, burner size, mounting, and cooling method (water-cooled, self-cooled, etc.), among others. In most cases, a new design is targeted at a specific application in a specific industry, although it may be applicable to other applications as well. The overall goal may be to modify an existing design, to start from scratch with a completely new design, or somewhere in between where significant changes are made to an existing design. The extent of the design changes often has an influence on what methods will be used in the design process. Simple changes may not require or warrant detailed analysis, while more complicated changes usually require detailed analysis.

7.3.1.2 Scaling

There are many instances where burner scaling is important. The most obvious is when burners are being used in processes of different capacities. For a given type of process, smaller furnaces typically require burners with lower firing rates, while larger furnaces use burners with higher firing rates. However, in larger furnaces, it may be desirable to simply use more burners with lower firing rates, than fewer burners with higher firing rates, depending on the application. In a given process, there may be several burners firing at different rates in different parts of the furnace. If the firing rate range is wide enough, different burner sizes may be required. The main concern in industrial combustors is often to maintain a certain temperature profile in the material being heated, which often equates to a specific temperature profile inside the combustor. It is often necessary to adjust the firing rate to meet the needs of a given application. For example, for an existing combustion system it may be desirable to increase the material processing rate, which normally means increasing the firing rate. The design question to be answered is by how much, because this is often not a linear relationship. Another example is the modification of a known heating process for higher or lower throughput rates. Again the question is how to do the scaling from the known design.

There are two important aspects to scaling. The first aspect is to determine what firing rate is needed for a given process. It may be possible to scale the rate in a given furnace based on empirical data. For example, data may be available on the firing rates required to produce 150 to 200 tons/day (140 to 180 m-tons/day) of product in a given furnace. Then, determining the firing rate for 175 tons/day (160 m-tons/day) may be a simple extrapolation. However, scaling to a higher production rate may or may not be as simple because the processes are often nonlinear. Therefore, computer modeling of a burner and the whole process can be helpful in those instances. Some examples of burner modeling are presented in this chapter. In Chapters 8 through 10, examples of CFD use in the steel, aluminum and glass, respectively, are given. Typically, process modeling techniques require knowledge of the burner details.

The second aspect of scaling is that once the firing rate has been determined, how is the burner scaled from known sizes to the new size (assuming the new size does not already exist)? There are many possible ways to scale up a burner according to changes in the firing rate that is the primary characteristic of interest in most industrial applications. Spalding (1963),[28] Beér and Chigier (1972),[29] and Damköhler (1936)[30] looked at numerous dimensionless groups based on considerations of the momentum, energy, and mass balances. Some of these groups include the Reynolds, Froude, and Damköhler numbers. However, it is not possible to maintain all of the dimensionless groups constant simultaneously.

The two most common methods used to scale industrial burners are constant velocity and constant residence time. Constant velocity scaling is by far the most popular. The burner thermal input (Q) can be calculated using:

$$Q_0 = K\rho_0 u_0 d_0^2 \tag{7.2}$$

where ρ_0, u_0, and d_0 are the inlet average fluid density, characteristic fluid velocity, and characteristic diameter, respectively, and K is a proportionality constant. Assuming that the inlet fluid density is constant, and the characteristic inlet fluid velocity u_0 is held constant, then the constant velocity scaling law can then be written as:

$$d_1 \propto \left(Q_1\right)^{0.5} \tag{7.3}$$

where the new characteristic burner diameter d_1 is proportional to the square root of the new firing rate. This law simply says that the outlet area of the burner is directly proportional to the firing rate for a constant velocity scaling law. For round outlets, the area is proportional to the square of the diameter.

The principle for constant residence-time scaling is to maintain the ratio of d_0/u_0, which has the units of time (typically seconds). This scaling law, sometimes known as the inertial or convective time scales, can then be written as:

$$d_1 \propto \left(Q_1\right)^{0.33} \tag{7.4}$$

This approach is not as commonly used because it leads to very low windbox pressures for smaller burners and excessive pressures for larger burners.[31] However, simple velocity or residence-time scaling laws often do not apply to complicated combustion problems.[32] This is where computational fluid dynamics can be most useful.

7.3.2 Design Based on Empiricism and Experience

This strategy for designing burners relies primarily on past experience and field operating data. The designer is knowledgeable about a wide range of burners in a wide range of applications and can apply that knowledge to design new or modify existing burners. Before the advent of advanced computational methods, this may have been the most common design tool. The success of the method relies heavily on the experience of the designer. This makes it difficult for someone inexperienced to effectively design new burners. The relatively few calculations that may be made are usually only to determine simple average velocities through various components in the burner. The rest of the design process involves educated "guesses" for where the fuel and oxidizer should go and how they should mix to achieve the desired results. The designs must be tested to get empirical data on the performance. This technique can be effective when relatively minor changes are needed or if the burner design is not a significant departure from the experience of the designer. The more radical the change or design, the less confidence there is in the result. This technique may be the only choice for the designer who does not have access to computational fluid dynamics modeling or to an appropriate pilot-scale furnace for testing and development. New burner designs can be tested in full-scale production furnaces. This may not be an option in all cases as it may be difficult to convince plant personnel to test an unproven device that could jeopardize their production. In other cases, the initial test may be limited to a single burner replacement in a multi-burner system. While that can give some insight into the new design, it often is difficult to predict the impact of complete replacement.

While this method can be inexpensive and fast, assuming that only a few iterations are needed to get to the end result, there are also many potential problems. It may be difficult for the experienced designer to transfer knowledge to someone less experienced as much of the design process may be based on gut feel and instinct developed over time. Another problem is that it may be difficult to understand and explain the results, even when the design does meet the objectives. Many iterations may be needed to get to an optimized design; but without the benefit of some type of computational tool, one may never know if a given design is truly optimized or not. It may also be more difficult to accurately scale new designs to other firing rates, without the benefit of some type of analysis. This method does not offer a straightforward way to estimate the performance of other burner sizes, or of the performance of the burners in different types of furnaces. In most cases, the performance can only be determined by testing. Another significant problem is that it may be very difficult and time-consuming to prove a design, in an actual industrial heating process, which has never been tested or modeled before. This method is now typically used only in very limited circumstances where the chances of success are very high and the risk is very low. This means that only relatively minor changes in existing designs would be applicable to this method.

7.3.3 Laboratory Tests and Analysis

This method is more sophisticated than the previous method as a new burner design is tested and demonstrated in some type of laboratory environment prior to full-scale field testing. For this method, there is essentially no limit to how radical the design departs from previous experience. Given the right type of laboratory equipment, this method can be relatively fast and inexpensive. The closer the lab furnace simulates actual full-scale production furnaces, the more believable the data and the more likely the design will be accepted in industry. This method also dramatically reduces the risk for the end user because performance data from the lab is available. Radical designs can be tested because failure does not jeopardize production as it would in a full-scale industrial furnace. It is often easy to make changes to the design, which can be then tested quickly to determine how well the burner performs. This makes it easier to optimize a design compared to the previous method.

There are also some potential problems with this method. First, the testing may be done on a smaller scale than an actual production furnace. This means there may be scaling issues that can only be tested and proven by field testing. Even when the testing is done at full scale, it is normally done for only a single burner. Many industrial processes have multiple burners, so the lab testing does not provide any information about the interactions between adjacent and opposing burners. Laboratory conditions are often much better than field conditions. For example, production furnaces often have high air infiltration rates compared to lab furnaces. Lab operators may be highly trained and skilled compared to production operators. This can also translate into significant differences between lab and field results. Because burner operation is a complex process in difficult industrial environments, lab experiments may never fully reflect burner performance in a real furnace. It is often much easier to control the operating conditions in a lab than it is in a full-scale production process. Labs may have highly precise equipment compared to the equipment on a production furnace. This may mean that the lab performance is significantly better than the field performance.

7.3.4   Field Validation

In the case of either of the previous burner design methods, field validation is essential. While some end users may be willing to experiment with an unproven burner design, most will not use unproven equipment because of the risk of lost production. While a design may seem valid in theory or on paper, it must ultimately be proven in a real furnace. The more radical the concept, the more difficult to convince an end user to test it. These types of projects must often be funded by outside agencies to minimize the risk for the end user. In many cases, the end user may need to modify existing equipment to even test the new design. It may be as simple as doing some minor re-piping to as complicated as moving the existing burner locations, which can mean moving structural beams on the furnace and changing the insulation inside the furnace. Proving the design in one furnace does not always mean the design will work in other furnaces, depending on how different the furnaces and their operation procedures are. This is another instance where CFD modeling can be effective in applying new designs to other conditions.

7.4   CFD AS A TOOL FOR BURNER DEVELOPMENT AND OPTIMIZATION

It must be stated up front that, given the current state of technology, radical new burners can not be designed solely with CFD. CFD should be used in conjunction with lab and field testing to validate any new designs. In the remainder of this chapter, CFD is presented as an additional tool for burner design. Some representative examples are given from the authors' past experience.

7.4.1   One More Step in Burner Development

With the increasing power of computers, it has become possible to develop quite sophisticated mathematical models using numerical techniques. One can write a program to solve a set of partial differential equations describing a process of interest. Usually programming by itself, and the debugging of the program, take quite an effort. Many industrial companies would rather outsource such developments. This has led to the formation of software companies specializing in computer modeling techniques. Some of these companies devoted to developing general-purpose programs that could be used for different applications. The programs are known under the general names of Computational Fluid Dynamics (CFD), Computational Chemistry, Stress Analysis, and others. As a rule, the codes are conveniently put together as a package with a pre-processor, solver, and post-processor and can be used by industrial companies to speed up the model building process. The existing packages may have limited physics, but they can be successfully used for different applications such as process study, new technology development, and equipment design.

7.4.2 MODELING STEPS

At least three main modeling steps or approaches can be used for a burner development process. The first one is a scoping-type computation where the problem is simplified to study many parameters quickly, without having to do comprehensive analysis. This is often done with a 2-D or 2-D axisymmetric model of the combustion process with a simplified burner geometry. This approach provides a rough, but very quick evaluation of different ideas and scenarios. It helps to restrict the number of evaluated parameters and combinations. It also helps to check the capabilities of the CFD code and the available physical models, and to debug the model for future use in more complex forms.

Once the scope of the initial work is narrowed down to several of the most important parameters, the second step of the modeling process can be implemented. Usually it involves a more complex and precise burner geometry with possibly more accurate physics and significantly increased number of gridpoints. Despite the increased complexity of the model, this is a widely used approach and still relatively fast. The models are usually 2-D or simplified 3-D. The aim of this modeling step is to come up with the two or three most critical parameters for the final and more accurate third-step evaluation.

The third step is a complex 3-D CFD model with the full spectrum of available necessary physics, complete burner geometry details, and a large number of gridpoints. It may be a significant effort to create such a model. The effort is a function of the required geometry details and grid complexity. The structure and grid size should be chosen very carefully because the numerical solution depends on it. In the case of a large geometry ratio (max/min dimensions), the number of gridpoints can increase drastically and the model might become too "bulky" and difficult to work with. The latter is especially tough for structured grids. In the case of unstructured grids, the geometry ratio problem is easier to resolve but solution convergence is typically slower. Sometimes, a full spectrum of physical models might not be available in the CFD code with an unstructured grid because this grid type is a fairly recent implementation. The restrictions on some physics with unstructured grids are only a temporary problem, but it may be a problem.

Substantially more time is required to obtain a solution from a full 3-D model. Therefore, it is usually a good idea to incorporate all possible geometry variations into one model. Indeed, if that extra work is done at the very beginning of the modeling process, then later when it is necessary, the changes to the model are usually easier to make and the previous solution can be used as the initial guess to obtain a new solution. A good initial guess significantly speeds up the solution convergence process and sometimes is the only possible way to obtain a converged solution.

7.4.3 COMPLEXITY OF THE PHYSICAL MODELS

In the case of burner development, one would need a wide spectrum of physical models to be available in the chosen CFD code. Usually, it is a turbulent chemically reacting flow of different species. Therefore, it is necessary to have a variety of models describing heat transfer, turbulence, chemical reaction, etc. The appropriate models should be used to obtain reliable modeling results. For example, a highly-ε turbulent swirling gas flow can be represented best by the RNG k-ε model.[33-35] A chemical reaction can be represented with a finite kinetics model[36] or so called probability density function (PDF) model.[37-39] In the case of the FLUENT code (see Chapter 3), the PDF model is a mixture fraction equilibrium or partial equilibrium chemistry model. The choice of physical models typically depends on the available empirical or experimental data and the expected accuracy of the solution.

A complex heat and mass transfer problem should be handled in the fluid and solid regions as well as at their interface. Radiation modeling is essential in the majority of cases. The challenge is to describe optical properties of the flames and radiation transfer through gaseous, liquid, and solid media as well as at their interface. There are several radiation models that are widely used in the CFD industry today. Some of these models were described in detail in previous chapters.

Only the most common ones are mentioned here. Models such as discrete ordinates (DORM),[40-44] discrete transfer (DTRM),[45-47] and flux (FRM)[48-54] radiation models are widely used. A graybody assumption is usually incorporated in the listed models. For some applications, this limitation might lead to a substantial error in the solution. Therefore, a selective radiation model is often required. A wide-band radiation model[55,56] is the most commonly used. It provides higher accuracy than graybody models, but it requires much more computer power and substantially delays the final solution.

7.4.4 Assumptions and Simplification

When a problem is modeled with CFD, some assumptions and simplifications should be applied to it. In most cases, it is impossible to exactly describe a real process mathematically. There may not be enough knowledge of the phenomena or a lack of mathematical techniques to simulate the phenomena. The numerical code may not have the necessary physics built in yet. Therefore, some assumptions and simplifications are required.

Restrictions are often applied at each modeling step. When a 2-D model is created, it is assumed that a 2-D slice, taken through the domain of interest, is representative of the process. It is also assumed that the boundaries of the real process can be considered far enough and not affecting the process in the section of interest. For example, gas flow from the flat flame Cleanfire® HR™ burner[57] can be initially represented by a 2-D model with a vertical slice taken through the middle section of the burner shown in Figure 7.4. In this specific case, the main assumption is that the burner discharge slot is wide enough and there are no significant changes in the discharge flow along the width of the burner. Such a simplification is helpful for the velocity field optimization near the burner tip and in the burner tile. It is very practical to use a 2-D model to obtain a number of quick solutions with different fuel and oxidizer discharge velocities. The next modeling step can be used when the discharge velocity range is already defined. In the case of the Cleanfire® HR™ burner, the next step is a full 3-D burner geometry (Figure 7.5). Such a model can be used for the internal burner geometry optimization. The optimization is based on the optimal flow distribution at the nozzle and on the pressure level at the back of the burner.

Another example is a "pipe-in-a-pipe" type burner. A 2-D axisymmetric model can be used to represent this burner type. The main assumption is that there are no changes in the angular direction of the domain. By definition, this model type has some physics restrictions. For example, gravity can be included only if it applies along the symmetry line. In the case of combustion, gravity can play an important role, and therefore, such a simplification of ignoring gravity could provide a misleading solution. On the other hand, this model type could still be useful for the burner geometry optimization. For example, the gravity force can be disregarded within the length of the jet which has strong enough momentum. Therefore, the solution can be valid in the vicinity of the burner tip, inside the burner tile, and in the furnace combustion space close to the burner tile.

The next modeling step is more complex but also somewhat limited. A three-dimensional part of the domain bounded by symmetry planes can be used to capture more burner details (Figure 7.6). This type of a model also includes some restrictions on the gravity force. Moreover, a process is assumed to be symmetric across the symmetry boundaries. Despite the restrictions, this approach saves a lot of time and effort at this stage of the model design and later, when obtaining the final solution. An example of such a model is given below.

7.5   EXAMPLES

Two examples will be presented in this section and CFD modeling use in burner development will be demonstrated.

FIGURE 7.4 Photo of Cleanfire® flat flame burner: (a) Cleanfire HR burner body, and (b) HR burner mounted together with burner block. (Courtesy of Air Products and Chemicals, Inc., Allentown, PA. With permission.)

7.5.1 Oxy/Fuel Burner Development for High-Pressure Vessels

7.5.1.1 Introduction and Modeling Procedure

This example involves a high-pressure reactor vessel used to neutralize waste materials. The vessel contained a molten metal under a syn gas environment, for which the composition is given below. A number of observation ports were located at the vessel headspace. Optical instruments were mounted a few feet from the hot face. The channel from the instrument to the hot face, known as headspace penetration, was several inches in diameter and refractory-lined. Recycled syn gas or natural gas was used to purge the optics. Splashed molten metal could condense near the port

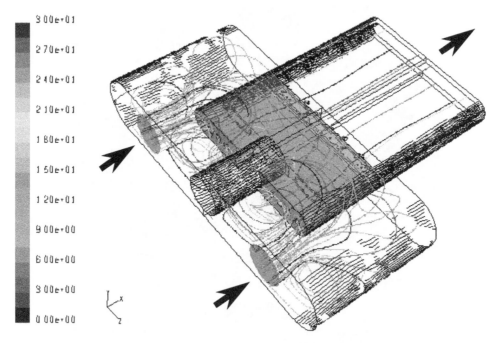

FIGURE 7.5   Flat flame Cleanfire® HR™ model with flowfield inside.

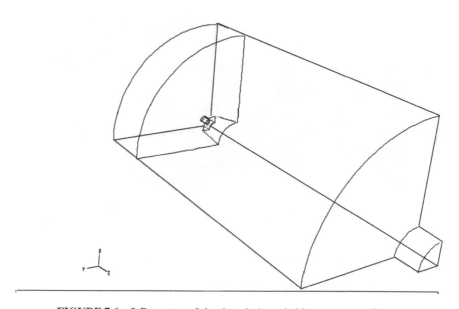

FIGURE 7.6   3-D quarter of the domain bounded by symmetry planes.

opening and form a "skull" to block the view of the instrument. An oxidant had to be injected inside the penetration to create a combustion zone, so that the temperature near the hot face would be above the melting point of the metal. The combustion took place inside a very constricted space; therefore, it was important that the refractory was not overheated. The injector had to be simple, accommodate large variations in the purging gas flow rate, and could not block the view of the instrument.

A design model was developed for the oxidizer injectors. The most critical factors included jet penetration distance in the cross-flow, maximum theoretical reaction temperature at any given fuel and oxidizer concentration, the main pipe diameter, the number of the holes in the injector, the injector hole diameter, and the oxidizer injection angle. A parametric study with immediate output was conducted with the help of the design model. The design of the injectors for two penetration diameters was obtained for two different installations. The developed injector designs were validated using detailed CFD analysis. The temperature distribution was calculated at the refractory surface of the penetrations. The predicted refractory temperature distribution of the penetrations confirmed the validity of the design model, which can be used for future injector design work. Nitrogen dilution of the fuel and oxidizer is recommended for installation #1 and installation #2, respectively.

7.5.1.2 Problem Definition

The end user was looking for a set of burners that would be installed in each penetration of the headspace of its furnaces that could be activated periodically to melt down any skull build-up in the penetration. The end user planned to use syn gas as the fuel in installation #1 and methane as the fuel in installation #2.

The requirements for the burners are listed below:

Installation #1	Operating pressure	135 psi (max.)
	Metal melting point	2650°F
	Surface temperature of the refractory	2800°F (3200°F max.)
	Refractory type	High alumina or silicon carbide
	Chemical composition of the syn gas	CO: 93.15% wt
		H_2: 6.58% wt
		H_2O: 0.27% wt
	Diameter of the penetrating pipe	10, 8, 6, 4 in.
	Range of syn gas flow rates	From: 1 SCFM for 4-in. pipe
		Up to: 47 SCFM for 10-in. pipe
Installation #2	Operating pressure	135 psi (max.)
	Metal melting point	2000°F
	Surface temperature of the refractory	2800°F (3200°F max.)
	Refractory type	High alumina or silicon carbide
	Chemical composition of the fuel	CH_4 + any amount of N_2 if needed
	Diameter of the penetrating pipe	6.5 in.
	Expected energy release	100,000 Btu/hr

The expected region of the skull build-up was between 1" and 13" from the internal surface of the vessel refractory for both installations. The goal was to combust the syn gas or methane to prevent any skull buildup and at the same time to not overheat the refractory.

7.5.1.3 Design Tool and Problem Solution Strategy

Originally, oxy/fuel combustion was anticipated for both installations because of the high flame temperatures often used for melting metals. It was decided to use the original pipe of the penetration as a burner body and to design an injector that provided the oxidizer for combustion inside the penetration. Because the oxidizer injector design had too many potential variables to consider, a simple design model was necessary for a fast parametric study. The goal of the parametric study was to narrow down the range of the variables and to optimize the injector design. Once this step was completed, a CFD model was used to validate the obtained injector design and to check the temperature profile at the penetration refractory surface.

The injector design model was built using an Excel spreadsheet. The model estimated the required oxidizer velocity, the number of oxidizer holes, and the oxidizer jet penetration distance for a nonconfined jet in the cross-flow with the jet angle effect included. One penetration diameter was chosen for each installation to build the appropriate injectors. A parametric study was performed using the developed design model. As a result of the study, injector designs were proposed. To validate the obtained injector designs, the following three steps were completed:

1. Run a 2-D axisymmetric CFD analysis for each penetration diameter with the design injectors to confirm the predictions from the design model, and to obtain the temperature distribution inside the penetration and at the refractory surface.
2. If necessary, run a 3-D CFD analysis of a 1/16 section of each penetration to confirm the predictions from the design model and the 2-D axisymmetric CFD model; obtain the temperature distribution inside the modeled penetration segment and at the refractory surface (basis: 2-D axisymmetric CFD model analysis).
3. If the results of the Excel spreadsheet model were confirmed by 2-D and/or 3-D CFD models, use the design model to obtain the rest of the injector designs.

7.5.1.4 Solution Procedure

Two cases were studied using the design model. The oxidizer injection rings were designed for both projects as suggested by the design model. The oxidizer flow rate was obtained for both cases. The criteria for choosing the right oxidizer flow rate was the jet penetration distance. The assumption was that the jet penetration distance should be equal to, or slightly more than the penetration diameter. The following flow rates of fuel and oxidizer were obtained:

Installation #1	Fuel	47 SCFM of syn gas (100% CO)
	Oxidizer	16.5 SCFM of 35% enriched oxygen stream, or 25 SCFM of air (may not attain sufficient temperature)
Installation #2	Fuel	120 SCFH (102 SCFH of CH_4 + 18 SCFH of N_2)
	Oxidizer	73.1 SCFH of 100% oxygen

It is important to note that the fuel:oxidizer ratio should stay the same for any fuel flow rate and the injector design should not be changed.

A thermal equilibrium code was used to predict the theoretical peak temperature for the specified fuel type and oxygen purity in the oxidizer. An example of the obtained peak reaction temperature distribution for methane/oxygen combustion is shown in Figure 7.7. The equivalence ratio (fuel:oxidizer mixture ratio) was determined from the calculation based on the desired maximum flame temperature. Finally, the appropriate oxidizer content was obtained to satisfy the refractory and metal freezing temperature limitations.

The 2-D axisymmetric CFD model was built using FLUENT to confirm the validity of the designed injector rings and to obtain a temperature profile at the burner refractory surface. The following assumptions and boundary conditions were used in this model:

1. No heat losses from the burner outer wall
2. Area of eight injector holes was represented by a continuous slot
3. k-ε turbulence model was used
4. PDF model was used with the following species:

Installation #1	CO, CO_2, O_2, O, N_2
Installation #2	CH_4, O_2, CO, CO_2, H_2O, O, H, H_2, N_2

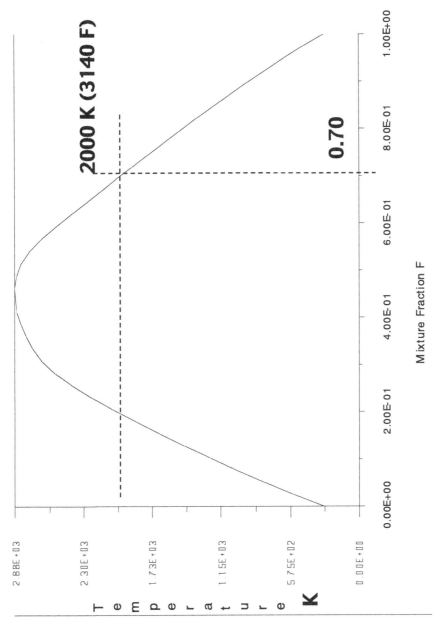

FIGURE 7.7 Peak flame temperature distribution of the equivalence ratio.

Several cases with different oxidizer velocities, mass flow rates, and O_2 content were calculated. In all the calculated cases, the reaction was pushed toward the burner refractory wall. This phenomenon was considered to be a limitation of the 2-D axisymmetric model. The following conclusions were made from the analysis of the results obtained from a 2-D axisymmetric CFD model:

1. The developed design model was sufficient as a tool for preliminary oxidizer injector selection.
2. Approximately 35% enriched oxygen stream should be used as an oxidizer in the case of syn gas fuel (Installation #1); the maximum temperature of the reaction significantly exceeds the specified limits for the refractory if the oxidizer oxygen content is more then 40%; in that case, the refractory is most likely will be overheated.
3. In the case of methane being used as a fuel, the injector with pure oxygen streams was designed; the maximum refractory temperature in this case was not expected to exceed the refractory limits (Installation #2).
4. The 2-D axisymmetric model was not sufficient and was not capable of predicting the correct local refractory temperature or the correct refractory profile temperature; a 3-D model was needed.

A 3-D CFD model was built as suggested above. The model calculated a 1/16 segment of a burner. The same assumptions and boundary conditions were used for the 3-D model as indicated above for the 2-D axisymmetric model. The main difference between the 3-D and the 2-D models was the correct oxidizer injection through each of the oxidizer holes and the absence of the continuous slot. In the 3-D model, the main flow in the penetration was not forced to go just through the center of the pipe being squeezed by the streams from the injector as it was in the 2-D axisymmetric model. So, the expected result was the correct flow and the temperature field distributions in the vicinity of each oxidizer jet stream.

The outline of the 3-D model of 1/16 of the pipe zoomed in the area of the oxidizer injector is shown in Figure 7.8. An example of the temperature distribution for the specified penetrations is shown in Figure 7.9 for Installation #1. Another example of the flow field distribution in the vicinity of the oxidizer injection hole is shown in Figure 7.10.

The results of the 3-D model show that for both Installation #1 and Installation #2, the suggested designs of the injectors developed with the help of the design model were satisfactory. In both cases, the maximum burner refractory temperature did not exceed the specified limits. At the same time, the burner refractory temperature at the expected penetration depth (12 in. to 13 in.) was above the metal freezing point. In the case of periodic use of the burners with a skull already built up inside the penetration, the high temperature from the flame core along the penetration centerline should help to melt down the plug.

7.5.1.5 Conclusions

1. An injector design model was built for an oxidizer injector. The model can be used to conduct fast parametric studies to narrow down the range of the parameters and for injector design optimization.
2. A 2-D axisymmetric CFD model was built to estimate the temperature profile and the concentrations of fuel and oxidizer for the designed burners. It was shown that the 2-D axisymmetric model was not sufficient and could not predict the correct flow field and temperature distribution inside the penetration and at the penetration refractory. A 3-D CFD model was required. A 3-D model was built for a 1/16 pipe segment. The model predicted the flow field and the temperature distribution inside the designed burners and the burner refractory.

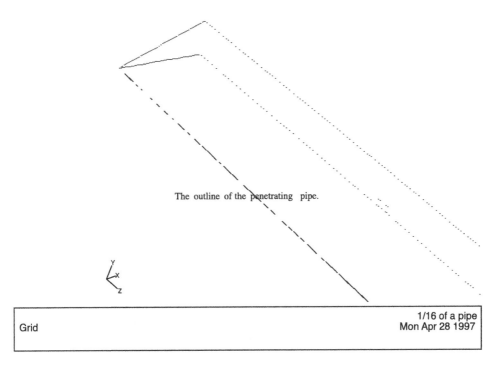

The outline of the penetrating pipe.

Grid

1/16 of a pipe
Mon Apr 28 1997

FIGURE 7.8 One-sixteenth of the penetration pipe shown with half of an oxidant injection nozzle zoomed in the nozzle region.

3. The results of the 3-D CFD model showed that the oxidizer injectors designed for both installations kept the penetration refractory temperature at the specified level, above the metal freezing point and below the refractory limiting temperature. The fuel:oxidizer ratios and the dilution of nitrogen in the fuel and the oxidizer were obtained for Installation #1 and Installation #2.

4. The developed design model was validated by the 3-D CFD model and is recommended as a tool for general design of the injectors in the cross-flow configuration.

7.5.2 AIR-OXY/FUEL BURNER COMPUTATIONAL ANALYSIS

7.5.2.1 Introduction and Problem Description

The evaluation of an air-oxy/fuel (AOF) burner design is described in this section. Different burner concepts were studied using CFD modeling techniques. Some modeling predictions were later confirmed by the results from laboratory tests. The goal was to understand the AOF burner design concept which provides an increase in melting efficiency and simultaneously generates less NOx. The results of this work helped to finalize and patent[58] the AOF burner design.

As discussed previously, it usually makes the most sense to develop a burner for a specific application. Therefore, an existing aluminum melter was chosen. The goals were to improve the melter performance by increasing the furnace production and by simultaneously reducing the NOx emissions. An adequate burner concept and design were defined as the primary targets because both of the goals are a function of the burner performance.

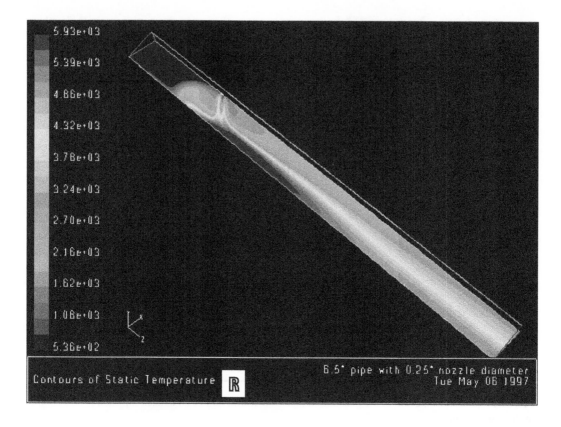

FIGURE 7.9   Example of temperature distribution inside the penetration.

The idea of the AOF burner is not new and is well-defined in the literature.[59] The challenge was to create an AOF burner, which not only helps to boost furnace production but also complies with NOx legislation.

Initially, two different burner concepts were evaluated: the low-NOx air/fuel (AF) burner, and the high-efficiency oxy/fuel (OF) burner. The idea was to merge two burners together in order to combine their best features. A preliminary AOF burner design was developed as a result of the merger. An outline of this original AOF burner is shown in Figure 7.11. Natural gas was introduced through the central burner pipe. Oxygen was introduced through the annulus around the fuel pipe. Air was introduced through the equal-area slots located at the vertical wall of the burner tile. The burner was designed for a maximum firing rate of 17.5×10^6 Btu/hr and was later tested in a semi-industrial experimental furnace. The secondary oxidizer ports were designed to accommodate the air stream as an oxidizer. It will be shown later that an understanding of the mixing and, therefore, the speed of chemical reaction obtained from this example led to a patented burner design.

The AOF burner was built based on the concept of the low NO_x AF burner. This concept assumes that nitrogen, which is introduced with the air stream, is diluted by the products of combustion. The dilution occurs due to the extensive entrainment of furnace gases into the flame and the burner tile region. The AOF burner was expected to have a higher flame temperature than the AF burner. Therefore, the following questions had to be answered:

1. What is the NOx level with the AOF burner?
2. What is the efficiency of the AOF flame compared to the AF flame?
3. What is the mechanism of the flame temperature dilution by the furnace gases for both the AOF and the AF burners?

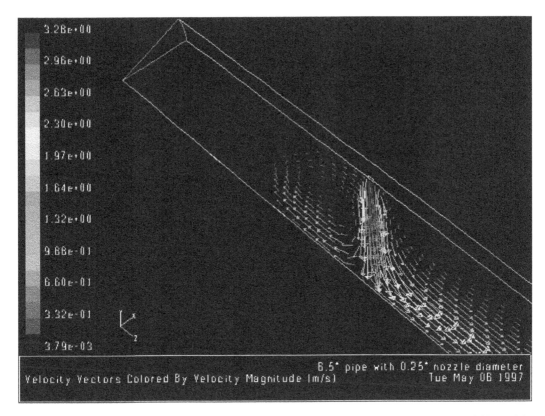

FIGURE 7.10   Example of the velocity distribution in the vicinity of the injector nozzle in 3-D CFD model.

The FLUENT CFD code was used to develop a model that helped to answer these and some other questions.

7.5.2.2   Model Description, Assumptions, and Boundary Conditions

An outline of the model is shown in Figure 7.11. The burner was installed in a refractory-lined, semi-industrial circular furnace. The furnace characteristics were used to build the model. The outline of the burner mounted on the furnace wall is shown in Figure 7.11a. It was possible to model only one quarter of the furnace due to its cylindrical shape. This simplification helped to reduce computational time. A quarter of the domain was built in such a way that a quick modification of the burner's geometry was easy if necessary.

The following boundary conditions were used:

- Pressure outlet for the furnace exhaust was located symmetrically opposite to the burner wall.
- Symmetry planes bounded the domain from the sides.
- A conducting wall was defined around the burner.
- A constant external heat transfer coefficient was defined at the furnace walls.
- Velocity components and species mole fractions were used at the inlets for natural gas, oxygen, and air.

The following cases were simulated:

- The low-NOx AF burner
- The AOF original burner design (simple merger of the OF and the AF burners) with total oxygen enrichment of 35% (two oxidizer streams)

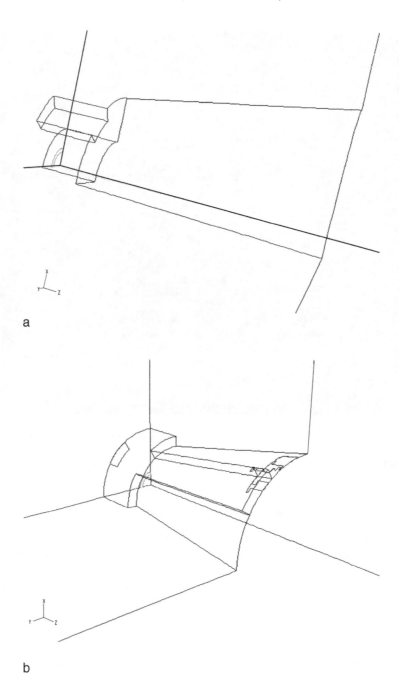

a

b

FIGURE 7.11 The outline of the model: (a) original air-oxy/fuel burner design, and (b) burner with air discharged from the ports located at the end of the tile.

- AOF original burner design with a 30% oxygen-enriched air stream (one oxidizer stream)
- Modified AOF burner design with the air stream introduced through the ports at the end of the burner tile (total oxygen enrichment or TOE of 35% was specified)

All the cases were calculated for the same burner firing rate of 17.5×10^6 Btu/hr. A 35% TOE (Equation 7.1) translates into a 50%-50% oxygen split, which means that 50% of the required oxygen comes through the oxygen annulus and another 50% comes with the air stream. An air preheat temperature of 900°F was used in all the reacting flow cases.

7.5.2.3   Results and Discussion

The flow field distribution is shown in Figures 7.12 through 7.18 at the different slices, changing from a 0 to a 45° angle and zoomed in the vicinity of the burner tile. Each figure shows both the AF and the AOF cases, so they can easily be compared. A 0° angle slice (Figure 7.12) was taken at the symmetry plane right in between two air ports. The slice at a 45° angle (Figure 7.18) was taken through the center of an air port. The slice at a 32.5° angle (Figure 7.17) was taken at the very edge of an air port.

The flow field comparison of both cases shows that more furnace gases are entrained into the burner tile area in the case of the AF firing mode. Here, the furnace gases are entrained into the burner tile within the 40 degree angle segment between the two air ports (Figure 7.12a to Figure 7.16a). No furnace gases are entrained into the burner tile region starting from the slice taken at a 20° angle from the symmetry plane, Figure 7.16a, and further toward the air port, Figure 7.17a and Figure 7.18a.

In the case of the AOF firing mode, the segment with the furnace gases entrained into the burner tile is much smaller. It narrows down to only about a 20° to 25° segment between the air ports, as compared to a 40° segment in the AF firing mode (Figure 7.12b to Figure 7.14). Moreover, in the case with the AOF firing mode, the entrainment is also limited by a vortex created by natural gas and oxygen inside the burner tile in the smaller segment. The vortex enhances mixing between the natural gas and oxygen, and speeds up the chemical reactions. Because of that, a significant temperature rise should be expected inside the burner tile area.

The temperature distribution for both the AF and the original AOF firing burners are shown in Figures 7.19 and 7.20. The same scale is used to plot the temperature field for both cases. As expected, the flame from the AOF firing burner has a significantly higher temperature as compared to the one from the AF burner. Also, this high-temperature region is located inside the burner tile and the air stream is discharged right into it. Therefore, the results suggest that a higher NOx level should be expected in the case of the AOF firing burner than in the AF case. At the same time, the flame efficiency should be higher in the AOF case rather than in the AF case. The calculated heat flux to the furnace wall for both cases is shown in Table 7.1.

The analysis above showed that the original AOF burner design did not satisfy the low-NOx criterion. Therefore, some burner modifications were necessary to help reduce the NOx and to keep the efficiency advantage of the AOF firing. Provided by the modeling results, the insights of the original AOF burner design suggested two burner modifications. The first one was a reduction of the total oxygen enrichment (TOE) up to only 30%. This approach was based on the AF burner design and premixed oxygen in the air stream. The maximum 30% oxygen enrichment is recommended based on safety guidelines.[60] The burner efficiency in this case was expected to be somewhat better when compared with the AF burner. The second modification was to change the location of the air ports. Moving the air ports to a different location would help to avoid direct air discharge into the hot oxy/fuel flame region. One of the possible air port locations was at the end of the burner tile (Figure 7.11b). Both suggested cases were modeled and compared to the original AOF burner design.

The temperature distribution for the modified cases is shown in Figures 7.21 and 7.22. The scale in these figures is kept exactly the same as in Figures 7.19 and 7.20 to make the visual comparison easier. Comparison of the temperature fields for all four cases, Figures 7.19 to 7.22, suggested that the most efficient case is the one with the AOF firing burner with the air ports located at the end of the tile. This burner also provided the highest average furnace gases entrainment inside the burner tile region, which was expected to provide lower NOx. The second most efficient case was the one with the enriched oxygen air stream. The calculated heat flux to the furnace wall is presented in Table 7.1 for both modified cases.

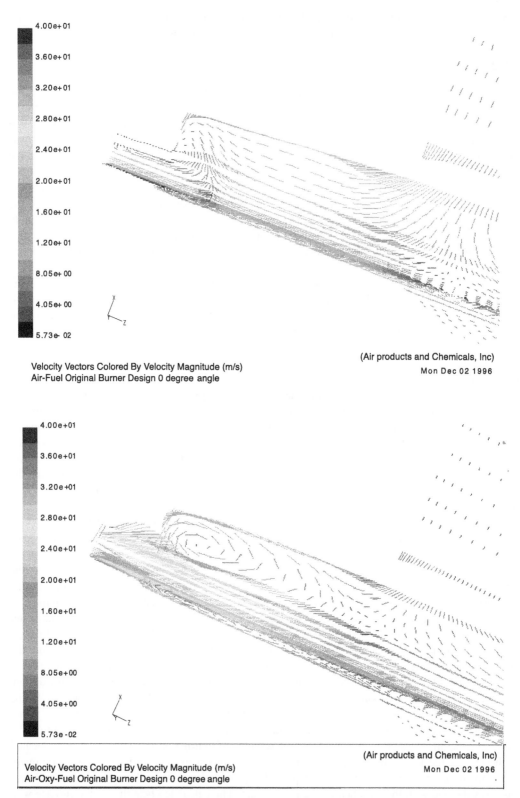

FIGURE 7.12   Velocity distribution in the vicinity of the burner tile region at 0° angle: (a) air-fuel firing mode, and (b) air-oxy/fuel firing mode.

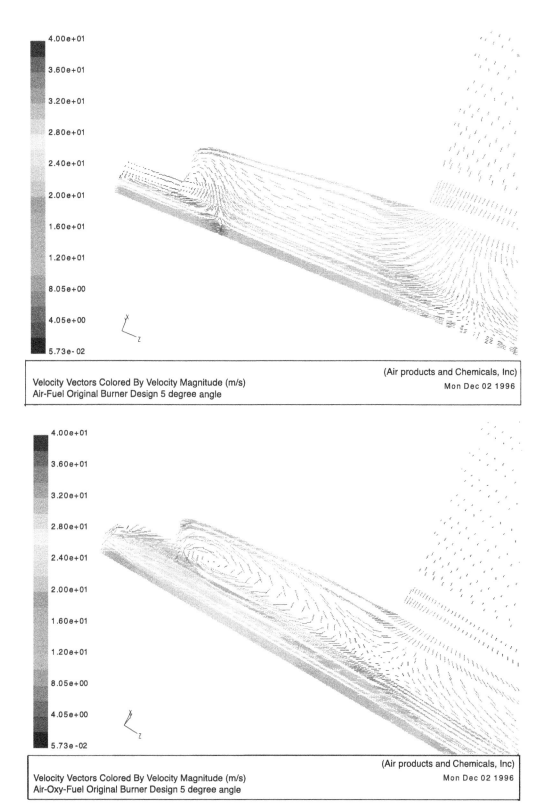

FIGURE 7.13 Velocity distribution in the vicinity of the burner tile region at 5° angle: (a) air-fuel firing mode, and (b) air-oxy/fuel firing mode.

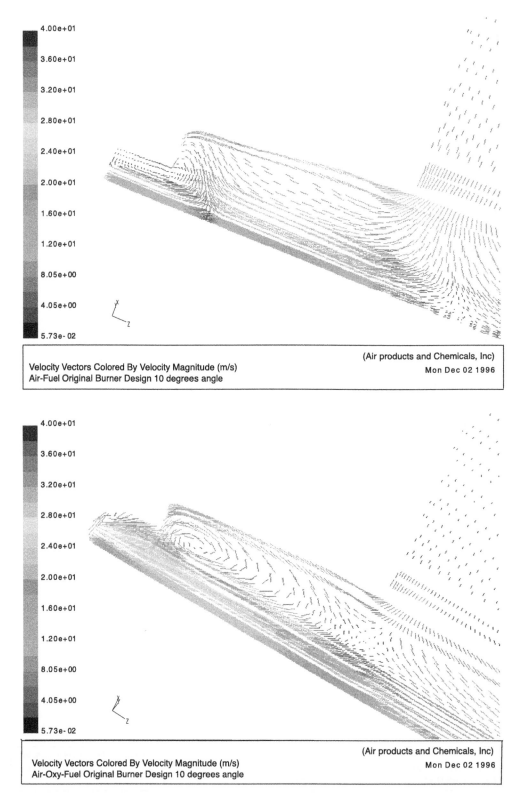

FIGURE 7.14 Velocity distribution in the vicinity of the burner tile region at 10° angle: (a) air-fuel firing mode, and (b) air-oxy/fuel firing mode.

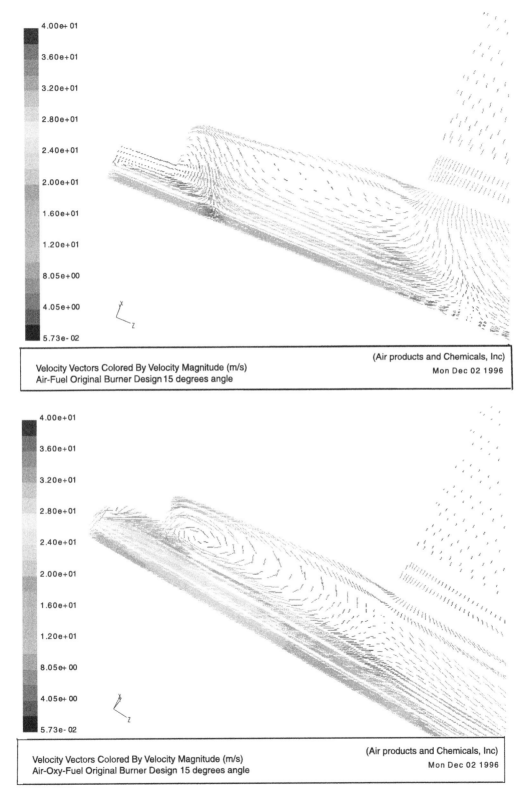

FIGURE 7.15 Velocity distribution in the vicinity of the burner tile region at 15° angle: (a) air-fuel firing mode, and (b) air-oxy/fuel firing mode.

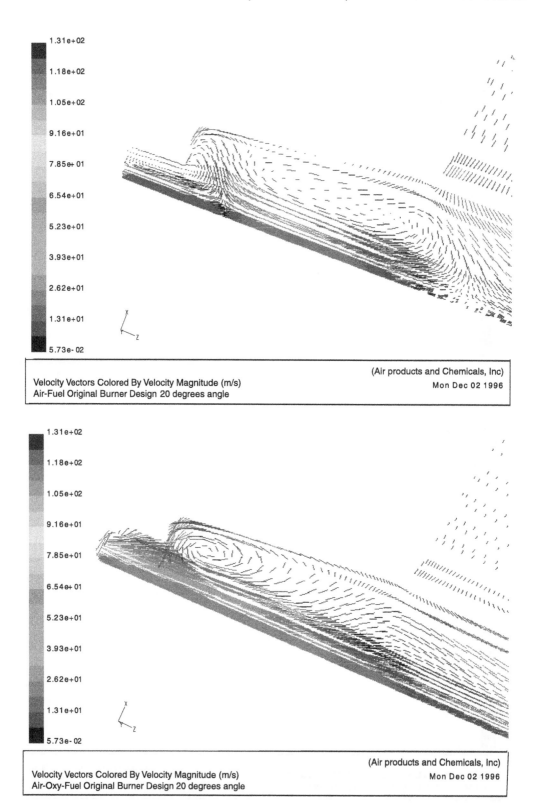

FIGURE 7.16   Velocity distribution in the vicinity of the burner tile region at 20° angle: (a) air-fuel firing mode, and (b) air-oxy/fuel firing mode.

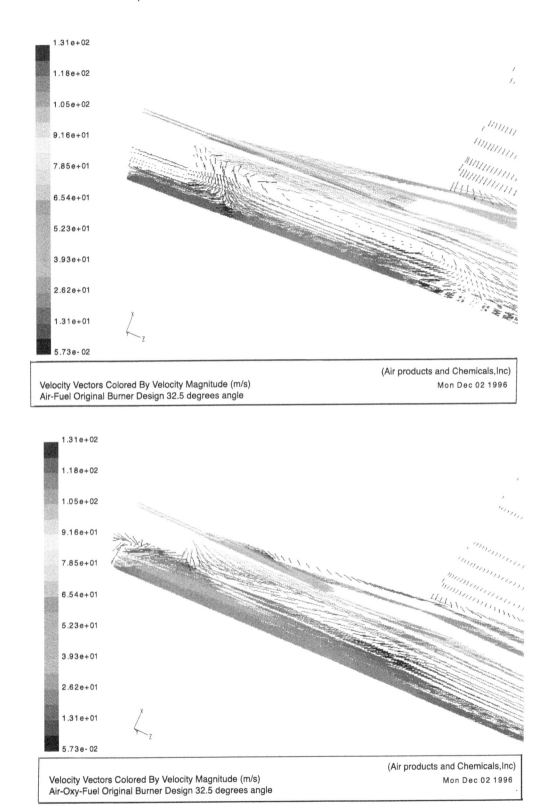

FIGURE 7.17 Velocity distribution in the vicinity of the burner tile region at 32.5° angle: (a) air-fuel firing mode, and (b) air-oxy/fuel firing mode.

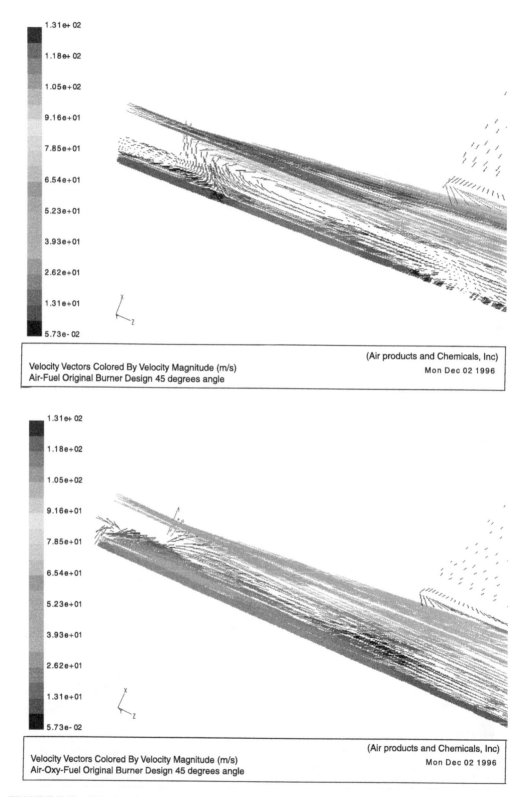

FIGURE 7.18   Velocity distribution in the vicinity of the burner tile region at 45° angle: (a) air-fuel firing mode, and (b) air-oxy/fuel firing mode.

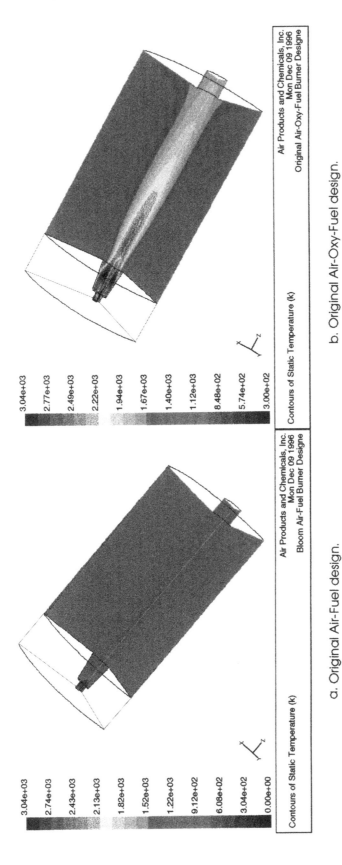

a. Original Air-Fuel design.

b. Original Air-Oxy-Fuel design.

FIGURE 7.19   Temperature distribution at the slices between the air ports for (a) the AF and (b) the original AOF designs.

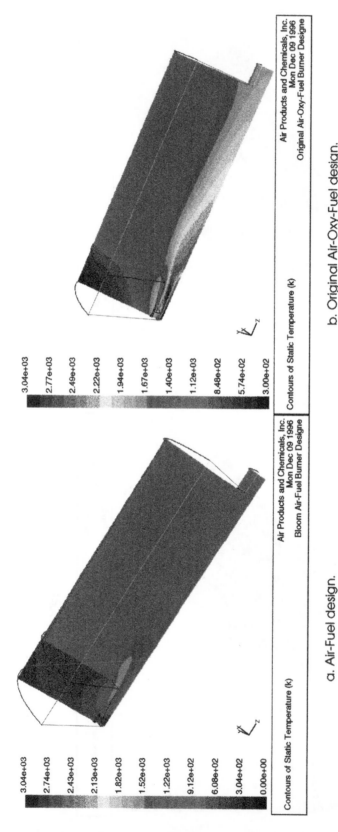

FIGURE 7.20 Temperature at the slices through the air ports for the AF and the original AOF designs: (a) 30% enriched oxygen air-oxy/fuel case, and (b) air-oxy/fuel case with air discharged at the end of the burner tile.

TABLE 7.1
Calculated Heat Flux to the Furnace Wall

Firing Mode	AF	AOF	AOF with 30% O_2 Enrichment	AOF with Air Port at the End of the Tile
Heat Flux to the Wall (kW)	3.3	10.9	11.3	23.0

Compared to the original AOF burner design, the one with the enriched oxygen air stream had a lower flame peak temperature and the flame hot spot was shifted toward the furnace combustion space. A lower NOx was expected to be in the furnace with modified burners compared to the furnace with the original AOF burner design.

A velocity flow field distribution is shown in Figure 7.23 for the modified AOF case with 30% oxygen-enriched air stream. The only difference between the two is in the velocity magnitude. This difference is especially pronounced at the air port discharge zone. The information on the velocity is helpful in understanding gas mixing and therefore, the chemical reaction rate and the NOx formation. Unfortunately, the NOx model available in FLUENT CFD code at the time of these calculations, had certain limitations and could not be used with the existing mesh. Considering these limitations, it is difficult to evaluate the difference in NOx between all the calculated cases. Once the NOx model is modified, it can be used in this particular application. Until then, even the NOx trend can only be evaluated indirectly based on the information on the flame temperature, velocity distribution, mixing, and species concentrations.

7.5.2.4 Conclusions

The most important outcome of the CFD model was that it provided: (1) an understanding of the reacting flow behavior in the calculated domain with different burner designs; (2) a possibility to evaluate a large number of burner modifications rapidly and inexpensively; and (3) a reduction in the number of experimental tests, which take time and may be costly. The following conclusions can be drawn from the AOF burner design evaluation:

- The AOF original burner design is more efficient than the AF burner, but creates substantially more NOx. This prediction was validated by the results obtained from the experimental semi-industrial furnace.
- The furnace melting cycle can be significantly reduced if the AOF burner is used with the air ports at the end of the burner tile. The NOx issue is yet to be evaluated for this case.
- The original AF burner design with a 30% enriched-oxygen air stream can be recommended as a second choice burner to help reduce the furnace melting cycle.

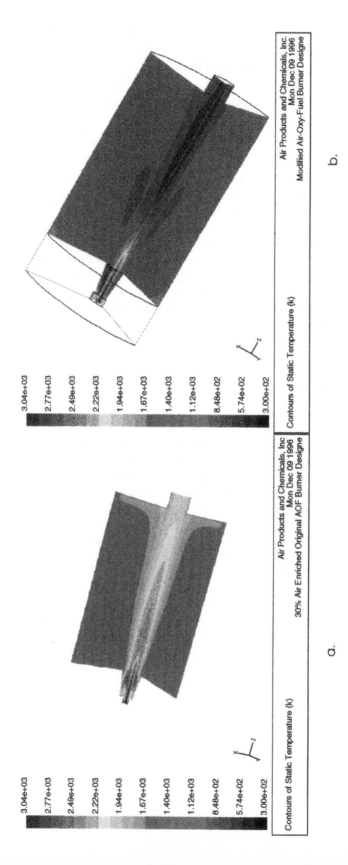

FIGURE 7.21 Temperature distribution at the slices between the air ports for the original and modified AOF designs: (a) 30% enriched oxygen air-oxy/fuel case, and (b) air-oxy/fuel case with air discharged at the end of the burner tile.

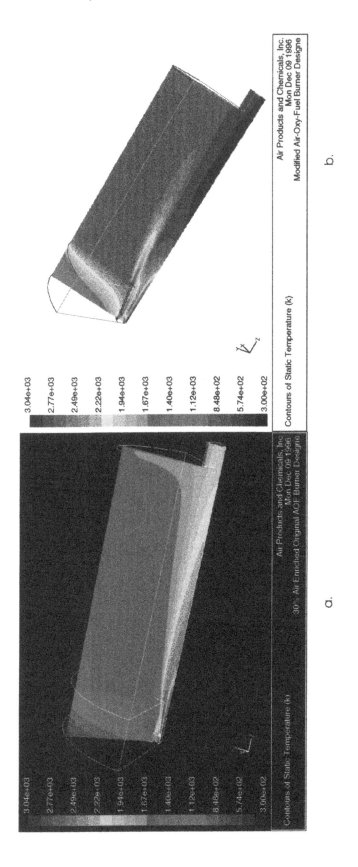

FIGURE 7.22 Temperature at the slices through the air ports for the original and modified AOF designs.

a. 30% enriched oxygen Air-Oxy-Fuel case.
b. Air-Oxy-Fuel case with air discharged at the end of the burner tile.

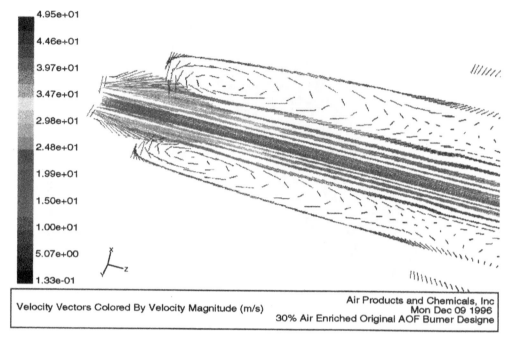

FIGURE 7.23 Velocity distribution in the vicinity of the burner tile region at 0° angle for the AOF firing burner with 30% oxygen enrichment of the air stream.

REFERENCES

1. A.M. Lankhorst and J.F.M. Velthuis, Ceramic recuperative radiant tube burners: simulations and experiments, in *Tranport Phenomena in Combustion*, Vol. 2, S.H. Chan, Ed., Taylor & Francis, Washington, D.C., 1996, 1330-1341.
2. F. Mei and H. Meunier, Numerical and experimental investigation of a single ended radiant tube, in *ASME Proceedings of the 32nd National Heat Transfer Conf.*, Vol. 3: Fire and Combustion, L. Gritzo and J.-P. Delplanque, pp. 109-118, ASME, New York, 1997.
3. H. Ramamurthy, S. Ramadhyani, and R. Viskanta, Development of fuel burn-up and wall heat transfer correlations for flows in radiant tubes, *Num. Heat Transfer, Part A*, 31, 563-584, 1997.
4. S. Bortz and A. Hagiwara, Inviscid model for the prediction of the near field region of swirl burners, in *Industrial Combustion Technologies*, M.A. Lukasiewicz, Ed., American Society of Metals, Materials Park, OH, 1986, 89-97.
5. Y. Tsujimoto and N. Machii, Numerical analysis of pulse combustion burner, *Twenty-first Symposium (International) on Combustion*, The Combustion Institute, Pittsburgh, PA, 1986, 539-546.
6. T.W. Tong, S.B. Sathe, and R.E. Peck, Improving the performance of porous radiant burners through use of sub-micron size fibers, in *Heat Transfer Phenomena in Radiation, Combustion, and Fires*, R.K. Shah, Ed., ASME, New York, HTD-106, New York, 1989, 257-264.
7. S.B. Sathe, R.E. Peck, and T.W. Tong, A numerical analysis of combustion and heat transfer in porous radiant burners, *Int. J. Heat Mass Transfer*, 33(6), 1331-1338, 1990.
8. S.H. Chan and K. Kumar, Analytical investigation of SER recuperator performance, in *Fossil Fuel Combustion Symposium 1990*, S. Singh, Ed., ASME PD-Vol. 30, New York, 1990, 161-168.
9. P.-F. Hsu, J.R. Howell, and R.D. Matthews, A Numerical Investigation of Premixed Combustion Within Porous Inert Media, *J. Heat Transfer*, 115(3), 744-750, 1993.

10. R. Mital, J.P. Gore, R. Viskanta, and S. Singh, Radiation efficiency and structure of flames stabilized inside radiant porous ceramic burners, M.Q. McQuay, W. Schreiber, E. Bigzadeh, K. Annamalai, D. Choudhury, and A. Runchal, Eds., *ASME Proceedings of the 31st National Heat Transfer Conf.*, Vol. 6, ASME, New York, HTD-328, New York, 1996, 131-137.

11. C.L. Hackert, J.L. Ellzey, and O.A. Ezekoye, Numerical simulation of a porous honeycomb burner, in *ASME Proceedings of the 32nd National Heat Transfer Conf.*, Vol. 3: Fire and Combustion, L. Gritzo and J.-P. Delplanque, Eds., ASME, New York, 1997, 147-153.

12. M.M. Sidawi, B. Farouk, and U. Parekh, A numerical study of an industrial hydrogen sulfide burner with air- and oxygen-based operations, in *Heat Transfer in Fire and Combustion Systems — 1993*, B. Farouk, M.P. Menguc, R. Viskanta, C. Presser, and S. Chellaiah, Eds., ASME, New York, HTD-250, New York, 1993, 227-234.

13. G.W. Butler, J. Lee, K. Ushimaru, S. Bernstein, and A.D. Gosman, A numerical simulation methodology and its application in natural gas burner design, in *Industrial Combustion Technologies*, M.A. Lukasiewicz, Ed., American Society of Metals, Materials Park, OH, 1986, 109-116.

14. A. Schmücker and R.E. Leyens, Development of the delta burner using computational fluid dynamics, *Proc. of 1998 International Gas Research Conf.*, Vol. V: Industrial Utilization, D.A. Dolenc, Ed., Govt. Institutes, Rockville, MD, 1998, 516-526.

15. B. Schmidt, B. Spiegelhauer, N.B. Kampp Rasmussen, and F. Giversen, Development of a process adapted gas burner through mathematical modelling and practical experience, *Proc. of 1998 International Gas Research Conf.*, Vol. V: Industrial Utilization, D.A. Dolenc, Ed., Govt. Institutes, Rockville, MD, 1998, 578-584.

16. J.R. Howell, M.J. Hall, and J.L. Ellzey, Combustion within porous media, in *Heat Transfer in Porous Media*, Y. Bayazitoglu and U.B. Sathuvalli, Eds., ASME, HTD-Vol. 302, New York, 1995, 1-27.

17. M. Perrin, P. Lievoux, R. Borghi, and M. Gonzalez, Utilization of a Numerical Model for the Design of a Gas Immersion Tube, in *Industrial Combustion Technologies*, M.A. Lukasiewicz, Ed., Amer. Soc. Metals, Materials Park, OH, 1986, 127-134.

18. C.L. Hackert, J.L. Ellzey, and O.A. Ezekoye, Combustion and heat transfer in model two-dimensional porous burners, *Comb. Flame*, 116, 177-191, 1999.

19. X. Fu, R. Viskanta, and J.P. Gore, Modeling of thermal performance of a porous radiant burner, in *Combustion and Radiation Heat Transfer*, R.A. Nelson, K.S. Ball, and Z.M. Zhang, Eds., Proceedings of the ASME Heat Transfer Division — 1998, Vol. 2, ASME, New York, HTD-361-2, New York, 1998, 11-19.

20. R. Weber, A.A. Peters, P.P. Breithaupt, and B.M.V. Visser, Mathematical modeling of swirling pulverized coal flames: what can combustion engineers expect from modeling?, *Amer. Soc. of Mech. Eng. (ASME) FACT*, 17, 71-86, 1993.

21. C.E. Baukal, Ed., *Oxygen-Enhanced Combustion*, CRC Press, Boca Raton, FL, 1998.

22. E.R. Bazarian, J.F. Heffron, and C. E. Baukal, Method for Reducing NOx Production During Air-Fuel Combustion Processes, U.S. Patent 5,308,239, 1994.

23. G.M. Gitman, Method and Apparatus for Generating Highly Luminous Flame, U.S. Patent 4,797,087, 1989.

24. C.E. Baukal, Basic Principles, in *Oxygen Enhanced Combustion*, C.E. Baukal, Ed., CRC Press, Boca Raton, FL, 1998, 1-44.

25. S. J. Williams, L. A. Cuervo, and M. A. Chapman, High-Temperature Industrial Process Heating: Oxygen-Gas Combustion and Plasma Heating Systems, Gas Research Institute Report GRI-89/0256, Chicago, IL, July 1989.

26. C.E. Baukal, P.B. Eleazer, and L.K. Farmer, Basis for enhancing combustion by oxygen enrichment, *Industrial Heating*, LIX(2), 22-24, 1992.

27. K.R. Benedek and R.P. Wilson, The Competitive Position of Natural-Gas in Oxy-Fuel Burner Applications, Gas Research Institute Report No. GRI-96-0350, Chicago, IL, September 1996.

28. D.B. Spalding, The Art of Partial Modeling, *Ninth Symposium (International) on Combustion*, Academic Press, New York, 1963, 833-843.

29. J.M. Beér and N.A. Chigier, *Combustion Aerodynamics*, Applied Science Publishers, London, 1972.

30. G. Damköhler, Elektrochem, 42, 846, 1936.

31. C.J. Lawn, T.S. Cunningham, P.J. Street, K.J. Matthews, M. Sarjeant, and A.M. Godridge, in *Principles of Combustion Engineering in Boilers*, C.J. Lawn, Ed., Academic Press, New York, 1987.

32. R. Weber, Scaling characteristics of aerodynamics, heat transfer, and pollutant emissions in industrial flames, *Twenty-Sixth Symposium (International) on Combustion*, The Combustion Institute, Pittsburgh, PA, 1996, 3343-3354.

33. V. Yahot and S.A. Orsag, Renormalization group analysis of turbulence. I. Basic theory, *J. of Sci. Comput*, 1(1), 1-51, 1986.

34. D. Choi, J.S. Sabnis, and T.J. Barber, Application of an RNG k-ε Model to Compressible Turbulent Shear Layers, AIAA Paper 94-0188, 1994.

35. D. Choudhury, Introduction of the Renormalization Group Method and Turbulence Modeling, Fluent, Inc., TM-107, 1993.

36. B.F. Magnussen and B.H. Hjertager, On mathematical models of turbulent combustion with special emphasis on soot formation and combustion, *Sixteenth Symp. (International) on Combustion*, Cambridge, MA, Aug. 15-20, 1976.

37. Y.R. Sivathann and G.M. Faeth, Generalized state relationships for scalar properties in non premixed hydrocarbon/air flames, *Combust. Flame*, 82, 211-230, 1990.

38. W.P. Jones and J.H. Whitelaw, Calculation methods for reacting turbulent flows: a review, *Combust. Flame*, 48, 1-26, 1982.

39. R.W. Bilger, *Turbulent Reacting Flows, Turbulent Flows with Nonpremixed Reactants*, P.A. Libby and F.A. Williams, Eds., Springer-Verlag, Berlin, 1980, chap. 3.

40. S. Chandrasekhar, *Radiative Transfer*, Dover Publications, New York, 1960.

41. W.A. Fiveland, Discrete-ordinates solutions of the radiative transport equation for rectangular enclosures, *J. Heat Transfer*, 106, 699-706, 1984.

42. A.S. Jamaluddin and P.J. Smith, Predicting radiative transfer in rectangular enclosures using the discrete ordinates method, *Comb. Sci. Tech.*, 59(4-6), 321-340, 1988.

43. A.S. Jamaluddin and P.J. Smith, Predicting radiative transfer in axisymmetric cylindrical enclosures using the discrete ordinates method, *Comb. Sci. Tech.*, 62(4-6), 173-186, 1988.

44. W.A. Fiveland and A.S. Jamaluddin, Three-dimensional spectral radiative heat transfer solutions by the discrete-ordinates method, in *Heat Transfer Phenomena in Radiation, Combustion, and Fires*, R.K. Shah, Ed., ASME, New York, HTD-106, New York, 1989, 43-48.

45. N.G. Shah, New Method of Computation of Radiative Heat Transfer in Combustion Chambers, Ph.D. thesis, Imperial College, London, 1979.

46. F.C. Lockwood and N.G. Shah, A new radiation solution method for incorporation in general combustion prediction procedures, *Eighteenth Symp. (International) on Combustion*, The Combustion Institute, Pittsburgh, PA, 1981, 1405-1414.

47. P. Docherty and M. Fairweather, Predictions of radiative transfer from nonhomogeneous combustion products using the discrete transfer method, *Combust. Flame*, 71, 79-87, 1988.

48. H.C. Hamaker, Philips Research Reports 3, 103, 112, and 142, 1947.

49. A.D. Gosman and F.C. Lockwood, Incorporation of a flux model for radiation into a finite difference procedure for furnace calculations, *Fourteenth Sym. (International) on Combustion*, the Combustion Institute, Pittsburgh, PA, pp. 661-671, 1973.

50. S.V. Patankar and D.B. Spalding, Simultaneous predictions of flow pattern and radiation for three-dimensional flames, *Heat Transfer in Flames*, N.H. Afgan and J.M. Beer, Eds., Scripta Book Co., Washington, D.C., 1974, 73-94.

51. T.M. Lowes, H. Bartelds, M.P. Heap, S. Michelfelder, and B.R. Pai, Prediction of radiant heat flux distributions, Int'l Flame Research Found. Report GO2/A/26, IJmuiden, The Netherlands, 1973.

52. W. Richter and R. Quack, A mathematical model of a low-volatile pulverised fuel flame, in *Heat Transfer in Flames*, N.H. Afgan and J.M. Beer, Eds., Scripta Book Co., Washington, D.C., 1974, 95-110.

53. R.G. Siddall and N. Selçuk, Two-flux modelling of two-dimensional radiative transfer in axi-symmetrical furnaces, *J. Inst. Fuel*, 49, 10-20, 1976.

54. R.G. Siddall and N. Selçuk, Evaluation of a new six-flux model for radiative transfer in rectangular enclosures, *Trans. IChem*, 57, 163-169, 1979.

55. D.K. Edwards, Molecular gas band radiation, in *Advances in Heat Transfer*, T.F. Irvine and J.P. Hartnett, Eds., Vol. 12, Academic Press, New York, 1976, 115-193.

56. A.T. Modak, Exponential wide band parameters for the pure rotational band of water vapor, *J. Quant. Spectosc. Radiat. Transfer*, 21(2), 131-142, 1979.

57. W. J. Horan, A.G. Slavejkov, and L.L. Chang, Heat Transfer Optimization in TV Glass Furnace, *Fifty-sixth Conference on Glass Problems,* IL, Oct. 24-25, 1995, 141-151.

58. C.E. Baukal, V.Y. Gershtein, J.F. Heffron, R.C. Best, and P.B. Eleazer, Method and Apparatus for Reducing NOx Production During Air-Oxygen/Fuel Combustion, U.S. Patent 5,871,343, issued on 16 February 1999.

59. D.J. Krichten, W.J. Baxter, and C.E. Baukal, Oxygen enhancement of burners for improved productivity, *EPD Congress 1997, Proceedings of the 1997 TMS Annual Meeting,* B. Mishra, Ed., February 9-13, Orlando, FL, 1997, 665-672.

60. M.A. Niemkiewicz, Safety Overview, in *Oxygen-Enhanced Combustion*, C.E. Baukal, Ed., CRC Press, Boca Raton, FL, 1998, 261-278.

8 CFD Modeling for the Steel Industry

Xianming Jimmy Li

CONTENTS

0-8493-2000-X/00/$0.00+$.50
© 2001 by CRC Press LLC

8.1   INDUSTRY OVERVIEW

Steel is a metal composed of iron plus varying amounts of carbon and other elements such as silicon, manganese, chromium, nickel, etc. Carbon, the most common alloying element, is the basis for primary classification. The carbon content is less than 0.10% in low-carbon steel, between 0.1 and 0.3% in medium carbon steel, between 0.30 and 2% in high carbon steel, and between 2.5 and 6% in cast iron. These classifications follow the iron-carbon phase equilibrium discussed later.

The 26th element in the periodic table, iron is one of the most common metals in the earth's crust. It can be found almost anywhere, combined with many other elements, in the form of ore. Pure iron has molecular weight 56. At room temperature, iron is a solid with a density of 7860 kg/m³.⁵ As temperature increases, iron goes through various transformations as shown in Table 8.1. Furthermore, these transformation points change dramatically with the level of carbon and other alloying elements. Steel, perhaps the most widely used material in the world, is a very complex material indeed.

Different types of steel — that is, steel with different properties and characteristics — are produced by adjusting the chemical composition and adapting any of the different stages of the steelmaking process, such as rolling, finishing, and heat treatment. As each of these factors can be modified, potentially there is virtually no limit to the number of different steels that can be made. Currently, there are over 3000 catalogued grades available, excluding custom formulas, ranging from basic grades for railway tracks to sophisticated high-alloy and stainless grades for specialized applications.

The steel industry worldwide produces over 750 million tons* of crude steel each year. The largest steel-producing countries are China, Japan, and the United States, each producing around 100 million tons per annum. Russia, Germany, and Korea each produce around 40 to 50 million tons.

Stainless steel is the generic name for a number of different steels that contain more than 10% chromium with or without other alloying elements. Stainless steel production has climbed rapidly in the past decade: over 16 million tons of finished stainless steel were produced in 1997, compared to 13 million tons in 1988 — an increase of more than 23%. Production has increased by more than 50% in Europe and Asia in that period. The largest stainless steel-producing countries are Japan (nearly 4 million tons), the United States (2 million tons), and Germany, Korea, Italy and France, each of which produced over 1 million tons of finished stainless steel in 1997.

Steel use in any country is closely linked to its economy, with the largest consumption in the wealthiest countries of the world. Consumption of finished steel products ranges from approximately 20 kilograms (kg) per capita each year in Africa to around 340 kg in Europe, 420 kg in North America, and 635 kg in Japan. However, the largest consumers are in Asia: Singapore (1200 kg per capita), Taiwan (over 970 kg), and Korea (830 kg). Per-capita consumption is climbing rapidly in Asia due to investments in industry, transport infrastructure, construction, and overall improved standards of living. For example, in the past decade, per-capita consumption has risen by nearly 470% in Malaysia, 240% in Korea, and nearly 80% in China.

Recycled steel (scrap) is a required and essential component of new steel, making steel a naturally environmentally responsible material. In basic oxygen steelmaking, scrap represents up to 30% of the raw materials charged into the furnace. It represents between 90 and 100% of the charge in electric arc furnace (EAF) production, which is also the principal route for stainless steel production. Steel is 100% recyclable and, with modern practices, can be used over and over again with no downgrading to a lower quality product. Steel's magnetic properties make it simple to

---

* Note that in this chapter, all tons are short tons unless otherwise stated: 1 ton = 2000 lb. = 0.90718 metric ton.

TABLE 8.1
Key Properties for Pure Iron

Temperature					Enthalpy of Transformation	
(K)	(°C)	(°F)	Transformation	Key Change	(cal/mole)	(kJ/kg)
1042	769	1416	Magnetic	Loss of magnetic properties	326	24.4
1184	911	1671	Polymorphic $\alpha \rightarrow \gamma$	bcc to fcc, or ferrite to austenite	215	16.1
1665	1392	2537	Polymorphic $\gamma \rightarrow \delta$	fcc to bcc again	165	12.3
1809	1536	2796	Melting point	Solid to liquid	3670	274.4
3135	2862	5183	Boiling point	Liquid to vapor	83712	6265.0

From Lankford, Jr., W.T., Samways, N.L., Craven, R.F., and McGannon, H.E., *The Making, Shaping and Treating of Steel,* 10th ed., United States Steel, 1985; Sargent-Welch Scientific Company, *Periodic Table of the Elements,* Skokie, IL, 1980; Sargent-Welch Scientific Company, *Table of Periodic Properties of the Elements,* Skokie, IL, 1980. With permission.

extract from other materials for recycling. Worldwide, approximately 350 million tons of steel scrap are recycled each year.

There is no one factor to explain the many environmental improvements achieved by the steel industry in recent years. One trend has become clear however: the steel industry has shifted its focus from end-of-pipe collection of emissions to considering improvements at every single stage of the steelmaking process. Steelmakers have therefore achieved even further emissions reductions by investing in overall cleaner production, better maintenance, and improved practices. New technologies, operating practices, employee education, and management attention have all been important.

Many of the improvements have resulted from very heavy investment programs. It is estimated that at least 10% of all steel industry capital expenditures have been specifically on environment improvement, or more than U.S.$ 20 billion in the last 10 years alone. This is almost certainly an underestimate, because it does not include investment in new steelmaking processes — such as continuous casting, thin slab casting, or coal injection — which enable much cleaner technologies to be introduced.

Steel production in the world has risen by approximately 30% in the past 25 years. In the same period, estimated employment in the major steel-producing countries (excluding China) has fallen from around 2.5 million to 1.3 million people. This enormous reduction has been the result of major investments by the world's steelmakers in modern steelmaking processes and technologies. The new technologies not only improve productivity but also increase yield efficiency while reducing resource consumption and the environmental impact of the steelmaking process.

In the U.S., steel is a $57 billion industry employing over 235,000 workers. It is the fourth largest energy-consuming industry in the U.S. and generates 3 million tons of solid waste. The U.S. steel industry competes in an environment where world capacity, some 900 million tons, exceeds actual annual production by nearly 150 million tons. Central to sustaining the industry's competitiveness are efforts to reduce costs and improve quality through fewer processing steps, better yields, greater energy efficiency, and better environmental performance.

EAF carbon steel producers, or "mini-mills," which are now among the most competitive steelmakers in the world, will continue to lead and grow due to their performance in such areas as worker productivity, cost savings, energy efficiency, shorter production times, expansion of product lines, and environmental improvements. The competitiveness of these companies will continue to generate improvements in other segments of the steel industry.

Contrast this with the steel industry of the 1970s and early 1980s. Steel was perceived to be in a perpetual, inescapable downturn. Many steel companies had gone bankrupt or underwent reorganization. Others faced severe financial difficulty. The U.S. steel industry, in fact, reduced

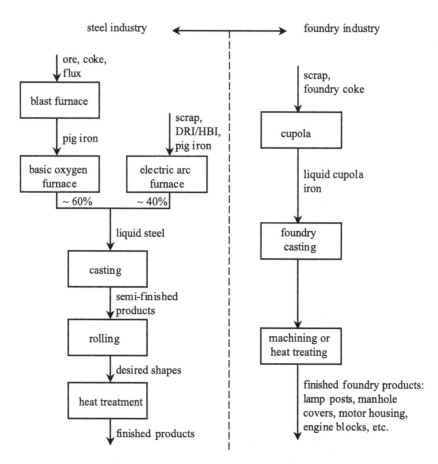

FIGURE 8.1 Key process steps of steelmaking.

total output by one third in the period between 1975 and 1985. During this same chaotic period, these "mini-mills" actually doubled their output, using EAF and continuous casting technology to make steel products efficiently from a feedstock of steel scrap. In 1970, EAF carbon steel companies accounted for only 15% of U.S. steel production. In 1999, the EAF carbon steel producers were approaching half of the steelmaking capacity in the U.S.

8.2 OVERVIEW OF STEEL MANUFACTURING PROCESSES

Although numerous variations exist, the overall steelmaking process can be conceptually described in several key steps, as shown in Figure 8.1. The majority of steel products start with iron ore that is converted in a blast furnace to hot metal or pig iron. The carbon content in the blast furnace hot metal is usually more than 3.5%, which must be reduced in a basic oxygen furnace (BOF). In the BOF, additional metallurgical treatments may be included to produce steel of a desired property. Alternatively, electric arc furnaces are used to melt mostly scraps to produce liquid steel, thus bypassing the blast furnace step. This approach eliminates the dusty operations of coke ovens and iron ore handling, reduces the capital investment considerably, and has been widely used by the "mini-mills" since the 1970s.

The liquid steel from either process (i.e., BOF or EAF) is sent to a continuous caster where it is carefully poured into water-cooled molds and solidified into semi-finished products. Next in the process are the rolling and finishing steps. Here, depending on the product specifications, many steps may be required. But the general process flow is to reheat the semi-finished, continuously

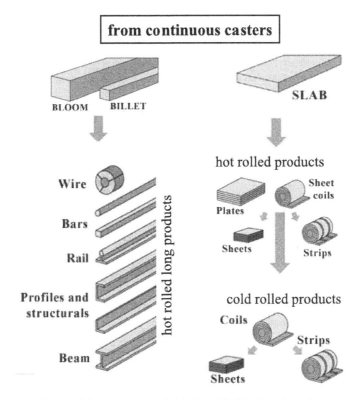

FIGURE 8.2 Major semi-finished and finished steel products.

cast products to sufficient temperature in reheat furnaces, roll them into flat or long products, and heat-treat the products to desired properties. A simplistic summary of the products in connection with the process steps is shown in Figure 8.2.

8.3 THERMAL PROPERTIES OF IRON AND STEEL

Properties relevant for thermal (rather than metallurgical) analysis are density, specific heat, thermal conductivity, and liquid iron viscosity. Because of the great variety of iron and steel products, these properties vary over a wide range. Most properties presented here are for pure iron, with only a few examples for alloys. For the most definitive and comprehensive source of data, the reader should consult References 1, 2, and 3.

The standard density for solid iron is 7860 kg/m³ and, for all practical purposes, is a constant in the solid phase. Iron has a linear thermal expansion coefficient of approximately 1.1 × 10⁻⁵ m/m/K.[4] For liquid iron, the density-temperature relation is approximately[5]

$$\rho = 8528 - 0.836T \tag{8.1}$$

where density is in kg/m³, and T in Kelvin.

The specific heat of steel has a considerable variation with temperature, as seen in Figure 8.3. Together with the data in Table 8.1, the enthalpy of pure iron can be computed and plotted with respect to temperature, as shown in Figure 8.4.

Thermal conductivity for pure iron is shown in Figure 8.5. At low temperatures, iron shows tremendous variability in thermal conductivity. But at room temperature and above, its thermal conductivity is relatively constant. Figure 8.6 shows thermal conductivity of pure iron, typical carbon

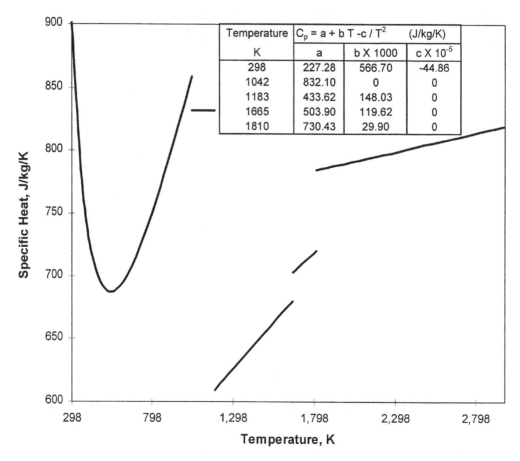

FIGURE 8.3   Specific heat of pure iron at various phases.

steels, and high alloy steels at higher temperatures. It is seen that pure iron has the highest thermal conductivity of these three groups. As temperature increases, however, the differences diminish.

Liquid iron viscosity is about 5 times higher than that of water at room temperature, as shown in Figure 8.7. There is considerable variability in the literature, depending on composition on the viscosity data. Figure 8.7 shows the upper and lower bounds in the form of the following correlations

$$\mu = \frac{2250}{T} - 3.475, \text{ lower bound}^{6} \tag{8.2}$$

$$\mu = 1.573 \times 10^{-8} T^{2} - 6.739 \times 10^{-5} T + 0.07741, \text{ upper bound}^{7} \tag{8.3}$$

where μ is in kg/m/s, and T is in Kelvin.

Because carbon is such an important ingredient in steel, its effect on liquid iron viscosity is shown in Figure 8.8[7] for a constant temperature of 1873K. Pure iron has the highest viscosity. Viscosity drops almost linearly with carbon content up to about 0.6%, then stays essentially constant for carbon content up to about 2.5%, and finally drops linearly again with carbon content.

Iron and steel surface emissivities vary greatly with product type, surface finish, impurity level and oxidation state. They vary from highly reflective (0.05–0.07) for highly polished or electrolytic surfaces, to almost black (0.95) for heavily oxidized cast iron.[8] Figure 8.9 shows the spectral

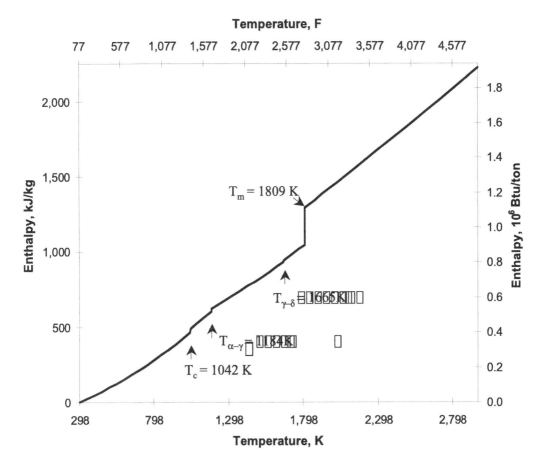

FIGURE 8.4 Pure iron (Fe) enthalpy-temperature relationship. Note that iron enthalpy is zero at 298.15K.

emissivity of polished iron at room temperature. Typically, a value of 0.4 can be used in the absence of additional information regarding surface conditions.

The melting point of steel has a strong dependence on carbon content. Figure 8.10 shows schematically the effect of carbon on iron melting point. Pure iron melting point is 1809K (1536°C). Up to 4.27% carbon level, the melting point decreases almost linearly to 1425K (1152°C). Beyond that, the melting point stays constant as carbon precipitates out in the form of graphite. Above 1425K, liquid and solid iron co-exist at even lower carbon content. See iron-carbon phase equilibrium diagram for more details.[5] Furthermore, similar behavior is observed for other alloying elements such as sulfur.

8.4 CONVENTIONAL FURNACES AND THEIR UNIQUE MODELING CHALLENGES

8.4.1 BLAST FURNACE

The modern blast furnace is an impressive structure, the largest being up to 12 m (39 ft) in diameter and over 50 m (164 ft) tall, producing up to 12,000 tons of hot metal per day, and involving hundreds of millions of dollars of capital investment. It is a refractory-lined steel vessel that is also referred to as a "shaft furnace." Currently in the U.S., there are approximately 55 million tons of blast furnace hot metal produced per year, requiring about 23 million tons of coke. Figure 8.11 shows a typical

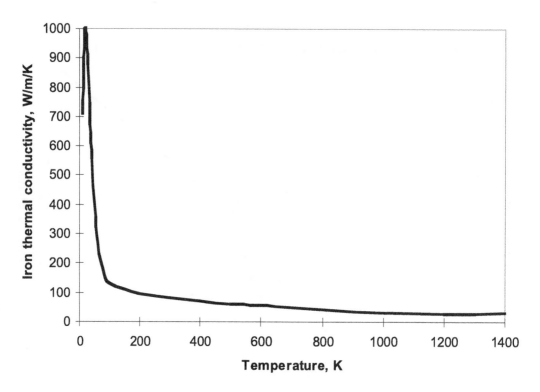

FIGURE 8.5   Thermal conductivity of pure iron. (From Perry, R.H. and Green, D., *Peery's Chemical Engineer's Handbook,* 6th ed., McGraw-Hill, New York, 1984.

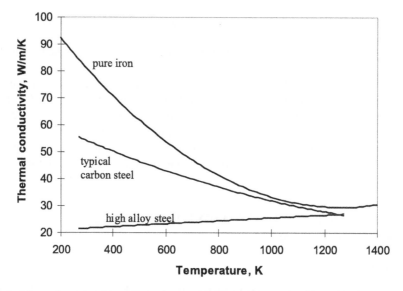

FIGURE 8.6   Thermal conductivity of pure iron and various alloy steels. (From Air Force Materials Laboratory, *Aerospace Structural Methods Handbook,* Wright-Patterson Air Force Base, Ohio, 1969; ASM, *Metals Handbook,* 8th ed., ASM, 1961.

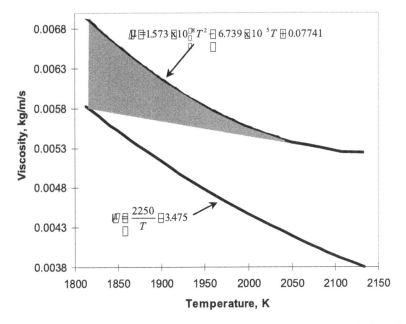

FIGURE 8.7 Viscosity of liquid iron (the equations describe the bounds, μ in kg/m/s, *T* in K). (From Cavalier, D., *Proc. Nat. Phys. Lab.,* 2(9), 40, 1958; Barfield, R.N. and Kitchener, J.A., *J. Iron and Steel Institute,* 180, 324-329, 1955.)

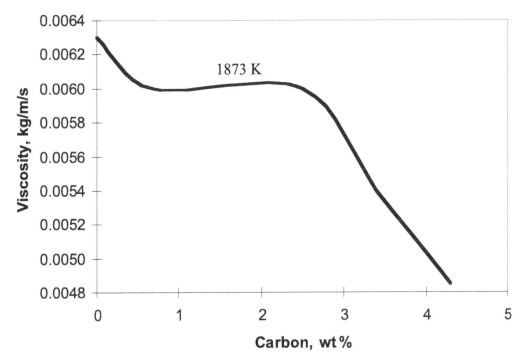

FIGURE 8.8 Effect of carbon content on viscosity of liquid iron at a constant temperature. (From Barfield, R.N. and Kitchener, J.A., *J. Iron and Steel Institute,* 180, 324-329, 1955.)

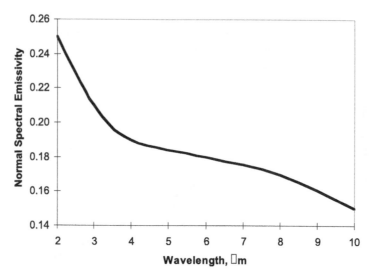

FIGURE 8.9   Spectral emissivity of polished iron at room temperature. (From Seigel, R. and Howell, J.R., Thermal Radiation Heat Transfer, 2nd ed., McGraw-Hill, New York, 1981.)

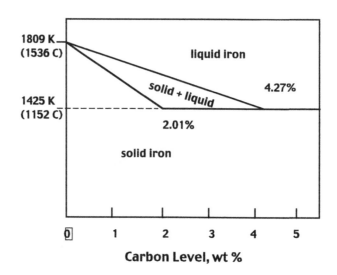

FIGURE 8.10   Schematic diagram showing the effect of carbon content on iron melting point.

material balance around a blast furnace that produces 3300 tons per day of hot metal. Iron ore is combined with flux materials (mainly limestone) and coke and charged into the furnace, layer by layer, from the top of the furnace. Preheated air at up to 1370K (2000°F) from regenerative hot-air stoves enters the furnace in the bottom part through injectors called tuyeres and supplies oxygen and a substantial portion of the process energy requirement. Oxygen reacts with coke in the charge to form CO, a reducing gas, and the reaction generates additional energy. CO reacts with iron ore (Fe_2O_3) in several steps to produce metallic iron (Fe), which melts and picks up considerable amounts of carbon, manganese, phosphorus, sulfur, and silicon. The principal reactions in a blast furnace are shown in Table 8.2. The gangue (mostly silica) of the iron ore and the ash in the coke combine with the limestone to form the blast furnace slag. The hot metal and slag are tapped periodically from the iron notch and the cinder notch, respectively.

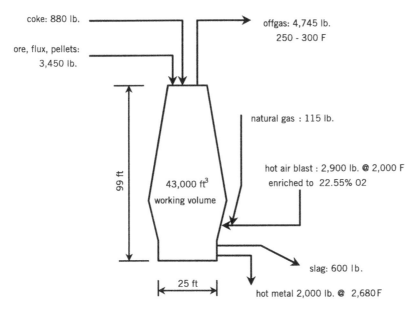

coke: 880 lb.

offgas: 4,745 lb.
250 - 300 F

ore, flux, pellets:
    3,450 lb.

natural gas : 115 lb.

99 ft

43,000 ft³
working volume

hot air blast : 2,900 lb. @ 2,000 F
enriched to 22.55% O2

25 ft

slag: 600 lb.

hot metal 2,000 lb. @ 2,680 F

FIGURE 8.11   Typical material and energy balance for a 3300 tons per day blast furnace (material streams are per ton of hot metal).

TABLE 8.2
Principal Reactions in a Blast Furnace

Blast Furnace Reaction	ΔH_{298K}, cal/mole[a] (negative sign for exothermic reaction)
$C + \frac{1}{2} O_2 \rightarrow CO$	−26,416
$3\ Fe_2O_3 + CO \rightarrow 2\ Fe_3O_4 + CO_2$	−12,636
$Fe_3O_4 + CO \rightarrow 3\ FeO + CO_2$	+8,664
$FeO + CO \rightarrow Fe + CO_2$	+4,136
$3\ Fe_2O_3 + H_2 \rightarrow 2\ Fe_3O_4 + H_2O$	−2,800
$Fe_3O_4 + H_2 \rightarrow 3\ FeO + H_2O$	+18,500
$FeO + H_2 \rightarrow Fe + H_2O$	+5,700
$CO_2 + H_2 \rightarrow CO + H_2O$	+9,836
$H_2O + C \rightarrow CO + H_2$	+31,380
$CO_2 + C \rightarrow 2\ CO$	+41,220
$FeO + C \rightarrow Fe + CO$	+38,575
$CH_4 \rightarrow C + 2\ H_2$	+17,900
$CH_4 + \frac{1}{2} O_2 \rightarrow CO + 2\ H_2$	−8,522
$C_{(coke)} \rightarrow C_{(dissolved\ in\ iron)}$	

[a] Data from Zumdahl, S.S., *Chemical Principles*, D.C. Heath and Company, Lexington, MA, 1992; Wark, K., *Thermodynamics*, 4th ed., McGraw-Hill, New York, 1983.

In ironmaking, coke is a unique, necessary ingredient that fulfills three functions. First, as stated above, it is partially combusted with oxygen (yielding CO) to provide energy. Second, it provides the structural support for the charge in the blast furnace, giving the necessary charge-bed porosity through which the gases can flow. Third, the coke provides the reductant that reacts with, and strips away, the oxygen from the iron in the ore, thereby yielding the metallic iron. The charge in the blast furnace in fact has several zones with a layered structure. The coke layer is considerably

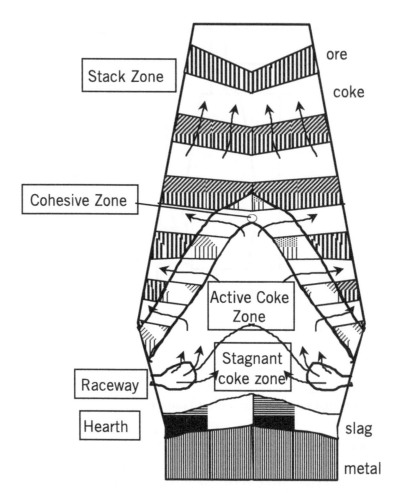

FIGURE 8.12   Various zones in the blast furnace.

porous so that the blast air can go through. A schematic diagram of the charge inside a blast furnace is shown in Figure 8.12. For thermal analysis purposes, the blast furnace charge is a gas-liquid-solid three-phase system with chemical reactions and heat transfer.

In addition to the blast furnace proper, many support facilities around a blast furnace also expend considerable amount of energy, such as hot air stoves and coke ovens. Downstream processing of the hot metal also involves extensive combustion, such as basic oxygen furnaces, electric arc furnaces, soaking pits, annealing furnaces, and reheat furnaces. Some of these combustion applications are discussed in later sections.

Besides energy efficiency and metal flow pattern, modeling plays an important role in technology development for the blast furnace, already under siege by alternative production technologies. One recent application is coke replacement with natural gas. As coke production is environmentally disruptive because of its heavy emissions, few new coke ovens are currently being built, and aging coke ovens gradually retire out of service. As blast furnace coke rates decrease and productivity increases, the limits of current understanding of blast furnace operations are exceeded. In a government-industry collaboration sponsored by the U.S. Department of Energy, the steel industry in the U.S. produced a "Steel Industry Roadmap" document that clearly identified a "comprehensive blast furnace model (including fluid flow and kinetics)" as a technical challenge for the industry.[9]

A step back from the overall blast furnace flow and kinetics is to understand the changes introduced by natural gas injection in the raceway. How does natural gas injection affect blast air

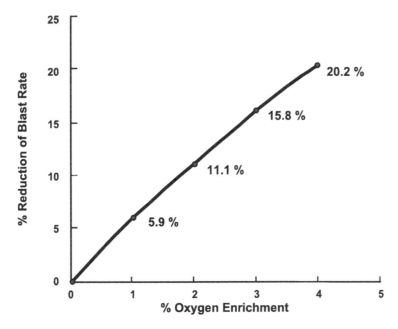

FIGURE 8.13 Oxygen enrichment reduces blast air requirement.

temperature in the raceway? Does soot formation occur significantly to affect porosity of the furnace charge? What is the time-temperature history of natural gas as it exits the injector? How should a natural gas injector be designed for optimum effectiveness and operation?

Another application of modeling is oxygen enrichment of blast air. Because of the large amount of air supply, slight enrichment with oxygen of about 1 to 2% can be economically attractive, as shown in Figure 8.13. But how does the elevated oxygen content interact with the blast furnace operation? Will it affect reactivity of the charge? Will it affect the quality of the hot metal? How does one quantify the benefit of oxygen enrichment? These questions can be answered with modeling.

Because of the complex nature of the blast furnace, no comprehensive model with fluid flow and kinetics currently exists, although considerable effort has been invested in this area as reviewed in a recent article.[10] Current models include global material and energy balance (zero dimensional), one-dimensional or zonal models,[11] and two-dimensional models.[12] Wang, et al.[13] attempted to develop a comprehensive model for the cohesive zone in the blast furnace starting from the fundamental transport equations. With more sophisticated multiphase reacting flow platforms available commercially,[14] better understanding of the kinetics, and field data from more reliable sensors, there is opportunity for progress in developing a comprehensive blast furnace model.

8.4.2 BASIC OXYGEN FURNACE

The basic oxygen furnace (BOF) produces about 60% of the liquid steel in the U.S. and Canada. Although this share may decrease due to the increasing use of mini-mills with electric arc furnaces (EAF), the BOF will continue to be a major source of steel for many years to come. On average, 74% of the BOF charge is liquid hot metal from blast furnaces, and the remainder is recycled scrap. The advantages of BOF are high production rates and the tapping of low residual element liquid steel.

In the BOF, hot metal at about 1770K (1500°C) is charged, together with some recycled scrap. The enthalpy contained in the hot metal accounts for about 60% of the total energy required by the process. Flux materials, such as burnt lime (CaO, 2–5% of charge weight), burnt dolomite (CaO-MgO, 0.7–0.8% of charge weight), and an even smaller amount of fluorspar (a mixture of mainly CaF_2 and $CaCO_3$), are also charged to aid in the refining process and to protect the furnace lining.

Oxygen at 99.5% or higher purity is injected supersonically into the melt through a water-cooled lance with multiple converging-diverging nozzles at its tip. The oxygen jets penetrate the slag and foam on the surface, stir up the hot metal, and react with carbon, silicon, phosphorus, manganese, sulfur, and other elements, as well as some iron, in the hot metal. These oxidation reactions provide about 35% of the energy requirement. Slag-making reactions supply the remaining energy requirement in a traditional BOF process.

Flow and heat transfer modeling of the BOF is important to process understanding, product quality, and furnace lining life. Modeling challenges include compressibility, chemical reactions, turbulence, and heat transfer. Furthermore, BOF again involves a possibly three-phase system: liquid hot metal, gas, and solid scrap.

Recent applications of CFD modeling include post-combustion and supplemental fuels to extend the use of hot metal, increase steel production without investing in new hot metal capacity, and increase the use of recycled scrap. CFD was also used to aid in the visualization of the BOF where measurements are very difficult.[26] Dispersed bubble flow from the bottom stirring equipment is handled using a "mixture model" or the volume of fluid (VOF) approach. Experimental data from water models were used to validate the simulations. CFD predictions show reasonable agreement with measurements and observations in the water model. Modeling as a visualization tool was investigated by Guthrie[16] for the BOF.

8.4.3 Electric Arc Furnace

Although electricity has been used for steelmaking since the late 1800s when Dr. Paul Heroult developed and patented the first alternating current direct-arc electric furnace (EAF), widespread application of electric steelmaking did take off until the 1970s. In 1970, electric arc furnace steel represented only 15% of the U.S. steel production. In 1999, it accounted for about half of the total capacity. Low investment costs, low man-hours per ton of steel produced, low scrap prices, and favorable electricity rates contributed to this dramatic increase. Inherently, EAF is flexible, highly efficient, and less capital-intensive.

A complete batch of steel in an EAF, from charging to tapping, is called a "heat." Each heat includes the charging, meltdown, oxidation, composition and temperature adjustment, and tapping periods. During charging, a combination of scrap, scrap substitutes (direct reduced iron or hot *briquette* iron), and pig iron is charged through the swung-open roof to form a scrap pile, although the charge can be 100% steel scrap. Carefully portioned flux materials are added to the charge. The roof is swung back into place and the electrodes are lowered to within inches of the scrap pile. An arc is struck and the meltdown begins. Current flows from one electrode to the scrap through an electric arc, then from the scrap to another electrode through a second arc to complete the circuit. The intense heat from the arc melts the scrap underneath the electrodes, and the electrodes are gradually lowered to follow the scrap pile. The molten iron starts to heat the scrap pile from below through radiation and contact. In this meltdown period, a significant amount of electrode consumption occurs due to direct oxidation in air.

Oxidation occurs concurrently with meltdown when molten metal begins to form. Phosphorus, silicon, manganese, carbon and iron combine with oxygen and supply about 35% of the total energy requirement for the "heat." Oxygen comes from the furnace atmosphere, direct oxygen injection, and/or cinder, ore, and scale. During oxidation, violent gas bubbles stir the molten metal, flush out impurities, and make the liquid more uniform in composition and temperature.

After the charge is melted, a steel sample is taken for composition analysis. Several temperature measurements are also taken. Based on these data, the required composition adjustments are made and the calculated ingredients are added. When the metallurgy is correct, the electrodes are raised, and the furnace is tilted to pour the liquid metal through a tapping spout into a ladle where further metallurgical refinements can be made.

Since 1970, EAF technology has improved substantially: the average tap-to-tap time decreased from 150 minutes to only 55 minutes, electricity consumption decreased from 550 to 375 kWh per ton of liquid, and electrode consumption decreased from 6 to 1.8 lb. per ton of liquid (2.72 kg to 0.82 kg).[9] Supplemental oxy-fuel melting is one such improvement where natural gas supplements electricity as a heat source and excess oxygen accelerates the oxidation process.

Like the BOF, the EAF involves very complex flow and heat transfer phenomena in addition to metallurgical processes. It is also a batch operation with inherently transient variations. A complete model describing the start-to-finish process does not exist today. However, snapshots of the process can be analyzed and appropriate assumptions can be made to extract useful information. For example, in the supplemental oxy-fuel melting application, the placement of burners will affect furnace life and productivity. For optimizing burner placement, the scrap shape can be approximated during a snapshot of the process when the supplemental fuel is at its maximum rate. Post-combustion of CO inside the EAF is another new development that captures the combustion energy of CO for electricity savings and CO emission reduction.[17] A detailed flow and heat transfer model, including the arc geometry, was presented by Szekely et al.[18] Gittler et al.[19] presented a study of EAF using the Euler-Lagrange approach for particle transport.

8.4.4 Continuous Reheat Furnace

Reheat furnaces are used to raise the temperature of semi-finished products, such as billets, slabs, and blooms, to a level suitable for hot-rolling toward more finished products. Depending on the product flow pattern, continuous reheat furnaces can be pusher, rotary-hearth, walking-beam, walking-hearth, or roller-hearth type. Several passes of reheat may be required before the final product is finished. As shown in Section 8.2, essentially all steel products must go through some reheat step, whereas semi-finished steel products may come from the blast furnace or the EAF route. Reheat operations consume 0.233 to 2.33 GJ ($0.221–2.21 \times 10^6$ Btu) energy per ton of product heated, depending on the charging temperature. They are the second largest energy consumer in steelmaking, next only to melting, and they account for about one third of the total NOx emissions of the steel industry.[20,21]

A typical walking-beam continuous reheat furnace can be 20 m (60 ft) long and 9 m (30 ft) wide. It usually has several heating zones powered by natural gas burners (or other available fuels) for better temperature control and product uniformity. Semi-finished steel products are charged into the furnace at one end, and progress through the furnace on walking beams, then discharge from the opposite end after attaining the desired temperature. Depending on the product being heated, the residence time can range from minutes for thin sheets to hours for blooms, billets, and slabs.

Although natural gas is a preferred fuel for reheat furnaces, coke oven gas, blast furnace top gas, or liquid fuel can also be used, depending on availability and economics. Almost all reheat furnaces use air as the oxidant, although some low-level enrichment can be beneficial for productivity boost. Usually, a recuperator or a regenerator is employed to preheat combustion air for this relatively high temperature process (1370–1590K, or 2000–2400°F).

Continuous reheat furnaces are perhaps the most amenable to CFD analysis among the conventional steel industry furnaces because the process is approximately steady state, and no phase change is involved. They are usually "clean" in that very little dust, foam, or slag is present. Yet their operation can impact the overall plant economics because of its large energy consumption and environmental emissions. Section 8.7 presents a case study on a continuous reheat furnace. Other studies in the literature include a burner development effort for reheat furnaces by IFRF,[22] rapid heating furnace for billets,[23] and NOx emission analysis for a gas fired reheat furnace.[24]

8.4.5 Batch Reheat and Other Furnaces

Batch reheat furnaces are the older type of furnaces used to reheat all sizes and all grades of steel. Soaking pits are used for ingots and rolls of coils. The products being heated remain stationary in these type of furnaces. A key difference between a batch reheat furnace and a soaking pit is perhaps the way the product is charged and withdrawn. In a soaking pit, usually the furnace roof — or for some types, the roof and the side walls — is lifted for charging and discharging. By contrast, products are typically charged and withdrawn from doors in a batch reheat furnace.

In either case, steel products are charged and the doors or other furnace sections are in place. Burners using liquid or gaseous fuel are ignited. A deliberate hot gas flow pattern is established that heats the products gradually to the desired temperature without oxidation or heat damage. Because of the unusual size of the charge, the heating time can vary tremendously, sometimes requiring several days for one charge to be completed.

Key considerations for batch reheat furnaces and soaking pits, as far as combustion modeling is concerned, are temperature uniformity, process thermal efficiency, and emissions. For these considerations, the batch furnace can be treated as steady state, especially when the heating time is very long. Obviously, the temperature distribution in the heated product is time-varying; thus, analysis of the thermal history of the heated products should be transient. Like continuous reheat furnaces, batch furnaces and soaking pits are "clean" processes without multiphase transitions. They are more amenable to analysis than BOF or other melting processes.

8.5   EMERGING TECHNOLOGIES IN IRON AND STEELMAKING

The desire and necessity of utilizing lower-grade ore and fuel unsuitable for blast furnace has stimulated the development of new processes for ironmaking. For example, there are so-called "direct reduction" processes that produce metallurgical iron units by reacting solid iron ore with gas-phase reductants at temperatures below the melting point of iron. The product of these direct reduction processes is called direct reduced iron (DRI), and is used as a scrap substitute for the electric arc furnace. The relatively pure DRI dilutes unwanted contaminants such as nickel, copper, and tin present in the steel scrap and improves steel quality. If the process exceeds the iron melting point and results in a liquid iron product, it is usually called a direct smelting process. Midrex is the major player in direct reduction, with about 22 million tons of DRI produced in 1995; the coal-based COREX process for direct smelting process is another commercial alternative to the blast furnace. AISI Direct Smelting (U.S.), DIOS (Direct Iron Ore Smelting, Japan), HIsmelt (Australia), ROMELT (Russia), and Hoogovens CCF (Cyclone Converter Furnace, The Netherlands) are major competing development efforts for COREX.[9] Ultimately, the most desirable process would be one using coal and ore fines to directly produce liquid iron. Liquid iron is preferred because it contains less gangue and is more amenable to the traditional BOF process. Coal is preferred over coke or natural gas because of its abundance and lower cost. Fines are preferred because no agglomeration costs are incurred. Many of the direct smelting processes promise a revolutionary change in ironmaking by replacing the blast furnace.

Since the revolutionary change from integrated mills to mini-mills where EAF and continuous casting are central players, the current steelmaking technology developments are more evolutionary rather than revolutionary. They involve process improvements, rather than radical redesigns, for productivity and efficiency gains. These incremental changes in the long run can also be significant.

CFD modeling has been used for new technology development for HIsmelt,[25] iron-bath reactors,[26] and the Hoogovens CCF development,[27] among others. With increasingly more powerful software and hardware, realistic physical phenomena can be included, and CFD will likely play a more important role in new technology development.

8.6   OTHER MODELING APPLICATIONS IN THE STEEL INDUSTRY

By far, the most applications of CFD modeling in the steel industry are for metallurgical studies, rather than for combustion, for good reason. For steel producers, product quality and production intensity are the ultimate measures of success. Working directly with process metallurgy seems the most appropriate. In connection with the mini-mill revolution, a substantial amount of work in modeling has related to understanding of the flow pattern and temperature distribution in the tundish,[28-32] liquid steel flow and heat transfer in ladles before and during teeming to a caster,[33] as well as the overall casting process.[34] Advances in EAF required deeper understanding of magnetohydrodynamics.[35] Improvements in the BOF need to start from further understanding of heat and mass transfer in the bottom blown process,[36] and overall steel quality improvement depends on the quality of desulfurization during inductive stirring,[37] sulfur[38] and inclusion removal,[39] novel treatment techniques such as pulsation,[40] etc. More precise control on steel composition has resulted in greater emphasis on ladle metallurgy with gas stirring,[41-44] vacuum degassing,[45] and new developments such as composition adjustment by sealed argon bubbling[46] and gas-injected iron bath.[47] Comprehensive reviews on CFD modeling for process metallurgy including flash smelting, electromagnetic stirring, gas-stirring in ladles, and solidification, are available in the literature.[48,49]

8.7 CASE STUDY: PRODUCTION INCREASE AND NOx REDUCTION FOR A CONTINUOUS REHEAT FURNACE

8.7.1 CASE BACKGROUND

Productivity increase and NOx control in reheat furnaces are two drivers for new technology introduction. In 1996, a U.S. Dept. of Energy program was initiated to investigate an advanced steel reheat furnace concept. The basic idea was to use oxygen enrichment for enhanced heat transfer and fuel reburn for NOx control, with over-fire air to control CO emission. A schematic diagram of this idea is shown in Figure 8.14. The heating zones (top and bottom) are oxygen-enriched to enhance heat transfer. Additional natural gas is injected downstream of the heating zones to destroy NOx via the reburn mechanism.[50] Over-fire air (OFA) is injected further downstream to destroy CO and recover fuel energy. A comprehensive discussion on oxygen-enhanced combustion — including theory, application, equipment, and safety — is available elsewhere.[51] The goal is to achieve up to 50% reduction in NOx, and 20% increase in throughput. CFD is used to analyze the flow pattern and heat transfer rates in a "typical" reheat furnace. The results are inputs for the reburn design, which is verified with CFD in an iterative procedure. This case study

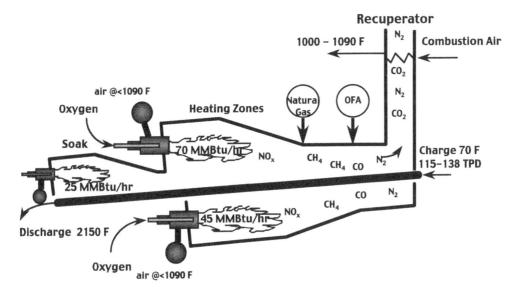

FIGURE 8.14 Typical reheat furnace modified for heat zone oxygen enrichment and gas reburn.

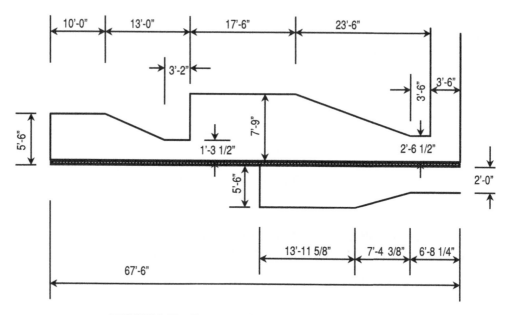

FIGURE 8.15   Geometry of the reheat furnace as modeled.

documents a process-level analysis followed by CFD modeling of the reheat furnace. The process-level analysis provides a broader perspective to the technology development effort.

8.7.2   METHODS

8.7.2.1   Geometry and Computational Grid

A market survey indicated that a pusher-type billet reheat furnace, as shown schematically in Figure 8.15 for the purpose of CFD analysis, is representative of a modern, continuous reheat furnace. It is 67.5 ft. (20.6 m) long, 30 ft. (9.1 m) wide, with three zones (soak, top, and bottom). The furnace is fired with natural gas with NO. 2 oil as backup. Steel billets are end-charged and side-discharged. This furnace was designed in 1992 and on-stream in June 1993. It is a modern furnace with computer controls and high fuel efficiency. The design production rate of this furnace is 115 tons/hr, and the steel discharge temperature is 1450K (2150°F). The furnace is equipped with a recuperator that preheats combustion air to 811K (1000°F). The total fuel firing rate at design condition is 41.1 MW (140 × 10⁶ Btu/hr), or approximately 1.29 GJ (1.22 × 10⁶ Btu) per ton of steel reheated.

Burners in this furnace are from a commercial supplier. The type, quantity, and key geometry information are summarized in Table 8.3. The overall dimension is large compared to the details

TABLE 8.3
Burner Specifications

Burner Type	Quantity	Zone	Air Duct Diameter (in.)	Main Gas Orifice (in.)	4 Gas Lances Location (Diameter, in.)	Orifices (in.)
S-1070-080 FTR	25	Soak	9.25	1.625	n/a	n/a
S-1070-160 FTR	8	Top	18.00	3.500	1.000	0.120
S-1070-125 FTR	7	Bottom	16.00	3.000	1.000	0.104

of the burners, which are critical to establishing the correct flames in the furnace. This large difference in geometric scale poses a challenge in the numerical solution because it results in a large computational grid. Overall, a 131 × 112 grid is used. This grid conforms to the irregular boundaries of the geometry. Denser grid points are used where high gradients are expected. Exploratory calculations show that the solution is independent of further grid refinement.

8.7.2.2 Basic Assumptions

For the purpose of this analysis, an average condition was sought that can be treated as steady state, which represents a steady production rate and a time-averaged turbulence field. Changes that are intentionally implemented, such as those resulting from the implementation of the reburn technology, are captured. The steady Reynolds-averaged Navier-Stokes equations, together with heat transfer, mass transfer, and turbulence models, will be solved.

The furnace is assumed to be wide so that conditions across the width of the furnace are constant; thus, it can be modeled as 2-D. Another implication is that the heat loss from the side walls is neglected. In actuality, total wall losses account for about 4% of gross fuel energy input, and the side walls are small as compared to the roof and bottom walls. Therefore, the 2-D assumption is reasonable. A third implication has to do with energy distribution by radiation as a result of the side walls. As far as radiation is concerned, a 2-D model is really 3-D with unit depth. The boundaries of the third coordinate direction are perfectly reflective. Because the refractory walls have very little heat loss, they are close to the reflective walls assumed in the 2-D model; therefore, the 2-D result will be representative of the true situation except, perhaps, in a small volume immediately next to the side walls.

Because of the 2-D assumption, the bottom zone must have a separate flue exit. In reality, the furnace width increases under the chimney, a feature known as "ears" in the industry. The ears allow flue gas from the bottom zone to join the rest of the flue and leave the furnace. And because such a flow pattern is truly 3-D, a simplification must be made. Without the flue gas from the bottom zone, the average velocity above the billets in the charging section of the furnace would be smaller, thus increasing the residence time. This condition is better than actual for CO burnout.

The fuel consists of 90% CH_4, 5% C_2H_6, and 5% N_2 by volume. The molecular weight is 17.3. Its gross heating value (HHV) is 997.2 Btu/scf at 60°F (50.75 MJ/kg), and the air:fuel mass ratio is 15.64.

There is speculation regarding soot in the flame and its influence on radiative heat transfer. Soot formation is a complex topic and, currently, only qualitative models are available for prediction. The accuracy of such a prediction depends critically on the flow and temperature fields, which in turn are intimate with mixing patterns. Because the 2-D model cannot predict the 3-D mixing pattern in the furnace, even qualitative trends of soot may not be available. For this reason, it is assumed that soot contribution to heat transfer is unchanged between the baseline and any future modification; therefore, it is not included in the model.

8.7.2.3 Fluid Flow and Heat Transfer Equations

With the assumptions listed above, the steady Reynolds-averaged Navier-Stokes equations assume the following form:

$$\frac{\partial}{\partial x_i}\left(\rho u_i\right) = 0 \tag{8.4}$$

$$\frac{\partial}{\partial x_i}\left(\rho u_i u_j\right) = -\frac{\partial p}{\partial x_j} + \frac{\partial}{\partial x_i}\left[\mu_t\left(\frac{\partial u_j}{\partial x_i} + \frac{\partial u_i}{\partial x_j}\right) + \frac{2}{3}\mu_t\frac{\partial u_k}{\partial x_k}\right] \tag{8.5}$$

Here, ρ is the mixture density, u is the velocity, x is the spatial coordinate variable, p is the pressure, g is the gravitational acceleration, and μ_t is the turbulent viscosity that is to be described by the turbulence model. The energy equation is written in terms of enthalpy of the mixture, defined as:

$$h = \sum_l Y_l \left(h_l + \Delta h_l^0 \right) \tag{8.6}$$

where Y_l is the mass fraction, and Δh_l^0 is the enthalpy of formation for species l:

$$\frac{\partial}{\partial x_i}\left(\rho u_i h \right) = \frac{\partial}{\partial x_i}\left(\frac{\mu_t}{\sigma_h}\frac{\partial h}{\partial x_i} \right) - \frac{\partial q_{r,i}}{\partial x_i} \tag{8.7}$$

where σ_h is the turbulent Prandtl number for the energy Equation ($\sigma_h = 0.85$), and the source term $\frac{\partial q_{r,i}}{\partial x_i}$ is due to radiation heat transfer, as described later.

8.7.2.4 Turbulence

Although more elaborate options such as the RNG-k-ε model and the full Reynolds stress model[52] are available, the standard k-ε two-equation turbulence model is used in this analysis as a compromise between speed and accuracy:

$$\frac{\partial}{\partial x_i}\left(\rho u_i k \right) = \frac{\partial}{\partial x_i}\left(\frac{\mu_t}{\sigma_k}\frac{\partial k}{\partial x_i} \right) + G_k + G_b - \rho\varepsilon \tag{8.8}$$

$$\frac{\partial}{\partial x_i}\left(\rho u_i \varepsilon \right) = \frac{\partial}{\partial x_i}\left(\frac{\mu_t}{\sigma_\varepsilon}\frac{\partial \varepsilon}{\partial x_i} \right) + C_{1\varepsilon}\frac{\varepsilon}{k}\left[G_k + \left(1 - C_{3\varepsilon}\right)G_b \right] - C_{2\varepsilon}\rho\frac{\varepsilon^2}{k} \tag{8.9}$$

Here, k and ε are the turbulent kinetic energy and its dissipation rate, respectively. $C_{1\varepsilon}$, $C_{2\varepsilon}$, σ_k, and σ_ε are model constants (1.44, 1.92, 1.0 and 1.3, respectively). $C_{3\varepsilon}$ is 0.8, according to Rodi.[53] G_k is the rate of production of turbulent kinetic energy by means of shear rate in the mean flow:

$$G_k = \mu_t \frac{\partial u_j}{\partial x_i}\left(\frac{\partial u_j}{\partial x_i} + \frac{\partial u_i}{\partial x_j} \right) \tag{8.10}$$

and G_b is turbulence generation due to buoyancy, which implies that density gradients in the opposite direction of gravity cause fluid motion, and thus amplifies turbulence:

$$G_b = -g_i \frac{\mu_t}{\rho\sigma_h}\frac{\partial \rho}{\partial x_i} \tag{8.11}$$

With k and ε known, the turbulent viscosity can be computed that ties other transport equations together:

$$\mu_t = \rho C_\mu \frac{k^2}{\varepsilon} \tag{8.12}$$

where C_μ is a model constant ($C_\mu = 0.09$).

The literature suggests "standard" values for the model constants as listed above in parentheses. Because no additional data are available to support or refute these values, they are used as suggested.

8.7.2.5 Chemistry and Turbulence–Chemistry Interaction

With CO emission prediction as a key objective, the following species list is the required minimum: CH_4, C_2H_6, O_2, CO_2, H_2O, CO, and N_2. This mixture is assumed to obey the ideal gas law. The viscosity, thermal conductivity, and specific heat of the mixture are computed from individual species properties, which are functions of temperature (as described in the JANAF tables). Experience shows that accurate physical properties are a prerequisite of CO emission predictions. The reaction mechanism considered in this study is the following:

$$CH_4 + 1.5\ O_2 \rightarrow CO + 2\ H_2O \tag{8.13}$$

$$C_2H_6 + 2.5\ O_2 \rightarrow 2\ CO + 3\ H_2O \tag{8.14}$$

$$CO + \tfrac{1}{2}\ O_2 \rightarrow CO_2 \tag{8.15}$$

Although the flow in the furnace is highly turbulent, chemical reactions still take place much more rapidly than the rate of mixing. Therefore, the reaction rate will be mixing limited. In the context of the Magnussen-Hjertager model,[54] the kinetic rates for these two reactions are thus deliberately set very high so that turbulent mixing is guaranteed to be the controlling rate. Mathematically, these statements translate into the following equation:

$$\dot{R}_{i,k} = \min\left\{\frac{Y_F}{v_{Fk}M_F}, A\frac{Y_O}{v_{Ok}M_O}, AB\frac{\sum Y_P}{\sum v_{Pk}M_P}\right\}M_i\rho\frac{\varepsilon}{k} \tag{8.16}$$

where $\dot{R}_{i,k}$ is the mass production rate for species i due to reaction k, Y is mass fraction, v is a molar stoichiometric coefficient, A and B are empirical constants, M is molecular weight, and ρ is the mixture density. Subscripts O, F, and P denote oxygen, fuel, and product, respectively. The key observation here is that the reaction rates are proportional to the ratio $\frac{\varepsilon}{k}$ with various proportionality constants. The rates ($\dot{R}_{i,k}$) are used in the transport equation for species mass fraction:

$$\frac{\partial}{\partial x_j}\left(\rho u_j Y_i\right) = \frac{\partial}{\partial x_j}\left(\frac{\mu_t}{\sigma_s}\frac{\partial Y_i}{\partial x_j}\right) + \sum_k \dot{R}_{i,k} \tag{8.17}$$

where Y_i is the mass fraction of species i, and σ_s is turbulent Schmidt number.

8.7.2.6 Radiation Treatment

The gases inside the furnace are radiatively participating, that is, the gases absorb and emit radiation. Figure 8.16 is a schematic that shows how this process takes place for a typical pencil of radiative energy through an infinitesimal distance. Mathematically, the process is described by the radiative transfer equation:[55]

$$\frac{dI}{ds} + \left(a + \sigma_s\right)I = a\frac{\sigma T^4}{\pi} + \frac{\sigma_s}{4\pi}\int_0^{4\pi} I(s,\omega')\Phi(\omega,\omega')d\omega' \tag{8.18}$$

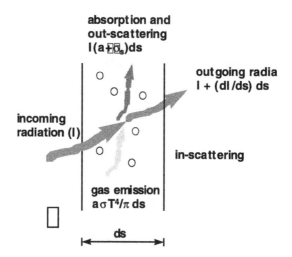

FIGURE 8.16   Radiation in participating medium.

where I is the radiation intensity, a is the gas absorption coefficient, σ_s is the scattering coefficient, ω is the solid angle, and Φ is the scattering phase function. The discrete transfer radiation model[56] solves this radiative transfer equation numerically for each ray. The rays start from a surface element, go through the gas medium, and terminate at another surface element in the optical line-of-sight. Every surface element emanates a specified number of rays, equally divided in the hemispherical solid angle. The number of rays determines the accuracy of the numerical solution, much as grid density controls the accuracy of the flow solution. Note that dI/ds can also be written as $\omega \cdot \nabla I$ to show the angular dependence explicitly. Computation of radiative transfer requires angular discretization in addition to spatial discretization already implemented for the flowfield. Once the radiation intensity is available, the source term in the energy equation is readily calculated:

$$-\frac{\partial q_{r,i}}{\partial x_i} = a \int_{4\pi} I d\omega - 4a\sigma T^4 \tag{8.19}$$

In this gas-fired reheat furnace, the predominant mode of radiation participation is through absorption, and the scattering coefficient is zero for practical purposes. The gas absorption coefficient depends on the pressure, temperature, concentrations of CO_2 and H_2O, the major contributors of thermal radiation, and soot concentration:

$$a = f\left(c_s, x_w, x_c, T, p\right) \tag{8.20}$$

where x_w and x_c are the water and carbon dioxide mole fractions in the furnace gas. The relationship above is, in general, very complicated. A compromise between an oversimplified gray gas model and a narrow-band model is the weighted-sum-of-gray-gases (WSGG) model.[55,57] As the name suggests, the WSGG model is a curve-fit to experimental data with the assumption that the mixture emissivity can be approximated with a number of fictitious gray gases, each well-representing a section of the spectrum. This idea is similar to a polynomial curve-fit except that every term in this fit is a fictitious gray gas. The fitting coefficients used in FLUENT[58] are from Taylor and Foster.[59,60] The (WSGG) model is used to compute the absorption coefficient as the solution evolves.

8.7.2.7 NOx Estimate

Three mechanisms of NOx formation have been identified in the literature: thermal, prompt, and fuel.[61] Among the three, thermal NOx is the most important in industrial furnaces and is well-described by the extended Zeldovich mechanism.[62,63] Prompt NOx usually plays a minor role, and it forms mainly in low-temperature, fuel-rich regions.[64] Fuel NOx can be the predominant contributor if there is a significant amount of nitrogen-containing organic compounds in the fuel.[65] Note that it is nitrogen compounds (such as HCN and amines) rather than the molecular form of N_2 in the fuel that is important, because the high-energy triple bond present in N_2 is already broken. In natural gas combustion systems, NOx production is almost solely due to thermal and prompt mechanisms — even with high-nitrogen natural gas — because most nitrogen found in natural gas is still in molecular form, similar to that found in combustion oxidant.

Although all three mechanisms are available in FLUENT, only thermal and prompt mechanisms will be used in the calculation. Specifically, partial oxygen equilibrium and a probability density function based on temperature for turbulence–chemistry interaction are used in the model. One transport equation for the NO mass fraction is solved using the converged flow, species, and temperature fields.

8.7.2.8 Enrichment Estimate and Process Model

When a fraction of total fuel input is diverted for reburn, oxygen is introduced in the primary combustion zone to compensate for the reduced primary zone firing rate so that a specified production rate is maintained or even exceeded. Assuming a constant production rate, the amount of oxygen required depends on the amount of fuel diverted. The amount of reburn fuel, in turn, depends on NOx reduction objectives. It is evident that by changing the amount of reburn fuel, various levels of NOx reduction can be achieved for a fixed production rate. Alternatively, varying amounts of oxygen can achieve different levels of production increase for a given level of NOx reduction. Or, by changing the amounts of reburn fuel and oxygen enrichment, a specified NOx reduction and a specified production rate can be achieved.

The "available heat" concept can be used to estimate the oxygen requirement. Because results from the CFD model can estimate the fraction (η) of total fuel energy that goes to the steel, the useful heat rather than the available heat can be calculated as:

$$q_a = \eta q_{HHV} \tag{8.21}$$

where q_{HHV} is the higher (gross) heating value of the fuel. With oxy-fuel, nitrogen that is normally associated with air is no longer present. The mass flow rates before and after oxygen enrichment are listed in Table 8.4. Thus, the sensible energy loss by nitrogen through the flue is captured:

$$q_o = q_a + r_a\left(1 - Y_{oa}/Y_o\right)c_p\left(T_{flue} - T_{amb}\right), \quad \text{or} \quad q_o/q_a = 1 + s - s/\theta \tag{8.22}$$

where r_a is the (actual, rather than stoichiometric) air:fuel mass ratio, c_p is the specific heat for nitrogen, Y_o is the mass fraction of oxygen in the oxidant after enrichment, Y_{oa} is the oxygen mass fraction in air, and

$$\theta = Y_o/Y_{oa}, \quad s = r_a c_p\left(T_{flue} - T_{amb}\right)/q_a \tag{8.23}$$

Suppose the primary zone is enriched to Y_o level, and the over-fire air is enriched to $Y_{o,ofa}$. Further, suppose that α fraction of the baseline fuel is diverted for reburn, of which γ fraction is

TABLE 8.4

Mass Flow Rates Before and After Oxygen Enrichment

	Air-Fuel (AF)	Oxygen-Enhanced Combustion (OEC)	Difference (AF-OEC)
O_2 mass fraction	Y_{oa}	Y_o	
Fuel rate	1	1	0
Oxidant rate	r_a	$r_a Y_{oa}/Y_o$	$r_a(1 - Y_{oa}/Y_o)$
O_2	$r_a Y_{oa}$	$r_a Y_{oa}$	0
N_2	$r_a(1 - Y_{oa})$	$r_a Y_{oa}(Y_o^{-1}Y_o - 1)$	$r_a(1 - Y_{oa}/Y_o)$
O_2 from ASU of purity Y_v (mass frac.)	0	$r_a Y_{oa}\dfrac{Y_o - Y_{oa}}{Y_v - Y_{oa}}$	

recovered by post-combustion with over-fire air. To meet a production change ς, the energy balance requires that

$$(1 + \varsigma)\dot{m}_f^0 q_a = \dot{m}_f q_o + \alpha \dot{m}_f^0 q_{o,ofa}\gamma \tag{8.24}$$

where $\dot{m}_f^0$ is the baseline fuel rate. Define

$$\theta_r = Y_{o,ofa}/Y_{oa}, \ \theta_v = Y_v/Y_{oa}, \ p = \dot{m}_f/\dot{m}_f^0 \tag{8.25}$$

and substitute definitions of q_o and $q_{o,ofa}$ into the energy balance, to obtain:

$$\varsigma = p(1 + s - s/\theta) + \alpha\gamma(1 + s - s/\theta_r) - 1 \tag{8.26}$$

Typically, the production rate ς, enrichment level θ, and the fuel rate p are the undetermined variables. Thus, the last terms are known quantities denoted as $B = \alpha\gamma(1 + s - s/\theta_r)$. The energy balance now gives an expression for the production increase rate:

$$\varsigma = p(1 + s - s/\theta) + B \tag{8.27}$$

The economic impact of the modifications consists of benefits from production increase (R \$/ton of steel), less the cost of oxygen (C_o \$/ton of oxygen) and additional fuel (C_f \$/ton of fuel):

$$\Gamma = \dot{m}_s^0\varsigma R - C_f\dot{m}_f^0(p + \alpha - 1) - C_o\dot{m}_f^0 pr_a Y_{oa}\frac{Y_o - Y_{oa}}{Y_v - Y_{oa}} - C_o\dot{m}_f^0 \alpha r_a Y_{oa}\frac{Y_{o,ofa} - Y_{oa}}{Y_v - Y_{oa}} \tag{8.28}$$

where $\dot{m}_s^0$ is the baseline production rate. Let $\omega = \dot{m}_f^0/\dot{m}_s^0$ be the baseline "fuel efficiency,"

$$D = C_o \alpha r_a Y_{oa}\frac{\theta_r - 1}{\theta_v - 1}, \ E = \frac{C_o r_a Y_{oa}}{\theta_v - 1} \tag{8.29}$$

Then rearrange the cost equation and spread the impact over the total production $\dot{m}_s^0(1 + \varsigma)$. The impact in dollars per ton of steel is:

$$I = \left\{ \zeta R - \omega \left[C_f (p + \alpha - 1) + E(\theta - 1)p + D \right] \right\} / (1 + \zeta), \quad \$/\text{ton of steel} \tag{8.30}$$

For dollars per incremental ton of steel, the relationship is:

$$I' = \left\{ \zeta R - \omega \left[C_f (p + \alpha - 1) + E(\theta - 1)p + D \right] \right\} / \zeta, \quad \$/\text{ton of steel incremental} \tag{8.31}$$

The total oxygen requirement per ton of steel is:

$$\Delta = \frac{\omega r_a Y_{oa}}{(\theta_v - 1)(1 + \zeta)} \left[p(\theta - 1) + \alpha (\theta_r - 1) \right], \quad \text{ton of } O_2 \text{ per ton of steel} \tag{8.32}$$

Three scenarios are envisioned for technology implementation. First, the flue volume is limited to some factor of the baseline by fan or pollution control equipment capacity. Assume the flue temperature and molecular weight are essentially the same before and after the modification; the mass flow rate after the modification must be related to the baseline through the "flue volume factor" f:

$$(1 + r_a) \dot{m}_f^0 f = \dot{m}_f \left(1 + r_a \frac{Y_{oa}}{Y_o} \right) + \alpha \dot{m}_f^0 \left(1 + r_a \frac{Y_{oa}}{Y_{o,ofa}} \right) \tag{8.33}$$

$$(1 + r_a) f = p \left(1 + r_a / \theta \right) + \alpha \left(1 + r_a / \theta_r \right) \tag{8.34}$$

so that the primary fuel factor is:

$$p = \frac{A\theta}{r_a + \theta} \tag{8.35}$$

where $A = ((1 + r_a)f - \alpha(1 + r_a/\theta_r))$.

The second scenario is due to firing rate limitations. Let the new firing rate be F times the baseline:

$$p + \alpha + F, \quad \text{so that } p = F - \alpha \tag{8.36}$$

The third scenario is a specified production increase ζ_c that must be achieved:

$$\zeta = \zeta_c \tag{8.37}$$

In all three scenarios, there is enough information to determine the relationship between enrichment rate and profit impact, as summarized in Table 8.5.

To obtain an estimate of the NOx trend for the purpose of economic evaluation, consider the effects of oxygen enrichment and gas reburn. NOx increases with oxygen purity of the combustion oxidant initially due to increased flame temperature, reaches a peak somewhere midway, and drops at high purity due to lack of nitrogen. Gas injection alters the overall furnace stoichiometry to fuel-rich as NOx is generally lower at fuel-rich conditions. Over-fire air is injected sufficiently post-flame where the flue gas temperature is high enough for CO oxidation but low enough to prevent

TABLE 8.5
Summary of Economic Impact under Three Scenarios

Scenario	Relations
Flue volume limited to a factor f relative to baseline	$p = \dfrac{A\theta}{r_a + \theta}, \quad \zeta = p(1 + s - s/\theta) + B,$
	$I = \left\{ \zeta R - \omega \left[C_f(p + \alpha - 1) + E(\theta - 1)p + D \right] \right\} \Big/ (1 + \zeta)$
Firing rate limited to a factor F relative to baseline	$p = F - \alpha, \quad \zeta = p(1 + s - s/\theta) + B,$
	$I = \left\{ \zeta R - \omega \left[C_f(p + \alpha - 1) + E(\theta - 1)p + D \right] \right\} \Big/ (1 + \zeta)$
Production increase specified at ζ_c over baseline	$\zeta = \zeta_c, \quad p = (\zeta - B)/(1 + s - s/\theta),$
	$I = \left\{ \zeta R - \omega \left[C_f(p + \alpha - 1) + E(\theta - 1)p + D \right] \right\} \Big/ (1 + \zeta)$

TABLE 8.6
Curve-Fitting Coefficients ($R^2 = 0.97$))

Coefficient	Value	Standard Error
a0	−2.911	1.334
a1	−112.2	11.219
a2	238.6	25.422
a3	−128.4	16.393
a4	0.09120	1.332
a5	168.5	11.198
a6	−314.6	25.375
a7	152.2	16.363

NOx formation. Thus, for the purpose of NOx estimate, post-combustion with over-fire air is practically irrelevant. The overall oxidant-fuel ratio of the furnace (excluding over-fire air) becomes:

$$f = \frac{\dot{m}_f r_a}{\dot{m}_f + \alpha \dot{m}_f^0} = \frac{r_a}{1 + \alpha/p} \tag{8.38}$$

NOx generation for CH_4 at 811K (1000°F) preheat oxidant temperature was calculated assuming adiabatic equilibrium. Oxygen mole fraction in the oxidant (x) varied from air-fuel to pure oxy-fuel, and stoichiometry ($y = r/r_{st}$, where r_{st} is the stoichiometric air:fuel ratio) varied between 0.9 and 1.1. Overall, 1050 points were evaluated. The resulting data were fitted to the following formula, for which the coefficients are listed in Table 8.6:

$$z = a_0 + a_1 x + a_2 x^2 + a_3 x^3 + a_4 y + a_5 xy + a_6 x^2 y + a_7 x^3 y, \quad lb_{NO_2}/MMBtu \tag{8.39}$$

Note that 1 $lb_{NO_2}/MMBtu = 0.451$ kg_{NO_2}/GJ. In the above curve fit, NOx emission is assumed to follow a cubic polynomial with oxygen mole fraction, whereas it is linear with stoichiometry. Suppose the baseline NOx is Z_0; then, the changes due to modification are:

TABLE 8.7
Cost of NOx Reduction Using Gas Reburn

	Total Fuel Rate Is Fixed	Production Rate Stays Constant
No oxygen enrichment	$x = 0.21, \theta = \theta_r = 1$	$x = 0.21, \theta = \theta_r = 1$
Given condition	$p + \alpha = 1 \rightarrow p = 1 - \alpha$	$\zeta = 0$
Energy balance	$\zeta = -\alpha(1 - \gamma)$	$p = 1 - \alpha y$
Stoichiometry $y = \dfrac{y_0}{1 + \alpha/p}$	$y = (1 - \alpha)\, y_0$	$y = \dfrac{y_0}{1 + \alpha/(1 - \alpha y)}$
NOx change z/z_0	$z/z_0 = 1 - G\alpha,\ G = 3.0995,$ (based on curve-fit)	$z/z_0 = 1 - Ky_0 + ky,\ K = 0.5367,$ (based on curve-fit)
NOx reduction cost, \$/ton NO_2	$C_n = \dfrac{2000}{e}\dfrac{(1 - \gamma)R}{G - 1 + \gamma},$ (independent of α)	$C_n = \dfrac{2000}{e}\dfrac{(1 - \gamma)C_f \omega}{Ky_0 - 1 + \gamma},$ (independent of α)

lb_{NO_2}/MMBtu basis: $\qquad\qquad Z/Z_0 - 1$ $\qquad\qquad\qquad\qquad\qquad\qquad$ (8.40)

lb_{NO_2}/ton$_{steel}$ basis: $\qquad\qquad \dfrac{Z}{Z_0}\left(\dfrac{p + \alpha}{1 + \zeta}\right) - 1$ $\qquad\qquad\qquad\qquad$ (8.41)

lb_{NO_2}/year basis: $\qquad\qquad Z/Z_0(p + \alpha) - 1$ $\qquad\qquad\qquad\qquad\qquad$ (8.42)

Suppose the actual baseline emission is e lb_{NO_2}/ton$_{steel}$. The cost of NOx reduction in dollars per ton of NO_2 removed is

$$C_n = \frac{2000I}{e\left[z(p + \alpha)/z_0(1 + \zeta) - 1\right]}, \quad \text{\$/ton of } NO_2 \text{ removed} \qquad (8.43)$$

When no oxygen enrichment is used, such as when only gas reburn is applied for NOx reduction, the cost of NOx reduction in dollars per ton of NO_2 removed is independent of the level of reburn fuel. Instead, this cost is a function of process efficiency, fuel cost, and product revenue, as shown in Table 8.7. This conclusion is a result of the assumption that NOx emission is linear with stoichiometry. The implication is that the process efficiency — and thus cost — is linearly affected by the amount of reburn, and so is the amount of NOx removed. Thus, on a cost-per-ton-of-NO_2 basis, the effects cancel. In reality, some variation in the cost is expected as the level of reburn is increased. But the estimate here is expected to be reasonably representative, as verified by the good curve-fitting statistics.

For a given oxygen level θ in the furnace, the mass fraction of fuel, δ, that burns with paid oxygen is such that:

$$q_0 = \delta q_v + (1 - \delta)q_a \qquad (8.44)$$

where q_v denotes the useful heat due to combustion with paid oxygen source. Substitute the relationships for useful heat to obtain

$$\delta = \frac{1 - \theta^{-1}}{1 - \theta_v^{-1}} \tag{8.45}$$

When the flow rate changes, the effectiveness of the recuperator changes, as does the preheat air temperature. To estimate the magnitude of the increase in air preheat temperature, consider the recuperator as a counterflow heat exchanger. Because both streams in the heat exchanger are gaseous, the following relationships hold:

$$C_{max} \equiv \dot{m}_h c_{p,h} \geq \dot{m}_c c_{p,c} \equiv C_{min}, \quad C_{min}/C_{max} \approx 1 \tag{8.46}$$

Then, the heat exchanger effectiveness is simply[66]:

$$\varepsilon \equiv \frac{\dot{m}_c c_{p,c}\left(T_{c,out} - T_{c,in}\right)}{C_{min}\left(T_{h,in} - T_{c,in}\right)} = \frac{T_{c,out} - T_{c,in}}{T_{h,in} - T_{c,in}} = \frac{T_p - T_{amb}}{T_{flue} - T_{amb}} = \frac{N}{N+1} \tag{8.47}$$

where N is the number of transfer units, defined as the ratio of the overall heat transfer conductance (UA) to the minimum fluid capacity rate (C_{min}):

$$N \equiv UA/C_{min} \tag{8.48}$$

When the heat exchanger device is fixed, the number of transfer units is approximately constant. This is so because the heat transfer coefficient depends on the Reynolds number to the power of 0.8 for turbulent flows,[67] and the minimum capacity rate depends on the flow rate. The two effects approximately cancel each other. In the reheat furnace recuperator, the specific heat of the gas is approximately fixed; and the heat exchanger effectiveness is related to the flue mass ratio f:

$$\varepsilon = \frac{N^o}{N^o + f^{0.2}} \tag{8.49}$$

where N^o is the number of transfer units evaluated at the baseline flow rates. The new preheat temperature is simply:

$$T_p = T_{amb} + \frac{N^o}{N^o + f^{0.2}}\left(T_{flue} - T_{amb}\right) \tag{8.50}$$

Basic assumptions regarding the reheat process are summarized in Table 8.8. For a given production increase target and reburn fuel usage, new preheat temperature, enrichment level, and burner velocities are calculated. These quantities are used as the starting boundary condition inputs for the new arrangement. Ideally, the actual production increase should be determined by iterating the amount of oxygen enrichment to match the target steel discharge temperature (1450K or 2150°F). In this case, the steel discharge temperature is allowed to float. The actual production increase is determined by comparing the amount of heat absorbed by the steel with that of the baseline.

8.7.2.9 Boundary Conditions and Treatment of Steel Billets

The furnace walls have several layers of material with different thermal properties. Based on the construction of the furnace, the material properties, and the expected hot-face and cold-face temperatures, the heat transfer coefficient is 3.97 W/m²/K (0.7 Btu/hr/ft²/°F). Note that this is an

TABLE 8.8
Assumptions Regarding the Steel Reheat Process

Description	Symbol	Value	Units	Value	Units
Oxygen in air	Y_{oa}	23.3%	Mass fraction	21.0%	Mole fraction
Oxygen source purity	Y_n	95.6%	Mass fraction	95.0%	Mole fraction
Stoichiometric air:fuel mass ratio	r_{st}	15.64			
Actual air:fuel mass ratio	r_a	17.20			
Nitrogen specific heat	c_p	1000	J/kg/K		
Flue gas temperature to recuperator	T_f	1144	K	1600	°F
Ambient temperature	T_a	294	K	70	°F
Base heat to steel as fraction of HHV	η	56%			
Natural gas HHV	q_{HHV}	50.76×10^6	J/kg	997.2	Btu/ft³ (at 60°F)
Recovery efficiency	γ	75%			
Equilibrium NOx at baseline		8.176	$lb_{NO_2}/10^6$ Btu	3.515	kg_{NO_2}/GJ
Actual baseline NOx		0.400	$lb_{NO_2}/10^6$ Btu	0.172	kg_{NO_2}/GJ
Total baseline firing rate		140×10^6	Btu/hr	40.03×10^6	W
Baseline production rate		115	tons/hr		

average over all walls (roof, bottom, side) and does not account for any crevices or imperfections in the walls. The cold-face temperature is assumed to be at 450K (350°F). All walls are non-slip, impermeable to gases, and are assumed to be gray and diffuse with an emissivity of 0.8.

The furnace hearth is packed with billets, 5.5 in. × 5.5 in. × 28 ft (140 mm × 140 mm × 8.53 m) each, with negligible spacing between billets. For the purpose of analysis, the billets are assumed to be a continuous slab, 5.5 in. (140 mm) thick, 28 ft (8.53 m) wide, that move at a speed consistent with the production rate. For 115 ton/hr, the speed is 7.4 in. (188 mm) per minute, or a total residence time of 106 minutes, assuming the specific gravity of steel at 7.8. The steel participates in the heat transfer analysis as conducting solids so that the discharge temperature of the steel will be calculated, rather than specified as a boundary condition. For that purpose, the thermal conductivity of steel is taken to be 40 W/m/K (23.1 Btu/hr/ft/°F), and the specific heat of steel is averaged to be 692 J/kg/K (0.165 Btu/lb./°F). Note that the specific heat already includes the effect of latent heat of polymorphic phase transformations. The theoretical energy requirement for heating steel from ambient temperature (294K, 70°F) to 2150°F is 0.688×10^6 Btu/ton (0.80 MJ/kg).

There is water cooling for the skid pipes above the bottom zone. There are six rows of pipes, each row has five risers and one horizontal pipe. The risers are 2.5 in. (63.5 mm) XX-strong pipes with 1.771 in. (44.98 mm) I.D., and the horizontal pipe is 3.0 in. (76.2 mm) XX-strong with 2.300 in. (58.42 mm) I.D. Based on a design water velocity of 7 ft/s (2.13 m/s), the total flow rate is 2160 gpm (0.1363 m³/s). When the insulation on the pipes is new, the water temperature rise is about 10°F (5.56 K), and the energy loss due to water cooling is 11×10^6 Btu/hr (3.22 MW). When the insulation is old, heat loss easily doubles. For the purpose of analysis, 15×10^6 Btu/hr (4.40 MW) loss is used. This heat loss is assumed to be distributed uniformly in the section above the bottom zone as a heat sink of 41,740 Btu/hr/ft³ (0.432 MW/m³).

The firing rate and its distribution under different operating conditions are listed in Table 8.9. At part load, the industry practice is to maximize firing capacity in the soak zone and prorate the remaining capacity between the top and bottom zones. The water flow rate stays constant. Because the temperature set-points must be the same for reheat quality, heat loss due to water cooling is the same as the full load. Proportionally, water cooling represents even more energy loss at part load. At idle condition, the furnace is fired with 10% of the design fuel flow with 15% of the design air flow (extra fuel lean). Although the cooling water flow rate stays the same, the temperature set-points are lower, and so is energy loss. The "baseline" is defined as current practice at full load.

TABLE 8.9
Firing Rates at Various Operating Modes

Load Level	Units	Full (baseline)	Part	Idle
Production rate	tons/hr	115	70	0
Soak zone	10^6 Btu/hr (MW)	25 (7.33)	25 (7.33)	2.5 (0.733)
Top zone	10^6 Btu/hr (MW)	70 (20.52)	40 (11.72)	7 (2.05)
Bottom zone	10^6 Btu/hr (MW)	45 (13.19)	19 (5.67)	4.5 (1.32)
Total firing rate	10^6 Btu/hr (MW)	140 (41.03)	84 (24.62)	14 (4.10)

In this analysis, radiation loss through openings and flue is assumed to be zero (emissivity equals 0.0 at openings). Although the opening areas are substantial — air passage in burners can be up to 18 in. (457 mm) in diameter, and the flue duct is 3.5 ft (1.07 m) wide — most radiation loss through them is recaptured. Loss through burner openings is returned in the form of preheat. The flue is well-insulated, steel is cold in that section, so that most energy is transferred to the steel. If the radiation loss is set at maximum (emissivity equals to 1.0 at openings), the steel discharge temperature can be 50°F (27.8K) lower. The actual condition is somewhere in between.

8.7.2.10 Numerical Issues

The governing equations for the conservation of mass, momentum, energy, and chemical species are solved with the FLUENT software package.[58] It uses a control volume-based finite-difference scheme where nonlinear variations are included inside each control volume, similar to the concept of a shape function in a finite-element scheme. This method is a variation of the original approach by Patankar.[68] This formulation ensures the balances of mass, momentum, energy, and species locally (within each control volume) to achieve physically realistic results — even on coarse grids. To further ensure the accuracy of the solution, a second-order discretization scheme[69] is used. The solution is allowed to iterate until the residuals are reduced by at least 5 orders of magnitude. More importantly, field variables are monitored to ensure that they do not vary with further iterations, and the overall mass, species, and energy balances are satisfied.

8.7.2.11 Analysis Plan

A baseline case was established for which many exploratory computations were carried out. The objective was to assess the adequacy of the grid size, the model sensitivities, and the solution approach. This baseline was chosen to be the current furnace design at full load. Once established, the baseline results are compared with other variations, a number of which are reported here in detail. Others are reported as sensitivities to production increase. In all CFD cases, 10% of the total furnace firing rate is assumed to be diverted for reburn (14×10^6 Btu/hr, or 4.10 MW), and that the total fuel rate in the furnace (including reburn gas injection) stays at the baseline level.

8.7.3 RESULTS

Process model results and first-order economics are reviewed first. CFD solutions are then presented for a few selected conditions, each with a summary of key results at the beginning, followed by details of temperature, flow, and other relevant quantities. Other variations are reported as sensitivities to production increase rate.

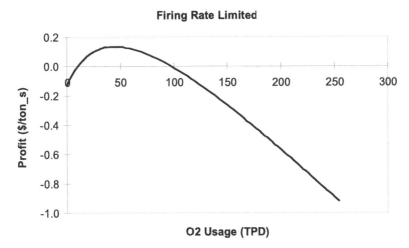

Firing Rate Limited

FIGURE 8.17   Impact of oxygen enrichment (in tons/day) and gas reburn on operating profit, assuming the total furnace fuel rate is fixed at 140×10^6 Btu/hr (41.0 MW), 10% of which is reburn fuel.

8.7.3.1   Process Model and First-Order Economics

The process model as described in Section 8.7.2.8 (enrichment estimate) can be used to perform gross estimates on the process performance. The assumptions regarding the steel reheat process are summarized in Table 8.8. The assumptions regarding economics are:

1. Natural gas is at \$2.5/MCSF (~\$2.5 per 10^6 Btu, \$2.37/GJ).
2. Oxygen is at 12.4 cents per CSCF (\$30/ton).
3. Revenue from steel production is \$5/ton.
4. Capital costs are not included.

Large deviations from these cost figures are common, and must be dealt with on a case-by-case basis. Typical range of fuel cost is between \$2 and \$5, typical O_2 cost is from 10 cents to 50 cents, and typical steel revenue can be from 0 to \$40 or more.

The first scenario analyzed assumes that the total fuel rate in the furnace is fixed at the baseline level (140×10^6 Btu/hr, or 41.0 MW). 10% of that total (14×10^6 Btu/hr, or 4.1 MW) is diverted for gas injection. The economic impact (\$/ton of steel throughput) of such a change is seen in Figure 8.17 where the optimum economic point corresponds to about 50 TPD O_2 usage. As seen in Figures 8.18 and 8.19, this level of oxygen enrichment corresponds to about 12% production increase, and 10% NOx reduction on a lb./ton of steel basis. But on a total emission basis ("bubble"), NOx is approximately unchanged. Figure 8.18 further suggests that to achieve 20% production without using additional fuel would require about 100 TPD of oxygen usage, and economically, operating profit is even slightly favorable. Figure 8.19 suggests that to achieve 50% NOx reduction as stated in the project objectives may require more than 14×10^6 Btu/hr (4.1 MW) reburn gas injection.

The second scenario analyzes the situation when the production increase is fixed at 20%. The total fuel for the furnace is allowed to change. Again, 14×10^6 Btu/hr (4.1 MW) of fuel is used for reburn. With the oxygen price, the best economics as seen in Figure 8.20 is somewhere in the 20 to 30 TPD O_2 enrichment range. That means using 10 to 15% more fuel to make up the available heat (Figure 8.22). The NOx bubble is roughly the same, or about 15% reduced on a lb./ton of

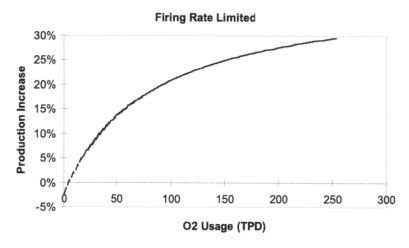

FIGURE 8.18   Possible production increase at different oxygen enrichment levels, assuming the total furnace fuel rate is fixed at 140×10^6 Btu/hr (41.0 MW), 10% of which is reburn fuel.

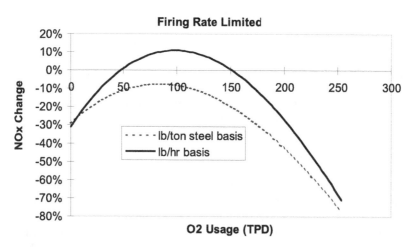

FIGURE 8.19   NOx reduction potential, assuming the total furnace fuel rate is fixed at 140×10^6 Btu/hr (41.0 MW), 10% of which is reburn fuel.

steel basis (Figure 8.21). Figure 8.22 further suggests that at about 80 to 100 TPD O_2 usage level, about 5% additional fuel must be spent to meet the production goal, and that economically the modification is still profitable.

When the production is fixed at baseline level and gas reburn is used solely for the purpose of NOx reduction (no oxygen is used), the cost is about $1097/ton of NO_2 removed. This figure compares favorably with existing NOx reduction technologies for regenerative reheat furnaces,[70] as shown in Table 8.10. This suggests that even if no production increase is desired, the gas reburn technology is still cost competitive.

Although cost competitive without production increase, it is the production increase that pays. This is evident from Table 8.11 where NOx reduction on a total emission basis is examined with various levels of production increase. For example, suppose the plant wants to maintain the same total emission. At 10% production increase, 6.5% of the baseline fuel is needed for NOx reduction, and the operation is $1.93 in profit per ton of steel incremental. At 20% production, the benefit gets better although more fuel is required for gas reburn to control NOx. These are optimum

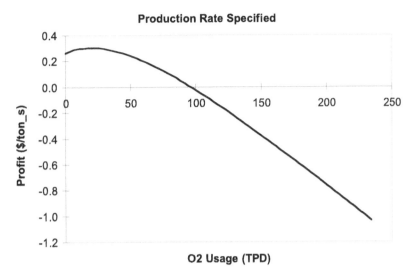

FIGURE 8.20 Profit impact of oxygen enrichment and gas reburn, assuming the production increase is fixed at 20%, and 14×10^6 Btu/hr (4.1 MW) of fuel is used for gas reburn.

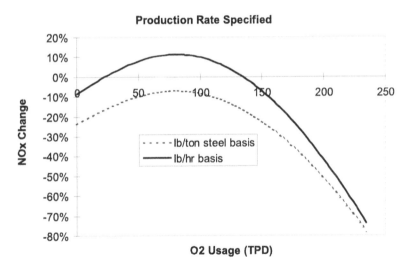

FIGURE 8.21 NOx reduction potential, assuming the production increase is 20%, and 14×10^6 Btu/hr (4.1 MW) of fuel is used for gas reburn.

economic figures as defined by the peak in curves similar to Figure 8.17. It is clear that only a few percent of production increase is required to make the operation profitable. It is also evident that the amount of reburn fuel can be tailored to meet the required level of NOx reduction for a particular customer. Thus, the combination of oxygen enrichment and gas reburn is a scalable technology.

8.7.3.2 Baseline CFD Results

Overall, about 55% of the fuel energy (HHV 140×10^6 Btu/hr or 41.0 MW) is absorbed by the steel to reach an average discharge temperature at 2120°F (1433 K). Water cooling amounts to 11% loss, and refractory loss is 4%. At an average exhaust temperature of 1725°F (1214K), flue loss amounts to 27% of total input. The flue temperature is in agreement with the recuperator design

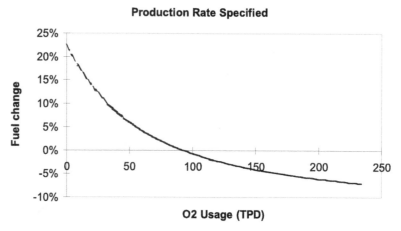

FIGURE 8.22 Fuel change, assuming the production increase is 20%, and 14 × 10⁶ Btu/hr (4.1 MW) of fuel is used for gas reburn.

TABLE 8.10
Cost Comparison with Competing NOx Reduction Technologies

Technology	Operating Cost ($/ton of NOx removed)
Gas reburn (current)	1097
Low NOx burners	260–510[a]
Low NOx burners plus flue gas recirculation	340–620[a]
Low excess air	1300–2100[a]

[a] 1994 dollars. Size dependent.

TABLE 8.11
Profitability Analysis for NOx Reduction on a Total Emission Basis

Scenario		NOx on Total Emission (lb. NO₂/yr) Basis			
		Same	−10%	−30%	−50%
Same production	$/ton of NOx	n/a	1097	1097	1097
	Reburn fuel[a]	n/a	3.5%	10.5%	17.5%
+10% Production	$/ton of steel[b]	1.93	1.65	1.17	0.62
	Reburn fuel[a]	6.5%	10.0%	16.0%	23.0%
+20% Production	$/ton of steel[b]	1.82	1.71	1.43	1.19
	Reburn fuel[a]	10.0%	13.0%	20.0%	26.0%

[a] Reburn fuel as percent of baseline total energy input.
[b] Dollars per ton of steel beyond the baseline production.

inlet temperature of 1600°F (1144K). The average gas velocity at the nose is 9.8 ft/s (3.0 m/s), and the average gas temperature is 2166°F (1459K). For comparison, note that the furnace designer suggested that the nose gas temperature is about 2300°F (1533K).

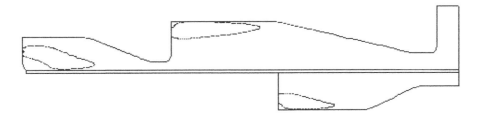

FIGURE 8.23   Flame pattern in the reheat furnace visualized by an iso-contour of CO at 200 ppm.

The flames in the top and bottom zones tend to attach themselves to the refractory surfaces, as visualized by iso-contours of CO at 200 ppm in Figure 8.23. This phenomenon occurs despite the fact that the burner has a 7° downward angle in the top zone. Such a flame pattern is desirable because more energy of the flame is directed to the refractory, which provides indirect heating of the steel. The result is more even heating of the steel rather than hot strips. With oxygen enrichment, higher flame temperature may result in refractory overheating with this flame pattern. Thus, it is important to consider this factor in the enrichment implementation. Further tilting/moving the burner downward or locally fuel-rich firing can be used to overcome this problem. Note that in this model, the effect of gravity is not included; therefore, the flame behavior is unrelated to buoyancy. This phenomenon, known as the Coanda effect, has to do with the burner placement relative to the refractory walls. The presence of a wall interferes with the entrainment process of a jet, resulting in a negative pressure on the wall and the jet bends toward it. Note that because the burner is placed midway on the vertical wall in the soak zone, the flame does not attach to the ceiling refractory.

To aid in quick evaluation of the temperature pattern, "one-dimensional" profiles along the furnace are plotted for the maximum and mass-averaged temperatures at each cross-section. These calculations are done separately for the bottom zone. Results are shown in Figure 8.24. For comparison, the steel surface temperature and bulk temperature are also displayed.

The flow pattern is shown in Figure 8.25 with velocity vectors. Overall, the flow is relatively slow in the furnace (on the order of 10 ft/s or 3.05 m/s); thus, the contribution to steel heating by convective heat transfer is minor (about 3% only). Low furnace gas velocity also implies that injection velocities for fuel reburn need not be too high.

NOx production rate in this furnace is estimated at 1.9 lb. per 10^6 Btu (0.817 kg/GJ) or 2.3 lb. (1.04 kg) per ton of steel processed. This value is an overestimate by at least several times,[70,71] but it will serve as the basis for comparison when firing patterns are modified.

8.7.3.3   Oxygen Enrichment in Top Zone, 50% Heat Recovery

Suppose 10% (14 × 10^6 Btu/hr, or 4.1 MW) of the total firing rate is diverted for reburn, and oxygen enrichment is simply via premixing in the air supply duct for the top zone only. No modifications are made for the bottom and soak zones. Total fuel rate in the furnace including reburn gas injection stays at 140 × 10^6 Btu/hr (41.0 MW). Furthermore, suppose 50% (7 × 10^6 Btu/hr, or 2.05 MW) of the reburn fuel energy is recovered inside the furnace, and 20% production increase is desired. Then, the new top zone firing rate is 56 × 10^6 Btu/hr (16.4 MW). The O_2 level in the top zone is 67% by volume. The total oxygen requirement is about 98 tons/day. The preheat temperature increases 21°F (11.7 K) because of the flue volume decrease. Because such a high O_2 level is not expected to be used, the authors use, for the purpose of analysis, an intermediate enrichment level at 79 TPD O_2, or 42% O_2 mole fraction. The theoretical production increase is expected to be about 18%. The actual production increase will be determined by the amount of heat absorbed.

Two approaches can be used to approximate the effect of energy recovery for the reburn implementation. First, the recovered energy is directly added to the steel as heat sources or a raised

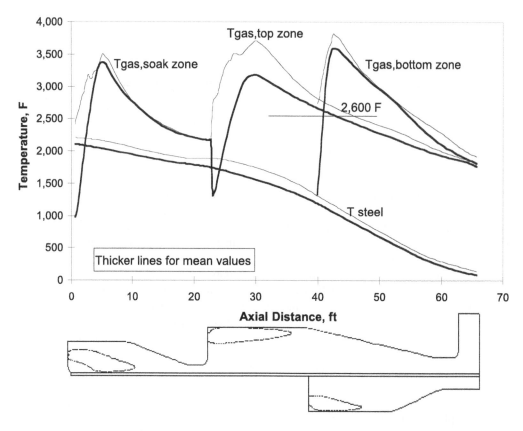

FIGURE 8.24   Baseline axial flue gas and steel temperatures (mean and maximum values).

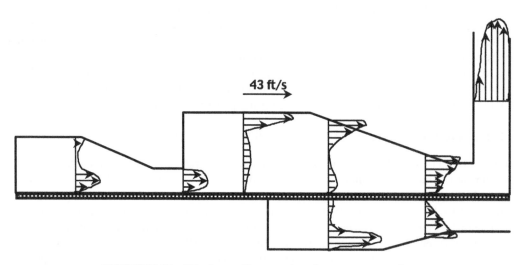

FIGURE 8.25   Velocity profiles at various furnace cross-sections.

charge temperature. Second, the recovered fuel is fired through the top zone burners. Both methods can account for the overall energy balance correctly, and neither account for the distribution correctly. The first approach is used for the results reported here.

The CFD model shows that the average steel discharge temperature is 1939°F (1333K), substantially lower than the target because less oxygen was being applied. The average flue temperature

TABLE 8.12
Summary of Results for the Three Casesalready

Description	Units	Baseline	Enrichment	Reburn	Design Data
Total firing rate	MMBtu/hr	140	140	140	140
Target production rate	tons/hr	115	138	138	
Achieved production rate	tons/hr	113.3	125.2	127.2	
Oxygen used	tons/day	0	79	79	
O_2 level in heating zones	mole fraction	21%	29%	29%	
Reburn fuel	MMBtu/hr	0	14	14	
Recovery efficiency			50% (specified)	77% (calc.)	
Steel discharge temperature	°F	2120	1939	1959	2150
Flue gas temperature	°F	1725	1528	1615	1600
Flue velocity	ft/s	13.9	8.2	11.8	
Nose temperature	°F	2166	2113	2097	2300
Nose velocity	ft/s	9.8	9.6	9.5	
Peak refractory temperature	°F	2363	2563	2471	
Peak flame temperature	°F	3679	4370	4166	
NOx emission	kg/GJ	0.817	1.07		0.215
NOx emission	kg/ton steel	1.04	1.18		

is 1528°F (1104K), lower than its design value. 61% of fuel energy (HHV) is now absorbed by the steel, and the flue loss decreases to 20%. NOx production in this case increases to 2.5 lb. per 10^6 Btu (1.07 kg/GJ), or 32% more than the baseline on net emission basis (350 lb./hr vs. 266 lb./hr, or 159 kg/hr vs. 121 kg/hr). On the basis of tons of steel processed, it is 2.6 lb. (or 1.18 kg) per ton. It is expected that reburn injections can be tailored to destroy more than the increased NOx, so that the overall NOx emission will be lowered.

At the nose, the average velocity is 9.6 ft/s (2.9 m/s), and the average gas temperature is 2113°F (1429K). Because the soak zone firing pattern has not changed, these changes are mainly due to lower steel temperature going into the soak zone. Although the peak flame temperature is considerably higher (691°F or 384K higher,), the peak ceiling refractory temperature increases more modestly from 2363°F (1568K) to 2563°F (1679K) — a 200°F (111K) increase. Table 8.12 summarizes these key facts. The steel temperature in both cases show a similar trend; but because the steel movement is 20% faster through the furnace in the enrichment case, it has a lower discharge temperature despite the increased heat transfer to the steel. With oxygen enrichment, the top zone has smaller flow rate, so that it picks up less of the furnace gas from the soak zone. A comparison of bulk gas and steel temperatures are shown in Figure 8.26 to illustrate the impact of oxygen enrichment.

8.7.3.4 Oxygen Enrichment in Both Zones with Reburn and Over-Fire Air Injections

Based on the results of the baseline and the oxygen enrichment cases, gas reburn and over-fire air injections are placed as illustrated in Figure 8.27. The gas injection is placed where the top zone starts to slope down. The injection angle is 30° counterflow at 1000 ft/s (305 m/s). It is assumed to come from a slot whose width is 42.9 μm so that the total gas flow corresponds to 14×10^6 Btu/hr (4.1 MW). Four feet (1.22 m) downstream, over-fire air is injected at 1000°F (811K), 20 ft/s (6.1 m/s), and normal to the sloped wall (~13°). The slot width is 55.2 mm (2.17 in.) so that the total air flow corresponds to about 10% excess for the gas reburn fuel. The grid was enlarged to 176×113 to accommodate the small injector slots.

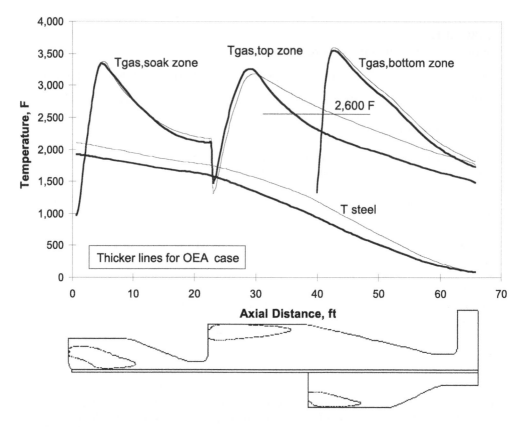

FIGURE 8.26   Axial flue gas and steel temperatures at baseline and oxygen-enriched conditions.

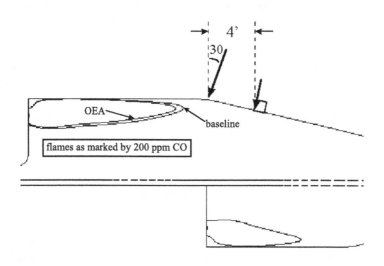

FIGURE 8.27   Locations for gas reburn and over-fire air injections.

The fuel rates at the top and bottom zones are reduced proportionally to divert the reburn fuel. The oxidant is modified so that approximately 98 tons/day of contained oxygen is used in the furnace. No assumption is necessary regarding the recovery efficiency of the reburn fuel because

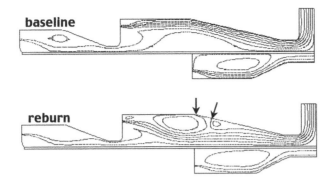

FIGURE 8.28   Flow pattern is visibly different before and after the injections.

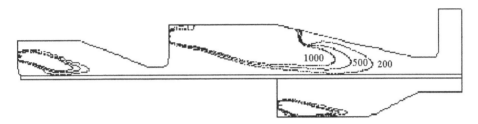

FIGURE 8.29   The proposed injection scheme allows complete burnout of CO within the furnace. However, high CO concentrations near the steel surface might impact scale formation. Shown are CO contours in ppm on a wet volume basis.

the post-combustion process is being modeled directly. For comparison purposes, the authors use only 79 TPD O_2, which corresponds to 29% O_2 mole fraction in the oxidant.

The resultant flowfield due to over-fire air injection is shown in Figure 8.28. The flame in the top zone is pushed downward toward the steel by the air jets. A large recirculation zone sets up between the flame and the roof where no such flow pattern was present in the baseline. It should be noted that because the air jets are modeled as slots in this 2-D analysis, flue gases have no way of going around the jets. In reality, over-fire air directly below the nozzles should penetrate farther into the flue gases, whereas between nozzles the flue gases may slip away and experience smaller disturbance. The results shown here are an "averaged" scenario.

The flame in the top zone is being displaced by the air jets toward the steel surface, as illustrated in Figure 8.29. The peak refractory temperature is 2471°F (1628K), in between the previous two scenarios (see Table 8.12). Cooler refractories mean less radiant heat transfer to the steel, an undesirable consequence. The steel surface temperature is still cool enough not to cause localized overheating even where the flame seems to touch the steel. Because of fuel injection later in the furnace, substantial CO concentrations persist much further downstream as compared to the baseline case. It is clear that most CO will be burned out within the furnace, which is good for thermal efficiency and emission compliance. It is also clear that the CO concentration near the steel surface is much different from the baseline. This difference might cause changes in the scale formation process.

8.7.3.5   Sensitivities of the CFD Results

8.7.3.5.1   Steel surface emissivity

As mentioned earlier in the report, the surface emissivity in the CFD model is 0.8. How does this model parameter impact performance predictions? The case discussed in Section 8.7.3.4 was re-examined with a steel surface emissivity of 0.9 while everything else remained unchanged. The results showed that 12.6% production gain is possible rather than 10.6% as predicted with the lower emissivity.

8.7.3.5.2   Enrichment arrangement

Intuition suggests that putting all the oxygen in a single zone will result in a very hot flame with a relatively small volume. Distributing the oxygen to multiple zones will result in relatively hot flames in a larger combined volume. Because heat transfer to steel is mostly radiative, larger flame volumes in multiple zones should be advantageous. The CFD model confirmed this suggestion. As compared to the case in Section 8.7.3.4, putting all the oxygen in the top zone reduces the production increase from 10.6% to 9.7%, almost a full percentage point.

8.7.3.5.3   Accuracy of the radiation model

Just as the number of gridpoints affects the accuracy of the flow solution, the number of rays affects the radiation model accuracy. All computations so far were done with 16 rays (4 in the circumferential φ direction, 4 in the polar θ direction). When the number of rays increases to 64, the production increase goes from 10.6% to 11.2%, a significant amount. But further increase in the number of rays results in excessive computational burden, and therefore was not pursued.

8.7.4   Discussion

Combustion is a complex phenomenon that presents great challenges for numerical modeling. Some of the fundamental aspects involved in combustion, such as turbulence, chemistry, radiation, and soot are not even well-understood today. Therefore, the results presented in this report are only approximate, consistent with the accuracy of the inputs and that of the physical models. However, experience shows that the results are helpful in predicting trends, thus helpful for engineering design decisions.

The assumptions detailed in Section 8.7.2.2 should be considered carefully. The most fundamental assumption in this model is the 2-D geometry, which simplifies the problem enormously but has limited applicability in the flow pattern near the exhaust region and on realistic enrichment options. The hope is that these calculations can provide enough guidance in the technology development phase, and the true impact of various implementations will be assessed during field trials.

The sensitivity analysis in Section 8.7.3.5 indicates that there are uncertainties in the model results, as expected. However, the CFD predictions do form a consistent trend. It is clear that holding the total fuel rate constant in the furnace is not economical because the amount of oxygen required to achieve a 20% production increase is too large. This observation is consistent with the process model results (Figures 8.17 and 8.18) in Section 8.7.3.1. Industry practice also agrees that a small amount of fuel is typically added in production boost applications.

Several alternatives exist for oxygen enrichment, such as premix in the air main, alternating air-fuel and oxy-fuel burners, oxygen lances, or specialized air-oxy-fuel burners.[51] Clearly, additional analysis must be carried out to examine their relative merits. One immediate comment is that the premix option, although simple and low cost, may not work in the scenarios reported here because the enrichment level is too high. However, for customers with smaller production increase, the lower level of oxygen enrichment may merit its use.

In this case study, oxygen enrichment for production increase and gas reburn for NOx reduction are field-tested technologies with a proven record of good results. However, application of these technologies together to steel reheat furnaces has not been done. To assess the process and economic impact, analyses were carried out first on a process level, then in detail with computational fluid dynamics tools. The analyses show that the proposed technology is feasible both technically and economically. It is a scalable technology so that particular customer needs can be met without sacrificing performance. Some specific conclusions are listed below.

1. The expected oxygen usage level is about 80 to 100 TPD for a mid-size reheat furnace with 115 tons of steel per hour throughput.
2. It is more economical to use on the order of 5% additional fuel as compared to the baseline in conjunction with oxygen enrichment.
3. It is thermally more efficient to enrich multiple zones rather than a single zone. However, this practice must be balanced with capital.
4. The process might modify CO concentrations near the steel surface. Its impact on scale formation is uncertain.

8.22 CONCLUSIONS

Steel is a fascinating material that is too often taken for granted. The steelmaking process is technologically sophisticated, and advances are made everyday. CFD modeling has been used widely to aid in the technology development and efficiency improvement. Specifically, CFD in combustion related processes for the steel industry has been widely applied, although major opportunities for improvement in modeling still exist for blast furnace and other multi-phase systems. Substantial gains in productivity are at stake. As the steel industry enters the new century, improvements on productivity, cost effectiveness, environmental protection and product quality to meet new material challenges such as polymers and composites are imperative. Computational fluid dynamics will most likely find more applications in the steel industry's effort to meet these challenges.

ACKNOWLEDGMENTS

The author wishes to thank the following individuals of Air Products and Chemicals, Inc.: Mr. Michael Lanyi for valuable discussions and mentoring; Mr. P. Buddy Eleazer, III, for introducing the application that resulted in the case study; and Dr. Vladimir Gershtein for encouragement in times of difficulty.

REFERENCES

1. Elliott, J.F. and Gleiser, M., *Thermochemistry for Steelmaking,* Vol. 1, The American Iron and Steel Institute, Addison-Wesley, Massachusetts, 1960.
2. Elliott, J.F., Gleiser, M., and Ramakrishna, V., *Thermochemistry for Steelmaking, Thermodynamic and Transport Properties,* Vol. 2, The American Iron and Steel Institute, Addison-Wesley, Massachusetts, 1963.
3. Wicks, C.E. and Block, F.E., Thermodynamic Properties of 65 Elements — Their Oxides, Halides, Carbides, and Nitrides, Bulletin 605, Bureau of Mines, U.S. Government Printing Office, Washington, D.C., 1963.
4. Avallone, E.A. and Baumeister III, T., *Marks' Standard Handbook for Mechanical Engineers*, 9th ed., McGraw-Hill, New York, 1987.
5. Lankford, Jr., W.T., Samways, N.L., Craven, R.F., and McGannon, H.E., *The Making, Shaping and Treating of Steel*, 10th ed., United States Steel, 1985.
6. Cavalier, D., *Proc. Nat. Phys. Lab.,* 2(9), 40, 1958.
7. Barfield, R.N. and Kitchener, J.A., *J. Iron and Steel Institute*, 180, 324-329, 1955.
8. Siegel, R. and Howell, J.R., *Thermal Radiation Heat Transfer,* 2nd ed., McGraw-Hill, New York, 1981.
9. Steel Industry Technology Roadmap, Office of Industrial Technologies, U.S. Department of Energy, March 1998.
10. Yogi, J., *Iron Steel Inst. Jpn. Int.,* 33, 619-639, 1993.
11. BFPCT — Blast Furnace PC Tool, Copyrighted by CCTT, Toronto, 1995.

12. Lanyi, M.D., Kurunov-Yaschenko Blast Furnace Model, personal communication, 1999.

13. Wang, G.X., Chew, S.J., Yu, A.B., and Zulli, P., Modeling the discontinuous liquid flow in a blast furnace, *Metallurgical and Materials Transactions,* 28B, 333-343, 1997.

14. Fluent, Inc., *Fluent 5 Users' Guide,* Fluent, Inc., Lebanon, NH, July, 1998.

15. Schulte, H.B.M., Development of a CFD-model for gas and fluid flow in the steel converter, Final Report to Directorate-General, Science, Research and Development, European Communities, 1999.

16. Guthrie, R.I.L., Scientific visualization of heat, mass and fluid flow phenomena in metallurgical reactor systems — a case history and future challenges, in *Quantative Description of Metal Extraction Processes,* N.J. Themelis and P.F. Duby, Eds., The Minerals, Metals & Materials Society, Canada, 1991.

17. Story, S.R. and Fruehan, R.J., Modeling post combustion in the EAF, *Proc. Materials 98: The Biennial Conference of the Institute of Materials Engineering,* Australasia, Ltd., July, 1998, 77-82.

18. Szekely, J., McKelliget, J., and Choudhary, M., Heat-transfer fluid flow and bath circulation in electric-arc furnaces and dc plasma furnaces, *Ironmaking Steelmaking,* 10(4), 169-179, 1983.

19. Gittler, P., Kickinger, R., Pirker, S., Fuhrmann, E., Lehner, J., and Steins, J., Applying computational fluid dynamics for the optimization of steelmaking processes, *BHM (Berg-Huettenmaenn Monatsh),* 144(5), 161-169, 1999.

20. Agarwal, J.C. and Jessiman, N.S., Nitrogen oxide emissions in the steel industry, *Iron and Steelmaker,* 19(8), 23-24, 1992.

21. *U.S. Shipments of Steel Mill Products — 1990–1994,* Metal Statistics 1996, 88th ed., American Metal Market, Chilton Publications, New York, 1996, 35.

22. Breithaupt, P.P. and Roberts, P.A., Improvement of burners for reheating furnaces with respect to heat transfer and emissions, Proc. *Eur. Steelmaking Developments and Perspectives in Rolling and Reheating,* Luxembourg, 1-2 Feb., 1995.

23. Sparrow, R., Clements, B., and Macfadyen, N., The development of a rapid heating billet furnace, *The Future and Changing Role of Combustion in Canada: Efficiency and the Environment,* CANMET Energy Technology Centre, Natural Resources Canada, Ottawa, Ontario, K1A 0G1, 1996, 1-12.

24. Byrnes, M.A., Chester, I., Foumeny, E.A., and Mahmum, T., Computational modeling of a gas fired steel reheating furnace, *Proc. 1995 ICHEME Research Event/First European Conference,* 1995, 475-477.

25. Bates, C.P., HIsmelt — A new approach to ironmaking, Australian Academy of Technological and Sciences and Engineering, 1998 Symposium.

26. Panjkovic, V., Truelove, J.S., and Ostrovski, O., Computational fluid dynamics for modeling the fuel efficiency of an iron-bath reactor, *Proc. Materials 98: The Biennial Conference of the Institute of Materials Engineering,* Australasia, Ltd., July 1998, 27-32.

27. Stroomer, P.J. and Bernard, H.K.A., Design of a cyclonic melter as part of the CCF ironmaking process, *Proc. Combustion and Heating. 18,* German-Dutch Flame Day, 1997.

28. Computer models fluid flow in the steel industry, *Steel Times,* April, 1989, 190.

29. Tsai, M.C., Parallel CFD model and its application in the design of a continuous casting tundish, *ASME FED,* 156, 1-7, 1993.

30. Yeh, J.-L., Hwang, W.-S., and Chou, C.-L., Three-dimensional mathematical modeling of fluid flow in slab tundishes and its verification with water model experiments, *J. Mater Eng. Perform.,* 1(5), 625-636, 1992.

31. Sahai, Y. and Ahuja, R., Fluid dynamics of continuous casting tundishes — physical modeling, *Steelmaking Proceedings,* Vol. 69, April 1986.

32. Szekely, J., Ilegbusi, O.J., and El-Kaddh, N., Mathematical modeling of complex fluid flow phenomena in tundishes, *PCH PhysicoChem. Hydrodyn.,* 9(3-4), 453-472, 1987.

33. Chakraborty, S. and Sahai, Y., Effect of slag cover on heat loss and liquid steel flow in ladles before and during teeming to a continuous casting tundish, *Metallurgical Transactions, B,* 23(2), 135-151, 1992.

34. Tavares, R.P. and Guthrie, R.I.L., Computational fluid dynamics applied to twin-roll casting, *Canadian Metallurgical Quarterly,* 37(3-4), 241-250, 1998.

35. Fujisaki, K., Sawada, K., Wajima, K., and Ueyama, T., Application of electromagnetic field field techniques to steelmaking processes, *Nippon Steel Technical Report,* 74, 29-35, 1997.

36. Sinha, A.K. and Lahiri, A.K., Mass transfer between gas and liquid in bottom blown process, *ISIJ Transactions,* 27(7), 560-562, 1987.

37. Alexis, J. and Jonsson, L., Modeling of desulphurization during inductive stirring, *The Steelmaking Conference Proceedings*, 81, 235-246, 1998, The Iron and Steel Society.

38. Jonsson, L., Sichen, D., and Jonsson, P., A new approach to model sulfur refining in a gas-stirred ladle — a coupled CFD and thermodynamic model, *ISIJ Int.*, 38(3), 260-267, 1998.

39. Takatani, K., Shirota, Y., Higuchi, Y., and Tanizawa, Y., Fluid flow, heat transfer and inclusion behavior in continuous casting tundishes, *Sumitomo Search*, (55), 56-62, 1994.

40. Shish, Y.I., Rubin, L.V., Volkov, L.G., Rod, A.G., Gladilin, Y.I., and Braginets, Y.F., Features of hydrodynamic processes in installation for pulsating treatment of steel in ladle, *Steel USSR*, 18(4), 158-159, 1988.

41. Nelson, L.R., Simulation of two-phase fluid flow in bottom, gas-stirred ladles, *S. Afr. Inst. Min. Metall.*, 2, 49-72, 1994.

42. Mazumdar, D. and Guthrie, R.I.L., The physical and mathematical modeling of gas stirred ladle systems, *ISIJ Int.*, 35(1), 1-20, 1995.

43. Jonsson, L. and Jonsson, P., Modeling of fluid flow conditions around the slag/metal interface in a gas-stirred ladle, *ISIJ Int.*, 36(9), 1127-1134, 1996.

44. Yokoya, S., Takagi, S., Souma, H., Iguchi, M., Asako, Y., and Hara, S., Removal of inclusion through bubble curtain created by swirl motion in submerged entry nozzle, *ISIJ Int.*, 38(10), 1086-1092, 1998.

45. Kocatulum, B., Mathematical modeling of two-phase flouid flow in vacuum degassing of steel, *ASME FED*, 238, 255-261, 1996.

46. Pan, Y., Guo, D., Ma, J., Wang, W., Tang, F., and Li, C., Mixing time and fluid flow pattern of composition adjustment by sealed argon bubbling with ladles of large height/diameter ratio, *ISIJ Int.*, 34(10), 794-801, 1994.

47. Turkoglu, H. and Farouk, B., Numerical computations of fluid flow and heat transfer in gas injected iron baths, *ISIJ Int.*, 30(11), 961-970, 1990.

48. Sahai, Y. and Guthrie, R.I.L., Recent advances in the hydrodynamics of metallurgical processing, *Adv. Transp. Processes*, 4, 1-48, 1986.

49. Jokilaakso, A., Computational fluid dynamics in process metallurgy, *Proc. Second Colloquium on Process Simulation*, Helsinki University of Technology, Finland, June 1995, 233-255.

50. Zamaski, V.M., Maly, P.M., and Seeker, W.R., Advanced Reburning Methods for High Efficiency NOx Control, U.S. Patent 5,756,059, May 26, 1998.

51. Baukal, C.E., Ed., *Oxygen-Enhanced Combustion*, CRC Press, Boca Raton, FL, 1998.

52. Launder, B.E., Reece, G.J. and Rodi, W., Progress in the development of a Reynolds-stress turbulence closure, *J. Fluid Mechanics*, 68(3), 537-566, 1975.

53. Rodi, W., *Turbulence Models and Their Application in Hydraulics*, 2nd ed., International Association for Hydraulic Research, Delft, The Netherlands, 1984.

54. Magnussen, B.F. and Hjertager, B.H., On mathematical models of turbulent combustion with special emphasis on soot formation and combustion, *Proc. 16th Symposium (Int.) on Combustion*, Cambridge, MA, August 15-20, 1976.

55. Siegel, R. and Howell, J.R., *Thermal Radiation Heat Transfer*, 2nd ed., McGraw-Hill, New York, 1981.

56. Shah, N.G., A New Method of Computation of Radiant Heat Transfer in Combustion Chambers, Ph.D. dissertation, Imperial College of Science and Technology, London, England, 1979.

57. Smith, T.F., Shen, Z.F., and Friedman, J.N., Evaluation of coefficients for the weighted sum of gray gas model, *J. Heat Transfer*, 104, 602-608, 1982.

58. FLUENT 5 User's Guide, 1998, Fluent, Inc., Lebanon, NH.

59. Taylor, P.B. and Foster, P.J., The total emissivities of luminous and non-luminous flames, *Int. J. Heat Transfer*, 17, 1591-1605, 1974.

60. Taylor, P.B. and Foster, P.J., Some gray gas weighting coefficients for CO_2-H_2O soot mixtures, *Int. J. Heat Transfer*, 18, 1331-1332, 1974.

61. Flagan, R.C. and Seinfeld, J.H., *Fundamentals of Air Pollution Engineering*, Prentice-Hall, NJ, 1988.

62. Zeldovich, Y.B., Sadovnikov, P.Y., and Frank-Kamenetskii, D.A., *Oxidation of Nitrogen in Combustion*, M. Shelef, Trans., Academy of Sciences of USSR, Institute of Chemical Physics, Moscow-Leningrad, 1947.

63. Hanson, R.K. and Salimian, S., Survey of rate constants in H/N/O system, *Combustion Chemistry*, W. C. Gardiner, Ed., Springer-Verlag, New York, 1984, 361.

64. Fenimore, C.P., *Proceedings of the Thirteenth Symposium (Int.) on Combustion*, The Combustion Institute, Pittsburgh, PA, 1971, 373.

65. De Soete, G.G., Overall reaction rates of NO and N_2 formation from fuel nitrogen, *Proc. Fifteenth Symp. (Int.) on Combustion*, The Combustion Institute, Pittsburgh, PA, 1975, 1093.

66. Kays, W.M and London, A.L., *Compact Heat Exchangers*, 3rd ed., McGraw-Hill, New York, 1984.

67. Kays, W.M. and Crawford, M.E., *Convective Heat and Mass Transfer*, 2nd ed., McGraw-Hill, New York, 1980.

68. Patankar, S.V., *Numerical Heat Transfer and Fluid Flow*, Hemisphere, Washington, D.C., 1980.

69. Leonard, B.P., A stable and accurate convective modeling procedure based on quadratic upstream interpolation, *Comput Methods Appl. Mech. Eng.*, 19, 59-98, 1979.

70. U.S. EPA, Alternative control techniques document — NOx emissions from iron and steel mills, EPA Report EPA-453/R-94-065, Office of Air and Radiation, Research Triangle Park, NC, 1994.

71. Gas Research Institute, NOx controls for gas-fired industrial boilers and combustion equipment: A survey of current practices, Topical report, GRI-92/0374, Chicago, IL, 1992.

72. *Periodic Table of the Elements,* Sargent-Welch Scientific Company, Skokie, IL, 1980.

73. *Table of Periodic Properties of the Elements,* Sargent-Welch Scientific Company, Skokie, IL, 1980.

74. Zumdahl, S.S., *Chemical Principles,* D.C. Heath and Company, Lexington, MA, 1992.

75. Wark, K., *Thermodynamics,* 4th ed., McGraw-Hill, New York, 1983.

76. Perry, R. H. and Green, D., *Peery's Chemical Engineers' Handbook,* 6th ed., McGraw-Hill, New York, 1984.

77. *Aerospace Structural Methods Handbook,* Air Foce Materials Laboratory, Wright-Patterson Air Force Base, Ohio, 1969.

78. *Metals Handbook,* 8th ed., ASM, 1961.

9 Aluminum Industry

Vladimir Y. Gershtein and Charles E. Baukal, Jr.

CONTENTS

9.1 ALUMINUM AND ALUMINUM INDUSTRY PROCESSES

Aluminum is a silver-white metal listed in Group 3 of the periodic table. It is the most abundant metal element on Earth. The first impure metallic aluminum was isolated in 1824 in a chemical laboratory. The present industrial method of aluminum production was discovered simultaneously in France and Ohio. The production and consumption of aluminum by country are given in References 1 and 2, respectively. The U.S. has the greatest potential supply of aluminum, which is approximately equal to 4.56×10^6 tons (4.14×10^6 m-tons).[3,4]

In recent years, the aluminum market growth has been defined by different industries. For example, in the building and construction industry, alloys such as 3003, 3004, and 3005 series are widely employed for residential and industrial siding, roofing, gutters, and down-spouts. A 6063 alloy is often used for fabrication of doors and windows. Different alloys are widely used for commercial and military aircraft,[5] and in the automotive industry. The demand for aluminum, especially the secondary aluminum segment, is predicted to grow significantly.[6] This reemphasizes the need for optimization of the processes used for aluminum production.

Aluminum alloys are available in a very wide variety and are produced by rolling, extrusion, drawing, or forging.[7] For economic reasons, aluminum ingots are typically alloyed and cast at smelting plants using primary or secondary metal. An overview of these processes is given in Reference 8. The processes vary from scrap treatment such as shredding, de-lackering, and preheating, to heat transfer in the melting and holding furnaces, and molten aluminum degassing and alloying. Transportation of the molten metal inside a crucible, for example to a casting machine or to another holding furnace, and casting by itself are widely used industrial processes in the aluminum industry.

This chapter is primarily devoted to the aluminum secondary furnaces and the processes inside them. Other processes, like aluminum bath demagging with the use of chlorine,[9,10] can be found in the literature and are beyond this discussion. A brief overview of the aluminum industry furnaces is given below to introduce a general concept of the aluminum melting process.

9.1.1 ROTARY FURNACES

Rotary furnaces have traditionally been used for melting dross. The furnace sizes range from 8000 lb to 30,000 lb (3600 to 14,000 kg), with fuel consumption of 1800 to 2400 Btu/lb (0.77 to 1.0 J/kg). A typical combustion system for rotary furnaces can be a single or double pass. The single-pass system has a burner at one end of the furnace and a flue at the other, as shown in Figure 9.1. The combustion gases flow straight through the combustion space, exiting through the flue. With the double-pass system, the flue is located on the same side of the furnace as the burner, as shown in Figure 9.2. The combustion gases make a "U" turn inside the furnace, which increases the gases residence time and improves furnace efficiency. This furnace type is often referred to as a concrete mixer type furnace.

Rotary furnaces can also be divided into fixed axis or tilters. A bottom taphole is required for the fixed axis furnaces to pour the molten aluminum. A delivery trough is needed for molten metal transport to a crucible, to the sow molds, or to a casting furnace. The fixed axis furnace also requires a procedure for saltcake removal. A towmotor with a rake or boom assembly mounted on the front is typically used to extract the saltcake. Such a procedure is time-consuming and also can damage the furnace refractory because of the raking process. The alternative is to remove the saltcake through the taphole when the "wet process" is used.

In tilting furnaces, the aluminum is poured from the furnace front, once the melting process is completed (Figure 9.3). The back of the furnace is tilted up gradually until all the metal is poured

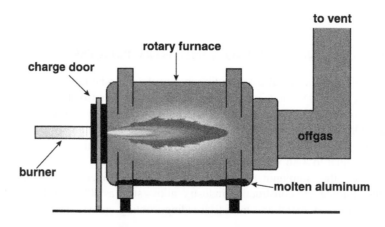

FIGURE 9.1   A single-pass rotary furnace.

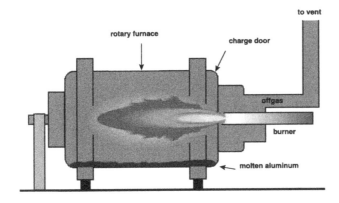

FIGURE 9.2 A double-pass rotary furnace.

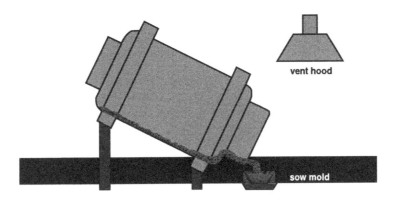

FIGURE 9.3 Pouring aluminum from a tilting furnace.

from the furnace lip. When the metal is removed completely, the furnace is rotated again and tilted more to remove the saltcake. The benefits of tilting furnaces are their ability to remove contaminants at any given time, to control the chemistry of the melt, and to effectively remove the metal. The melt and saltcake removal processes reduce the pouring time, increase furnace production, and reduce maintenance cost.

Another significant distinction of the rotary furnaces is the melting process selection. The melting process can be "wet" or "dry." The wet process uses twice as much salt compared to the dry one. It also requires more energy because the flux has to be molten throughout the process. When the dry process is used, a saltcake is formed above the molten metal bath. This saltcake has to be removed from the furnace at the end of the melting cycle by tilting the furnace or raking it out. The removal procedure depends on the type of furnace. More dross can be processed per cycle in the rotary furnaces using the dry process because less flux is used.

Flux is used for aluminum melting in both the wet and dry processes. It serves several purposes. First, the flux melts at a lower temperature than the aluminum. Therefore, it is used as an oxidation protective blanket over the metal surface. Second, the salt flux acts as a separation agent between the molten metal and the oxides. A typical salt flux consists of sodium chloride (NaCl) and potassium chloride (KCl). Sometimes, Cryolite is added to the flux to reduce its melting temperature.

9.1.2 SIDE-WELL REVERBERATORY FURNACES

Reverberatory furnaces, often referred to as reverbs, have a much wider use in the aluminum industry than rotary furnaces. These furnaces are typically used for primary or secondary aluminum

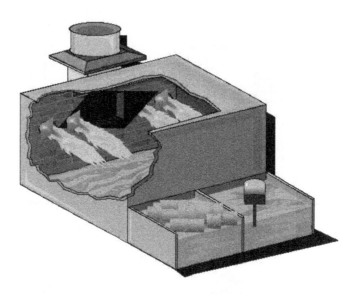

FIGURE 9.4   A schematic of a typical side-well reverberatory furnace.

melting. The operation procedures for both rotary and reverb furnaces are similar. Side-well
reverberatory furnaces are discussed next.

A typical side-well reverberatory furnace is shown schematically in Figure 9.4. The furnace
consists of a number of burners typically firing parallel to the hearth, against the furnace hot wall
or against the furnace door. A charging well and a pump well are attached to the furnace hot wall
on the outside of the furnace. Both wells are connected to each other and with the furnace hearth
by the arches, which are used for aluminum circulation in the furnace. Two types of arches are
used between the wells and the hearth. The first type is the tall arch. These are built to stretch from
the furnace bottom to well above the metal line. These arches are typically used in side-well furnaces
with no metal pump[11] and, therefore, no pump well. The second type is the submerged arch. These
arches are built from the furnace floor to a level below the metal line. A typical height of this type
of arch varies from 10 to 15 in. (25 to 38 cm), depending on the furnace design. Reverb furnaces
with submerged arches typically use a metallic pump.

Material is charged into the furnace charging (side) well, which already has a certain level of
molten aluminum in it. The material melts into the existing molten metal in the charge well. The
molten aluminum circulates from the hearth into the charging well and back into the hearth for
reheating. As the metal level reaches its maximum, the furnace is tapped into a casting line, crucible,
ladle, or into a holding furnace. Often, the same melting furnace is used for degassing and alloying.
In that case, the total furnace cycle consists of two sub-cycles: the melting cycle and the holding
cycle. A typical holding cycle time varies from 25% to 50% of the total furnace cycle time.

The efficiency of the side-well furnaces is determined by the ability of the molten metal to
receive the energy from the combustion chamber. There are several important factors influencing
heat transfer inside the aluminum melter. First, the layer of dross at the molten bath surface inside
the furnace hearth can be viewed as a thermal insulation, restricting heat transfer between the
combustion space and the melt. At the same time, a thin layer of dross is essential to prevent
excessive aluminum oxidation at the gas-melt interface. Second, the furnaces with a metallic pump
are more efficient than the ones without a pump. The melt circulation helps to enhance heat transfer
to the melt compared to the process with no melt circulation. Convective heat transfer is added to
conductive heat transfer in the case with a non-stagnant melt (the furnace with the pump). Third,
flame characteristics have a great influence on the heat transfer to the melt. For example, high-
momentum air/fuel burners with relatively transparent (nonluminous) flames deliver less energy to

the melt compared to highly luminous, high-temperature oxygen-enriched flames. This is discussed in more detail later.

Furnace design and operating practices are of great importance for aluminum producers. An optimized ratio of the furnace length to its width, molten bath height, combustion space height, shape and cross-sectional area of the arches, burner firing rates, and burner position and orientation usually determine the furnace performance. Until recently, furnace designs and operating practices were based on the experience and limited field data available from aluminum producers. CFD modeling is an excellent tool that offers significant help to both furnace designers and operators. Examples of furnace modeling are presented in Section 9.3.

9.1.3 DIRECT CHARGED FURNACES

Another type of a reverberatory furnace is a direct charged furnace.[12,13] A photograph of a typical direct charged furnace is shown in Figure 9.5. This furnace type differs from the side-well furnace by the absence of a side-well. The metal is charged directly into the furnace hearth where it is exposed to the open flames. Therefore, three phases are present in this melting process. Combustion gases, molten aluminum, and the solid metal are all together inside the furnace. The metal recovery of such a system is usually lower compared to the side-well operation but, nevertheless, these furnaces are widely used in the aluminum industry.

The metal is charged into the furnace through the furnace door (see Figure 9.6) and piled in the furnace hearth. Once the charging procedure is completed, the furnace door is closed and the burners are turned on full-fire. The melting process begins. As the solid melts, bath height increases while the solid pile height decreases. The heat transfers into the solid metal from both the furnace

FIGURE 9.5 Photograph of a typical direct charged furnace.

FIGURE 9.6   Metal charged into the furnace through the furnace door.

combustion space and from the molten phase. Once the top of the charged pile is flush with the molten bath surface, the charging procedure starts again until the metal level reaches its maximum. The dross is skimmed from the furnace hearth through the door, in between the charges and at the end of the melting cycle (see Figure 9.7).

The melting efficiency in these furnaces is typically high, due to simultaneous exposure of the charged material to the hot furnace gases and to the molten aluminum. A significant disadvantage of the direct charged furnace compared to the side-well furnace may be lower overall furnace efficiency due to excessive energy loss during the charging and the dross skimming operations. The refractory life of the direct charged furnaces is usually shorter compared to side-well furnaces because of the possible mechanical damages and the extensive thermal shocks. The process is more difficult to model due to its dynamic nature and the presence of the three phases. Despite this, however, modeling can still help the furnace designers and operators. For the latter, the modeling support helps to optimize furnace operation and suggests desired modifications during shutdowns. An example of such a model is presented in Section 9.3.

9.2   CHALLENGES IN MODELING OF ALUMINUM PROCESSES

9.2.1   MODELING GOALS

A typical goal of modeling an existing furnace is to optimize its operation. This means defining burner placement, firing rate, firing profile, comparing different firing technologies, and optimizing the components of the furnace geometry. All these can be achieved based on better understanding

FIGURE 9.7 Dross skimming process with rake through the furnace door.

of the process, which can be provided by the model. Therefore, each model should be constructed according to the initially defined specific goal. Different models might be required to answer different questions. Simulating a dynamic three-phase aluminum melting process in a rotary or in a direct charged furnace is a very challenging problem. The system complexity can be overcome by introducing certain simplifications to the process and dividing it into stages. Each stage should be a simpler snapshot than the overall complex process. For example, a two-phase steady-state model with a flat bath could be assumed, to compare different firing technologies and to estimate oxygen availability at the melt interface in a direct charged furnace. The model will not provide any information about the dynamic solid-phase melting process or furnace gas behavior with the pile of solid material present. It will not take into the account the dynamic behavior of the molten bath either. At the same time, the model can compare different firing technologies, accounting for the refractory and melt temperature distribution, flow pattern of the furnace gases, and species distribution. Therefore, the information obtained from such a model helps to answer some of the questions defined initially. Another example is a side-well furnace with different burner orientations. A common goal is to optimize burner position and orientation to achieve the maximum furnace efficiency and metal recovery rate. Only a furnace combustion space model is needed to complete this task and to answer the stated questions. Certain boundary conditions can be assumed at the gas-melt interface, which helps to avoid the extra effort of calculating the molten bath region.

These examples demonstrate the importance of having well-defined modeling goals. This ensures that the right problem is solved and that the appropriate level of detail is included in the simulation. In addition, any modeling work assumes that an engineer has good knowledge of the process.

9.2.2 Different Approaches in Furnace Modeling

As discussed, the modeling approach depends on the questions that need to be answered. For example, only a small portion of the furnace combustion space needs to be simulated[14] if the goal is to determine refractory temperature on the furnace wall opposite the burner. This approach is very helpful when hot spots are anticipated on the furnace wall, especially when different burners[15-18] are evaluated for the job.

When the goal is to analyze just temperature distribution and the hot spots on the furnace refractory, a full 3-D combustion space model can be used.[19-23] The model can be simplified by excluding the turbulent reacting flow with multiple species. Instead, high-temperature turbulent jets, representing the flames from the burners, can be used. The accuracy of such a simplification can be fair as long as the flame representation is realistic. An additional 2-D burner model can be used in conjunction with the simplified 3-D combustion space model. In that case, the results from the 2-D models can be directly incorporated into the 3-D furnace model to increase the accuracy of the flame energy release output.

A more complex 3-D combustion space model can be used when the modeling goal includes species distribution throughout the furnace space. In the aluminum melting process, the oxygen distribution, for example, can play a critical role in dross formation and also impacts NOx and CO emissions. This modeling approach requires the input of at least some, if not all of the burner details and typically impacts the grid size of the model. Often, the grid becomes too large, which influences the time required to solve the problem. This could be a serious problem, assuming limited computer power available for engineers in the majority of industrial companies. On the other hand, reducing the grid density could result in inaccurate, and sometimes misleading results. Therefore, the use of full 3-D furnace models is restricted by available hardware and project time.

Changes in the molten metal bath may require only a melt model. Assuming a specified energy input, the melt model can be constructed with all the geometric details. Typically, these models are much easier to solve because they do not include reacting flow. The biggest downside of such a model is the questionable accuracy of the energy distribution at the boundaries. Even a small inaccuracy in assumptions and boundary conditions can generate misleading results.

Another approach is a complex 3-D furnace model that consists of two separate models. The model includes both the combustion space and the molten bath models.[24,25] Typically, the output from one of the models is used as an input to the other, and vice versa. The iterative solution conversion procedure between both models takes some time, but may not be a problem if enough computing power is available. This approach requires an additional program that links the model parts together and allows the parts to communicate with each other. This small inconvenience can be avoided if all the parts are included into one model.[26] The advantage of this approach is that no assumptions or data transfer between models are required between the phases. It is also convenient to use the exact furnace geometry during the presentation of results and it simplifies the communication process between the modeler and the end user. The disadvantage is the increased number of gridpoints. The last two approaches are somewhat similar; and in the majority of cases, it is a matter of preference for the modeler as to which approach to use.

9.2.3   Common Modeling Assumptions

Nearly all mathematical models have some assumptions. The assumptions help to reduce the model complexity, to restrict the number of parameters influencing the process, and to focus on the most important issues of the specific process stage. For example, in the case of a reverberatory side-well aluminum furnace, the metal level increases constantly due to the melting process in the side-well. Strictly speaking, the process is time dependent and, in general, should be represented by a transient model. In fact, the melting rate is quite low compared to the capacity of the furnace and, therefore, a melt level increase of 0.5 ft (0.15 m) could take hours. If the goal is to find the temperature and species distribution throughout the furnace hearth, the melt bath can be fixed at a certain level and the distribution of the temperature, species, velocity, etc. can be obtained for the specified furnace conditions. The results can be viewed as the immediate or the steady-state solution for a specified level of the melt. Thus, the steady-state model assumption is reasonable and provides reliable results. The results from several steady-state models with different metal levels can later be combined into one broader time-dependent analysis that can indirectly account for the dynamic nature of the process. The same procedure can be applied, for example, to a variable burner firing rate. If the changes in the firing rate do not occur with a high frequency compared to the process itself, a sequence of steady-state models with different firing rates can be implemented to capture all variations inside the furnace.

Another assumption is the grid size. It is well-known that a significantly higher grid density is required for a combustion problem than, for example, for a nonreacting flow. The reason is the influence of mixing on the final solution. The rate of mixing typically controls the chemical reaction and thus greatly influences the energy release, temperature distribution, species concentration, and other parameters. The more accurately the mixing is calculated, the more accurate the final solution becomes, assuming that the other physical models are within the specified accuracy. High-density grids are especially necessary in places with high gradients. In the past, when only structured grids were available, the need for a high grid density sometimes led to a high grid aspect ratio. The high grid aspect ratio, by itself, can cause a solution convergence problem or can lead to misleading results. The problem became less important with the implementation of unstructured grids. However, the grid size and its structure can influence the solution convergence and the accuracy of the final results.

The next very important assumption is the type of physical models chosen for the problem description. For example, consider turbulence models. A number of models are well-known and available.[27-32] The choice depends on the process type. The standard κ-ε model is quite adequate for the majority of engineering applications. However, the major limitation of this model is that the turbulence viscosity μ_t is isotropic. Therefore, the standard κ-ε model can produce physically incorrect results for a highly swirling flow. The alternative is the Reynolds stress model (RSM). This model provides the flexibility of the directional variation of both the velocity and the length scale in the momentum equation. Another alternative is the RNG-based κ-ε turbulence model. This model can be successfully implemented in the regions where, for example, the jet is impinging on a wall (this example by all means does not limit the application of the model). The RNG model accounts for the normal stresses near the impingement point and does not overpredict the turbulent viscosity and, thus, the mixing rate downstream.

The chemical reaction model should be chosen in the case of a detailed combustion model that requires species distribution and accounts for chemical reactions. Different physical models are available.[33-35] The "eddy break-up" model is usually useful for premixed and diffusion flames. It can also be used for partially premixed reacting flow. A typical limitation of this model is the number of chemical reactions that can be used. The literature suggests that at least a two-step chemical reaction of methane and oxygen should be used. The intermediate reaction with carbon monoxide production is critical for valid temperature predictions. The number of reactions, in general, can be unlimited. The problem again is in the speed of the calculations and the available computer power.

Another approach, called the PDF/mixture fraction model, has became quite popular in recent years.[36] The model is accurate and requires less computational time once the lookup tables are generated. The accuracy depends on the choice of species. In the case of premixed flames, the combustion takes place in a very thin layer. There are a number of models that were specifically developed for flame front tracking[37,38] under premixed conditions.

Another assumption is the choice of a radiation model. A number of models are available.[39-50] The final results of a CFD simulation can differ significantly with the use of different radiation models. Therefore, it is very important to choose the appropriate radiation model suitable for the application. For a gas firing aluminum furnace, good results can be achieved using discrete transfer (DTRM) or discrete ordinate (DORM) radiation models. These models are suitable for both semi-transparent and transparent media and have proven to be reliable in these kinds of applications. The downside of these models is the high demand for computer memory and storage space. Solution convergence can decrease (up to 4 times) when the DTRM is used for a 3-D furnace combustion space with a high-density grid. Spectral radiation models could also be used for improved accuracy. It has yet to be shown that this added accuracy is required in aluminum applications.

The degree of complexity of a CFD model can vary and depends on the simulation objectives. A typical combustion space or molten bath model of an aluminum furnace requires the input of the physical properties as a function of temperature. Some of the physical properties of the melt can be found in References 51 and 52. Physical properties of the gases can be found in a number of references.[53-57]

9.3   EXAMPLES OF CFD MODELS FOR THE ALUMINUM INDUSTRY

Several examples are presented in this section, showing how CFD has been used to help understand, optimize, and change the aluminum melting process in reverb furnaces. An oxygen-based firing technology will be compared to a conventional air/fuel firing technology for the side-well and direct charged aluminum furnaces.

Both air-oxy/fuel and oxy/fuel technologies are presently used by the aluminum industry all over the world. One of the tools that helps in understanding a process is CFD modeling. It is shown here how this tool can be used in process design, especially when a new technology is being implemented in an industrial aluminum furnace.

9.3.1   REVERB FURNACE WITH AIR-, AIR/OXY-, AND OXY/FUEL BURNERS

A CFD model was created to simulate a furnace at Roth Bros. Corporation (now Wabash Aluminum) in East Syracuse, NY. The model solves two sets of equations simultaneously: one for the turbulent reacting gas flow in the combustion space, the other for the laminar flow of the liquid aluminum bath. Such an approach provides the flexibility to evaluate furnace performance for a variety of production and firing rates. The model was built in such a way that the burner firing configuration could easily be changed. Three furnace firing modes — air/fuel, air-oxy/fuel, and oxy/fuel — were modeled.

The main objective of this work was to evaluate the furnace's ability to produce aluminum with different burner types and to compare the advantages and disadvantages of each firing technology. The furnace production rate and molten bath height were fixed. In addition, the firing rate of the burners and their orientation were also kept the same for all three cases. The following three cases were compared: air/fuel (AF), air-oxy/fuel (AOF) with 35% total oxygen enrichment (TOE), and oxy/fuel (OF).

Comparison of the three cases showed that the specified firing rate (14×10^6 Btu/hr, or 4.1 MW) was not high enough to melt aluminum for the AF case at the chosen production rate (12,000 lb/hr or 5,400 kg/hr). For both the AOF and OF cases, the furnace was able to produce molten metal at the specified rate without the need to increase the burner firing rate and, therefore, the natural gas consumption. The results predicted the refractory overheating in the OF case, meaning that the burner firing rate would need to be reduced for pure oxygen operation (OF). The metal temperature in the OF case was shown to be very close to that temperature in the AOF case. This may mean that reducing the firing rate in the OF case might result in the melt temperature dropping below the required level. The AOF case provides the same benefits as the OF case, but uses less oxygen during the melting cycle. In addition, the AOF-fired burner does not use oxygen while in the low firing mode. The low firing mode is mostly used during the degassing and alloying cycle, when not much energy is needed to hold the metal bath at the required temperature.

The flue gas volume is dramatically reduced as the TOE increases. If the goal is to reduce the gas volume through the furnace flue, then switching to a higher TOE-firing technology may be very important. The ability of the AOF burner to fire in the AF mode as well as in the AOF mode with TOE up to 60% makes it a very attractive option.

9.3.1.1 Problem Description

Oxygen-based combustion was tried in the aluminum industry to increase furnace production as long as 40 years ago. These experiments gained a lot of negative publicity. Most of the trials ended with a high rate of aluminum oxidation and furnace refractory damage. The loss in yield and expensive furnace relining forced the aluminum industry to reject the use of pure oxygen. Today, the industry is reevaluating the concept of oxygen firing in aluminum furnaces because of the growing demand to increase production without investing significant capital and because of advancements in oxygen-enhanced combustion technology.[58]

The project described here was sponsored by the U.S. Dept. of Energy, Office of Industrial Technology (DOE/OIT). The objective was to help aluminum makers increase the production of reverb furnaces at low cost, simultaneously addressing environmental issues such as NOx.

Air Products and Chemicals, Inc. (APCI) was chosen by the DOE as the project managing organization because of its significant experience in oxy/fuel combustion for aluminum and other industries. The scope of the project included the development and subsequent demonstration of a low-cost, high-efficiency, and low-NOx technology for the aluminum industry. Roth Bros. Smelting Corp. (RBSC) was chosen as the host site for implementation of the new technology.

A schematic of the furnace is shown in Figure 9.4. This furnace consists of three main elements: furnace hearth with burners, pump well, and charging well. Aluminum scrap is constantly delivered into the charging well during the melting cycle. A pump circulates molten metal from the furnace hearth through the pump and charging wells, back into the furnace hearth. The speed of the metal circulation and the metal temperature dictate the charging rate. The desired average metal temperature at the inlet to the pump well is 1000 to 1035K (1350 to 1400°F). The melt temperature changes in the charging well as a function of the amount of scrap charged. Thus, rapid temperature recovery in the furnace hearth is of great importance and the rate of energy delivered to the melt is directly linked to the production achieved.

The dross formation rate is determined by oxygen availability at the melt interface. Therefore, another important parameter is the oxygen concentration throughout the combustion space of the furnace. As long as this concentration is kept low, excess dross is not formed and, as a result, yield and heat transfer are maximized inside the furnace.

9.3.1.2 Model Description

The CFD model was developed using FLUENT. The model solves two sets of equations simultaneously: one for the turbulent reacting flow of gases in the combustion space, the other for the laminar flow of the molten aluminum. The interactions of furnace production, melt rate, metal pump speed, metal level, burner orientation, and firing mode can be studied with this model. For example, a change in the melt height automatically recalculates velocity, temperature, and species concentration everywhere in the domain.

An outline of the model is shown in Figure 9.8. The model did not include the pump well and the charging well. The metal circulation through the wells and the pump speed were approximated using the average inlet velocity of the molten metal through the inlet arch. The temperature at this location was calculated based on an energy balance around the well. Except for the above assumption, no other assumptions were made for the furnace geometry. The elimination of the well in the model helped to reduce the computational domain and to obtain faster results without a significant sacrifice in accuracy. It would be a simple matter to include the well if necessary.

All three cases were built around the same 90,000-cell grid. The only difference was the number of cells assigned to represent the burners. For example, the AF burner had the highest number of cells because the burner had the largest cross-sectional area. The OF burner had the smallest number of cells. The inlet velocity and temperature were specified for each burner for natural gas, oxygen, and air. The firing rate of each burner was the same for each case. This firing rate was 3.5×10^6 Btu/hr

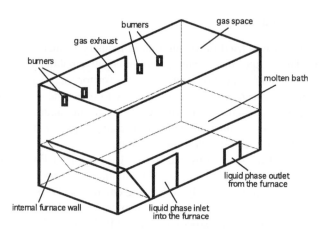

FIGURE 9.8   Model outline.

(1.0 MW), so the total firing rate for the furnace was 14×10^6 Btu/hr (4.1 MW), regardless of the firing mode (AF, AOF, or OF).

The molten aluminum bath height was chosen as 25 in. (64 cm) from the furnace floor. This metal height represents an intermediate point during the furnace melting cycle. It is possible to change the metal line and study the process at different metal levels. Ideally, this approach gives the flexibility to simulate quasi-steady-state solutions, the combination of which should represent the overall melting cycle. For a 25-in. (64 cm) metal level, the pump speed was chosen to circulate 8800 lb (4000 kg) of aluminum per minute. A scrap charging rate of 12,000 lb/hr (5400 kg/hr) was specified. Therefore, the aluminum mass and energy balance through the furnace wells was defined. The following boundary conditions were applied at the arch between the charging well and the furnace hearth: velocity of 0.062 m/s (0.20 ft/s), temperature of 976K (1297°F).

An important part of the modeling work was model validation. The model temperature prediction was compared with the temperature measured in the pump well, in the flue, and in the roof refractory. The agreement of the temperatures was within 10% based on the measurements in the melt. A relative comparison of the cases is reasonable because all the cases were built with the same grid and assumptions.

9.3.1.3   Results and Discussion

Some results of the model predictions are shown in Figures 9.9 through 9.20. Each figure has three sub-figures (a, b, and c), representing the AF, AOF, and OF cases, respectively. The same color scheme ranges are used for each sub-figure. This way, it is possible to make a visual comparison without closely focusing on the numbers displayed on the left-hand side of each sub-figure. However, if more accurate information is needed, it can be obtained from the scale provided. It is also important to note that the same colors are used regardless of the parameters presented (temperature, mole fraction, velocity, etc.). It is important to look at the values and the range of the displayed parameter. For example, compare Figures 9.9 and 9.12. The temperature field in Figure 9.9 is presented in K with a temperature variation of 1500K (2240°F) from 1000K (1340°F) to 2500K (4040°F). The temperature distribution inside the melt (Figure 9.12), is also presented in K, but the temperature range variation is only 7K (13°F) from 975K (1296°F) to 982K (1308°F). The white spots indicate that the values are outside the chosen range, for example, the hot flame temperature in Figures 9.10c and 9.11c.

The temperature distribution in the furnace combustion space is shown in Figures 9.9 through 9.11. The OF case shows the hottest calculated flame temperature with T_{max} of about 2760K

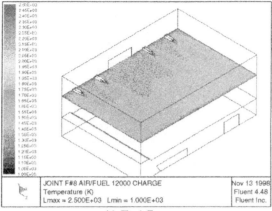

a. Air/Fuel Case

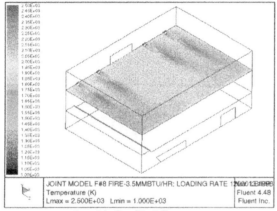

b. Air-Oxy/Fuel Case

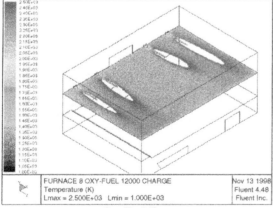

c. Oxy/Fuel Case

FIGURE 9.9 Horizontal temperature distribution through burner level: (a) air/fuel case; (b) air-oxy/fuel case; and (c) oxy/fuel case.

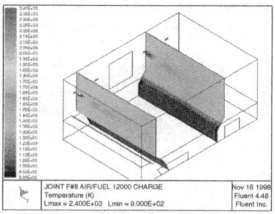

a. Air/Fuel Case

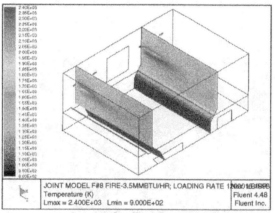

b. Air-Oxy/Fuel Case

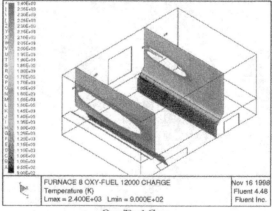

c. Oxy/Fuel Case

FIGURE 9.10 Vertical temperature distribution through 1st and 3rd burner location: (a) air/fuel case; (b) air-oxy/fuel case; and (c) oxy/fuel case.

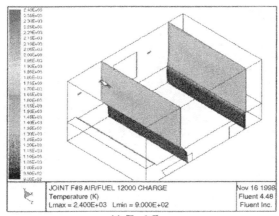

a.   Air/Fuel Case

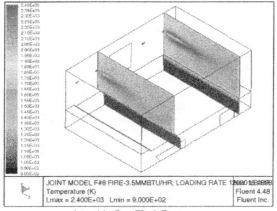

b.   Air-Oxy/Fuel Case

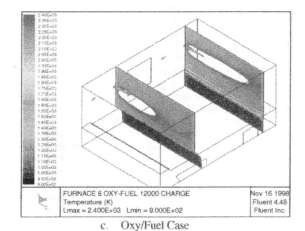

c.   Oxy/Fuel Case

FIGURE 9.11 Vertical temperature distribution through 2nd and 4th burner location: (a) air/fuel case; (b) air-oxy/fuel case; and (c) oxy/fuel case.

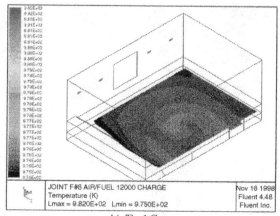

a. Air/Fuel Case

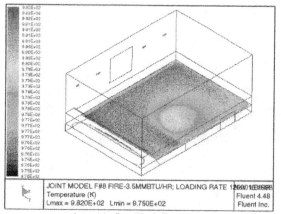

b. Air-Oxy/Fuel Case

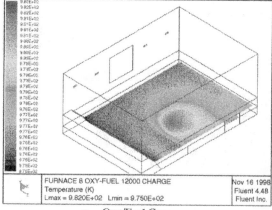

c. Oxy/Fuel Case

FIGURE 9.12   Horizontal temperature distribution through the melt: (a) air/fuel case; (b) air-oxy/fuel case; and (c) oxy/fuel case.

(4510°F). The AOF case has about a 260K (470°F) lower calculated peak flame temperature, and calculated flame T_{max} in the AF case is only about 1680K (2565°F). It is also shown that the calculated average gas temperature in the furnace combustion space in the OF case is significantly hotter than for the AF case. The calculated average gas temperature in the AOF case is shown to be much closer to the OF case than to the AF case. From these results, one can expect the following.

1. Because the AOF case provides just enough energy to melt 12,000 lb (5400 kg) of the charged scrap, the AF case does not supply enough energy to the melt at the same firing rate. A significantly higher firing rate would be required for the AF operation. At the same time, it is not obvious what firing rate the furnace can sustain without overheating or excessive refractory wear.

2. Slightly more than the required energy is supplied to the melt in the OF case; however, the refractory is a concern in this case because the temperature exceeds 3000°F (1900K) in several places. Additional work is needed to understand if a lower firing rate can provide enough energy to the melt with a simultaneous reduction of the refractory temperature.

The melt temperature distribution is shown in Figures 9.10 through 9.12. It is difficult to compare the temperatures in the melt as it is presented in Figures 9.10 and 9.11 because the temperature range was chosen to emphasize the differences in the combustion space. By focusing only on the melt temperature (Figure 9.12), it is possible to narrow the temperature range. An analysis of the results shown in Figure 9.12 leads to the conclusion that the melt temperature at the pump door in both the AOF and OF cases is sufficient to melt the scrap charged at the specified rate of 12,000 lb/hr (5400 kg/hr) with the melt circulation of 8800 lb/min (4000 kg/min). The AF case shows that the melt temperature is actually lower at the pump-well arch than at the charging-well arch. This is obviously impossible in the actual furnace. However, in the model, this leads to an important conclusion. One should recognize that the boundary conditions at the charging-well arch are fixed. Then, the temperature distribution shows that the melt temperature cannot be recovered to melt the aluminum scrap at the charging rate of 1200 lb/hr (540 kg/hr) with the AF firing rate of 3.5 × 10⁶ Btu/hr (1.0 MW) per burner. Consequently, the firing rate needs to be increased in the AF case to be able to melt the scrap. The only question is what the firing rate should be. Additional study is recommended to answer this question. Increasing the firing rate will obviously increase the natural gas consumption. This should be taken into consideration when the economics of all three technologies are evaluated. A field trial has confirmed that an increase in firing rate is required for the AF case compared to the AOF case. The burners were fired 30 to 40% higher in the AF case, but the same production as with the AOF burners was not achieved.

A hot spot is shown in the melt in Figure 9.12 for the AOF and OF cases. This hot spot is more pronounced in the OF case. Once again, the scrap charging rate was kept the same for all three cases. These results lead to the conclusion that even a slight increase in the firing rate in the OF case might overheat the aluminum surface. Such overheating can result in faster aluminum surface oxidation, yield loss, excessive drossing, and reduced aluminum alloy quality. It is predictable that the hot-spot location is at the metal stagnation point if it exists in the melt. The temperature distribution pattern does not change with the melt depth, but the melt temperature rises toward the melt surface. The melt temperature in the side-well furnace is usually kept around 1000 to 1035K (1350 to 1400°F) at the pump-well arch. The deeper the melt bath, the less efficient the furnace operation. Some furnaces do not have a molten metal pump. A "puddler" is used instead to stir the metal in the furnace charging well. A furnace of this kind is less efficient than a furnace that operates with a metal pump. Heat transfer in a molten aluminum bath occurs due to conduction through the metal depth and convection, which is defined by the melt motion. Therefore, the melt circulation is critical for overall furnace thermal efficiency. At the same time, any stagnation zone could cause surface overheating.

The metal flow field distribution is shown in Figure 9.13 at the horizontal plane, 12 in. (30 cm) above the furnace floor. This plot confirms that the melt hot spot is located at the stagnation zone, where the melt is recirculating between two submerged furnace arches. Because the pump speed was kept the same, one would expect the same melt flow pattern for all three cases. Figure 9.13 suggests that the attempt to move the hot spot toward the pump-well arch would lead to an increase in furnace production.

Figures 9.14 through 9.20 address important issues of the process and furnace control. The placement of a thermocouple in the roof is very important, especially if the roof temperature distribution is not uniform. The roof thermocouple is used to prevent the furnace refractory from overheating and should be installed at the anticipated hot spot. The hot face roof temperature distribution is shown in Figure 9.14. The roof temperature in the AF case, Figure 9.14a, is significantly below the refractory temperature limit, which is approximately 1900K (3000°F). However, as previously discussed, the AF case cannot melt aluminum at a 12,000 lb/hr (5400 kg/hr) charging rate with four burners firing only at 3.5×10^6 Btu/hr (1.0 MW) each. The calculated furnace roof temperature in the AOF case is within the specified refractory limits Figure 9.14b. In the OF case (Figure 9.14c), the roof temperature is approximately 2150K (3400°F), which exceeds the temperature limit by 250K (400°F). The burner firing rate should be reduced in the OF case, to bring the refractory temperature to the limit that might cause the melt temperature to be at a lower level than required. This leads to the conclusion that the OF case, production-wise, may not be more attractive than the AOF case. Another option is to charge more metal into the furnace to provide a better heat sink. However, there is no guarantee that the refractory temperature would be reduced enough. Additional work should be done to fully evaluate this question.

One can argue that the reduction of the firing rate in the OF case would result in fuel savings compared to the AOF case. In practice, however, fuel savings are offset by higher oxygen consumption, leading to higher overall cost. The cost of oxygen is always higher in the OF case than it is in the AOF case. The oxidizer stream consists of pure oxygen for the OF firing burner. For the AOF burner, the total oxidizer stream is a mixture of pure oxygen and air. For example, even a split of oxygen between the air stream and the high-purity oxygen stream provides 35% TOE. In that case, the cost of the oxidizer for the OF technology is twice as high as that for AOF. In addition, for the OF case, oxygen is consumed during the alloying and degassing cycle when the burner firing rate is reduced to low fire. In the case of AOF technology, the burners maintain melt temperature at a reduced firing rate using AF only (no oxygen). This flexibility provides better economics for the AOF firing technology than for the OF one.

As discussed above, AF technology requires a higher firing rate to melt the same amount of aluminum. An increased firing rate will produce furnace refractory hot spots. The nonuniform temperature distribution at the furnace side walls can be seen in Figure 9.15. One should expect a higher refractory temperature at the wall opposite to the burners, as shown in Figure 9.15a. In both the AOF and OF cases, the refractory temperature distribution is fairly uniform, as shown in Figures 9.15b and c. In fact, the largest predicted temperature gradient along the roof hot surface is 5K (9°F) in the OF case. The roof thermocouple placement in these cases is not as important. The nonuniform temperature distribution in the AF case is caused by the gas flow pattern shown in Figures 9.16 through 9.18. The white spots in the vector fields represent velocities of higher magnitude than the specified range. The gas velocities in the AF case are significantly higher compared to the AOF and the OF cases. Impingement of the high-temperature gases on the refractory (Figure 9.16a through Figure 9.18a) could cause local refractory overheating (see Figure 9.15a).

Higher energy loss and lower furnace thermal efficiency are expected in the AF case, compared to the AOF and OF cases. This can be explained by a higher volume of combustion gases leaving the furnace through the flue (Figure 9.16), despite the fact that the flue gas temperature in the AF case is lower than that in the AOF and OF cases (Figure 9.19).

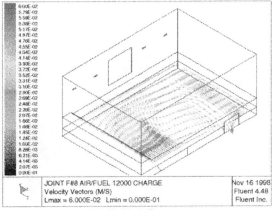

a.   Air/Fuel Case

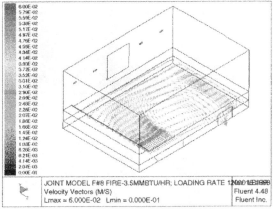

b.   Air-Oxy/Fuel Case

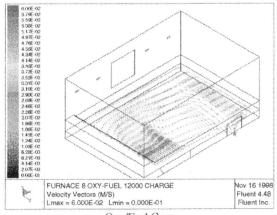

c.   Oxy/Fuel Case

FIGURE 9.13   Horizontal velocity vector field in the melt: (a) air/fuel case; (b) air-oxy/fuel case; and (c) oxy/fuel case.

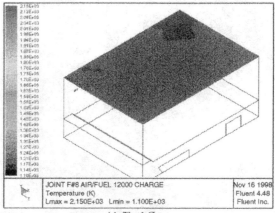

a. Air/Fuel Case

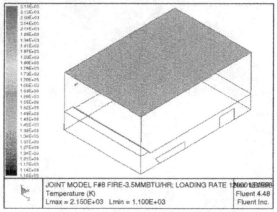

b. Air-Oxy/Fuel Case

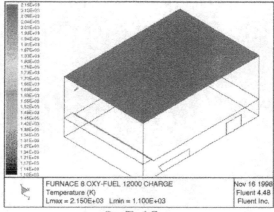

c. Oxy/Fuel Case

FIGURE 9.14   Furnace roof temperature distribution: (a) air/fuel case; (b) air-oxy/fuel case; and (c) oxy/fuel case.

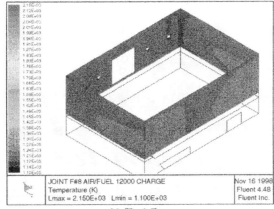

a.   Air/Fuel Case

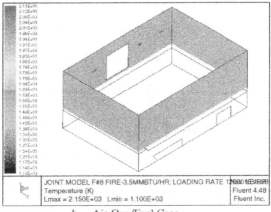

b.   Air-Oxy/Fuel Case

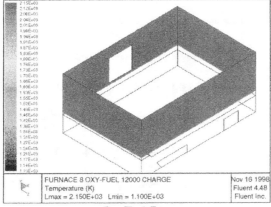

c.   Oxy/Fuel Case

FIGURE 9.15   Temperature distribution at the furnace walls in the combustion space: (a) air/fuel case; (b) air-oxy/fuel case; and (c) oxy/fuel case.

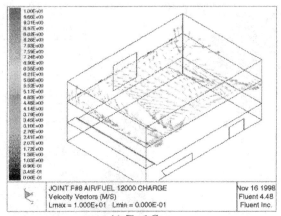

a.   Air/Fuel Case

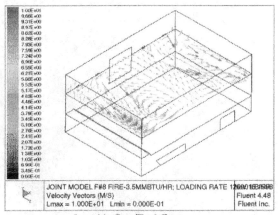

b.   Air-Oxy/Fuel Case

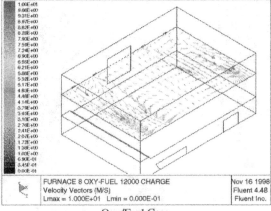

c.   Oxy/Fuel Case

FIGURE 9.16   Gas velocity vectors at the burner level: (a) air/fuel case; (b) air-oxy/fuel case; and (c) oxy/fuel case.

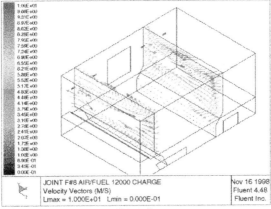

a. Air/Fuel Case

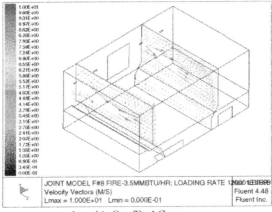

b. Air-Oxy/Fuel Case

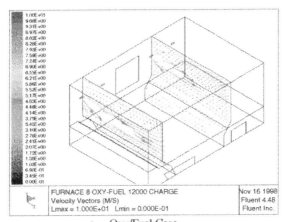

c. Oxy/Fuel Case

FIGURE 9.17 Vertical slices of the velocity field through the location of the 1st and 3rd burners: (a) air/fuel case; (b) air-oxy/fuel case; and (c) oxy/fuel case.

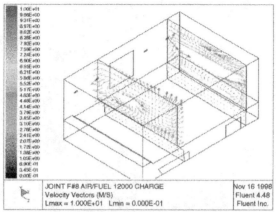

a. Air/Fuel Case

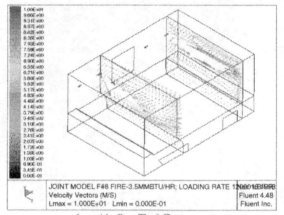

b. Air-Oxy/Fuel Case

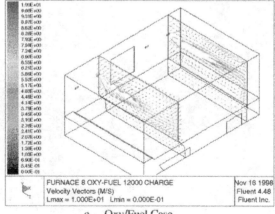

c. Oxy/Fuel Case

FIGURE 9.18 Vertical slices of the velocity field through the location of the 2nd and 4th burners: (a) air/fuel case; (b) air-oxy/fuel case; and (c) oxy/fuel case.

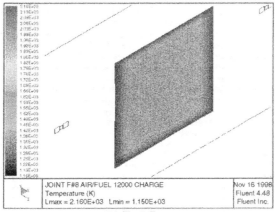

a. Air/Fuel Case

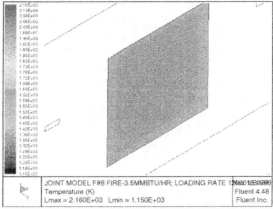

b. Air-Oxy/Fuel Case

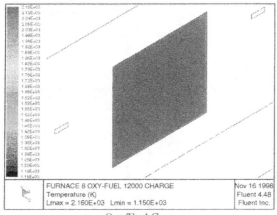

c. Oxy/Fuel Case

FIGURE 9.19 Temperature distribution at the furnace exhaust: (a) air/fuel case; (b) air-oxy/fuel case; and (c) oxy/fuel case.

One of the most important issues is the oxidation of the aluminum at the gas-melt interface. In general, it is very difficult to measure the oxygen concentration at the interface on a regular basis. Thus, it is difficult to judge which technology is more favorable in terms of the dross formation and yield until actual field data are obtained. The ability of the model to predict species concentration distribution at any chosen location in the furnace helps in evaluating all three technologies. For example, the oxygen concentration distribution, in mole fraction, is shown in Figure 9.20 at the aluminum surface. In comparing all three cases, the lowest O_2 concentration is predicted for the AOF case (Figure 9.20b). The highest oxygen concentration at the aluminum surface is predicted for the OF case (Figure 9.20c). This is another concern for the OF technology. The available oxygen at the melt interface depends on the burner and furnace designs, firing mode, oxygen staging, stoichiometry, velocity field distribution in the combustion space, and burner orientation. The evaluated AOF burner has the advantage of providing lower oxygen concentration at the melt surface than the specified flat flame OF burner and the AF burner.

9.3.1.4 Conclusions

A modeling study was conducted to evaluate different firing technologies using a continuously-charged aluminum furnace: air/fuel, air-oxy/fuel, and oxy/fuel firing. As a result, the following was found:

1. The AF technology could not melt 12,000 lb/hr (5400 kg/hr) of aluminum scrap with a total firing rate of 14×10^6 Btu/hr (4.1 MW) and a metal circulation rate of 8800 lb/min (4000 kg/min). For this case, a higher firing rate is required.
2. Refractory hot spots are anticipated in the AF case at the furnace wall opposite the burner wall and at the furnace roof. The hot spots are the result of hot gas impingement on the wall, due to an increase in the burner firing rate.
3. The AOF case provides enough energy to melt aluminum at the specified conditions. The refractory temperature did not exceed the refractory temperature limit.
4. The OF case provides enough energy to the melt, but overheats the furnace refractory at the evaluated conditions. A reduction in the burner firing rate in the OF case to satisfy the refractory temperature limit could cause a drop in the melt temperature, which would affect furnace production. An additional study will be conducted.
5. It is easy to overheat and to oxidize the melt surface when the OF firing technology is used. As a result, yield loss, excessive dross formation, and a decrease in alloy quality are expected. The OF case, production-wise, is less attractive than the AOF case and uses more oxygen.
6. The AOF case is more fuel cost attractive due to its flexibility, especially if the full furnace production cycle is considered. A VSA-integrated oxygen supply system is recommended, based on field test results obtained in April 1998 from RBSC.

9.3.2 DIRECT CHARGED FURNACE WITH AIR/OXY- AND OXY/FUEL BURNERS

An evaluation of a direct charged aluminum furnace was done on the computer, using the CFD commercial code FLUENT. The goal was to compare two different firing systems: air-oxy/fuel (AOF) and oxy/fuel (OF).

Initially, two cases were simulated. Case one has an AOF burner built into the model. In case two, the AOF burner was exchanged for a large flat flame OF burner. Both cases were constructed with the assumption that the aluminum bath does not contain any solid scrap and the metal is fully molten. This assumption represents the end of the furnace melting cycle. The burner firing rate of 1.75 MW (6×10^6 Btu/hr) was kept the same in both cases. This firing rate matches the low firing rate of the AF burner originally installed in the furnace. The modeling results showed that the

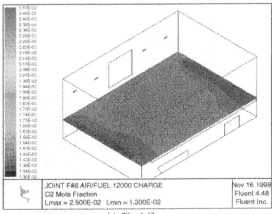

a. Air/Fuel Case

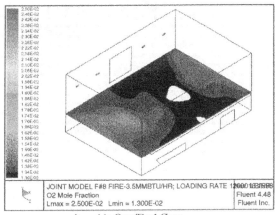

b. Air-Oxy/Fuel Case

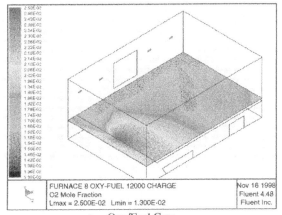

c. Oxy/Fuel Case

FIGURE 9.20 Oxygen mole fraction distribution at the melt surface: (a) air/fuel case; (b) air-oxy/fuel case; and (c) oxy/fuel case.

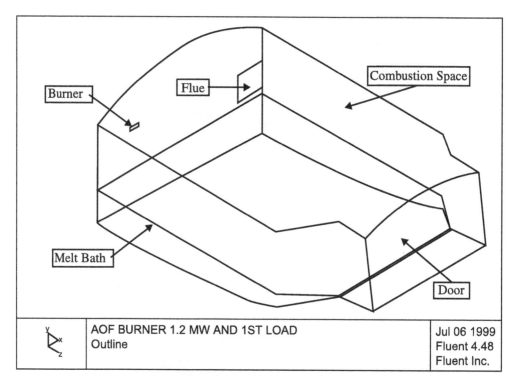

FIGURE 9.21 Furnace outline.

specified firing rate would overheat the furnace in both the AOF and OF cases. Therefore, two additional cases were calculated.

Both additional cases were calculated with the energy sink uniformly distributed throughout the molten bath. Such an assumption represents the melting process taking place inside the aluminum bath. The firing rate was reduced to a 1.2 MW (4×10^6 Btu/hr) in both the AOF and the OF cases. The Large OF flat flame burner geometry was changed for the Small OF flat flame burner geometry. The results showed that the AOF case has an advantage over the OF case. The model predictions were later confirmed by field tests. An analysis of the results suggests that the AOF technology is the preferred technology in the case of a direct charged furnace.

9.3.2.1 Process Description

The goal of the modeling work was to compare two different firing technologies: AOF and OF. Also, the modeling results were expected to help identify the maximum burner firing rate and to suggest the size of the OF flat flame burner for the field test. A direct charged furnace was chosen for the test. The furnace geometry and its refractory specifications were used to construct a computer model and to conduct a computer evaluation of the two firing technologies.

The furnace outline is shown in Figure 9.21. The total furnace capacity is 25 tons (23 m-tons) of molten metal. The aluminum scrap is charged into the furnace through the furnace door as shown in Figure 9.6. After the furnace is full and the furnace door is closed, the molten metal is poured into the furnace from the side (Figure 9.22). The melting process starts. Approximately half of the metal comes as scrap and the other half comes as molten metal. The scrap is usually charged into the furnace two or three times during the melting cycle. This cycle takes approximately 8 to 10 hours and then degassing and alloying cycle take place for approximately another 2 to 3 hours. As the metal is ready, it is tapped from the furnace into the ladle or into the casting machine (Figure 9.23) and the full furnace cycle starts again.

FIGURE 9.22 Pouring molten metal from the ladle into the furnace.

The furnace is equipped with one burner. Traditionally, it is a high-momentum AF burner. The maximum designed firing rate for the AF burner is 2.2 MW (7.5×10^6 Btu/hr). The burner is tilted toward the furnace centerline and downward to the melt.

The melt level rises inside the furnace during the melting cycle as the scrap turns from a solid into a molten state. There is no forced circulation of the melt inside the bath, which suggests that conduction is the primary heat transfer mechanism inside the molten aluminum. A furnace operator skims the melt surface two or three times during the melting cycle, removing dross from the melt surface. This skimming practice is very important; it helps to clean the aluminum surface, increases heat transfer between the melt and the furnace combustion space, and therefore increases the furnace thermal efficiency.

The process description of the direct charged furnace leads to the conclusion that the furnace production increase is defined by the length of the melting cycle. Therefore, a direct charged furnace throughput can be increased if the melting rate is improved by switching from the traditional AF technology to an oxygen-based firing technology.

Yield is also one of the most important production factors. The metal could be excessively oxidized while exposed to the open flame inside the furnace. Oxidation happens because of flame impingement on the aluminum scrap, and because of excessive free oxygen availability at the melt interface during the flat bath mode. A typical yield is approximately 89% to 92% for this type of furnace with the AF burner, and varies with the type of scrap used. An extra amount of metal loss could be expected as the flame temperature increases with the use of an oxygen-based technology. Therefore, the oxygen-based firing technology should be carefully evaluated. Such an evaluation can be performed with the help of a CFD model.

a. Pouring into the ladle

b. Pouring into the casting machine

FIGURE 9.23   Pouring the metal out of the furnace: (a) pouring into the ladle, and (b) pouring into the casting machine.

9.3.2.2   Model Description

A model was built using a CFD commercial code from FLUENT with 215,000 nodes. An example of the grid distribution is shown in Figure 9.24 for the constructed domain. The model geometry was built based on the furnace drawings provided by the plant. The same model was used for both the AOF and OF cases. The only differences between the cases were the burner geometry and the number of cells representing the burner.

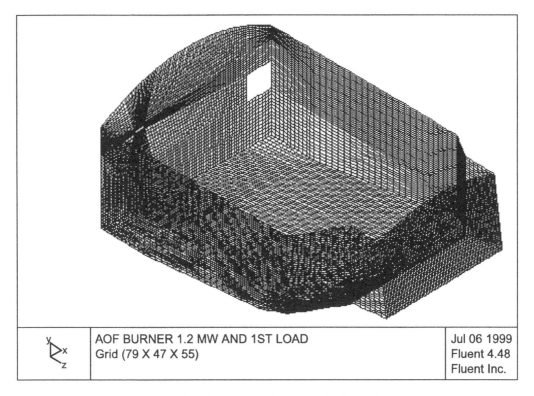

y⤢x z	AOF BURNER 1.2 MW AND 1ST LOAD Grid (79 X 47 X 55)	Jul 06 1999 Fluent 4.48 Fluent Inc.

FIGURE 9.24   An example of the grid distribution.

The same boundary conditions and assumptions were used in both cases. All the furnace walls were represented by boundaries with zero velocity and with an appropriate external heat transfer coefficient. These coefficients were calculated based on the wall thickness and thermal conductivity of each refractory layer. The refractory specifications were provided by the plant. The furnace exhaust was introduced as a pressure boundary and the burners were represented with multiple inlets. The inlets for natural gas, oxygen, and air were defined as separate streams with appropriate species. The velocity magnitude, velocity components, and species mass fractions were defined at each inlet to represent the corresponding firing rate and the firing technology. The combustion reactions were solved by using a mixture fraction (PDF) approach. Ten species were used in each case to represent the reactants and the products of combustion.

The model was constructed to account for only two phases: the combustion space and the molten bath. The molten bath height was fixed at 70 cm (27 in.) and the melt surface was assumed flat. Such an assumption provided the process information at the end of the melting cycle and reduced the model complexity. The model would be very complex at the beginning of the melting cycle because the beginning of the cycle has three phases: combustion gases, solid scrap, and molten bath. Intuitively, the beginning of the cycle requires intensive convection in the combustion space in order to heat the scrap as uniformly as possible. At the same time, the process requirements are not obvious at the end of the cycle. For example, what mechanism of the heat transfer is the most important? What is the temperature distribution at the refractory surfaces and at the melt? What is the dross formation rate due to oxygen concentration at the melt surface? Therefore, the flat bath assumption will help with the process clarification at the end of the melting cycle and will simultaneously reduce the model complexity.

The first two cases were constructed with the assumption of no solid scrap left in the melt. Thus, no heat sink was specified inside the furnace bath. The burners were defined to fire at a constant firing rate of 1.75 MW (6×10^6 Btu/hr). The total oxygen enrichment (TOE) for the AOF case was specified as 35%, which corresponds to a 50%-50% oxygen split between the air and oxygen streams. The large flat flame OF burner geometry was built into the model for the OF case with 30% of oxygen flow coming through the burner staging. A 95% oxygen purity was assumed in both cases.

The additional two cases have been constructed after the results of the first two cases were analyzed. The burner firing rate was reduced to a 1.2 MW (4×10^6 Btu/hr) for both the AOF and OF cases. The heat sink of 3.5×10^4 W/m^3 (3400 Btu/hr-ft^3) was assumed in the melt. Such a heat sink corresponds to a gradual scrap loading procedure. It was assumed that 12 tons (11 m-tons) of the aluminum scrap was loaded into the furnace during the time period of 10 hours. Obviously, the latter assumption is a big simplification, but it should help to provide insight into the process at the end of the melting cycle.

9.3.2.3   Results and Discussion

All the results are shown in gray scale. The appropriate gray range of each presented parameter was chosen to be the same for corresponding figures in both calculated cases, so that both cases can be visually compared. Every figure in this chapter section is plotted in the same manner. As in the previous example, the same gray scheme is also used to represent different parameters. For example, compare the temperature distribution shown in Figure 9.25 with the oxygen concentration shown at the melt level in Figure 9.29.

9.3.2.3.1   High fire, 1.75 MW, at the end of the furnace cycle

The results of the first two cases are presented in Figures 9.25 through 9.29. The AOF and the OF cases were computed with no heat sink in the melt. The large flat flame OF burner geometry was specified for the OF case. The firing rate of 1.75 MW (6×10^6 Btu/hr) was assumed in both cases.

The temperature distribution through the combustion space of the furnace is presented in Figures 9.25 and 9.26. It is shown that the combustion gases are hotter inside the combustion space in the OF case. The average gas temperature is about 80 to 120K (140 to 220°F) higher in the OF case compared to the AOF case. At the same time, the OF case is thermally more efficient due to the lower flow rate of gases through the furnace exhaust. The total energy loss is a function of both the exhaust temperature and the gas flow rate through the furnace exhaust. The volumetric gas flow rate in the OF case is 52% lower compared to the AOF case. This suggests that the energy losses through the furnace exhaust are significantly lower in the OF case, even with the average exhaust temperature up to 120K (220°F) higher compared to the AOF case. The flame temperature distribution shown in Figure 9.26 leads one to believe that more metal losses should be expected in the OF case due to possible excessive overheating of the scrap surface.

The melt temperature is shown in Figure 9.27. The average melt temperature is higher and more uniform in the OF case compared to the AOF case. In both cases, it is significantly above the required maximum melt temperature. The required average melt temperature should be around 1005 to 1060K (1350 to 1450°F) vs. 1650 to 1850K (2510 to 2870°F), as shown in Figure 9.27. This fact suggests that the burner firing rate should be gradually reduced in both the AOF and the OF cases as the pile of scrap in the furnace melts down and the melting cycle approaches its end. Also, a significant melt overheating suggests that the firing rate of a 1.75 MW (6×10^6 Btu/hr) might be required (if at all) only in the very beginning of the melting cycle when the scrap is piled up inside the furnace and can yet receive a substantial amount of energy. Still, the danger of using such a high firing rate, even at the very beginning of the melting cycle, is that the flame temperature is significantly higher for both the AOF and the OF cases compared to the AF case. The increased flame temperature and its direct impingement on the scrap will cause excessive metal oxidation, which reduces the yield.

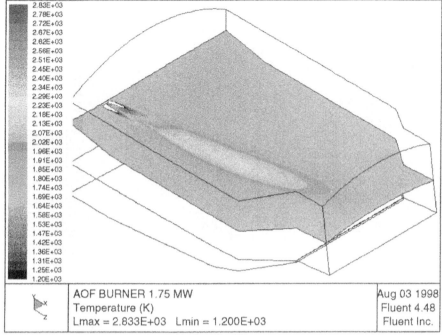

Air-Oxy/Fuel Burner

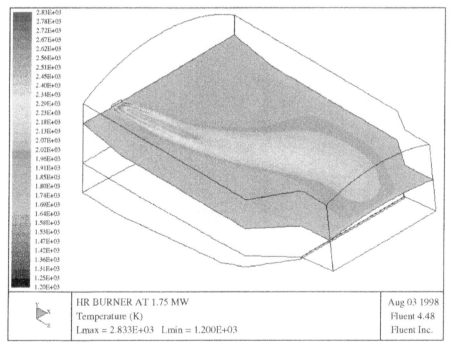

HR Oxy/Fuel Burner

FIGURE 9.25 Temperature distribution through the furnace combustion space: (a) air-oxy/fuel burner, and (b) HR oxy/fuel burner.

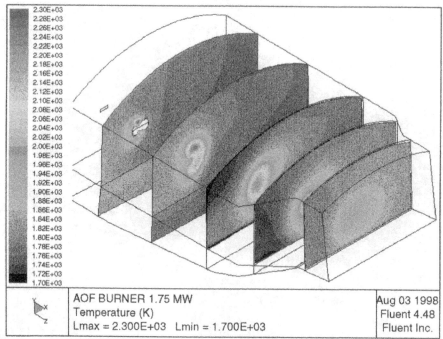

Air-Oxy/Fuel burner

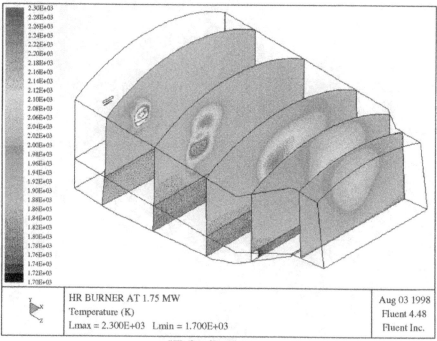

HR Oxy/Fuel burner

FIGURE 9.26 Vertical temperature distribution through the furnace combustion space (no heat sink): (a) air-oxy/fuel burner, and (b) HR oxy/fuel burner.

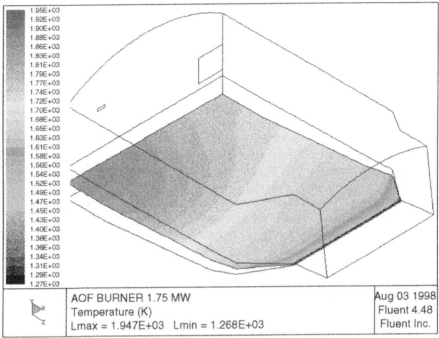

Air-Oxy/Fuel Burner

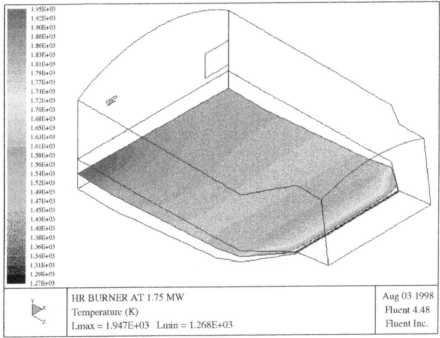

Oxy/Fuel burner

FIGURE 9.27 Temperature distribution through the middle of the melt (no heat sink): (a) air-oxy/fuel burner, and (b) HR oxy/fuel burner.

The temperature distribution at the furnace roof (see Figure 9.28) is in the range of 1592 to 1890K (2406 to 2942°F). As shown in Figure 9.28, the hot face of the roof refractory has a higher and more uniform temperature distribution in the case with the OF burner. At the same time, the maximum refractory temperature is the same for both cases and located at the same upper-left corner inside the furnace. The calculated refractory temperature is shown to be within the maximum refractory limit, which is typically 1920K (3000°F). Therefore, one can conclude that if the burner firing rate is reduced, the refractory temperature is expected to be substantially below its maximum limit. In many operations, the temperature set point at the furnace roof is defined at 1500K (2250°F). Once the roof thermocouple reaches the set point, the firing rate is automatically reduced to a "low fire" mode. Refractory thermal shock is more likely to occur if the burner is overpowered and changes its firing rate too often. Therefore, the high firing rate, such as 1.75 MW (6×10^6 Btu/hr), could cause refractory thermal shock. This suggests that, for both cases, the burner firing rate should be reduced to minimize refractory damage.

Oxidation of the aluminum surface is a function of the oxygen concentration at the metal level. Thus, the formation of dross and also the metal recovery (yield) will be influenced by free oxygen availability at the metal line when the scrap pile of scrap forms a flat surface. The oxygen distribution is shown in Figure 9.29 at the metal line. The OF case with the large flat flame OF burner has a well-pronounced high oxygen region at the metal line. Twice as much oxygen is available in this region in the OF case, compared to the AOF case. Distribution of the oxygen concentration in Figure 9.29 leads to the conclusion that more dross formation, lower heat transfer, and decreased yield should be expected in the OF case compared to the AOF case. Unfortunately, the model does not account for the dross formation and, therefore, for appropriate reduction of the heat transfer from the combustion space to the melt.

9.3.2.3.2   Low fire, 1.2 MW, with heat sink in the melt

It was shown previously that a firing rate of 1.75 MW could be too high for the described process. Thus, two additional cases were built and calculated. In both of these new AOF and OF cases the firing rate was reduced to 1.2 MW (4×10^6 Btu/hr). The heat sink of 3.5×10^5 W/m³ (3.4×10^4 Btu/hr-ft³) was assumed in the melt, while the melt line was still flat and was kept at the same height as in the previous two cases. Such a heat sink corresponds to a gradual scrap loading procedure assuming that 10 hours are needed to melt 12 tons (11 m-tons) of scrap. A small flat flame OF burner was used for the OF case. The small flat flame burner is required now so the burner low firing rate can be reduced to a low firing mode of 0.3 MW (1×10^6 Btu/hr) during the furnace holding time.

Vertical slices of the temperature distribution are shown in Figure 9.30 through the furnace combustion space and the melt depth. As before, the white spots fall out of the specified temperature range. A comparison of Figures 9.30a and b shows that the average temperature of the furnace combustion gases is about 15 to 20K (27 to 36°F) higher in the AOF case than in the OF case. It is opposite to what was found in the two previous cases. This should not come as any surprise if one considers that the heat transfer is more efficient in the OF case compared to the AOF case and that the specific heat of the furnace gases is also higher in the OF case as shown in Figure 9.31. Although the average furnace gas temperature is higher in the AOF case, the flame temperature is still significantly higher in the OF case. The same calculated flame temperature difference of 270K (490°F) was found in the additional two cases as in the two previously calculated cases. The heat sink, the velocity distribution in the furnace combustion space (Figure 9.32), and the specific heat of the furnace gases (Figure 9.31) are the major contributing factors defining the average furnace gas temperature.

Figures 9.33 through 9.35 show the melt temperature; the melt temperature is shown at two different elevations inside the melt and at its surface. The melt surface temperature is shown in Figure 9.33. The melt surface is overheated near the burner region and at the shallow end of the furnace in the OF case. The overall surface temperature is not as uniform in the OF case as it is in the AOF case. A similar temperature pattern is shown in the middle of the melt depth (Figure 9.34)

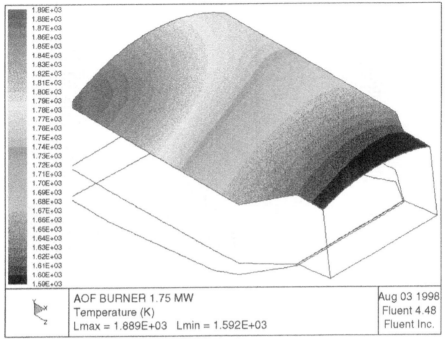

Air-Oxy/Fuel burner

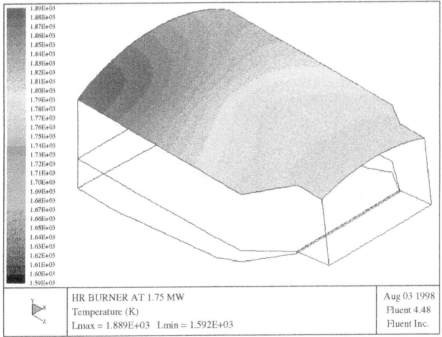

HR Oxy/Fuel burner

FIGURE 9.28 Temperature distribution at the furnace roof (no heat sink): (a) air-oxy/fuel burner, and (b) HR oxy/fuel burner.

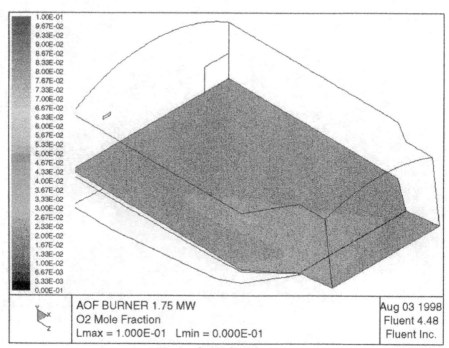

Air-Oxy/Fuel burner

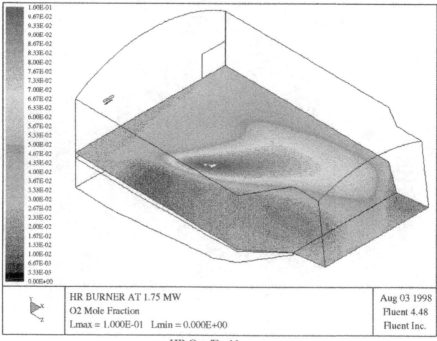

HR Oxy/Fuel burner

FIGURE 9.29 Oxygen distribution next to the melt surface (no heat sink): (a) air-oxy/fuel burner, and (b) HR oxy/fuel burner.

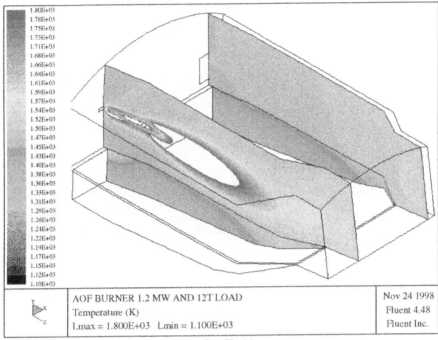

a.   Air-Oxy/Fuel burner

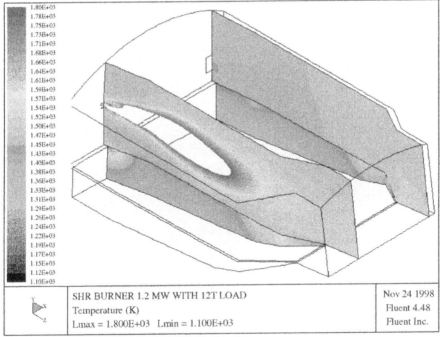

b.   Oxy/Fuel Burner

FIGURE 9.30   Temperature distribution at vertical slices: (a) air-oxy/fuel burner, and (b) HR oxy/fuel burner.

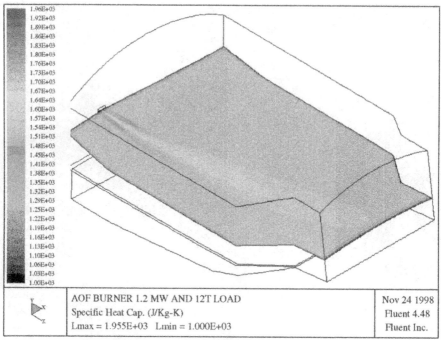

a.   Air-Oxy/Fuel burner

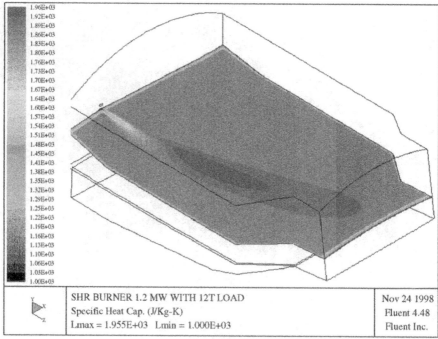

b.   Oxy/Fuel Burner

FIGURE 9.31   Specific heat of furnace gases in the combustion space: (a) air-oxy/fuel burner, and (b) HR oxy/fuel burner.

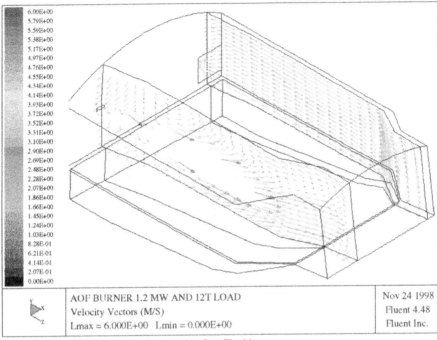

a.   Air-Oxy/Fuel burner

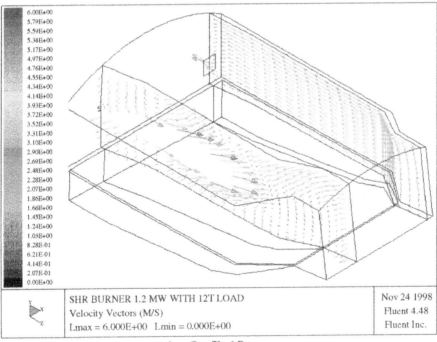

b.   Oxy/Fuel Burner

FIGURE 9.32   Velocity distribution in the furnace combustion space: (a) air-oxy/fuel burner, and (b) HR oxy/fuel burner.

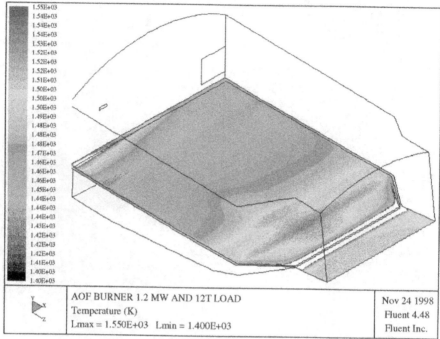

a.   Air-Oxy/Fuel burner

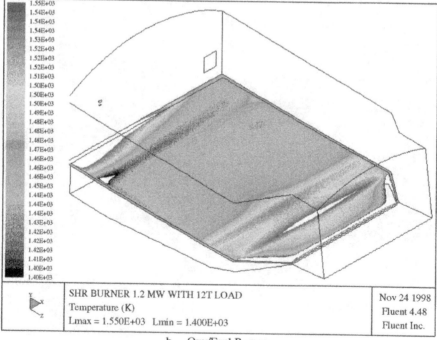

b.   Oxy/Fuel Burner

FIGURE 9.33   Temperature distribution at the melt surface: (a) air-oxy/fuel burner, and (b) HR oxy/fuel burner.

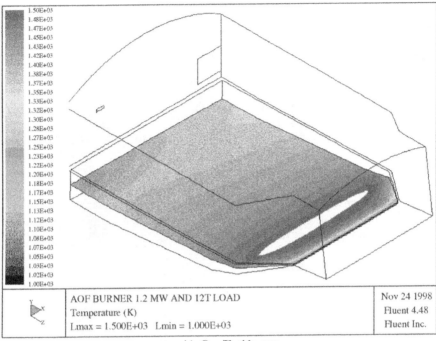

a.   Air-Oxy/Fuel burner

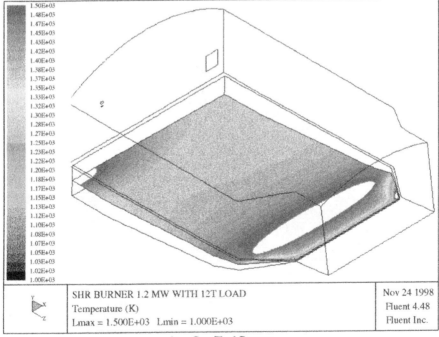

b.   Oxy/Fuel Burner

FIGURE 9.34   Temperature distribution through the middle of the melt: (a) air-oxy/fuel burner, and (b) HR oxy/fuel burner.

and at the slice next to the furnace bottom (Figure 9.35). The melt temperature in both cases is still higher than is required, 1300-1350K (1880-1970°F). The results suggest either that the model assumption of the scrap loading procedure was not accurate or the firing rate of 1.2 MW (4 × 10^6 Btu/hr) is still too high with the flat bath model. In any case, it is shown that the OF firing technology is more likely to overheat the molten bath locally, which would lead to greater metal loses, more dross formation at the melt surface, and reduced metal quality.

Figures 9.36 and 9.37 show the calculated refractory temperature. The furnace roof temperature is shown in Figure 9.36, and the temperature at the furnace side walls is shown in Figure 9.37. A quick visual comparison can be deceiving. The temperature difference between the roof's hottest spots in the AOF and the OF cases is only 15K (27°F), as shown in Figure 9.36. This temperature difference is even smaller, 10K (18°F), at the furnace door as shown in Figure 9.37. The maximum refractory temperature does not exceed the refractory temperature limit and is about 1510K (2260°F), which is usually used as a thermocouple set point to switch from a "high" to a "low" firing mode during furnace operation. A significant temperature reduction at the refractory face suggests that the furnace can be operated more efficiently without frequent changes in the firing modes. This will also lead to fuel savings during the furnace melting cycle.

9.3.2.4   Conclusions

A CFD model was built for a direct charged furnace. The model was used to analyze the furnace operating conditions and the expected melting results with different firing technologies and different maximum burner firing rates. The following conclusions were drawn based on analysis of the modeling results:

1. The burner firing rate should be gradually reduced in both the AOF and the OF cases as the scrap pile melts down and the melting cycle approaches its end.
2. The OF case is more thermally efficient compared to the AOF case.
3. The yield is expected to decrease in the OF case. This can be explained by the increased flame temperature and direct impingement on the pile of scrap which would cause excessive metal oxidation.
4. The oxygen concentration at the melt level suggests that more dross formation, possible reduction in heat transfer, and a decrease in the yield is expected in the OF case compared to the AOF case. It is also shown that it is easy to overheat the molten bath locally with the OF firing technology. Local metal overheating will result in the metal quality reduction.
5. The burner firing rate should be reduced significantly compared to AF operation. The maximum recommended burner firing rate is about 1.2 MW (4 × 10^6 Btu/hr). Refractory thermal shock is expected to be reduced with either the AOF or the OF technology if the firing rate is properly defined.
6. The small flat flame burner OF should be used if OF technology is chosen.
7. The AOF technology with air-oxy/fuel burner is recommended due to the burner's flexibility, expected increase in yield, expected decrease in fuel consumption, and reduced dross formation.

9.3.3   Performance of Newly Designed Air/Fuel reverb Furnace

9.3.3.1   Objective

A new side-well aluminum furnace was designed for an aluminum smelting company (RBSC) by a third party. The new furnace design assumed a 50% production increase compared to the existing side-well aluminum furnace. To achieve the scheduled production increase, a new furnace was

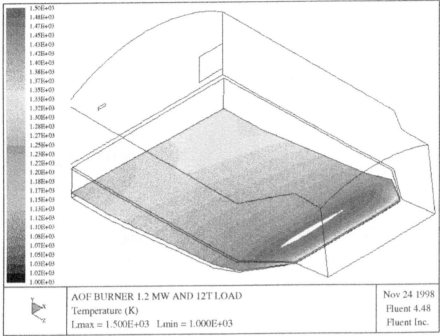

a.   Air-Oxy/Fuel burner

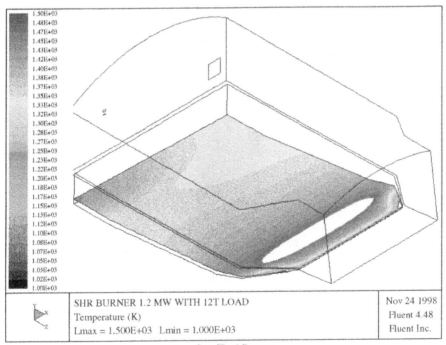

b.   Oxy/Fuel Burner

FIGURE 9.35   Temperature distribution through the melt, next to the furnace bottom: (a) air-oxy/fuel burner, and (b) HR oxy/fuel burner.

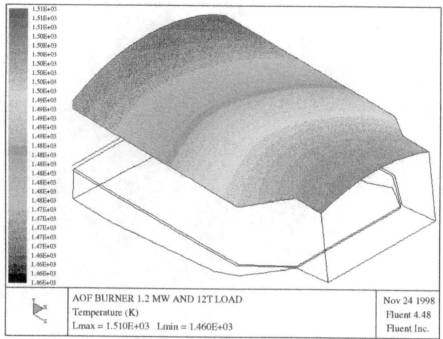

a.　Air-Oxy/Fuel burner

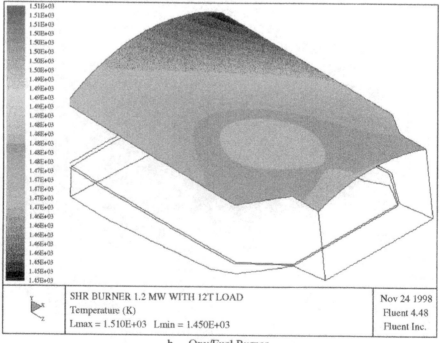

b.　Oxy/Fuel Burner

FIGURE 9.36　Temperature distribution at the furnace roof: (a) air-oxy/fuel burner, and (b) HR oxy/fuel burner.

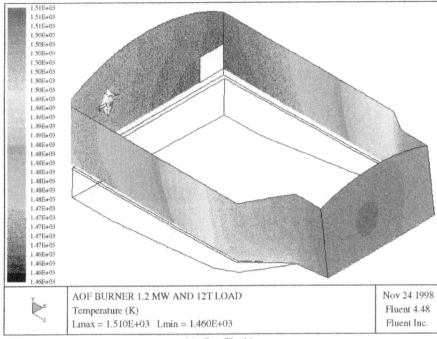

a. Air-Oxy/Fuel burner

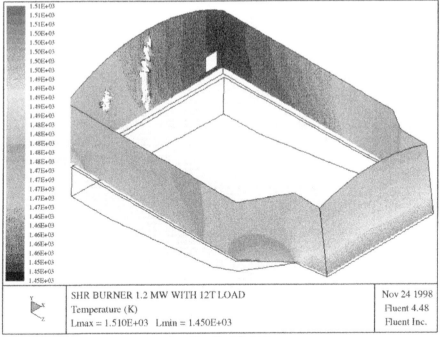

b. Oxy/Fuel Burner

FIGURE 9.37 Temperature distribution at the furnace walls in the combustion space: (a) air-oxy/fuel burner, and (b) HR oxy/fuel burner.

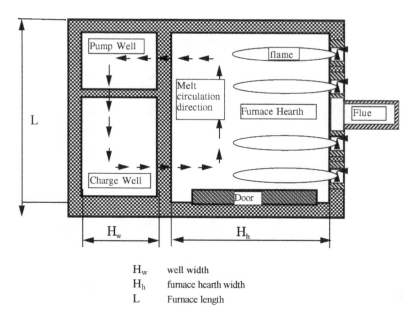

H_w	well width
H_h	furnace hearth width
L	Furnace length

FIGURE 9.38 Schematic of the furnace design.

designed with 50% more aluminum bath surface area. There were some questions about the new furnace design, because RBSC had no experience operating such a large furnace. Air Products was asked to evaluate the new furnace design and to predict its production rate. Assuming that the total firing rate in the furnace of 28×10^6 Btu/hr (8.2 MW) was evenly distributed among four air/fuel burners, the questions were:

- Would the new furnace handle the scheduled production increase?
- What burner angles should be used?

To help answer these questions, Air Products proposed an evaluation of the new furnace design using CFD modeling. A CFD model was used to compare an existing furnace with the newly designed bigger furnace. A relative comparison of the furnaces provided enough information to answer the above questions.

9.3.3.2   Model Description

Two models were built to evaluate both the existing and the new side-well aluminum furnaces. The general furnace design is the same for both furnaces and is shown in Figure 9.38. The new furnace had a 50% longer side wall. Both furnaces had side-wells and aluminum pumps that were not modeled. The aluminum bath inside the furnace was also not included. For modeling purposes, the same refractory material was assumed for both furnaces. With this information, the boundary conditions at each wall were specified with the appropriate external heat transfer coefficient (EHTC). To calculate the EHTC for each furnace wall, an electric analogy was used. The expression for the EHTC has the form:

$$ h = \frac{1}{\displaystyle\sum_{i=1}^{n} \frac{\delta_i}{\lambda_i}} \tag{9.1} $$

where h = EHTC

δ_i = Thickness of the ith refractory layer

λ_i = Thermal conductivity of ith refractory layer

The same EHTC boundary condition was used for the aluminum surface in both cases. Here, the thickness of the aluminum layer was included. The molten aluminum circulation was not included. The final results should be corrected for this assumption, at least qualitatively.

The exhaust port and four air/methane firing burners were placed at the furnace burner wall, as shown in Figure 9.38. The exhaust port has the same dimensions for both cases. The elevation of the burners and the exhaust port over the melt line was also kept the same. The outline of the models is shown in Figure 9.39.

The total firing rate from four air/methane burners in the existing furnace was specified at 16 × 10^6 Btu/hr (4.7 MW). This total firing was assumed to be evenly distributed among the burners, so each burner was firing at 4 × 10^6 Btu/hr (1.2 MW). The same assumption was used for the new furnace. However, the total firing rate in the new furnace was increased to 28 × 10^6 Btu/hr (4.7 MW). In both cases, all four burners were tilted down toward the aluminum surface at a 10° angle. In both cases, the side burners near the furnace door were also tilted toward the center of the furnace hearth at a 7° angle. The gas velocity from the burner tip was changed from 143 ft/s (43.6 m/s) in the existing furnace, to 250 ft/s (76.2 m/s) in the new furnace. These velocities were used as the boundary conditions at the inlets representing the burners. The same discharge cross-section of the burner tip was used in both cases. The main parameters of the compared furnaces are summarized in Table 9.1.

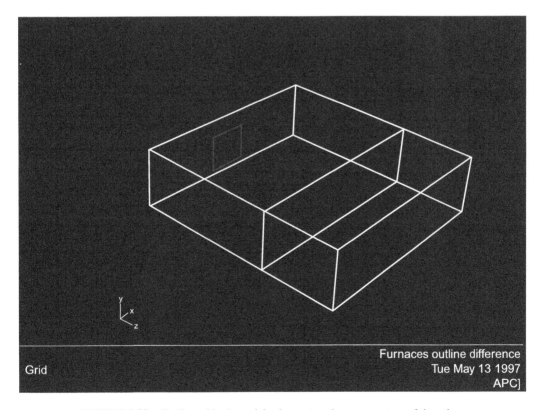

Grid

Furnaces outline difference
Tue May 13 1997
APC]

FIGURE 9.39 Outline of both models shown together, one on top of the other.

TABLE 9.1
A Summary of Furnace Parameters

Furnace Parameters	Units	Old Furnace	New Furnace
Length	ft	L	L
Width	ft	W	$1.54 \times W$
Number of burners		4	4
Firing rate per burner	10^6 Btu/hr	FR	$1.75 \times FR$
Burner elevation from Al	ft	h	h
Burner tilt down	degree (°)	10	10
Side burner tilt	degree (°)	7	7
Exhaust dimension	ft × ft	Ex	Ex
Gas velocity at burner tip	ft/s	V	$0.57 \times V$
Gas inlet temperature	°F	80	80
Refractory	type	A	A

FUENT/UNS CFD was used to construct the model. The following physical models were used:

1. Discrete transfer radiation model (DTRM), to account for radiative heat transfer inside the furnace combustion space
2. Standard κ-ε turbulence model
3. Mixture fraction (PDF) chemical reaction model with local equilibrium
4. Six chemical species for combustion reaction: CH_4, O_2, CO, CO_2, H_2O, and N_2

9.3.3.3 Results and Discussion

Because there was no way to validate the model results for the newly designed furnace, the existing furnace was modeled first. Once the modeling results were obtained and validated for the existing furnace, the same modeling techniques were used to obtain the results for the new furnace. Both results were then compared. Therefore, the validity of the modeling results for the new furnace were obtained based on the experience of operating the existing furnace. The temperature level and the temperature distribution at the internal refractory walls are indicators of furnace performance. The furnace thermal efficiency can also be used as a comparison criterion. In general, the furnace thermal efficiency can be defined as the ratio of the useful energy needed to melt the aluminum to the total energy introduced into the system, and can be written in the following form:

$$\text{Furnace thermal efficiency} = \frac{\text{Useful energy}}{\text{Total energy}} \tag{9.2}$$

In these cases, the "useful energy" in Equation (9.2) takes into account only the heat transferred through the aluminum bath and the furnace floor refractory to the outside. It does not consider the energy that is carried away with the molten aluminum into the side well. Additional model complexity should be added to account more accurately for the latter portion of energy. To avoid increasing both the model set-up time and the computational time, the additional model complexity was not used here, in particular because the model predictions were compared on a relative basis. If more rigorous analysis is needed, the molten aluminum bath can be included in the model and the aluminum flow rate through the side-well would then be taken into account.

The two cases were computed. The first case was the existing furnace, with four air/fuel burners firing at 4×10^6 Btu/hr (1.2 MW) each. The second one was the new furnace with a 50% larger aluminum surface area with the same four burners, but firing at 7×10^6 Btu/hr (2.1 MW) each. In both cases, all four burners were firing downward at a 10° angle to the melt surface. Also, the side burner near the furnace door was tilted toward the furnace hearth at a 7° angle in both cases.

The model predictions are shown in Figures 9.40 through 9.63. The same color scale is used in each figure for ease of visual comparison. Each figure consists of two pictures: a and b. Picture "a" presents the model predictions for the existing furnace, and picture "b" presents the results for the newly designed furnace.

The temperature distribution is shown in Figures 9.40 through 9.46 at different slices inside both furnaces. Figure 9.40 shows the temperature distribution inside the combustion space at a slice along the burner firing direction. Figure 9.41 shows the temperature distribution at the horizontal slice taken at the burner elevation. Figure 9.42 and Figure 9.43 present the temperature distribution at the vertical slices through each burner along the furnace side walls. Figure 9.44 and Figure 9.45 show the temperature distribution at the roof and the floor of the furnaces, respectively. Figure 9.46 presents the temperature distribution at the vertical furnace walls.

A comparison of the results presented in Figures 9.40 through 9.46 leads to the conclusion that the new furnace, at a total firing rate of 28×10^6 Btu/hr (8.2 MW), will be at a higher temperature than the existing furnace, at a total firing rate of 16×10^6 Btu/hr (4.7 MW). For the new furnace, the maximum refractory temperature increase is about 40°F (22K) compared to the same temperature of the refractory of the existing furnace. A comparison of the calculated average temperatures at different refractory surfaces for both furnaces is shown in Table 9.2.

The temperature distribution in the region near the furnace side wall, opposite the furnace door, is very different for the new and the existing furnaces, as shown in Figure 9.40. The firing direction of the side-wall burner is shown tilted toward the furnace's center plane in Figure 9.40a. The burner is firing straight along the furnace side wall, exactly the same way as it is shown in Figure 9.40b. Because the flame of the side burner is tilted into the furnace combustion space, the temperature gradient is well-defined between the flame and the furnace side wall, as it is shown in Figure 9.40a. This prevents the furnace side wall from overheating. There was no flame tilt at the furnace side wall in the case of the new furnace design. Thus, the wall could be easily overheated.

The reason for different flame behavior from the side burner in the new and the existing furnaces is the flow patterns near the side wall. In Figure 9.47, the velocity vectors of the internal flowfield are shown for the same slice as in Figure 9.40. The hot jet from the side burner propagates all the way down to the front wall, impinging on it in the case of the existing furnace design. This jet creates a recirculation zone that keeps the flame away from the wall. In the case of the new furnace design, the front wall of the furnace is too far away from the burner nozzle and does not actively influence the jet propagation. Figures 9.47 through 9.51 show the furnace gas flow distribution at different slices through the furnace combustion space. These figures reinforce the conclusion that the influence of the furnace hot wall on the flow pattern is much stronger inside the combustion space of the existing furnace, compared to the new furnace design. Thus, the distance between the furnace burner wall and the furnace front wall is important.

A comparison of the temperature and velocity distribution inside the existing (old) and new furnaces leads to the conclusion that the burner orientation inside the furnace is of great importance and influences furnace efficiency and production. To evaluate the influence of burner orientation on the furnace efficiency and productivity, two additional cases (cases 3 and 4) were constructed. These cases were calculated for the new furnace only.

In case 3, all the burners were still firing downward at a 10° angle. Both side burners were tilted toward the furnace center plane (Figure 9.52a) at a 4.5° angle. In case 4, the burners on the sides remained at a 4.5° firing angle toward the center of the furnace hearth, as in case 3. The difference between the cases is in the downward firing angle. This angle is changed from 10° to 5° for all four burners. No other changes were made. The goal is to concentrate the hottest spot near the furnace front wall in the vicinity of the most intensive aluminum motion in the molten bath.

Cases 3 and 4 were calculated, and the results are shown in Figures 9.52 through 9.63. The results of the cases 3 and 4 are shown in pictures "a" and "b," respectively. Three cases of the new

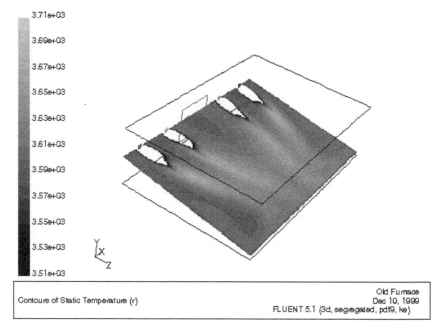

a. Old furnace

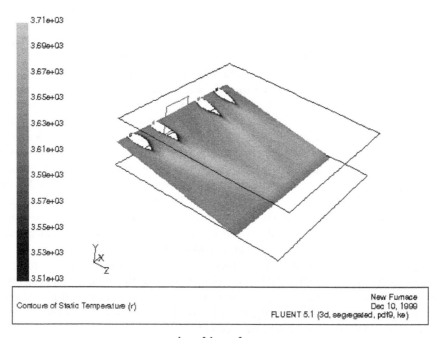

b. New furnace

FIGURE 9.40 Temperature distribution at the slice along the burner firing direction: (a) old furnace, and (b) new furnace.

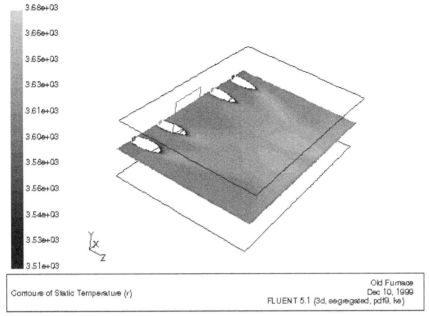

a. Old furnace

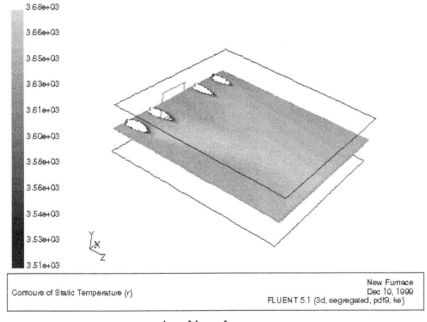

b. New furnace

FIGURE 9.41   Horizontal temperature distribution at the burner level: (a) old furnace, and (b) new furnace.

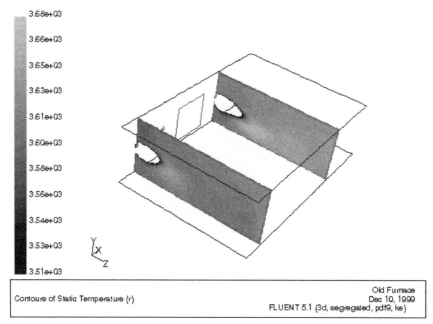

a. Old furnace

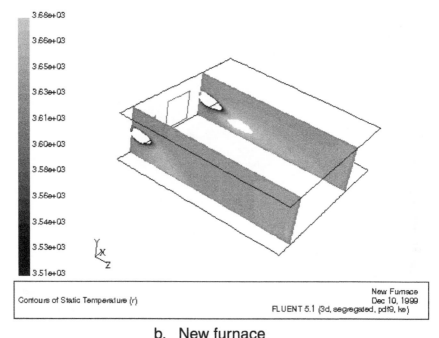

b. New furnace

FIGURE 9.42   Vertical temperature distribution at the 1st and 3rd burner location: (a) old furnace, and (b) new furnace.

furnace design with different burner orientations can now be compared. First, cases 3 and 4 are compared to choose the best. Then, the chosen case will be compared with case 2, presented above.

The temperature distribution is shown in Figure 9.52 at a slice taken at a 10° downward angle for both pictures, starting from the burner elevation level. One can see a significant difference in

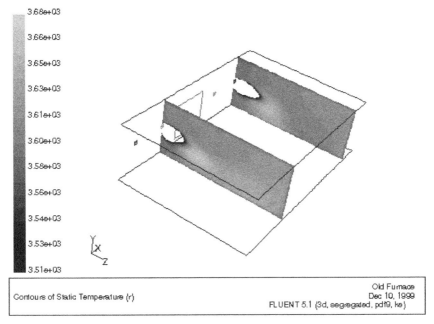

a. Old furnace

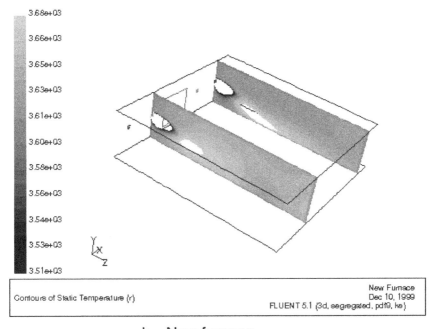

b. New furnace

FIGURE 9.43   Vertical temperature distribution at the 2nd and 4th burner location: (a) old furnace, and (b) new furnace.

the temperature profile of the two compared cases. The flame from the third burner, which is next to the side-wall burner, is shifted toward the furnace center as shown in Figure 9.54a. In the case with a 5° downward firing angle (Figure 9.52b), the third burner flame is also slightly shifted toward

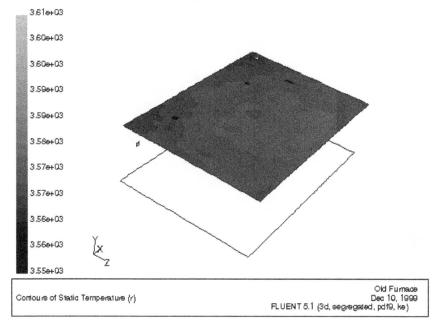

a. Old furnace

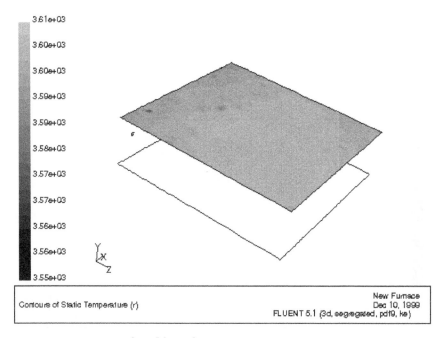

b. New furnace

FIGURE 9.44 Temperature distribution at the furnace roof: (a) old furnace, and (b) new furnace.

the center of the furnace combustion space, but not as dramatically as in the case with a 10°
downward firing angle.

The temperature distribution at the horizontal plane through the burner level is shown in
Figure 9.53. Here, the hot spot is more pronounced near the furnace side wall (Figure 9.53b) for

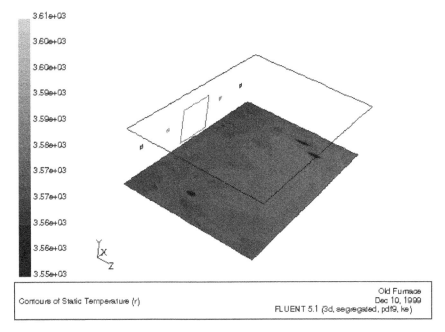

a. Old furnace

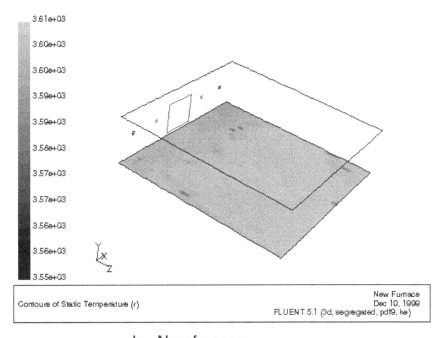

b. New furnace

FIGURE 9.45 Temperature distribution at the melt level: (a) old furnace, and (b) new furnace.

the case with a 5° downward firing angle. The case with a 10° downward firing angle has the hot spot shifted toward the furnace center plane. These results lead to the conclusion that the temperature profile at the aluminum surface should be better in the case with a 10° downward burner firing angle.

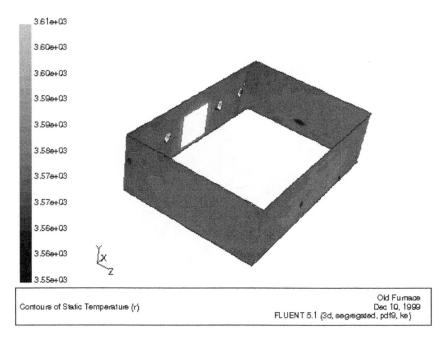

a. Old furnace

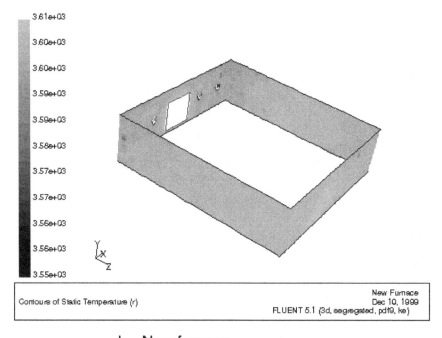

b. New furnace

FIGURE 9.46 Temperature distribution at the furnace side walls: (a) old furnace, and (b) new furnace.

The temperature distribution at the vertical planes through the burner positions (Figures 9.54 and 9.55) also shows the differences in the temperature profile for the two studied cases.

The temperature distributions at the furnace roof and the aluminum surface are shown in Figures 9.56 and 9.57. The results confirm the conclusion drawn before that the case with the

TABLE 9.2
Calculated Average Refractory Temperature Difference

Internal Furance Surface	Average Calculated Temperature Increase, °F	
	Old Furnace	New Furance
Roof	Tr	Tr + 21
Aluminum	Ta	Ta + 32
Side walls	Ts	Ts + 29
Exhaust	Tf	Tf + 29

downward firing burner angle of 10° should give better results in terms of production. The location of the hot spot at the aluminum surface is shifted toward the furnace hot wall and simultaneously toward the furnace melt entrance arch (Figure 9.56a). The temperature distribution at the furnace vertical walls, presented in Figure 9.58, supports the latter observation. Moreover, the temperature hot spot on the furnace side wall is located approximately in the middle of the wall and is evenly distributed between the aluminum surface and the furnace roof, as shown in Figure 9.58b. For the case with a 10° firing angle, the hot spot is significantly smaller in size and is shifted down toward the aluminum surface (Figure 9.58a). The furnace hot wall has also shifted the hot spot to the corner where the molten aluminum enters the furnace chamber in the case with a 10° firing angle. This ensures faster temperature recovery by the molten bath and, therefore, more efficient furnace operation.

The flow fields inside the furnace combustion space, for the cases with different downward firing angles, are shown in Figures 9.59 through 9.63. The velocity vectors shown at different slices and locations reveal the cause of the different temperature distributions. The interaction of the hot jets with the furnace gases recirculated inside the combustion space and with the gases leaving the furnace is shown in Figures 9.59, 9.60, and 9.63. The influence of the aluminum surface on the velocity profile at different firing angles is shown in Figures 9.61 and 9.62.

A comparison of the two cases with the same side burner angles and different downward burner firing angles leads to the following conclusion: a 10° downward firing angle is preferable to a 5° angle for air/fuel firing burners. Such an angle helps to establish a better temperature profile at the aluminum surface and at the furnace side walls. It also ensures higher furnace efficiency and production.

Now, cases 2 and 3 can be compared. Both cases have a 10° downward burner firing angle. In case 2, the side burner at the furnace door is tilted into the furnace combustion space at a 7° angle, and the other side burner is firing parallel to the furnace side wall. In case 3, both side burners are tilted toward the furnace center plane at a 4.5° angle. As an additional reminder, the results for case 2 are shown in Figures 9.40b through 9.51b, and the results for case 3 are shown in Figures 9.52a through 9.63a.

The most significant outcome from the results presented for cases 2 and 3 is that both cases are quite similar and can be used as a firing profile for the new furnace. At the same time, case 3 seems to be preferable. Comparison of the temperature distribution at the furnace roof and at the aluminum surface (Figure 9.44b vs. Figure 9.56a, and Figure 9.45b vs. Figure 9.57a) shows the well-defined hot spots in case 2. In case 3, these hot spots are not well-pronounced. This observation indicates that, at the same firing rate, it is much easier to overheat the surfaces with a burner orientation like in case 2.

A comparison of the temperature distribution at the furnace vertical walls, shown in Figures 9.46b and 9.58a, also indicates that the burner orientation with both side burners tilted into the furnace combustion space is preferable. The hot spot at the furnace side wall in the case 3 is significantly smaller compared to case 2. The hot spot at the furnace front wall is shifted toward the molten aluminum entrance, which should help to heat up the melt faster and to increase the thermal efficiency and the production rate of the furnace.

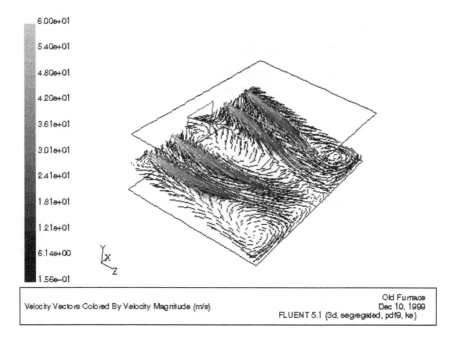

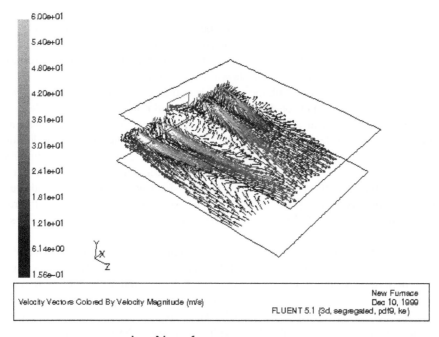

FIGURE 9.47 Velocity vectors at the slice along the burner firing direction: (a) old furnace, and (b) new furnace.

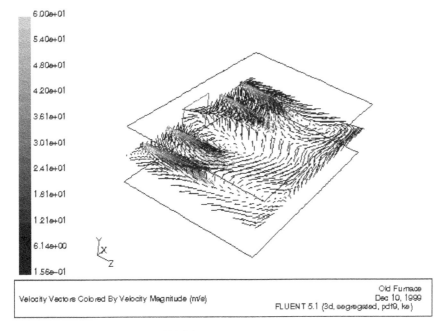

a. Old furnace

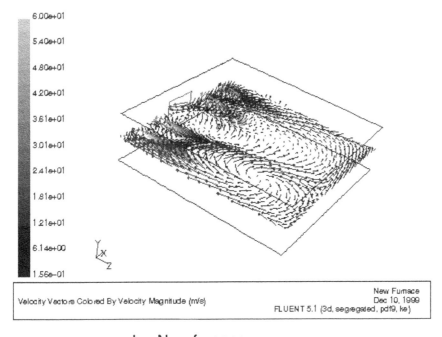

b. New furnace

FIGURE 9.48 Velocity vectors distribution at a horizontal slice through the burner level: (a) old furnace, and (b) new furnace.

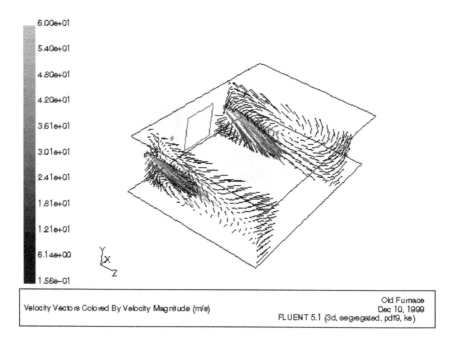

a. Old furnace

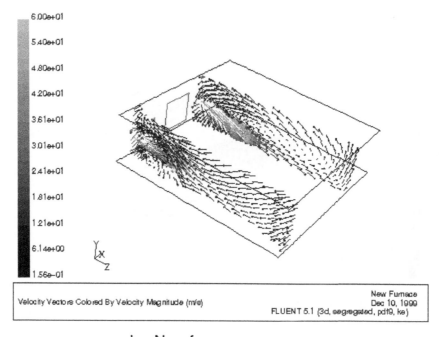

b. New furnace

FIGURE 9.49 Velocity vectors at a slice through the 1st and 3rd burner location: (a) old furnace, and (b) new furnace.

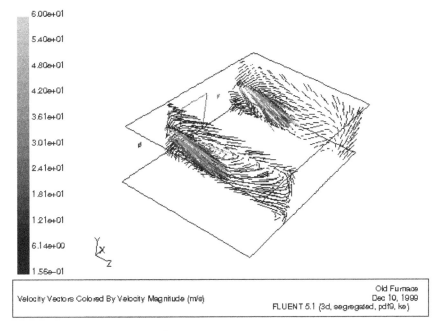

a. Old furnace

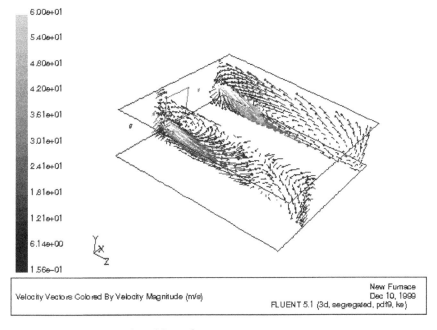

b. New furnace

FIGURE 9.50 Velocity vectors at a slice through the 2nd and 4th burner location: (a) old furnace, and (b) new furnace.

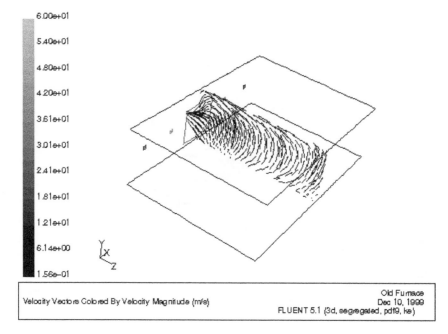

a. Old furnace

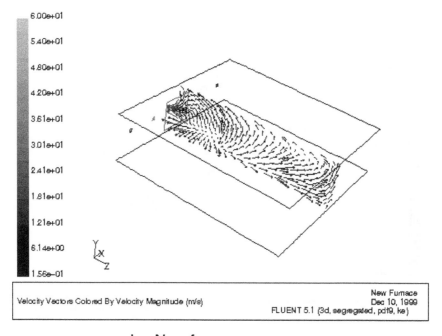

b. New furnace

FIGURE 9.51 Vertical slice of the velocity vectors through the furnace exhaust port: (a) old furnace, and (b) new furnace.

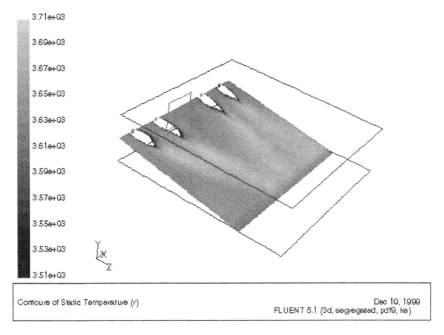

a. Old furnace

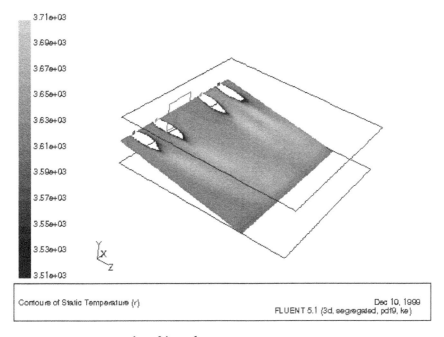

b. New furnace

FIGURE 9.52   Temperature distribution inside the furnace at 10° slice toward the melt: (a) new furnace with burner angled downward 10°, and (b) new furnace with burner angled downard 5°.

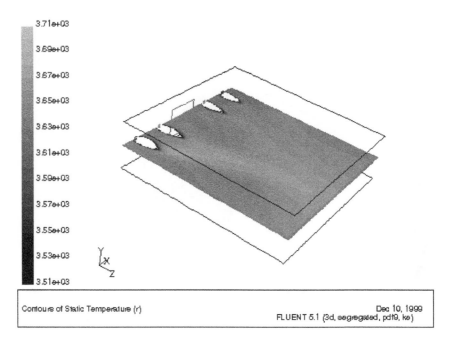

a. Old furnace

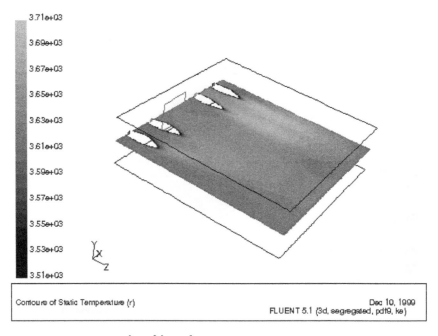

b. New furnace

FIGURE 9.53 Temperature distribution at the horizontal slice through the burner level: (a) new furnace with burner angled downward 10°, and (b) new furnace with burner angled downard 5°.

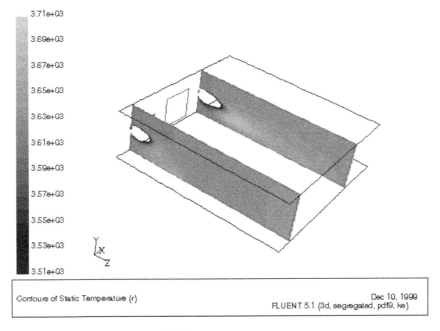

a. Old furnace

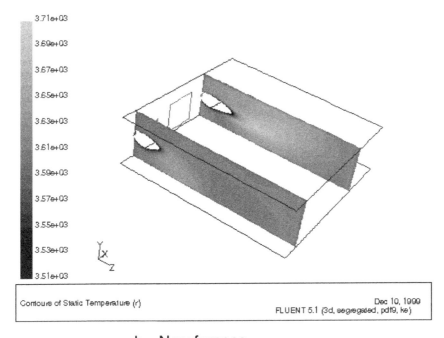

b. New furnace

FIGURE 9.54 Temperature distribution through slices at the 1st and 3rd burner location: (a) new furnace with burner angled downward 10°, and (b) new furnace with burner angled downard 5°.

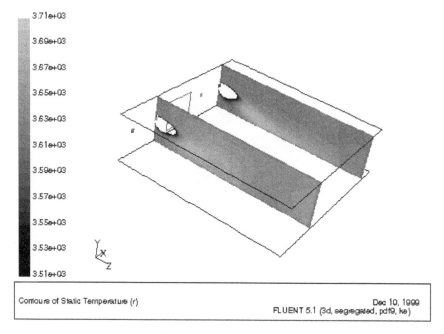

a. Old furnace

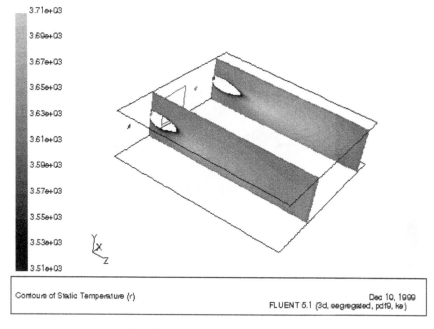

b. New furnace

FIGURE 9.55 Temperature distribution through slices at the 2nd and 4th burner location: (a) new furnace with burner angled downward 10°, and (b) new furnace with burner angled downard 5°.

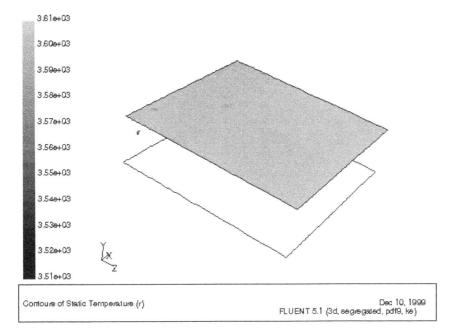

a. Old furnace

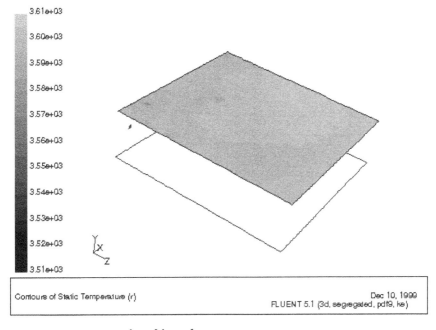

b. New furnace

FIGURE 9.56 Temperature distribution at the furnace roof: (a) new furnace with burner angled downward 10°, and (b) new furnace with burner angled downard 5°.

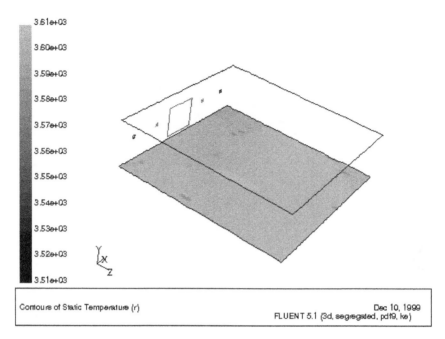

a. Old furnace

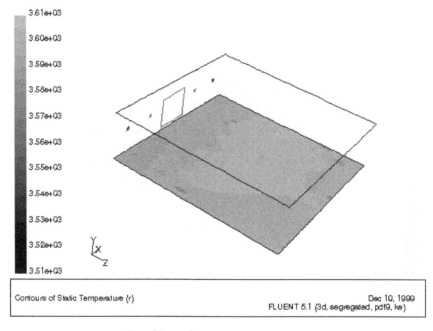

b. New furnace

FIGURE 9.57 Temperature distribution at the furnace melt surface: (a) new furnace with burner angled downward 10°, and (b) new furnace with burner angled downard 5°.

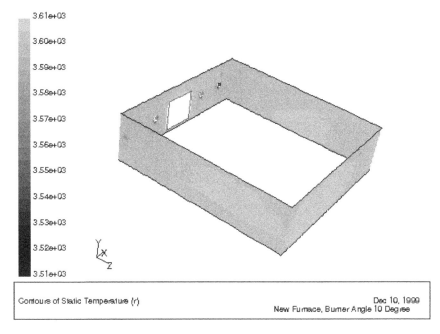

a. Old furnace

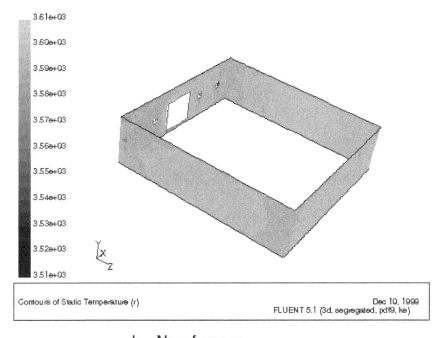

b. New furnace

FIGURE 9.58 Temperature distribution at the furnace side walls: (a) new furnace with burner angled downward 10°, and (b) new furnace with burner angled downard 5°.

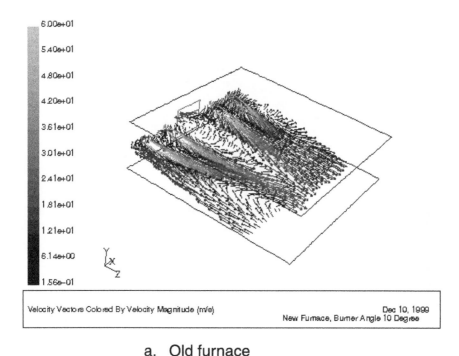

a. Old furnace

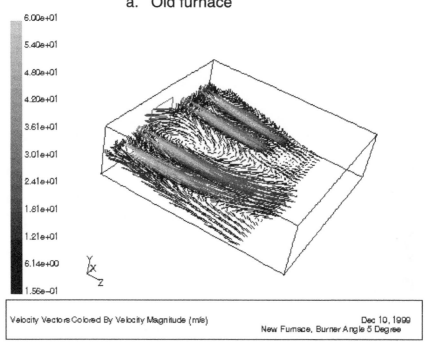

b. New furnace

FIGURE 9.59   Velocity vectors at a slice at a 10° angle toward the melt: (a) new furnace with burner angled downward 10°, and (b) new furnace with burner angled downard 5°.

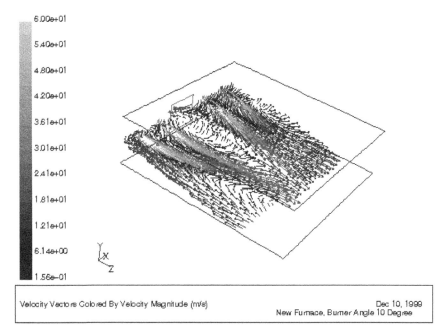

a. Old furnace

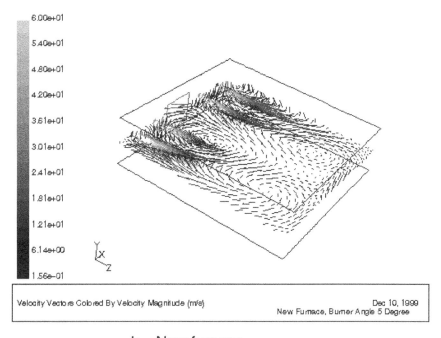

b. New furnace

FIGURE 9.60 Velocity vectors at a horizontal slice through the burner level: (a) new furnace with burner angled downward 10°, and (b) new furnace with burner angled downward 5°.

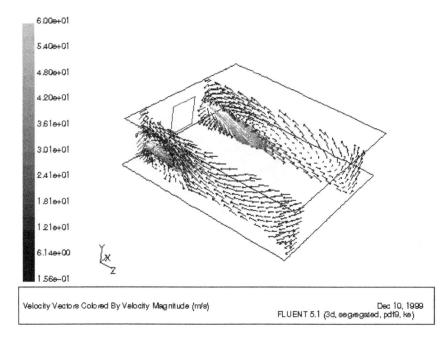

a.　Old furnace

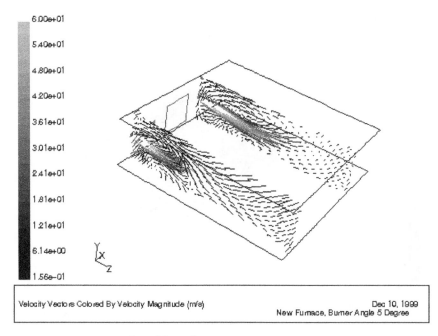

b.　New furnace

FIGURE 9.61　Velocity distribution at slices through the 1st and 3rd burner location: (a) new furnace with burner angled downward 10°, and (b) new furnace with burner angled downward 5°.

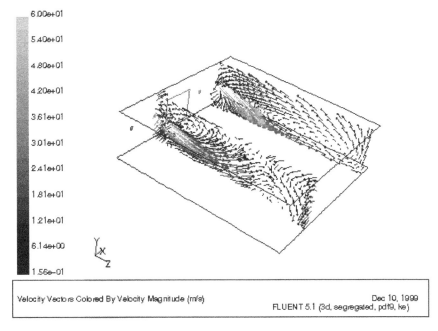

a. Old furnace

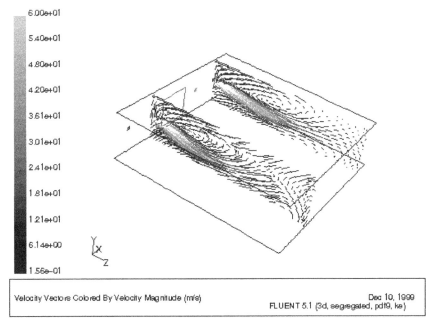

b. New furnace

FIGURE 9.62 Velocity distribution at slices through the 2nd and 4th burner location: (a) new furnace with burner angled downward 10°, and (b) new furnace with burner angled downard 5°.

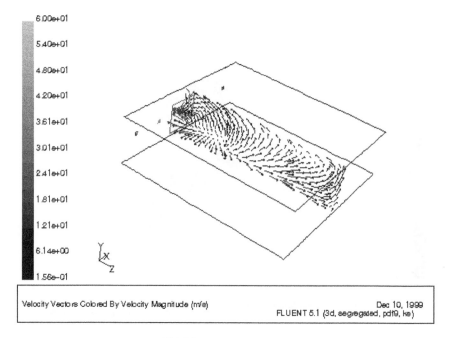

a. Old furnace

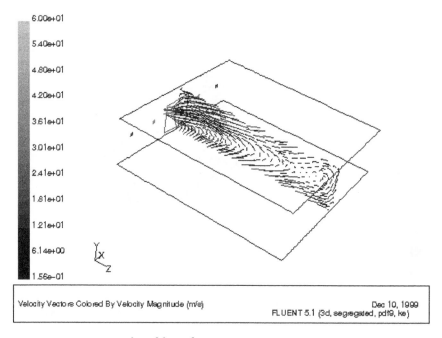

b. New furnace

FIGURE 9.63 Velocity vectors at a vertical slice through the furnace exhaust port: (a) new furnace with burner angled downward 10°, and (b) new furnace with burner angled downard 5°.

9.3.3.4　Conclusions

A new aluminum side-well furnace design was evaluated and compared with an existing furnace at RBSC. A comparison was conducted for the air/methane firing burners. The following conclusions can be drawn from the comparison:

1. The new furnace design should be able to handle a production increase of 50% compared to the old furnace design. This production increase was achievable with a total firing rate of 28×10^6 Btu/hr (8.2 MW).
2. The furnace thermal efficiency was found to be about 4% lower for the new furnace design than for the existing furnace.
3. The burner orientation plays a significant role in the optimization of the furnace efficiency, production rate, and operation. Small changes in the burner firing orientation may result in perceptible changes of the temperature profile at the internal surfaces of the furnace combustion space.
4. The case with the following burner orientation is recommended as the best out of the three evaluated scenarios: all four burners firing downward at a 10° angle toward the melt surface and both side burners inclined toward the center of the furnace hearth.
5. More rigorous analysis can simulate the molten aluminum flow rate, aluminum circulation pattern, and the melt temperature distribution. Such analyses will require more problem set-up time and computational time. At the same time, the model with both the combustion space and the molten bath will provide a more accurate evaluation of the furnace production rate as a function of any changes in the combustion space. Thus, full furnace optimization would be possible.
6. The flow pattern of each furnace design is unique and requires separate study to achieve the desirable results. As shown, a small increase in one of the furnace dimensions leads to significant changes in the velocity and temperature distribution inside the furnace. One must be careful to utilize the operating practice from one existing base case furnace in designing new furnaces.

REFERENCES

1. *Metal Statistics 1976*, 69th anual ed., American Metal Market, New York, 1976.
2. *Aluminum Statistical Review 1976*, The Aluminum Association, New York, 1976.
3. *Nonferrous Metal Data 1976*, American Bureau of Metal Statistics, New york, 1976.
4. Spector, S.R., *Survey of Free World Primary Aluminum Capacity, 1974–1979*, Oppenheimer & Co., Inc., Mar. 1975.
5. Bossing, E., Aluminum aerospace casting — 25 years in review, *Proceedings of International Molten Aluminum Processing*, City of Industry, CA, Feb. 17-18, 1986.
6. Novelli, L.R., Secondary aluminum smelting, *Scrap*, November/December 1997, 49-58.
7. *Aluminum-Fabrication and Finishing*, Vol. 3, American Society for Metals, Metals Park, OH, 1967.
8. Kearney, A.L., Furnace characteristics and selection, *CMI/AFS Aluminum Conference*, September 25, 1979.
9. Mangalick, M.C., Demagging aliminum, *Die Casting Engineer*, January–February, 1974.
10. Kellogg, H. H., Thermodynamic relationships in chlorine metallurgy, *J. Metals*, 188, 862-872, 1950.
11. Neff, D.V., The use of gas injection pumps in secondary aluminum metal refining, *Proc. AIME, Metallurgical Society*, Florida, December 1-4, 1985.
12. West, J.S., U.S. experience with high efficiency aluminum melting techniques, *IHEA–AGA Energy Seminar*, October 4, 1983.
13. Becker, J.S. and Heffron, J.F., The changing role of oxygen-based combustion in aluminum melting, *Light Metal Age*, June 1994.

14. Gershtein, V.Y., Glass furnace combustion space modeling using FLUENT Version 3.03 and FLUENT Version 4.11, *Proc. FLUENT Users' Group Conference,* Burlington, VT, October 5-7, 1993.
15. Baukal, C.E., Gershtein, V.Y., Heffron, J.F., Best, R.C., and Eleazer, P.B., Method and Apparatus for Reducing NOx Production During Air-Oxygen/Fuel Combustion, U.S. Patent 5,871,343, issued 16 February 1999.
16. Gershtein, V.Y., Baukal, C.E., and Kiczek, E.F., High efficiency, high productivity, low NOx aluminum melting, *Proc. Aluminum Association, Prevention Pollution Workshop,* IL, October 29, 1998.
17. Slavejkov, A.G., Gosling, T.M., and Knorr, R.E., Low-NOx Staged Combustion Device for Controlled Radiative Heating in High Temperature Furnaces, U.S. Patent, 5,611,682, March 18, 1997.
18. Slavejkov, A.G., Gosling, T.M., and Knorr, R.E., Method and Device for Low-NOx High Efficiency Heating in High Temperature Furnaces, U.S. Patent, 5,575,637, Nov. 19, 1996.
19. Gershtein, V.Y., 3D combustion space model of the glass furnace, *Russian J. Eng. Thermophys.,* 4(3), 1994.
20. Gershtein, V.Y., Mathematical model of complex heat transfer in the industrial furnace, *Thermophysics Aeromechanics,* 2(2), 1995.
21. Carvalho, M.G., Computer Simulation of a Glass Furnace, Ph.D. Thesis, London University, 1983.
22. Carvalho, M. G., Oliveira, P., and Semiao, V.A., Three-dimensional modeling of an industrial glass furnace, *J. Inst. Energy,* Sept., 1988.
23. Urgan, A. and Viskanta, A., Three-dimensional numerical modeling calculation of heat transfer in a glass melting tank, *Glastech. Ber.,* 60, 71-78 and 115-124, 1987.
24. Hoke, B.C. Jr. and Gershtein, V.Y., Coupling combustion space and glass melt models improves predictions, *The American Ceramic Society,* 74(11), November 1995.
25. Hoke, B.C. Jr. and Gershtein V.Y., Improved predictive results from coupling the gas space and glass bath models for a glass melter, *Proc. 3rd (Int.) Seminar on Mathematical Simulation in Glass Melting,* Ostrava, Cz Republic, May 26-28, 1995.
26. Gershtein, V.Y., Baukal, C.E., and Hewertson, R.J., Oxygen-enrichment of side well aluminum furnaces, presented at *TMS Conference,* February 1999.
27. Yahot, V. and Orsag, S.A., Renormalization group analysis of turbulence, I. Basic theory, *J. Sci. Comput.,* 1(1), 1-51, 1986.
28. Jones, W.P. and Lander, B.E., The prediction of laminarisation with a two-equation turbulence model, *Int. J. Heat Mass Transfer,* 15, 301, 1972.
29. Launder, B.E., Reece, G.L., and Rodi, W., Progress in the development of a Reynolds-stress turbulence closure, *J. Fluid Mech.,* 68(3), 537-566, 1975.
30. Rodi, W., *Turbulence Models and Their Application in Hydraulics,* 2nd ed., International Association for Hydraulic Research, Delft, The Netherlands, 1984.
31. Choi, D., Sabnis, J.S., and Barber, T.J., Application of an RNG κ-ε Model to Compressible Turbulent Shear Layers, AIAA Paper 94-0188, 1994.
32. Choudhury, D., Introduction of the Renormalization Group Method and Turbulence Modeling, Fluent, Inc., TM-107, 1993.
33. Magnussen, B.F. and Hjertager, B.H., On mathematical models of turbulent combustion with special emphasis on soot formation and combustion, *Sixthteenth Symp. (Int.) on Combustion,* Cambridge, MA, Aug. 15-20, 1976.
34. Dryer, F.L., The phenomenology of modeling combustion chemistry, in *Fossil Fuel Combustion — A Source Book,* Batrok W. and Serofim, A. F., Eds., John Wiley & Sons, New York, 1991, chap. 3.
35. Spalding, D.B., Mixing and Chemical Reaction in Steady Confined Turbulent Flames, *Thirteenth Symp. (Int.) on Combustion,* The Combustion Inst., Salt Lake City, UT, Aug. 23-29, 1970.
36. Sivathann, Y.R. and Farth, G.M., Generalized state relationship for scalar properties in non-premixed hydrocarbon/air flames, *Combust. Flam.,* 82, 211-230, 1990.
37. Kerstein, A., Ashurst, W., and Williams, F., Field evaluation for interface propagation in an unsteady homogeneous flow field, *The American Physical Society,* 37(7), 2728, 1987.
38. Yakhot, V., Scale invariant solution of the theory of thin turbulent flame propagation, *Combust. Sci. Tech.,* 62, 127-129, 1988.
39. Chandrasekhar, S., *Radiative Transfer,* Dover, New York, 1960.
40. Fiveland, W.A., Discrete-ordinates solutions of the radiative transport equation for rectangular enclosures, *J. Heat Transfer,* 106, 699-706, 1984.

41. Jamaluddin, A.S. and Smith, P.J., Predicting radiative transfer in rectangular enclosures using the discrete ordinates method, *Comb. Sci. Tech.*, 59(4-6), 321-340, 1988.
42. Jamaluddin, A.S. and Smith, P.J., Predicting radiative transfer in axisymmetric cylindrical enclosures using the discrete ordinates method, *Comb. Sci. Tech.*, 62(4-6), 173-186, 1988.
43. Fiveland, W.A. and Jamaluddin, A.S., Three-dimensional spectral radiative heat transfer solutions by the discrete-ordinates method, in *Heat Transfer Phenomena in Radiation, Combustion, and Fires*, R.K. Shah, Ed., ASME, New York, HTD-Vol. 106, pp. 43-48, New York, 1989.
44. Shah, N.G., New Method of Computation of Radiative Heat Transfer in Combustion Chambers, Ph.D. thesis, Imperial College, London, 1979.
45. Lockwood, F.C. and Shah, N.G., A new radiation solution method for incorporation in general combustion prediction procedures, *Eighteenth Symp. (Int.) on Combustion*, the Combustion Institute, Pittsburgh, PA, 1981, 1405-1414.
46. Gosman, A.D. and Lockwood, F.C., Incorporation of a flux model for radiation into a finite difference procedure for furnace calculations, *Fourteenth Symp. (Int.) on Combustion*, The Combustion Institute, Pittsburgh, PA, 1973, 661-671.
47. Patankar, S.V. and Spalding, D.B., Simultaneous predictions of flow pattern and radiation for three-dimensional flames, *Heat Transfer in Flames*, N.H. Afgan and J.M. Beer, Eds., Scripta Book Co., Washington, D.C., 1974, 73-94.
48. Siddall, R.G. and Selçuk, N., Evaluation of a new six-flux model for radiative transfer in rectangular enclosures, *Trans. IChem*, 57, 163-169, 1979.
49. Edwards, D.K., Molecular gas band radiation, in *Advances in Heat Transfer*, T.F. Irvine and J.P. Hartnett, Eds., Vol. 12, Academic Press, New York, 1976, 115-193.
50. Modak, A.T., Exponential wide band parameters for the pure rotational band of water vapor, *J. Quant. Spectosc. Radiat. Transfer*, 21(2), 131-142, 1979.
51. Kearney, A.L., Properties of Cast Aluminum Alloys, *Metals and Alloys,* 154-177, 1992.
52. Kelly, K.K., *Selected Values of the Thermodynamic Properties of Metals and Alloys,* John Wiley & Sons, New York, 1963.
53. Reid, R.C., Prausnitz J.M., and Poling, B.E., *The Properties of Gases & Liquids,* 4th ed., McGraw-Hill, New York, 1986.
54. Din, F., Ed., *Thermodynamic Functions of Gases,* Vol. 3, Butterworth, London, 1961.
55. Vargaftic, N.B., *Tables on the Thermophysical Properties of Liquids and Gases,* 2nd ed., Hemisphere, Washington, D.C., 1975.
56. Dul'nev, G.N. and Zarichnyak Y.P., *Thermophysical Properties of Substances and Materials,* 3rd issue, Standards Pablications, Moscow, 1971, 103.
57. L'Air Liquid, *Gas Encyclopedia,* Elsevier Science Publishers B.V., The Netherlands, 1976.
58. Saha, D. and Baukal, C.E., Nonferrous metals, in *Oxygen-Enhanced Combustion*, C.E. Baukal, Ed., CRC Press, Boca Raton, FL, 1998, 181-214.

10 CFD Modeling for the Glass Industry

Bryan C. Hoke, Jr. and Petr Schill

CONTENTS

10.1   GLASS MELTING FURNACES

There are several types of furnaces for melting glass. The type used typically depends on the kind and quantity of glass being produced, and the local fuel and utility costs. While there are exceptions, the following discussion describes the primary furnace types and the glass segments that most commonly use each style. CFD modeling has been performed for all types of glass melting furnaces.

Figure 10.1 shows a schematic of a glass tank. The main parts of the glass tank are the melter section, throat, refiner, and forehearth. The throat connects the melter section to the refiner and the forehearths transport the glass to the forming machines. For some glasses (e.g., fiberglass), the refiner is absent and the glass moves directly from the melter section through a throat to the forehearths.

For flat glass production, the design (and terminology) is slightly different, as shown in Figure 10.2. The melter and refiner sections are in the same room. The melter section is distinguished from the refiner by the presence of combustion above the glass. After the refiner, the waist section leads to the working end and finally to forming.

A greater variety of heating systems are found in glass manufacture. These types are discussed in the Sections 10.1.1 through 10.1.5.

10.1.1   UNIT MELTER

The term "unit melter" is generally given to any fuel-fired glass melting furnace that has no heat recovery device. Typically, the air/fuel unit melters are relatively small in size and are fired with 2 to 16 burners. Furnaces range in production from as large as 40 tons (36 m-tons) of glass per day to as small as 500 lb. (230 kg) of glass per day. Larger air/fuel unit melters are found in areas where fuel is extremely cheap. Frit, tableware, ophthalmic glass, fiberglass, and specialty glasses with highly volatile and corrosive components are produced in unit melters.

Generally, one is referring to an air/fuel-fired furnace when using this furnace term. However, most full oxy/fuel furnaces have no heat recovery system and are, therefore, technically "unit melters."

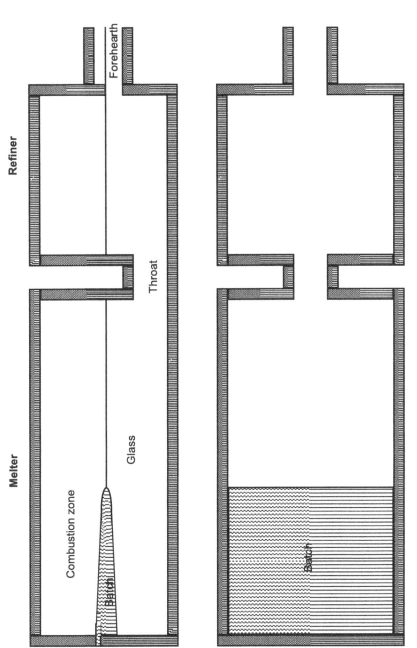

FIGURE 10.1 Schematic of glass tank geometry: longitudinal view (top) and plan view (bottom).

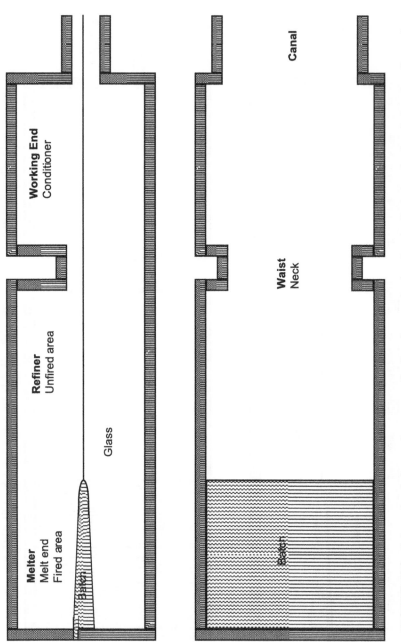

FIGURE 10.2 Schematic of glass tank geometry for flat glass production: longitudinal view (top) and plan view (bottom).

10.1.2 Oxy/Fuel Melter

Oxy/fuel melters are furnaces in which the air for combustion has been replaced by oxygen. Depending on the oxygen supply system, the concentration of oxygen can vary from 90% to nearly 100%. Cryogenically produced oxygen provides concentrations near 100% while vacuum swing adsorption (VSA) technology produces oxygen concentrations closer to 90%.

Oxy/fuel melters have been applied to all of the glass segments. The frit industry was the first to convert during the 1970s and 1980s. During this time, Corning also converted many of its smaller specialty glass furnaces, primarily unit melters, to oxy/fuel firing.[1]

A full size range of oxy/fuel melters have been built with pull rates greater than 500 tons (460 m-tons) per day for float glass production, down to 500 pounds (230 kg) of glass per day for art glass.

Many oxygen suppliers have developed capabilities for modeling oxy/fuel furnaces in an effort to support customers when they convert from air/fuel to oxy/fuel.[2,3]

Other types of melters described in the following sections have been modified to include the use of oxygen for enhancing combustion. Oxygen-enhanced combustion is described in greater detail elsewhere.[4] Oxygen suppliers have also developed modeling capabilities to support the use of these oxygen technologies.

10.1.3 Recuperative Melter

A recuperative melter is a unit melter equipped with a recuperator. Typically, the recuperator is a metallic shell and tube-style heat exchanger that preheats the combustion air to 1000 to 1400°F (540 to 760°C). The furnace is fired with 4 to 20 individual burners. These furnaces range in size from as large as 280 tons (250 m-tons) per day of glass, to as small as 20 tons (18 m-tons) per day of glass. These furnaces are common in fiberglass production but can also be used to produce frit. Some recuperative furnaces are used in the container industry, although this is not common. A typical recuperative melter is shown in Figure 10.3.

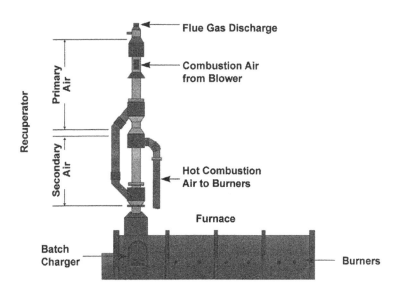

FIGURE 10.3 Typical recuperative melter. (From Baukal, C.E., Ed., *Oxygen-Enhanced Combustion*, CRC Press, Boca Raton, FL, 1998, Figure 7.1.)

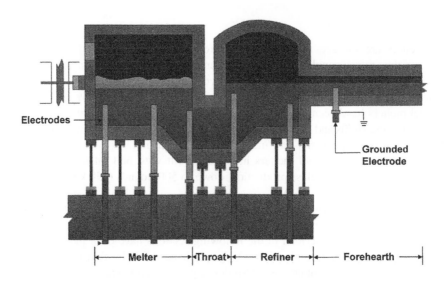

Electrodes

Grounded
Electrode

|←———— Melter ————→|←Throat→|←——— Refiner ——→|←——— Forehearth ——→|

FIGURE 10.4 Typical electric melter.

10.1.4 ALL ELECTRIC MELTER

As the name implies, all-electric melters receive all of the energy for glass melting through electric heating. Electric current is passed through the glass by means of electrodes. Because of the electrical resistance of the glass, the glass is heated by Joulean heating. Electrodes are typically made of molybdenum; however, tin oxide, platinum, graphite, and iron have also been used.[5] The electrodes are usually rod- or plate-type and can be located in the melter side walls or bottom.

The refractory tends to degrade much faster in all-electric furnaces, resulting in very short furnace campaigns, typically less than 2 years. The glass temperature and velocity are typically higher near the electrodes and this promotes localized erosion. Models are important for analyzing this effect. Most of these furnaces are less than 40 tons (36 m-tons) of glass per day; however, furnaces as large as 200 tons (180 m-tons) per day have been built.[6] A typical electric melter is shown in Figure 10.4.

10.1.5 REGENERATIVE OR SIEMANS FURNACE

The regenerative furnace was patented in the U.S. by Siemans Corporation in the late 19th century. While some design evolution has occurred, the basic concept has remained unchanged. In a regenerative furnace, air for combustion is preheated by being passed over hot regenerator bricks, typically called checkers. This heated air then enters an inlet port to the furnace. Using one or more burners, fuel is injected at the port opening, mixes with the air, and burns over the surface of the glass. Products of combustion exhaust out of the furnace through a non-firing port and pass through a second set of checkers, thereby heating them. After a period of 15 to 30 minutes, a reversing valve changes the flow and the combustion air is passed over the hot checkers that were previously on the exhaust side of the process. The fuel injection system also reverses. After reversing, the exhaust gases pass through and heat the checkers that had previously heated the combustion air.

The Siemans furnace is the workhorse of the glass industry. Most flat glass and container glass are produced in this furnace type. Regenerative furnaces are also used in the production of TV products, tableware, lighting products, and sodium silicates. There are two common variants of the Siemans furnace: the end-port regenerative melter and the side-port regenerative melter.

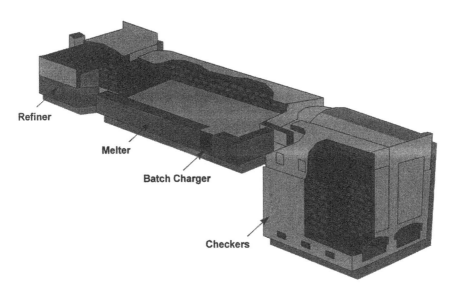

FIGURE 10.5 Typical end-port regenerative furnace. (From Baukal, C.E., Ed., *Oxygen-Enhanced Combustion*, CRC Press, Boca Raton, FL, 1998, Figure 7.3.)

10.1.5.1 End-port Regenerative Furnace

End-port regenerative furnaces are typically used for producing less than 250 tons (230 m-tons) of glass per day. In an end-port furnace, the ports are located on the furnace back wall. Batch is charged into the furnace near the back wall from one or both of the side walls. Figure 10.5 shows the layout of a typical end-port furnace. These furnaces are commonly used for producing container glass, but are also used for producing tableware and sodium silicates.

10.1.5.2 Side-port Regenerative Furnace

Side-port regenerative furnaces have ports located on the furnace side walls. Batch is charged into the furnace from the back wall. Figure 10.6 shows the layout of a typical side-port furnace. Side-port regenerative furnaces are typically used for producing greater than 250 tons (230 m-tons) of glass per day. A side-port furnace for float glass commonly produces 500 to 700 tons (460 to 630 m-tons) of glass per day. For container glass, side-port furnaces ordinarily produce between 250 and 350 tons (230 and 320 m-tons) of glass per day. These furnaces are normally used in container and float glass production, but are also used for the production of tableware and sodium silicates.

10.2 INTRODUCTION TO THE THEORY OF THE GLASS MELTING PROCESS

Glass melting is a process that includes all modes of transport phenomena: fluid flow, heat transfer, and mass transfer. Glass batch, originally a mixture of glass forming compounds like quartz, soda ash, limestone, etc., responds to heating by producing new solid and liquid phases and by releasing gases.

Before the glass batch melts, it undergoes a number of processes, such as drying, removal of chemically bound water, burning and decomposition of organics, chemical reaction in the solid state, reaction in the presence of molten salts, production of glass forming melts, and redox

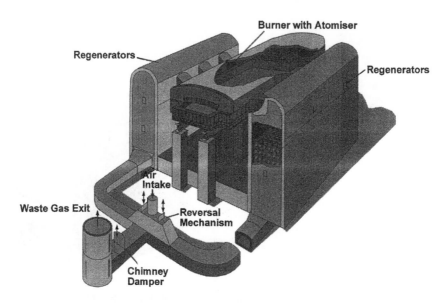

FIGURE 10.6   Typical side-port regenerative furnace. (From Baukal, C.E., Ed., *Oxygen-Enhanced Combustion,* CRC Press, Boca Raton, FL, 1998, Figure 7.4.)

reactions. The molten glass contains solid and gaseous inclusions and its composition is spatially nonuniform. The solid particles must be dissolved, bubbles removed, and chemical homogeneity established. After this process occurs, solid particles, bubbles, and inhomogeneities can be reintroduced due to interaction of the glass with the furnace walls, heating electrodes, and the combustion atmosphere.

The glass melting process can be divided into the following stages:

1. Initial stage before a melt appears: drying, removal of chemically bound water, hydrothermal reactions, crystalline inversions, and solid-state reactions
2. Vigorous chemical reactions between batch components
3. Dissolution of solid grains in the primary forming melt: refractory particles (e.g., SiO_2, MgO) with a melting point higher than the maximum temperature used to produce glass are dissolved
4. Refining stage: removal of gases from the melt
5. Homogenization of the glass melt: smoothing local (microscale) differences in concentration

These processes occur in the glass melter and are influenced by the flow and temperature patterns in the melter. Flow and temperature patterns can be predicted using CFD modeling and, consequently, knowledge of the temperature history of each glass particle passing from the furnace can be deduced. This time–temperature history can then be related to the rates of elementary processes such as dissolution, refining, homogenization, etc. with the aim of establishing the optimum conditions for their completion without unwarranted over-processing.[7]

10.3   GLASS MELT FLOWS IN THE MELTING FURNACE

Most of the glass produced on a large scale is made in continuous melters. During the melting process, intensive flow patterns occur naturally. The positive effect of the glass motion is mixing of the melt, which accelerates the processes of heat and mass transfer. The negative effect is increased corrosion/erosion of the refractory wall materials that make up the tank and possible carry-over of unmelted batch into the refining and working end.

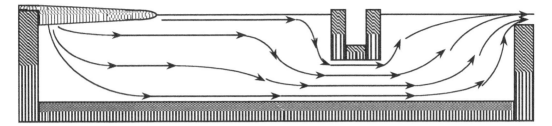

FIGURE 10.7 Longitudinal cross-section of a glass melting tank showing throughput flow.

As described by Hlavač,[7] two principal types of glass flow in continuous melters can be distinguished: (1) throughput flow (i.e., movement of glass from the batch charging end to the working end), and (2) circulating or secondary flow. Both flow types are laminar because the Reynolds number is less than 1. Throughput flow is based on forced convection, and circulating flow is based on natural convection.

10.3.1 THROUGHPUT FLOW

Throughput flow results from the difference in pressure caused by charging the batch and withdrawing (pulling) the melt. Flowlines are principally parallel, as shown in Figure 10.7. The velocity is zero at the bottom and side walls, and reaches a maximum on the melt surface in the vicinity of the longitudinal axis of the melting tank.

10.3.2 CIRCULATING FLOW

Natural convection-induced circulating flow is caused by density gradients resulting from temperature variations in the melt. Typical operation of tanks where the melt is heated from above by combustion prescribes the maximum temperature (hot spot) at roughly half the furnace length. Consequently, the density of the glass is relatively lower at mid-furnace length. Hydrostatic pressure differences give rise to upward flow in the middle of the tank. At the end walls, the temperature is lower and the density is higher, leading to downward flow at the walls. Because of heat losses from the walls, the glass near the walls is cooled, increasing the local glass density and further enhancing the downward flow. Downward flow along the side walls produces a dual corkscrew-like flow in the melter.

Combining circulating flow with throughput flow results in the flow pattern shown schematically in Figure 10.8. This figure also shows the return current from the refiner section back to the melter.

Return flow from the refiner back to the melter depends on the pull rate, the throat height, and the density difference of the glass on either side of the throat. Return flow can be important for glass quality and has been a topic of study for CFD modeling.[8]

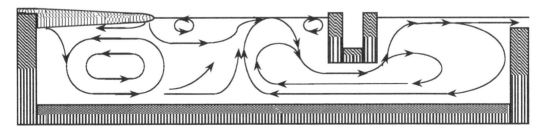

FIGURE 10.8 Longitudinal cross-section of a glass melting tank showing circulating flow.

The flow pattern of the glass melt depends strongly on the maximum temperature and temperature distribution in the furnace. As described above, the glass rises in the hot-spot region. This flow can be reinforced through the use of forced bubbling and/or electric heating.

10.4   GLASS FURNACE MODELING

There is a lot of activity worldwide in the field of glass furnace modeling, as illustrated by the number of members of the Technical Committee 21, Modeling of Glass Melts, of the International Commission on Glass.[9] Committee participants and others are listed in Table 10.1.

Works on mathematical modeling of flow, heat transfer, and electric heating of glass furnaces have been published since the 1970s. Initially, 2-D simulations were performed. One of the earliest works, by Mase and Oda,[17] solved 2-D flow and temperature fields for the glass melt. A constant batch velocity was used over the melt. The Hottel zone method was used to prescribe the combustion chamber temperature.

Recent 3-D numerical simulations reported in the literature have revealed that the 2-D simulations are not very realistic. Flows in a glass tank are basically 3-D in nature because of the batch blanket, the presence of sidewalls, throat, or waist; the presence of obstructions such as dams, steps, or weirs; and other nonsymmetrical flow-altering causes such as bubblers and electric boosting.

Development of 3-D models continues. Works by TNO,[18,19] Glass Service,[20,21,28,34] Choudhary,[22] Ungan,[23] and Carvalho[24] are commonly cited. Most of these works are based on finite-difference techniques and use the common procedures SIMPLE, SIMPLER, etc. of Patankar.[25]

The finite-element approach is seldom used for glass melt furnace modeling but has been applied.[26,27]

Although there have been some studies to improve the convergence rate using multigrid techniques,[28] most of the development effort has been in the development of submodels and coupling to combustion space models.

10.4.1   GLASS MELT SUBMODELS

Various phenomena in the glass melting process require specialized submodels, which have been described in the literature. The physical processes and quantities include:

1. Glass flow and temperature (including refractory materials)
2. Batch melting (sometimes together with submodel 1)
3. Electric boost (Joulean heat and electric currents)
4. Redox state and refining in the glass melt
5. Particle concentration, often including nucleation (sand, bubbles, etc.)
6. Additional momentum sources (forced air bubbling, mechanical stirring)

In most cases, submodel calculations are based on the solution of the general transport equation.

10.4.1.1   Transport Phenomena

The general form of the transport equation can be expressed by:

$$\frac{\partial}{\partial t}(\rho\Phi) + \rho\vec{V}\cdot grad\Phi = div(\Gamma grad\Phi) + Q \tag{10.1}$$

where the common notations for time, t, dependent variable, Φ, diffusion coefficient, Γ, source term, Q, and velocity vector, $\vec{V}$, are used. The meaning of the quantities Γ and Q depend on the particular

TABLE 10.1

Organizations Active in Glass Furnace Model Development and Use

Organization	Country	Glass Melt Model	Combustion Model
Technical University (RWTH) Aachen	Germany	Self-developed	
Battelle Pacific Northwest Laboratory	U.S.	Self-developed	
Air Products and Chemicals, Inc.[10]	U.S.	Glass Service	FLUENT based
Thomson Consumer Electronics	U.S.	TNO	TNO
Corning, Inc.[11]	U.S.	Self-developed and FLUENT based	Self-developed and FLUENT based
Institute of Chemical Technology (VSCHT)[77]	Czech Republic	FLUENT based	FLUENT based
Institute of Inorganic Chemistry ASCR	Czech Republic	Glass Service	Glass Service
Nippon Electric Glass	Japan	Self-developed	
TNO TPD	Netherlands	Self-developed	Self-developed
Instituto Superior Tecnico[12]	Portugal	Self-developed	Self-developed
Pilkington	England, U.S.	Self-developed and FLUENT based	Self-developed and FLUENT based
Asahi[13]	Japan	Self-developed	STAR-CD based
Schott Glaswerke	Germany	TNO	TNO
Sisecam Research Center	Turkey	Self-developed based on Ungan	Self-developed
Jenaer Schmelztechnik Jodeit GmbH	Germany	Self-developed	
Nikolaus Sorg GmbH	Germany	TNO	
Battelle Ingenieurtechnik	Germany	Self-developed	
Glass Service Ltd.	Czech Rep.	Self-developed	Self-developed
Uni Louvain la Neuve	Belgium	Self-developed	
St. Gobain Conceptions Verrieres	France	Self-developed	Self-developed
Samsung Corning[14]	Korea	FLUENT based	FLUENT based
BOC	U.S.	FLUENT based	FLUENT based
Hankuk Electric Glass	Korea	Glass Service	Glass Service
Hanglass	Korea	Glass Service	Glass Service
PPG Industries	U.S.	TNO	TNO
L'Air Liquide	France	TNO	Self-developed
AFG	U.S.	Glass Service	Glass Service
Glaverbel	Belgium	Self-developed	Self-developed
Visteon Glass Systems	U.S.	Self-developed	Self-developed
		TNO	TNO
		FLUENT based	FLUENT based
		BYU[15]	BYU
Brigham Young University[15]	U.S.	Fluent based with self-developed batch model	FLUENT based
Purdue University[16]	U.S.	Self-developed	Self-developed
Johns Manville	U.S.	FLUENT based	
Owens Illinois, Inc.	U.S.	Glass Service	Glass Service

meaning of Φ. In the case of the energy equation, Φ is temperature, $\Gamma = \lambda/c$, where λ is the thermal conductivity and c is the specific heat; and $Q = q/c$, where q is the specific volumetric heat source. In the case of the flow calculation, the transport equation must be specified for each velocity component i. Here, Φ represents the velocity component V_i, $\Gamma = \eta$, where η is the dynamic viscosity, and the source term is a combination of two parts: the pressure derivative, $\partial p/\partial x_i$, and the body force component $\rho g + B_i$, where B_i is an external force such as is imposed by bubbling or stirring, and g is the gravity force in the appropriate direction. The resulting source term for the flow equation can be written $Q = \rho \vec{g}_i + B_i - \partial p/\partial x_i$. The general transport Equation (10.1) can also be used to describe other physical phenomena such as conservation of chemical species, turbulence kinetic-energy, etc.

The density, ρ, can be related, via an equation of state, to other variables such as mass fraction, pressure, and temperature. Further, the flowfield must satisfy mass conservation, the so-called continuity equation:

$$\frac{\partial \rho}{\partial t} + div\left(\rho \vec{V}\right) = Q_\rho \tag{10.2}$$

where Q_ρ is a mass source term.

The general transport equations (10.1) and (10.2) are usually solved by numerical methods using discrete variables located on computational grids. The numerical method generally consists of two steps: (1) discretization, and (2) solving the set of algebraic difference equations. Discretization means that the nonlinear partial difference equations (10.1) and (10.2) are converted into linear difference equations. The conversion is based on applying the Gauss divergence theorem to a grid cell having finite volume, v, and surface s, with the result

$$\int_s \left(\rho \vec{V} \cdot \Phi - \Gamma grad\Phi\right) \vec{n} \cdot ds = \int_v Qdv \tag{10.3}$$

where $\vec{n}$ is the outer normal.

The finite difference approximation of the partial derivatives and numerical integration transforms Equation (10.3) into an algebraic equation of the form

$$\Phi_p = \left[\sum_m \left(A_m \Phi_m\right) + Q_p\right] \Bigg/ \sum_m A_m \tag{10.4}$$

This equation is valid for each grid cell, p. The index m indicates neighbor cells surrounding the central cell p. The formulation of the differential equation must be done carefully with respect to the magnitude of Peclet or Reynolds numbers.[25]

The resulting set of algebraic equations (10.4) together with the discrete form of the continuity equation, are solved by one of various smoothing methods. Smoothing is based on the calculation of nodal updates during the iterative procedure. Convergence and stability are sensitive to the iterative process, which requires relaxation. There are many calculation methods, such as segregated, coupled, and multigrid. The calculation can be applied on rectangular grids or on curvilinear grids (body-fitted coordinates), and the solvers can work with primitive variables (temperature, velocity components, etc.) or with indirect variables such as flow potentials and streamfunctions.

Most of the smoothing techniques are based on a finite-difference approach and use the common procedures SIMPLE, SIMPLER, etc. of Patankar.[25] This classical computational technique segregates the iterative procedure into two parts and an additional pressure correction equation. This pressure equation is created from the continuity equation. The numerical process has the disadvantage in decoupling the momentum equation. The decoupling may be responsible for instabilities, especially for complex geometries and may increase the computational time needed for reaching a converged solution.

A different approach, based on symmetrical coupled Gauss-Seidel method (SCGS) was applied by Schill.[21,28] This method is based on coupling the solution of all the transport difference equations inside each computational cell. The iterative procedure sweeps the whole computational domain and new variable updates are calculated by passing through all the cells. Because the pressure is one of the regular variables, no pressure corrections are used. This coupled technique is advantageous for calculating circulating flows and complicated furnace geometries.

Various calculation methods applied in glass furnace modeling can be found in the literature.[16,23]

10.4.1.1.1 Energy transport

Energy transport in the glass melt is realized by three different mechanisms: radiation, conduction, and convection. Glass is optically thick, meaning that it is a highly absorbing medium for radiation. This is fortunate because it permits the solution of the radiative transfer phenomena as a diffusion process. This is the so-called Rosseland approximation and is typically used.[20,22-24,29,30] In the energy equation, an effective thermal conductivity, λ_{eff}, is used which consists of two contributions: the Rosseland radiation, λ_r, and conduction, λ_c. The effective thermal conductivity can then be written as

$$\lambda_{eff} = \lambda_c + \lambda_r = \lambda_c + \frac{16n^2\sigma T^3}{3\kappa_R} \tag{10.5}$$

where n is the refractive index, σ is Stefan-Boltzmann constant, and κ_R is the Rosseland absorption coefficient. The resulting effective thermal conductivity is a strong function of temperature, which is usually estimated experimentally in laboratories by various methods such as direct measurements of absorption, indirect measurements in an experimental furnace, etc. The effective thermal conductivity has been described mathematically in a number of ways, including a piecewise quadratic function[20] and cubic function[30] of temperature.

The optically thick assumption seems to be well-justified for colored glass, but lower precision is expected for clear glasses, especially near boundaries.[31] However, the accuracy, even for clear glasses, is reported to be within engineering accuracy.[32]

The energy equation, derived from the general transport Equation (10.1), is:

$$\frac{\partial T}{\partial t} + \vec{V} \cdot gradT - div\left(a_{eff} gradT\right) = Q_e \tag{10.6}$$

where $a_{eff} = \lambda_{eff}/\rho c$ is the effective thermal diffusivity, and $Q_e = q/(\rho c)$ is the heat source. A set of thermal boundary conditions must be specified on all of the edges of the computational domain. The numerical procedures allow application of different kinds of boundary conditions (Dirichlet, Neumann, heat transfer, etc.) on separate boundary planes. The numerical solution of this model Equation (10.6) is based on the general principles described in Section 10.4.1.1.

10.4.1.1.2 Momentum transport and continuity

Glass melt flows are calculated by solving the transport equation for momentum, together with the continuity equation. The resulting set of partial differential equations consists of three equations for velocity components U, V, W in the directions x, y, z:

$$\frac{\partial}{\partial t}\left(\rho U\right) + \rho\vec{V} \cdot gradU - div\left(\eta gradU\right) + \frac{\partial P}{\partial x} = B_x \tag{10.7a}$$

$$\frac{\partial}{\partial t}\left(\rho V\right) + \rho\vec{V} \cdot gradV - div\left(\eta gradV\right) + \frac{\partial P}{\partial y} = B_y \tag{10.7b}$$

$$\frac{\partial}{\partial t}\left(\rho W\right) + \rho\vec{V} \cdot gradW - div\left(\eta gradW\right) + \frac{\partial P}{\partial x} = B_z + \rho g \tag{10.7c}$$

and the continuity equation

$$\frac{\partial \rho}{\partial t} + div(\rho \vec{V}) = Q_p \qquad (10.8)$$

These are the so-called Navier-Stokes equations, which can be derived from general transport Equation (10.1). In these equations, the z-direction is taken to be aligned with gravity.

The glass melt is considered a Newtonian fluid with strongly temperature-dependent viscosity $\eta(T)$. The effects of density variation $\rho(T)$ are assumed to be important only in the buoyancy term ρg and the Boussinesq approximation can be used. In this case, the right-hand side of Equations (10.7a), and (10.7b) involve only common body force-components B_x and B_y. Assuming that the z-coordinate is in the same direction as gravity, g, and the temperature function of density is linear $\rho(T) = \rho_R[1 - \beta(T - T_R)]$, then the right side of Equation (10.7c) includes the common body force-components B_z and the non-zero buoyancy term of the form $-g\beta\rho_R(T - T_R)$. The value of ρ_R is the reference density at reference temperature T_R.

A set of flow boundary conditions must be specified on all of the edges of the computational domain. The main boundary conditions are inputs, outputs, the slip condition on the free surface, and non-slip conditions on the walls. The numerical solution of this set of model equations (10.7) and (10.8) is based on general principles described in Section 10.4.1.1.

10.4.1.2 Batch

Raw materials (batch) and cullet (recycled solid glass) for the glass making process are typically charged to the furnace on the top of the glass melt. The batch can be introduced from the back of the furnace or from one or both sides. The partially submerged batch floats on the surface and can form islands or a continuous blanket. Because the batch ordinarily enters at room temperature, it acts as a large heat sink to the combustion chamber as well as the glass melt.

The batch blanket is a very important driving force for natural convection in the tank because the temperature difference between the glass under the batch and the hot spot is about 400 to 500°C. Because of the batch's influence on the convective melt flow and inlet glass flow distribution, batch modeling is one of the most important sub-models for glass tank models. For an accurate furnace simulation, one of the most important results from the batch sub-model is the shape of the batch region on top of the glass melt. The shape of the batch will be a function of material feed rate, charging temperature, thermophysical properties, cullet ratio, batch feeder type, and heating system (burner locations, electrode distribution, etc.).

Some assumptions used for modeling batch were tested experimentally.[33] In that study, it was found that the primary melt did not form along any specific isotherm. Therefore, the concept of a uniform melting temperature, which is used by several theoretical models, was not confirmed.

Most batch models developed by glass companies or glass consulting companies are proprietary and details are not published. Some general descriptions of batch sub-models are given by Glass Service[34] and TNO,[35] but their approaches are not fully disclosed — to prevent duplication by others. Detailed batch sub-model descriptions developed at universities include those of Ungan,[23] Carvalho,[24] and Wang.[36]

10.4.1.2.1 Glass Service

In Glass Service's model,[34] batch is converted to melted glass inside the batch region, depending on the temperature and a macroscopic "degree of conversion" of the batch into melt. Fully three-dimensional temperatures and flows inside the batch region are coupled with the glass flows in the tank, allowing the interface between the batch and glass to be calculated automatically. To achieve a transition between batch and glass, the degree of conversion in each node is calculated based on the state of adjacent nodes. An overall energy balance is calculated in the batch area and a correction scheme is used for updating the local heat consumption. The batch shape and local batch thickness is calculated automatically according to the thermal boundary conditions on batch surfaces and energy and mass balances inside batch blanket.

The degree of conversion is defined as the fraction of the total gas released from the batch and is determined experimentally by thermogravimetric measurements. The degree of conversion is zero when the batch is first introduced to the furnace. Water evaporates and transformation reactions begin at about 77°C. All the chemical reactions are completed (and all the gases evolved) at about 1000°C, where the degree of conversion is 1.

The batch behavior depends on thermophysical properties and additional kinetic properties, which are in turn dependent on the degree of conversion. The batch thermophysical properties are taken to be a function of temperature and degree of conversion and continuously approach the glass property functions as the degree of conversion approaches 1. The batch becomes glass and enters the melt space when the degree of conversion is 1. Proper specification of properties is important for realistic batch melting behavior.

The energy consumed by the batch melting process, h[W/m^3], is calculated in each control volume and depends on the heat of reaction, H_R, the cullet ratio, C_u, and experimental functions of the degree of conversion, α:

$$h = \rho \frac{d\alpha}{dt} H_R \cdot \left(1 - C_u\right) = \rho \cdot f_\alpha(T) \cdot K_\alpha \cdot H_R \cdot \left(1 - C_u\right) \tag{10.9}$$

where $d\alpha/dt$ is the time derivative of α, $f_\alpha(T)$ is the temperature derivative of α, and K_α is rate of batch heating. An example measured degree of conversion function and its temperature derivative is plotted in Figure 10.9 as a function of temperature for a heating rate of 0.06 K/s. The quantity h is used as a heat sink term in the model energy equation.

The same degree of conversion approach has been attempted in the general-purpose CFD code FLUENT, but no results were published.[40]

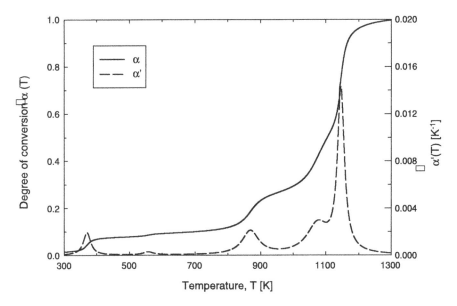

FIGURE 10.9 Degree of conversion functions for the Glass Service batch model.

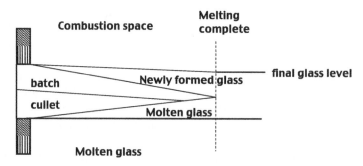

FIGURE 10.10 Geometry of batch region for the Carvalho batch model. (From Carvalho, M.G., Computer Simulation of a Glass Furnace, Ph.D. thesis, The University of London, 1983.)

10.4.1.2.2 TNO

The TNO batch blanket model is based on heat transfer equations and chemical kinetics, describing the melting or dissolving of the sand grains of the batch. The interaction of the blanket and the molten glass has been described using "apparent viscosities" for the batch floating on the melt.[37]

10.4.1.2.3 Carvalho

Carvalho[24] modeled the batch region as two-dimensional flow where three distinct subregions are present in the batch region: virgin raw materials (batch), cullet (recycled glass), and molten glass. The batch is placed on top of the cullet and newly melted glass is formed above the batch subregion and below the cullet subregion, as shown in Figure 10.10. The batch and cullet are treated as solids and a conventional viscosity law is used for the molten glass forming over the batch blanket. The momentum equation in the streamwise direction is solved throughout the batch flow domain and a uniform velocity profile is ensured in the solid-phase regions by setting a large value of viscosity.

Specific heat, density, and thermal conductivity are specified as linear functions of temperature. Correction terms for the density and thermal conductivity are added to account for air voids in the batch region.

The raw batch materials are treated as inert solid up to 850°C, at which temperature it reacts to form glass. The endothermic reaction and phase change are assumed to occur over a small finite temperature range. The cullet is treated similarly and changes phase to become liquid glass at a prescribed temperature, again 850°C.

Viscous dissipation is neglected, although the Prandtl number has a high value, because the velocities and velocity gradients in the batch are very low.

Uniform inlet boundary conditions are prescribed for velocity and temperature. A zero shear boundary condition is applied between the batch region and the molten glass and on the top surface. The front edge of the batch region is found when all the temperatures through the depth of the batch are greater than both the temperature of reaction and the melting temperature (850°C).

10.4.1.2.4 Wang

In the batch model by Wang[36] (a student of Carvalho), it is assumed that all chemical reactions take place at a specified temperature (for normal Na-Ca-Si glasses, 850°C is used), consuming heat and producing gas, which escapes from the top surface of the batch during the melting process. The batch becomes molten glass and melting down when its temperature arrives at the batch melting temperature, which is 1423K for normal Na-Ca-Si glasses. It was also assumed that the melting mass factor (fraction of raw materials evolved as gases) is constant for a given glass batch. The thickness of the batch is a function of the temperature, density, and fraction of unmelted batch.

Wang's batch melting model is a quasi-three-dimensional model, consisting of a series of thermally unconnected, two-dimensional "lanes" extending from the doghouse toward the throat. Batch inlet velocities from the batch charger are calculated according to the pull rate of the furnace,

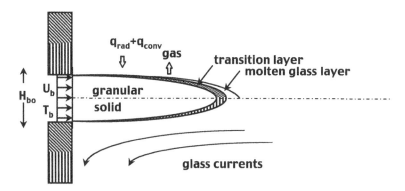

FIGURE 10.11 Schematic of axially charged loose batch blanket for the Ungan batch model. (From Ungan, A., Three-dimensional Numerical Modeling of Glass Melting Process, Ph.D. thesis, Perdue University, 1985.)

and the variation in batch feed along the width of the tank is approximated from observations of the batch in the modeled furnace. Based on observation, the batch-occupied area on the melt surface was divided into two zones: the continuous batch blanket and batch islands. In the continuous batch blanket zone, the batch is assumed to occupy 100% of the melt surface. In the batch islands zone, the batch is assumed to occupy a predetermined fraction of the melt surface area and molten glass is assumed to occupy the remainder. The boundary line between the two zones is based on observation and experience.

10.4.1.2.5 Ungan

Ungan[23] describes a "loose batch blanket" model. A schematic of the physical situation considered is shown in Figure 10.11. There are some similarities between batch melting and solid-liquid phase change heat transfer problems generally known as Stefan problems. However, phase transformation in batch melting is associated with sequential chemical reactions that start at a certain temperature and continue over a temperature range.

Major assumptions of the model are:

1. The processes for batch melting are time independent (steady state), and governed by two-dimensional flow and heat transfer. Diffusion in the horizontal direction is negligible in comparison to the transverse direction so that the upstream values are independent from those downstream.

2. A lumped velocity distribution is used for the batch velocity field. It is assumed that each grain within the batch takes on a horizontal velocity, which is constant and equal to the charging speed throughout the transformation process. The vertical movement at any axial location is derived from Newton's first law of motion by ignoring all of the internal shear stresses and acceleration in the vertical direction.

3. Phase transition of the batch takes place over a temperature range. The first-order approximation employed by Pugh[38] and Mase and Oda[17] is used. At a specified temperature, the silica grains are assumed to be reduced in size so that for all practical purposes, the mixture can be considered to be liquid batch or glass melt and the viscosity approaches the function for glass.

In the mathematical formulation of the batch model, two phases are considered: solid-liquid and gas. The velocity and temperature of the gas phase may be different from the solid-liquid phase in the same region. Each phase is therefore taken into account separately.

As the batch temperature increases, chemical reactions take place, forming products according to conservation equations. Reaction rates are described by a modified form of the Arrhenius equation.

Cullet is treated as a subcooled liquid without reaction, and chemically nonactive species are transformed as a function of temperature without solving additional species conservation equations.

Based on the simplified chemical reaction model used, only carbon dioxide is produced in the gas phase and is described by the mass conservation equation. Only the vertical component of the gas is considered and is calculated based on Darcy's law.[39]

At the combustion space-batch interface, a zero shear boundary condition is assumed and the heat flux is given by the radiation heat flux from the combustion model. Species and gas temperature gradients are taken to be zero.

At the batch-glass interface, the species gradients and gas pressure gradient are taken to be zero. The axial velocity component is equal to the feed velocity and the interface temperature is specified.

The approach described by Ungan has been implemented in the general-purpose CFD code FLUENT[40] and results were published showing general agreement with the original work.

10.4.1.3 Bubbling

Bubbling is an operation where air (or other gas) is introduced at the bottom of the melt through a small tube. The purpose is to significantly affect the flow pattern in the melter by introducing an upward force in the location of the bubblers. The frequency of bubbles is typically on the order of 10 to 100 bubbles per minute, and jetting flow is normally avoided. The size of the bubbles ranges from 0.06 m to 0.2 m.

Proper use of bubbling can significantly improve melting in a furnace. Bubbling lifts cold glass from the bottom of the melter to the surface and returns hotter glass from the surface to the bottom. Higher bottom temperatures result, which can improve melting but can also decrease the minimum residence time because the viscosity of the glass is reduced at these higher temperatures. Bubbling significantly changes the flowfield in the melt and can have both positive and negative impacts. A study to examine the effect of bubbling and optimize the number of bubblers to use is reported by Hoke.[41]

Two approaches for implementing bubbling into glass melt models are described by Franěk.[42] The first approach is to describe bubbling in the flow-temperature equations as an additional force acting on the glass melt. The second is to directly solve the flow equations around bubbles and impose a velocity on the glass melt into the flow-temperature equations. Most often, the first approach is used and is described here. The reader is referred to Franěk[42] for a description of the alternative approach, where a solution for spherical bubbles is given.

According to the law of action and reaction, the buoyancy force acting from a bubble on the glass melt is the same as the drag force of the glass acting on the bubble but with opposite sign. This force is $F = \rho V g$, where V is the volume of the bubble, ρ is the glass melt density, and g is the acceleration due to gravity. The gravity force is neglected since the density of the gas is very small relative to the density of the glass.

Measurable characteristics of bubbling on an existing tank are the volumetric flow rate, Q, and the frequency of the bubbles, Ω. From these, the volume of an individual bubble can be calculated by $V = Q/\Omega$. The total force acting from one bubble is $F_d = Q\rho g/\Omega$.

At any time, there will be several bubbles existing in the bubble column with a distance D between them. The distance will depend on Ω and the velocity of the bubbles U as $D = U/\Omega$. The number of bubbles in the column is the depth of the glass, H, divided by the distance between bubbles, D. The total force acting in the column is $F = Q\rho g H/\Omega D = Q\rho g H/U$, and the average force per unit volume is $F_V = F/HS$, where S is the cross-sectional area of the bubble column. The total average force per unit mass of the glass melt is $F_M = F_V/\rho = Qg/US$. This force is applied in a cylinder with axis intersecting the bubbler nozzle.

10.4.1.4 Electric Boost

Electric boost or direct electric heating in glass furnaces is achieved by immersing electrodes into the glass melt and connecting the electrodes to an electric power source, either single- or three-phase transformers. Molybdenum or tin oxide are commonly used as electrode materials. The electrodes are usually rod- or disk-type. AC sinusoidal electric power is usually applied. Because the glass melt is an electrically conductive medium, an electric current field and electric potential fields are created and Joulean heat is generated. The Joule heat distribution is used as the source term in energy equation. An overview of electric melting of glass is given in the monograph by Staněk.[43]

Complete models calculate the electric fields in the glass melt as well as in refractories and insulation materials. In this way, the influence of the electric fields on glass convection due to buoyancy and corrosion due to electrochemical phenomena can be investigated.

The first models of electrical boosting in glass-melting furnaces were created by Curran,[44] Austin and Bourne,[45] and Mardorf and Woelk.[46] These models were further enhanced by Hofman and Hilbig,[47] Choudhary,[48] Ungan and Viskanta,[49] and Schill.[50,51] The basic approach for electric boost calculations is the same.

The glass melt and refractories are considered an electrically linear medium where non-transverse linear electrical fields occur. Only the temperature dependence of electric conductivity is considered. The model describing the electrical field leads to the solution of a direct boundary problem, described by the differential equation of conservation of electric potential, ϕ,

$$div(\sigma grad\phi) = 0 \tag{10.10}$$

The instantaneous local specific Joule heat source, q, emitted at the certain point is given by

$$q = \vec{E}\vec{J} = \sigma\left(\vec{E}\right)^2 = \sigma\left[grad\phi\right] \tag{10.11}$$

where the electric current, $\vec{J}$, and electric intensity, $\vec{E}$, are related by

$$\vec{J} = -\sigma grad\phi = \sigma\vec{E} \tag{10.12}$$

To properly describe the physical situation, the temperature dependence of electrical conductivity, $\sigma(T)$, must be included to involve the coupling of glass temperature and Joule heating. In practice, it is usual to express the quantities by effective values. For the Joulean power output, the integral time mean values are used so that the local specific voluminous Joule heat source, q_v, is expressed by

$$q_v = \frac{1}{t_{p0}}\int^{t_p}\sum \sigma(x_i, T)\left(\frac{\partial\phi}{\partial x_i}\right)^2 \tag{10.13}$$

where t_p is time period of the AC current and $\phi(x_i)$ is electric potential given by solving Equation (10.10).

Using the superposition principle, the entire potential in each location is given by the superposition of partial electric potentials induced on each electrode. These electrode potentials are constant as are the electric boundary conditions. According to the superposition principle, only one resulting sinusoidal potential field in the glass melt will be generated by applying several power sources, independent of the single- or three-phase sinusoidal form of the power source.

This resulting sinusoidal potential field $\phi(x_i, T, t)$ has the form

$$\phi(x_i, t) = A(x_i)\sin\left[wt + \phi(x_i)\right] \tag{10.14}$$

with circular frequency $w = 2\pi/t_p$ const. The complex potential method is usually used because of the position dependency of amplitude, $A(x_i)$, and phase shift, $\phi(x_i)$. This method is based on an equivalent description using the complex potential,

$$\phi(x_i,t) = \hat{A}x_1(x_1)\exp\{iwp\} \tag{10.15}$$

where the complex phasor, $\hat{A}(x_i)$, is expressed by its real and imaginary parts

$$\hat{A}(x_i) = PR(x_i) + iPI(x_i) \tag{10.16}$$

Applying the conservation Equation (10.10) to these real and imaginary components, one needs to solve two Laplacian equations

$$div\left[\sigma(x_i,T)gradPR(x_i)\right] = 0 \text{ and } div\left[\sigma(x_i,T)gradPI(x_i)\right] = 0 \tag{10.17}$$

to get the distribution of real and imaginary potentials in the glass melt and in the refractories. The amplitude, $A(x_i) = |\hat{A}(x_i)|$, and phase shift, $\phi(x_i)$, can be stated by the relations

$$A(x_i) = \sqrt{\left[PR^2(x_i) + PI^2(x_i)\right]} \text{ and } \phi(x_i) = arctg\left[PR(x_i)/PI(x_i)\right] \tag{10.18}$$

Using Equation (10.13) the mean specific voluminous Joule heat source, $q_v(x_i)$, is then given by:

$$q_v(x_i) = \left[\sigma(x_i,T)/2\right]\hat{A}(x_i)\cdot\hat{A}^*(x_i) \tag{10.19a}$$

$$q_v(x_i) = \left[\sigma(x_i,T)/2\right]\left\{|gradPR(x_i)|^2 + |gradPI(x_i)|^2\right\} \tag{10.19b}$$

Finally, the Joulean heat generation is included as a source term in the steady energy equation, and has the form

$$Q(x_i) = q_v(x_i)/2\rho c \tag{10.20}$$

10.4.1.5   Mechanical Stirring

Glass mixing can be intensified by the use of mechanical stirrers. Mechanical stirring is applied in a furnace to support homogenization of the glass. The stirrers may be of various forms and they are normally made from refractory. Stirrers are typically located in the working end, or in the waist of the float furnace. Stirrers are also used in the gathering section of hand-made crystal glass furnaces.

Calculation of the stirring effect is based on an empirical formula that modifies the basic flow by means of an additive velocity or by an additive body force. The parameters of the empirical formula are calculated from experimental measurements from liquid models using the common theory of similarity.[52,53]

10.4.1.6   Redox and Fining

In the glass melt, gases are released during chemical reactions of the batch materials, resulting in a large amount of bubbles (commonly called seeds) in the glass melt. These bubbles are considered defects if they are in the final glass product. The process of removing the bubbles from the melt is called fining of the glass melt. In industrial glass furnaces, chemical agents are used to enhance the fining process. These chemical fining agents consist of oxides of polyvalent ions that are introduced in their most oxidized form into the batch charge. At high temperatures, the fining agents are reduced and oxygen or sulfur dioxide gas is released. This fining gas diffuses into existing bubbles and causes them to grow. These enlarged bubbles have a higher rising velocity and are able to move to the melt surface and leave the melt more quickly.

The most common fining agents are $NaCl$, Na_2SO_4, As_2O_4, and Sb_2O_5. The latter three agents undergo a redox reaction in the molten glass where the highest valency is shifted to a lower valence state and the gases O_2 and/or SO_2 are released in the glass melt. Redox reactions also occur when the glass melt contains other polyvalent elements, such as iron or chromium, but the valence state of these ions influence the color of the melt.

The temperature-dependent equilibrium constants of the various redox reactions are of great importance for the course and optimization of the fining process. These constants also depend on glass composition and can be measured by various methods. Two methods, square-wave voltammetry and oxygen equilibrium pressure measurements, are used most often. A detailed review is provided by Wondergem-de Best.[54] Contributors to the theory and experimental techniques include Ruessel,[56] Johnston,[55] Beerkens,[58,62,63] Němec,[59-61,64] and Kawachi.[66]

The overall reaction scheme in a general form[55] is described by

$$M^{(x+n)+} + (n/2)O^{2-} \leftrightarrow M^{x+} + (n/4)O_2(g) \tag{10.21}$$

where the polyvalent ions, M, are able to convert from one valence state to another

$$M^{(x+n)+} + n\,e^- \leftrightarrow M^{x+} \tag{10.22}$$

(n denotes the number of transferred electrons) and the electrons can be provided or taken up by oxygen:

$$O_2 + 4\,e^- \leftrightarrow 2O^{2-} \tag{10.23}$$

The reaction equilibrium constant K*(T) can be defined by

$$K^*(T) = \left\{ a\left(M^{x+}\right) \cdot \left[a\left(O_2\right)^{n/4}\right] \right\} \Big/ \left\{ a\left(M^{(x+n)+}\right) \cdot \left[a\left(O^{2-}\right)^{n/4}\right] \right\} \tag{10.24}$$

where $a(i)$ is the activity of component i. The ratio of the activities of polyvalent ions is generally approximated by the ratio of their concentrations, C(M), which are usually low. The activity of O^{2-} is constant[55] and the activity of O_2 is replaced by the partial equilibrium pressure, p_{O_2}. The modified equilibrium constant can be expressed as

$$K(T) = C\left(M^{x+}\right) \cdot \left(p_{O_2}\right)^{n/4} \Big/ C\left(M^{(x+n)+}\right) \tag{10.25}$$

All the quantities in this form of K(T) are measurable.

The temperature dependency of K(T) can be theoretically expressed as a function of enthalpy, H, and entropy, S, with gas constant, R, which leads to common exponential function

$$K(T) = \exp\left\{ \frac{-H}{RT} + \frac{S}{R} \right\} = \exp\left\{ C_1 + \frac{C_2}{(T + C_3)} \right\} \tag{10.26}$$

where coefficients C_1, C_2, and C_3 are fitted parameters.

The interaction of more polyvalent elements is based on same approach but is more compli-cated.[56] Pelton[57] gives an interesting thermodynamic approach for estimating activities and equi-librium constants.

To simulate the fining process in a glass melting furnace, the bubble behavior and redox state of the glass melt must be solved. The bubble behavior is controlled by glass-melt convective flow, buoyancy forces (radius dependent), and diffusion of gases to and from the bubble. Diffusion depends on temperature, diffusion coefficients, and gas concentration in the glass melt. Gas con-centration can be estimated by so-called REDOX models.[58-60]

REDOX models calculate the distribution of all redox species (ions and gases) in the glass melt based on a general assumption of dynamic and chemical equilibria at each point. The species concentrations, $C(i)$, for each reaction are calculated from the equations:

$$K(i) = \frac{\prod_j C(j)}{\prod_k C(k)}; \quad K_i(T) = \exp\left\{ C_{1i} + \frac{C_{2i}}{(T + C_{3i})} \right\}$$ (10.27a)

$$\nabla(D(i)\nabla C(i)) - \vec{V}\nabla C(i) = \omega(i)$$ (10.27b)

where j represents products, k represents reactants, $D(i)$ is the diffusion coefficient, ω_i is the rate of concentration change of species brought about the shifting of chemical equilibrium in the oxidation-reduction reaction, and $\vec{V}$ is the vector of glass melt velocity. The system of the model Equations (10.27) are solved by special numerical methods [a combination of algebraic nonlinear equations (10.27a) and partial differential equations (10.27b)) and must be coupled to the solution of the flow and temperature fields in the glass melt. One numerical approach is described by Rakova and Němec[61] and is used in the Glass Service model.

The fining calculation based on bubble behavior has been calculated by various models devel-oped by Beerkens and colleagues,[62,63] Němec and colleagues,[59-61,64,65] and Kawachi and Kawase.[66] The latest development is dealing with the calculation of the sulfate layer that can form on bubble surfaces. This model was created by the group of Němec.[67] All of the models are based on solving equations for diffusion mass flow of each gas component i on the boundary of a rising bubble, expressed as

$$\frac{dm_i}{d\tau} = 4\pi a^2 \cdot \frac{0.381(m_{ib} - m_{ia}) \cdot D_i^{2/3} g^{1/3} \rho^{1/3}}{\eta^{1/3}}$$ (10.28)

where a is the bubble radius, m_{ib} and m_{ia} are the mass concentrations of i-th gas in the glass melt and on the bubble boundary, respectively, D_i is the appropriate diffusion coefficient, ρ is glass density, and η is the glass viscosity. The value of gas concentration in glass, m_{ib}, is estimated by solution of the REDOX model. If the sulfate layer is precipitated, more complicated equations must be used.[65,67] The diffusion equation must be solved for each gas together with the Stoke's equation for the bubble rising velocity, and

$$p_{ex} + \rho g Z + \frac{2\sigma}{a} = \sum_{i=1}^{n} p_i$$ (10.29)

where p_{ex} is the external pressure and σ is the surface tension of glass. Equations (10.28) and (10.29) are usually solved by a Runge-Kutta method for all gas species ($i = 1, n$). For the case of bubbles moving in the glass melt, the rising velocity and the influence of the glass flow velocity must be included. The fining model can be concluded by several methods[63,65-69]:

1. Solving Equations (10.27), (10.28), and (10.29) using direct experimental data of bubble growth without calculation of a REDOX model, and tracking each individual bubble in glass melt.
2. Solving Equations (10.27), (10.28), and (10.29) using concentration fields precalculated (or coupled) with a REDOX model and tracking each individual bubble in glass melt.
3. Solving Equations (10.27), (10.28), and (10.29) using method (1) or (2) and tracking individual bubbles generated with given frequency.
4. Solving Equations (10.27) and (10.29) using direct calculation of the bubble concentration field without tracking of individual bubbles.

10.4.1.7 Glass Quality Indicators

Glass quality depends strongly on the performance of the glass melting processes in the furnace. Glass quality is defined in terms of glass homogeneity (cord = inhomogeneity), the presence of gas bubbles (seeds), and the presence of unmelted materials (stones) in the product. Temperatures, convection currents, and redox state influence the magnitude of these glass defects.

Temperature and flowfield results calculated by CFD models are simply not sufficient for making decisions regarding furnace function. Great effort has gone into post-processing models to try to describe indicators of glass quality. These post-processing models can describe the dissolution of sand grains, the distribution of refining species (redox state), the growth, shrinkage, and movement of gas bubbles, or diffusion of inhomogeneities for known temperatures, flows, and redox conditions.

10.4.1.7.1 Massless particle tracking

Massless particle tracking (also referred to as particle tracing and particle trajectory calculation) is a relatively simple procedure that is included as a post-processor in most glass melt models. It is based on the calculation of moving an imaginary point inside the three-dimensional flow field $\vec{V}$ of the glass melt. The resulting path is called a trajectory. The trajectory location is calculated by a sequence of position updates, Δr, according to a defined time increment Δt, whereby

$$\Delta \vec{r} = \vec{V} \cdot \Delta t \tag{10.30}$$

Some special numerical treatment must be used to interpolate velocity values between grid nodes. Proper selection of the time increment is also important. A typical tracking procedure follows the steps: (1) specify starting positions (individual point locations, planes, or volumes); (2) specify target positions (typically, the glass outlet plane or glass surface in case of bubbles tracking); (3) set the number of starting points (up to several thousands) into starting positions with a defined distribution (regular, random, etc.); (4) calculate trajectories; (5) process and summarize the results. The trajectory results can be processed in a number of ways to obtain the following results:

- Minimum, maximum, and average temperature of each trajectory to determine the possible influence of furnace function on fining or sand dissolution.
- Plot of temperature of trajectories as a function of time as an indication of fining or sand dissolution.
- Minimum, maximum, and average residence time for the particles traversing from the starting position to the target position (the minimum trajectory time can be a good estimation of minimum residence time of the melter).
- Residence time distribution (the number and position of peaks on the so called E-curve[70] indicates mixing ability of glass melt flow in the furnace).
- Estimation of deadwater regions in furnace,[70] which is an indication of stagnant or recirculating cells in the melt.

Using a massless particle tracking procedure permits investigation of the mixing ability of the furnace and flow character changes after geometry changes, operational changes, adding/removing/changing electric boost and/or forced air bubbling, etc.

10.4.1.7.2 Bubbles tracking

Bubbles tracking is calculated by a combination of principles described in the Sections 10.4.1.6 (redox and fining) and 10.4.1.7.1 (massless particle tracking). Typical results include bubble trajectories with various distributions such as residence time, bubble radius, bubble concentration, bubble gas composition, etc. Using bubble tracking, various phenomena can be investigated, such as fining and refining, bubble nucleation, foam formation, gas released from refractories, electrodes, etc.

10.4.1.7.3 Sand dissolution and tracking

Calculation of sand particle dissolution in the glass melt is based on the studies of Hixson and Crowell,[71] Hrma,[72] Cooper,[73] Choudhary,[74] and Němec and Muehlbauer.[75,76]

The dissolution process (changing of sand particle sphere radius, r, with time, t) can be described by a differential equation involving the mass transfer coefficient, α,

$$\frac{dr}{dt} = \alpha\left(w_m - w_e\right) \tag{10.31}$$

where w_m, and w_e denote mass fractions of dissolved SiO_2 in glass (e means equilibrium value). The value of w_m can be expressed in terms of w_t, the total mass fraction of SiO_2 in the mixture of glass and sand, and of w_s, the mass fraction of undissolved SiO_2 in the mixture, as:

$$w_t = w_s + \left(1 - w_s\right) \cdot w_m \tag{10.32}$$

Then, Equation (10.31) can be rewritten:

$$\frac{1 - w_s}{w_s\left(w_e - 1\right) + w_t - w_e}\, dr = \alpha \cdot dt \tag{10.33}$$

The mass transfer coefficient can be approximated by various formulae,[75,76] as

$$\alpha(T,t) = \frac{\alpha_0 \alpha_f(T)}{\alpha_0 + \left[\alpha_f(T) - \alpha_0\right] \cdot \exp\{-\theta(T)t\}} \tag{10.34}$$

where the rate of chemical reaction on the particle surface, α_0, the rate of particle a dissolution at steady state, $\alpha_f(T)$, and the rate of reaching steady state, $\theta(T)$, must be measured in laboratory.

According to experiments, it is assumed that the dissolution velocity of sand particles (the rate of radius decrease dr/dt) is independent of their size (on actual radius r), which implies that the value of w_s for polydisperse sand at time t may be given by the formula

$$w_s(\tau) = \frac{4\pi\rho_s}{3\rho_g}\, N_{so} \int_0^{r_{max}} f\left(r + r_{max0} - r_{max}\right) r^3\, dr \tag{10.35}$$

where ρ_g and ρ_s are the densities of glass and sand, respectively, N_{so} is initial number of sand particles in the unit volume of the glass-sand mixture, $f(r)$ is probability density function of sand particles, and r_{max0} and r_{max} are the radii of maximum sand particles at time zero and t, respectively. The values of N_{so} and $f(r)$ can be estimated from a sieve analysis of a given batch sand.

The simultaneous solution of differential Equation (10.31) and integral Equation (10.35), taking into account the time-temperature history of the maximum sand particle on its trajectory, yields the value of sand concentration, w_s, at any time t and location.

Various quantities are obtained from sand particle trajectory calculations. The distribution of undissolved sand concentration, w_s, in the melt can be displayed by graphical procedures. The intermediate and final radius of the sand particle with initial maximum radius can be obtained from the trajectory history. And the distribution of remaining sand particles in the target plane (usually furnace exit plane) can be obtained and is an important indicator of glass quality.

10.4.1.7.4 Homogenization

The process of glass homogenization[37,77] has been evaluated using post-processing methods that rely on tracking techniques. Using particle tracking, the melting, refining, and homogenization behavior of representative small elements of molten glass are investigated.

Particle tracking is a very powerful technique for determining the melting history of glass, bubbles, sand grains, and other inhomogeneities. The residence time distribution and various melting or refining indexes that can be derived from particle tracking, are also important for evaluating the furnace and operating conditions. Particle tracking can be used to locate dead zones in the tank. The knowledge of dead spaces helps furnace designers improve the tank geometry to eliminate stagnation zones.

The mixing behavior in a glass tank can be characterized by the residence time distribution (RTD), which is an indicator of the history of different paths of the glass melt. The shape the RTD curves[70] indicates the portion of glass having lower or higher residence time with respect to the mean residence time (which is equal to the volume of the glass in the furnace divided by the pull rate). The shape of the RTD curve can indicate the presence of short-circuiting in the tank as well as deadwater regions. Moreover, the number and position of peaks on the so-called E-curve[70] indicates number of recirculating regions in the furnace. A very low minimum residence time (estimated from a set of many trajectories) could indicate that part of the molten glass may not have had enough time to melt completely or may not be well-degassed.

The glass quality is determined not only by residence time but also by temperature history. The dissolution of defect particles (originated from refractories, for example) depends on the melting characteristic of the furnace. Therefore, a melting index has been defined,[37] which combines temperature, viscosity, and time. High temperature, low viscosity, and long time support the melting of many impurities. The melting index, G_i, for molten glass on trajectory i, starting from the batch and finishing at the glass output plane is defined by the formula

$$G_i = \int_0^{t_i} \frac{T_i(t)}{\eta_i(t)} \, dt \tag{10.36}$$

where $T_i(t)$ and $\eta_i(t)$ are time courses of temperature and dynamic viscosity on trajectory i, and t_i is the time when the trajectory reaches the target position. Similarly, a refining factor is defined by:

$$R_i = \int_0^{t_i} \frac{T_i(t)}{\eta_i(t)} \, dt \tag{10.37}$$

where $T_i(t)$ is zero when the temperature is below a user-defined firing temperature, and a temperature index is defined:

$$\Psi_i = \int_0^{t_i} T_i(t)dt \tag{10.38}$$

The melting index distribution (MID) is derived in the same way as the RTD, and various types of MID curves can be obtained.

Kasa et al.[77] recently proposed relations to describe deformation of inhomogeneities (cords) and concentration change due to diffusion.

10.4.2 THERMOPHYSICAL PROPERTIES

The precision of model results is highly dependent on the quality of physical property data of the glass melt and refractory building materials. Obtaining good thermophysical data is very difficult and expensive. Approximating real glass and refractory materials using literature or theoretical data could cause differences between the model and the real furnace. Laboratory data obtained for the real glass and/or refractory materials could similarly be in error because of differences between conditions in a laboratory crucible and in a real furnace. Good modeling results can be obtained only by using clever approximations of a combination of literature and measured data, which involves tuning the data by technological aspects (furnace design, operation conditions, time of furnace life, etc.).

Error in the models is often due to lack of availability of good physical property data.

10.4.2.1 Thermophysical Properties of Glass

Important thermophysical properties of glass include the thermal conductivity, density, viscosity, specific heat, and electrical conductivity.

10.4.2.1.1 Thermal conductivity

Thermal conductivity is the most important glass property because it dramatically influences the calculated glass temperatures and flows calculated by the model. Radiation transfer in the glass melt can be treated as a diffusion process. This is because radiation can only propagate a short distance before being absorbed due to a sufficiently high absorption coefficient of the glass melt caused by a small but significant metal content. The effective thermal conductivity, k_{eff}, can be expressed as $k_{eff} = k_c + k_r$, where k_c is the contribution due to conduction and k_r is the contribution due to radiation. The conductive part can be simply measured at low temperatures by classical methods and extrapolated to high temperatures. The part due to radiation can be theoretically expressed by

$$k_r = 16n^2\sigma T^3/3\Gamma_r \tag{10.39}$$

where Γ_r is Rosseland absorption coefficient, n is the refractive index, and σ is the Stefan-Boltzmann constant.

The radiation part, k_r, can be determined by two different methods: high temperature absorption coefficient measurements and direct measurements in an experimental furnace.[78,79]

10.4.2.1.2 Density

The high temperature dependence of glass density is very important because it is responsible for buoyancy forces in the melt. Buoyancy glass influences the main natural circulation loops and consequently the global glass mixing. Glass density is approximated by a linear function and fits available data well. Because of the Bousinesq approximation, which is used in most models, the following form is frequently used:

$$\rho(T) = \rho_R\left[1 - \beta\left(T - T_R\right)\right] \tag{10.40}$$

where ρ_R is reference density at the reference temperature, T_R.

10.4.2.1.3 Viscosity

Glass viscosity has a very strong temperature dependence and therefore viscosity has an important influence on the glass flows in the furnace. Glass viscosity is typically approximated by an exponential function that fits data very well. There are various forms of this function depending on units and tradition. A common form is:

$$\eta(T) = \exp\left[A + B/(C + T)\right] \tag{10.41}$$

10.4.2.1.4 Specific heat

Glass specific heat has a weak temperature dependence. Specific heat is normally fitted by a quadratic function:

$$C(T) = A_1 + A_2 T + A_3 T^2 \tag{10.42}$$

10.4.2.1.5 Electrical conductivity

Electrical conductivity increases rapidly with increasing temperature. This is important because electrical conductivity influences Joulean heating inside the glass during electric boosting. As the glass temperature rises due to Joulean heating, the electrical conductivity increases and electric current moves more easily through the hotter glass. It has been observed with bottom electrodes that the electric current can find a path of least resistance by flowing a longer distance through the hot surface glass instead of a shorter path through cold bottom glass.[80] Glass electrical conductivity is typically approximated by an exponential function. There are various forms, but the common form is:

$$\sigma(T) = \exp\left[A + B/(C + T)\right] \tag{10.43}$$

10.4.2.2 Physical Properties of Materials

Important thermophysical properties of furnace building materials include the thermal conductivity and electrical conductivity.

10.4.2.2.1 Thermal conductivity

For modeling, the most important property of the materials of furnace construction is thermal conductivity. It dramatically influences heat losses and consequently the overall heat balance. Unfortunately, it is often the case that the values and temperature functions taken from the refractory product literature are not very accurate. The problem is that the materials are not homogeneous because of small holes and other impurities, which cause variation in the thermal conductivity values. Another problem arises due to material changes during the furnace lifetime due to chemical reactions within the refractory material. The temperature functions used in models are generally linear or quadratic forms.

10.4.2.2.2 Specific heat

Specific heat is important only for transient modeling.

10.4.2.2.3   Electrical conductivity

Electrical conductivity is only important for electric boost calculations and, moreover, in special cases when the electrical conductivity of the refractory material is similar to the glass electrical conductivity because of Joule heating inside the refractory. This occurs in borosilicate or E-glasses. The temperature function has the same form as glass.

10.4.3   COMBUSTION SPACE MODELS

Combustion space models are able to simulate turbulent fluid flow, heat transfer, combustion species concentrations, and often pollutants formation. Heat transfer, fluid flow, and combustion reactions are calculated by solving the governing partial differential equations (Navier-Stokes equations), which can be found in standard textbooks.

10.4.3.1   Turbulence

Most models of the glass furnace combustion space[3,10,12,13,15,91,111] use the well-established standard k-ε model.[81] Wall functions are sometimes included. The standard k-ε model is a two-equation model and is the simplest "complete model" of turbulence in which the solution of two separate transport equations allows the turbulent velocity and length scales to be independently determined. Robustness, economy, and reasonable accuracy for a wide range of turbulent flows explain its popularity in industrial flow and heat transfer simulations. It is a semi-empirical model, and the derivation of the model equations relies on phenomenological considerations and empiricism.

10.4.3.2   Combustion Reactions

Two approaches have been used to describe the combustion reactions in a glass furnace: (1) a generalized finite rate formulation, and (2) a mixture fraction/pdf formulation.

The finite rate formulation is based on the solution of species transport equations for reactants and product concentrations, with various chemical reaction rates. The reaction rates appear as source terms in the species transport equations and are computed from Arrhenius rate expressions or by using the eddy dissipation concept of Magnussen and Hjertager.[82] Models of this type are suitable for a wide range of applications, including premixed, partial premixed, and non-premixed combustion. Multiple inlet streams are allowed and are not limited with regard to concentrations of fuels or oxidants. Although the combustion of fuel can be described by a large number of elementary chemical reactions, single- or two-step reactions are often used for economy of computing resources. The finite rate formulation has been used by a number of investigators.[10,13,91]

In the mixture fraction/pdf formulation, individual species transport equations are not solved. Instead, transport equations for one or two conserved scalars (mixture fractions) are solved and individual component concentrations are derived from the calculated mixture fraction distribution. Reaction mechanisms can be treated using infinitely fast chemistry ("mixed-is-burnt" approach), chemical equilibrium calculations, or nonequilibrium (flamelet) calculations.

This approach has been specifically developed for the simulation of turbulent diffusion flames and similar reaction processes in which turbulent mixing is the limiting reaction rate. The mixture fraction method offers several benefits over the finite rate formulation. The method allows intermediate species formation, dissociation effects, and coupling between turbulence and chemistry in a rigorous way. And, the method is computationally efficient because it does not require the solution of a large number of species transport equations.

There are, however, limitations to the mixture fraction approach. It can only be applied to diffusion reaction systems; it is not applicable to premixed or partially premixed systems. Further, the number of fuel and oxidant streams is limited. FLUENT is a three-stream model and is limited to systems with one "fuel," one "oxidant," and one secondary stream (another fuel or oxidant, or a nonreacting stream). Other three-stream models have been reported.[12,83] The mixture fraction/pdf approach has been used by a number of investigators for modeling combustion in a glass furnace.[12,15,83]

A well-written, well-organized, and more detailed description of the pdf approach can be found in the FLUENT User's Manual.[84]

10.4.3.3 Radiation

Radiative heat transfer between superstructure sidewalls, crown, batch, and glass surfaces, as well as radiative absorption and emission by H_2O, CO_2, and soot is the dominant mode of heat transfer in the combustion space. It is therefore a very important sub-model for accurate simulation of the combustion space.

The discrete transfer method (DTM), credited to Lockwood and Shah,[85,86] is commonly used to model radiation in a glass furnace combustion space.[10,12] The method is based on a direct solution of the radiative transfer equation. For each wall point, the solid angle is discretized into a number of beams. The incident heat flux on a wall point coming from opposing wall points after crossing the non-transparent gas mixture is determined by summation of the contributions for each beam by performing ray tracing. The gas mixture extinction coefficient is calculated from the emissivities of H_2O, CO_2, and soot.

The P-1 radiation model, which is valid for optically thick media, has also been used.[10] The P-1 method is the simplest case of the more general P-N model, which is based on the expansion of the radiation intensity into an orthogonal series of spherical harmonics.[87,88] For the P-1 model, the radiative transfer equation is a diffusion equation, which is easy to solve with minimal CPU demand.

The finite-volume scheme[89,90] of the discrete ordinates model (DOM) was used by Wang et al.[15] and Önsel and Eltutar.[91] The DOM solves the radiative transfer equation for a finite number of discrete solid angles, each associated with a vector direction fixed in the global Cartesian system. The fineness of the angular discretization can be varied by the modeling engineer. Unlike the DTM, however, the DOM does not perform ray tracing.

10.4.3.4 Soot

Soot formed in flames promotes radiation and therefore the efficiency of heat transfer from the flame to the glass melt. To calculate the impact of soot, the amount and distribution in the flame have to be known. The amount of soot formed in a flame depends strongly on the fuel and mixing of the fuel with oxidant (staging). Relatively little soot is formed in methane, ethane, and natural gas flames compared to acetylene or fuel oils. Accurate simulation of soot requires solving a transport equation for the soot concentration with appropriate source terms.

Most combustion models applied to glass furnaces[12,91,110] use the empirically based Khan and Greeves[92] soot generation model and the Magnussen and Hjertager[82] soot oxidation model. The Khan and Greeves soot generation model is a single-step model and solves a single transport equation for the soot mass fraction. The rate of soot formation is given by a simple empirical Arrhenius-type rate expression. The Magnussen and Hjertager soot oxidation model is an eddy-breakup model and assumes that turbulence decay controls the rate of soot oxidation. The rate is computed as the minimum of two expressions: one for the regions where the local mean soot concentration is low compared to the oxygen concentration, and the other applicable to regions where oxygen concentration is low and limits the rate of soot combustion.

FLUENT[84] offers a two-step soot formation model in addition to the single-step model. The Tesner model[93] predicts the generation of radical nuclei and then computes the formation of soot on these nuclei. Transport equations are solved for two scalar quantities: the soot mass fraction and the normalized radical nuclei concentration.

The soot formation models that have been applied to glass furnaces are empirically based, approximate models of the soot formation process in combustion systems. The detailed chemistry and physics of soot formation are quite complex and can only be approximated by these models. The models provide only a qualitative indicator of the impact of soot. The empirical constants were derived from air/fuel combustion data and care must be taken to extend these models for oxy/fuel combustion.

10.4.3.5   NOx

Environmental regulations have become a key driver for changes in glass manufacturing. As a result, mathematical modeling is being used to provide insight into and guidance about emissions resulting from the glass melting process.

NOx is a precursor to photochemical smog, contributes to acid rain, and causes ozone depletion. Thus, NOx is a pollutant. The principal constituent of NOx is nitric oxide (NO), although minor amounts of nitrogen oxide (NO_2) and nitrous oxide (N_2O) may also be present.

Resultant NOx formation from combustion can be attributed to four distinct chemical kinetic processes: thermal NOx formation, prompt NOx formation, fuel NOx formation, and reburning. Thermal NOx is formed by the oxidation of atmospheric nitrogen present in the combustion air. Prompt NOx is produced by high-speed reactions at the flame front, and fuel NOx is produced by oxidation of nitrogen contained in the fuel. The reburning mechanism reduces the total NOx formation by accounting for the reaction of NO with hydrocarbons.

At the high temperatures encountered during combustion, most of the NOx is thought to be thermal NOx, formed by the so-called Zeldovich mechanism. The principal reactions governing the formation of thermal NOx from molecular nitrogen are:

$$O + N_2 \xrightleftharpoons{k_1} N + NO \qquad\qquad (10.44)$$

$$N + O_2 \xrightleftharpoons{k_2} O + NO \qquad\qquad (10.45)$$

and

$$N + OH \xrightleftharpoons{k_3} H + NO \qquad\qquad (10.46)$$

Various rate constants have been proposed for the forward and reverse reactions, and have been evaluated by Baulch et al.[94] and Hanson and Salimian.[95]

The prompt NOx mechanism was first identified by Fenimore[96] and is promoted by low-temperature, fuel-rich conditions and short residence times. Staged combustion systems, which are used in glass furnaces, can create such conditions.

The accepted reaction route for prompt NOx is:

$$CH + N_2 \leftrightarrow HCN + N \qquad\qquad (10.47)$$

$$CH_2 + N_2 \leftrightarrow HCN + NH \qquad\qquad (10.48)$$

$$N + O_2 \leftrightarrow NO + O \qquad\qquad (10.49)$$

$$HCN + OH \leftrightarrow CN + H_2O \qquad\qquad (10.50)$$

$$CN + O_2 \leftrightarrow NO + CO \qquad\qquad (10.51)$$

Reaction rate parameters needed for the prompt NOx mechanism have been proposed by DeSoete,[97] Backmier et al.,[98] and DuPont et al.[99]

Fuel NOx is formed by oxidation of nitrogen-containing organic compounds present in liquid or solid fossil fuel. Fuel NOx is important when the fuel is a residual fuel oil or coal, which can contain 0.3 to 2% nitrogen by weight. Fuel NOx has not received attention for glass furnace modeling.

The NO reburning mechanism is a reaction pathway whereby NO reacts with hydrocarbons and is subsequently reduced. The reburn reactions include:

$$CH + NO \xrightarrow{k_1} HCN + O \tag{10.52}$$

$$CH_2 + NO \xrightarrow{k_2} HCN + OH \tag{10.53}$$

$$CH_3 + NO \xrightarrow{k_3} HCN + H_2O \tag{10.54}$$

Rate constants for the reburn reactions can be found in Bowman.[100]

Some glass furnace combustion models have included only the thermal NOx mechanism[12] while others (typically based on FLUENT[84]) have included all of the mechanisms.[22,101,126] The reader is referred to the FLUENT manual[84] for more details of NOx formation models.

10.4.3.6 Particulate Emissions

Particulates are small particles that exit in the stack gases. Environmental regulations limit emissions of particulates of less than 10 microns in diameter from glass manufacturing facilities. Particulate emissions are a result of two general mechanisms. The first mechanism is volatilization of lead, alkalis, borates, and fluorides from the glass melt in the hot furnace and later reaction and/or condensation of these species in the cooler furnace exhaust. The second mechanism is carryover of fine batch materials by entrainment in high-velocity furnace gases. Most of the particulate emissions are believed to come from the volatilization mechanism.

In addition to contributing to particulate emissions, alkali volatilization is important because it can also lead to corrosion attack of furnace refractory in the combustion space.

As a result of the importance of volatilization of glass species, some models have been developed to examine the impact of furnace atmosphere (oxy/fuel vs. air/fuel) and flows in the furnace on the rate of volatilization.

A model for alkali volatilization and subsequent particulate formation for a soda-lime glass was reported by Kobayashi et al.[102] The model assumes that the most important mechanism for NaOH volatilization is due to water vapor reacting with Na_2O in the melt:

$$Na_2O(l) + H_2O(g) \rightarrow 2NaOH(g) \tag{10.55}$$

NaOH can also be formed from the batch by the reaction

$$Na_2CO_3(s) + H_2O(g) \rightarrow 2NaOH(g) + CO_2(g) \tag{10.56}$$

but this was not included in the model.

As the furnace gases exit the furnace and cool in the regenerator and flue ducts, NaOH vapor reacts with SO_2 and O_2, also in the flue gas, to form Na_2SO_4, which subsequently condenses to form sub-micron size particles. Total particulate emissions are evaluated assuming that all of the NaOH vapor is converted to form Na_2SO_4 particles.

The volatilization model is a post-processing model. This approach assumes that the NaOH volatilization has negligible effects on the overall mass and energy balance of the furnace. The mass transfer equation is evaluated over glass surface zones, using local glass surface temperatures and gas velocities from the combustion space model. The surface concentration of NaOH is defined based on the assumption of thermodynamic equilibrium at the surface (i.e., mass transfer resistance in the glass is neglected). Mass transfer in the gaseous boundary layer above the melt is calculated by the traditional approach.[103]

Results from an air/fuel and oxy/fuel comparison study for a 300 ton/day flint container furnace were presented. The model showed that the average NaOH concentration for the oxy/fuel furnace were 4 times higher than the air/fuel furnace. Na_2SO_4 particulate emissions were predicted to be the same for both furnace types.

Jurcik et al.[104] derived models for the volatilization mechanism and carryover mechanism to compare oxy/fuel and air/fuel furnaces. The model gives the ratio of the amount of NaOH volatilized for an air/fuel vs. an oxy/fuel furnace. Like the previous approach, the volatilization mechanism is based on the reaction between NaO and H_2O to form volatile NaOH at the glass surface. Based on the increased water concentration for the oxy/fuel combustion atmosphere, they derived that the partial pressure of NaOH in the oxy/fuel furnace is 4.2 times the partial pressure in an air/fuel furnace. The mass transfer resistance in the glass appears to be neglected. Then the ratio of mass transfer resistance in the combustion space, based on film theory, is calculated from CFD models of oxy/fuel and air/fuel furnaces. Integrating over the melt surface and combining the two effects, partial pressure differences and mass transfer resistances, the relative mass flux of NaOH for air/fuel and oxy/fuel are calculated. For the example furnace, the calculations show that the volatilization of NaOH in an air/fuel furnace will be a factor of 1.6 to 2.0 larger than that of an oxy/fuel furnace.

10.4.4   COUPLED COMBUSTION SPACE AND GLASS MELT MODELING

Coupling combustion space and glass melt models is now common practice[11,13-15,91,105-109] and the importance of coupling the combustion space and the glass melt space is well-documented.

Kim and Yoon[14] have identified four basic approaches for transferring energy balance boundary condition information between the combustion space and glass melt. In method 1, the temperature is read from the combustion space and applied to the glass melt, and the heat flux is read from the glass melt and applied to the combustion space. Method 2 is the opposite of method 1. In method 3, the heat flux is read from both the glass melt and combustion space, compared, and a new temperature profile is applied to both the glass melt and the combustion space. In method 4, the temperature is read from both the glass melt and combustion space, compared, and a new heat flux profile is applied to both the glass melt and combustion space.

Eltutar and Önsel[108] and Wang et al.[15] apply a heat flux boundary condition calculated from the combustion space to the glass melt and batch, and a temperature boundary condition calculated from the glass melt to the combustion space.

Hoke and Gershtein[106,109] showed significant improvement to combustion modeling results by coupling to a glass melt model. Hoke and Marchiando[107] showed the influence of the combustion model on glass melt model results. Bottom temperatures, spring zone location, residence times, and other important flow features were shown to be significantly affected by the use of the coupled model approach. In the analysis of how the batch blanket influences model results, Eltutar and Önsel[108] concluded that combined glass and combustion space modeling is required for proper simulation of the glass furnace. The combustion space model influences the temperature, surface area, and shape of the batch blanket, and these are important parameters which affect the success of the model.

Coupling combustion space and glass melt regions obviously increases the time for calculation considerably. Therefore, it is important to optimize the method of coupling for speed to convergence, as was done by Kim and Yoon.[14] Four coupling approaches were tested for speed to convergence. The approach used was shown to have little impact on the speed to convergence but, surprisingly, the coupling approach slightly influenced the solution results.

10.4.5 VALIDATION

Trust in modeling results is key to the acceptance of conclusions and recommendations formulated from a modeling study. The validity of the simulation depends on two parts: the numerical solver software (numerical procedure, sub-models, etc.) and the formulation of the model (geometry, boundary conditions, physical properties, assumptions, etc.) by the modeling engineer.

Several papers have been written about validation of CFD models where model results are compared to furnace data.

Validation studies have been performed as part of the Technical Committee 21 of the International Commission on Glass (ICG). Results from modeling a former Ford Nashville float glass furnace have been published by some of the participants.[110,111] Details of the operating conditions of the furnace can be found in Lankhorst et al.[110]

Temperature measurements in the combustion space and glass melt were available for this furnace. Crown temperatures were measured by optical pyrometer. Glass temperatures at various glass depths along the furnace length were measured by thermocouple.

Figure 10.12 shows modeling results compared with measured crown temperatures. Crown temperatures are plotted as a function of furnace length. Predicted temperatures are within 50°C of measured data for both models. In-glass predictions and measured values are shown in Figures 10.13 to 10.16. Appropriate trends are observed, and accuracy is generally within 50°C.

Hoke[111] presents results from seven different furnace simulations and compares those results to measured glass melt and combustion space data to validate his solver software and model formulation. For combustion models, different boundary condition approaches for the glass/batch surface are examined and shown to impact the accuracy of the model. Two separate boundary condition regions — one for the glass and one for the batch — are required when modeling the combustion space without the glass melt space. The greatest accuracy is shown for cases where

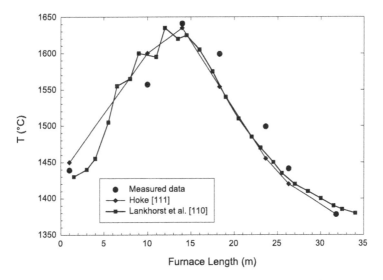

FIGURE 10.12   Centerline crown temperature for Ford float furnace.

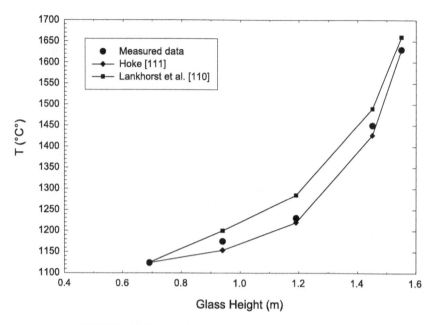

FIGURE 10.13   In-glass temperature at center of port 4.

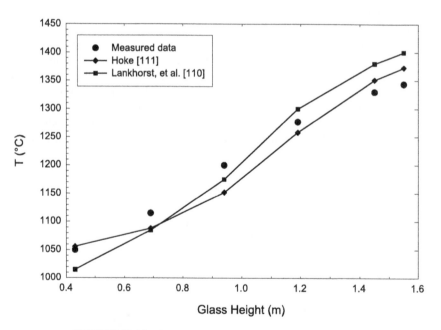

FIGURE 10.14   In-glass temperature 5 ft upstream of waist.

the combustion space and glass melt models are coupled. Validation of an oxy/fuel coupled model shows results for both the combustion space and glass melt.

These examples illustrate the accuracy that can be obtained from CFD models.

In addition to comparing CFD results with furnace data, CFD models have been validated against physical models.[112] A comparison study was made of a scaled glycerin model[113] and of a mathematical simulation. Qualitative and quantitative agreement was found. The same general flow

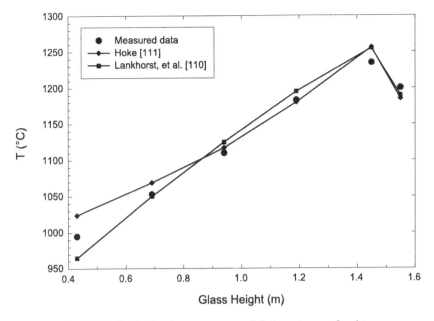

FIGURE 10.15   In-glass temperature 5 ft downstream of waist.

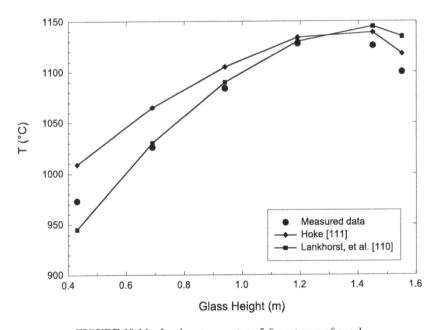

FIGURE 10.16   In-glass temperature 5 ft upstream of canal.

pattern was seen in both models, as were Bernard roll cells under the batch region. Quantitative agreement between velocity profiles determined by probes in the glycerin model numerical tracking in the mathematical model was obtained.

Validation with physical models is instructive. However, the physical model is also an approximation of the real process. As discussed earlier in Section 10.4.1.2, the batch behavior is very important to the accuracy of the model and the batch may not be well-represented in physical models.

10.4.6   Industrial Applications of CFD Modeling

CFD models have been applied to investigate a number of issues. Production increase, glass quality improvement, furnace design, melting capability improvement for batch changes, furnace efficiency improvement, pollution (NOx and particulate) decrease, sensor placement ... have all been studied. For the conversion from an air/fuel to an oxy/fuel furnace, it is typical to conduct studies to evaluate oxy/fuel burner placement, firing rate distribution, exhaust number and placement. Depending on the objectives of the study, the researcher must determine whether a coupled solution of the combustion space and glass melt is required or whether a model of the combustion space alone or glass melt alone will be sufficient.

10.4.6.1   Production Increase

Önsel and Eltutar[114] used modeling to investigate using electric boost to successfully increase the furnace capacity by 20% on a end-fired container furnace with bubbling and a dam. They studied placing electric boost under the batch and in the hot-spot region. The study showed that electric boosting in the hot spot would be more efficient for increasing furnace capacity or glass quality, but that electric boosting under the batch would also provide sufficient benefits to meet the capacity increase.

Chmelař et al.[115] illustrated a number of applications of CFD models.

1. CFD modeling was used to evaluate a flow dam and the location of the hot spot. The model showed that the flow and temperature fields were better with the hot spot moved further down tank and installing a flow dam just after the hot spot. The spring becomes more intensive and the general circulation in the melter becomes stronger.
2. A refiner was optimized to reduce the backward current between the refiner and the melter. The model suggested that the best way to suppress the backward flow was by reducing the refiner glass depth.
3. Electric boost was added to increase production by 25% in a container furnace. The boost location and power was optimized in an effort to minimize dam corrosion and establish a stable batch pattern.

10.4.6.2   Furnace Design and Operation

Intensification of glass melting using a mechanical barrier (dam) and furnace insulation was studied by Wagnerova et al.[116] The position and height of the dam was varied in the study and heat losses were varied to simulate the effect of insulation changes. Some 92 variations were conducted as part of this study. Acceptable furnace performance was based on the resulting throat temperature, whether the minimum glass melt temperature was lower than the liquidus temperature, and general flow pattern causing the bulk of glass to flow through a high-temperature region. Of the 92 variants, 17 variants satisfactorily met the set criteria.

Oxy/fuel combustion generally provides a higher rate of heat transfer to the glass compared to air/fuel combustion with lower combustion space temperatures. It is therefore possible to increase the production rate in a furnace when converting an air/fuel furnace to oxy/fuel. Hoke and Marchiando[8] presented results from three mathematical simulations, showing that although enough energy can be transferred to the glass using oxy/fuel, other furnace design features, such as throat geometry, must be considered for higher production rates. The most important feature identified by the model was correlation between glass quality and the existence of return flow from the refiner back into the melter.

Hoke[41] evaluated bubbling as a means to intensify glass melting so that batch composition could be altered without adversely affecting glass quality. Three cases were presented: no bubblers, 6 bubblers, and 16 bubblers. The study suggested that six bubblers provided the best results.

Although 16 bubblers intensified the mixing in the melter the most and raised the bottom temperatures, glass quality parameters such as the melting index, minimum residence time, and mean residence time were compromised.

In another forced-air bubbling study, Aume and Hong[117] analyzed the impact of bubblers on a CRT panel furnace. The model predicted a 50 to 60% decrease in viscous knot inclusions when bubbling was used, and a similar result was observed for the furnace. In this paper, the forced-air bubbling sub-model was validated against a physical model.

Two operating conditions that produced very different glass quality results were modeled by Hoke and Marchiando[118] using a coupled model of the combustion space and glass melt. The pull rate, total firing rate, and bubbler flow rate were the same for both cases. The firing rate distribution between burners was different for each case. Results from the models were compared to explain the change in glass quality with regard to operating conditions. The case corresponding to improved glass quality showed longer minimum and mean residence times, a higher temperature on the critical trajectory, and spring zone and turning point location that are moved back toward the batch when compared to the other case.

The objective of the study by Bauer and Lankhorst[119,120] was to evaluate the use of oxy/fuel in a Sorg LoNOx® Melter design. Oxy/fuel has typically been applied only to traditional furnace designs. The study compared the standard air/fuel design with a side-port-fired tube-in-tube type oxy/fuel burners, and front end-wall-fired Cleanfire®* HR™ oxy/fuel burners.[121] The study concluded that the best results were obtained using the Cleanfire® HR™ oxy/fuel burners where the highest thermal efficiencies were obtained (71%). The air/fuel recuperative LoNOx® Melter was predicted to have an efficiency of 52% and the side-port oxy/fuel LoNOx® Melter an efficiency of 68%.

Schill and Vlček[122] reported a furnace design study for radioactive waste glass. In this study, the effect of the tank bottom slope, electrode dimensions and locations, and electrode voltages and phasing were evaluated. This work was done under U.S. Department of Energy Contract DE-AC06-76RLO 1830.

Sensors are a very important aspect of furnace design and operation. By properly locating sensors, a furnace operator can be notified of furnace upsets sooner. This can potentially lead to higher overall yields if appropriate actions are taken. Proper placement of sensors can also help improve yield for the planned changes in the furnace operation. Fletcher et al.[123] performed a 3-D time-transient simulation of an air/fuel recuperative fiberglass furnace, subject to a process change involving 3% increase in total fuel. The modeling results helped them to recommend the optimum temperature sensor locations in the furnace.

In a study that is typical for industrial gas suppliers, Schnepper et al.[124] confirmed that a proposed oxy/fuel setup provided glass temperature and circulation profiles very similar to those resulting from the regenerative air/fuel furnace case. This capability is important for oxygen suppliers to support their customers converting to oxy/fuel combustion.

Hoke et al.[127,128] evaluated air/fuel backup of an oxy-fuel furnace using Cleanfire® AOF technology. Modeling was used to confirm suitable flame interaction and furnace temperatures during the air/fuel backup mode.

10.4.6.3 Pollution Abatement

To meet government regulations on pollution, many technologies have been developed to reduce NOx. One NOx reduction technology for the glass industry is oxygen-enriched air staging (OEAS).[125] OEAS is a technique whereby the primary combustion is fired without excess air, which reduces NOx formation but increases CO. A secondary stream of oxidant is added in a second combustion zone to burn the CO before the exhaust gases leave the furnace. In a 3-D, single port-to-port model of a regenerative furnace combustion space, Li[126] evaluated four strategies for

---

* Cleanfire is a registered trademark of Air Products and Chemicals, Inc.

injection into the second combustion zone: injection from inside of the exhaust port using two nozzles, injection from the crown with a single nozzle, and injection from under the exhaust port using one or two nozzles. All of the cases predicted effective burnout of the CO without overheating the furnace refractory, and without fuel penalty. Under-port lancing with two nozzles was predicted to be the best choice in terms of CO destruction, impact on flow above the glass, and furnace thermal efficiency.

Carvalho et al.[12] examined NOx reduction by various oxygen-enhanced enrichment techniques. Twelve cases were reported, which included a baseline air/fuel case, various oxy/fuel boost conditions, oxygen lancing, and full oxy/fuel. Cases were evaluated based on predicted natural gas consumption, NOx emission, particle emission, and furnace fuel efficiency.

REFERENCES

1. Brown, J.T., 100% oxygen-fuel combustion for glass furnaces, in *Collected Papers from the 51st Conf. Glass Problems*, The American Ceramic Society, Westerville, OH, 1991, 202-217.
2. Shamp, D.E., Slavejkov, A.G., and Joshi, M.L., Oxy-fuel firing for emissions control on a fiberglass melter, *53rd Conf. Glass Problems*, The American Ceramic Society, Westerville, OH, 1992, 87-102.
3. Schnepper, C., Modeling oxy-fuel combustion space for glassmelting tanks, *The Glass Researcher: Bulletin of Glass Science and Engineering*, New York State College of Ceramics at Alfred University, 8(1), 1998.
4. Baukal, C.E., *Oxygen-Enhanced Combustion*, CRC Press, Boca Raton, FL, 1998.
5. Tooley, F. V., Ed., *The Handbook of Glass Manufacture*, 3rd ed., Vol. I, Ashlee Publishing, 1984, 400-9 to 400-15.
6. Moore, R.D. and Davis, R.E., Electric furnace application for container glass, *47th Conf. Glass Problems*, The American Ceramic Society, Westerville, OH, Nov. 1986.
7. Hlaváč, J., *The Technology of Glass and Ceramics: An Introduction*, Elsevier Scientific, Amsterdam, 1983.
8. Hoke, B.C. and Marchiando, R.D., Using computational fluid dynamics models to assess melter capacity changes when converting to oxy-fuel, *Proc. 18th Int. Congress on Glass*, San Francisco, July 1998.
9. Muschick, W. and Muysenberg, E., Round robin for glass tank models — report of the International Commission on Glass (ICG) Technical Committee 21, Modeling of Glass Melts, *Glastechn. Ber.*, 71(6), 153-156, 1998.
10. Hoke, B.C., Validation of CFD models for glass furnaces, *Proc. 5th Int. Semin. Mathematical Simulation in Glass Melting*, Horní Bečva, Czech Republic, 1999, 16-25.
11. Bergman, R. and Tselepidakis, D., Simulation of TV glass refiner using coupled combustion and glass flow models, *Advances in Fusion and Processing of Glass II*, The American Ceramic Society, Westerville, OH, 1998, 239-244.
12. Carvalho, M.G., Speranskaia, N., Wang, J., and Nogueira, M., Modeling of glass melting furnaces: applications to control, design and operation optimization, *Advances in Fusion and Processing of Glass II*, The American Ceramic Society, Westerville, OH, 1998, 109-135.
13. Takamuku, H., Ozeki, Y., Kudo, E., and Miyamoto, T., Evaluation of combustion chamber in furnaces by numerical simulation, *ICG Int. Symp. Glass Problems*, Istanbul, Turkey, 1996, 392-397.
14. Kim, D.Y. and Yoon, S.K., The coupling technique studies for glass furnace simulation, *Proc. 5th Int. Semin. Mathematical Simulation in Glass Melting*, Horní Bečva, Czech Republic, 1999, 138-145.
15. Wang, J., Brewster, S., Webb, B.W., McQuay, M.Q., and Bhatia, K., A coupled combustion space/batch/melt tank model for an industrial float glass furnace, *Proc. 5th Int. Semin. Mathematical Simulation in Glass Melting*, Horní Bečva, Czech Republic, 1999, 84-93.
16. Viskanta, R., Review of three-dimensional mathematical modeling of glass melting, *Journal of Non-Crystalline Solids*, 177, 347-362, 1994.
17. Mase, H. and Oda, K., Mathematical model of glass tank furnace with batch melting process, *Journal of Non-Crystalline Solids*, 38, 39, 807-812, 1980.
18. De Waal, H., Mathematical modeling of the glass melting process, *2nd International Conference on Advances in the Fusion and Processing of Glass*, Dusseldorf, 1990, 1-18.

19. Simonis, F. De Waal, H., and Beerkens, R., Influence of furnace design and operation parameters on the residence time distribution of glass tanks predicted by computer simulation, *Collected Papers from the XIV Int. Congress on Glass*, Vol. III, New Delhi, 1986, 118-127.

20. Schill, P., Simulation of 3-dim. temperatures and flows in glassmelt via mathematical modeling, *1st Int. Semin. Mathematical Simulation in Glass Melting*, Valašské Meziříčí, Czechoslovakia, 1991, 1-13.

21. Schill, P., Models of glass melting furnaces, *Proc. II Semin. Mathematical Simulation in Glass Melting*, Vsetin, Czech Republic, 1993, 102-116.

22. Choudhary, M.K., A modelling study of flows and heat transfer in an electric melter, *J. Non-Crystalline Solids*, 101, 41-53, 1988.

23. Ungan, A.,Three Dimensional Numerical Modeling of Glass Melting Process, Ph.D. thesis, Purdue University, 1985.

24. Carvalho, M.G., Computer Simulation of a Glass Furnace, Ph.D. thesis, University of London, 1983.

25. Patankar, S.V., *Numerical Heat Transfer and Fluid Flow*, McGraw-Hill, New York, 1980.

26. Vanandruel, N. and Deville, M., Unsteady numerical simulation of circulation and heat transfer in a glass melting tank, *Proc. 2nd ESG Conf. Glass*, 1993, 85.

27. Luňáček, K., Numerical simulation and optimization of processes in glass furnace, *Proc. III Int. Semin. Mathematical Simulation in the Glass Melting*, Horní Bečva, Czech Republic, 1995, 62-69.

28. Schill, P., Calculation of 3D glassmelt flow in large furnaces via two grid method, *Glastechn. Ber.*, 63K, 39, 1990.

29. Carvalho, M.G. and Nogueira, M., Modelling of glass melting industrial process, *J. de Physique IV*, 1357-1366, 1993.

30. Choudhary, M.K., Mathematical modeling of flow and heat-transfer phenomena in glass furnace channels and forehearths, *J. Am. Ceramic Soc.*, 74, 3091-3099, 1991.

31. Viskanta R. and Anderson, E.E., Heat transfer in semi-transparent solids, *Advances in Heat Transfer*, Vol. 11, T.F. Irvine, Jr. and J.P. Hartnett, Eds., Academic Press, New York, 1975, 317.

32. Curran, R.L., Mathematical model of an electric glass furnace: effects of glass color and resistivity, *IEEE Trans. on Ind. Appl.*, IA-9, 348, 1973.

33. Conradt, R., Suwannathada, P., and Pimkhaokham, P., Local temperature distribution and primary melt formation in a melting batch heap, *Glastechn. Ber.*, 67, 103, 1994.

34. Schill, P., Batch melting in mathematical simulations of glass furnaces, *Proc. III Int. Semin. Mathematical Simulation in the Glass Melting*, Horní Bečva, Czech Republic, 1995, 97-101.

35. Bauer, R.A., Op den Camp, O., Simons, P., Verheijen, O., Noot, M., and Lankhorst, A.M., Advanced possibilities of the TNO glass tank model and future developments, *Proc. Opening TNO Glass Technology Representative Office, North America*, 1998, 71-91.

36. Wang, J., Three-dimensional Mathematical Model of Thermal Phenomena Occurring in Industrial Glass Melting Tanks, Ph.D. thesis, Instituto Superior Tecnico, Lisbon, Portugal, 1998.

37. Beerkens, R.G.C., Van der Heijden, T., and Muysenberg, E., Possibilities of glass tank modeling for the prediction of the quality of melting processes, *Ceram. Eng. Sci. Proc.*, 14(3-4), 139-160, 1993.

38. Pugh, A.C.P., A method of calculating the melting rate of glass batch and its use to predict effects of changes in the batch, *Glastek. Tidskr.*, 23, 95-104, 1968.

39. Bear, J., *Dynamics of Fluids in Porous Media*, Elsevier, New York, 1972.

40. Prasad, R.O.S., Mukhopadhyay, A., and Dutta, A., Implementation of a glass batch melting model in the general purpose three-dimensional CFD code FLUENT, *Proc. 5th Int. Semin. Mathematical Simulation in Glass Melting*, Horní Bečva, Czech Republic, 1999, 43-51.

41. Hoke, B.C., Application of glass melt modeling for examining forced bubbling design, *Proc. 5th ESG Conf. Glass Science and Technology for the 21st Century*, Prague, Czech Republic, 1999.

42. Franěk, A., Implementation of bubbling into the glass-furnace model, *Proc. II Semin. Mathematical Simulation in the Glass Melting*, Vsetin, Czech Republic, 1993, 117-121.

43. Staněk, J., *Electric Melting of Glass*, Elsevier Scientific, 1977.

44. Curran, R.L., Use of mathematical modeling in determining the effects of electrode configuration on convection currents in an electric glass melter, *IEEE Trans. Ind. Gen. Appl.*, IGA-7, 116-129, 1971.

45. Austin, M.J. and Bourne, D. E., A mathematical model of an electric glass furnace, *Glas. Technol.*, 14(3), 78-84, 1973.

46. Mardorf, L. and Woelk, G., Calculation of the behaviour of an electrically heated glass-melting tank with the aid of a mathematical model, *Glastechn. Ber.*, 56(4), 73-84, 1983.

47. Hofmann, O.R. and Hilbig, G., Calculation of the Joule effect heat liberated directly in electric glass melting, *Glastechn. Ber.,* 54(2), 36-43, 1981.

48. Choudhary, M.J., A three-dimensional mathematical model for flow and heat transfer in electrical glass furnaces, *IEEE Trans. Ind. Appl.,* IA-22(5), 912-921, 1986.

49. Ungan, A. and Viskanta, R., Three-dimensional numerical simulation of circulation and heat transfer in an electrically boosted glass melting tank, *IEEE Trans. Ind. Appl.,* IA-22(5), 922-933, 1986.

50. Schill, P., Electric fields in close vicinity of electrodes, *Proc. VIIth Conf. Electric Melting of Glass,* Prague, 1986, 74-80.

51. Schill, P., Mathematical modeling of complex shapes of glass tank, *Proc. XVI Int. Congress on Glass,* Madrid, Boletin de la Sociedad Espanola de Ceramica y Vidrio, 6, 31-38, 1992.

52. Cooper, A.R., Cable, M., and Bradford, I.T., Model study of the mixing of viscous liquids in a crucible, *Glass Technol.,* 8(2), 48-54, 1967.

53. Spremulli, P.F., Physical modelling of flow behaviour in a stirred glass system, *Glastechn. Ber. Glass Sci. Technol.,* 70(2), 41-51, 1997.

54. Wondergem-de Best, A., Redox Behaviour and Fining of Molten Glass, Ph.D. thesis, Technische Universiteit Eindhoven, The Netherlands, 1994.

55. Johnston, W.D., Oxidation-reduction equilibria in molten $Na_2O \cdot SiO_2$ glass, *J. Am. Ceramic Soc.,* 48(4), 184-190, 1965.

56. Ruessel, C., Polyvalent ions in glass melts, *Glastechn. Ber.,* 63K, 197-211, 1990.

57. Pelton, A.D., Thermodynamic calculations of chemical solubilities of gases in oxide melts and glasses, *Glastechn. Ber. Glass Sci. and Technol.,* 72(7), 214-226, 1999.

58. Beerkens, R., Heutige moeglichkeiten zur modellierung von glasschmelzoefen voraussagen zur qualitaet des glasschmelzprozesses, *Glastechn. Ber. Glass Sci. and Technol.,* 71(4), N35-N47, 1998.

59. Kloužek, J., Němec, L., Schill, P., and Ullrich, J., Refining under real conditions of glassmelting furnaces, *Proc. Fundamentals of Glass Science and Technology,* Vaexjoe, Sweden, 1997, 340-347.

60. Němec, L., Some critical points of the glassmelting process, *Ceramics-Silikaty,* 39(4), 121-160, 1995.

61. Rakova, M., and Němec, L., The significance of the redox state of glass on the bubble behaviour in the isothermal glass melts, *Ceramics-Silikaty,* 41(3), 81-87, 1997; 42(1), 1-7, 1998.

62. Beerkens, R.G.C., Chemical equilibrium reactions as driving forces for growth of gas bubbles during refining, *Glastechn. Ber.,* 63K, 222-242, 1990.

63. Beerkens, R.G.C., van der Heijden, T., and Muijsenberg, E., Possibilities of glass tank modeling for the prediction of the quality of melting process, *Ceram. Eng. Sci. Proc.,* 14(3-4), 139-160, 1993.

64. Němec, L., The behaviour of bubbles in glass melts, *Glass Technol.,* 21(3), 134-144, 1980.

65. Němec, L., Kloužek, J., and Ullrich, J., The technological significance of interactions between gases and glasses, *Ceramics-Silikaty,* 42(4), 186-198, 1998.

66. Kawachi, S. and Kawase, Y., Evalution of bubble removing performance in a TV glass furnace, *Glastechn. Ber. Glass Sci. and Technol.,* 71(4), 83-91, 1998.

67. Němec, L. and Ullrich, J., Calculations of interactions of gas bubbles with glass liquids containing sulphates, *J. Non-Crystalline Solids,* 238, 98-114, 1998.

68. Ungan, A. and Balkani, B., Numerical simulation of bubble behaviour in glass melting tanks. Part 4. Bubble number density distribution, *Glass Technol.,* 37(5), 164-168, 1996.

69. Johnson, W., Bubble concentration — redox model, *Proc. 5th Int. Semin. Mathematical Simulation in Glass Melting,* Horní Bečva, Czech Republic, 1999, 105-111.

70. Levenspiel, O., *Chemical Reaction Engineering,* John Wiley & Sons, New York, 1962.

71. Hixson, A.W. and Crowell, J.H., Dependence of reaction velocity upon surface and agitation, *Ind. Engin. Chem.* 23, 923-930, 1931.

72. Hrma, P., A kinetic equation for interaction between grain material and liquid with application to glass melting, *Silikaty,* 24(1), 7-16, 1980.

73. Cooper, A.R., *Chem. Eng. Sci.,* 21, 1095, 1966.

74. Choudhary, M.K., The effect of free convection on the dissolution of a spherical particle in a viscous melt, *Glass Technol.,* 29(3), 100-102, 1988.

75. Němec, L. and Muehlbauer, M., The concentration field of sand in the glass tank furnace, *Glastechn. Ber.,* 56K(1), 82-87, 1983.

76. Muehlbauer, M. and Němec, L., Dissolution of glass sand, *Ceramic Bull.,* 64(11), 1471-1475.

77. Kasa, S., Jandeček, P., and Wagnerova, S., Studies on homogenizing process by the 3D numerical model, *Proc. 5th ESG Conf. Glass Science and Technology for the 21st Century*, Prague, Czech Republic, 1999.
78. Blažek, A., Endrys, J., Kada, J., and Staněk, J., *Glastechn. Ber.*, 49(4), 75-81, 1976.
79. Endrys, J., Blažek, A., and Ederova, J., *Glastechn. Ber.*, 66(6/7), 151-157, 1993.
80. Hoke, B.C., Jr., Customer modeling study, 1997.
81. Launder, B.E. and Spalding, D.B., The numerical computation of turbulent flows, Imperial College of Science and Technology, London, England, NTIS N74-12066, January, 1973.
82. Magnussen, B.F. and Hjertager, B.H., On mathematical models of turbulent combustion with special emphasis on soot formation and combustion, *16th Symp. (Int.) on Combustion*, The Combustion Institute, 1976.
83. Boerstoel, G.P., Lankhorst, A.M., and Muysenberg, H.P.H., Numerical study on the effects of oxy-boosting in a float furnace, *Proc. IV Int. Semin. Mathematical Simulation in Glass Melting*, Horní Bečva, Czech Republic, 1997, 130-139.
84. FLUENT User's Guide, Fluent, Inc., Lebanon, NH, July 1998.
85. Lockwood, F.C. and Shah, N.G., A new radiation solution method for incorporation in general combustion prediction procedures, *18th Symp. (Int.) on Combustion*, The Combustion Institute, 1981, 1405-1414.
86. Shah, N.G., A New Method of Computation of Radiant Heat Transfer in Combustion Chambers, Ph.D. dissertation, Imperial College of Science and Technology, London, 1979.
87. Cheng, P., Two-dimensional radiating gas flow by a moment method, *AIAA J.*, 2, 1662-1664, 1964.
88. Siegel, R. and Howell, J.R., *Thermal Radiation Heat Transfer*, Hemisphere, Washington, D.C., 1992.
89. Chui, E.H. and Raithby, G.D., Computation of radiant heat transfer on a non-orthogonal mesh using the finite-volume method, *Numerical Heat Transfer, Part B*, 23, 269-288, 1993.
90. Raithby, G.D. and Chui, E.H., A finite volume method for predicting a radiant heat transfer in enclosures with participating media, *J. Heat Transfer*, V. 112, 1990, pp. 415-423.
91. Önsel, L. and Eltutar, Z., Sisecam mathematical model for glass furnaces, personal communication, Sept. 1999.
92. Khan, I.M. and Greeves, G., A method for calculating the formation and combustion of soot in diesel engines, *Heat Transfer in Flames*, Afgan, N.H. and Beer, J.M., Eds., Scripta Book Co., Washington, D.C., 1974, 391-402.
93. Tesner, P.A., Snegiriova, T.D., and Knorre, V.G., Kinetics of dispersed carbon formation, *Combustion Flame*, 17, 253-260, 1971.
94. Baulch, D.L., Drysdall, D.D., Horne, D.G., and Lloyd, A.C., *Evaluated Kinetic Data for High Temperature Reactions*, Vols. 1, 2, and 3, Butterworth, 1973.
95. Hanson, R.K. and Salimian, S., Survey of rate constants in H/N/O systems, *Combustion Chemistry*, W.C. Gardiner, Ed., 1984, 361.
96. Fenimore, C.P., *13th Symp. (Int.) on Combustion*, The Combustion Institute, 1971, 373.
97. De Soete, G.G., Overall reaction rates of NO and N_2 formation from fuel nitrogen, *15th Symp. (Int.) on Combustion*, The Combustion Institute, 1975, 1093.
98. Backmier, F., Eberius, K.H., and Just, T., *Comb. Sci. Technol.*, 7, 77, 1973.
99. DuPont, V., Porkashanian, M., Williams, A., and Woolley, R., Reduction of NOx formation in natural gas burner flames, *Fuel*, 72(4), 497-503, 1993.
100. Bowman, C.T., Chemistry of gaseous pollutant formation and destruction, *Fossil Fuel Combustion*, Bartok and Sarofim, Eds., J. Wiley & Sons, Canada, 1991.
101. Wang, J., Webb, B.W., McQuay, M.Q., and Bhatia, K., Numerical simulation of an oxy-fuel-fired float glass furnace by means of a model coupling the combustion space and the glass tank, in *Collected Papers from the 60th Conference on Glass Problems*, The American Ceramic Society, Westerville, OH, 1999.
102. Kobayashi, H., Wu, K.-T., and Richter, W., Numerical modeling of alkali volatilization in glass furnaces and applications for oxy-fuel furnace design, *Glastechn. Ber. Glass Sci. Technol.*, 68(C2), 119-126, 1995.
103. Bird, R.B., Stewart, W.E., and Lightfoot, E.N., *Transport Phenomena*, John Wiley, New York, 1960.
104. Jurcik, B., Philippe, L., Wayman, S., and Ruiz, R., How oxy-fired glass furnaces reduce particulate emissions, *Glass Industry*, 14-23, May 1997.

105. Muysenberg, H.P.H., Simonis, F., and Van der Heijden, T.C.M., The coupling between a glass tank and combustion chamber model, *Proc. II Int. Semin. Mathematical Simulation in Glass Melting*, Vsetin, Czech Republic, 1993, 175-181.

106. Hoke, B.C. and Gershtein, V.Y., Improved predictive results from coupling the gas space and glass bath models for a glass melter, *Proc. III Int. Semin. Mathematical Simulation in the Glass Melting*, Horní Bečva, Czech Republic, 1995, 82-96.

107. Hoke, B.C., Jr. and Marchiando, R.D., Using combustion modeling to improve glass melt model results, *Proc. IV Int. Semin. Mathematical Simulation in Glass Melting*, Horní Bečva, Czech Republic, 1997, 87-95.

108. Eltutar, Z. and Önsel, L., The analysis of the batch blanket and its characteristics for a realistic simulation of glass furnaces, *Proc. 5th Int. Semin. Mathematical Simulation in Glass Melting*, Horní Bečva, Czech Republic, 1999, 57-67.

109. Hoke, B.C., Jr. and Gershtein, V.Y., Coupling combustion space and glass melt models improves predictions, *Am. Ceramic Soc. Bull.*, 74(11), 75-78, 1995.

110. Lankhorst, A.M., Boerstoel, G.P., Muysenberg, H.P.H., and Koram, K.K., Complete simulation of the glass tank and combustion chamber of the former Ford Nashville float furnace including melter, refiner and working end, *Proc. IV Int. Semin. Mathematical Simulation in Glass Melting*, Horní Bečva, Czech Republic, 1997, 77-86.

111. Hoke, B.C., Jr., Validation of CFD models for glass furnaces, *Proc. V Int. Semin. Mathematical Simulation in Glass Melting*, Horní Bečva, Czech Republic, 1999, 16-25.

112. Schill, P., Verification of mathematical model and glycerine model, *Proc. III Int. Semin. Mathematical Simulation in the Glass Melting*, Horní Bečva, Czech Republic, 1995, 70-75.

113. Curlet, N.W.E., Experimental and Numerical Modeling of Three Dimensional Natural Convection in an Enclosure, Sc.D. thesis, MIT, 1976.

114. Önsel, L. and Eltutar, Z., Modelling of a container furnace for electric boosting evaluations, *Proc. IV Int. Semin. Mathematical Simulation in Glass Melting*, Horní Bečva, Czech Republic, 1997, 108-116.

115. Chmelar, J., Schill, P. and Casteckova, A., Practical use of models, *Proc. II Int. Semin. Mathematical Simulation in Glass Melting*, Vsetin, Czech Republic, 1993, 160-169.

116. Wagnerova, S., Kasa, S., Jandacek, P., and Paur, F., Influence of intensifying means upon technological characteristic of glass melting furnaces, *Proc. IV Int. Semin. Mathematical Simulation in Glass Melting*, Horní Bečva, Czech Republic, 1997, 117-129.

117. Aume, V.O. and Hong, B., Application of bubbling in a CRT panel furnace, *Advances in Fusion and Processing of Glass II*, The American Ceramic Society, Westerville, OH, 1998, 251-256.

118. Hoke, B.C., Jr. and Marchiando, R.D., Using computational fluid dynamics models to explain operating condition effects on glass quality, *Advances in Fusion and Processing of Glass II*, The American Ceramic Society, Westerville, OH, 1998, 137-142.

119. Bauer, R.A. and Lankhorst, A.M., Advanced furnace design for energy efficient glass melting, *Proc. V Int. Semin. Mathematical Simulation in Glass Melting*, Horní Bečva, Czech Republic, 1999, 68-73.

120. Bauer, R.A. and Lankhorst, A.M., Advanced furnace design for energy efficient glass melting, *Proc. 5th ESG Conf. Glass Science and Technology for the 21st Century*, Prague, Czech Republic, 1999.

121. Slavejkov, A.G, Gosling, T.M., and Knorr, R.E., Jr., Low NOx Staged Combustion Device for Controlled Radiative Heating in High Temperature Furnaces, U.S. Patent 5,611,682, March 18, 1997.

122. Schill, P. and Vlcek, P., Mathematical model of radioactive waste glass melter, *Proc. V Int. Semin. Mathematical Simulation in Glass Melting*, Horní Bečva, Czech Republic, 1999, 112-118.

123. Fletcher, J.P., Peters, G., and Shome, B., Time-transient mathematical modeling of a fiber glass furnace, *Advances in Fusion and Processing of Glass II*, The American Ceramic Society, Westerville, OH, 1998, 177-185.

124. Schnepper, C., Jurcik, B., Champinot, C., and Simon, J.-F., Coupled combustion space-glass bath modeling of a float glass melting tank using full oxy-combustion, *Advances in Fusion and Processing of Glass II*, The American Ceramic Society, Westerville, OH, 1998, 219-224.

125. Khinkis, M.J. and Abbasi, H.A., Oxygen-enriched Combustion Method, U.S. Patent 5,203,859, April 20, 1993.

126. Li, X., Effect of oxygen enriched air staging on CO emission for an Owens-Brockway side-port regenerative container glass furnace, in *Final Report — Demonstration of Oxygen-Enriched Air Staging at Owens-Brockway Glass Containers*, GRI-97/0292, Gas Research Institute, Chicago, IL, October 1997.

127. Hoke, B.C., Jr., D'Agostini, M.D., Slavejkov, A.G., Lievre, K.A., and Horan, W.J., Goodbye LOX? *Ceramic Industry,* April, 2000, 71-78.

128. Hoke, B.C., Jr., D'Agostini, M.D., Slavejkov, A.G., Lievre, K.A., and Horan, W.J., Burner design developments for oxy/fuel furnaces, *Proc. Int. Commission on Glass Annual Meeting 2000, Glass in the New Millennium: Challenges and Break-through Technologies,* Amsterdam, May 2000.

11 Modeling Impinging Flame Jets

Charles E. Baukal, Jr.

CONTENTS

0-8493-2000-X/00/$0.00+$.50
© 2001 by CRC Press LLC

11.1　INTRODUCTION

Impinging flame jets have been extensively studied because of their importance in a wide range of applications. Figure 11.1 shows a flame impinging normal to a water-cooled plate.[1] The nomenclature is given at the end of this chapter. Early experiments were used to simulate the extremely high heat fluxes encountered by space vehicles re-entering the Earth's atmosphere. These heat levels

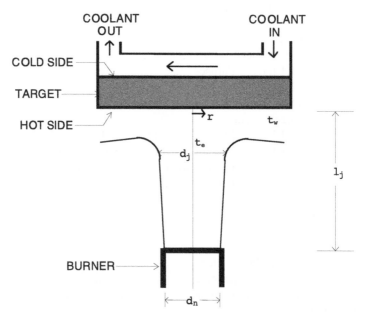

FIGURE 11.1 Flame impingement normal to a cooled target. (Adapted from C.E. Baukal, L.K. Farmer, B. Gehhart, and I. Chan, *1995 Int. Gas Research Conf.,* Vol. II, D.A. Dolenc, Ed., Government Institutes, Rockville, MD, 2277-2287, 1996.)

are caused by the hypersonic flow impact velocities that ionize the highly shocked atmospheric gases.[2] Subsequent studies have since investigated the heat fluxes attainable using high-intensity combustion, with pure oxygen instead of air, to increase metal heating and melting rates.[3] High-intensity impinging jet flames have been used in recent years to produce synthetic diamond coatings by chemical vapor deposition.[4] Supersonic-velocity, high-intensity flames have been used in a process known as thermal spallation. In this process, the impinging jet bores through rock by causing it to fragment, due to the large thermal stresses arising from the high heat fluxes on a cold surface.[5] This may be more rapid and economical than traditional mechanical rock drilling, depending on the rock type. High-velocity flames impinging on structural elements have been used to simulate large-scale fires caused by ruptured piping in the chemical process industry.[6] Low-intensity impinging flame jets have been used in safety research to quantify the heating rate caused by buoyant fires impinging on walls and ceilings.[7] Eibeck et al. (1993) studied the impact of pulse impinging flames on the heat transfer to a target.[8] Zhang and Bray (1999) studied the effects of the burner nozzle exit velocity and the distance between the impinging flame and a water-cooled plate positioned directly above the burner nozzle exit.[9] As shown in Figure 11.2, five different flame shapes were identified: (1) ring flame, (2) conic flame, (3) disc flame, (4) envelope flame, and (5) cool central core flame.

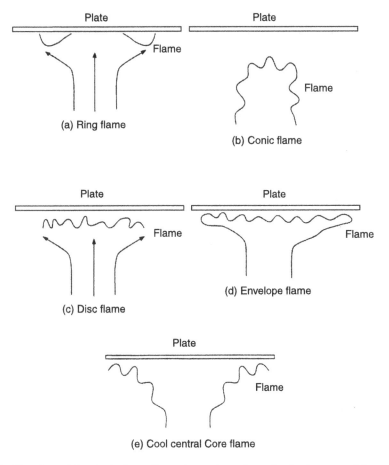

FIGURE 11.2 Sketches of five typical reacting patterns for a flame impinging normal to a water-cooled plate. (a) ring flame, (b) conic flame, (c) disc flame, (d) envelope flame, and (e) cool centrla core flame. (From Y. Zhang and K.N.C. Bray, *Comb. Flame,* 116, 672, 1999. With permisison.)

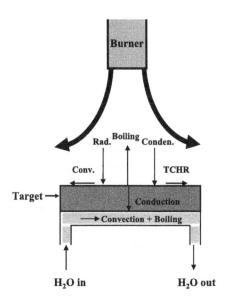

FIGURE 11.3   Potential heat transfer mechanisms in flame impingement on a water-cooled target. (From C.E. Baukal, *Heat Transfer in Industrial Combustion,* CRC Press, Boca Raton, FL, 2000.)

Most previous research has concerned air/fuel combustion. In these lower intensity flames, the predominant heat transfer mechanism is usually forced convection. Much work has also concerned fuels combusted with pure oxygen. These high-intensity flames produce significant amounts of dissociated species (e.g., H, O, OH, etc.) and uncombusted fuel (e.g., CO, H_2, etc.). These reactive gases then impact on a relatively low-temperature target surface. As these species cool, they exothermically combine into products such as CO_2 and H_2O, which are more thermodynamically stable at lower temperatures. This chemical heat release is sometimes referred to as "convection vivre," or live convection.[10] Here, that process is referred to as thermochemical heat release or TCHR. This mechanism may be comparable in magnitude to the forced convection heat transfer at the surface.[11] Such high-intensity flames require much more complicated analysis. Several experiments have used fuels combusted with an oxidizer having an oxygen content between that of air and pure oxygen. This process is sometimes referred to as oxygen-enriched air combustion.[12] These medium-intensity flames have received much less attention. They are not currently used in industrial applications.

Depending on the specific problem, many different types of heat transfer may be important. Six heat transfer mechanisms have been identified in previous flame impingement studies[1]:

- Convection (forced and natural)
- Conduction (steady-state and transient)
- Radiation (surface, luminous, and nonluminous)
- Thermochemical heat release (equilibrium, catalytic, and mixed)
- Water vapor condensation
- Boiling (internal and external).

These are shown schematically for a water-cooled target in Figure 11.3. All of the mechanisms are not usually present simultaneously and depend on the specific problem.

11.2   HEAT TRANSFER MECHANISMS

Three of the six heat transfer mechanisms identified in previous flame impingement studies are briefly considered here because of their importance in flame impingement modeling. These mechanisms include: forced convection, thermochemical heat release (TCHR), and thermal radiation.

11.2.1 FORCED CONVECTION

The impinging flame is an external flow with several flow regions including the free jet, the stagnation and the wall jet regions. Within a given region there may be several sub-regions. In the free jet region, the flow may be developing or fully developed depending on the distance from the nozzle exit. Ambient fluid entrainment is an important consideration, since there is commonly a large temperature difference between the flame jet and the ambient gases. In the stagnation zone, there are very large pressure gradients caused by the rapid deceleration of the impinging jet. In the boundary layer flow of the wall jet region, there may be laminar, transition, and turbulent subregions.

In many conventional furnace heating processes, forced convection is often only a small fraction of the total heat transfer to the product. Most of the heating comes from the radiation from the hot refractory walls. However, in flame impingement with no furnace enclosure, forced convection may be 70–90% of the total heat flux.[10,41] For flame temperatures up to about 1700K (2600°F), forced convection is the dominant mechanism in flame impingement heat transfer.[32]

For low temperature flames common in air/fuel combustion systems, forced convection has generally been the only mechanism considered. In highly dissociated oxygen/fuel flames, a large fraction of the heat release is from exothermic reactions. However, even for those flames, forced convection is still an important contributor to the overall heat transfer to the target.

The turbulence level directly affects the importance of forced convection. The flow regime is determined by the Reynolds number. There are many possible choices for the characteristic length l used to calculate the Reynolds number, including:

- the burner nozzle diameter (d_n),
- the axial distance from the nozzle exit to the target (l_j),
- the width of the jet at the edge of the stagnation zone (d_j), or
- some dimension of the target like the diameter (d_b) for a disk or cylinder or the radial distance from the stagnation point (r) for a plane surface.

Laminar flames have been used in many studies. Many of the semi-analytic heat transfer equations developed for flame impingement heating are for laminar flow normal to the stagnation point of an axisymmetric, blunt-nosed target (discussed later). Turbulent flames have also been commonly used. Some of the empirical heat transfer correlations incorporate the turbulence intensity Tu (discussed later).

11.2.2 THERMOCHEMICAL HEAT RELEASE (TCHR)

This mechanism refers to the energy release that occurs when hot and dissociated gaseous species cool down and exothermically recombine into more thermodynamically stable molecules. This process has been given many names including *chemical recombination* or simply *recombination*, *convection vivre* (which means live convection), *aerothermochemistry*, and the *exothermic displacement of equilibrium*. Here, the process is referred to as *thermochemical heat release* or TCHR. This name includes the aspects of thermodynamics, chemical reactions, and exothermic energy release. The other names do not explicitly indicate the importance that the exothermic heat release has on the heat transfer.

The exhaust products of combustion processes may contain dissociated species. The level of such dissociation increases with the flame temperature. When a flame impinges on a cool surface, these dissociated species diffuse in the direction of the concentration gradients toward the lower temperature regions. As the gases cool, the dissociated species exothermically combine with each other to form more stable molecules. The new components are thermodynamically favored at lower temperatures. For example, when CH_4 is combusted adiabatically with pure O_2, significant amounts of unburned fuel in the form of CO (16 vol%) and H_2 (7 vol%) are produced, along with radicals like O (4 vol%), H (5 vol%), and OH (9 vol%). As these combustion products cool to temperatures below about 1600K (2400°F), they react to form CO_2 and H_2O, while simultaneously releasing energy. However, when CH_4 is combusted with air, the adiabatic equilibrium combustion products are nearly all CO_2, H_2O, and N_2. There are only trace amounts of any dissociated species, due to the lower flame temperature. The large concentration of N_2 acts as a heat sink, resulting in a lower flame temperature.

TCHR becomes important when high temperature dissociated gases contact cooler bodies. One example is the catalytic reaction of hydrogen atoms to form stable H_2 molecules:

$$H + H + M \rightarrow H_2 + heat$$

where M is a third body. Giedt et al. (1960) studied fuel rich, turbulent, O_2/C_2H_2 flames flowing parallel to a flat plate.[2] It was estimated that H atom recombination increased the heating rate by 30 to 90% compared to forced convection alone. In high temperature flame impingement, the combustion products diffuse through the boundary layer to the colder surface where they exothermically react and form new species. Giedt found that two chemical mechanisms initiate the thermochemical heat release: equilibrium and catalytic. Nawaz (1973) referred to a third mechanism as *mixed flow*, which is a mixture of equilibrium and catalytic chemistries.[44]

11.2.2.1 Equilibrium TCHR

Equilibrium TCHR has also been referred to as a *homogeneous* effect. The gas phase chemical reactions occur in the boundary layer. Unstable species collide in the gas phase with other atoms or molecules that are the third bodies that initiate the reactions. The reaction time is much less than the time required for the gases to diffuse to the surface. Free radicals enter the laminar boundary layer by molecular diffusion. The diffusion rate is small compared to the chemical reaction rate. Therefore, there is a higher probability of homogeneous free radical chemical reactions because of the fast kinetics. Because of higher levels of dissociation, TCHR is very significant when O_2, instead of air, is used as the oxidizer. Kilham and Purvis (1978) tested mostly fuel rich, laminar, O_2/CH_4 and C_3H_8 flames impinging normal to a refractory plate.[74] Heat flux gages made of silicon carbide (nearly non-catalytic) and platinum (highly catalytic) were used to try to measure TCHR effects. No difference in heating rates was found, which led to the conclusion that the TCHR process occurred in the boundary layer, before the impinging gases reached the surface, so that it was an equilibrium effect. Many semi-analytic solutions for the heat transfer from the flame to the target have been recommended for this type of TCHR.[11] Some examples are given later.

11.2.2.2 Catalytic TCHR

Catalytic TCHR has also been called a *heterogeneous* effect. It involves chemical diffusion reactions at the target surface. The chemical reaction times are much greater than the transit time for the diffusing species to reach the surface. There is insufficient time for the radical species to react before reaching the surface. Turbulence and catalytic surface materials may accelerate these surface reactions. Some semi-analytic solutions for the heat transfer from high temperature gases impinging on a target have been given for this type of TCHR.[11]

11.2.2.3 Mixed TCHR

Mixed TCHR was a mechanism suggested by Nawaz[44] and is a combination of equilibrium and catalytic TCHR. Some of the dissociated species in the flame may react within the boundary layer prior to reaching the surface (equilibrium TCHR). Some of the species may react catalytically upon contact with the cool surface (catalytic TCHR). Some of the dissociated species may also remain unreacted, which would occur if either some of the gases did not reach the surface, or if the surface was not perfectly catalytic. No semi-analytical heat transfer solutions have been recommended for this type of TCHR.[11]

11.2.3 RADIATION HEAT TRANSFER

Nonluminous, luminous, and surface emission radiation have been considered in previous flame impingement studies and are briefly discussed next.

11.2.3.1 Nonluminous Radiation

The complete combustion of hydrocarbon fuels produces, among other things, CO_2 and H_2O. These gaseous products generate gaseous or nonluminous radiation, which is highly wavelength dependent and has been extensively studied. This heat transfer mode depends on the gas temperature, the partial pressure and concentration of each species, and the molecular path length through the gas. In some studies, this mode has been significant. Kilham (1949) tested nearly stoichiometric, laminar, air/CO flames impinging normal to an uncooled refractory cylinder.[31] Flame radiation was 5 to 16% of the total heat flux. Jackson and Kilham (1956) also tested laminar flames impinging normal to refractory cylinders.[32] A variety of fuels, oxidizers and stoichiometries were tested. The measured flame radiation was up to 5% of the total heat flux. Dunham (1963) tested nearly stoichiometric, laminar, air/CO flames impinging normal to the nose of a hemi-nosed cylinder.[124] The estimated nonluminous radiation was up to 13% of the total heat flux. Ivernel (1979) tested nearly stoichiometric, natural gas flames impinging normal to the nose of a hemi-nosed cylinder.[46] The calculated nonluminous flame radiation was up to 34% of the total heat flux.

11.2.3.2 Luminous Radiation

Luminous radiation is important when soot particles in the flame radiate approximately as blackbodies. This heat transfer mechanism is usually important when liquid and solid fuels, like oil and coal, are used. It is usually not significant for gaseous fuels like natural gas. You (1985) measured the radiation from pure diffusion flames ($\phi = \infty$). Radiation to the stagnation point was 13 to 26% of the total heat flux. Much of this radiation was likely to have been luminous for these very fuel rich flames. Radiation was negligible at the edge of the plate (R = 7.3). Hustad and Sønju (1991) calculated the radiant flux to be 7 to 14% and 20 to 40% of the total flux for CH_4 and C_3H_8 pure diffusion flames ($\phi = \infty$), respectively.[34] The majority of that radiation was likely luminous due to the very fuel rich flame conditions. Hustad assumed that these long flames were optically thick radiating cylinders. The convective flux was calculated using correlations for flow over a cylinder and was subtracted from the measured total flux to determine a calculated radiant flux. This calculated radiation compared favorably with the measured radiation.

11.2.3.3 Surface Radiant Emission

Beér and Chigier (1968) studied flame impingement inside a furnace and experimentally determined that radiation was at least 10% of the total heat flux to the target.[41] Although it was not specified how much of that radiation came from the hot walls and how much came from the flame, it is likely that much of that radiation was from the walls (surface radiation). Vizioz and Lowes (1971)[43] and Smith and Lowes (1974)[55] studied flame impingement on a flat plate in a hot furnace. It was determined that radiation and forced convection were of comparable magnitudes. Vizioz measured radiation to be 4 to 100% of the total heat flux. Smith calculated the surface radiant emission to the target, using Hottel's zone method, to be 30 to 43% and 10 to 17% of the total heat transfer to water-cooled and air-cooled flat plates, respectively. Matsuo et al. (1978) studied turbulent, preheated air/coke oven gas flames impinging on a metal slab inside a hot furnace.[48] The top of the slab was exposed to the impinging flame. The rest was exposed to the radiation from the furnace walls. Furnace radiation was the dominant mechanism for: 1) large L, 2) high t_w, and 3) large R. Ivernel (1979) calculated the radiation from hot furnace walls to a hemi-nosed cylinder to be as much as 42% of the total heat flux for impinging O_2/natural gas flames.[46] You (1985) measured the convective and total heat flux using gages plated with gold and black foils, respectively.[7] By subtracting the convection from the total heat flux, radiation was calculated to be up to 35% of the convective flux. Van der Meer (1991) estimated that the radiation from the hot inner refractory wall of a tunnel burner was up to 15% of the total heat flux to the target.[71]

11.3   INDUSTRIAL APPLICATIONS

Some of the more common industrial flame impingement applications, which have been modeled with CFD and are discussed later, are briefly presented next.

11.3.1   FLAME-JET DRILLING

There is often a need to get to resources (oil, natural gas, water, etc.) located well below the surface of the Earth. The most common way to do this has been with mechanical drilling equipment. While this has been successful in reaching the desired resources, the costs can be very high due to wear on the drill bit and the associated rig piping, as well as the large capital expense associated with the drilling rig and ancillary equipment. The harder the rock, the more costly the drilling due to more rapid bit wear and the more time it takes to bore the hole. While more advanced materials like synthetic diamonds and tungsten carbide have improved drilling rates and reduced bit wear, mechanical drilling in hard rock remains slow and costly.

Thermal drilling methods, sometimes referred to as thermal spallation, use high heat flux rates to cause rock to fracture so that the fragments can be easily removed. A variety of techniques have been used, including lasers, electron beams, plasmas, and flames. An important advantage to these techniques is that there is no direct contact between the heating equipment and the rock, so there is no mechanical wear of the components due to grinding on rocks as there is in mechanical drilling. Thermal drilling methods are particularly advantageous compared to mechanical methods when drilling through hard rock. If the hard rock properly fractures under high heat loads, then thermal drilling may be much less costly and be much faster than using drill bits that wear out rapidly. While there is still normal wear and tear of the equipment, thermal drilling rigs may last much longer than mechanical drilling rigs so there is often less downtime and maintenance. However, an important challenge for thermal drilling methods is the transport of the rock fragments out of the hole. In conventional mechanical drilling, water is often used to both flush out the fragments and cool the drill bit. The fragment transport can be a challenge in thermal methods for several reasons. Water cannot practically be used because of the danger of using it in proximity to electricity used in some of the thermal drilling techniques. Another problem is that it defeats the purpose of the high heating rates because it is cooling the rock while the thermal heating system is simultaneously trying to rapidly heat the rock. Because the thermal drilling system is not in direct contact with the rock, the water can act as a barrier between it and the rock, which can seriously reduce the drilling performance. Gaseous transport is more effective because it is safer, much less cooling on the rock, and much less of a barrier between the heat source and the rock. The flame-jet boring method automatically incorporates a gaseous transport system. Another challenge of some of the thermal methods can be the high operating cost for the power source.

A thermal drilling method that has low operating costs and a built-in rock transport system is referred to as flame-jet drilling, shown in Figure 11.4. In this technique, a very-high-intensity flame, often at supersonic velocities, impinges directly on the rock. The bulk of the rock is relatively cool, while the surface is very hot in the vicinity of the impinging flame. This produces very high thermal stresses in the rock, causing it to fracture. No separate fluid transport system is needed because the naturally produced combustion products carry away the fractured or spalled rock fragments.

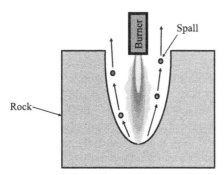

FIGURE 11.4   Elevation view of the thermal spallation drilling process.

Because of thermal expansion, the hot combustion products maintain high velocities that improve the effectiveness of the spall transport. This method is still relatively new, so research is needed to determine what drilling rates are possible through the various types of rock and to better understand the physics of the process so it can be optimized. CFD modeling can be an effective tool to help understand and optimize flame-jet drilling.

11.3.2   CHEMICAL VAPOR DEPOSITION

Chemical vapor deposition (CVD) is a process used to grow a wide variety of materials such as thin films, coatings, wafers, or optical fibers on various substrates. The films may be semiconductors, optical fibers, or synthetic diamonds. The substrate is often a flat disk that may be cooled and may be rotating to achieve more uniform deposition. The appropriate gases are impinged on the substrate, where the resulting chemical reactions in the presence of heat produce surface deposition. The heat for the reactions usually comes from either a heated substrate or from exothermic reactions in the gaseous phase, prior to impingement on the substrate. The latter is of interest here, where a flame impinges on the substrate that is often cooled.

An example of a CVD flame process is synthetic diamond production.[4,13,14] Thick diamond wafers can be used as infrared windows. Diamond coatings are used to improve the wear charac-teristics of machinery and equipment such as drill bits or cutting blades. The flames in the process are commonly produced by combusting hydrocarbons (often acetylene mixed with H_2) with pure oxygen under fuel-rich conditions, often at low pressures (<100 torr). The pure oxygen, instead of air, produces much higher flame temperatures that aid in the chemical reactions. A premixed flame from a welding torch is often used. The combustion products are impinged on a water-cooled substrate. There are many factors that affect the quantity and quality of the deposited film; these include the fuel flow rate, the ambient pressure, the substrate temperature, the distance between the burner and the substrate, the fuel composition, the mixture ratio between the fuel and the oxidizer, and the burner geometry.

Another example of a flame CVD process is in the production of optical fibers manufactured from high-temperature gas-phase synthesis techniques.[15] The optical fiber is drawn from silica preforms produced in particle deposition processes known as outside vapor deposition (OVD) and vapor axial deposition (VAD). A hydrocarbon fuel such as CH_4 is combusted with an oxidizer, often pure O_2, in the presence of silicon-containing gases like $SiCl_4$. This is different from the diamond synthesis process in that the goal is to produce silicon particles, not carbon-containing particles that would contaminate the optical fiber. Therefore, the combustion process here is designed to completely combust the fuel so no soot particles are generated and so that the heat can be used to promote the production of SiO_2 particles for deposition on the substrate. The major mechanism controlling the rate of deposition on the substrate is the thermophoretic effect.[16,17]

There are two major concerns in the CVD process: (1) the deposition rate, which is related to the process economics, and (2) the properties of the deposited film. The two are interrelated. If the deposition rate is too high, then the quality of the film often suffers. At slower deposition rates, the film quality may improve, but the process economics suffer. The film properties include things like the thickness, composition homogeneity, and morphology. The gas-phase transport is a large factor in determining the quality of the deposited film.

Flame CVD processes are very complicated and difficult to experimentally study due to the high temperatures and high gradients. They include fluid dynamics in free space and in boundary layers, thermophoretic particle deposition, forced convection and radiation heat transfer to a surface, thermal conduction through the film and the substrate, multiphase flow of particles generated in the flame and combustion product gases, and numerous chemical reactions. CFD modeling can be an effective tool for improving the understanding of this complicated process and for optimizing it by maximizing the deposition rate with acceptable film properties.

11.3.3 METAL REHEATING

One common use of impinging flames is to heat metal parts. This process is important in rolling mills and forge shops. Metal is heated to the point where it becomes plastic enough that it can be rolled or forged into the desired shape.[18] The raw materials are supplied as large castings or ingots, usually in the form of bars or slabs. In a rolling mill, these heated castings are commonly rolled into sheet products. In a forge shop, the heated castings are machine-worked into the final product.

There have been two conventional methods for reheating metals. The first is commonly referred to as a reheat furnace. Here, the parts pass through a refractory-lined furnace. The parts are heated primarily by radiation from the furnace walls. These walls are heated by burners, using a fossil fuel combusted with air. The heating costs are generally low. A wide range of part sizes and shapes can be heated. However, the thermal efficiency is low. Also, the times, to both start-up the furnace and to heat the parts, are long. The second method is referred to as electric induction heating. Here, the parts pass through electric coils. A magnetic field is produced around the parts. The induced electrical current heats the parts. Electric induction heating is known for very fast heating times. No time is required to heat a furnace. However, both the equipment and the operating costs are high. Also, the heating system must be readjusted whenever the part shape changes, to optimize the magnetic field.

The ultimate goal of any heating process is to heat the load — not the furnace. Therefore, it is desirable to use direct flame impingement on the load. In some cases, this is not possible. The flame may contaminate or damage the product. However, in metal heating, this is usually not the case. The metal can withstand direct flame impingement, as long as the surface temperature does not exceed the melting point of the metal. A schematic of direct flame impingement in a continuous reheat furnace is shown in Figure 11.5.

British Gas has studied rapid metal heating for some time (e.g., Masters et al., 1971[19]). It has funded most of the work done in the U.K. related to this technology. Many benefits were identified: product quality improved; throughputs increased; less material was at risk during an unscheduled stoppage; and less floor space was required for the heating equipment. Hara et al. (1979) showed that flame impingement heating may eliminate low-temperature stripes, known as skid marks.[20] These are caused by the contact of the hot metal slabs with the cooled transfer skids. Rapid heating allows more flexibility in switching to different types of slabs. This is because the delay time in the reheat furnace is greatly reduced. Jayaraman et al. (1984) discussed the application of flame impingement heating in a strip heating furnace.[21] Compared to conventional reheat furnaces, the heat transfer rate increased by 60%, at a 10% higher line speed using direct flame impingement. The fuel consumption was reduced by 36%. Galvin (1991) stated that lower fuel consumption and

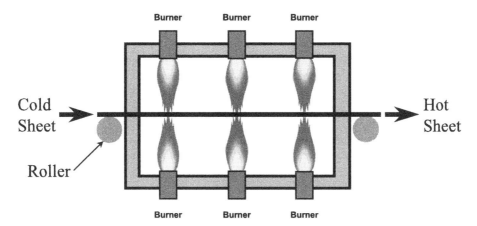

FIGURE 11.5 Elevation view of flame impingement inside a hot continuous reheat furnace.

higher thermal efficiency result with rapid heating.[22] Farmer et al. (1994) noted some additional benefits for direct flame impingement, rapid heating systems, including less maintenance on the rolls in the rolling mill, lower noise, and a more compact size.[23]

The Gas Research Institute (Chicago, IL) funded a project to develop a rapid metal heating system.[24] The process involved the direct impingement of oxygen/natural gas flames onto metal parts. The goal was to have the fast heating times of electrical induction, with the low costs of fossil-fuel-fired reheat furnaces. This a challenging new technology that would benefit from the use of CFD modeling to better understand the physics in this high-temperature and high-intensity heating process.

11.3.4 FLAME IMPINGEMENT ON A TANK

Fires are potentially very serious threats in chemical plants and petrochemical refineries that contain large quantities of flammable liquids and gases. Therefore, much effort has been devoted to preventing these fires with a combination of experimental and numerical simulations. One area of interest is the effect of flame impingement on surrounding flammable storage tanks that could lead to an explosion. The impinging flame could cause the unwanted rupture of a pipe and the subsequent ignition of flammable liquids or gases flowing through that pipe. To protect against such a potentially catastrophic event, sufficient models need to be developed to predict the consequences of a flame impinging on a flammables storage tank.

Cowley and Pritchard (1990) studied the heat transfer from large-scale natural gas flames impinging on structures.[25] Propane and natural gas flames were studied with firing rates of up to 1100 and 560 MW (3.8 and 1.9×10^9 Btu/hr), respectively. Flames were impinged on a 2 m (6 ft) diameter cylindrical tank and a 900 mm (35 in.) diameter pipeline section. Flame shape, heat flux, and radiation were studied to determine their impact on the tank and pipe section. The heat flux distribution was found to be complex and dependent on the flame type and relative position. Cracknell et al. (1995) studied the heat flux from a large-scale natural gas flame impinging on a cylindrical tank.[26] Four different types of flames were tested, ranging from subsonic to supersonic gas release velocities. Two tank diameters of 0.94 and 2.17 m (3.1 and 7.12 ft) were used as targets. The distance from the gas discharge point to a target ranged from 9 to 28 m (30 to 92 ft). Temperatures and heat fluxes were measured at the tank targets. Empirical correlations were developed for the heat flux distribution on the tanks. The predicted results compared favorably with the experimental data. Goose (1995) discussed the development of a model to predict explosions caused by jet-flame impingement on a tank containing flammable fluids.[27] The model used data from the literature for the heat fluxes produced by large-scale jet flames impinging on vessels.

11.4 IMPINGING FLAME-JET STUDIES

Many important parameters arise in flame-jet impingement processes.[28] The first and most important aspect is the overall geometric configuration. This includes the target shape and its orientation relative to the burner. These operating conditions strongly influence the heat transfer intensity. They also determine which mechanisms will be most important. These conditions include the oxidizer composition, the fuel composition, the equivalence ratio, and the Reynolds number at the nozzle exit. Other factors commonly have secondary influences on the heat transfer processes. These include the burner design and the position relative to the burner. The characteristics of the target also influence the heat transfer. These include the dimensions, material composition, surface treatments or coatings, and the surface temperature. The above parameters are briefly discussed below. A more comprehensive discussion is given elsewhere.[29]

11.4.1 GEOMETRIES

There are several aspects of the geometry that are important. The first is the surrounding environment: inside a furnace or in the ambient environment. The next is the target geometry, which specifies the relative orientation of the target surface and the burner. This is often the most important consideration for the designer or researcher.

11.4.1.1 Environment

Industrial flame impingement studies have been typically done in three types of environments. The first is inside a hot, refractory-lined furnace. The second is in an ambient environment with no confinement. The third type is in a combination of the two where there is some confinement that influences the flame shape and gas flow. Other flame impingement studies, to simulate the processes in internal combustion engines, have been done in pressurized environments. These are not discussed here as they are not considered to be industrial combustion processes. The environment can have a significant impact on the predominant mechanisms in a given flame impingement process. The environments are briefly discussed next.

11.4.1.1.1   Furnace

Impinging flames located inside a furnace are significantly impacted by both the confinement of the enclosure and by the radiation from the hot refractory walls. A cartoon of this is shown in Figure 11.6. This type of environment is common in certain types of metal reheating furnaces where

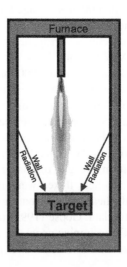

FIGURE 11.6   Elevation view of flame impingement inside a hot furnace.

FIGURE 11.7 Flame impinging normal to a cylinder in crossflow. (From C.E. Baukal, *Heat Transfer in Industrial Combustion,* CRC Press, Boca Raton, FL, 2000.)

products are heated by direct flame impingement (see Figure 11.7). Some of these have been modeled and are discussed later. The gas flow from the burner to the target can be influenced by currents in the furnace caused by multiple burners exhausted to a common outlet. The burners closest to the outlet see all of the exhaust products from the rest of the burners, while the burners farthest from the outlet may be little disturbed. The impinging flames often entrain some of the surrounding combustion products, which can improve the heating rate and thermal efficiency caused by the internal furnace gas recirculation.

The hot refractory walls in the furnace may play an important role in the heat transfer to the target. For flames in open air, there is essentially no contribution to the heat transfer from the surrounding environment. In impinging flames in a furnace, the radiant heat transfer from the walls to the target may be significant, depending on the wall temperature and how well the target sees the walls. This complicates the analysis and computer modeling.

11.4.1.1.2 No enclosure

Many industrial impinging flames have no enclosure and are in ambient air. One example is the simulation of flame impingement from a ruptured pipe onto surrounding pipes or vessels in a chemical plant. Another example is flame impingement in a large room or building onto a wall or ceiling caused by an unwanted fire. These geometries are generally simpler to analyze because it is assumed there is no disturbance of the gas flows in the vicinity of the burner and target and because there is no heat transfer contribution from the surrounding environment. The main influence from the ambient environment is by entrainment into the impinging flame, which cools the flame and can quench some of the chemical reactions on the periphery of the flame. This geometry has been the most commonly studied experimentally. Some CFD modeling has been done for this type of environment and is discussed later.

11.4.1.1.3 Partial confinement

The main industrial example of partial confinement of impinging flames is in thermal spallation used to bore holes in rock, as shown in Figure 11.4. The target is the rock, which is in the general shape of a cylinder closed at one end with a curvature away from the impinging flame. The diameter of the hole is a function of the burner design. To further complicate this problem, the rock chips are carried away with the combustion exhaust products. The surfaces in the hole are not uniform due to the nature of rock. The bottom of the hole is moving as the rock is fragmented and carried away. The gas flows are strongly influenced by the confinement from the hole and there is some radiation from the side walls. This geometry has been modeled with CFD and is discussed later.

11.4.1.2 Targets

The four most common geometric configurations in flame jet experiments have been flames impinging (1) normal to a cylinder in crossflow, (2) normal to a hemi-nosed cylinder, (3) normal to a plane surface, and (4) parallel to a plane surface. Other configurations have been tested, including flames at an angle to a plane surface and flames normal to and around the circumference of a large cylindrical furnace.[30] These are not common geometries and little information is available on them. Flame impingement on rock in a bore hole is a specific geometry that has been studied for thermal spallation drilling. It is not specifically discussed in this section, but is treated later.

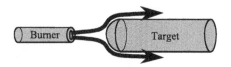

FIGURE 11.8 Flame impinging normal to a hemi-nosed cylinder. (From C.E. Baukal, *Heat Transfer in Industrial Combustion,* CRC Press, Boca Raton, FL, 2000.)

11.4.1.2.1   Flame normal to a cylinder in crossflow

In this configuration, shown in Figure 11.7, the cylinder axis is perpendicular to the burner axis. It applies to industrial processes such as heating round metal billets, and in fires impinging on pipes in chemical plants. Some of the earliest studies investigated impingement on refractory cylinders.[31,32] A pipe was the most common target for this geometry (see Reference 33, for example). In some studies, the local heat flux was measured at the forward stagnation point of the pipe (e.g., References 6, 34, and 35). In many studies, the pipe was cooled with water circulating through it.[36] The average heat flux, over the entire surface, was calculated from the sensible energy gain of the cooling water.

11.4.1.2.2   Flame normal to a hemispherically-nosed cylinder

For this geometry, shown in Figure 11.8, the cylinder axis is parallel to the burner axis. The flame impinges on the end of the cylinder, which is hemispherical. These tests have been very important in aerospace applications. This is a relatively uncommon geometry for industrial application, but it is important because the heat transfer results can be directly compared to some of the analytic solutions (see Reference 37) derived for aerospace applications, such as rockets and missiles. In this configuration, the heat flux has been measured only at the forward stagnation point. Most such studies have concerned laminar flames.

11.4.1.2.3   Flame normal to a plane surface

This configuration, shown in Figure 11.9, has received the most attention, because it has been widely used in many industrial processes. It has also been used for the widest range of operating and target surface conditions. Shorin and Pechurkin (1968)[38] and Kremer et al. (1974)[39] also investigated the effect of flame impingement, at an angle between parallel and normal, onto plane surfaces. Only the results for flame impingement normal to a plane surface are included here. The reader is referred to the literature for further information on angled jets.

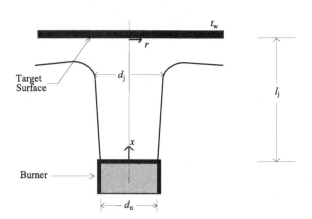

FIGURE 11.9 Flame impinging normal to a plane surface. (From C.E. Baukal, *Heat Transfer in Industrial Combustion,* CRC Press, Boca Raton, FL, 2000.)

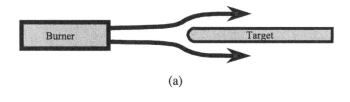

(a)

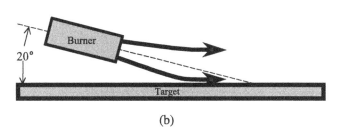

(b)

FIGURE 11.10 Flames impinging (a) parallel and (b) oblique to a plane surface. (From C.E. Baukal, *Heat Transfer in Industrial Combustion*, CRC Press, Boca Raton, FL, 2000.)

11.4.1.2.4 Flame parallel to a plane surface

This configuration, shown in Figure 11.10, has been the least studied. However, it is very important for flight applications. It simulates the heat transfer processes on airfoil surfaces. The studies by Giedt and co-workers (1960, 1966)[2,40] investigated flow across the top and bottom of a plane surface. Neither of these studies investigated the heat flux at the leading edge of the target. Beér and Chigier (1968)[41] studied impingement only on one side of a plane surface, which was the hearth of a furnace. The flame was inclined at 20° above the horizontal (see Figure 11.10b). Mohr et al. (1996) studied a special configuration for flames parallel to a plane surface, which they called radial jet reattachment flames as shown in Figure 11.11.[42] This type of flame promises to give more uniform heating of the surface compared to flames impinging normal to a surface.

11.4.2 OPERATING CONDITIONS

The operating conditions have been found to strongly influence the heat transfer intensity. The conditions include the oxidizer and fuel composition, the flame equivalence ratio and firing rate, the Reynolds number at the nozzle exit, the burner type, the nozzle diameter, and the location of the target with respect to the burner.

The most important variable, after the physical configuration, is the oxidizer composition. The oxygen mole fraction in the oxidizer, Ω, has a very large influence on heat transfer intensity. Almost all of the previous studies used either air ($\Omega = 0.21$) or pure oxygen ($\Omega = 1.0$) as the oxidizer. This affects both the flame temperature and the amount of dissociation in the combustion products. As an example, the adiabatic flame temperature for methane combusted stoichiometrically with air and with pure oxygen is 2220K and 3054K (3537 and 5038°F), respectively. The products of combustion for a stoichiometric air/CH_4 adiabatic flame contain essentially no unreacted fuel or dissociated species, except at very high flame temperatures approaching adiabatic conditions. However, the adiabatic equilibrium combustion products of an O_2/CH_4 flame contain nearly 23 vol.% unreacted fuel (CO and H_2), and over 18 vol.% dissociated species (H, O, and OH). As these products cool down in the boundary layer along the target surface, they exothermically release

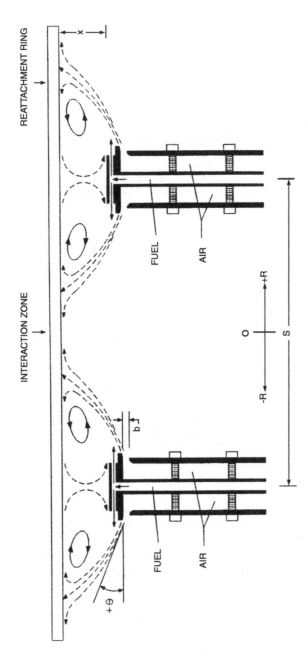

FIGURE 11.11　Pair of radial jet reattachment flames (courtesy of ASME, New York). (From J.W. Mohr, J. Seyed-Yagoobi, and R.H. Page, in Fire and Combustion, *ASME Proc. 31st Natl. Heat Transfer Conf.*, Vol. 6, McQuay et al., Eds., HTD-Vol. 328, ASME, New York, 1996, 17. With permission.)

heat. Some studies have used oxygen concentration levels between those of air and pure oxygen.[36,41,43-49]

Another parameter of interest is the fuel composition. Only studies using gaseous fuels are considered here. Natural gas and methane, the main constituent in natural gas, have been the most widely used. Some of the uncommon fuels that have been used include acetylene,[50] MAPP® gas,[51] coke oven gas,[41] town gas,[38,52] and a mixture of propane and butane in unspecified proportions.[38]

The equivalence ratio directly affects both the sooting tendency and the level of dissociation in the combustion products. Fuel rich flames ($\phi > 1$) produce a combination of both luminous and nonluminous thermal radiation. The combustion products of these flames may also contain unreacted fuel components, due to insufficient oxygen. Fuel-lean flames ($\phi < 1$) normally do not produce luminous thermal radiation, due to the absence of soot particles. These flames seldom produce significant quantities of unreacted fuel species, unless the flame temperature is high enough to produce dissociation. Flames at or near stoichiometric ($\phi = 1$) produce the highest flame temperatures, due to complete combustion. They also generally produce only nonluminous radiation, because no soot is generated. Most of the studies used stoichiometric mixtures. Some have used equivalence ratios as fuel-lean as $\phi = 0.5$.[52-54] Other studies have used pure fuel jets ($\phi = \infty$).[6,7,10,34,35] Many studies have investigated a range of equivalence ratios, in order to determine the resulting effects on heat transfer to the target.

The firing rate or gross heat release (q_f) of the flames has ranged from 0.3 to 3000 kW (1×10^3 to 1×10^7 Btu/hr). Large industrial-scale flames have been used by researchers at the International Flame Research Foundation (IFRF) in IJmuiden, The Netherlands.[41,43,55,56] Many of the studies (e.g., Reference 57) used torch tips, firing at under 50 kW (170,000 Btu/hr).

The Reynolds number at the burner nozzle, Re_n, varied from 50 to 330,000. Both laminar (e.g., Reference 45) and turbulent (e.g., Reference 58) flow conditions arose. The Reynolds number varies directly with the burner diameter. It is also influenced by the burner design. In partially or fully premixed flames, the combustion products leave the burner at an elevated temperature. However, in diffusion flames, the gases leave the burner at essentially ambient conditions. Because the gas viscosity increases with temperature, Re_n is generally lower if the gases have been heated at the burner exit. In some studies, turbulence effects were analyzed (e.g., Reference 59). However, in most studies, the turbulence effects were not assessed.

Many different types have been used, ranging from fully premixed to diffusion mixing downstream of the burner exit. In fully premixed burners (e.g., Reference 60), the fuel and oxidizer mix prior to reaching the nozzle exit. The resulting flame in premix burners may have either a uniform or a nonuniform velocity profile, depending on the nozzle design. It also depends on the distance between the ignition point and the exit. In most studies using this burner type, both the temperature and the composition of the combustion products at the burner exit were approximately uniform. The tunnel burner is a common, fully premixed burner. The gases are mixed and ignited inside the burner and travel through a refractory-lined chamber before leaving the burner. The temperature and composition are fairly uniform at the exit. The velocity profile may not be uniform and may be approximately developed pipe flow, depending on the downstream length of the equilibration chamber. In partially premixed burners, the fuel and oxidizer mix prior to reaching the nozzle exit. However, only a portion of the stoichiometric amount of oxygen is supplied through the burner. The remainder is provided by mixing with the surrounding ambient air, which is entrained into the flame. At the nozzle exit, the velocity profile is commonly nonuniform. Both uniform and nonuniform outlet temperature profiles and compositions have been reported. In diffusion-mixing burners, the fuel and oxidizer begin to mix at the nozzle exit, where the velocity is often nonuniform. An example is a glass working torch.[61] In diffusion burners, the exit temperature field is commonly homogeneous and equal to ambient conditions. The gas composition at the exit is pure fuel and pure oxidizer, with no combustion products. If the oxidizer is not supplied through the burner, a pure diffusion flame results. The oxygen is provided for combustion by ambient air entrainment into the flame.

The exiting flame shape was round in all but two studies[31,36] where the flames were slot-shaped. The burner nozzle diameter, d_n, ranged from 0.53 to 152 mm (0.02 to 6.0 in.). This dimension was not given in some of the studies. In many studies (e.g., Reference 3), the burner outlet consisted of a nested bundle of small tubes or orifices, arranged in a circular pattern. This is common practice for nozzles used on cutting and welding torches.

11.4.3   STAGNATION TARGETS

The important features of the targets include the dimensions, composition, surface conditions, and temperature. These properties vary widely among the studies reviewed here. In some studies, these were varied to examine the effect on the heat transfer mechanisms.

11.4.3.1   Size

The targets ranged in size from a 0.56 cm (0.22 in.) o.d. disk to a 200 × 625 cm (6.6 × 20.5 ft) furnace hearth. Most of the cylindrical targets have been hollow pipes, while some have been solid refractory rods[31,32] or solid steel cylinders.[6,34,35] All of the hemi-nosed cylinders have been under 22 mm (0.87 in.) o.d., except in one study[62] which were from 50 to 150 mm (2 to 6 in.). Most of the plane surfaces have been disks of over 16 mm (0.63 in.) in diameter, except for the 0.56 mm (0.022 in.) o.d. disks used by Horsley et al. (1982).[63,64] In addition to disks, many studies used square and rectangular plates as the target surfaces.

11.4.3.2   Target Materials

The most commonly used materials were aluminum, copper, brass, and stainless steel. For example, Posillico (1986) studied flames impinging normal to uncoated aluminum disks.[65] Aluminum, copper, and brass have very high thermal conductivities. Therefore, they are easier to water-cool, because the heat from the impinging flame is quickly conducted away. However, these materials also have relatively low melting points. For that reason, they have not been used at very high surface temperatures. Refractories, including alumina (Al_2O_3) and sillimanite (Al_2SiO_5), were also used.

11.4.3.3   Surface Preparation

Most of the target surfaces were untreated. However, in some experiments, the surfaces were treated or coated to study a specific surface effect. Kilham[31] and Jackson and Kilham[32] coated the surface of the refractory cylinders with different oxides. The objective was to estimate the emissivities of the coatings. The surface temperature was approximately determined using a thermocouple imbedded inside the cylinder body. Radiation from the surface was measured with a thermopile. Using the temperature level and the radiation, along with an energy balance on the cylinder, the emissivity of the coatings was calculated as a function of temperature.

Four studies used coatings to determine the effects of surface catalysis on the heat transfer from an impinging flame. Giedt et al.[2] used both uncoated and porcelain-coated iron plates. These surfaces were considered to be catalytic and noncatalytic, respectively. The measured heat flux for these two surface conditions was about the same. It was concluded that surface chemical recombination effects were negligible for those test parameters. Kilham and Dunham (1967)[66] used three different surface conditions to study the effects on air/CO flames impinging on a hemi-nosed cylinder. The results indicated that the lowest heat flux was for the National Bureau of Standards Coating A418. This coating is a mixture of eight oxides. It was assumed to be noncatalytic. On the other hand, the highest flux was found for a platinum coating, which was considered to be perfectly catalytic. The heat flux to an uncoated copper heat flux gage surface was intermediate between the results for the A418 and platinum coatings. However, the differences were small. It was concluded the bulk of the chemical reactions occurred in the boundary layer. In addition, as

L increased, the heat flux for all three surface conditions converged to the same value. Conolly and Davies[3,67] found no difference in heat flux between a surface coated with highly catalytic platinum and a surface coated with SiO_2, which was expected to be noncatalytic. This again led to the conclusion that the recombination of radical species occurs in the boundary layer, prior to reaching the surface. Baukal and Gebhart (1997) also found no difference in heat flux to surfaces coated with alumina or platinum.[68]

You[7] used a high-emissivity (0.96) paint, to maximize the amount of radiant heat absorbed by the target. Three interrelated studies[69-71] used polished copper surfaces to reduce the effects of flame radiation. In some studies, the heat flux gage surface was coated or treated.

11.4.3.4   Surface Temperatures

Target surface temperatures ranged from 290 to 1900K (63 to 2960°F). For many measurements, the target temperature, t_w, was maintained below 373K (212°F) using water-cooled targets. In some studies (e.g., Hargrave and Kilham, 1984[59,72]), the surface temperature was slightly above 373K (212°F). This eliminated the possibility of combustion products condensing on the target. Ethylene glycol was used as the target coolant because it has a higher boiling temperature than water. Beér and Chigier[41] and Vizioz and Lowes[43] measured surface temperatures above 1200K (1700°F) for refractory targets. In some studies, the surface temperature level was actually for the heat flux gage, and not the target. For example, Fairweather et al.[45] reported a maximum surface temperature of 1600K (2400°F). However, the target was made of brass, which melts at about 1300K (1900°F). A stainless steel heat flux gage, imbedded in the brass target, was used to measure the heat flux. Stainless steels have a melting point of about 1700K (2600°F).

11.5   SEMI-ANALYTICAL HEAT TRANSFER SOLUTIONS

Heat transfer from high-temperature gases to axisymmetric and blunt-nosed bodies has been studied for many years. These processes are very important in aerospace applications. Aerospace vehicles, such as rockets and missiles, travel at high supersonic velocities. Commonly, the nose of these surfaces is axisymmetric, blunt, and rounded. Gas shock-waves are produced as the vehicles travel through the atmosphere. The resulting temperatures at the stagnation point are generally high enough to cause the atmospheric gases to dissociate into many chemical species. Very high heat fluxes arise in that region. Several semi-analytic solutions have been proposed for calculating such fluxes. Those solutions were derived from the laminar, two-dimensional, axisymmetric, boundary layer equations applied in the stagnation region. The equations were simplified using similarity flow formulations. In the resulting heat flux solutions, a constant was numerically determined. Therefore, the solutions are referred to as semi-analytic.

The heat transfer from impinging, chemically active flames has also been extensively studied. The experimental conditions and measurements for those studies have been previously reviewed[73] and were discussed above. In twelve of those studies,[3,44-46,53,57,59,62,63,66,71,74] the measured heat flux was compared against one or more semi-analytic solutions. The stagnation body has commonly been a hemispherically nosed cylinder (see Figure 11.12). This geometry has been used because of its similarity to the shape of aerospace vehicles. Then, the semi-analytic heat transfer solutions derived for aeronautical applications were used to model the measured heat fluxes in energetic flame impingement. The applicability of those equations to flame heating applications has been determined. In aerospace applications, the vehicle moves through stagnant atmospheric gases. In impinging flames, the combustion products move around a stationary target. Therefore, the relative motion is similar in both applications.

Chen and McGrath (1969)[75] reviewed impingement heat transfer from combustion products containing dissociated species. Sample heat transfer calculations were given for a stoichiometric O_2/C_3H_8 flame impinging on a 2 cm (0.8 in.) o.d. sphere. The equations recommended by McAdams

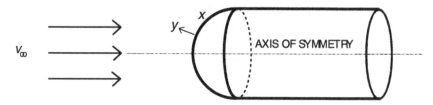

FIGURE 11.12 Stagnation flow around a hemi-nosed, axisymmetric, body of revolution. (From C.E. Baukal, *Heat Transfer in Industrial Combustion,* CRC Press, Boca Raton, FL, 2000.)

(1954),[76] Altman and Wise (1956),[77] and Rosner (1961)[78] were used. The Lewis number, Le, is the ratio of the mass diffusion rate to the thermal diffusion effect. The results were evaluated for both Le = 1 and Le > 1. It was concluded that, for Le = 1, the existing information was sufficient to adequately predict heat transfer in chemically reacting systems. However, for flows in which Le > 1, more analytical and experimental work was recommended. Such systems are important for high-temperature flames, where considerable dissociation occurs.

Two types of heat transfer behavior have been compared with experimental data from the studies considered here. The first was forced convection with no chemical dissociation. This is applicable in lower temperature-level flame impingement. The second was forced convection with dissociation. The heat transfer relations for the second type have been variations of those recommended for the first type. These equations are useful in predicting the heat transfer in the absence of experimental data.

11.5.1 Equations

The thermophysical properties and the stagnation velocity gradient (β_s) have been used in all of the semi-analytic solutions. Many methods have been used to calculate these parameters. Those methods are discussed here, prior to presenting the semi-analytic solutions.

This chapter section discusses the early semi-analytic solutions for the heat transfer in stagnation flows. Sibulkin (1952),[37] Fay and Riddell (1958),[79] and Rosner (1961)[78] developed equations to compute the heat flux at the stagnation point of an axisymmetric body in a uniform, external, steady flow. Radiation effects were ignored. Sibulkin's equation included only the forced convection effect. Fay and Riddell and Rosner developed solutions for both equilibrium and catalytic TCHR.

11.5.1.1 Sibulkin's Equation

The heat transfer at the forward stagnation point of a body of revolution was considered (see Figure 11.13). The flow was uniform, except in the boundary layer. The flow around the body, in the boundary layer, was assumed to be laminar, incompressible, axisymmetric, and of low speed. Using the axisymmetric boundary layer equations, the following relation for the local surface heat transfer was given[37]:

$$q_s'' = 0.763\left(\beta_s \rho_e \mu_e\right)^{0.5} \Pr_c^{-0.6} c_{p_e}\left(t_e - t_w\right) \tag{11.1}$$

This applies for $0.6 < \Pr_e < 2.0$. This formulation was actually developed for the hypersonic velocities common to space vehicle re-entry. It was assumed that the velocity is very low behind the bow shock-wave near the stagnation point, due to boundary layer flow conditions. The constant 0.763 was determined numerically. All other semi-analytic solutions presented here are based on Equation (11.1).

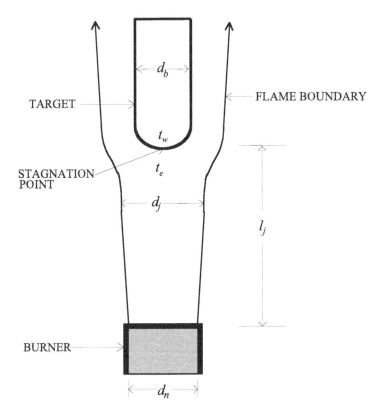

FIGURE 11.13   Flame impingement on a hemi-nosed cylinder. (From C.E. Baukal, *Heat Transfer in Industrial Combustion,* CRC Press, Boca Raton, FL, 2000.)

11.5.1.2   Fay and Riddell's Equation

Fay and Riddell used the same axisymmetric boundary layer equations as Sibulkin. However, chemical dissociation was included. The driving force for heat transfer was enthalpy, instead of temperature. A factor was added, that contains the ratio of $\rho\mu$, evaluated at the wall and at the edge of the boundary layer. The heat transfer at the stagnation point for equilibrium TCHR was given as[79]:

$$q''_s = 0.76\left(\beta_s\rho_e\mu_e\right)^{0.5}\left(\frac{\rho_w\mu_w}{\rho_e\mu_e}\right)^{0.1}\mathrm{Pr}_e^{-0.6}\left[1+\left(\mathrm{Le}_e^{0.52}-1\right)\frac{h_e^C-h_w^C}{h_e^T}\right]\left(h_e^T-h_w^T\right) \tag{11.2}$$

The heat transfer at the stagnation point for catalytic TCHR was given as:

$$q''_s = 0.76\left(\beta_s\rho_e\mu_e\right)^{0.5}\left(\frac{\rho_w\mu_w}{\rho_e\mu_e}\right)^{0.1}\mathrm{Pr}_e^{-0.6}\left[1+\left(\mathrm{Le}_e^{0.63}-1\right)\frac{h_e^C-h_w^C}{h_e^T}\right]\left(h_e^T-h_w^T\right) \tag{11.3}$$

The only difference between the equations is the exponent of the Lewis number, Le_e.

11.5.1.3   Rosner's Equation

In Rosner's formulation, the boundary layer equations were not solved directly, as had been done by Sibulkin and by Fay and Riddell. Instead, Fay and Riddell's equations were modified to include the effects of chemical dissociation. The resulting equation was then the sum of a forced convection term, from Sibulkin, and a diffusion-chemical reaction term. Two different forms of the solution were given, depending on the nature of the thermochemical heat release.[78] For equilibrium TCHR, the following form was recommended:

$$q_s'' = 0.763(\beta_s \rho_e \mu_e)^{0.5} \Pr_e^{-0.6} \left[1 + (Le_e - 1)\frac{h_e^C - h_w^C}{h_e^T - h_w^T}\right]^{0.6} (h_e^T - h_w^T) \tag{11.4}$$

For catalytic TCHR, the recommended form was:

$$q_s'' = 0.763(\beta_s \rho_e \mu_e)^{0.5} \Pr_e^{-0.6} \left[1 + (Le_e^{0.6} - 1)\frac{h_e^C - h_w^C}{h_e^T - h_w^T}\right] (h_e^T - h_w^T) \tag{11.5}$$

For equilibrium TCHR, Equation (11.4) differs significantly from that recommended by Fay and Riddell in Equation (11.2). For catalytic TCHR, Equation (11.5) is similar to that recommended in Equation (11.3). It has commonly been assumed that the driving force for energy transport is the total enthalpy difference across the boundary layer. Rosner noted that this is a common misconception. It applies only for $Le_e = 1$, because the terms inside the square brackets above also contain the total and chemical enthalpy differences across the boundary layer. For $Le_e = 1$, Equations (11.4 and 11.5) yield the same result. For the O_2/H_2 system considered by Rosner, the calculated heat flux was very similar using either equation, across a realistic range of values. The application was rocket motors. For that system, it was not important whether the TCHR was equilibrium or catalytic. Rosner showed that the factor $\frac{h_e^C - h_w^C}{h_e^T - h_w^T}$ becomes more important as the target temperature approaches the flame temperature ($t_w/t_e \rightarrow 1$). In that case, assuming $Le_e = 1$ may seriously underestimate the actual heat flux.

11.5.2 COMPARISONS WITH EXPERIMENTS

In the studies using semi-analytical solutions, the measured heat flux rates were compared against the rates predicted using one or more forms of a semi-analytic solution. This comparison and the detailed forms of the semi-analytical solutions have been given by Baukal and Gebhart (1996)[11] and are not repeated here for the sake of brevity.

There have been two types of experiments. The first was for lower temperature flames where forced convection was the dominant mechanism and TCHR was negligible. The second type was for higher temperature flames where forced convection and TCHR were both important. Both types are further classified into laminar and turbulent flow regimes. Representative examples are given next.

11.5.2.1 Forced Convection (Negligible TCHR)

For air/fuel flames, the heat transfer was almost exclusively by forced convection, with negligible TCHR. In some studies,[3,59,62] the driving potential was taken as $(h_e^T - h_w^T)$. For air/fuel flames, this is essentially the same as $(h_e^S - h_w^S)$. Horsley et al.[63] recommended the following modification of equation (11.1):

$$q_s'' = 1.67(\beta_s \overline{\rho \mu})^{0.5} \overline{\Pr}^{-0.6} (h_e^S - h_w^S) \tag{11.6}$$

This applies to $7050 \leq \text{Re}_n \leq 16200$. These values of Re_n are generally considered to be turbulent,[73] although the flames were described as having a "laminar appearance."

Horsley also inserted metal grids into industrial burners to promote turbulence.[63] A similar modification of equation (11.1) was determined as,

$$q_s'' = 1.12\left(\beta_s \overline{\rho\mu}\right)^{0.5} \overline{\text{Pr}}^{-0.6}\left(h_e^S - h_w^S\right)$$
(11.7)

This also applies to $7050 \leq \text{Re}_n \leq 16200$. The burner was located at the axial position, L, which yielded the maximum surface heat flux.

11.5.2.2 Forced Convection with TCHR

The combination of forced convection and TCHR has been most important in O_2/fuel flames. The total enthalpy difference has been used as the driving potential. Some of the equations also included the effect of Le.

Kilham and Purvis[57,80] measured the total heat flux from laminar, mostly fuel-rich flames. The data was compared to three different equations using the recommendation by Fay and Riddell[79] for $\rho_e\mu_e$. Only forced convection heat transfer was assumed in the first equation:

$$q_s'' = 0.763\left(\beta_s \rho_e \mu_e\right)^{0.5}\left(\frac{\rho_w \mu_w}{\rho_e \mu_e}\right)^{0.1} \overline{\text{Pr}}^{-0.6}\left(h_e^S - h_w^S\right)$$
(11.8)

Equilibrium TCHR, with no Le augmentation (Le = 1), was assumed in the second equation:

$$q_s'' = 0.763\left(\beta_s \rho_e \mu_e\right)^{0.5}\left(\frac{\rho_w \mu_w}{\rho_e \mu_e}\right)^{0.1} \overline{\text{Pr}}^{-0.6}\left(h_e^T - h_w^T\right)$$
(11.9)

Equilibrium TCHR, with Le augmentation (Le > 1), was assumed in the third equation:

$$q_s'' = 0.763\left(\beta_s \rho_e \mu_e\right)^{0.5}\left(\frac{\rho_w \mu_w}{\rho_e \mu_e}\right)^{0.1} \overline{\text{Pr}}^{-0.6}\left[1 + \left(\text{Le}_{e,H} - 1\right)\frac{h_{e,H}^C - h_{w,H}^C}{h_e^T - h_w^T}\right]\left(h_e^T - h_w^T\right)$$
(11.10)

Equation (11.10) is a form of equation (11.8). Kilham showed that equation (11.8) could be simplified by calculating Le_e and $(h_e^C - h_w^C)$, based only on H atom recombination. Equation (11.8) underpredicted the experimental data by 24 to 42%. Equation (11.9) underpredicted the data by 3 to 7%. Equation (11.10) overpredicted the data by 2 to 10%.

Ivernel and Vernotte[46] tested air-O_2/natural gas ($\Omega = 0.25 - 0.90$) and O_2/natural gas flames. Although the flow regime was not specified, it appears to have been turbulent, based on a comparison with other studies.[73] The following equation was given:

$$q_s'' = 0.763\left(\beta_s \rho_w \mu_w\right)^{0.5}\frac{\overline{\text{Pr}}^{-0.4}}{\text{Pr}_w}\left(h_e^T - h_w^T\right)$$
(11.11)

This differs from the other semi-analytic solutions because most of the properties are evaluated at the wall temperature. Equation (11.11) overpredicted the experimental data by up to 68%.

11.6 EMPIRICAL HEAT TRANSFER CORRELATIONS

Empirical heat transfer correlations for impinging flame jets have been generally given in two forms.[81] The first uses the Nusselt number, a dimensionless heat transfer coefficient, which has appeared as:

$$Nu \sim a\,Pr^b\,Re^c$$

where a, b, and c are constants. The second form has been directly in terms of the heat flux, q''. These two forms are related as follows:

$$q'' = (k/d)\{Nu\}\,\Delta t$$

where k is the thermal conductivity of the fluid and d is a characteristic dimension.

For consistency, the correlations given here are written in terms of q''. This makes it easier to directly compare equations. Also, the driving force, or the potential for heat transfer, is explicitly given in the second form. In some correlations, this potential is the temperature difference. In others, it is the enthalpy difference. In the first form, the Nu formulation, the potential is not explicitly given. For the correlations converted from the first form to the second form (i.e., Nu to q''), the Nu relationship is shown inside curly brackets { }.

Baukal and Gebhart (1996)[81] have extensively reviewed the empirical correlations used in impinging flame experiments. The correlations were arranged according to the geometry, the types of heat transfer, and the flow type (laminar or turbulent) in each experiment. The heat transfer types have included convection, TCHR, radiation, and combinations of these. Some representative examples of those correlations are given next.

Hargrave et al. (1987)[82] studied the effects of turbulence. The following correlation was determined for the forced convection heat transfer (with no TCHR) from flames impinging normal to the stagnation point of a cylinder (Figure 11.7):

$$q''_{s,conv} = \frac{\overline{\overline{k}}}{d_b}\left\{\overline{\overline{Re}}_b^{0.5}\left[1.071 + 4.669\left(\frac{\overline{\overline{Tu\,Re}}_b^{0.5}}{100}\right) - 7.388\left(\frac{\overline{\overline{Tu\,Re}}_b^{0.5}}{100}\right)^2\right]\right\}(t_e - t_w) \qquad (11.12)$$

This correlation underpredicted the data by up to 14%.

Ivernel and Vernotte[46] determined the following correlation for forced convection with TCHR for the heat transfer from turbulent flames impinging normal on the stagnation point of a hemi-nosed cylinder (Figure 11.8):

$$q''_{s,conv+TCHR} = \frac{k_w}{r_b}\left\{0.853 Re_{r_b,w}\,\overline{Pr}^{0.4}\right\}\frac{h_e^T - h_w^T}{c_{p_w}} \qquad (11.13)$$

This was valid for $1000 < Re_{r_b,w} < 45000$. Equation (11.13) correlated the experimental data within 10%. Although the flow conditions were not given, they are believed to have been turbulent. This is based on a comparison of the reported velocities with those of other studies (see Baukal and Gebhart[73]). Radiation to the target was calculated. It was specifically excluded from the above correlation.

You (1985)[7] studied pure diffusion flames impinging on a plate (Figure 11.9). The fuel flow from the burner nozzle was laminar. However, the buoyant plume impinging on the plate was turbulent. It was found that the average convective heat flux in the stagnation zone was essentially constant:

$$q''_{b,conv} = 3.12\left(q_f/l_j^2\right)Ra_e^{-1/6}\,Pr_e^{-3/5}$$

(11.14)

for R < 0.16. The flux decreased with R in the wall jet region:

$$q''_{b,conv} = 1.46R^{-1.63}\left(q_f/l_j^2\right)Ra_e^{-1/6}\,Pr_e^{-3/5}$$

(11.15)

for R > 0.16. In both cases, $10^9 < Ra < 10^{14}$ and $Pr \cong 0.7$. The Rayleigh number was defined as:

$$Ra = g\tilde{\beta}_e q_f\, l_j^2/\rho_e c_{p_e} v_e^3$$

(11.16)

No correlations were given for the measured radiation heat flux. This accounted for up to 26% of the total heat flux. The radiant flux was specifically excluded from the above correlations.

Beér and Chigier (1968)[41] studied flames impinging at a 20° angle from the furnace hearth (see Figure 11.10b). The correlation for the total heat flux (forced convection + radiation) as a function of axial distance from the leading edge of the plate was given as:

$$q''_{r,conv+rad} = \left(k_{max}/r\right)\left\{0.13\,Re_{r,max}^{0.8}\right\}\left(t_{max} - t_w\right)$$

(11.17)

The velocity in the Reynolds number was the maximum velocity measured at the axial distance from the point of impingement. The temperature, t_{max}, was the maximum temperature measured at the axial distance from the point of impingement. A Prandtl number of 0.7 was assumed.

11.7   CFD MODELING OF IMPINGING FLAMES

Numerous studies have been done for CFD modeling of impinging, nonreacting jets. Two examples illustrate those studies. Alkam and Butler (1994) used a finite difference technique to model the transient laminar flow of a jet impinging on a disk.[83] Dianat et al. (1996) applied a Reynolds stress closure model to turbulent, axisymmetric impinging jets.[84]

Relatively few studies of CFD modeling of impinging flame jets have been performed. The studies related to industrial combustion studies are discussed below.

11.7.1   CHALLENGES

Studying fundamental flame-jet impingement problems has many unique challenges, in addition to those faced in studying any type of flame. The flames are often turbulent and in close proximity to a stagnation surface, which complicates the geometry in modeling and limits access in experimental measurements. At a minimum, the flames are at least two-dimensional. Modeling the stagnation target increases the complexity of the problem.

Industrial flame impingement further complicates analysis. Industrial heating processes nearly always have multiple burners where there is often flame interaction. For experimental measurements, the environment is much less controllable than in the laboratory, which complicates the model validation. The flames are often located inside an insulated chamber, which can dramatically increase the size and complexity of a model. The target is usually moving, unlike laboratory studies where the target is nearly always stationary. Although the movement adds little to the complexity of the model, it significantly complicates measurements on the target, which can affect validation.

The scale is often much larger than in the laboratory, which may mean that more nodes and longer computation times are needed to characterize the system. Along with the transient nature of the turbulent flames, the overall operating conditions of the heating system may also be transient to adjust for different part shapes, sizes, and production rates. The heating system may be dependent on both upstream and downstream processes in the plant, which makes it difficult to independently control its operating conditions to make measurements for validating the model.

11.7.2   FUNDAMENTAL INVESTIGATIONS

This chapter section is not intended to be exhaustive, but merely representative of the fundamental studies on CFD modeling of impinging flame jets. It should be noted that two types of stagnation flames are commonly used in fundamental studies: opposed jet and jet impingement.[85] Only the latter is considered here. The two geometries considered here are flames impinging normal to a flat plate and normal to a hemi-nosed cylinder.

11.7.2.1   Flame Impingement on a Flat Plate

Some fundamental investigations have been done to numerically study flame jets impinging on flat plates, sometimes referred to as a stagnation point geometry. Bray et al. (1992, 1994) developed an asymptotic approach for modeling turbulent premixed flame impingement on an adiabatic flat plate.[85,86] The flow was divided into three regions: the viscous sublayer next to the wall, the shear layer outside the viscous sublayer, and the turbulent stream of reactants approaching the wall. The reaction zone was assumed to only be present in the first two regions, closest to the wall. Favre-averaging was used for the turbulence parameters. Asymptotic equations and boundary conditions were developed for the three regions, with appropriate matching conditions between each region. The numerical results were in qualitative agreement with previous modeling results, but were not compared with any experimental data.

Park and Kim (1996) modeled a pure, round propane jet, issuing into and combusting in open air, impinging on a flat plate.[87] Velocities, temperatures, and species were reported for the flow in the gas space. Heat fluxes up to about 60 kW/m^2 (19,000 Btu/hr-ft^2) were reported at the impingement surface.

Ito et al. (1996) modeled preheated, premixed flat flames, stabilized near and impinging on a stagnation surface.[88] The effects of the stagnation surface temperature, Lewis number of the reactant mixture, and mixture preheat temperature were studied. Heat fluxes to the surface were reported, among other things.

Popp and Baum (1997) modeled laminar, stoichiometric air/methane flames impinging on and being quenched by an inert wall.[89] Wall temperatures ranging from 300 to 600K (80 to 620°F) were investigated, using one-dimensional fluid mechanics with detailed chemical kinetics, including radical recombination, to study the quenching by and heat flux to the wall. A direct numerical simulation code was used.[90] Reported results included wall heat flux, flame thickness, total reaction rate, flame wall distance, maximum reaction rate, species mass fractions, and net heat release as functions of position and wall temperature.

Lindstedt and Váos (1998) numerically studied stoichiometric, premixed, turbulent, air/ethylene flames impinging normal to a flat plate.[91] The focus of the study was to investigate turbulent premixed combustion using a full second-moment closure for both velocity and scalar turbulent transport, which included extended variable density forms for the pressure redistribution, pressure scrambling, and dissipation generation terms. A uniform stream of reactants was assumed at the burner outlet. The flame was stabilized far enough from the plate so that the turbulence quantities were not strongly influenced by the plate in the flamelet region. The simulations were two-dimensional and transient. Comparison of the numerical results with the experimental measurements by Cheng and Shepherd (1991)[92] showed good agreement of both mean and turbulent quantities.

Comparison of the second-moment method with gradient-diffusion closure showed that the second-moment method was in better agreement with the experimental data. According to the authors, this demonstrates the superiority of second-moment methods over gradient-diffusion closure methods.

Treviño (1999) numerically studied catalytic ignition from stagnation flow of combustible gases on a flat plate.[93] Catalytic combustion presents some unique modeling challenges due to the heterogeneous chemical reaction mechanisms. The ignition process is typically numerically studied using either elementary chemistry or an elementary one-step overall reaction mechanism with a large activation energy, asymptotic analysis. Treviño used the latter technique. The problem was modeled using 2-D boundary layer equations and a heterogeneous reaction model. The equations and boundary conditions were nondimensionalized and simplified using an asymptotic analysis. The numerical results for the ignition temperature of an air/CO flame fell directly between two sets of experimental data available in the literature.

11.7.2.2 Flame Impingement on a Hemi-nosed Cylinder

While this is not a particularly common geometry in industry (but is in aerospace), the results of this work were extremely useful in validating the semi-analytical equations that have been used to simulate industrial processes. The earliest numerical modeling of impinging flames was done by Conolly and Davies (1971, 1972).[3,67] They numerically and experimentally investigated a wide range of flames impinging normal to a 12.7 mm (0.5 in.) diameter hemi-nosed sphere. Stoichiometric O_2/CH_4, O_2/H_2, O_2/C_2H_4, O_2/C_3H_8, O_2/CO, and air/natural gas flames impinged on a blunt-nosed surface at a temperature of 400K (260°F). The combustion reactions were not directly computed, but were simulated by calculating weighted average transport properties and then using them in the axisymmetric boundary layer equations. Only the flow in the boundary layer was simulated. The following boundary layer partial differential equations were used:

$$\frac{\partial}{\partial x}(\rho u r) + \frac{\partial}{\partial y}(\rho v r) = 0 \tag{11.18}$$

where x is the distance from the stagnation point measured along the surface, ρ is the gas density, r is the radius, u is the gas velocity in the x-direction, y is the direction normal to the surface, v is the velocity in the y-direction. The conservation of momentum was given by:

$$\rho u \frac{\partial u}{\partial x} + \rho v \frac{\partial u}{\partial y} = -\frac{\partial p}{\partial x} + \frac{\partial}{\partial y}\left(\mu \frac{\partial u}{\partial y}\right) \tag{11.19}$$

where p is the pressure and μ is the gas viscosity, and

$$\frac{\partial p}{\partial y} = 0 \tag{11.20}$$

The conservation of energy equation was given by:

$$\rho u \frac{\partial I}{\partial x} + \rho v \frac{\partial I}{\partial y} = \frac{\partial}{\partial y}\left(k_{eq}\frac{\partial t}{\partial y}\right) + \frac{\partial}{\partial y}\left\{\frac{\mu}{2}\left(1 - \frac{1}{Pr}\right)\frac{\partial u^2}{\partial y}\right\} \tag{11.21}$$

where I is the fluid enthalpy including kinetic energy, k_{eq} is the gas equilibrium thermal conductivity, and Pr is the gas Prandtl number. These equations were transformed into ordinary differential equations using the transformations proposed by Lees (1956)[94]:

TABLE 11.1

Measurements and Predictions for Flames Impinging Normal to a Hemi-nosed Cylinder

Parameter	O_2/CH_4	O_2/H_2	O_2/C_2H_4	O_2/C_3H_8	O_2/CO	Air/N.G.
$t_{adiabatic}$ (K)	3052	3073	3216	3097	3004	2224
$t_{measured}$ (K)	3023	3040	3144	3072	2957	2200
Measured heat transfer coeff. (W/m²-K)	521	611	526	465	256	147
Numerically predicted heat transfer coeff. (W/m²-K)	544	585	520	493	301	152

Adapted from R. Conolly and R.M. Davies, *Int. J. Heat Mass Transfer*, 15, 2155-2172, 1972.

$$u(\bar{x}, \eta) = u_e(\bar{x}) f'(\eta)$$

$$I = I_e g(\eta)$$

$$\eta = \frac{\rho_e u_e r}{(2\bar{x})^{0.5}} \int_0^y \frac{\rho}{\rho_e} dy \qquad (11.22)$$

$$\bar{x} = \int_0^x \rho_e u_e^2 r^2 dx$$

where the subscript *e* refers to the edge of the boundary layer. The velocity at the edge of the boundary layer was taken as:

$$u_e = \frac{3}{2} U_\infty \sin \frac{x}{r_b} \qquad (11.23)$$

where U_∞ is the free stream velocity and r_b is the radius of the hemi-nosed cylinder. The transformed equations were solved with finite-difference methods. The primary parameter of interest was the heat flux to the stagnation point. As shown in Table 11.1, the computed values were within 6% of the measured values, except for the O_2/CO flames where the numerical results overestimated the measured value by 18%.

Fairweather et al. (1984) modeled turbulent, premixed, air/methane flames impinging normal to a hemi-nosed cylinder.[58] Radiative heat transfer was ignored as it was argued to be negligible due to the low emissivity of the flames. Fairweather et al. used similar transformed boundary layer equations as Conolly and Davies, except for the addition of turbulence. The momentum equation was modified as:

$$\rho u \frac{\partial u}{\partial x} + \rho v \frac{\partial u}{\partial y} = -\frac{\partial p}{\partial x} + \frac{\partial}{\partial y}\left((\mu + \varepsilon)\frac{\partial u}{\partial y}\right) \qquad (11.24)$$

The conservation of energy equation was modified as:

$$\rho u \frac{\partial I}{\partial x} + \rho v \frac{\partial I}{\partial y} = \frac{\partial}{\partial y}\left[\left(\frac{\mu}{Pr_{eq}} + \varepsilon\right)\frac{\partial I}{\partial y}\right] + \frac{\partial}{\partial y}\left\{\frac{\mu}{2}\left(1 - \frac{1}{Pr_{eq}}\right)\frac{\partial u^2}{\partial y}\right\} \qquad (11.25)$$

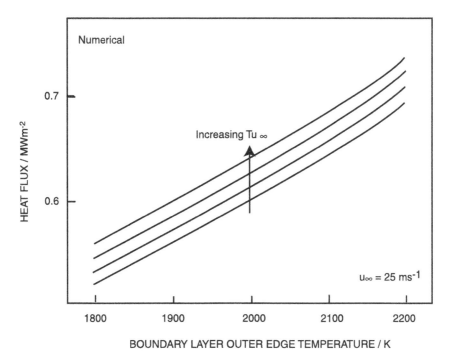

FIGURE 11.14 Predicted heat flux to the stagnation point of a hemi-nosed cylinder as a function of the boundary layer edge temperature and turbulence intensity (Tu$_\infty$ = 0, 5, 10, 15%) for an air/methane flame (ϕ = 0.957). (From M. Fairweather, J.K. Kilham, and A. Mohebi-Ashtiani, *Comb. Sci. Tech.*, 35, 231, 1984. With permission.)

The following eddy law was used:

$$\varepsilon = C\rho Tu_\infty u_\infty y \tag{11.26}$$

where ε is the eddy viscosity, C is a constant (0.1 was used), Tu$_\infty$ is the turbulent intensity in the free stream, and u_∞ is the free stream velocity (25 m/s in this case). Because of the assumption of unity turbulent Lewis and Prandtl numbers, only eddy viscosity needed to be modeled. Figure 11.14 shows that the predicted heat flux increases with both the turbulence and the boundary layer, outer edge gas temperature. It was noted that increases in free stream turbulence have little effect on the heat flux rates at low gas velocities. The numerical predictions severely underpredicted the experimental data when the hemi-nosed cylinder was in the reaction zone. The agreement was much better as the reaction zone moved upstream, as shown in Figure 11.15.

11.7.3 INDUSTRIAL INVESTIGATIONS

Related studies not considered here are for flames impinging on cylinder walls in an internal combustion engine, where the flames are confined and the gases are at elevated pressures, so they are not particularly relevant to industrial flame impingement processes.

11.7.3.1 Flame Impingement in Thermal Spallation

Thermal spallation is a technique used to drill through rock with high-temperature impinging flames, as shown in Figure 11.16.[95] It is designed to compete with mechanical drilling where rapid bit wear and frequent bit replacement can be problems, and where the drilling costs often increase exponentially with depth.[96] The flames rapidly heat the rocks, which fracture due to the extremely high temperature

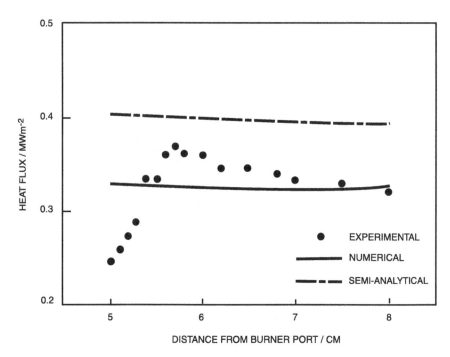

FIGURE 11.15   Predicted and measured heat flux to the stagnation point of a hemi-nosed cylinder as a function of the distance from the burner for a turbulent air/methane flame ($\phi = 0.943$)· (From M. Fairweather, J.K. Kilham, and A. Mohebi-Ashtiani, *Comb. Sci. Tech.*, 35, 231, 1984. With permission.)

gradients near the surface of the rock. Such gradients cause the rock to fracture due to the high thermal stresses. Localized heating is preferred to increase the temperature gradients that cause the rock to fragment. This technique has no bits to wear out or be replaced. It also provides a ready means for rock removal where the high-velocity combustion products carry the rock fragments or spalls out of the hole. The fluid flow in the vicinity of the rock surface and the thermal gradients through the rock surface are both important in determining the effectiveness of thermal spallation.

Rauenzahn and Tester (1991) numerically[95] and experimentally[97] studied supersonic flames impinging on rock in a bore hole. For the modeling, the reactant gases were assumed to be at equilibrium; thus, the reactions were not modeled, based on previous analyses that showed the reactions to be over 95% completed at the exit of the burner. The effect of the rock chips on the flowfield was ignored, so that only single-phase, compressible, axisymmetric, transient fluid flow was modeled. The 2-D equations were discretized and solved using a finite-difference formulation and an upwind differencing technique. Turbulence was modeled using a combined k-ε technique in the bulk flow and the traditional law-of-the-wall in the boundary layer region.[98] The boundary layer equations were modified for compressibility according to the recommendations of Van Driest (1951).[99] The rock surface temperature was calculated using Weibull's statistical failure theory.[100,101] The effects of the standoff distance between the nozzle and the rock and the inlet pressure ratio (ratio of the jet pressure at the nozzle outlet to the ambient pressure) on the heat transfer and the hole radius were studied. As expected, decreasing the distance between the nozzle and the rock significantly increased the heat transfer coefficient to the rock. Increasing the pressure ratio decreased the heat transfer coefficient. The drilling effectiveness increased as the standoff distance decreased at moderate pressure ratios. However, at higher pressure ratios, the trend actually reversed, so that the drilling effectiveness decreased with decreasing standoff distances. Under those conditions, the drilling effectiveness actually improved by moving the nozzle away from the rock surface. This was explained by the formation of a flow reversal that caused a sort of stagnation "bubble,"

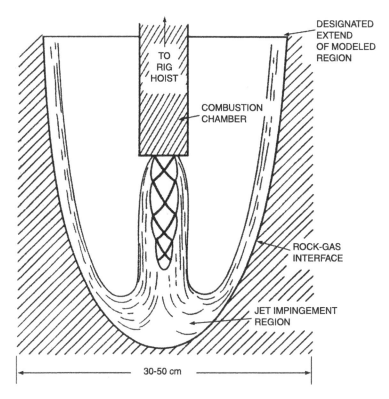

FIGURE 11.16 Thermal spallation drilling through rock. (From R.M. Rauenzahn and J.W. Tester, *Int. J. Heat Mass Transfer,* 34(3), 795, 1991. With permission.)

previously unreported in the literature. It was expected that there would be an optimum standoff distance, beyond which the drilling effectiveness would again decrease.

Wilkinson and Tester (1993) numerically[102] and experimentally[103] studied thermal spallation drilling to further improve on the understanding of this phenomenon by continuing the work of Rauenzahn and Tester described above. The rock surface temperature was modeled using Dey's thermal spallation rock mechanics model,[104] which is based on a failure mechanism characterized by Weibull statistics that requires some empirical parameters. The primary objective was to study the drilling rate and hole diameter as a function of various parameters, using improved empirical parameters derived from experiments.[103] The 2-D, compressible, axisymmetric fluid flow and energy equations were discretized using the finite-volume formulation. The chemistry was not directly modeled but was simulated by varying the gas properties. Turbulence was modeled using a combined k-ε technique in the bulk flow and the traditional law-of-the-wall in the boundary layer region. Artificial dissipation was included to stabilize the solution algorithm. As shown in Figure 11.17, there was good agreement between the predicted and measured heat transfer coefficients, where the model incorporated variable heat capacity and mass injection. Further experiments designed to expand the thermal drilling database were recommended.

11.7.3.2 Flame Impingement in CVD Processes

Dong and Lilley (1992) modeled round, turbulent, fuel-rich O_2/C_2H_2 flames impinging normal to a substrate to simulate the chemical vapor deposition (CVD) process used to make synthetic diamonds.[105] The parameters of primary interest were the distance between the burner nozzle and the substrate, the nozzle diameter, the overall equivalence ratio, and the total flow rate of reactants.

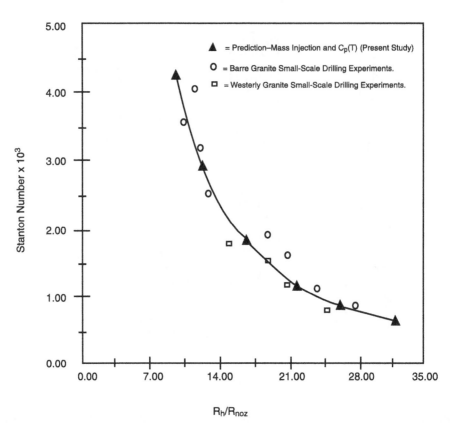

FIGURE 11.17 Comparison of the predicted and measured Stanton number vs. hole radius for the model incorporating variable heat capacity and mass injection. (From M.A. Wilkinson and J.W. Tester, *Int. J. Heat Mass Transfer,* 36(14), 3474, 1993. With permission.)

The system was modeled using 2-D axisymmetric finite-difference equations solved using an iterative relaxation solution technique known as SIMPLE[106] (semi-implicit method for pressure linked equations). By tripling the distance from the nozzle to the substrate, the predicted surface temperatures decreased by more than 50%. Keeping all other conditions constant and varying the nozzle diameter between 4.8 and 7.2 mm (0.19 and 0.28 in.) had relatively little effect on the substrate temperature profile. Increasing the equivalence ratio from 2.5 to 3.0 reduced the substrate temperatures by less than 10%. Increasing the reactant flow rate by 340% increased the substrate temperature by as much as 50%.

Tsai and Greif (1995) studied thermophoretic transport in CVD processes used in the fabrication of optical fibers and semiconductor devices.[107] Two different geometries were modeled. The first was for the 3-D flow for a flame impinging on a stationary disk, where the centerline of the burner was not aligned with the center of the disk, based on the experimental study by Hwang and Daily (1994).[108] The second was for the 2-D flow of a flame impinging normal to a rotating disk, where the centerlines of the burner and disk were aligned, which was based on the experimental study by Tsai and Greif (1994).[109] The equations were discretized using a finite-volume technique. The burner consisted of eight concentric rings, where the rings were at different distances from the substrate. The outermost ring was the longest and closest to the substrate. The innermost ring was the shortest and furthest from the substrate. The centermost ring carried a mixture of O_2 and $SiCl_4$. The innermost annulus, next to the center ring, carried N_2. The six outer rings carried methane and O_2 alternately. Silica particles are produced in the flame reactions and thermophoretically deposited on the substrate. The flow was assumed to be laminar and soot radiation was neglected. The particles were assumed to be smaller than 1 μm in diameter and the concentration was dilute, so particle-particle and particle-wall interactions were neglected. The chemical kinetics were simulated with the following two equations:

$$CH_4 + 2O_2 \rightarrow CO_2 + 2H_2O$$

$$SiCl_4 + O_2 \rightarrow SiO_2 + 2Cl_2$$

The discretized equations were solved using the Modified Strongly Implicit (MSI) procedure.[101] The diameter of the disk (4.5 to 9 cm or 1.8 to 3.5 in.), the rotation speed of the disk (0 to 1200 rpm), and eccentricity for the nonrotating disk (0 to 2/3 of the disk radius, where 0 = aligned) were the parameters varied. The numerical results showed that the deposition efficiency increased significantly with the diameter of the substrate, decreased only slightly with eccentricity, and increased only slightly with rotation speed.

11.7.3.3 Flame Impingement in a Furnace

Chapman et al. (1992, 1994) modeled impinging flame jets in a direct-fired continuous metal strip reheating furnace.[111,112] The objective of the study was to determine the parameters that significantly impact the performance of the system, which has previously been studied empirically. Natural gas flames impinge on the metal strip from above, and in some cases also from below. The 3-D system was modeled with a 2-D steady-state simulation, using the finite volume SIMPLER algorithm. Turbulence was modeled with the k-ε approximation. The combustion reactions were modeled using a one-step global reaction:

$$Fuel + Oxidant \rightarrow Products$$

A probability density function (pdf) was used to predict the fuel and oxygen mass fraction distributions in the reaction zone.[113] Radiation was modeled with the S_4 discrete ordinates method[114,115] and the weighted-sum-of-gray-gases (WSGG) model for the gaseous radiation.[116] In the actual furnace being simulated, the metal enters at 310K (99°F) and exits at 1000K (1300°F). For the steady-state model, an average metal temperature of 650K (710°F) was used. Three parameters were studied: firing rate, refractory emissivity, and load emissivity. The parametric investigations were compared by examining the radiative and convective heat transfer rates to the load, the refractory temperature, and the fuel efficiency. The base-case analysis showed that 60 to 70% of the heat flux to the load was from radiation from the refractory walls; 20 to 30% came from gaseous radiation; and only about 10% came from forced convection from the gases, depending on the position in the furnace. The heat transfer to the load increased and the thermal efficiency decreased as the firing rate increased. Figure 11.18 shows that the system thermal efficiency was the lowest for a refractory emissivity of 0.3 and increased at both higher and lower emissivities. The system thermal efficiency increased rapidly with the emissivity of the load. A number of weaknesses in the model were noted. The turbulence model for jet impingement needs improvement. The assumption of fast chemistry may be invalid in the stagnation zone. The model results could not be validated because of insufficient experimental data. Turbulence/radiation interactions, interactions between the load and the combustion space, and the third spatial dimension should be included in future models.

11.7.3.4 Flame Impingement on a Cylindrical Tank

Johnson et al. (1997) (Shell Research) modeled high-velocity, pure propane diffusion flames impinging on a 2 m (7 ft) diameter cylindrical tank, 9 m (30 ft) downstream of the burner.[117] The purpose was to simulate a hydrocarbon jet fire impinging on a tank to assess the safety risks of a

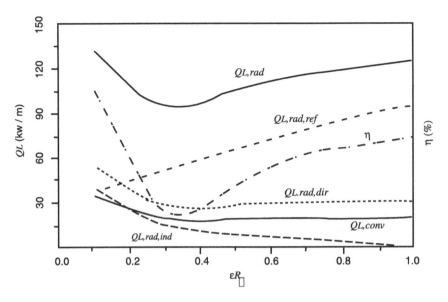

FIGURE 11.18 Effect of fuel firing rate on heat transfer to the load and thermal efficiency. (From K.S. Chapman, S. Ramadhyani, and R. Viskanta, *Comb. Sci. Tech.,* 97, 115, 1994. With permission.)

fire caused by a ruptured pipe in petrochemical plants. They used the commercial CFD code CFX-FLOW3D with an extension of new physical sub-models for turbulent combustion, soot formation, and radiative heat transfer.[118] The high-velocity lifted flames were modeled in two stages. The first stage was modeled using an extension of the work by Bradley et al. (1988)[119] to determine the lift-off height using a PDF premixed laminar flamelet model that employed a mean turbulent volumetric heat release, separately calculated using the PREMIX code from Sandia.[120] The flame lift-off position was defined as the position where the mean turbulent heat release reached 10 MW/m³ (1 × 10⁶ Btu/hr-ft³). After the lift-off point, the CFX-FLOW3D code was run with the combustion switched on to calculate the 3-D flame structure, using an assumed PDF and strained laminar diffusion flamelet model similar to that in Reference 118. Radiative heat transfer was calculated as a post-process using the following formula to determine the diffusion flamelet temperature and including radiative heat loss:

$$T(f) = T_{ad}(f)\left[1 - \chi\left(\frac{T_{ad}(f)}{T_{ad}^4}\right)\right]$$ (11.27)

where χ is a constant (0.15) throughout the flame, f is the mixture fraction, and T_{ad} is the adiabatic flame temperature. Modifications were made to the combustion model to include the radiative loss term and the laminar diffusion flamelet model based on the local turbulent mean Eulerian strain rate based on the Taylor microscale. A strain rate of 500 s⁻¹ gave the best agreement between the measured and predicted temperatures and heat fluxes (measured with 40 calorimeters on the tank surface). Further work was recommended to validate the turbulence sub-models for predicting the convective heat transfer for flames impinging on curved surfaces.

11.8 RECOMMENDATIONS FOR FURTHER RESEARCH

In a report prepared for the U.S. Dept. of Energy, direct flame impingement has been identified as a critical combustion technology.[121] Further research and development of the technology for use in industry was recommended. In a review, Viskanta (1991) indicated that very little study has concerned heat transfer from flames impinging on solids.[122] Carefully controlled experiments were recommended to improve the understanding of convective and radiative heat transfer in impinging flame jets. This represents an extremely fruitful area of research. In a more recent review, Viskanta (1993) indicated that there is a lack of fundamental information on the relative importance of both radiation and convection in flame impingement heating.[123] This is needed to understand the flame structure and to predict the heat transfer to the target. Detailed CFD modeling can greatly aid in the understanding of this complex phenomenon.

Although flame impingement heat transfer has been studied for nearly 50 years, much still remains to be done. Very little CFD modeling has been done compared to the many experimental studies available in the literature. Detailed modeling needs to be done for other geometries, fuel compositions, target types and surface conditions, burner designs, and oxidizer compositions. Parametric studies of the influence of the spacing between the burner and the target, the target emissivity and temperature, fuel and oxidizer compositions, and burner conditions (firing rate and equivalence ratio) will help equipment designers to optimize the heating process.

New CFD models for flame impingement should include more comprehensive chemistry, especially in the higher temperature systems using oxygen-enhanced combustion. Detailed spectral modeling could be done to improve the radiation modeling. Coupled modeling between the gas phase and the heat conduction transfer through the solid target is needed for the high heat flux systems. In many of those systems, the flame must be turned on and off to avoid overheating and melting the target surface and to allow the energy received at the surface to be conducted into the target.[23,24] The heating sequence varies, depending on the system. Modeling would help to optimize and adapt the system for new part sizes, shapes, and compositions.

CFD modeling results need to be compared to semi-analytical and empirical equations to find their range of applicability. CFD can also be used to determine why those equations break down under certain conditions, and may be useful to either extend their ranges or to help develop new simplified correlations. CFD modeling can also be used to determine the importance of various heat transfer mechanisms such as forced convection and TCHR. This is particularly important for TCHR because there is no easy method for measuring this mechanism by itself. This information will be useful for both finding what mechanisms need to be included in correlations, as well as for optimizing the heating system. For example, if it is found that radiation is important under certain conditions, then increasing the flame luminosity or the target emissivity may be useful. It is expected that new system designs would result from CFD modeling.

11.9 CONCLUSIONS

Flame impingement heating is an important and complicated process used in industrial heating applications. Much fundamental research has been done experimentally, while relatively little research has been done with computational fluid dynamics to study this phenomenon. Those few CFD studies have been considered here. They vary considerably in complexity. Some have not modeled the chemical reactions at all. Others have modeled the reactions indirectly by including their effects by modifying transport properties. Others have used simplified reduced chemical mechanisms to simulate complex chemical reaction systems. Various models have been used to simulate the radiant heat transfer, ranging from zone models to more sophisticated radiant transport models. No attempts have been made to study TCHR, vapor condensation, or boiling with CFD modeling.

Because of the complexity, high gradients, harsh conditions, and relatively small scale, it is difficult to make measurements on flame impingement systems. Although advanced laser diagnostic techniques will help this in the future, modeling still promises to provide more information, especially with dramatic advances in hardware and software. While some modeling work has been done, much more needs to be done. CFD modeling promises to dramatically improve the understanding of flame impingement heating, which should be useful for improving the performance of these systems in industrial combustion applications.

REFERENCES

1. C.E. Baukal, L.K. Farmer, B. Gebhart, and I. Chan, Heat transfer mechanisms in flame impingement heating, in *1995 Int. Gas Research Conf.*, Vol. II, D.A. Dolenc, Ed., Govt. Institutes, Rockville, MD, 1996, 2277-2287.
2. W.H. Giedt, L.L. Cobb, and E.J. Russ, Effect of Hydrogen Recombination on Turbulent Flow Heat Transfer, ASME Paper 60-WA-256, New York, 1960.
3. R. Conolly and R.M. Davies, A study of convective heat transfer from flames, *Int. J. Heat Mass Trans.*, 15, 2155-2172, 1972.
4. M.A. Cappelli and P.H. Paul, An investigation of diamond film deposition in a premixed oxyacetylene flame, *J. Appl. Phys.*, 67(5), 2596-2602, 1990.
5. R.M. Rauenzahn, Analysis of Rock Mechanics and Gas Dynamics of Flame-Jet Thermal Spallation Drilling, Ph.D. thesis, Massachusetts Institute of Tech., Cambridge, MA, 1986.
6. J.E. Hustad, M. Jacobsen, and O.K. Sønju, Radiation and heat transfer in oil/propane jet diffusion flames, *Inst. Chem. Eng. Symp. Series*, 10(129), 657-663, 1992.
7. H.-Z. You, Investigation of fire impingement on a horizontal ceiling. 2. Impingement and ceiling-jet regions, *Fire & Materials*, 9(1), 46-56, 1985.
8. P.A. Eibeck, J.O. Keller, T.T. Bramlette, and D.J. Sailor, Pulse combustion: impinging jet heat transfer enhancement, *Comb. Sci. Tech.*, 94, 147-165, 1993.
9. Y. Zhang and K.N.C. Bray, Characterization of impinging jet flames, *Comb. Flame*, 116, 671-674, 1999.
10. A. Milson and N.A. Chigier, Studies of methane-air flames impinging on a cold plate, *Comb. Flame*, 21, 295-305, 1973.
11. C.E. Baukal and B. Gebhart, A review of semi-analytic solutions for flame impingement heat transfer, *Int. J. Heat Mass Trans.*, 39(14), 2989-3002, 1996.
12. C.E. Baukal, Ed., *Oxygen-Enhanced Combustion*, CRC Press, Boca Raton, FL, 1998.
13. L.M. Hanssen, W.A. Carrington, J.E. Butler, and K.A. Snail, Diamond synthesis using an oxygen-acetylene torch, *Materials Lett.*, 7(7-8), 289-292, 1988.
14. K.V. Ravi, D.S. Olson, and C.A. Koch, High Rate, High quality, diamond synthesis by the combustion flame process, *Mat. Res. Soc. Symp. Proc.*, 349, 373-383, 1994.
15. J. Hwang and J.W. Daily, Characterization of a vapor axial deposition (VAD) flame impinging on a disk for a study of electric field enhanced deposition, *Exper. Thermal Fluid Sci.*, 8, 58-66, 1994.
16. L. Tailbot, R.K. Cheng, R.W. Schefer, and D.R. Willis, Thermophoresis of particles in a heated boudary layer, *J. Fluid Mech.*, 101(4), 737-758, 1980.
17. A.D. Eisner and D.E. Rosner, Experimental and theoretical studies of submicron particle thermo-phoresis in combustion gases, *PhysicoChem. Hydrodyn.*, 7, 91-100, 1986.
18. W.T. Lankford, N.L. Samways, R.F. Craven, and H.E. McGannon, *The Making, Shaping and Treating of Steel*, 10th ed., Herbick and Held, Pittsburgh, PA, 1985.
19. J. Masters, C.J. Towler, and P.F.Y. Wong, The development and exploitation of rapid heating tech-niques, *Inst. Gas Eng. J.*, 11(10), 695-717, 1971.
20. H. Hara, T. Ohkohchi, and T. Ono, Application of Jet Impinging Heating to Slabs in a Continuous Reheating Furnace, Inter. Flame Research Found. document F 29/la/12, IJmuiden, The Netherlands, 1979.
21. V. Jayaraman, F.J. Koinis, M.A. Gipko, and G.W. Whiteman, Jet impingement radiation (JIR) heating of steel strip, *Iron & Steel Eng.*, 61(1), 43-46, 1984.
22. N. Galvin, Choosing direct fired burners, *Heat Treating*, February, 30-31, 1991.
23. L.K. Farmer, I.S. Chan, and J.G. Nelson, Development of rapid gas heating process for semifinished steel products, *Iron Steel Engineer*, 71(9), 11-13, 1994.
24. J.S. Becker and L.K. Farmer, Rapid fire heating system uses oxy-gas burners for efficient metal heating, *Industrial Heating*, 62(3), 74-78, 1995.
25. L.T. Cowley and M.J. Pritchard, Large-scale natural gas and LPG jet fires and thermal impact on structures, *Proc. of Gastech 90 Conf. & Exhibition*, Amsterdam, Paper 3.6, 1990.
26. R.F. Cracknell, J.N. Davenport, and A.J. Carsley, A model for heat flux on a cylindrical target due to the impingement of a large-scale natural gas jet fire, *IChemE Symp. Series*, (139), 161-175, 1995.
27. M.H. Goose, Recent developments with ALIBI, a model for site specific prediction of LPG tank BLEVE frequency, *IChemE Symp. Series*, (139), 327-333, 1995.

28. C.E. Baukal and B. Gebhart, A review of flame impingement heat transfer studies. 1. Experimental conditions, *Comb. Sci. Tech.,* 104, 339-357, 1995.

29. C.E. Baukal, *Heat Transfer in Industrial Combustion,* CRC Press, Boca Raton, FL, 2000.

30. K.H. Hemsath, A novel gas fired heating system for indirect heating, in *Fossil Fuel Combustion Symposium 1990,* S. Singh, Ed., ASME PD-Vol. 30, New York, 1990, 155-159.

31. J.K. Kilham, Energy transfer from flame gases to solids, *Third Symposium on Combustion and Flame and Explosion Phenomena,* Williams and Wilkins, Baltimore, MD, 1949, 733-740.

32. E.G. Jackson and J.K. Kilham, Heat transfer from combustion products by forced convection, *Ind. Eng. Chem.,* 48(11), 2077-2079, 1956.

33. J.E. Anderson and E.F. Stresino, Heat transfer from flames impinging on flat and cylindrical surfaces, *J. Heat Trans.,* 85(1), 49-54, 1963.

34. J.E. Hustad and O.K. Sønju, Heat transfer to pipes submerged in turbulent jet diffusion flames, in *Heat Transfer in Radiating and Combusting Systems,* Springer-Verlag, Berlin, 1991, 474-490.

35. J.E. Hustad, N.A. Røkke, N.A., and O.K. Sønju, Heat transfer to pipes submerged in lifted buoyant diffusion flames, in *Experimental Heat Transfer, Fluid Mechanics, and Thermodynamics, 1991,* J.F. Keffer et al., Ed., Elsevier, New York, 1991, 567-574.

36. D.R. Davies, Heat Transfer from Working Flame Burners, B.S. thesis, University of Salford, Salford, U.K., 1979.

37. M. Sibulkin, Heat transfer near the forward stagnation point of a body of revolution, *J. Aero. Sci.,* 19, 570-571, 1952.

38. S.N. Shorin and V.A. Pechurkin, Effectivnost' teploperenosa na poverkhnost' plity ot vysokotemperaturnoi strui produktov sjoraniya razlichnykh gazov, *Teoriya i Praktika Szhiganiya Gaza,* 4, 134-143, 1968.

39. H. Kremer, E. Buhr, and R. Haupt, Heat transfer from turbulent free-jet flames to plane surfaces, in *Heat Transfer in Flames,* N.H. Afgan and J.M. Beér, Eds., Scripta, Washington, D.C., 1974, 463-472.

40. L.W. Woodruff and W.H. Giedt, Heat transfer measurements from a partially dissociated gas with high Lewis number, *J. Heat Trans.,* 88, 415-420, 1966.

41. J.M. Beér and N.A. Chigier, Impinging jet flames, *Comb. Flame,* 12, 575-586, 1968.

42. J.W. Mohr, J. Seyed-Yagoobi, and R.H. Page, Heat transfer from a pair of radial jet reattachment flames, in *Combustion and Fire, ASME Proc. 31st Natl. Heat Transfer Conf.,* Vol. 6, M. McQuay, K. Annamalai, W. Schreiber, D. Choudhury, E. Bigzadeh, and A. Runchal, Eds., HTD-Vol. 328, New York, 1996, 11-17.

43. J.-P. Vizioz and T.M. Lowes, Convective Heat Transfer from Impinging Flame Jets, Int. Flame Research Found. Report F 35/a/6, IJmuiden, The Netherlands, 1971.

44. S. Nawaz, Heat Transfer from Oxygen Enriched Methane Flames. Ph.D. thesis, The University of Leeds, Leeds, U.K., 1973.

45. M. Fairweather, J.K. Kilham, and S. Nawaz, Stagnation point heat transfer from laminar, high temperature methane flames, *Int. J. Heat Fluid Flow,* 5(1), 21-27, 1984.

46. A. Ivernel and P. Vernotte, Etude expérimentale de l'amélioration des transferts convectis dans les fours par suroxygénation du comburant, *Rev. Gén. Therm., Fr.,* Nos. 210-211, 375-391, 1979.

47. K. Kataoka, H. Shundoh, and H. Matsuo, Convective heat transfer between a flat plate and a jet of hot gas impinging on it, in *Drying '84,* A.S. Majumdar, Ed., Hemisphere/Springer-Verlag, New York, 1984, 218-227.

48. M. Matsuo, M. Hattori, M., T. Ohta, and S. Kishimoto, The Experimental Results of the Heat Transfer by Flame Impingement, Int. Flame Research Found. Report F 29/1a/1, IJmuiden, The Netherlands, 1978.

49. C.E. Baukal and B. Gebhart, Heat transfer from oxygen-enhanced/natural gas flames impinging normal to a plane surface, *Exp. Therm. & Fluid Sci.,* 16(3), 247-259, 1998.

50. E.M. Schulte, Impingement heat transfer rates from torch flames, *J. Heat Trans.,* 94, 231-233, 1972.

51. R.H. Fay, Heat transfer from fuel gas flames, *Welding J.,* Research Supplement, 380s-383s, 1967.

52. R.M. Davies, Heat transfer measurements on electrically-boosted flames, *Tenth Symp. (Int.) on Combustion,* The Combustion Institute, Pittsburgh, PA, 1965, 755-766.

53. R.A. Cookson, An Investigation of Heat Transfer from Flames, Ph.D. thesis, The University of Leeds, Leeds, U.K., 1960.

54. R.A. Cookson and J.K. Kilham, Energy Transfer from Hydrogen-Air Flames, *Ninth Symp. (Int.) on Combustion,* Academic Press, New York, 1963, 257-263.

55. R.B. Smith and T.M. Lowes, Convective heat transfer from impinging tunnel burner flames — A short report on the NG-4 trials, Int. Flame Research Found. Report F 35/a/9, IJmuiden, The Netherlands, 1974.

56. J.B. Rajani, R. Payne and S. Michelfelder, Convective heat transfer from impinging oxygen-natural gas flames — Experimental results from the NG5 trials, Int. Flame Research Found. Report F 35/a/12, IJmuiden, The Netherlands, 1978.

57. J.K. Kilham and M.R.I. Purvis, Heat transfer from hydrocarbon-oxygen flames, *Comb. Flame*, 16, 47-54, 1971.

58. M. Fairweather, J.K. Kilham, and A. Mohebi-Ashtiani, Stagnation point heat transfer from turbulent methane-air flames, *Comb. Sci. Tech.*, 35, 225-238, 1984.

59. G.K. Hargrave and J.K. Kilham, The effect of turbulence intensity on convective heat transfer from premixed methane-air flames, *Inst. Chem. Eng. Symp. Ser.*, 2(86), 1025-1034, 1984.

60. C.C. Veldman, T. Kubota, and E.E. Zukoski, An Experimental Investigation of the Heat Transfer from a Buoyant Gas Plume to a Horizontal Ceiling. 1. Unobstructed Ceiling, National Bureau of Standards Report NBS-GCR-77-97, Washington, D.C., 1975.

61. T.B. Reed, Heat-transfer intensity from induction plasma flames and oxy-hydrogen flames, *J. Appl. Phys.*, 34(8), 2266-2269, 1963.

62. A.O. Hemeson, M.E. Horsley, M.R.I. Purvis, and A.S. Tariq, Heat transfer from flames to convex surfaces, *Inst. of Chem. Eng. Symp. Series*, 2(86), 969-978, Inst. of Chem. Eng., Rugby, U.K., 1984.

63. M.E. Horsley, M.R.I. Purvis, and A.S. Tariq, Convective heat transfer from laminar and turbulent premixed flames, in *Heat Transfer 1982*, U. Grigull, E. Hahne, K. Stephan, and J. Straub, Eds., Hemisphere, Washington, D.C., 3, 409-415, 1982.

64. A.S. Tariq, Impingement Heat Transfer From Turbulent and Laminar Flames, Ph.D. thesis, Portsmouth Polytechnic, Hampshire, U.K., 1982.

65. C.J. Posillico, Raman Spectroscopic and LDV Measurements of a Methane Jet Impinging Normally on a Flat Water-Cooled Boundary, Ph.D. thesis, Polytechnic Institute of New York, New York, 1986.

66. J.K. Kilham and P.G. Dunham, Energy Transfer from Carbon Monoxide Flames, *Eleventh Symp. (Int.) on Combustion*, The Combustion Institute, Pittsburgh, PA, 1967, 899-905.

67. R. Conolly, A Study of Convective Heat Transfer from High Temperature Combustion Products, Ph.D. thesis, University of Aston, Birmingham, U.K., 1971.

68. C.E. Baukal and B. Gebhart, Surface condition effects on flame impingement heat transfer, *Therm. & Fluid Sci.*, 15, 323-335, 1997.

69. C.J. Hoogendoorn, C.O. Popiel, and T.H. van der Meer, Turbulent heat transfer on a plane surface in impingement round premixed flame jets, *Proc. 6th Int. Heat Trans. Conf.*, Toronto, Ontario, Canada, 4, 107-112, 1978.

70. C.O. Popiel, T.H. van der Meer, and C.J. Hoogendoorn, Convective heat transfer on a plate in an impinging round hot gas jet of low Reynolds number, *Int. J. Heat Mass Trans.*, 23, 1055-1068, 1980.

71. T.H. van der Meer, Stagnation point heat transfer from turbulent low Reynolds number jets and flame jets, *Exper. Ther. Fluid Sci.*, 4, 115-126, 1991.

72. G.K. Hargrave, A Study of Forced Convective Heat Transfer from Turbulent Flames, Ph.D. thesis, University of Leeds, Leeds, U.K., 1984.

73. C.E. Baukal and B. Gebhart, A review of flame impingement heat transfer studies. 2. Measurements, *Comb. Sci. Tech.*, 104, 359-385, 1995.

74. J.K. Kilham and M.R.I. Purvis, Heat transfer from normally impinging flames, *Comb. Sci. Tech.*, 18, 81-90, 1978.

75. D.C.C. Chen and I.A. McGrath, Convective heat transfer in chemically reacting systems, *J. Inst. Fuel*, 42(336), 12-18, 1969.

76. W.H. McAdams, *Heat Transmission*, 3rd ed., McGraw-Hill, New York, 1954, chap. 10.

77. D. Altman and H. Wise, Effect of chemical reactions in the boundary layer on convective heat transfer, *Jet Propulsion*, 26(4), 256-269, 1956.

78. D.E. Rosner, Convective Heat Transfer with Chemical Reaction, Aeron. Res. Lab. Rept. ARL 99, Part 1, AD269816, 1961.

79. J.A. Fay and F.R. Riddell, Theory of stagnation point heat transfer in dissociated air, *J. Aero. Sci.*, 25, 73-85, 1958.

80. M.R.I. Purvis, Heat Transfer from Normally Impinging Hydrocarbon Oxygen Flames to Surfaces at Elevated Temperatures, Ph.D. thesis, The University of Leeds, Leeds, U.K., 1974.

81. C.E. Baukal and B. Gebhart, A review of empirical flame impingement heat transfer correlations, *Int. J. Heat Fluid Flow*, 17(4), 386-396, 1996.

82. G.K. Hargrave, M. Fairweather, and J.K. Kilham, Forced convective heat transfer from premixed flames. 2. Impingement heat transfer, *Int. J. Heat Fluid Flow*, 8(2), 132-138, 1987.

83. M.K. Alkam and P.B. Butler, Transient conjugate heat transfer between a laminar stagnation zone and a solid disk, *J. Thermo. Heat Trans.*, 8(4), 664-669, 1994.

84. M. Dianat, M. Fairweather, and W.P. Jones, Reynolds stress closure applied to axisymmetric, impinging turbulent jets, *Theoret. Comput. Fluid Dynamics*, 8, 435-447, 1996.

85. K.N.C. Bray, M. Champion, and P.A. Libby, Flames in stagnating turbulence, in *Turbulent Reacting Flows*, P.A. Libby, Ed., Academic Press, London, 1994, 573-607.

86. K.N.C. Bray, M. Champion, and P.A. Libby, Premixed flames in stagnating turbulence. III. The $\tilde{k}$-$\tilde{\varepsilon}$ theory for reactants impinging on a wall, *Comb. Flame*, 91, 165-186, 1992.

87. Y.Y. Park and H.Y. Kim, The characteristics of laminar diffusion flame impinging on the wall, *Trans. Korean Soc. Mech. Eng.*, 20(3), 979-987, 1996.

88. S. Ito, K. Asato, T. Kawamura, and Y. Ito, Characteristics of combustion and heat transfer of premixed flat flames with heat loss stabilized in a stagnation flow, *JSME Int. J.*, 39B(4), 852-858, 1996.

89. P. Popp and M. Baum, Analysis of wall heat fluxes, reaction mechanisms, and unburnt hydrocarbons during the head-on quenching of a laminar methane flame, *Comb. Flame*, 108, 327-348, 1997.

90. P. Popp, M. Hilka, M. Baum and T. Poinsot, Using Direct Numerical Simulation to Study Turbulent Combustion, Centre de Recherche sur la Combustion Turbulente, 1995, 67-95.

91. R.P. Lindstedt and E.M. Váos, Second moment modeling of premixed turbulent flames in impinging jet geometries, *Twenty-Seventh Symposium (International) on Combustion*, The Combustion Institute, Pittsburgh, PA, 1998, 957-962.

92. R.K. Cheng and I.G. Shepherd, The influence of burner geometry on premixed turbulent flame propagation, *Comb. Flame*, 85, 7-26, 1991.

93. C. Treviño, An asymptotic analysis of catalytic ignition in a stagnation-point flow, *Combust. Theory Modelling*, 3, 469-477, 1999.

94. L. Lees, Laminar heat transfer over blunt-nosed bodies at hypersonic flight speeds, *Jet Propulsion*, 26(4), 259-274, 1956.

95. R.M. Rauenzahn and J.W. Tester, Numerical simulation and field testing of flame-jet thermal spallation drilling. 1. Model development, *Int. J. Heat Mass Trans.*, 34(3), 795-808, 1991.

96. J.W. Tester and H.J. Herzog, The economics of heat mining: an analysis of design options and performance requirements of hot dry rock (HDR) geothermal power systems, *Energy Systems and Policy*, 15(1), 33-63, 1992.

97. R.M. Rauenzahn and J.W. Tester, Numerical simulation and field testing of flame-jet thermal spallation drilling. 2. Experimental verification, *Int. J. Heat Mass Trans.*, 34(2), 809-818, 1991.

98. B.E. Launder and D.B. Spalding, The numerical computation of turbulent flows, *Comput. Meth. Appl. Mech. Engng.*, 3, 269-289, 1974.

99. E.R. Van Driest, Turbulent boundary layer in compressible fluids, *J. Aero. Sci.*, 18, 145-160, 1951.

100. R.M. Rauenzahn and J.W. Tester, Rock failure mechanisms of flame-jet thermal spallation drilling: theory and experimental testing, *Int. J. Rock Mech. Geomech. Abstr.*, 26, 381-399, 1989.

101. R.M. Rauenzahn, Analysis of Rock Mechanics and Gas Dynamics of Flame-Jet Thermal Spallation Drilling, Ph.D. thesis, MIT, Cambridge, MA, 1986.

102. M.A. Wilkinson and J.W. Tester, Computational modeling of the gas-phase transport phenomena during flame-jet thermal spallation drilling, *Int. J. Heat Mass Trans.*, 36(14), 3459-3475, 1993.

103. M.A. Wilkinson and J.W. Tester, Experimental measurement of surface temperatures during flame-jet induced thermal spallation, *Rock Mech. Rock Engng.*, 26(1), 29-62, 1993.

104. T.N. Dey, More on Spallation Theory, Los Alamos National Laboratory Internal Memorandum No. ESS-3-286-84, 1984.

105. M. Dong and D.G. Lilley, Combustion Flowfield Prediction for CVD Diamond Synthesis, *Proc. 1992 ASME Int. Computers in Engineering Exhibition*, 1, 661-668, 1992.

106. S.V. Patankar, *Numerical Heat Transfer and Fluid Flow*, Hemisphere, New York, 1980.

107. H.C. Tsai and R. Greif, Thermophoretic transport for a three-dimensional reacting flow impinging on a disk with burner misalignment, *J. Materials Processing Manuf. Sci.*, 3, 217-242, 1995.

108. J. Hwang and J. Daily, Characterization of a vapor axial deposition (VAD) flame impinging on a disk for a study of electric field enhanced deposition, *Exper. Therm. Fluid Sci.*, 8, 58-66, 1994.

109. H.C. Tsai and R. Greif, Thermophoretic transport with application to external vapor deposition processes, *Int. J. Heat Mass Trans.*, 37, 257-268, 1994.

110. G.E. Schneider and M. Zedan, A modified strongly implicit procedure for the numerical solution of field problems, *Num. Heat Trans.*, 4, 1-19, 1981.

111. K.S. Chapman, S. Ramadhyani, and R. Viskanta, Modeling and parametric studies of heat transfer in a direct-fired furnace with impinging jets, *Proc. Heat and Mass Transfer in Fire and Combustion Systems,* P. Cho and J. Quintiere, Eds., ASME, New York, HTD-223, New York, 1992, 63-71.

112. K.S. Chapman, S. Ramadhyani, and R. Viskanta, Two-dimensional modeling and parametric studies of heat transfer in a direct-fired furnace with impinging jets, *Combust. Sci. Tech.*, 97, 99-120, 1994.

113. W.P. Jones, Models for turbulent flows with variable density and combustion, in *Prediction Methods for Turbulent Flows*, W. Kollmann, Ed., Hemisphere, Washington, D.C., 1980, 379-422.

114. W.A. Fiveland, Discrete-ordinates methods for radiative heat transfer in isotropically and anisotropically scattering media, *J. Heat Transfer*, 109, 809-812, 1987.

115. J.S. Truelove, Discrete-ordinates solutions of the radiative transport equation, *J. Heat Trans.*, 109, 1048-1051, 1987.

116. J.S. Truelove, A Mixed Grey Gas Model for Flame Radiation, Report for the United Kingdom Atomic Energy Authority, No. AERE-R-8494, 1976.

117. A.D. Johnson, A. Ebbinghaus, T. Imanari, S.P. Lennon, and N. Marie, Large-Scale Free and Impinging Turbulent Jet Flames — Numerical Modelling and Experiments, *IChemE Symp. Series*, No. 141, 113-126, 1997.

118. P.W.H. Barker, A.D. Johnson, and N. Goto, CFD Calculation of the Combustion and Radiation Properties of Large-Scale Natural Gas Jet Flames, Major Hazards Onshore and Offshore II, *IChemE Symp. Series* 139, Manchester, U.K., October 1995.

119. D. Bradley, L.K. Kwa, A.K. Law and M. Missaghi, Laminar Flamelet Modelling of Recirculating Premixed Methane and Propane-Air Combustion, *Comb. Flame*, 71, 109-122, 1988.

120. R.J. Kee, J.F. Grcar, M.D. Smooke, and J.A. Miller, A Fortran Program for Modelling Steady One-Dimensional Premixed Flames, Sandia Labs Report SAND85-8240.UC-401, 1991.

121. A.S. Chace, H.R. Hazard, A. Levy, A.C. Thekdi, and E.W. Ungar, Combustion Research Opportunities for Industrial Applications. Phase II, U.S. Dept. of Energy Report DOE/ID 10204-2, Washington, D.C., 1989.

122. R. Viskanta, Enhancement of Heat Transfer in Industrial Combustion Systems: Problems and Future Challenges, *Proc. ASME/JSME Thermal Engineering Joint Conf.,* J.R. Lloyd and Y. Kurosaki, Eds., 5, 161-173, ASME, New York, 1991.

123. R. Viskanta, Heat transfer to impinging isothermal gas and flame jets, *Exper. Therm Fluid Sci.*, 6, 111-134, 1993.

NOMENCLATURE

Symbol	Description	Units
c_p	Specific heat	J/kg-K or Btu/lb-°F
d	Diameter	m or ft
D	Dimensionless diameter $= d/d_n$	dimensionless
$D_{i\text{-mix}}$	Mass diffusivity of component i in mix	m²/s or ft²/s
f	Mixture fraction	dimensionless
g	Gravitational constant	m/s² or ft/s²
h	Convection heat transfer coefficient	W/m²-K or Btu/hr-ft²-°F
h^C	Chemical enthalpy	J/kg or Btu/lb
h^S	Sensible enthalpy $= \int c_p dt$	J/kg or Btu/lb
h^T	Total enthalpy $= h^C + h^S$	J/kg or Btu/lb
I	Fluid enthalpy	J/kg or Btu/lb
k	Thermal conductivity	W/m-K or Btu/hr-ft-°F
k	Turbulent kinetic energy	J or Btu
l	Length	m or ft
L	Distance between the burner and the target $= l_j/d_n$	dimensionless
Le	Lewis number $= \rho c_p D_{i\text{-mix}}/k$	dimensionless
Nu	Nusselt number $= hd/k$	dimensionless
p	Pressure	Pa or lb/ft³
Pr	Prandtl number $= c_p \mu/k$	dimensionless
q''	Heat flux	kW/m² or Btu/hr-ft²
q_f	Burner firing rate	kW or Btu/hr
r	Radial distance from the burner centerline	m or ft
R	Dimensionless radius $= r/d_n$	dimensionless
Re	Reynolds number $= \rho v d/\mu$	dimensionless
Ra	Rayleigh number	dimensionless
t	Temperature	°C or °F
T	Absolute temperature	K or °R
Tu	Turbulence intensity $= \sqrt{(v')^2}/\bar{v}$	dimensionless
u	Velocity in the x-direction	m/s or ft/s
v	Velocity	m/s or ft/s
	Velocity in the y or r direction	m/s or ft/s
v'	velocity fluctuation	m/s or ft/s
x	Coordinate direction	m or ft
$\bar{x}$	Transformation variable	dimensionless
y	Coordinate direction	m or ft

Greek Symbols	Description	Units
β_s	Velocity gradient stagnation $= \left(\dfrac{\partial u}{\partial x}\right)_{x=0,\,y=\delta}$	s^{-1}
$\tilde{\beta}$	Volumetric coefficient of expansion	K^{-1} or $^\circ R^{-1}$
ε	Eddy viscosity	m^2/s or ft^2/s
η	Transformation variable	dimensionless
μ	Absolute or dynamic viscosity	kg/m-s or lb/ft-s
Ω	Oxidizer composition $= \dfrac{O_2 \text{ volume in the oxidizer}}{O_2 + N_2 \text{ volume in the oxidizer}}$	O_2 volume in the oxidizer
ϕ	Equivalence ratio $= \dfrac{\text{Stoichiometric oxygen/Fuel volume ratio}}{\text{Actual oxygen/Fuel volume ratio}}$	stoichiometric oxygen/fuel volume ratio
ρ	Density	kg/m^3 or lb/ft^3

Subscripts	Description
ad	Adiabatic
b	Stagnation body or target
conv	Convective heat transfer
e	Edge of boundary layer
eq	Equilibrium
∞	Free stream conditions
j	Jet
H	H atom
max	Maximum
n	Burner nozzle
r	Radial direction
rad	Thermal radiation
s	Stagnation point
TCHR	Thermochemical heat release
w	Wall (target surface)

Overbar	Description
$\bar{p}$	$\bar{p} = p[(t_e + t_w)/2]$ = property evaluated at average temperature between the edge of the boundary layer and the wall
$\bar{\bar{p}}$	$\bar{\bar{p}} = \dfrac{\int_{t_w}^{t_e} p\,dt}{t_e - t_w}$ = property evaluated at the integrated average temperature between the edge of the boundary layer and the wall

12 Gas Turbines

D. Scott Crocker and Clifford E. Smith

CONTENTS

12.1 INTRODUCTION

A gas-turbine combustor is located between the compressor and turbine in a gas turbine engine, and performs the function of burning fuel that releases heat and increases the temperature of the working fluid (air plus products of combustion). The combustion process occurs at nearly constant pressure. Various fuels are used, including gaseous fuels (e.g., natural gas, syngas, etc.) and liquid fuels (e.g., JetA, JP fuels, diesel, etc.).

The modeling of gas-turbine combustors is extremely difficult because of the many physical processes that occur and interact with each other, such as turbulence, combustion, spray, heat transfer, radiation, etc. Until the advent of CFD analysis, combustors were designed using past experience, simple correlations, and the "build and bust" testing approach. Needless to say, this design approach was not very efficient. It was not uncommon to test 30 to 50 combustor designs. There was little innovation because of the excessive test costs that were required to prove a new design.

In the late 1970s, CFD analysis and the use of CFD analysis in the design process were first realized. Mongia et al.[1] was perhaps the first to utilize three-dimensional (3-D) Reynolds-averaged Navier-Stokes (RANS) CFD analysis in the design process. Although their TEACH-type code with Cartesian grids was very crude by today's CFD standards, they showed CFD could be used as a tool to increase understanding of the complex physical features of combustor flowfields. NASA, in the Hot Section Technology (HOST) Program, helped to assess the viability and shortcomings of gas-turbine combustor CFD analysis. Results from HOST Phase I[2-4] determined TEACH-type CFD codes were lacking in many areas, including:

1. Numerical differencing schemes
2. Physical models of turbulence and turbulence/chemistry interaction
3. Lack of benchmark experimental data to assess the physical models

HOST Phase II addressed some of these issues,[5] providing spray data and dome airflow measurements, and developing advanced differencing schemes. After HOST Phase II, combustor CFD code improvements were left mainly to in-house developments of gas turbine engine companies. The only commercial code that had a significant impact on combustor analysis and design was the FLUENT code. Developed in the mid-1980s, early versions of FLUENT were of the TEACH-type, and they became popular because of user friendliness and availability.

As CFD analysis became more popular in the 1990s, limitations soon surfaced that rendered TEACH-type codes obsolete. Complex 3-D combustor geometries needed curvilinear grids and solutions with accurate discretization schemes for highly skewed grid cells. The fuel nozzle and swirler boundary conditions into the combustor were shown to be very important,[6] so the computation needed to start inside the fuel nozzle and swirler, even upstream of the fuel nozzle and swirler. In order to get grid independent solutions of internal combustor flowfields, millions of grid cells were required.[7] To reduce the number of cells required, advanced software codes have been developed that utilized multi-block grids with arbitrary interfaces. Such codes are now routinely used at most gas-turbine manufacturers (e.g., CFD-ACE+, FLUENT, STAR-CD, CFX, etc.), although single-block codes (sometimes with cartesian grids) are still used in some day-to-day combustor analyses.

Other improvements have been realized in recent years. Unstructured solvers are becoming more and more popular. Most grid generators (e.g., CFD-GEOM, ICEM, Gridgen, TrueGrid, GridPro, etc.) can now import CAD geometries and produce multi-block grids or unstructured grids in a few days or less. More advanced physical models have been developed for turbulence, kinetics, turbulence-combustion interaction, turbulence-spray interaction, and radiation. To predict dome and liner temperatures, conjugate heat transfer capability has been added to many CFD codes, in which the temperature is calculated in both the gas path and solid material. To get more accurate predictions, it has become necessary to simulate not just the combustor internal flowfield, but also the external flowfield, going from the compressor discharge to the turbine inlet. Visualizing and processing the large data sets produced by CFD analysis can now be easily accomplished using advanced software such as CFD-VIEW, Fieldview, Tecplot, Ensight, etc.

In the chapter sections that follow, the methods and issues important in modeling gas-turbine combustors are discussed. The authors have attempted to present a practical approach to modeling gas-turbine combustors, emphasizing the models and techniques that are commonly used by combustor designers in real-world applications. Because the authors routinely design and analyze fuel nozzles/combustors, the discussion that follows incorporates many practical aspects that have been accumulated through direct experience and usage of CFD.

12.2 GAS-TURBINE COMBUSTOR MODELING

12.2.1 Modeling Overview

CFD modeling of combustors and fuel nozzles has progressed to the extent that CFD simulations are frequently included in the design process. Thus far, CFD models have generally been limited to isolated parts of the combustion system. Most models include only the reacting flow inside the combustor liner with assumed profiles and flow splits at the various liner inlets. Carefully executed models of this type can provide valuable insight into mixing performance, pattern factor, emissions, and combustion efficiency. Some insight into liner wall temperatures can also be indirectly derived from the near-wall gas temperatures. Perhaps the most important benefit of CFD modeling is the innovative ideas that CFD solutions sometime inspire. A typical annular combustor and the terminology used to describe gas turbine combustors is illustrated in Figure 12.1. In general, airflow discharges from the compressor, flows around the combustor, and enters the combustor through the fuel nozzle, dome swirler, dome cooling, liner cooling, primary holes, and dilution holes. Typical gas-turbine combustors are axial through-flow annular combustors (Figure 12.1), reverse-flow annular combustors (Figure 12.2),[8] and can-annular combustors (Figure 12.3).

There are numerous examples of combustor CFD solutions in the literature, of which Lawson,[9] Fuller and Smith,[10] and Lai[11] are a representative sample. Lawson was able to successfully match exit radial profile experimental data and then use the anchored CFD model to predict the radial profile that resulted from different cooling and dilution air patterns. Lawson used a one-dimensional code to predict flow splits and a two-dimensional CFD model to predict the flow profile at the exit plane of the swirl cup. This profile was then applied as a boundary condition in the three-dimensional model. Fuller and Smith were able to predict exit temperature profiles of an annular through-flow combustor that were in fairly good agreement with measurements. They also used a two-dimensional model to provide boundary conditions at the exit plane of the swirl nozzle. They did introduce the use of multi-block grids, which are essential for modeling complex geometries.[10] Lai was able to

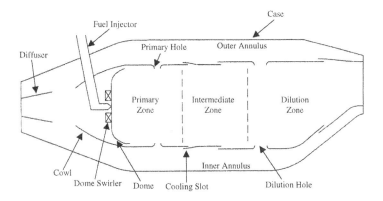

FIGURE 12.1 Main components of an axial through-flow annular combustor.

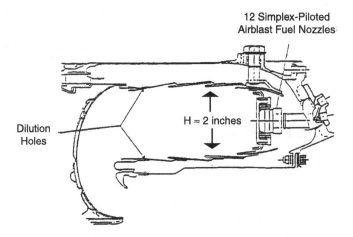

FIGURE 12.2   Schematic of a reverse-flow annular combustor. (From Crocker, D.S., Smith, C.E., and Myers, G.D., Pattern Factor Reduction in a Reverse Flow Gas Turbine Combustor Using Angled Dilution Jets, ASME Paper 94-GT-406, 1994. With permission.)

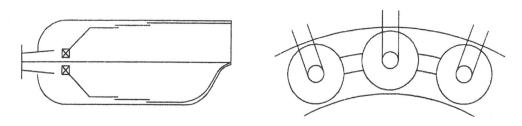

FIGURE 12.3   Schematic of can-annular combustor.

predict hot spots based on near-wall gas temperatures that corresponded to locations in the combustor that had experienced deterioration. Lai included the swirler passages in his model, which is an important step in reducing the uncertainty of prescribed boundary conditions.[11] Although Lai modeled the swirler passages and the rest of the combustor with a single-block grid, it is generally much easier and more efficient to use multi-block grids.

CFD models of the diffuser and annulus region have also been performed such as those of Srinivasan et al.,[12] Karki et al.,[13] and Little and Manners.[14] All three of these efforts at modeling the flowfield outside of the combustor liner were reasonably successful in predicting velocity profiles with relatively coarse two-dimensional and three-dimensional models. However, Mongia[7] points out that prediction of pressure losses is more difficult and the results are relatively poor.

A logical next step is to model the entire flowfield from the compressor diffuser to the turbine inlet with the flow inside and outside the combustor liner fully coupled. There are at least two important reasons to attempt such an ambitious task: (1) flow splits and boundary conditions for the combustor liner inlets are modeled and no longer need to be approximated; and (2) liner wall temperatures can be predicted when the flowfields on both sides of the liner walls are modeled in a coupled fashion. An example of this type of analysis is illustrated in Figure 12.4.[15]

One-dimensional annulus codes have traditionally been utilized to provide a reasonable prediction of flow splits, but they cannot capture important two- and three-dimensional flow features that can significantly affect total pressure losses. Nonuniform profiles of flow conditions, including velocities, jet angles, and turbulence properties, at inlet boundaries can also have a significant influence on the flowfield in a combustor. Turbulence properties at inlets are particularly difficult

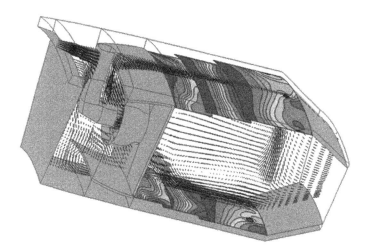

FIGURE 12.4 Conjugate heat transfer CFD analysis of full combustor from compressor exit to diffuser inlet. (From Crocker, D.S., Nickolaus, D., and Smith, C.E., *J. Engr. for Gas Turbine and Power,* 121, 89-95, 1999. With permission.)

to estimate. McGuirk and Spencer[16] discuss the sensitivity of discharge coefficient and jet trajectory of dilution holes to small changes in the dilution hole geometry. They emphasize the need to provide realistic profiles at dilution hole inlets.

Another important issue is accurate definition of the flow through the dome/fuel injector air swirlers. Fuller and Smith[10] demonstrated the sensitivity of the overall solution to the details of the swirler air flow. The aerodynamics of recent lean dome combustor designs are dominated by the large percentage of air that is admitted through the dome swirlers. Crocker et al.[17] discusses the importance of detailed prescription of fuel nozzle/dome swirler boundary conditions and describes an approach for accurate prediction of flow splits and effective area. Fuller and Smith[10] modeled the air swirler and combustor flow fields in a decoupled manner that may be appropriate in some cases. In general, however, it is necessary to couple the combustor solution and at least part of the swirler passages because of strongly recirculating flow in the region of the swirler. Another important consideration is the effect of nonuniform feed pressure at the inlet of the dome swirlers. The advent of multi-block and many-to-one grid topology, discussed later in the chapter, make it feasible to include a detailed model of the swirler passages in the overall combustor model.

The treatment of dome/liner cooling flow is usually of secondary importance in most analyses and is not modeled in much detail. Liner cooling airflow is typically introduced as film cooling at the tip of cooling louvers if conventional film cooling is employed, or distributed along the liner if effusion cooling is employed. Circumferentially preferred cooling is easily handled by tailoring the airflow distribution. The liner temperature is usually specified (typically 1600°F) or assumed adiabatic. However, the prediction of liner wall temperatures is becoming increasingly important as more aggressive cycles increase liner heat loads and low emissions designs reduce the available cooling air. Modeling liner wall temperatures requires appropriate grid resolution near the walls and cooling flow inlets. Although conjugate heat transfer approaches that couple heat transfer in solid material with the fluid flow solution have been available for some time, the authors are not aware of any attempt (in the open literature) to directly model gas-turbine combustor liner wall temperatures using CFD.

12.2.2 GRID GENERATION

Generation of a grid for complex combustor geometries in a timely manner and with sufficient grid resolution is a challenging task. Mongia,[7] in a review of combustion modeling, estimated that as

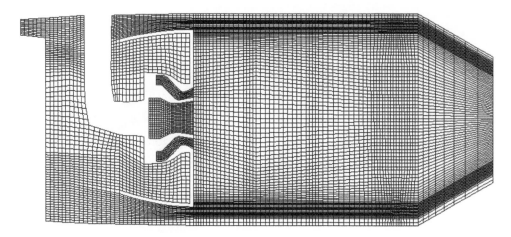

FIGURE 12.5   Centerline axial plane of fine grid (370,000 cells). (From Crocker, D.S., Nickolaus, D., and Smith, C.E., *J. Engr. for Gas Turbine and Power,* 121, 89-95, 1999. With permission.)

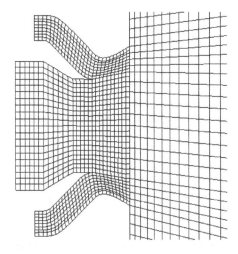

FIGURE 12.6   Centerline axial plane of fuel nozzle portion of grid showing many-to-one interface. (From Crocker, D.S., Nickolaus, D., and Smith, C.E., *J. Engr. for Gas Turbine and Power,* 121, 89-95, 1999. With permission.)

many as 12 million cells may be required to provide a near-grid-independent solution for a comprehensive one-nozzle sector combustor model. Such a model would include detailed modeling of swirler vane passages. Second-order numerical schemes can reduce that requirement to less than 5 million cells. Mongia's estimate is reasonable given the assumption of one-to-one grid cell interfaces between blocks. Many-to-one grid cell interfaces can dramatically reduce the total number of required cells. An axial cross-section of a typical 3-D combustor grid with 11 domains and 370,000 cells is shown in Figure 12.5.[15] Many-to-one is illustrated by a close-up of the fuel nozzle grid interfaced with the primary zone grid shown in Figure 12.6. Detailed structures, such as swirler passages and even swirler vanes, can be modeled with a fine grid without transferring the high-density grid into the rest of the combustor. The swirler passages shown in Figure 12.6 were modeled with 35,000 cells, which does not include swirler vanes. Swirler vanes could be included with approximately 100,000 additional cells.

The challenge in generating the grid is twofold: (1) it must be possible to generate a quality grid in a reasonable amount of time; and (2) the grid must have adequate resolution to capture the relevant physics but be small enough that approximately 24-hour execution time can be maintained (if the CFD model is to be used as a practical design tool). Most CFD codes that have adequate physical models for solving reacting flow with spray utilize a structured grid topology. It is possible to model virtually any geometry using a multi-block, many-to-one structured grid approach; however, the process is quite painful for complex geometries such as swirler vane passages. Unstructured grids are generally easier to generate for complex geometries and the potential for automation is greater. Automation of grid generation (e.g., swirler vanes, dilution holes, and cooling slots) is an important improvement for making comprehensive combustor modeling a more practical design tool. Unstructured grids also may require fewer grid cells because the grid packing can be isolated in the appropriate area (the many-to-one feature accomplishes the same thing to a lesser extent in structured grids.) Solution adaptive grids, which automatically refine the grid in regions where the gradient of key variables is high, is another promising approach for achieving grid-independent solutions with a reasonable number of grid cells. The need for mixed-element "brick" cells near walls is one of the primary difficulties. Physical models in unstructured grid solvers are generally less mature than for structured grid solvers, but should catch up in the relatively near future. One interesting approach is to use an unstructured solver for the nonreacting and generally more geometrically complex regions feeding the liner, and a more mature structured solver for the reacting region inside the liner in a fully coupled manner. An additional advantage to this approach is that the computational expense associated with reaction and radiation can be limited to the appropriate regions of the flow.

In the near- to mid-term future, it should be possible to perform fully coupled combustion system CFD calculations. Adequate grid resolution can be achieved with roughly 1 million cells using advanced grid generation techniques and second-order discretization schemes. Execution time for the 370,000 cell case illustrated in Figures 12.4 and 12.5 was about 48 hours (with radiation and with little optimization) on a 500-MHz DEC Alpha 500. If one assumes 4 times more execution time for a 1 million cell case, a factor of 8 improvement is needed to achieve one-day turnaround time. If history is a guide, such improvement may be realized for a single workstation within 5 years. Overnight execution can be achieved now with 10 to 15 workstations operating in parallel.

12.2.3 THE ROLE OF CFD IN GAS-TURBINE COMBUSTOR MODELING

CFD models are capable of providing the designer with detailed information about the behavior of the complex flowfield in and around the gas turbine combustor. The results of the analyses give the designer both a unique window into the mechanisms of the combustor and a quantitative prediction of combustor performance. Although the value of the former is more difficult to quantify, it is usually just as important as the latter for the experienced designer. The insight derived from detailed predictions of flow, temperature, and species fields enables the development of innovative designs and allows for more efficient optimization of promising concepts.

A typical CFD analysis provides the following predictive information:

1. **Harmonic temperature pattern factor.** This is similar to rig temperature pattern factor, except the CFD analysis is generally only performed for a one-nozzle segment (for an annular combustor), so the effect of nozzle-to-nozzle fuel flow or air flow variations cannot be assessed. The temperature pattern factor is calculated as the maximum temperature at the combustor exit, minus the average temperature at the combustor exit, divided by the temperature rise (average combustor exit temperature minus inlet temperature). Typical "good" rig pattern factors are 0.20 to 0.30 for axial through-flow combustors and 0.15 to 0.25 for reverse-flow combustors. CFD harmonic pattern factors should be less than rig pattern factors, and "good" values should range between 0.12 to 0.20.
2. **Radial temperature profile.** Turbine designers prefer a center-peaked radial profile. Radial temperature profiles can be accurately predicted using CFD, as long as the dilution

jets are accurately prescribed, and the combustor front-end heat release is well-predicted. For example, Mongia,[18] Anand et al.,[19] and Smiljanovski and Brehm,[20] all present good agreement between predicted radial temperature profiles and measurements.

3. **Combustion efficiency.** Normally, combustion efficiency is 99.9% or better for medium and full power operation. At idle conditions combustors typically have about a 1% inefficiency caused by excessive unburned hydrocarbons (UHC) and carbon monoxide (CO). This inefficiency is typically caused by "quenching" at the liner walls, or overly rich pockets in the primary zone that are not oxidized before exiting the combustor. CFD modeling usually attempts only to predict CO levels because realistic prediction of UHC requires detailed kinetics that are not generally available with a practical design analysis approach.

4. **NOx emissions.** NOx can be post-processed because changes in NOx have a relatively insignificant effect on heat release. Thermal NOx is comparatively easy to predict using the one-step Zeldovich mechanism. For most aero applications, only thermal NOx need be considered. However, for industrial gas turbines, in which NOx emission levels are less than 25 ppmv, prompt and nitrous mechanisms must also be considered. NOx is produced in high-temperature regions of the combustor, typically where stochiometric ($\phi = 1$) burning is occurring. In most engine applications, most of the NOx is formed at high power conditions.

5. **Liner temperature.** Liner temperatures are predicted using a conjugate heat transfer analysis that solves the internal and external combustor flowfields as well as the liner material (including thermal barrier coating, if used). The internal combustor flowfield must include convection and radiation effects, and the liner must be allowed to radiate to the engine case. For lean front-end combustors (common for low-emission designs), soot radiation can be ignored (although gas radiation should be included), but for rich and/or high-performance combustors, soot radiation may be the dominant heat load to the liner and should be included.

12.3   FUEL NOZZLE MODELING

12.3.1 INTRODUCTION

Design of the fuel nozzle for gas turbine engines is critical to the performance of the combustor. The fuel spray distribution affects performance parameters, including ignition, lean blow-out, pattern factor, gaseous and smoke emissions, combustion efficiency, and liner wall temperatures. The air flow through the fuel nozzle and/or dome swirler is at least partially, and often completely, responsible for determining the fuel atomization and distribution in the combustor primary zone. In most combustors, moderate to strongly swirling air is used to maintain a stable flame by creating a central recirculation zone that entrains hot combustion products and mixes them with unreacted fuel and air. The swirling flow is usually produced by some combination of radial vanes or axial vanes. Radial vanes generally have lower loss and can be packaged with shorter length. Axial vanes can usually be packaged in a smaller diameter and are necessary if fuel is introduced between two air passages. Specific design details of swirler vanes and downstream air passages vary widely.

CFD can be used as a tool for modeling and for increasing understanding of the air flowfield through the fuel nozzles and in the combustor primary zone. A basic characteristic of a fuel nozzle is its effective flow area (ACd). Accurate prediction of the ACd of a given fuel nozzle geometry can significantly improve design productivity, especially with the advent of lean injection in which a relatively large percentage of the total air flow enters the combustor through the fuel nozzle. It is now possible to produce a fuel nozzle with the correct ACd reasonably quickly using rapid prototyping techniques. Rapid prototyping and simple ACd measurements, however, will provide only limited understanding of the physics of the flow through the nozzle. Accurate CFD modeling can provide the designer with a level of understanding (i.e., the influence of design parameters) that cannot be practically achieved through experimental measurements. Greater understanding will inevitably lead to better designs produced in less time.

Currently, fuel nozzles/air swirlers are designed primarily through a combination of scaling from existing designs, trial-and-error flow tests, and past experience of the designer. CFD is just beginning to be used as an effective design tool. Numerous studies have been performed in which CFD analyses of fuel nozzles were compared with experimental velocity field data. Most of the work that has been done in modeling fuel nozzles has assumed two-dimensional axisymmetric flow. The quality of the simulations has been inconsistent at best. Mongia,[7] in a review of combustion modeling, admits that qualitative agreement with LDV measurements using simple profile boundary conditions and the standard k-ε model has been "less than successful." One of the most comprehensive studies covering a wide range of axial swirler design parameters was performed by Liley[21] and several of his students. Rizk,[76] Rizk and Chin,[22] and Rizk et al.[23] have analyzed various fuel nozzles. In these studies, the agreement between CFD results and experimental measurements for air flow (without fuel spray) was generally poor. The results were, however, somewhat better when adjustments were made to the air flow rates and/or passage flow splits. Fuller and Smith[6] showed good agreement with the experimental measurements of McVey et al.[24] for a research nozzle. Smith et al.[77] also demonstrated reasonably good comparison with experimental data for a dual filmer air blast nozzle.

Significant difficulties still remain in the application of CFD to the design of fuel nozzles. The most important of these difficulties are:

1. Rapid grid generation for complex passages
2. Timely solution convergence with adequate grid resolution
3. Adequate turbulence models
4. Definition of boundary conditions

Grid generation difficulties have been largely overcome with the availability of multi-domain topology for structured grids and easy to use graphic interfaces. A quality grid for a geometry with numerous complex passages can usually be generated in, at most, a few hours. Modified grid meshes for minor model variations can then be generated in minutes. Adequate grid resolution for two-dimensional axisymmetric models usually requires from 10,000 to 15,000 cells. Solution convergence for such a model can typically be achieved on state-of-the-art workstations in 1 or 2 hours.

Two-equation turbulence models, such as the standard k-ε model,[25,26] that employ an isotropic eddy viscosity, have been demonstrated to be less than completely adequate for swirling flows.[27] The turbulence models are most inadequate near the centerline of axisymmetric swirling flows. It is generally observed that the inadequacy of the standard model arises due to the lack of sensitivity of the dissipation rate to extra strains such as swirl. To remedy this, Yakhot et al.[28] proposed a modified k-ε model based on the renormalization group (RNG) theory. In their RNG k-ε model, the coefficient of production of dissipation varies as a function of the strain parameter η, which is defined as the ratio of turbulent to mean strain time scales. More advanced turbulence models, such as the Reynolds-stress transport models, have the potential to improve solution accuracy, but have not been demonstrated to justify the associated price of computational effort and convergence difficulty. Poor CFD results are frequently blamed on poor turbulence models and the blame is certainly justified to some extent. On the other hand, one must be careful not to falsely accuse turbulence modeling when other, more easily resolved sources of error may exist. The most prevalent source of error in modeling fuel nozzle flows, in the author's opinion, is inaccurate specification of boundary conditions. The most significant boundary conditions are the flow properties (velocity components and turbulence levels) where the swirler air is introduced to the computational domain, usually at the trailing edge of the swirl vanes. Specification of the velocities is not as easy as it may seem at first glance. The biggest difficulty is determining what percentage of the total flow goes through each passage. Even measurement of ACd for individual passages is generally not adequate because strong interactions between air passages may exist. Determination of flow splits requires accurate prediction of the total pressure drop through each air passage. The pressure drop through the swirl vanes is usually significant. A systematic method for correctly determining swirler boundary conditions is first presented below.

12.3.2 DETERMINATION OF SWIRLER BOUNDARY CONDITIONS

This discussion of swirler boundary conditions assumes that the computational domain begins at the trailing edge of the swirl vanes or the exit of angled holes or slots and that the CFD analysis is two-dimensional axisymmetric. For a given mass flow through a swirler, the velocity normal to the swirler exit plane (at a location just downstream of the exit plane) is determined from continuity as

$$V_n = W_{sw}/\left(A_{se}\rho\right) \tag{12.1}$$

W_{sw} is the swirler mass flow rate, A_{se} is the swirler exit area, and ρ is the air density. V_n is the axial velocity u for axial swirlers and the radial velocity v for radial swirlers. The velocity normal to the swirler walls (v for axial swirlers and u for radial swirlers) can generally be assumed to be zero. The only case where it might not be zero is when there is a strong gradient in the normal velocity V_n. The simplest way to determine the tangential velocity w is to assume that the flow exiting the swirler has the same angle as the swirl vanes. However, this assumption can lead to significant errors because small changes in the swirl angle of the air can result in large changes in the flow pattern. A better expression for the tangential velocity considers the effects of vane blockage b, slot discharge coefficient Cd_s, and vane turning efficiency η. The tangential air velocity is determined from

$$w = \eta V_{ns} \tan \alpha_v \tag{12.2}$$

where the turning efficiency is given by

$$\eta = \sin \alpha_a/\sin \alpha_v \tag{12.3}$$

and V_{ns} is the velocity normal to the swirler exit plane just before exiting the swirler:

$$V_{ns} = W_{sw}/\left(A_{se}\rho Cd_s(1-b)\right) \tag{12.4}$$

α_v is the swirler vane angle and α_a is the average air exit angle. The higher value of V_{ns} (compared to V_n) should be used to determine the tangential velocity because the tangential velocity is not significantly affected by the dump at the swirler exit. Halpin[29] calculated a maximum turning efficiency of 0.92 for radial swirlers and an even lower value when a vane loading parameter was increased. However, the turning efficiency was assumed to be 1.0 for the analyses given in the following sections with no apparent detrimental effect on the results. The vane blockage is typically about 0.1 for axial or radial swirlers with thin vanes. The blockage is sometimes higher for radial swirlers with straight slots or for angled holes. The discharge coefficient can be assumed to be very close to 1.0 when the flow area through the swirler contracts substantially, as it does for thin vane swirlers. If constant area holes or slots are used, the discharge coefficient can have a substantial effect. A value for Cd_s of 0.75 to 0.80 is a typical estimate for slots or holes with length/diameter greater than unity.[30] The swirl vane pressure loss discussed below should be set to zero for constant area slots or holes.

This discussion on setting the velocities (or swirl angle) at the computational inlets assumes that the velocities are uniform across the inlet. This will not be the case in general, especially for axial swirlers with a small hub diameter.[21] Even the swirl vane angle is frequently not constant for small hub diameter swirlers. For helical vanes, the tangential velocity is proportional to the tangent of the radius and decreases toward zero as the radius approaches zero. The axial velocity (for axial swirlers) also increases with increasing radius because of the swirl-generated radial pressure gradient. Velocity gradients at the swirler exit are important in some cases, but can often be neglected in practice for two reasons: (1) radial swirlers and most axial swirlers do not have a small hub diameter relative to the outer diameter; and (2) even when there is a small diameter axial swirler at the core of the nozzle, its mass flow contribution is generally quite small.

Turbulence levels at the inlet boundaries are generally not known, but they seem to have a relatively small effect if there is a confined passageway that extends for some length downstream of the swirler exit. Such is the case for most designs. A turbulence intensity of 10% and turbulence length scale of 10% of the swirl vane height are reasonable guesses.

The mass flow through each swirler is controlled by the total pressure drop through the fuel nozzle. Most of the pressure drop for a well-designed fuel nozzle is a result of accelerating the flow through the nozzle and then dumping into the combustor primary zone. The dump loss is determined from the CFD solution. The loss through the swirl vanes is also usually important and should not be neglected. The loss through swirl vanes can be estimated by the following equations, which approximate the data of Kilik[31] for vanes with aspect ratio of 0.4 and repeated by Lefebvre[32]:

Flat vanes:

$$\Delta P_{sw} = 0.5\rho V_i^2 \left(0.02625\alpha_v^2 - 2.1\alpha_v + 45.5\right), \quad \alpha_v \geq 40°$$

$$\Delta P_{sw} = 0.5\rho V_i^2 \left(0.0875\alpha_v\right), \quad \alpha_v < 40°$$

(12.5)

Curved vanes:

$$\Delta P_{sw} = 0.5\rho V_i^2 \left(0.0075\alpha_v^2 - 0.6\alpha_v + 12.5\right), \quad \alpha_v \geq 40°$$

$$\Delta P_{sw} = 0.5\rho V_i^2 \left(0.0125\alpha_v\right), \quad \alpha_v < 40°$$

(12.6)

V_i is the velocity normal to and upstream of the swirler inlet plane. These pressure loss curves were generated based on experimental data for axial swirlers, but they should also apply to radial swirlers as long as the ratio of the swirler inner radius to outer radius approaches unity. Equations (12.5) and (12.6) account for the minor dump loss that occurs at the swirl vane exit for typical thin vanes. If there is vane blockage much greater than 0.1, the vane exit dump loss may be significant and should be added to the swirler total pressure loss.

The mass flow rate must be determined by an iterative process in which the nozzle dump loss, and any other air passage losses, are calculated by the CFD solution and the swirler loss is calculated by Equations (12.5) and (12.6). The steps of the iterative procedure are as follows:

1. Estimate the flow rates, w_{sw}, for each air passage for the overall fuel nozzle total pressure drop, ΔP_{tot}.
2. Use the estimates for Step 1 to calculate the total pressure at the swirler exit (computational domain inlet) for each air passage.
3. Use the flow rate estimates from Step 1 to set the inlet velocities (or use the total pressures from Step 2 and set the inlet total pressures) and obtain a converged CFD solution.

4. Check the calculated total pressure at the swirler exit (CFD inlet) for each air passage and compare with the values calculated from Step 2 (or check the calculated mass flow for each air passage and compare with the values from Step 1).
5. Return to Step 1 and adjust the air flow rates as necessary and repeat until the total pressure from Step 2 and Step 4 agree (or until the mass flows from Step 1 and Step 4 agree).

The ACd for the fuel nozzle can be found from

$$\text{ACd} = W_{tot} / \sqrt{2\rho\Delta P_{tot}} \tag{12.7}$$

The result of the iterative process is an accurate prediction of the total effective flow area and individual passage flow splits for the fuel nozzle. The extra effort required for the iterative process is not excessive because only three or four iterations are usually required and the CFD solutions converge much more quickly after the initial run. Table 12.1 shows a comparison of measured and calculated ACd's (calculations performed at CFD Research Corporation) for several different fuel nozzles with a range of design characteristics and size. In most cases, the calculations were performed as part of the design process before the measurements were taken. It is evident from the results shown in Table 12.1 that it is possible to accurately predict fuel nozzle effective flow area.

12.3.3 EXAMPLE FUEL NOZZLE CALCULATION

A low-emissions fuel nozzle with both radial inflow and axial air swirlers was developed by CFD Research Corporation for Pratt & Whitney. Experimental tests at ambient conditions using full-scale prototype hardware were conducted by United Technologies Research Center (UTRC). A cross-section of the fuel nozzle is shown in Figure 12.7. The low-emissions fuel nozzle consisted of a central fuel injector with two axial swirlers. The fuel was injected radially outward through discrete holes and atomized by a combination of fuel pressure and the high-velocity air stream. A large percentage of the total air flow (82%) entered through the radial inflow swirler. The fuel was

TABLE 12.1
Comparison of Measured and Calculated ACd Values

Fuel Nozzle	ACd (cm²)	
	Measured	Calculated
Low emissions-1	15.1	15.2
Integrated combustor	2.45	2.42
Low emissions-2	7.2	7.0
Dual-spray airblast-1	0.49	0.48
Dual-spray airblast-2	0.37	0.37
LDI airblast-1	7.3	7.4
LDI airblast-2	6.8	7.4
LDI airblast-3	12.9	12.5
LDI high shear	14.8	15.3

From Crocker, D.S., Fuller, E.J., and Smith, C.E., Fuel Nozzle Aerodynamic Design Using CFD Analysis, ASME Paper 96-GT-127, 1996. With permission.

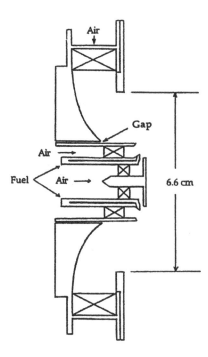

FIGURE 12.7 Low-emissions fuel nozzle cross section. (From Crocker, D.S., Nickolaus, D., and Smith, C.E., *J. Engr. for Gas Turbine and Power,* 121, 89-95, 1999. With permission.)

injected radially outward to provide better fuel/air uniformity. The radially outward fuel injection was the motivation behind the rather unique fuel nozzle design with the center pintle. Aerodynamically curved vanes with a 50° vane exit angle were employed for the radial inflow swirler and the outer axial swirler. The inner swirler had 45° straight helical vanes to simplify fabrication. Unswirled air was also admitted through a small gap as indicated in Figure 12.7. The diameter of the fuel nozzle at the exit plane to the combustor was 6.6 cm.

The experimental tests and the corresponding CFD analyses were performed for unconfined flow at atmospheric pressure and ambient temperature conditions. The total pressure drop through the fuel nozzle was 1.7%. Axial and tangential velocities at a position 0.76 cm downstream of the nozzle exit (the downstream face of the center pintle) were experimentally measured using a laser-Doppler velocimeter (LDV).

A portion of the two-dimensional axisymmetric, multi-block grid used for the CFD analyses is shown in Figure 12.8. The remainder of the grid extended 25 cm downstream in the axial direction and 15 cm in the radial direction. The model began at the trailing edge of the swirler vanes. The swirler locations have been included in Figure 12.8 for clarity. The exit boundary and the outer radial boundary were assumed to be fixed at atmospheric pressure. The swirler inlet boundaries were modeled as fixed mass inlets with air swirl angles of 48.0°, 52.8°, and 52.9° for the inner and outer axial swirlers and the radial inflow swirler, respectively. The mass flow rate for each passage was set using the iterative procedure described in the previous section. The air swirl angles were slightly higher than the vane angles because of the effect of vane blockage. The turbulent intensity was assumed to be 10% and the turbulent length scale was assumed to be 10% of the swirl vane height.

The predicted ACd values for the fuel nozzle and for each passage individually are shown in Table 12.2. The predicted ACd of 15.2 cm² (obtained before experimental measurement was performed) is almost identical to the measured ACd of 15.1 cm². Table 12.2 also shows the pressure drop through each swirler. A relatively large fraction of the 1.7% total pressure drop occurred in

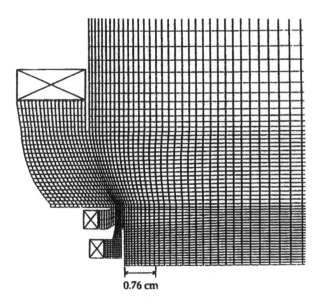

0.76 cm

FIGURE 12.8   Partial computational grid for low-emissions fuel nozzle. (From Crocker, D.S., Fuller, E.J., and Smith, C.E., Fuel Nozzle Aerodynamic Design Using CFD Analysis, ASME Paper 96-GT-127. With permission.)

TABLE 12.2
Low-Emissions Fuel Nozzle ACd and Swirler Pressure Drop

	Axial, Inner	Axial, Outer	Radial Inflow	Gap[a]	Total
ACd (cm²)	0.45	1.74	12.39	0.58	15.2
Swirler ΔP/P (%)	1.34	0.78	0.16	0.36	—

[a] Pressure drop based on gap length.

the axial swirlers because they were at a low radius, resulting in a high axial velocity through the swirler. The total pressure drop through the radial inflow swirler, where most of the air is admitted, was quite low.

Four CFD cases were performed and compared with experimental measurements:

Case 1: k-ε turbulence model, upwind differencing
Case 2.   RNG turbulence model, upwind differencing
Case 3.   RNG turbulence model, second-order upwind differencing
Case 4.   Low Reynolds number turbulence model, upwind differencing

Streamline contours for Case 3 are shown in Figure 12.9. A recirculation zone behind the center pintle is evident. Axial and tangential velocity profiles for each of the four cases are compared with the experimental results in Figure 12.10. All of the cases show reasonable agreement with the experimental data. The RNG turbulence model in Case 2 had slightly better agreement with the diameter of the jet (axial velocity near 3 cm radius) and the tangential velocity in the center (near 2 cm radius) of the jet flow. The second-order upwind differencing in Case 3 further improved the prediction of the jet width. All four cases predicted a secondary recirculation zone on the centerline at the face of the pintle. A secondary recirculation zone streamline is not captured in Figure 12.9 because it has a very low flow rate. The size of the secondary recirculation zone is over predicted by the first three cases, as evidenced by the increase in axial velocity near the centerline.

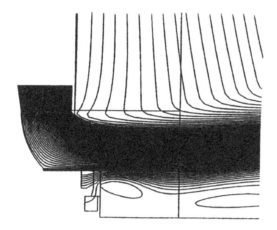

FIGURE 12.9 Streamline contours for low-emissions fuel nozzle discharging into ambient environment: Case 3. (From Crocker, D.S., Fuller, E.J., and Smith, C.E., Fuel Nozzle Aerodynamic Design Using CFD Analysis, ASME Paper 96-GT-127. With permission.)

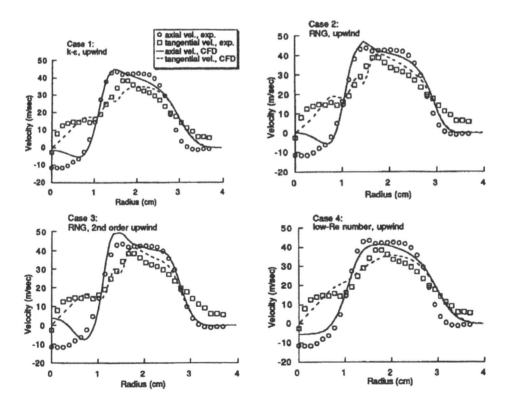

FIGURE 12.10 Velocity profiles at axial location 0.76 cm downstream of pintle face for low-emissions fuel nozzle. (From Crocker, D.S., Fuller, E.J., and Smith, C.E., Fuel Nozzle Aerodynamic Design Using CFD Analysis, ASME Paper 96-GT-127. With permission.)

The low Reynolds number turbulence model (Case 4) better matches the experimental data near the centerline. The low Reynolds number model[33] avoids difficulties with wall functions by permitting integration of the momentum and k-ε equations all the way to the wall. The low Reynolds number model probably performed better because the grid cells adjacent to the pintle and near the centerline were in the laminar sublayer. The elevated swirl velocity at the outer edge of the jet was also not predicted by any of the cases.

12.3.4   MODELING OF LIQUID SPRAY

Liquid spray in current industrial CFD applications is modeled as packets of droplets with common properties (size, velocity, temperature, etc.) that are tracked in a Lagrangian frame of reference. The droplets are coupled to the gas-phase equations in an explicit manner through mass, momentum, and energy source terms for the gas equations. Atomization models are generally not available, so appropriate assumptions must be made for initial droplet size/distribution, position relative to the injector exit, and velocity.

Droplet size can be estimated based on flow conditions and atomizer features using empirical data.[32] Experimental data helps to provide a starting point for the empirical relationships, but is usually limited to a few points at atmospheric pressure. More thorough measurements can certainly lead to better characterization of the injection conditions,[34] but the degree of overall solution improvement is not clear. At high power, where drop sizes are small and evaporation rates are high, the solution is relatively insensitive to drop size. NOx emissions are more likely to be affected because drop size strongly affects evaporation rate, which in turn affects fuel-air distribution. The converse is true at lower power conditions. A Rosin-Rammer drop size distribution is often assumed. A small number (<10) of size bins is generally adequate, although neglected large drops will be most likely to reach the liner walls, affecting CO emissions and carbon deposition.

Initial injection position and velocity magnitude/direction are not known and must be assumed based on quite rough guidelines. Unfortunately, the solution can be strongly sensitive to initial injection assumptions depending on the local aerodynamics of the case. Swirling flow in particular tends to magnify the variations in downstream drop trajectory. The sensitivity is greatest for low power conditions where the drops live longer. Guidelines for drop injection conditions must consider the following issues: (1) atomization mechanisms for pressure atomizers and airblast atomizers are fundamentally different, (2) droplets are formed with a significant distribution of position and velocity, and (3) spray models normally assume that droplet concentration is dilute (<10% by volume), which is usually not the case very close to the injection point.

12.4   JETS-IN-CROSSFLOW

Primary and dilution holes introduce airflow into the combustor, and can represent 10 to 40% of the combustor airflow in conventional combustors. In Rich burn-Quick mix-Lean burn (RQL) configurations used to reduce NOx emissions, dilution (quick mix) jets can represent up to 60 to 70% of the combustor airflow. Needless to say, it is very important to correctly predict the penetration as well as the mixing of primary/dilution jets. Based on experience (see Bain[35]), it is best to include external flow in the calculation domain if accurate predictions are required. In this way, the velocity profile entering the orifice and the turbulence quantities are numerically captured, and not "guessed." Otherwise, one usually assumes a uniform velocity profile at the orifice, with a flow angle predicted by a 1-D flow code (if such information is available). Also, if the external flow is not included, the orifice diameter must be reduced from its geometric dimension to account for the discharge coefficient, and the turbulence quantities are set to a turbulence intensity of 0.10 and length scale of 10% of the orifice diameter.

If the external flow is modeled, the jet flow conditions are determined from the actual CFD calculation. In this case, it is important to have a grid density that accurately captures the discharge coefficient. CFDRC has performed benchmark calculations and shown that a 6 × 6 grid of the orifice diameter, combined with five cells for the liner thickness, is sufficient to capture the discharge coefficient (with second-order upwind differencing on velocity and pressure). Figure 12.11 compares

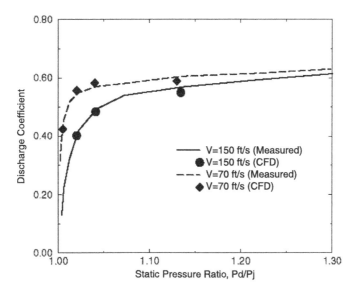

FIGURE 12.11   Comparison between predicted and measured discharge coefficients using 6 × 6 × 5 grid through orifice. (Measured from Dittrich, R.T. and Graves, C.G., Discharge Coefficients for Combustor-Liner Air-Entry Holes. I. Circular Holes with Parallel Flow, NACA Technical Note 3663, 1956.)

C_d predictions with measurements, and show that good agreement can be obtained with a 6 × 6 × 5 orifice grid.[36] It is also important to have smooth grid size transition axially through the hole from the annulus to the inside of the combustor. The k-ε turbulence model with Prandtl and Schmidt numbers of 0.25 give good agreement with measurements for combustor dilution zones. The Prandtl and Schmidt number of 0.25 is somewhat controversial (e.g., see Anand et al.[19] and Smiljanovski and Brehm,[20] who prefer to use 0.7), but there is substantial evidence in the literature for using 0.25 (e.g., He et al.[37] and Gulati et al.[38]). In reality, the value of 0.25 is really just a way to compensate for the inadequacy of the k-ε turbulence model, which tends to underpredict jet-in-crossflow mixing. When the external flow is modeled, it is important to check the pressure drop across the liner once the CFD analysis is converged. Because the airflow through the liner orifices is specified by the user (usually from 1-D flow codes with assumed C_d), it is relatively easy to get the incorrect pressure drop, which then translates into incorrect jet velocity and jet penetration. Normal procedure is to modify the hole diameter to get the correct pressure drop (especially in the design stage). It is important to know that the total pressure in the external flow is determined from a specified combustor exit pressure, and the external pressure has to "build up" through a flow restriction (liner orifices).

   As part of the NASA High Speed Research (HSR) program, research in jets-in-crossflow was performed applicable to low-emission RQL combustors. The NASA-sponsored work involved both experimental and computational research performed by university (University of California, Irvine), small business (CFD Research Corporation), and engine manufacturers (United Technologies and Rolls Royce-Allison). The mixing of multiple jets with a confined crossflow was studied for cylindrical ducts (Holdeman et al.[39]) and for rectangular ducts (Holdeman et al.[40]). Dilution jet penetration was shown to be the most important parameter in optimizing mixing for a given pressure drop. Two variables greatly influence jet penetration:

1. Jet-to-mainstream momentum-flux ratio, defined as:

$$J = \rho_j V_j^2 / \rho_\infty V_\infty^2$$

where J   = Momentum-flux ratio
     ρ_j = Density of jet
     V_j = Velocity of jet
     ρ_∞ = Density of mainstream
     V_∞ = Velocity of mainstream

2. Orifice spacing-to-duct height, defined as:

$$S/H$$

where S = Orifice spacing
     H = Duct height

For optimum penetration, it was observed that

$$C = (S/H)\sqrt{J}$$

where the constant C is approximately 2.5 for single-sided injection, 1.25 for double-sided injection with opposed rows of in-line orifices, and 5.0 for double-sided injection with opposed rows of staggered orifices. It was found that the optimum value for C was a function of downstream location (x/H); and for high mass-flow ratios (jet-to-mainstream >1), the optimum C appeared to be about twice that of low mass-flow ratios (< 0.5) for rectangular duct flows.

CFD analysis was used to assess various aspects of jet-in-crossflow mixing in rectangular ducts (see Figure 12.12 for the numerical mixing geometry studied). These included:

1. Orifice shape had little impact on overall mixing as long as optimum penetration occurred. This can be seen in Figure 12.13.[41] For optimum penetration, orifice spacing varied for different shapes, due to the size of the jet wake. The effect of orifice aspect ratio on jet wakes can be seen in Figure 12.14.[41]
2. For opposed rows of jets, staggered orifices (in the lateral direction) produced better downstream mixing than in-line orifices for high momentum flux ratios (J > 36), and worse mixing for low momentum flux ratios (J < 36). In all cases, in-line configurations produced better initial mixing. Figure 12.15 shows predictions of mixing for optimum inline and staggered arrangements for a J of 36 and a jet-to-mainstream mass-flow ratio of 2.0.[42] For this J, mixing at x/H of 1.5 is almost exactly the same for both configurations.

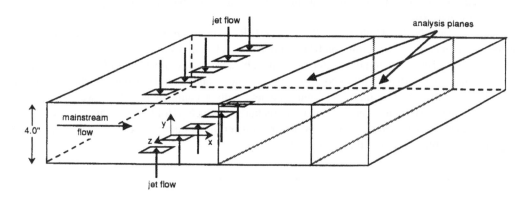

FIGURE 12.12   Schematic of numerical mixing model.

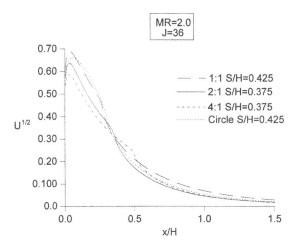

FIGURE 12.13 Effect of orifice aspect ratio on unmixedness (U); J = 36, MR = 2.0. (Data from Bain, D.B., Smith, C.E., and Holdeman, J.D., CFD Assessment of Orifice Aspect Ratio and Mass Flow Ratio on Jet Mixing in Rectangular Ducts, AIAA Paper 94-0218, 1994; also NASA TM 106434.)

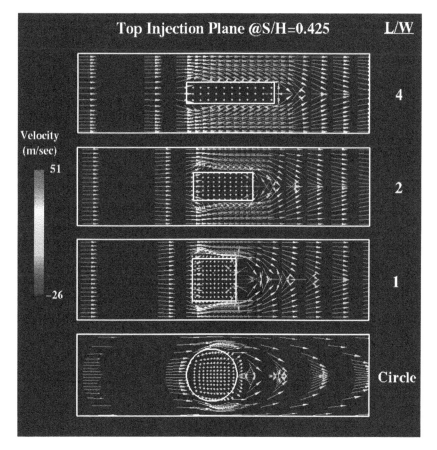

FIGURE 12.14 Effect of orifice aspect ratio on jet wakes; J = 36, MR = 2.0. (Data from Bain, D.B., Smith, C.E., and Holdeman, J.D., CFD Assessment of Orifice Aspect Ratio and Mass Flow Ratio on Jet Mixing in Rectangular Ducts, AIAA Paper 94-0218, 1994; also NASA TM 106434.)

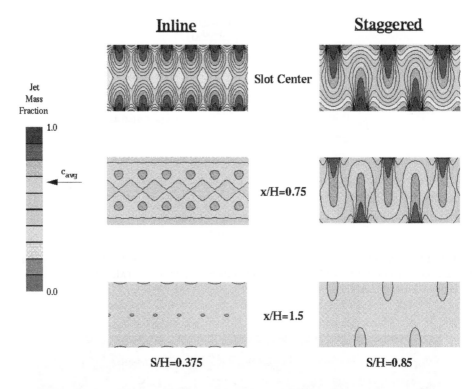

FIGURE 12.15 Effect of lateral arrangement on mixing; J = 36, MR = 2.0. (Data from Bain, D.B., Smith, C.E., and Holdeman, J.D., AIAA Paper 93-2044, 1995.)

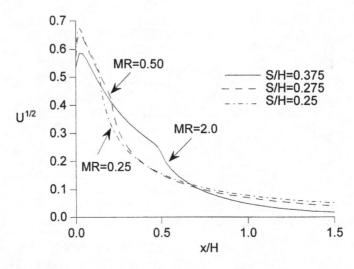

FIGURE 12.16 Effect of jet-to-mainstream mass flow ratio on unmixedness at optimum S/H; orifice aspect ratio = 4, J = 36. (Data from Bain, D.B., Smith, C.E., and Holdeman, J.D., CFD Assessment of Orifice Aspect Ratio and Mass Flow Ratio on Jet Mixing in Rectangular Ducts, AIAA Paper 94-0218, 1994; also NASA TM 106434.)

3. Jet-to-mainstream mass flow ratio (MR) affects penetration and mixing (at optimum penetration). Figure 12.16 presents mixing results for different mass flow ratios (at optimum spacing, S/H).[41] Note that the spacing varies for optimum mixing, indicating that there is a significant effect of mass flow ratio on optimum C.

Other CFD studies were performed for cylindrical ducts, but they are not discussed here. Further information on these studies can be found in Holdeman et al.[39]

12.5 TURBULENT/CHEMISTRY INTERACTION

Turbulence generally plays a dominant role in determining the overall rate of reaction and mean properties such as temperature, density, and species concentrations in gas-turbine combustor applications. The turbulent fluctuations in the reactant and product fluid dramatically influence mean chemical kinetic rates, particularly when reaction time scales are on the order of the turbulent mixing time scale or less. Most reactions for gas-turbine combustion lie in this critical area where the interactions between the chemical reaction and the turbulent mixing are crucial to the overall system behavior.

Turbulent fluctuations most strongly influence the mean reaction rate source term in the Reynolds- or Favre-averaged species conservation equations. Most reaction rates are highly nonlinear functions of temperature and species concentrations and, thus, mean reaction rates cannot simply be calculated from the time mean variable without introducing substantial error. Norris and Hsu[43] compared mean reaction rates for H_2O and CO_2 in a turbulent $CO/H_2/N_2$-air non-premixed flame using both a laminar chemistry approximation and an exact calculation. Table 12.3 shows these calculated rates. A comparison between the exact calculation and the laminar approximation showed that the CO_2 rate differed by an order of magnitude and the H_2O rate even differed in sign. This example demonstrates that a reasonable accounting for turbulent fluctuations in reacting flow is essential for obtaining a realistic solution.

A fundamental classification of turbulence-chemistry interactions is given by the Damkohler number Da, the ratio of a flow time to a chemical time. The flow time can be based on the integral or Kolmogorov time scale, corresponding to large and small eddies, respectively. The chemical time depends on the overall reaction rate of a particular reaction and was estimated by Correa[44] to vary between 50.0 and 0.0002 ms for gas-turbine combustion (CH_4 at 300K, air at 600K, 0.7 equivalence ratio, 10 atm, and 10 ms residence time). The estimated Damkohler numbers, depending on the reaction, ranged from the reaction-sheet regime (Da $\gg$ 1; H + O_2 = OH + O and CH_4 + OH = CH_3 + H_2O) to the distributed-reaction regime (Da $\rightarrow$ 0; N_2 + O = NO and CH_4 + M = CH_3 + H + M). Typically, most turbulent-chemistry interaction models are valid within a certain Damkohler number range, depending on the assumptions and complexity of the model. The following paragraphs describe several well-known turbulence-chemistry interaction models. During recent years, intensive research activity in the modeling of turbulent reacting flow has occurred due to the demand for higher performance gas-turbine combustors and the concern for pollutant emissions. These models can be classified into the following groups: mean flow methods, probability density function (PDF) methods, and large eddy simulation (LES) methods.

TABLE 12.3
Mean Reaction Rates in a Turbulent Flame Using a Laminar Chemistry Approximation and an Exact Calculation. Units Are in Mass Fractions per Second (Norris and Hsu, 1994)[43]

	dCO_2/dt	dH_2O/dt
Laminar Approximation	24.1	−0.62
Exact	4.77	0.77

12.5.1   MEAN FLOW METHODS

The simplest modeling approach is the mean flow (or first-moment) method, which is based on the conservation equations for the mean fields of velocity and composition. Turbulence-chemistry interaction models that are used with this method rely on mean flowfield variables and include the following: laminar chemistry, eddy breakup, eddy dissipation, and laminar flamelet.

12.5.1.1   Laminar Chemistry

The laminar chemistry approach approximates the mean reaction rate source term according to:

$$\langle S(\phi) \rangle = S(\langle \phi \rangle) \tag{12.8}$$

where ϕ is the composition vector. This approximation neglects the fluctuations in the composition and, because of the highly nonlinear reaction rates, it can be in error by orders of magnitude.

12.5.1.2   Eddy Breakup

The eddy breakup type models (Spalding,[45] Magnussen and Hjuertager[46]) assume that the chemical reaction rates are limited by mixing. This approach is valid for Da $\gg$ 1. Reactions between fuel and oxidizer occur quickly, but do not take place until the separate eddies in which the reactants are found dissipate and allow the reactants to mix. Also, in premixed systems, reaction does not take place until the eddies of hot products and cold reactant fluid mix. The eddy lifetime is represented by k/ε, where k is the turbulent kinetic energy and ε is its rate of dissipation. Some approaches also compare the turbulent mixing limited rate with a chemical reaction rate and use the slowest of the two. This algorithm is simple to code and computationally efficient; however, the approach is empirical and ignores the statistical nature of turbulence.

12.5.1.3   Eddy Dissipation

The eddy dissipation model[47] is related to the eddy breakup model in that reaction is limited, based on a relation between the kinetic time scale and a mixing time scale. The eddy dissipation model focuses on the fine eddy structures with characteristic dimensions on the order of the Kolmogorov microscale. The reactants are assumed to be homogeneously mixed within these fine structures, which can then be assumed to be well-stirred reactors. If the mixing time scale of the fine structures, determined from the available k-ε field, is less than the kinetic time scale, extinction of the reaction is assumed. The eddy dissipation model can be applied to cases in which detailed kinetics are required.

12.5.1.4   Laminar Flamelet

Laminar flamelet models cover a regime in turbulent combustion where chemistry (as compared to transport processes) is fast enough to occur in asymptotically thin layers (flamelets) embedded within the turbulent flowfield.[48,49] The layers are assumed to be one-dimensional so that they can each be studied separately with detailed chemistry. Laminar flamelet calculations typically require the setup of a pre-processed flamelet library. Features that are not accounted for in the laminar flamelet approach are the premixing that follows local extinction and the broadening of reaction zones due to slow pyrolysis or recombination reactions, which destroy the asymptotic thin-flame structure. The flamelet model can implicitly include complicated chemistry within the context of a two-variable description and thus there remains an incentive to develop this approach further. Laminar flamelet models can also be combined with prescribed PDF models.

12.5.2 PDF Methods

PDF methods have been used extensively for modeling turbulent combustion. PDF methods, unlike mean flow methods, require the solution of an additional scalar equation that describes the second moment (or variance), or higher moments of the scalar variable(s). PDF methods can be further grouped into prescribed PDF, composition PDF, and conditional moment closure (CMC) methods.

12.5.2.1 Prescribed PDF Method

Prescribed PDF methods have typically been used with success in modeling turbulent non-premixed flames. Once the local values for the mean and variance of a conserved scalar are obtained, one can calculate mean properties of other species or density by convoluting the instantaneous values of the scalar with the assumed shape of the conserved scalar PDF. As with the eddy breakup and flamelet models, the prescribed PDF methods work best when the chemical reaction rates are fast compared to the turbulent mixing. The prescribed PDF methods account for turbulent fluctuations in a more statistical manner than the eddy breakup type models. Typically, 1 or 2 reactive scalars can be considered within prescribed PDF methods. CFD Research Corporation[50] describes a 2 scalar (1 conserved, 1 reactive) prescribed PDF approach that has been used with success to model 3-D gas-turbine combustor geometries.

12.5.2.2 Composition PDF Method

An alternative to assuming the shape of the PDF is to solve for it from a PDF transport equation. A derived transport equation for the composition PDF can be obtained from the conservation equation for composition scalars.[51] The main advantage of the composition PDF method over the previously described methods is that chemical reactions of arbitrary complexity and their interactions with turbulence appear in closed form and do not need modeling. Because the joint composition PDF equation requires more independent variables than typical mean flow equations (i.e., now all the composition scalars are independent variables in full PDF methods), a more economical, Monte Carlo approach is used to solve the PDF transport equation. As first described by Pope,[52] rather than discretizing space, the fluid is discretized into a large number of representative or stochastic particles.

12.5.2.3 Conditional Moment Closure

This approach was developed by Bilger,[53] and requires that the species mass fractions and enthalpy are conditionally averaged over a fluctuating progress variable to account for the effects of turbulent fluctuations on the mean reaction rate. The conditioning variable is the mixture fraction for non-premixed flames. The fluctuations about the conditional mean are assumed to be small, thus limiting the range of application to flames that are far from extinction. The PDF shape of the conditioning variable must be assumed.

12.5.2.4 Application of PDF Methods

Biagoli[54] strongly recommends the use of the prescribed PDF method for design applications because comparable results are obtained with prescribed and composition PDF methods, and because the prescribed PDF method required 20 times less CPU time compared to the Monte Carlo method. An example of diffusion flame calculations with laminar chemistry, eddy breakup model, prescribed PDF, and composition PDF is shown in Figure 12.17.[55,56] As expected, pure laminar chemistry produces completely unrealistic results. The eddy breakup model is a considerable improvement, but still over predicts peak flame temperature by about 400K. The PDF approaches

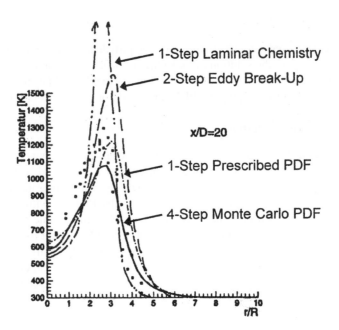

FIGURE 12.17 Comparison of turbulent-chemistry modeling approaches for a diffusion flame. (Data from Masri, A.R., Bilger, R.W., and Dibble, R.W., *Combust. Flame,* 71, 245-266, 1988; PDF calculations from Pfitzner, M., Mack, A., Brehm, N., Leonard, A., and Romaschov, I., *ASME Pressure Vessels and Piping Conf. on Computational Technologies for Fluid/Thermal/Thermal/Structural/Chemical Systems with Industrial Applications,* 1, 93-104, 1999.)

provide a reasonable approximation of the data. A useful approach for the design of gas-turbine combustor hardware has been in use at CFD Research Corporation. This approach consists of the prescribed PDF method with a one-step finite-rate reaction that proceeds to equilibrium products. The right side of the one-step reaction can include any number of species (rather than just CO_2 and H_2O), and the stoichiometric coefficients for each species are calculated as a function of fuel mixture fraction, enthalpy, and pressure. This method combines the excellent flame temperature prediction of an equilibrium model with finite-rate chemistry, and can be efficiently combined with a prescribed PDF. A pre-processed lookup table is created for the mean reaction rate and scalar properties to enhance computational efficiency. Oxidation of CO is also accounted for in the construction of the look-up table. The prescribed PDF calculation shown in Figure 12.17 was performed using this approach. Post-processing is also performed using appropriate finite-rate chemical kinetic mechanisms. NOx emissions are typically predicted within 10 to 20% of measured values and CO emissions are within 50% of measured values at combustor exit conditions. More quantitatively accurate predictions (<10% of measurements) of emissions are currently not possible with practical design-type (overnight turnaround) calculations. Despite this deficiency, qualitative information can be obtained as design perturbations are attempted.

12.5.3 Large Eddy Simulation

When unsteady, or transient effects in turbulent reacting flows are important, then large eddy simulation (LES) is typically required. LES is an approach that accounts for large-scale turbulent-chemistry interaction in a direct manner. Large eddy simulations, although computationally expensive, are now becoming a more practical method to resolve temporal and spatial scales. In the LES method, the transient forms of the conservation equations governing fluid flow, reaction, and heat transfer are solved. Improved numerical accuracy (compared to typical mean flow models), with a fine enough grid and appropriate time step are required for LES because it is used to fully resolve large-scale motions. Only the effects of smaller scales are modeled. Subgrid models are required to compute the effects of the small-scale turbulence on momentum, species, and enthalpy transport. The development and testing of turbulence-chemistry interaction models for the subgrid scale in LES methods is far less mature than the steady-state methods presented earlier. Nonetheless, several models have been proposed and developed for LES. These LES subgrid chemistry models have some of the same framework as mentioned earlier, with laminar chemistry, prescribed PDF, and full PDF methods. In addition, a different model, called the linear eddy model,[57] has also been developed for LES. Section 12.6 describes the LES approach in more detail in connection with modeling of combustion acoustic dynamics.

12.6   COMBUSTION INSTABILITY

Combustion instability is a serious challenge for designers of modern gas-turbine combustors. Typical combustion-driven oscillations involve a feedback cycle that converts chemical energy to oscillatory energy at a frequency corresponding to an acoustic resonant frequency. The well-known Rayleigh criterion must be satisfied for flame-driven instability. The current drive toward fuel-air premixing and lean front-end equivalence ratios in gas-turbine combustors have promoted the coupling of periodic heat release with resonant acoustics. Driving mechanisms that result in unsteady heat release include (1) fuel time lag causing variation in equivalence ratio, and (2) expansion and compression in the premix passage causing variation in the feed-rate of reactants. High-performance military combustors also have a strong potential for damaging pressure oscillations, caused by increased combustion intensities that accompany higher thrust-to-weight engines.

Time-accurate CFD analysis is required to accurately model combustion instability and understand the physical mechanisms that drive the instability. Unsteady FANS, in which the temporal terms are added to the Favre-Averaged Navier Stokes (FANS) equations, are sometimes used to capture the acoustic waves within the combustor and the coupling of the unsteady heat release with these waves. This methodology is adequate when the time scale of turbulence k/ε is significantly less than the acoustic time scale (1/frequency).

12.6.1   LES MODELING

A higher fidelity solution is attained when large eddy simulation (LES) is employed. In the LES method, large-scale motions are directly computed and the effects of smaller scales are modeled. Each variable in LES is decomposed into a large-scale component (indicated by an overbar) and a residual component (indicated by a prime):

$$f = \bar{f} + f' \tag{12.9}$$

The large-scale component is obtained by spatially averaging f with a filter function G,

$$\bar{f} = \int f(X) G(x, X) dX \tag{12.10}$$

where X is a dummy variable. The filter effectively eliminates fluctuations on scales smaller than a specified size.[58] The large-scale, or resolved, components are time dependent, in contrast to the average components in conventional (FANS) turbulence modeling.

The governing transport equations (continuity, momentum, energy, mass fraction) are density-weighted (Favre) filtered to give equations for the resolved quantities,

$$\frac{\partial \bar{\rho}}{\partial t} + \frac{\partial \bar{\rho}\tilde{u}_i}{\partial x_i} = 0$$

$$\frac{\partial \bar{\rho}\tilde{u}_j}{\partial t} + \frac{\partial \bar{\rho}\tilde{u}_i\tilde{u}_j}{\partial x_i} = \frac{\partial \tilde{\tau}_{ij}}{\partial x_i} - \frac{\partial T_{ij}}{\partial x_i}$$

$$\frac{\partial \bar{\rho}\tilde{E}}{\partial t} + \frac{\partial \bar{\rho}\tilde{u}_i\tilde{E}}{\partial x_i} = \frac{\partial \tilde{\tau}_{ij}\tilde{u}_j}{\partial x_i} - \frac{\partial \tilde{Q}_i}{\partial x_i}$$

$$\frac{\partial \bar{\rho}\tilde{\phi}_\alpha}{\partial t} + \frac{\partial \bar{\rho}\tilde{u}_i\tilde{\phi}_\alpha}{\partial x_i} = -\frac{\partial \tilde{J}_i^\alpha}{\partial x_i} - \frac{\partial M_i^\alpha}{\partial x_i} - \tilde{\dot{w}}_\alpha$$

(12.11)

which contain unknown terms such as τ_{ij}, that arise from filtering nonlinear terms and are known as the subgrid-scale (SGS) stresses. The SGS stresses account for the effect of small scales on large scales and need to be modeled to achieve closure. Similar methodology is employed for the energy (subgrid heat flux, Q) and mass fraction equations (subgrid scalar flux, J_i^α and filtered reaction rate, $\tilde{w}_\alpha$).

A finite-volume formulation (typically used in commercial codes) implicitly filters the equations by integrating over control volumes (grid cells), which is equivalent to using a filter function that is nonzero only within the control volume. The finite-volume approach gives algebraic equations for flow variables at discrete points, the center of each cell. The nonlinear terms in the transport equations yield the SGS stresses, which must be modeled. These terms are explicitly added to the difference equations for LES.

12.6.1.1   Subgrid Stress Modeling

The SGS modeling used in LES is typically based on an eddy viscosity model originally proposed by Smagorinsky.[59] The anisotropic part of the SGS is assumed to be proportional to the resolved strain rate:

$$\tau_{ij} - \delta_{ij}\frac{\tau_{kk}}{3} = -2\nu_t \bar{S}_{ij}$$

(12.12)

thereby defining a subgrid viscosity. The subgrid viscosity, by dimensional arguments, is the product of an appropriate length scale (usually the filter width) and velocity scale.

Advanced subgrid scale models have been shown to be more appropriate for a wide variety of flows, such as the dynamic model.[60] The dynamic subgrid model uses two filters to calculate the eddy viscosity dynamically, based on the flow at the smallest scales. This approach allows the Smagorinsky "constant" to vary with flow conditions and to vanish in laminar flows or near solid bodies. This model, unlike the simple Smagorinsky model, also predicts transfer of energy from the small to large scales.

12.6.1.2   Turbulent Chemistry Subgrid Modeling

As mentioned earlier, the turbulent fluctuations in the reactant and product fluid dramatically influence mean chemical kinetic rates, particularly when reaction time scales are on the order of the turbulent mixing time scale or less. Most reactions in combustion applications lie in this critical area, even at the subgrid scales that are not resolved in LES. A subgrid chemistry model is required to describe the effects of small-scale turbulent fluctuations on the filtered reaction rate source term. Two such models suitable for practical LES calculations are the linear eddy model (LEM) and the conditional moment closure (CMC) model. LEM, developed by Kerstein[55] and popularized by Menon,[61] is well-suited as a subgrid chemistry model for LES. The LEM provides an exact description of chemical kinetics and molecular diffusion at all length scales of the flow, while modeling the effects of turbulent advection. This is achieved by formulating the model in one spatial dimension. The reaction-diffusion equation for the species concentration is[62]:

$$\frac{\partial \phi}{\partial t} = D \frac{\partial^2 \phi}{\partial x^2} + \dot{w}_\phi \tag{12.13}$$

where ϕ is the species concentration, D is the molecular diffusion coefficient, and $\dot{w}_\phi$ represents the chemical source (sink) term. Turbulent stirring or convection is simulated by making random rearrangement events and the PDF for the size distribution (eddy sizes) must also be provided.

Recently, Klimenko,[63] Bilger,[53] and Bushe and Steiner[64] proposed a new approach called conditional moment closure (CMC). The CMC method employs the transport equations of conditionally averaged quantities instead of their spatially filtered counterparts. Variables on which the chemical reactions are known to depend on are chosen to be the conditioning variables. CMC allows the evaluation of the chemical source term in an affordable and sufficiently precise manner.

12.6.1.3   Other Issues

Future LES combustion codes will require other advanced features. Reduced chemical mechanisms from detailed kinetic schemes (using steady-state assumptions) need to be developed so that enough species are tracked to provide relevant information, but not so many species that storage requirements are excessive. Chen[65] has developed a general procedure to construct reduced mechanisms. Examples of mechanisms developed using this automated approach can be found in Bowman et al.[66] and Chen.[67] To reduce storage requirements and to optimize calculation times, fast, efficient lookup tables containing thermochemical information (reaction rate, source term, temperature, density, etc.) are required with *in situ* adaptive tabulation schemes.[68] Such tabulation schemes will permit the use of 15 to 20 scalars in a reasonable storage space (less than 1 gigabyte). Another tabulation scheme that shows much promise is artificial neural networks (ANN) used by Christo et al.[69] and Blasco et al.[70]

12.6.2   CFD Applications for Combustion Instability

Over the next 5 years, combustor 3-D LES calculation on clusters of parallel computers will become a reality. Such simulations will be able to accurately predict not only combustion instability, but also emissions, lean blowout, autoignition/flashback, etc. In recent years, some progress has been made to show the feasibility of using 2-D time-accurate analysis to predict combustion instability, realizing the limitations and inaccuracy that 2-D analysis brings.

A few applications are discussed below. In these cases, laminar chemistry was assumed (subgrid fluctuations on reaction rates were ignored) and a one-step finite kinetic rate with combustion products going to equilibrium was used. The reaction rate constants were tuned to laminar flow speeds and ignition delay times of (1) methane for natural gas and (2) propane for JETA fuel. Subgrid turbulence for LES calculations was modeled using either the Smagorinsky or Dynamic models. For unsteady FANS calculations, the RNG k-ε model was used. The second-order Crank-Nicholson scheme was used for temporal differencing, and the second-order upwind scheme was used for spatial differencing.

12.6.2.1   2-D Premixed Pipe Combustor

A numerical investigation of modeling combustion instability in the experimental combustor of Janus et al.[71] was performed by Cannon and Smith[72] using the unsteady FANS approach. The axisymmetric geometry for this combustor is shown in Figure 12.18. The supply of air and fuel was choked and mixed at the entrance point of the nozzle. This choked-flow inlet allowed the combustor to be

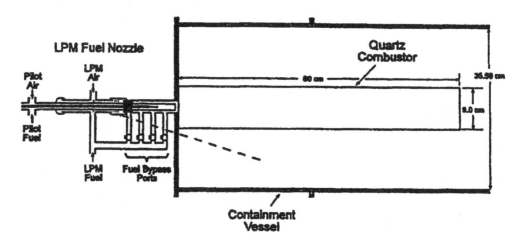

FIGURE 12.18 Schematic of fuel nozzle and premixed pipe combustor. (From James, M.C., Richards, G.A., Yip, M.J., and Robey, E.H., *Int. Gas Turbine and Aeroengine Congr. and Exhibition,* 97-GT-266, Orlando, FL, 1997. With permission.)

modeled with a fixed mass flow boundary condition and ensured that the local equivalence ratio throughout the combustor was constant. The driver of combustion instability in this experiment was the expansion and compression of gases in the premix fuel passage causing variation in the feed rate of reactants at the flame zone. The geometry of the combustor was such that longitudinal acoustic waves could be driven at or near the quarter wavelength frequency (f = C/4L).

The swirl vanes were modeled as source terms that modified the axial and tangential momentum equations. The initial numerical perturbation caused by starting from a steady-state solution was sufficient to allow large-amplitude pressure oscillations to develop if the unsteady heat release was in phase with an acoustic resonant frequency. Figure 12.19 shows comparisons between predicted and measured rms pressure as a function of the combustor inlet temperature. The correct trend of decreasing rms pressure with inlet temperature was predicted.

12.6.2.2   2-D Helmholtz Combustor

A numerical investigation of modeling combustion instability in the experimental combustor of Richards and Janus[73] was performed by Smith and Cannon[74] using the unsteady FANS approach and the LES approach. The driver of combustion instability in this experiment was believed to be a fuel time lag causing variation in equivalence ratio at the flame zone. The combustor included an uncooled ceramic plug that reduced flow area and produced a Helmholtz resonating frequency (see Figure 12.20). A grid of 21,290 computational cells was used in the simulation. The grid extended from the inlet plenum all the way to the exhaust duct valve.

Two baseline cases were first studied: one that experimentally produced instability and one that did not. Predicted and measured limit cycle combustor pressures are shown in Figure 12.21. Good overall agreement between the unsteady FANS predictions and measurements was obtained. The pressure amplitude was slightly higher than the measurements (9% vs. 6%), and the predicted frequency was 269 Hz vs. 225 Hz for the measurements. The unsteady case was repeated using the LES approach (using the Smagorinsky subgrid turbulence model). The predicted pressure limit cycle is shown in Figure 12.22. It can be seen that the predicted pressure amplitude with the LES methodology agrees much better with the measurements. The predicted frequency still did not match the measured frequency. It remains unclear why there is a difference in frequency; possible

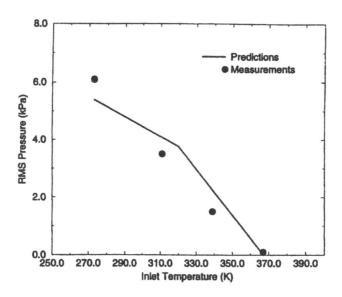

FIGURE 12.19 Comparisons of measured and predicted RMS pressure as a function of combustor inlet temperature.

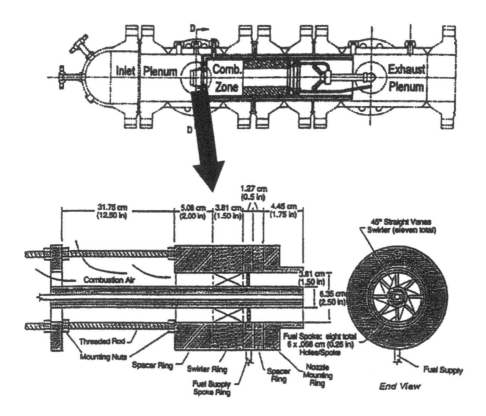

FIGURE 12.20 Features of Richards and Janus experimental rig. (From James, M.C., Richards, G.A., Yip, M.J., and Robey, E.H., *Int. Gas Turbine and Aeroengine Congr. and Exhibition,* 97-GT-266, Orlando, FL, 1997. With permission.)

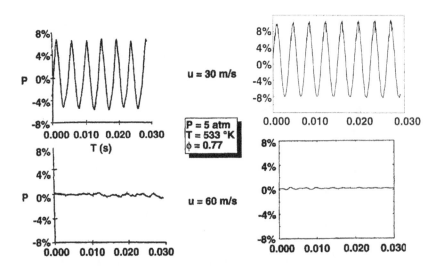

FIGURE 12.21   Comparison of limit cycle pressures (experiment vs. CFD analysis).

explanations could be the lack of modeling quench water injection in the exhaust duct or the inaccuracy of the chemistry model.

Time snapshots of the temperature field at various times in the limit cycle are shown in Figure 12.22. Predictions using both the LES approach and the unsteady FANS are shown. More detailed flowfield structure is seen for the LES approach, including the formation of multiple vortices in the combustor.

To further understand the effect of fuel time lag on instability, the fuel injection location was moved to various axial locations. Time histories of the pressure at the combustor mid-length are shown for eight cases in Figure 12.23. Fuel injection locations C, D, E, and F were all unstable, having maximum pressure amplitudes between 5 and 9%. Fuel injection locations A, B, G, and H produced stable flows. Active control (without feedback) was also investigated by pulsing the fuel at 300, 400, and 450 Hz (results are not shown in figure). The only frequency that damped the pressure osciallations was the 450-Hz modulation.

12.6.2.3   Liquid Fueled Flametube Combustor

Combustion dynamics of a Pratt & Whitney high-shear fuel injector, tested in a single-injector flametube combustor, has been studied using large eddy simulation (LES).[75] Independently measured data, including the magnitude and frequency of combustor pressure, were used to evaluate the model. A high-shear fuel injector was tested in a single-nozzle rig that mimics a high-performance combustor. Figure 12.24 shows the single nozzle rig that includes translatable inlet and exit boundary conditions to allow for acoustic tuning. Test conditions corresponded to:

$$P_3 = 174.7 \text{ psia}$$
$$T_3 = 769°F$$
$$W_3 = 3.91 \text{ lbm/s}$$
$$W_f = 346.7 \text{ lbm/hr}$$

Dynamic pressure measurements were obtained for several configurations, two of which were modeled. The baseline configuration included a 0.6-in. diffuser plenum length and exhibited a

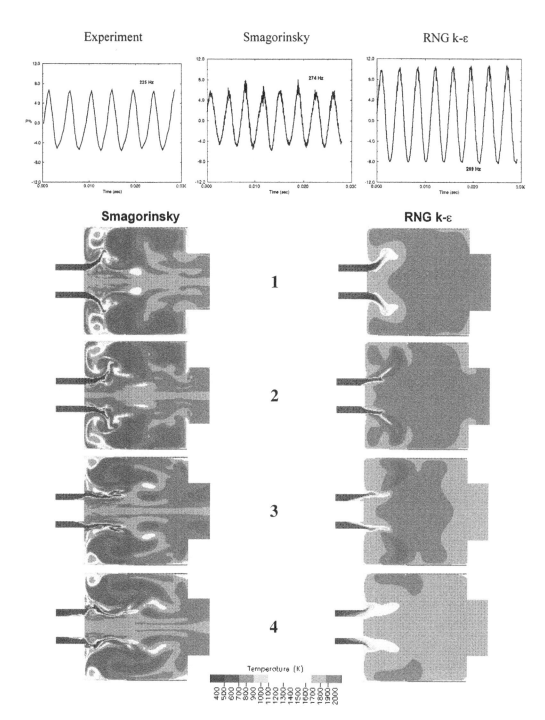

FIGURE 12.22 Comparison of predicted temperature contours during the unstable oscillation using LES and unsteady FANS.

moderate instability near 575 Hz. The extended configuration included a 19.25-in. extension to the diffuser plenum and measured pressure oscillations reached amplitudes of ±4% at a dominant frequency near 275 Hz.

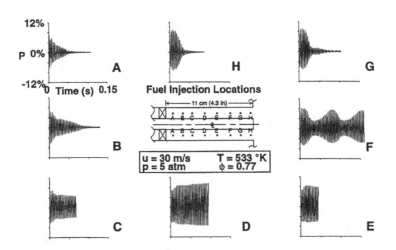

FIGURE 12.23   Time history of combustor pressure for different fuel injection locations.

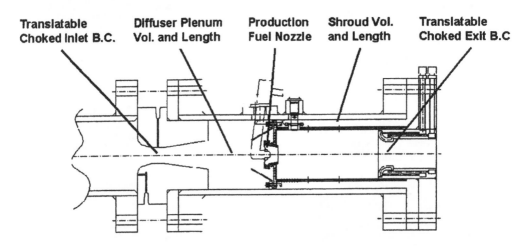

FIGURE 12.24   Single nozzle test rig.

Axisymmetric 2-D CFD calculations were performed using both LES methodology (Smagor-insky and dynamic models for subgrid turbulence) and unsteady FANS methodology (using RNG k-ε turbulence model). Swirl source terms were included within the radial swirlers and allowed the entire computational domain to extend from the known boundary conditions at the choked diffuser inlet to downstream of the choked combustor exit. Unsteady FANS was used in the external flowfield. Downstream of the swirler and in the combustor, the LES subgrid turbulence models were used where the grid was finer and flow structures that influence unsteady heat release were resolved. Multi-disciplinary computing environment (MDICE) allowed the separate simulations to communicate information at the appropriate interface boundaries at each time step.

Figure 12.25 shows the measured and predicted pressure history in the combustor for the extended case. The measured 275-Hz oscillation corresponds approximately to the longitudinal mode frequency through the diffuser, plenum, and combustor. The predictions also show high-amplitude pressure oscillations in the combustor at a frequency of 330 Hz (20% higher than measured). The pressure amplitude was best predicted using the dynamic model, although all three turbulence models gave good engineering results.

Snapshot images of the flame structure at various times during the oscillation cycle (six points marked in Figure 12.22) were captured and are shown in Figure 26. These temperature contours

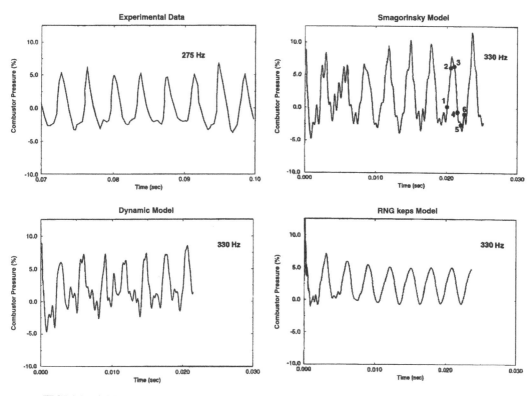

FIGURE 12.25   Measured and predicted combustor pressure for the 19.25-in. extension case.

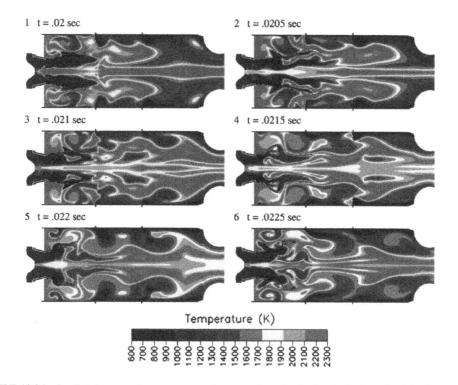

FIGURE 12.26   Predicted temperature contours and spray trajectories during the limit cycle of the Smagorinsky extension case.

and superimposed spray trajectories show that significant motion is occurring during the cycle. The amount of swirler air penetration into the combustor responds directly to the combustor pressure. It was interesting that hot gases were predicted on the fuel nozzle tip; experimental evidence showed a similar result.

Transient CFD calculations were also performed for the baseline case, where experimental data showed a much weaker instability ($\pm 0.2\%$) at about 570 Hz. Coupling did not occur during the predictions for the baseline case. The trend of going from a large amplitude (extension case) to a small amplitude (baseline case) pressure oscillation was captured with the transient CFD model.

REFERENCES

1. Mongia, H.C., Reynolds, R.S., Coleman, E., and Bruce, T.W. (1979), Combustor Design Criteria Validation, USARTL-TR-78-55A, 55B, and 55C.
2. Kenworthy, M.J., Correa, S.M., and Burrus, D.L. (1983), Aerothermal Modeling: Phase I Final Report, NASA CR-168243.
3. Srinivasan, R., Reynolds, R., Ball, I., and Berry, R. (1983), Aerothermal Modeling Program: Phase I Final Report, NASA CR-168243.
4. Sturgess, G.J. (1983), Aerothermal Modeling: Phase I Final Report, NASA CR-168202.
5. Nikjooy, M., Mongia, H.C., McDonnell, V.G., and Samuelsen, G.S. (1993), Fuel Injector–Air Swirl Characterization Aerothermal Modeling: Phase II Final Report, NASA CR-189193.
6. Fuller, E.J. and Smith, C.E. (1994), CFD Analysis of a Research Gas Turbine Combustor Primary Zone, Paper No. AIAA-94-2768.
7. Mongia, H.C. (1994) Combustion Modeling in Design Process: Applications and Future Direction, AIAA Paper 94-0466.
8. Crocker, D.S., Smith, C.E., and Myers, G.D. (1994), Pattern Factor Reduction in a Reverse Flow Gas Turbine Combustor Using Angled Dilution Jets, ASME Paper 94-GT-406.
9. Lawson, R.J. (1993), Computational Modeling of an Aircraft Engine Combustor to Achieve Target Exit Temperature Profiles, ASME Paper 93-GT-164.
10. Fuller, E.J. and Smith, C.E. (1993), Integrated CFD Modeling of Gas Turbine Combustors, AIAA Paper 93-2196.
11. Lai, M.K. (1997), CFD Analysis of Liquid Spray Combustion in a Gas Turbine Combustor, ASME Paper 97-GT-309.
12. Srinivasan, R., Freeman, W.G., Mozumdar, S., and Grahmann, J.W. (1990), Measurements in an Annular Combustor-Diffuser System, AIAA Paper 90-2162.
13. Karki, K.C., Oeclsle, V.L., and Mongia, H.C. (1992), A computational procedure for diffuser-combustor flow interaction analysis, *ASME Journal of Engineering for Gas Turbines and Power,* 114, 1-7.
14. Little, A.R. and Manners, A.P. (1993), Predictions of the Pressure Losses in 2-D and 3-D Model Pump Diffusers, ASME Paper 93-GT-184.
15. Crocker, D.S., Nickolaus, D., and Smith, C.E. (1999), CFD modeling of a gas turbine combustor from compressor exit to turbine inlet, *J. Engr. for Gas Turbine and Power,* 121, 89-95.
16. McGuirk, J.J. and Spencer, A. (1995), Computational Methods for Modeling port Flows in Gas-Turbine Combustors, ASME Paper 95-GT-414.
17. Crocker, D.S., Fuller, E.J., and Smith, C.E. (1996), Fuel Nozzle Aerodynamic Design Using CFD Analysis, ASME Paper 96-GT-127.
18. Mongia, H.C. (1998), Aero-thermal design and analysis of gas turbine combustion systems: current status and future direction, *34th AIAA Joint Propulsion Conference & Exhibit,* AIAA 98-3982.
19. Anand, M.S., Zhu, J., Connor, C., and Razdan, M.K. (1999), Combustor Flow Analysis Using an Advanced Finite-Volume Design System, ASME, 99-GT-273.
20. Smiljanovski, V. and Brehm, N. (1999), CFD Liquid Spray Combustion Analysis of a Single Annular Gas Turbine Combustor, ASME, 99-GT-300.
21. Liley, D.G. (1985), Swirling Flows in Typical Combustor Geometries, Paper No. AIAA-85-0184.
22. Rizk, N.K. and Chin, J.S. (1994), Comprehensive Fuel Nozzle Model, Paper No. AIAA-94-3278.

23. Rizk, N.K., Chin, J.S., and Razdan, M.K. (1997), Modeling of gas turbine fuel nozzle spray, *ASME Journal of Engineering for Gas Turbines and Power,* 119, 34-44.

24. McVey, J.B., Kennedy, J.B., and Russell, S. (1989), Application of advanced diagnostics to airblast injector flows, *ASME Journal of Engineering for Gas Turbines and Power,* 111, 53-62.

25. Launder, B.E. and Spalding, D.B. (1974), The numerical computation of turbulent flow, Comp. *Methods Appl. Mech. Engr.,* 3, 269.

26. Launder, B.E., and Spalding, D.B. (1974), *Comp. Methods Appl. Mech. Eng.,* 3, 269.

27. Sloan, D.G., Smoot, L.D., and Smith, P.J. (1985), Modeling of swirl in turbulent flow systems, *Central and Western States Sections of the Combustion Institute, 1985 Spring Technical Meeting.*

28. Yakhot, V., Orszag, S.A., Thangam, S., Gatski, T.B., and Speziale, C.G. (1992), Development of turbulent models for shear flows by a double-expansion technique, *Physics of Fluids,* 4, 1510-1520.

29. Halpin, J.L. (1993), Swirl Generation and Recirculation Using Radial Swirl Vanes, ASME Paper No. 96-GT-169.

30. Lichtarowicz, A., Duggins, R.K., and Markland, E. (1965), Discharge coefficients for incompressible non-cavitating flow through long orifices, *J. Mechanical Engineering Science,* 7(2), 210-219.

31. Kilik, E. (1976), The Influence of Swirler Design Parameters on the Aerodynamics of the Downstream Recirculation Region, Ph.D. thesis, School of Mechanical Engineering, Cranfield Inst. of Tech., England.

32. Lefebvre, A.H. (1983), *Gas Turbine Combustion,* Hemisphere, New York.

33. Chien, K.Y. (1982), "Predictions of Channel and Boundary-Layer Flows with Low-Reynolds Number Turbulence Model," *AIAA Journal,* 20, 33-38.

34. Benjamin, M.A. and Crocker, D.S. (1996), Spray Characterization of Relatively High Flow Simplex Atomizer Using Experiment and CFD, AIAA Paper 96-3165.

35. Bain, D.B., Smith, C.E., Liscinsky, D.S., and Holdeman, J.D. (1999), Flow Coupling Effects in Jet-in-Crossflow Flowfields, *Journal of Propulsion and Power,* AIAA 15(1), 10-16.

36. Dittrich, R.T. and Graves, C.G. (1956), Discharge Coefficients for Combustor-Liner Air-Entry Holes. I. Circular Holes with Parallel Flow, NACA Technical Note 3663.

37. He, G., Guo, Y., and Hsu, A.T. (1999), The Effect of Schmidt Number on Turbulent Scalar Mixing in a Jet-in-Crossflow, ASME 99-GT-137.

38. Gulati, A., Tolpadi, A., and VanDeusen, G. (1994), Effect of Dilution Air on Scalar Flowfield at Combustor Sector Exit, AIAA 94-0021.

39. Holdeman, J.D., Liscinsky, D.S., Oechsle, V.L. Samuelsen, G.S., and Smith, C.E. (1997), Mixing of multiple jets with a confied subsonic crossflow. I. Cylindrical Ducts, *Journal of Engineering for Gas Turbines and Power,* 119.

40. Holdeman, J.D., Liscinsky, D.S., and Bain, D.B. (1999), Mixing of multiple jets with a confined crossflow. II. Opposed rows of orifices in rectangular ducts, *Journal of Engineering for Gas Turbines and Power,* 121, 551-562.

41. Bain, D.B., Smith, C.E., and Holdeman, J.D. (1994), CFD Assessment of Orifice Aspect Ratio and Mass Flow Ratio on Jet Mixing in Rectangular Ducts, AIAA Paper 94-0218 (also NASA TM 106434).

42. Bain, D.B., Smith, C.E., and Holdeman, J.D. (1995), Mixing analysis of axially opposed rows of jets injected into confined crossflow, *Journal of Propulsion and Power,* 11(5), 885-893 (*see also* AIAA Paper 93-2044 and NASA TM 106179).

43. Norris, A.T. and Hsu, A.T. (1994), Comparison of PDF and moment closure methods in the modeling of turbulent reacting flows, *30th Joint Propulsion Conference-AIAA, ASME, SAE, and ASEE,* Indianapolis, IN, June 27-29.

44. Correa, S. (1989), Relevance of non-premixed laminar flames to turbulent combustion, *Proceedings of NASA Langley/ICASE Combustion Workshop,* 45-69.

45. Spalding, D.B. (1971), Mixing and chemical reaction in steady confined turbulent flame, *Proceedings 13th Symposium (International) on Combustion,* The Combustion Institute, 649-657.

46. Magnussen, B.F., and Hjertager, B.H. (1976), On mathematical modeling and turbulent combustion and special emphasis on soot formation and combustion, *16th Symposium (International) on Combustion,* The Combustion Institute, Pittsburgh, PA, 719-728.

47. Gran, I.R., Melaaen, M.C., and Magnussen, B.F. (1994), Numerical simulation of local extinction effects in turbulent combustor flows of methane and air, *Proc. of 25th Int. Symp. on Combustion,* 1283-1291.

48. Peters, N. (1986), Laminar Flamelet Concepts in Turbulent Combustion, *21st Symposium (International) on Combustion,* The Combustion Institute, Pittsburgh, PA, 1231-1250.

49. Mueller, C.M., Breitbach, H., and Peters, N. (1994), Partially Premixed Turbulent Flame Propagation in Jet Flames, *Proc. of 25th Int. Symp. on Combustion,* 1109-1106.

50. CFD Research Corporation, *CFD-ACE Theory Manual,* Ver. 4.0, Huntsville, AL, February 1998.

51. Pope, S.B. (1985), PDF methods for turbulent reactive flows, *Progress in Energy and Combustion Science,* 11, 119-192.

52. Pope, S.B. (1981), A Monte Carlo method for the PDF equations of turbulent reactive flow, *Combustion Science and Technology,* 25, 159-174.

53. Bilger, R.W. (1993), Conditional moment closure for turbulent reacting flow, *Phys. Fluid,* A5(2).

54. Biagioli, F. (1997), Comparison between presumed and Monte Carlo probability density function combustion model, *Journal of Propulsion and Power,* 13(1).

55. Masri, A.R., Bilger, R.W., and Dibble, R.W. (1988), Turbulent nonpremixed flames of methane near extinction: mean structure from raman measurements, *Combust. Flame,* 71, 245-266.

56. Pfitzner, M., Mack, A., Brehm, N., Leonard, A., and Romaschov, I. (1999), Implementation and validation of a PDF transport algorithm with adaptive number of particles in industrially relevant flows, *ASME Pressure Vessels and Piping Conference on Computational Technologies for Fluid/Thermal/Structural/Chemical Systems with Industrial Applications,* 1, 93-104.

57. Kerstein, A. R. (1988), A linear-eddy model of turbulent scalar transport and mixing, *Combustion Science and Technology,* 60, 391.

58. Piomelli, U. (1993), Applications of large eddy simulations in engineering — an overview, in *Large Eddy Simulation of Complex Engineering and Geophysical Flows,* Galperin, B. and Orszag, S.A., Eds., Cambridge University Press, Cambridge.

59. Smagorinsky, J. (1963), General circulation experiment with the primitive equations. I. The basic experiment, *Monthly Weather Review,* 91, 99-164.

60. Germano, M., Piomelli, U., Moin, P., and Cabot, W.H. (1991), A dynamic subgrid-scale eddy viscosity model, *Physics of Fluids A,* 3, 1760-1765.

61. Menon, S., Stone, C., Sankaran, V., and Sekar, B. (2000), Large-eddy simulations of combustion in gas turbine combustors, *38th AIAA Aerospace Sciences Meeting and Exhibit,* Reno, NV, AIAA 00-0960.

62. Frankel, S.H., McMurtry, P.A., and Givi, P. (1995), Linear eddy modeling of reactant conversion and selectivity in turbulent flows, *AICHE Journal,* Vol. 41, pp. 258-266.

63. Klimenko, A. Yu. (1990) Multicomponent diffusion of various admixtures in turbulent flows, *Fluid Dynamics,* 25, 327-334.

64. Bushe, K. and Steiner, H. (1999), Conditional moment closure for large eddy simulation of non-premixed turbulent reacting flows, *Phys. Fluids.,* 11, 1896-1906.

65. Chen, J.-Y. (1988), A general procedure for constructing reduced reaction mechanisms with given independent relations, *Combustion Science and Technology,* 57, 89-94.

66. Bowman, C.T., Hanson, R.K., Davidson, D.F., Gardner, W.C., Lissianski, V., Smith, G.P., Golden, D.M., Frenklach, M., and Goldenbert, M. (1996), http://www.me.berkeley.edu/gri_mech/.

67. Chen, J.-Y. (1997), Development of reduced mechanisms for numerical modeling of turbulent combustion, *Workshop on Numerical Aspects of Reduction in Chemical Kinetics,* CERMICS-ENPC Cite Descartes — Champus sur Marne, France, September 2.

68. Pope, S.B. (1997). Computationally efficient implementation of combustion chemistry using *in situ* adaptive tabulation, *Combustion Theory and Modeling,* 1, 41-63.

69. Christo, F.C., Masri, A.R., and Nebot, E.M. (1996), Artificial neural network implementation of chemistry with PDF simulation of H_2/CO_2 flames, Combustion and Flame, 106(4), 406-427.

70. Blasco, J.A., Fueyo, N., Dopazo, C., and Ballester, J. (1998), Modeling the temporal evolution of a reduced combustion chemical system with an artificial neural network, *Combustion and Flame,* 113, 38-52.

71. Janus, M.C., Richards, G.A., Yip, M.J., and Robey, E.H. (1997), Effects of ambient conditions and fuel composition on combustion stability, presented at the *International Gas Turbine and Aeroengine Congress and Exhibition,* 97-GT-266, Orlando, FL.

72. Canon, S.M. and Smith, C.E. (1998), Numerical modeling of combustion dynamics in a lean premixed combustor, presented at the *International Joint Power Generation Conference,* FACT-Vol. 22, 243-250, Baltimore, MD.

73. Richards, G.A. and Janus, M.C. (1997), Characterization of oscillations during premix gas turbine combustion, presented at the *International Gas Turbine and Aeroengine Congress and Exhibition*, 97-GT-244, Orlando, FL.

74. Smith, C.E. and Cannon, S.M. (1999), CFD assessment of passive and active control strategies for lean, premixed combustors, presented at the *37th AIAA Aerospace Sciences Meeting and Exhibit*, Reno, NV, AIAA 99-0714.

75. Cannon, S.M., Smith, C.E., and Lovett, J. (2000), LES Modeling of Combustion Dynamics in a Liquid-Fueled Flametube Combustor, *AIAA Joint Propulsion Conference, AIAA*, 2000-3126, July 2000.

76. Rizk, N.K. (1994), Model for Research Swirl Atomizers, Paper No. AIAA-94-2777.

77. Smith, C.E., Fuller, E.J., Crocker, D.S., Mekkes, L.T., and Sheldon, J.C. (1995), Dual-spray airblast fuel nozzle for advanced small gas turbine combustors, *J. of Propulsion and Power*, 11(2), 244-251.

13 CFD Modeling in the Petrochemical Industry

Mike Henneke

CONTENTS

13.1 INTRODUCTION

This chapter focuses on CFD simulations applied to furnaces and incinerators in the petrochemical industry. Petrochemical furnaces are typically gas- or oil-fired. In the U.S., nearly all petrochemical furnaces are gas-fired, but in other parts of the world, various grades of oil are used. The focus of this chapter is on gas-fired furnaces, with only brief discussion of oil-fired furnaces. This bias is reflective of the bias in the open literature, where little discussion of modeling full-scale, oil-fired furnaces can be found.

The reader will quickly observe that industrial-scale combustion modeling is an immature field (even after years of scientific effort), and may even wonder why industries invest any money into such efforts. The simple answer is that there is no better alternative. In the past, experimentalists have tried to simulate the flow patterns inside industrial furnaces using scaled-down plexiglass representations of the geometry. Although the complex physics of the real furnace (with combustion, radiation, buoyancy, etc.) is completely neglected, these experimentalists have met with some success. Even with its multitude of approximations about turbulence, radiation, and chemical reactions, a CFD model offers much more information about the furnace than a nonreacting, scaled-down plexiglass model.

Reacting flow experimentation on a scaled-down model is also problematic. Scaling combustion systems from the laboratory scale to industrial scale is very difficult. The problem is well-understood theoretically. To have "similarity" as a combustion system is scaled, the Reynolds' number and Damkohler numbers must be unchanged. This is practically impossible, so some sort of incomplete scaling is used. Typically, combustion systems are scaled using a constant-velocity or constant-momentum flux method.

Development of burners for petrochemical applications is usually done experimentally using a single burner at full scale. This method frequently produces a burner that performs well in the industrial setting; however, multiple-burner firing typically produces higher NO_x than single-burner firing.[1] Other flame interaction problems can arise as well. For example, the flames from the individual burners are occasionally observed to merge together and the length of the flame can increase significantly when the burner-to-burner spacing is not sufficient. These problems are especially significant for ultra-low-NO_x burners because these burners typically produce flame lengths significantly longer than a conventional burner. As the flame is "stretched out" by fuel staging strategies, it becomes more and more difficult for the burner designer to control the mixing between the fuel, air, and furnace gases (products of combustion). The stability of these flames becomes a significant and poorly understood problem. In addition, the dominant flow currents in an operating environment may be very different from those in the test environment. These operational problems provide an opportunity to use CFD analysis to solve problems that can otherwise only be solved by trial and error.

Given that these industrial furnaces frequently cost tens of millions of dollars, full-scale experimentation is prohibitively expensive. There is ample justification for the petrochemical industry to pursue CFD technology to better understand the performance of furnaces. Hopefully, this chapter will illustrate some of the potential value of CFD technology in petrochemical furnaces and stimulate studies to better understand the performance of CFD models in these applications.

This chapter takes a CFD code user approach to the discussion of CFD modeling. The intent is to provide an overview of how CFD modeling can be profitably used to diagnose and correct problems in petrochemical applications. Discussion of models is limited primarily to those that are available in commercial CFD software. Other models are briefly discussed, as needed, to indicate possible directions commercial CFD software may move in and to discuss needed improvements in modeling tools. In this author's experience, CFD analysis of full-scale combustion systems requires a large amount of user discretion. It is by no means a solved problem, and hopefully the contents of this chapter will clarify some of the limitations and illustrate how an understanding of these limitations allows the user to glean information from their CFD models.

13.2 MODELING APPROACHES

The current state-of-the-art approaches in modeling petrochemical furnaces are described in this section. This is not a comprehensive discussion of all models and algorithms available to the CFD analyst because not all models are commonly used in modeling furnaces. While there a large number of algorithms and models studied in academic circles, analyses of large-scale furnaces are done with commercial CFD packages. There are currently several commercial CFD packages with very similar modeling capabilities. This section discusses the modeling approaches used by these codes, as well as several models being studied at the research level that should lead to improved combustion modeling capability for industrial users.

There are a large number of approximations involved in modeling combustion processes in furnaces. Even in a gas-fired furnace, the multitude of important physics is daunting. The flow in the furnaces is turbulent flow with a very large integral length scale (the characteristic dimension of the furnace). The combustion chemistry in the furnace involves tens to hundreds of chemical species reacting with time scales from less than a microsecond to several seconds. Radiative transport from a non-gray gas (the products of combustion) to the furnace walls and tubes (with the process fluid flowing inside), whose emissivity is temperature dependent, is the primary mode of heat transfer. The interaction between these physical processes is of considerable importance. The turbulence-chemistry interaction has been well-studied for many years, particularly for non-premixed systems. More recently, the interaction between turbulence and radiative emission from a non-gray gas with properties has been studied.[2]

The geometries of the burners used in these furnaces are increasingly complex. The dominant factor in most burner designs is NO_x reduction, and this leads to burners that are more and more geometrically complex. In addition to the geometric complexity is chemical complexity. The dominant strategy for reducing NO_x emissions from gas-fired burners is to used staged fuel systems and use the fuel jets to entrain large amounts of the products of combustion into the flame zone. This means that in order to make a burner with lower NO_x emissions, one has to make a turbulent flame that is less stable. For the CFD analyst, modeling the stability of a flame or its lift-off height is at present a very imposing problem. For this reason, CFD predictions of NO_x emissions from ultra-low-NO_x burners are typically poor. However, even if the quantitative NO_x predictions are poor, frequently the qualitative information from a CFD calculation can be very useful. For example, the results of a CFD analysis can be used to study the entrainment of cooled furnace gases by fuel jets as well as the mixing of the fuel jets with combustion air. Combined with an analyst who has a thorough knowledge of what real flames do, these quantitative measures can be used (in conjunction with experimentation) to solve equipment problems. As an example, McDermott and Henneke[3] used an axisymmetric CFD model to design turning vanes in a premixed burner. The problem being addressed was flashback, which is very difficult to model in a CFD study. However, by combining CFD analysis with knowledge of what conditions allow flashback, the authors were able to design a series of turning vanes that eliminated flashback for a wide range of operating conditions.

13.2.1 COMPUTATIONAL ALGORITHMS

The equations governing a turbulent reactive flow include the continuity equation:[4]

$$\frac{\partial \rho}{\partial t} = \nabla \cdot (\rho \vec{v}) \tag{13.1}$$

and the momentum equation[4]:

$$\frac{\partial}{\partial t}(\rho \vec{v}) + \nabla \cdot (\rho \vec{v} \vec{v}) = -\nabla P + \nabla : \tau + \rho \mathbf{g} \tag{13.2}$$

where τ is the viscous stress tensor and g is the gravitational acceleration. The energy equation for an low Mach number flow is:

$$\rho C_p \left(\frac{\partial T}{\partial t} + \vec{v} \cdot \nabla T \right) = -\nabla \cdot \vec{q} + \tau : \nabla \vec{v}$$

$$+ \beta T \left(\frac{\partial P}{\partial t} + \vec{v} \cdot \nabla P \right) - \sum_{species} \dot{\omega}_i h_i M_i \tag{13.3}$$

The equation governing species transport is:

$$\frac{\partial}{\partial t}\rho Y_i + \nabla \cdot (\rho \vec{v} Y_i) = \nabla \cdot (\rho D_{im} \nabla Y_i) + M_i \dot{\omega}_i \qquad (13.4)$$

This section discusses the algorithms used to solve the Reynolds- (or Favre-) averaged Navier-Stokes equations, the Reynolds- (or Favre-) averaged energy and species equations, and the radiative transfer equation. No discussion is given here of large eddy simulations (LES) or direct numerical simulations (DNS). LES may become a viable option for CFD modeling of furnaces in the near future, but at present the increase in computational cost of a LES calculation is usually not justifiable. Pope[5] notes that the appeal of LES in nonreacting flows is the expectation that the small scales of turbulence are universally related to the large scales. In a reacting flow, there is no similar expectation. LES does have the advantage of resolving the large-scale structures that challenge the Favre-averaged models, but many difficult problems remain to be addressed before LES will be a useful tool. Bray[6] notes that despite the fact that LES faces current difficulties, it will be successfully developed and will be a useful tool for the combustion modeler. DNS can be categorically neglected for this class of problems because the computational demands are far in excess of current computational resources.[6,7]

13.2.2 DISCRETIZATION SCHEMES

There are many schemes used to discretize the partial differential equations of fluid flows onto different types of meshes. Because the primary focus of this chapter is applied CFD where mesh types by necessity include tetrahedra, the two important discretization schemes are the finite-volume method and the finite-element method. The finite-volume method is clearly the method of choice in the industry today for large-scale computations of turbulent flows. The dominant software products commercially available for these problems almost exclusively use the finite-volume method. There are occasions when other methods, such as the finite-difference method, are used. Most advanced combustion models are first implemented in academic CFD codes which, as mentioned in the introduction, are rarely intended to model complex geometry. For this reason, industrial CFD analysts rarely have access to advanced combustion models.

13.2.3 TURBULENCE MODELING

A turbulent flow is a flow with a wide range of temporal and length scales. Figure 13.1 is an example of a typical point measurement (e.g., pressure or velocity) within a turbulent flow. Within a turbulent flow, the quantities of interest such as pressure and velocity fluctuate in an apparently random fashion. Analysis reveals that these quantities are not truly random.[8] Information revealed by spectral analysis of point measurements reveals that there are ranges of temporal and length scales that contain significant energy (the large or integral scales), and smaller scales where this turbulent energy is dissipated by viscous processes. The energy cascade is the mechanism by which energy is moved from the large scales to the small scales. For more information on the physics of turbulent flows, the reader should refer to Libby.[8]

Prediction of turbulent flow from the Reynolds- or Favre-averaged conservation equations requires closure approximations. This is because the time-averaged conservation equations contain terms unclosed. In the case of the momentum equations, the time-averaging of the convection terms leads to the following Reynolds stresses: $\rho \overline{u'^2}$, $\rho \overline{v'^2}$, $\rho \overline{w'^2}$, $\rho \overline{u'v'}$, $\rho \overline{u'w'}$, $\rho \overline{v'w'}$. Closure approximations are required to solve the Reynolds-averaged conservation equations.

The workhorse turbulence model used in furnace simulations is the k-ε model.[9] The popularity of this model can be ascribed to its relative simplicity (compared to a Reynolds stress model, for

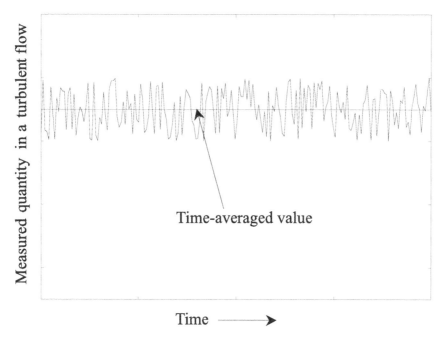

FIGURE 13.1   Graph illustrating measurements in a turbulent flow. The measurement of a flow variable (such as pressure or velocity) at a particular location will appear random.

example) and its good performance in a variety of engineering flows. Its weaknesses include its performance in unconfined flows, in rotating and swirling flows, and in flows with large strains, such as curved boundary layers. The Reynolds stress model (RSM) addresses some of these performance issues. RSM is much more computationally demanding because it involves seven extra partial differential equations (PDE) rather than the two of the k-ε model. However, in a typical combustion calculation, the number of PDEs solved is typically quite large, so adding five more may be easily justified if the quality of the predictions improve.

A number of variants of the classical k-ε model exist. The classical k-ε model uses a single eddy viscosity in all directions. The nonlinear k-ε model by Speziale[10] addresses this assumption, which is known to be poor even in relatively simple flows. Another development in k-ε modeling is the renormalization group (RNG) k-ε model of Yakhot et al.[11] Its performance in complex flows has been promising, so much so that several of the commercial CFD code vendors have implemented the RNG k-ε model. The realizable k-ε model represents yet another variant recently introduced.[12] The advantages and limitations of these turbulence models are discussed in more detail in Veersteeg and Malalasekera.[13]

13.2.4   RADIATION MODELING

Typical petrochemical furnaces consist of a radiant section and a convection section. These regions are so named because of the dominant mode of heat transfer. In the radiant section, refractory surface temperatures can be higher than 2200°F (1200°C). Radiant heat is incident on the process tubes both from the high-temperature surfaces and directly from the flame. Accurate modeling of the heat delivered to the process fluid requires an accurate prediction of the radiant intensity inside the furnace. In addition, accurate prediction of radiation from the flame is necessary to accurately predict emissions. For example, Barlow et al.[14] note that the different radiation models can affect NO_x predictions as much as the different turbulence-chemistry interaction models evaluated in that paper.

Thermal radiation transport presents a difficult problem because of the number of independent variables. The radiation transport Equation (RTE) describes radiation transport in absorbing, emitting, and scattering media. The equation is (notation of Modest[15]):

$$\frac{\partial I_\lambda}{\partial s} = \kappa_\lambda I_{b\lambda} - \beta_\lambda I_\lambda + \frac{\sigma_{s\lambda}}{4\pi} \int I_\eta\left(\bar{s}_i\right)\Phi\left(\bar{s}_i,\bar{s}\right)d\Omega_i; \tag{13.5}$$

Radiant intensity is a function of location (three coordinates in a 3-D problem), direction (two angular independent variables), and wavelength (one independent variable varying from 0 to infinity). This means that the problem of radiative transport is a six-dimensional problem. A common approach is to remove the wavelength dependence by making gray media approximations discussed below. The angular dependence can be treated by angular discretization, such as in the P-1 method or the discrete-ordinates method. Finite-difference and finite-volume methods discussed above treat the spatial dependence. Both the P-1 and discrete-ordinates methods approximate the angular dependence of the equation of transfer. The Monte Carlo method takes a much different approach. In the Monte Carlo method, individual photons of radiant energy are emitted, reflected, and absorbed by both solid surfaces and participating media using ray tracing algorithms. This method provides a very elegant approach to treating non-gray radiation, as well as the directional dependence of radiation. Its use is limited by its computational cost.

Siegel and Howell[16] and Modest[15] provide extensive discussion of the solution methods for radiation in participating media. These texts discuss the accuracy, computational effort, and limitations of the various models. The reader should consult these books for more discussion on these solution methods.

13.2.5 GAS RADIATION PROPERTIES

Molecular gas radiation is an important mode of heat transfer in gas-fired furnaces. Radiative emission from nonluminous hydrocarbon flames is mostly due to the H_2O and CO_2 species present in the products of combustion. Radiation from these gases is fairly well understood, but a rigorous treatment of this radiation requires significant computational resources. For instance, Mazumder and Modest[3] considered ten radiative bands in modeling emission from a hydrocarbon flame. This means that they solved the RTE for ten different intensities. In a large-scale furnace calculation, such a model would be extremely demanding computationally.

Quantum mechanics postulates that molecular gases emit and absorb gases only at distinct wavenumbers, called spectral lines. However, in reality, these distinct lines are broadened by several mechanisms, including collision broadening, natural line broadening, and Doppler broadening. These individual lines are characterized by a line strength and a line width. These lines are caused by quantum transitions in the vibrational or rotational state of a molecule. Frequently, vibrational and rotational transitions occur simultaneously, leading to a tightly clustered array of lines around a given vibrational transition. This subject is beyond the scope of the present chapter. The intent is to illustrate the complexity of modeling a radiating gas.

13.2.5.1 Weighted Sum of Gray Gases

The weighted-sum-of-gray-gases model[16] provides formulae for computing the emissivity of a gas volume as a function of its temperature and partial pressures of CO_2 and H_2O. The model assumes the gas is a mixture of radiating gases that is transparent between the absorption bands. The weighted-sum-of-gray gases model is probably the most widely used method to calculate radiation within combustion gases. Alternatives include band models (discussed below). The computational cost of radiation transport can be very high compared to the flow solver portion of a simulation because of the large number of independent variables in the RTE. In practice, it is usually reasonable to lag the calculation of the RTE for a number of flow solver iterations, with the actual number dependent on the solver in use and stability requirements.

13.2.5.2 Soot

The presence of soot in a flame can significantly increase the flame emissivity. Predicting soot formation within a flame is very difficult because soot is formed in fuel-rich regions of a flame when the temperature is high. Models such as those by Khan and Greeves[17] and Tesner[18] allow the prediction of soot concentrations, but these models are very empirical and can not be expected to provide quantitative results.

Soot within a flame is caused by the combustion of hydrocarbons under fuel-rich conditions. Soot is visually observed as a yellow-red brightness in the flame. C_2 hydrocarbons and higher have more tendency to soot, while methane does not normally produce a sooty flame. Soot has a strong impact on flame radiation. Radiative emission from soot in flames is frequently much larger than the gas radiation emitted by the flame.[15] In some applications (oil-firing, in particular), particulate emissions are regulated by environmental agencies. In flaring applications, smokeless (smoke results from unoxidized soot particles leaving the flame) operation is frequently guaranteed by the flare vendor for some range of conditions. In petrochemical applications, the gases flared are a wide range of hydrocarbons, typically ranging in molecular weight from 16 to 40. These gases may have components such as ethylene and acetylene, which are known precursors to soot formation. Current CFD codes (limited by physical model availability) cannot predict smoking from these large, buoyant flare fires, but current LES work in this area appears promising.

13.2.5.3 Turbulence-Radiation Interaction

The turbulence-radiation interaction plays an important role in predicting the radiative emission from a flame. Unfortunately, none of the available commercial products that this author is aware of attempt to model this interaction. To appreciate the significance of this issue, consider the time-averaged radiative transport equation in an absorbing/emitting media:

$$\frac{\partial \bar{I}_\lambda}{\partial s} = \overline{\kappa_\lambda \left(I_{b\lambda} - I_\lambda \right)} \tag{13.6}$$

Frequently, the time-averaged emission is computed as $\overline{\kappa_\lambda \, I_{b\lambda}} \approx \overline{\kappa_\lambda} \, \overline{I_{b\lambda}}$, which neglects correlations between κ_λ and T, as well as the effect of temperature fluctuations on the time-averaged emission (i.e., $\overline{T^4} \neq \bar{T}^4$). Mazumder and Modest[2] discuss the history of the turbulence-radiation interaction.

13.2.6 REACTION MODELING

This section discusses the modeling of combustion chemistry in petrochemical applications. The focus of this section is on methods for modeling the interaction of turbulence with combustion chemistry. This is an area of intense current research, and some of this research is briefly discussed as it pertains to current CFD calculations as well as near-future CFD calculations. There are several relatively new turbulence-chemistry interaction models (such as CMC and joint-PDF transport models) that are not currently available for use in any of the commercial CFD packages. One can hope that this situation will change soon and that these models will be available for more widespread use.

13.2.6.1 Regimes of Turbulent Combustion

Damkohler numbers are ratios of a fluid dynamical time scale to a chemical time scale.[19] In a turbulent flow, there are a variety of time scales, such as the integral scale (a convective scale) and the Kolmogorov scale (a viscous scale). There are also a variety of chemical time scales because of the many chemical reactions that accompany the combustion of even a simple molecule such as CH_4. Frequently, combustion problems are described as being in the high Damkohler or flamelet regime. The term "flamelet" is used because of the notion that within a turbulent non-premixed flame, the actual combustion reactions take place within small layers termed flamelets. These flamelets are so small that they are not affected by the turbulent motions within the fluid; instead, molecular diffusion effects dominate and the structure of the reaction zone is that of a laminar flame (albeit a strained laminar flame).

Following Bray,[6] one can define the Damkohler number as $Da = \frac{t_T}{t_L^0} = \frac{ku_L^0}{\varepsilon l_L^0}$, where the subscript L and superscript 0 refer to an unstretched laminar flame, and the subscript T refers to the scale of the turbulence. In cases where non-premixed combustion is studied, it is common to use the velocity and length scales (the laminar premixed flame speed and thickness) as representative of the relevant chemical scales. The Karlovitz number is $Ka = \frac{t_L^0}{t_k^0} = \frac{t_L^0}{\sqrt{v/\varepsilon}}$, where the subscript K refers to the Kolmogorov time scale. When the laminar flame time is less than the Kolmogorov scale (i.e., $Ka < 1$), the flame is considered to be a laminar flame convected and stretched by a turbulent flow. Combustion in this regime is referred to as flamelet combustion. When the Damkohler number is less than one, the time scale of larger turbulent eddies has become smaller than the chemical time scale. In these conditions, the combustion process is described as a well-stirred reaction zone. For intermediate values of Da and Ka, combustion is said to occur in distributed reaction zones. This term indicates that the turbulent flow can affect the structure of the reaction zone, in contrast to the flamelet regime, but the turbulent mixing is not so fast that the reaction can be considered to occur under well-stirred conditions.

13.2.6.2   Non-premixed Combustion

This section discusses modeling of non-premixed combustion systems. The notion of non-premixed combustion is an idealization. In real combustion systems, mixing occurs simultaneously with combustion, and to call the combustion process non-premixed implies that the combustion takes place much faster than the mixing and that the flame is not lifted off or near extinction at any location. Although it is an idealization, the assumption that combustion is non-premixed provides very useful insights into the combustion processes occurring in real systems.

There are a multitude of computational models for non-premixed (formerly called diffusion) flames. One of the earliest models to appear is the eddy breakup model of Spalding.[20] The model of Magnussen and Hjertager[21] limits the reaction rate according to the local mass fractions of the reactant concentrations or product concentrations. The ratio of the turbulent kinetic energy k to the dissipation rate ε is used as the time scale of the turbulent eddies controlling mixing. These models give physically reasonable predictions of species concentrations in non-premixed systems, but do not consider the important effect of turbulent fluctuations on reaction rates. The model can be extended to consider finite-rate chemistry, but the model is a moment model, using the time-averaged temperature in the Arrhenius rate expression. This limitation is severe in light of the large temperature fluctuations observed in flames.

The mixture fraction concept plays a central role in reducing a turbulent non-premixed flame to a mixing problem. The mixture fraction is a conserved scalar, meaning that it is convected and diffused by fluid motions and gradients, but it is neither created nor destroyed. The mixture fraction, f, represents the mass fraction of fluid at a particular location that originated with the fuel stream. The pure fuel stream then will have $f = 1$, while the oxidant stream will have $f = 0$.

In a turbulent flow, the mixture fraction f fluctuates at a given point with time. A probability density function (PDF) for these fluctuations can be defined so that the probability of f lying between some value x and $x + dx$ is $P(x)dx$. The PDF has some additional properties[8]:

$$\int_0^1 P(f)df = 1 \tag{13.7}$$

$$\langle f \rangle = \int_0^1 f\, P(f)df \tag{13.8}$$

$$\langle f'^2 \rangle = \int_0^1 \left(f - \langle f \rangle \right)^2 P(f)df \tag{13.9}$$

where the $\langle f \rangle$ notation indicates the expectation value (or ensemble average, equivalent to the time-average in a statistically stationary flow) of f, and f' is the turbulent fluctuation of f. $\langle f'^2 \rangle$ is the variance of f.

In CFD calculations of large-scale furnaces with non-premixed burners, the most common combustion model used is probably the assumed-PDF model with equilibrium chemistry. In this model, the shape of the PDF of f is assumed. The β-PDF is a commonly used function that describes the probability of finding the instantaneous fluid to have mixture fraction. The beta-function is given by:

$$P(\langle f \rangle) = \frac{\langle f \rangle^{\alpha-1} \left(1 - \langle f \rangle\right)^{\beta-1}}{\int_0^1 \langle f \rangle^{\alpha-1} \left(1 - \langle f \rangle\right)^{\beta-1} d\langle f \rangle} \tag{13.10}$$

where

$$\alpha = \langle f \rangle \left[\frac{\langle f \rangle \left(1 - \langle f \rangle\right)}{\langle f'^2 \rangle} - 1 \right] \tag{13.11}$$

$$\beta = \left(1 - \langle f \rangle\right) \left[\frac{\langle f \rangle \left(1 - \langle f \rangle\right)}{\langle f'^2 \rangle} - 1 \right] \tag{13.12}$$

Other PDF shapes such as a clipped Gaussian function and a double-delta function are discussed in Jones and Whitelaw.[22] The equilibrium chemistry assumption is poor in flames that are lifted or flames near extinction. Figure 13.2 shows the shape of the β-PDF for several values of $\langle f \rangle$ and $\langle f'^2 \rangle$.

An alternative to the equilibrium chemistry discussion in the previous paragraph is to use the laminar flamelet model. In this model, the relationship between the state of the mixture and the mixture fraction f is determined by a laminar diffusion flame calculation. Peters[23] introduced this idea, which assumes that the reaction length scale, L_R, is much smaller than the Kolmogorov length scale, L_K. Bilger[24] has criticized the classical flamelet method, claiming that for most nonpremixed flames of interest, the flamelet criterion, $L_R < L_K$, is violated. Bish and Dahm[25] discuss the concept further and attempt to eliminate what they view as a key limitation of the method: its assumption that the reaction layers are bounded by pure fuel on one side and pure oxidizer on the other. Their SDRL model is based on the one-dimensionality of the reaction layer, but does not assume the reaction layer to be thin relative to the dissipative scales.

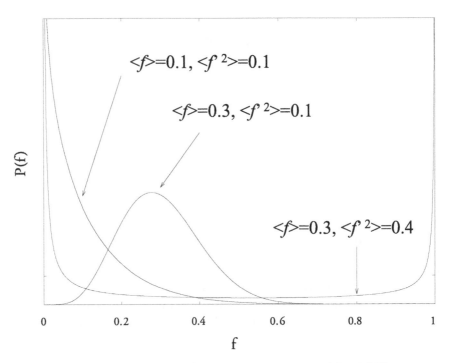

FIGURE 13.2   Plot of the β-PDF for several values of $\langle f \rangle$ and $\langle f'^2 \rangle$.

The classical flamelet model's assumption that the reaction zones are bounded by pure fuel on one side and pure oxidizer on the other is severe in light of the NO_x control strategies used in practical combustion systems. NO_x control is predicated on entraining cooled combustion products into the reaction zone, and the proportion of these gases entrained varies along the length of the flame. The effect of this flue gas entrainment is to reduce flame temperatures and dilute the reactants. Both of these effects are effective at reducing NO_x formation.

Research of models of non-premixed combustion continues at a fervent pace. Pope's[5] joint PDF methods appear promising because they have the ability to treat finite-rate kinetics and eliminate the closure problems. Bilger's (1993) conditional moment closure (CMC) method[26] is also a promising model for non-premixed combustion modeling. Both of these models are applicable to premixed combustion as well.[6] These models are still the subject of active research and academic debate and are not implemented in any of the commercial packages that this author is aware of.

13.2.6.3 Premixed Combustion

Most petrochemical applications use non-premixed combustion because of safety issues in premixed combustion. There are some important exceptions to this statement, however. One important class of premixed burners in the petrochemical industry are venturi-based radiant wall burners. These burners use high-pressure fuel to educt combustion air from the ambient environment. The fuel and combustion air are then mixed in a tube prior to the combustion zone.

Turbulent premixed flames have proven to be much more difficult to model than their non-premixed counterparts.[5] In a turbulent, mixing-limited, non-premixed flame, the flame structure is governed by turbulent mixing, a reasonably well-understood phenomena. The ideal turbulent premixed flame consists of a flame sheet propagating at some flame speed with respect to the fluid around it, which is itself undergoing turbulent motions. The consequence of superposing flame propagation and turbulent fluid motions is that premixed flame modeling is much more challenging than non-premixed flames.[7] For this reason, most commercial CFD codes only include limited support for premixed flame modeling.

The model of Magnussen and Hjertager[21] can be used to simulate a premixed flame. The model is unsatisfactory in many ways, however. It has no means of modeling the effect of temperature fluctuations on the reaction rate and no description of the turbulent flame as an ensemble of premixed flamelets.

A significant limitation of many of the flamelet models for premixed combustion (see Reference 7, for example) is that they assume the combustion process is adiabatic. In operating furnaces, heat losses from the flame to the load are an integral part of the process. The inability to adequately model premixed flames is a significant limitation. Methods such as PDF methods discussed by Pope[5] may allow improved simulations in the future. In addition, it may be possible to include heat losses in the flamelet models of premixed combustion.

13.2.7 POLLUTANT EMISSIONS

Pollutant emissions are among the most important drivers in the petrochemical industry, especially in the U.S. The EPA allowed levels of NO_x and SOx emissions from petrochemical plants and refineries continue to decrease. To respond to this challenge, burner manufacturers strive to develop burners that produce lower and lower emissions. In addition, furnace manufacturers and other vendors develop post-combustion technologies such as SCR (selective catalytic reduction) and SNCR (selective non-catalytic reduction) to deal with NO_x in the stack. Sulfur scrubbers are used to reduce SOx levels in stack gases.

A recent paper[14] evaluated NO_x predictions using two different models for the turbulence-chemistry interaction: the probability density function (PDF) model of Pope (see Reference 5, for example) and the conditional moment closure (CMC) model of Bilger.[26] The PDF model here is not the assumed-PDF discussed at length above. This method solves for the transport and production of the scalar joint PDF and is extremely computationally expensive because a Monte Carlo solution algorithm must be used. The particularly interesting thing about this article is the comment in the introduction that "a realistic target for agreement between experiment and prediction might be $\pm 20\%$ to $\pm 30\%$." The article goes on to discuss how sensitive NO_x predictions are to the radiation model used. The flame studied in that article is a simple diluted hydrogen jet flame. If the most sophisticated turbulence-chemistry models currently under research applied to a very simple flame in a very simple geometry can only be expected to yield an accuracy of $\pm 30\%$, then how accurately can one realistically expect to predict NO_x emissions?

13.2.8 COMPLEX GEOMETRY

The capability of a CFD package to treat complex geometries is an important consideration for industrial applications. The geometries encountered in low-emissions burners frequently employ complicated shapes and jet angles. The purpose of these geometries is to precisely control when and where the fuel is oxidized. These combustion control strategies are critical to the performance of the equipment. CFD models must be able to accurately capture the effect of these complex geometries in order to be useful. This means, as a practical matter, that cell types other than hexahedra are required. All of the current generation of commercial codes are compatible with a variety of cell types, including hexahedra, prisms, pyramids, and tetrahedra. Figure 13.3 shows a rendered view of a CFD model of a John Zink Co. hearth burner. The burner shown in Figure 13.3 is vertically fired, and the view shown is looking down at the burner from above. There are four gas tips: two in the primary combustion zone (inside the tile) and two firing secondary fuel (these tips are sitting on the tile). Each primary tip has five small orifices firing fuel in different directions, as shown in Figure 13.4. The orifice diameters range from 0.0625 in. to about 0.25 in. Each secondary tip has three fuel orifices of similar size. The figure illustrates the necessity of a CFD package to treat complex geometry in order to accurately model the performance of an industrial burner.

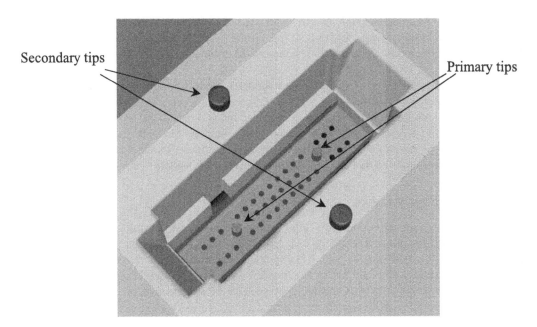

FIGURE 13.3 Rendered view of a CFD model of a John Zink Co. burner. This view illustrates the complex geometry, which necessitates a variety of cell types. This mesh consists of hexahedral, pyramid, and tetrahedral cell types. (Courtesy of John Zink Co., Tulsa, OK.)

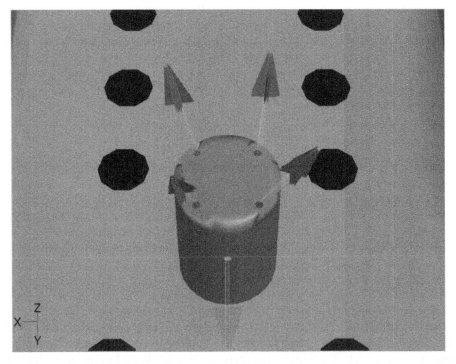

FIGURE 13.4 Close-up view of primary tip. This view reveals the five fuel jets (indicated by the arrows on the image) issuing from the primary tip.

Generating a computational mesh for these geometries is a well-known bottleneck in CFD analysis. Mesh generation can consume well over half of the time budgeted for a CFD project. Improvements in mesh generation technology greatly benefit industrial CFD users as they allow more and more of the actual geometry to be included in the CFD model. In addition, mesh generation improvements frequently simplify the process of modifying an existing geometry. Making geometric modifications to a geometry after the initial meshing can be nearly as time-consuming as generating the initial mesh.

New combustion and turbulence models are frequently developed only for simple Cartesian grids. Only much later, after considerable proof of concept in the simple geometries, are these models adapted for use on mesh types used in industrial analysis. In this author's opinion, the ability to treat complex geometry is the main reason that industries turn to commercial CFD vendors instead of using codes developed at national laboratories and universities.

13.3 POST-PROCESSING

A typical CFD simulation provides on the order of 10^6 to 10^8 discrete numerical outputs. For example, a simulation with 500,000 nodes and 11 variables per node (pressure, density, three velocity components, k, ε, temperature, mixture fraction, variance of mixture fraction, and irradiation) would generate 5,500,000 numbers. If the various chemical species are considered as well as the detailed results of a discrete ordinates model, the number of variables per node could easily exceed 50, leading to 25,000,000 numerical results. The generation of xy plots (for example, temperature vs. position along the burner centerline), contour plots, velocity vector plots, streamline plots, and combinations and animations of these outputs are necessary for the analyst to understand the results of a simulation. The production of these different sorts of outputs becomes very important in communicating the results of a simulation. This is especially true when the intended audience is not composed of CFD specialists. Current post-processing packages have the ability to add lighting to a model, which makes the images more realistic to the viewer. Figure 13.5 shows an example of using the rendering capabilities of a CFD package to generate an image with photorealistic qualities. Images such as Figure 13.5 can take anywhere from several seconds to several minutes for current-generation scientific workstations to render, depending on the number of lights applied, the number of surfaces in the scene, and the complexity of these surfaces. High-performance, virtual reality environments must be able to regenerate these scenes many times per second.

In addition to still images, animations can be effectively used to illustrate CFD results. Animated velocity vectors and streamlines illustrate very well the path of fluid flow in internal and external flow problems. Sweeping planes showing either velocity vectors or filled contour maps of a scalar result can quickly present information about an entire three-dimensional simulation.

Generating effective presentations, including still images and animations, is a time-consuming task. Creating a suitable image to make a specific argument frequently requires the analyst to look at and reject a large number of candidate images. It also requires significant expertise from the CFD analyst. It is certainly true that CFD results can be misinterpreted or misapplied to lead to an incorrect conclusion. In addition, in an industrial setting, the audience will frequently not have the expertise required to assess the quality of a simulation.

13.4 CASE STUDIES

This section describes several applications of CFD in petrochemical applications. CFD can address a wide variety of problems in the industry. The applications discussed here relate to fired heaters and incinerators. Many other applications, flare systems for example, exist in petrochemical plants where CFD analysis is valuable.

FIGURE 13.5   Rendered view inside an ethylene pyrolysis furnace showing flow patterns near the premixed radiant wall burners. (Courtesy of John Zink Co., Tulsa, OK.)

13.4.1   ETHYLENE PYROLYSIS FURNACE

Ethylene pyrolysis furnaces produce ethylene and propylene from feedstock containing ethane, propane, butane, and hydrocarbons through naptha. The process entails rapidly heating the feedstock for a short time (less than 1 second is typical) to a temperature of about 1600°F (870°C). The feed gases are then rapidly cooled and subjected to a number of separation processes.

This section focuses on modeling of the pyrolysis furnace. Typical pyrolysis furnaces are approximately 10 ft wide, 30 ft long, and 40 ft high. There are two rows of "flat flame" burners that directly fire onto the walls of the furnace. These fired walls then radiate heat to the process tubes in the center of the furnace. Figure 13.6 illustrates this geometry. In the figure, the process fluid tubes extend from the floor of the furnace to the roof of the radiant section. In the image, only the radiant section is shown because the radiant section is where the combustion occurs. In a production furnace, the products of combustion would leave the radiant section and enter a convection section where heat is recovered from the products of combustion.

Figure 13.3 shows a view of the burner geometry. The burner is a staged-fuel gas burner. This example illustrates the disparity in scales in a furnace analysis. The furnace has a height of approximately 30 ft, while the fuel orifices can be as small as 0.0625 in. in diameter. The ratio from the largest dimension to the smallest is then greater than 5000. In this example, a nonconformal mesh interface was used to reduce the cell requirements.

The CFD model of the ethylene pyrolysis furnace includes detailed information about all of the fuel jets in the burner. In this particular burner, there are five fuel jets on each of the two primary tips (Figure 13.3), and four fuel jets on each secondary tip.

Figure 13.7 shows the predictions of heat flux to the process tubes as a function of height above the furnace floor. These heat flux profiles drive the design of modern ethylene pyrolysis furnaces. CFD is being used in these designs more and more as the results of the model become better validated. These results have not been validated, and it seems unlikely that such data will become available,

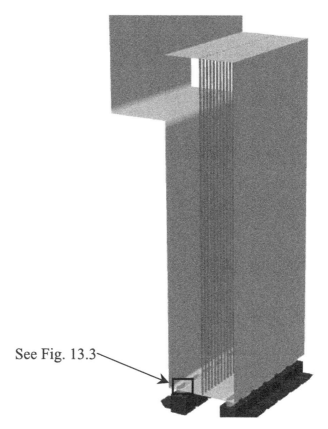

See Fig. 13.3

FIGURE 13.6 CFD model of an ethylene pyrolysis furnace. There are six burners shown in each row at the bottom of the furnace, and the tubes are approximately 35 feet long. (The endwalls are not shown in this image.)

given the difficulty of acquiring data in operating furnaces. Availability of data is a significant limitation in further use of CFD in petrochemical applications, as discussed in the Section 13.1.

13.4.2 Xylene Reboiler

The geometry of the vertical cylindrical furnace is shown in Figure 13.8. The small wall around the burners is a reed wall, and is used to heat the cold flue gases coming from the tubes. The periodicity of the furnace was used to simplify the model. The computational model of the vertical/cylindrical heater consisted of only one burner with periodic boundary conditions applied. This model has the shape of a very tall slice of pie. The images shown here are created by rotating the results about the vertical axis of the furnace.

The combustion model used in these calculations is an assumed PDF of mixture fraction. Because heat transfer to the tubes and furnace temperature are known to be important, a nonadiabatic mixture fraction table was constructed. The independent variables in the lookup table are mixture fraction, variance of mixture fraction, and enthalpy. Radiation was modeled using the discrete ordinates model with 32 ordinates. All solid surfaces were assumed to be radiatively black. Gas properties were computed using the weighted-sum-of-gray-gases method.

This study involves an operating furnace in a refinery. The problem observed in the furnace was that the flames from the ultra-low-NO_x burners were very long and had the potential to damage process tubes in the top of the furnace. This phenomena has been observed in several vertical/cylindrical furnaces with ultra-low-NO_x burners. The problem is related to the flow pattern within the

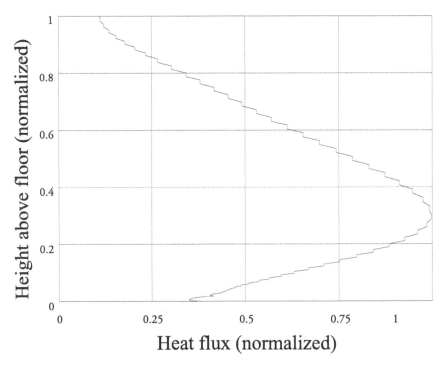

FIGURE 13.7 Plot showing heat flux to the process tubes in the modeled ethylene furnace as a function of height above the furnace floor.

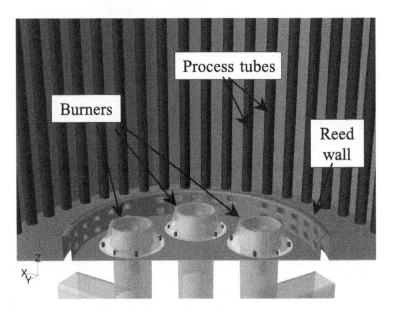

FIGURE 13.8 Geometry of xylene reboiler. This view shows half (sliced vertically) of the furnace. This view shows only three of the six burners at the bottom of the image.

furnace as it does not allow complete mixing of the combustion air with the fuel, but rather distorts the flame prior to burnout.

Figure 13.9 shows a CFD simulation of the burners as they were originally installed. The figure shows an isosurface of OH, which is a good indicator of flame shape in this case. The results reveal

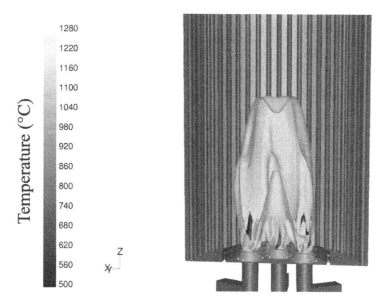

FIGURE 13.9 This view shows half of the furnace with unmodified burners. The "blob" in the furnace is the 50-ppm OH mole fraction isosurface. This surface is colored according to its temperature (°F).

that the flames from adjacent burners merge together to produce a single long flame, which is confirmed by observations of the operating furnace. This burner has two primary fuel tips that fire fuel inside the tile, and four secondary fuel tips. The solution to this flame interaction problem was to change the burner so that only three of the secondary tips actually fired. The CFD results for that configuration are shown in Figure 13.10. This solution was implemented and tested in the

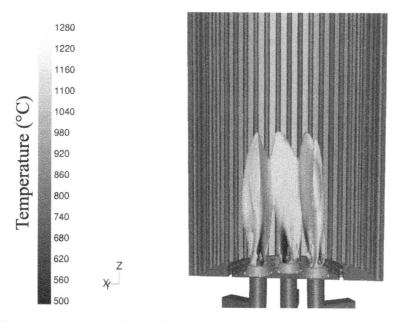

FIGURE 13.10 This view shows half of the furnace with the modified burners firing. The 50-ppm OH mole fraction isosurface is shown as an indicator of the flame shape. This surface is colored according to its temperature (°F).

operating furnace and found to yield qualitatively the same result: the flames became distinct and burned out at the appropriate height.

13.5 NEEDED FUTURE WORK

It seems fitting to summarize the needs for future work. Throughout this chapter, there have been discussions of model limitations and needs for better basic physical models. These needs are relatively clear and easily understood. The focus of this section will then be on issues that hinder CFD from becoming a valued design and troubleshooting tool in the petrochemical industry.

Given the importance of NO_x emissions in the installation of new combustion equipment, NO_x predictions may be the single most important improvement in the CFD analysis of industrial systems. Accurate NO_x predictions will continue to be a challenge due to the nature of the physical problem (i.e., the strong coupling between the relatively slow chemical reactions and the turbulence in the flame).

The ability of CFD models to treat stability problems in non-premixed combustion is another area needing improvement. As discussed above, the presumed PDF model with equilibrium chemistry is probably the most accessible model to the industrial analyst, yet significant departures from equilibrium are observed in flames near extinction.[7] Given the chemical complexity of the problem, finding a model that can be valid for a range of fuels will be a significant challenge, but such a model would benefit the industry. It seems likely that a model that can capture these effects is a prerequisite to improving NO_x predictions due to the strategies employed in industrial combustion systems for reducing NO_x.

The inability of current-generation commercial products to model premixed flames with heat losses is a significant limitation. Premixed combustion is significantly more challenging than non-premixed combustion because of the coupling of flame propagation and turbulent motions in the fluid. Improvement in models of premixed combustion that are applicable to petrochemical burners would benefit the industrial CFD user.

13.6 CONCLUSIONS

CFD modeling of industrial furnaces is a valuable tool that can be used profitably. CFD modeling can help identify the cause of problems and it can be used to test solutions. In addition, CFD modeling can be a valuable design tool for combustion equipment in the petrochemical industry. It is also clear that CFD has not achieved the status of stress analysis in terms of ease of use. In many cases, stress analysis of mechanical designs is done by engineers without advanced understanding of the physics and reasonable results are achieved. In CFD, especially the study of combustion systems, this is not the case. Understanding and interpreting the results of a CFD model requires a thorough understanding of the underlying physics. In a typical furnace model, the science involved is multi-disciplinary, involving heat transfer, fluid flow, and combustion kinetics.

REFERENCES

1. Davis, K.A., Bockelie, M.J., Smith, P.J., Heap, M.P., Hurt, R.H., and Klewicki, J.P., Optimized fuel injector design for maximum in-furnace NOx reduction and minimum unburned carbon, presented at the *First Joint Power and Fuel Systems Contractor Conference,* Pittsburgh, July 1996.
2. Mazumder, S. and Modest, M., Turbulence — Radiation Interactions in Nonreactive Flow of Combustion Gases, *Journal of Heat Transfer,* 121, 726-729, 1999.
3. McDermott, R. and Henneke, M.R., High capacity, ultra low NOx radiant wall burner development, *12th Ethylene Forum,* May 11-14, 1999. The Woodlands, TX.
4. Panton, R., *Uncompressible Flow,* Wiley-Interscience, New York, 1984.

5. Pope, S.B., Computations of turbulent combustion: progress and challenges, *23rd Symposium (International) on Combustion,* The Combustion Institute, 1990.
6. Bray, K.N.C., The challenge of turbulent combustion, *26th Symposium (International) on Combustion,* The Combustion Institute, 1996.
7. Warnatz, J., Mass, U., and Dibble, R.W., *Combustion: Physical and Chemical Fundamentals, Modeling and Simulation, Experiments, Pollutant Formation,* Springer-Verlag, Berlin, 1995.
8. Libby, P.A., *Introduction to Turbulence,* Taylor and Francis, New York, 1996.
9. Launder, B.E. and Spalding, D.B., *Mathematical models of turbulence,* Academic Press, London, 1972.
10. Speziale, C.G., On non-linear $k-l$ and $k-\varepsilon$ models of turbulence, *Journal of Fluid Mechanics,* 178, 459-478.
11. Yakhot, V. et al., Development of turbulence models for shear flows by a double expansion technique, *Physics of Fluids A,* 4(7), 1510-1520, 1992.
12. Shih, T.-H., Liou, W.W., Shabbir, A., and Zhu, J., A new $k-\varepsilon$ eddy-viscosity model for high Reynolds number turbulent flows — Model development and validation, *Computers Fluids,* 24(3), 227-238, 1995.
13. Versteeg, H. K. and Malalasekera, W., *An Introduction to Computational Fluid Dynamics. The Finite Volume Method,* Addison-Wesley Longman Limited, England, 1995.
14. Barlow, R.S. Nitric Oxide formation in dilute hydrogen jet flames: isolation of the effects of radiation and turbulence-chemistry submodels, *Combustion and Flame,* 117, 4, 1999.
15. Modest, M.F., *Radiative Heat Transfer,* McGraw-Hill, New York, 1993.
16. Siegel, R. and Howell, J., *Thermal Radiation Heat Transfer,* Hemisphere, Washington, D.C., 1992.
17. Khan, I.M. and Greeves, G., A method for calculating the formation and combustion of soot in diesel engines, in Afgan, N. H. and Beer, J. M., Eds., *Heat Transfer in Flames,* Scripta, Washington, D.C., 1974, chap. 25.
18. Tesner, P.A., Snegiriova, T.D., and Knorre, V.G., Kinetics of dispersed carbon formation, *Combustion and Flame,* 17, 253-260, 1971.
19. Williams, F.A., *Combustion Theory,* Addison-Wesley, 1985.
20. Spalding, D. B., Mixing and chemical reaction in steady confined turbulent flames, *13th Symposium (International) on Combustion,* The Combustion Institute, 1970.
21. Magnussen, B.F. and Hjertager, B.H., On mathematical modeling of turbulent combustion with special emphasis on soot formation and combustion, *16th Symposium (International) on Combustion,* The Combustion Institute, 1976.
22. Jones, W.P. and Whitelaw, J.H., Calculation methods for reacting turbulent flows: a review, *Combustion and Flame,* 48, 1-26, 1982.
23. Peters, N., *Progress in Energy and Combustion Science,* 10, 319, 1984.
24. Bilger, R.W., The Structure of Turbulent Nonpremixed Flames. *22nd Symposium (International) on Combustion,* The Combustion Institute, 1988.
25. Bish, E.S. and Dahm, W.J.A., Strained dissipation and reaction layer analyses of nonequilibrium chemistry in turbulent reaction flows, *Combustion and Flame,* 100, 3, 1995.
26. Bilger, R.W., *Physics of Fluids A,* 5, 436, 1993.

Section III

Advanced Techniques

14 Design Optimization

Richard W. Johnson, Mark D. Landon, Ernest C. Perry

14.1 INTRODUCTION

Design optimization has been practiced by humankind, in some form or other, for millennia, although the designers may not have recognized their efforts as such. Many would have said that they were simply looking for improvements in their canoe paddle or electric motor to enhance performance. Many would not have said that they were trying to obtain the best possible design to maximize the performance of their sailboat. In fact, even today, management does not often allow the design engineer to find the absolute best possible design to maximize performance. They want to sell intermediate designs that work well relative to competitors, the actual objective being to maximize sales. This is not a bad thing. Design optimization can be employed to find improved designs, which may be the best designs one can find, given the time and resources allocated to the problem. However, the authors believe that if the principles discussed below are applied, design engineers can find much better designs for given resources than would be found using traditional methods.

This chapter explains what is meant by design optimization, some basic mathematics, how to pose an optimization problem, some popular and useful optimization algorithms, optimization as applied to CFD, and also provides some examples.

14.2   THE PHILOSOPHY OF OPTIMIZATION

14.2.1 INTRODUCTION

While engaged in taming, harnessing, or dealing with the forces of nature, humankind has primarily been interested in wagon and boat designs that work (feasible designs), at least some of the time. One may have tried to find improvements that made the design work better or longer than the previous design. Now, of course, one lives in a competitive world where people not only want something that works, but works as well or better than the competition's. This means that most people are not satisfied simply with something that works some of the time. How does one improve designs? One might go and consult with Bill, the senior designer at the company (experience), or one might have one's own ideas, so one builds something and tests it (trial and error). One might do this over and over again, making improvements much of the time (exhaustive search). This is probably how the Phoenicians came up with sailing boat designs that allowed them to become expert sailors.

One wonders how the Phoenicians came up with the need for multiple sails, the shapes of the sails, the need for rudders and keels, and how to size the width vs. the length of the boat. Probably, many sailors lost their lives while ship designers tried to figure out these things. The number and shapes of the sails can be thought of as design variables (variables that can be modified to change performance), while ship stability can be thought of as a constraint. (If the boat goes down in moderate winds, much of the master's gold is going to the bottom). The objective of the ship designers may simply have been to build ships that transported cargo well most of the time. Of course, with the appearance of pirates, the objective may have changed to finding a design that would outrun the bandits. Then the bandits steal the new design and the design race is on. The objective (objective function) of the design is the central issue of the design problem. In the case of the Phoenician ship designers, the objective probably was not merely the speed or capacity of the ship, but the capability of the ship to provide maximum wealth to the owners. Once the overall objective has been defined, the next lower level of objectives can be stated: the speed and capacity of the ship, subject to the ship being stable in most weather situations and one that does not cost the owner most of his wealth.

The best ship designers were those who knew what the true objective was, who knew about everything that could affect the objective, who knew what the constraints were, and who had some idea of how to meet the objective. This is still true today. The designer who has a good starting design, who can identify all the design variables, who knows what the technical constraints are and is sensitive to the political ones, and who employs the most effective design tools will usually end up with the best design. There is still much need for design engineers with bountiful ingenuity.

14.2.2 ENGINEERING DESIGN OPTIMIZATION

Recent decades have witnessed the development of powerful engineering analysis tools, including computational fluid dynamics (CFD) codes that have allowed analysis of both natural phenomena and man-made equipment that humankind has never analyzed previously. These codes have been hailed as tremendous resource savers for design engineers because they are supposed to be able to analyze a new design much faster and much cheaper than building and testing a physical model. As is usually the case, reality has worked out a bit differently than early expectations predicted.

Design engineers must be careful before placing confidence in a numerical analysis of the performance of their design. They also realize the necessity of validating the predicted design performance by building a physical model and testing it. What are the possible sources of error that can corrupt a numerical analysis? The list of potential error sources includes:

1. A fine enough grid (grid dependence) was not used.
2. Inaccurate assumptions were made about the problem (the flow is really compressible, unsteady and/or turbulent).
3. The problem thought to be set up was not what was really set up (inexperience).
4. Inaccurate material properties (density, permeability, reaction rates) were used.
5. The computer was inadequate, requiring a coarse mesh (grid dependence).
6. There was insufficient time to get the model right.
7. Some important physics were neglected.
8. The physics models employed in the model were inappropriate for the problem (turbulence model, radiation heat transfer model, heat transfer coefficient, wrong set of reactions).

Remark. Some of the above may not be the fault of the design engineer, or may have only a small influence on the problem analysis; for example, the best turbulence model available was used, or the convective heat transfer coefficient is not so important because the heat is transferred mostly by radiation. Still, if a computer model is a reasonable representation of the physics, it can be very useful. It can be used to help identify which physics are most important and eliminate many poor designs. The alternatives to using a good computer model in engineering design are to do it the old-fashioned way, experimentally, (sometimes, this is still the only way to do it), find a similar problem in the literature and hope it is close, or imitate the competition's design.

It is assumed in this chapter that the design engineer has developed a reasonably accurate model of the physics of the component or system design that he/she wishes to improve and/or optimize. At this point, the typical approach is to begin tweaking the design variables and rerunning the numerical model to see if improvements can be made. This approach is illustrated in Figure 14.1 where design changes are generated by application of experience or simply trial and error.

While the time required for finalizing the design is shortened relative to the old-fashioned approach of building a new physical model and testing it each time, there is still a significant amount of time used to tweak and reanalyze. Also, this approach does not by any means guarantee that the optimal design will be found.

The application of engineering design optimization is the next step that should be employed if there are more than two design variables or the mathematics describing the physics are nonlinear. That is, computer-based design optimization represents the next step beyond application of analysis codes such as CFD codes. A computer code that incorporates optimization algorithms, which are used to search for optimal sets of design variables, are coupled to engineering analysis codes such that the search (optimization algorithm) and design testing (analysis code) are done automatically. Figure 14.2 illustrates this approach, called engineering design optimization. The starting point is an existing design, whose performance is measured by performing a numerical (CFD) simulation. The optimizer then determines how to search for improved designs based on the nature of the optimizing algorithm, usually employing the analysis code to perform sensitivity analyses. The

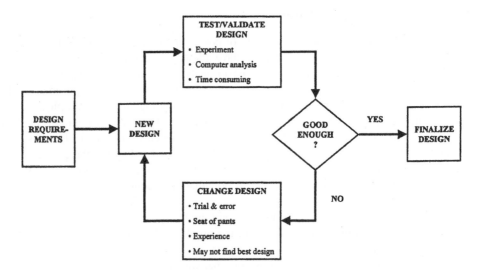

FIGURE 14.1 The traditional engineering design process.

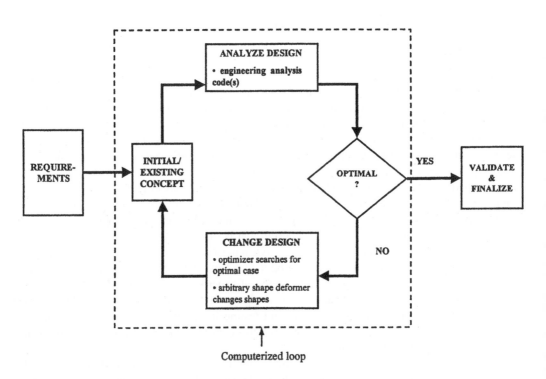

FIGURE 14.2 Computer-based engineering design optimization.

optimizer can then search for improvements by altering the collection of design variables subject to the specified constraints until no improvement can be found (an optimal design) or until allocated time or resources are exhausted. If the optimizer is stopped before the optimal design is found, the engineer can use the last feasible design as the final design. Often, the latter is the most practical option, for example, if each analysis requires a long time, or if improvements become miniscule. If geometric shape is important to the performance of the design, some means of changing the shape can be employed along with the optimizer and analysis code. Shape optimization is discussed in Section 14.5. Also, it is possible to employ multiple analysis codes in the optimization problem if multiple engineering fields are important (e.g., fluid dynamics, solid mechanics, etc.)

The rest of the present section is devoted to expanding on the details of design optimization, including discussing the basic concepts, explaining how to pose an optimization problem, and the application of (computer-based) optimization to design engineering.

14.2.3 BASIC CONCEPTS IN OPTIMIZATION

Suppose that an army research department is given the task to design the means of propulsion for a new armored personnel carrier (APC). The project head must begin the design process by first considering what the available options are in view of what the major performance requirements are. For example, the APC must be able to function in all kinds of weather, including very muddy conditions. It must be capable of functioning in 5 ft of water. It must use readily available fuel that is also cheap. It must be able to operate for 300 miles without refueling. The designers must now survey existing engines: gasoline, diesel, turbomachine, and others. Just for fun, one could add solar- and nuclear-powered engines, and engines that run on propane and hydrogen. Each of these engines can be considered to be a different **design concept** or **archetype**. Perhaps it is straightforward to select the best archetype for this design problem; perhaps it is not. Maybe an archetype needs to be developed to the point of practicality from amongst those currently in the research phase.

Now suppose that the designer has chosen an archetype for the APC, but wants to enhance the performance currently available in existing designs. Or perhaps the designer has selected two archetypes and wishes to optimize both to see which will be best overall. The designer must determine how to proceed with the optimization. There can be regular meetings, where each engineer on the project proposes ideas that are then implemented and tested on a physical prototype. This, however, is determined to be very time- and resource-consuming. The designer can have engineers develop mathematical models of the target engine(s) using sophisticated analysis software: structural analysis, thermal analysis, and computational fluid dynamics. After the models have been developed, staff meetings can be rescheduled to discuss how to improve the target engine based on the simulation results obtained. This approach is actually rather typical of design programs today. Unfortunately, the models that simulate the engine performance may be very nonlinear, with a high degree of coupling of physics of various engine components. It may be very difficult to conjure up changes that will ultimately lead to improved performance of the engine, especially if there are many design variables to be fine-tuned. The approach that is recommended here is the **engineering design optimization** approach. That is, let proven computer-based optimization algorithms go to work on the optimization process.

At this point, the terms used in engineering design optimization need to be defined. The mathematical or numerical models of the engine are simply called the **model**. It is computer-based and written in some scientific code. It may be written for a commercial code or an in-house code for which one has the source code. The input variables that basically define the design are the **analysis variables**. Some of these will be allowed to change to enhance the design performance; these are the **design variables**, a subset of the analysis variables. The design variables may be such things as the compression ratio, number of cylinders, engine displacement, etc. These design variables are inputs to the model. A unique set of values for the design variables constitutes a particular **design**. Outputs from the model, such as horsepower, torque, exhaust effluents, etc., are called **analysis functions**. One or more of the analysis functions can be chosen to be the objective of the optimization. If more than one is chosen, it will be necessary to determine how to weight each one as they are combined to form a single function. This analysis function (or collection of analysis functions) is the **objective function**. The objective function is sometimes called the cost function. One seeks to either minimize or maximize this function to perform the optimization. For example, one might want to maximize the horsepower of the engine.

Additional analysis functions can be defined for the purpose of constraining the design. That is, one can compute the fuel economy of the design and decide that it must be equal to or greater than a specific value in order to meet the goal of a 300-mile range for the APC. This is an example of an **inequality constraint** applied to an analysis function. Also, one can determine that for this APC, the exhaust temperature must be equal to a certain value, perhaps to avoid having an infrared signature, but yet to maximize fuel consumption. This would be an example of an **equality constraint** applied to an analysis function (the computed exhaust temperature). Some constraints may be required to ensure that unphysical results are avoided (infinite velocity at a point of zero cross-sectional area in a duct). Other constraints may be required to avoid impractical results (too expensive; need room for other components). (Note that the APC example is not intended to have any connection with a real design.)

We can represent the objective function f as a function of the design variables $x_1, x_2, \ldots, x_n$:

$$f\left(x_1, x_2, \cdots, x_n\right) \text{ or } f(\mathbf{x}) \tag{14.1}$$

where $\mathbf{x} = x_1, x_2, \ldots, x_n$. A bold letter is used to represent a vector quantity (or set of variables). The set of design variables $\mathbf{x}$ may vary, depending on which variables the designer chooses to identify as design variables. The objective function $f(\mathbf{x})$ may be a closed-form analytical function, but usually it is not. If employing CFD to define the model, the objective function will typically not be explicitly definable, but will be some kind of output from the CFD code. One will probably have to perform some post-processing operation to obtain a value for it (such as integrating to obtain a mass flux or drag coefficient).

The analysis functions that are chosen to be inequality constraints can be represented as:

$$g_i(\mathbf{x}) \le 0, \text{ for } i = 1, m \tag{14.2}$$

where there are 'm' inequality constraints. Using vector notation, one can instead write

$$\mathbf{g}(\mathbf{x}) \le \mathbf{0} \tag{14.3}$$

The equality constraints can be represented as:

$$h_j(\mathbf{x}) = 0, \text{ for } j = 1, k \tag{14.4}$$

where there are 'k' equality constraints, or more simply,

$$\mathbf{h}(\mathbf{x}) = \mathbf{0} \tag{14.5}$$

Note the use of zero on the right-hand sides above; one can subtract any constant from both sides to yield a zero on the right. Now the optimization problem can be stated as:

$$\text{Minimize or maximize } f(\mathbf{x})$$
$$\text{subject to} \quad\quad \mathbf{g}(\mathbf{x}) \le \mathbf{0}$$
$$\text{and} \quad\quad\quad\quad \mathbf{h}(\mathbf{x}) = \mathbf{0}$$

The design space is n-dimensional because there are n design variables.

In the case of **unconstrained optimization**, the extremum (minimum or maximum) will most likely occur interior to the domain at a "peak" or a "valley." There may be multiple places where a "peak" or "valley" occurs. The problem statement seeks to find the "global" optimum, as opposed to a local optimum. However, some optimization algorithms are gradient based and may only find a local optimum. This may not be a bad thing. A local optimum can still provide improved performance. Other non-gradient-based optimization algorithms may be able to find the global optimum, but may require hundreds of analyses to find it. This may not be practical if the time it takes to perform an analysis is large or if the allocated time and resources are insufficient. Optimization algorithms for unconstrained optimization are fairly simple and are discussed in Section 14.3.

It is not sufficient that the gradient of the objective function be zero to be able to identify an extremum. For example, the function $f(x) = x^3$ has a zero gradient at $x = 0$. However, this is an inflection point, not an extremum.[1] Also, the function $f(\mathbf{x}) = x_1 x_2$ has a zero gradient at $x_1 = x_2 = 0$. However, this is a saddle point. For the case of a single design variable or $n = 1$, it is sufficient to have $f'(x_o) = 0$ and $f''(x_o) > 0$ to have a minimum at x_o or $f'(x_o) = 0$ and $f''(x_o) < 0$ to have a maximum. There are analogous conditions for $n > 1$.[1] See the list of reference texts at the end of the chapter for additional details.

In the case of **constrained optimization**, the optimum will most likely lie on a line or surface defined by one or more constraints. When a constraint blocks the further minimization (maximizing) of the objective function, the constraint is said to be **active** or **binding**. This situation often occurs in engineering design optimization. For constrained optimization, different conditions must be met to ensure that an optimum has been found. Special algorithms are required to handle the constraints. Constraint-handling algorithms, along with the sufficient conditions for a constrained optimum, are discussed in Section 14.3.

For constrained optimization, the objective function can be improved by relaxing the constraint. Of course, permission may be needed from another design group that has responsibility for that particular constraint or from management. However, armed with some information about how much the objective could be improved by relaxing the constraint, one may be able to negotiate a change with management.

A design (set of values for the design variables) is called a **feasible design** if is does not violate any of the constraints. A particular optimization effort may not produce the true global optimum due to limited time and resource constraints, but it still may produce an improved design. Some optimizing algorithms will only allow feasible designs to be considered, while others may jump into an infeasible region during the process of finding the optimum. The designer should be cognizant of what the optimizing algorithm is doing in this regard.

14.2.4 POSING THE OPTIMIZATION PROBLEM

Section 14.2.3 provided the basics used to pose the optimization problem. This section focuses on the practical aspects and some of the pitfalls associated with posing the problem. Of course, the following discussion assumes that the paradigm for the optimization/design problem is computer based. One must be careful to adequately describe the problem to the computer in order to obtain meaningful results.

14.2.4.1 The Model

The model that is used is only an approximation to reality. It is important that it is a reasonable approximation to reality. The optimization process is only as good as the model. One must be able to identify which pieces of the physics are most important and which are not. One must be able to employ suitable mathematical descriptions of each important piece of the physics. For example: Is the turbulence model adequate? Have the important chemical reactions been included? Is the radiation heat transfer model good? Is the numerical grid fine enough? One sometimes has to put up with something less than satisfactory because the state-of-the-art is somewhat less than satisfactory.

In such cases, one should realize the inadequacy of the model. One will probably also want to validate the numerical results with an experimental test. The designer should also be aware that values for the material properties (e.g., density, viscosity, reaction rate coefficients) will affect the accuracy of the model.

14.2.4.2 The Objective Function

Surprisingly, it is sometimes difficult for the design project leader to properly define the objective function. Sometimes the objective function used on the computer is a translation of the actual objective, which may be to maximize company profits. Then the designer has to figure out what that means at the technical level. Or, the traditional approach to defining the objective may not really be the correct way. The objective function must be defined as something that the numerical code can calculate and produce a value for. Examples of objective functions are: minimize the carbon monoxide, maximize the thermal energy input to the scrap, or minimize the variation of temperature or species concentration in the melt. An objective function that could be used to minimize the carbon monoxide is the total mass of CO in the furnace as obtained from an integration of CO species concentration in the model for the furnace volume.

In the approach to optimization called *inverse design*, the objective function is defined in a particular way. Inverse design is related to solving an inverse problem, wherein the solution to a problem is known (e.g., temperature field, pressure field) but the boundary condition is not. Then one seeks to find the boundary condition that gives the desired solution. An example of inverse design methodology is to choose a particular pressure distribution one desires on an airfoil and then change the shape of the airfoil until the desired pressure distribution is achieved. The objective function would be to minimize the difference between the current pressure distribution and the desired one. This approach can work fine if one is confident that the desired distribution is the best one. A forward approach to the design of the airfoil would be to define the objective function to be the ratio of lift to drag and then seek to maximize this ratio.

14.2.4.3 Constraints

Constraints are very important. If the designer forgets to include an important constraint, the optimization algorithm may simply exploit this weakness and produce an optimum that is wholly impractical or impossible. Of course, this should alert the designer that something has been forgotten, so that the correction can be made. On the other hand, one does not want to add constraints that are inappropriate, or one may be removing the true optimum from the feasible design space.

14.2.4.4 Design Variables

The set of design variables is a subset of the analysis variables that jointly define the design. The designer might determine to choose different sets of design variables and then seek the optimum for each set. This may be because the designer is unsure which variables should be allowed to change, or there is some political reason to disallow some variables to change. The designer may then choose the overall best from the several optima that are found. In fact, the fewer the number of design variables, the better, because the dimension of the design space is smaller. This means that the optimization problem will require less time, sometimes much less time. The designer must weigh the desire to include all the analysis variables possible to find an overall optimum, versus the time and computer resources that it will take to do so.

14.2.5 USING OPTIMIZATION IN DESIGN ENGINEERING

Using computer-based engineering design optimization may seem difficult and expensive. The manager or even the engineer may not wish to incur the expense of yet another software license or another staff member with expertise in optimization. Or it may seem that the process is going to take a long time because multiple analyses will be required to find optimal designs. However, the thoughtful engineer/manager must compare the actual cost of doing design the old way with the new costs of optimization. One must consider all of the time used for trial-and-error testing. How many years did it take "old Bill" to figure out the current operating parameters used for the furnace? How much does it cost to tweak the furnace to see if a better operating condition can be found? What if one just makes no effort to try and optimize our furnace operation? Will the competition eventually leave us in the dust?

The application of engineering design optimization should be fairly transparent to a good engineer, even one who has very little experience with formal optimization. More than likely, this engineer has performed optimization, but in a less efficient way. There are commercial codes that can be used to jump-start the process. The process of thinking through what needs to be done to adequately pose the optimization problem is an excellent exercise that will lead the designer to finally ask and answer questions that should and need to be answered. The entire design process will be clearer. In the authors' opinion, the application of computer-based engineering optimization tools will actually **shorten** the design cycle and lead to **improved** designs that would never have been found otherwise.

14.3 OPTIMIZATION SEARCH ALGORITHMS

Optimization can be viewed as a search for improved or optimal designs. Although there have been many optimization algorithms developed for a variety of problems, the purpose here is to present a review of some basic nonlinear optimization as well as constraint-handling algorithms that the authors have found useful. The preferred algorithms used in engineering design optimization can be divided into gradient-based and non-gradient-based classes. Discussion focuses on algorithms used for unconstrained optimization and some constraint-handling algorithms for constrained problems, and references are provided where further details for these and other algorithms can be found.

14.3.1 GRADIENT-BASED UNCONSTRAINED OPTIMIZATION ALGORITHMS

Gradient based algorithms require, as expected, gradients. This implies the use of differentiable functions, or, at least, well-behaved functions which, when called with small perturbations in the design variables, will provide the appropriate change in the analysis function values. Most gradients are simply calculated with a forward difference operation that requires one analysis call for every design variable. The central difference operation can give a more accurate gradient, but it requires two analyses for every design variable. If one has the luxury of explicit gradient functions, this will eliminate the analysis calls for gradient calculations. It is rare to have these available when the analysis call involves a CFD code. Automatic differentiation codes are available that apply the chain rule of calculus to the original analysis software. These codes may produce new software that calculates the desired gradients as if they were explicit. (See the Argonne National Laboratory Web site at the end of the chapter for information on automatic differentiation software.)

Described here are the most common or basic gradient-based optimization methods available, including comments on their strengths and weaknesses. First derived are the basic unconstrained gradient-based methods, namely, the method of steepest descent, Newton's method, and the quasi-Newton method, followed by a discussion of constrained optimization methods. Because most design optimization problems of interest require constraints on analysis functions, it is important to understand constraint-handling algorithms. Two of these algorithms — the generalized reduced gradient method and the sequential quadratic programming method — are reviewed here. There are many derivatives and implementations of these methods, but only the basic derivation or outline of these methods is given here.

14.3.1.1 Method of Steepest Descent

The very basic method of steepest descent has two questions that must be answered at each iteration:

1. What should the search direction be?
2. How far should one go in that direction?

The answer to the first question is the negative of the gradient, $-\nabla f(x)$, when searching for a minimum. The answer to the second question is to perform a line search. One performs a line search because, usually, the objective function is not a nice function characterized by nice circular contours if one draws a contour map of it. A line search is a one-dimensional minimization search along the direction of the gradient until a minimum point is reached. The gradient calculation and line search is repeated until a minimum of the objective function is reached, or the given number of iterations have been preformed. The method of steepest descent has a major weakness in that its rate of convergence is very slow for very eccentric functions.

14.3.1.2 Newton's Method

Newton's method uses a quadratic approximation to a function at the current point in the design space. For each design iteration, an n-dimensional quadratic function fit is performed and its minimum is found. This quadratic approximation and minimization is repeated until the minimum of the actual function is reached. A Taylor series expansion of the objective function $f(x)$, where x is the vector of design variables, at the current point in design space gives

$$f(x) = f(x_i) + \nabla f(x_i)^T (x - x_i) + \frac{1}{2}(x - x_i)^T \nabla^2 f(x_i)(x - x_i) + \cdots \tag{14.6}$$

The gradient vector, $\nabla f(x)$, is a vector of constants; and the hessian, $\nabla^2 f(x)$ or H, is a matrix of constants. Defining $(x - x_i) = \Delta x$ and dropping high order terms gives

$$f(x) = f(x_i) + \nabla f(x_i)^T \Delta x + \frac{1}{2}\Delta x^T H(x_i)\Delta x \tag{14.7}$$

Taking the gradient of $f(x)$,

$$\nabla f(x) = \nabla f(x_i) + H\Delta x \tag{14.8}$$

Setting this to zero (because the min or max has a zero gradient) and solving for Δx, one obtains

$$\Delta x = -H^{-1}\nabla f(x_i) \tag{14.9}$$

Thus, one has a steplength and a direction for the quadratic approximation. Note that the right-hand side requires costly second-derivative calculations and a matrix inversion for H. Furthermore, this method only works if the initial guess is close enough to the true minimum that the quadratic approximation is good. Otherwise, the search direction, $-H^{-1}\nabla f(x_i)$, could well be in a direction of increasing f.

14.3.1.3 Quasi-Newton (Variable Metric) Method

The "quasi" comes from the fact that the hessian, H, is not actually calculated. Rather, one converges to it as the iterative search proceeds. The first iteration is started with $H = I$, the identity matrix.

This, in effect, is starting with the method of steepest descent because $\Delta x = -I\nabla f = -\nabla f$. H is then updated at the end of each design iteration. H is updated using the Broyden-Fletcher-Goldfarb-Shanno (BFGS) rank two update. The BFGS update is the most popular update because of its robustness.

If one defines $\Delta x_i = x_{i+1} - x_i$ and $\Delta y_i = \nabla f(x_{i+1}) - \nabla f(x_i)$, one form of the BFGS hessian update is given as:

$$H_{i+1} = \left(I - \frac{\Delta x_i \Delta y_i^T}{\Delta y_i^T \Delta x_i}\right) H_i \left(I - \frac{\Delta y_i \Delta x_i^T}{\Delta y_i^T \Delta x_i}\right) + \frac{\Delta x_i \Delta y_i^T}{\Delta y_i^T \Delta x_i} \tag{14.10}$$

The search direction is $d_i = -H_i \nabla f(x_i)$.

The BFGS update keeps the hessian matrix positive definite even for functions that are not approximated well by quadratics. This helps ensure convergence even when one starts the search far from the optimum. Practical numerical algorithms to perform the quasi-Newton method using the BFGS update are given in the *Numerical Recipes* series of texts. See the list of reference texts at the end of the chapter.

14.3.2   GRADIENT-BASED CONSTRAINED OPTIMIZATION ALGORITHMS

Most engineering design optimization problems are constrained at the optimum. It is therefore important to understand constraint-handling algorithms as well as the conditions that need to be satisfied to ensure an optimum for this case. Two of the most effective constraint-handling algorithms — sequential quadratic programming and generalized reduced gradient — are discussed.

The conditions needed to ensure that one has reached an optimum when one or more of the constraints are active, are called the Kuhn-Tucker conditions. Consider that there is one inequality constraint that prevents reaching a peak in a maximization problem. The optimum will lie on the boundary of this constraint, somewhere below the peak where the objective function is largest. One can define the gradient of the objective function $\nabla f(x)$, as well as the gradient of the constraint $\nabla g(x)$, where the constraint is given as $g(x) - b \leq 0$. The constraint gradient, $\nabla g(x)$, will point uphill toward infeasible space; the function gradient, $\nabla f(x)$, will also point uphill. Moving along the constraint boundary, one finds that the two gradients will align when the maximum value of $f(x)$ that is allowed by the constraint has been reached. thus,

$$\nabla f(x) = c\nabla g(x) \tag{14.11}$$

where c is a positive constant. If c is negative, then the function $f(x)$ could increase without violating the constraint (because the constraint gradient would be pointing downhill). For two binding constraints at the optimum, the maximum will have to be "contained" by the two constraint gradients such that

$$\nabla f(x) = c_1 \nabla g_1(x) + c_2 \nabla g_2(x) \tag{14.12}$$

where c_1 and c_2 are positive constants, later called Lagrange multipliers. In general, the Kuhn-Tucker conditions, which are sufficient for a *local* optimum, state that if x^* is a local maximum for the problem

$$\text{Maximize} \quad f(x)$$

$$\text{such that} \quad g_i(x) - b_i \leq 0 \quad i = 1, \ldots, k$$

$$g_i(x) - b_i = 0 \quad i = k+1, \ldots, m$$

or a local minimum for the problem

$$\text{Minimize} \quad f(\mathbf{x})$$

$$\text{such that} \quad g_i(\mathbf{x}) - b_i \geq 0 \quad i = 1, ..., k$$

$$g_i(\mathbf{x}) - b_i = 0 \quad i = k+1, ..., m$$

and the constraint gradients at the optimum, $\nabla g_i(\mathbf{x}^*)$, are independent, then there are Lagrange multipliers λ_i such that $\mathbf{x}^*$ and λ^* satisfy the following equations,

$$g_i(\mathbf{x}) - b_i \text{ is feasible} \qquad i = 1, ..., m \qquad\qquad (14.13a)$$

$$\nabla f(\mathbf{x}^*) - \sum_{i=1}^{m} \lambda_i^* \nabla g_i(\mathbf{x}^*) = 0 \qquad\qquad (14.13b)$$

$$\lambda_i^* \left[g_i(\mathbf{x}^*) - b_i \right] = 0 \qquad i = 1, ..., k \qquad\qquad (14.13c)$$

$$\lambda_i^* \geq 0 \qquad i = 1, ..., k \qquad\qquad (14.13d)$$

$$\lambda_i^* \text{ unrestricted} \qquad i = k+1, ..., m \qquad\qquad (14.13e)$$

Equation (14.13a) requires that the constrained optimum be feasible with respect to all constraints, and Equation (14.13b) ensures that there is no direction that will simultaneously improve the objective function and be feasible (satisfy the constraints). Equation (14.13c) requires that either an inequality constraint be binding or else the corresponding Lagrange multiplier be zero. Equation (14.13d) states that the Lagrange multipliers for the binding inequality constraints must be positive, and Equation (14.13e) states that the Lagrange multipliers for equality constraints can be either positive or negative.

14.3.2.1   Sequential Quadratic Programming

The sequential quadratic programming (SQP) method, as it is usually implemented, can be viewed as a generalization of the quasi-Newton method. The generalization is the addition of constraints to the quadratic approximation of the objective function, where the constraints are linear approximations of the constraint functions. SQP has been judged to be one of the best algorithms for finding a constrained optimum; it is very efficient in terms of minimizing the number of analysis calls that it must make to find the optimum. However, if the search is started with an initial guess that violates a constraint (an infeasible design), succeeding designs will also likely be infeasible until the optimum is reached. SQP converges by simultaneously improving the objective function and the feasibility of the constraints, when violated. Only the optimum is guaranteed to be feasible. This is a clear drawback for SQP because, for some particular problem, time and/or resources may expire before the optimum is reached, potentially leaving the engineer with an infeasible design.

To demonstrate SQP, solve the optimization problem

$$\text{Minimize} \quad f(\mathbf{x})$$

$$\text{such that} \quad g_i(\mathbf{x}) - b_i \geq 0 \quad i = 1, ..., k$$

$$\text{and} \quad g_i(\mathbf{x}) - b_i = 0 \quad i = k+1, ..., m$$

where there are n design variables x_i and m constraints with associated Lagrange multipliers, λ_i.

The SQP method proposes to solve the above minimization problem by satisfying the Kuhn-Tucker conditions given earlier in Equations (14.13a–e). Equation (14.13b) states that one is searching for a zero because the left-hand side of this equation will not add to zero until the optimum is found. One searches for a design vector comprised of the design variables x_i plus the Lagrange multipliers λ_i that yields a zero (vector). That is, one has now cast the problem into a zero-finding problem by seeking the satisfaction of the Kuhn-Tucker conditions.

One can apply the well-known Newton-Raphson zero-finding method to solve the Kuhn-Tucker conditions. However, the standard practice is to recast the problem into an equivalent minimum-finding problem (in contrast to a zero-finding problem) and use the quasi-Newton method described above to find the minimum. Powell[2-4] determined that the solution to the problem as stated by the Kuhn-Tucker conditions, Equations (14.13a–e), is the same as the solution to a quadratic programming problem given by

$$\text{Min} \qquad f(\Delta\mathbf{x}) = f\left(\Delta\mathbf{x}^0\right) + \nabla f(\Delta\mathbf{x})^T \Delta\mathbf{x} + \frac{1}{2}\Delta\mathbf{x}^T \nabla_x^2 L\left(\mathbf{x}^0, \lambda^0\right)\Delta\mathbf{x}$$

$$\text{such that} \qquad g_i\left(\mathbf{x}^0\right) + \nabla g_i\left(\mathbf{x}^0\right)^T \Delta\mathbf{x} \geq b_i \qquad i = 1, ..., k$$

$$g_i\left(\mathbf{x}^0\right) + \nabla g_i\left(\mathbf{x}^0\right)^T \Delta\mathbf{x} = b_i \qquad i = k+1, ..., m$$

The above function, f, is a quadratic approximation of the true function given by the left-hand side of Equation (14.13b) (including the binding constraints). This approximation is not valid far from the point of expansion. That is, one approximates the scalar function f defined by the left-hand side of Equation (14.13b) with a Taylor series, expanded from some point, $\Delta\mathbf{x}^0$, truncated after the quadratic term. The term $\nabla_x^2 L(\mathbf{x}^0, \lambda^0)$ is the Lagrangian hessian matrix. It is a hessian matrix that has been augmented by the terms containing the binding constraints with associated Lagrange multipliers. For an unconstrained problem, the problem stated above is the same as given earlier for the quasi-Newton method. To find the solution to the above quadratic programming problem, one applies the quasi-Newton method, obtaining the Lagrangian hessian eventually by using the BFGS update. The process is iterated until $\Delta\mathbf{x}$ approximates zero to some tolerance.

A summary of the steps taken in the SQP optimization and constraint-handling algorithm is given as follows:

1. Approximate the original problem with a quadratic approximation. For the first iteration, use the identity matrix as Lagrangian hessian.
2. Find the optimum to the quadratic approximation. Lagrangian multipliers are obtained and used to evaluate the Lagrangian gradient at the current point.
3. Perform a simple line search to get to the optimum of the quadratic approximation. If the penalty function (composed of the objective function and violated constraints, $P = f(\mathbf{x}) + \sum_{\text{violated}} \lambda_i |g_i(\mathbf{x})|$) is reduced, press on. If not, reduce the step size until the penalty function is reduced. Evaluate the Lagrangian gradient at the new point and the difference of the Lagrangian gradients for the current iteration.
4. Update the Lagrangian hessian with the BFGS update.
5. Repeat 1 through 4 until $\Delta\mathbf{x} \rightarrow 0$. When $\Delta\mathbf{x} \rightarrow 0$, the Kuhn-Tucker conditions are satisfied.

14.3.2.2   General Reduced Gradient

Reduced-gradient algorithms avoid using penalty functions by performing the search along the constraint curves to stay in (or near) the feasible design space. Standard reduced-gradient algorithms perform searches along a steepest-descent direction while generalized reduced-gradient (GRG) methods use a more sophisticated approach by keeping a BFGS update of the hessian. GRG methods may take more analysis calls to converge than SQP methods. This is outweighed by the strong advantage of remaining feasible once a feasible design is found. Therefore, once feasible, any subsequent design iteration will also be feasible. This is particularly important in CFD-based optimization because if the search is stopped due to exhaustion of resources, the latest design will be both improved and feasible. A GRG algorithm is described here that is an active constraint algorithm for a problem with only two design variables. The problem is posed as follows

$$\text{Min.} \qquad f(\mathbf{x}) \qquad \mathbf{x} = \left[x_1, x_2 \right]$$

$$\text{Subject to} \qquad g(\mathbf{x}) = 0$$

Suppose one starts at a feasible point, and wants to move to improve the objective function.

The gradient of f is:

$$df = \frac{\partial f}{\partial x_1} dx_1 + \frac{\partial f}{\partial x_2} dx_2 \qquad (14.14)$$

and one wants to stay on the constraint; that is, one wants

$$dg = 0 = \frac{\partial g}{\partial x_1} dx_1 + \frac{\partial g}{\partial x_2} dx_2 \qquad (14.15)$$

Solving for dx_2 gives

$$dx_2 = - \frac{\partial g / \partial x_1}{\partial g / \partial x_2} dx_1 \qquad (14.16)$$

Substituting dx_2 into df from above gives

$$df = \left[\frac{\partial f}{\partial x_1} - \frac{\partial f}{\partial x_2} \left(\frac{\partial g / \partial x_1}{\partial g / \partial x_2} \right) \right] dx_1 \qquad (14.17)$$

The term in square brackets is the *reduced gradient*, or

$$\left(\frac{df}{dx_1} \right)_{red} = \left[\frac{\partial f}{\partial x_1} - \frac{\partial f}{\partial x_2} \left(\frac{\partial g / \partial x_1}{\partial g / \partial x_2} \right) \right] \qquad (14.18)$$

If one substitutes Δx_1, for dx_1, then the equations are only approximate. The reduced gradient provided the direction to step tangent to the constraint in a direction that improves the objective function. This derivation can be expanded to n-dimensional problems. See the references for a more detailed formulation. Much of the above discussion on gradient-based constrained optimization algorithms is derived from Parkinson.[5-7]

14.3.3 Non-gradient-based Optimization Algorithms

Two of the most popular and successful non-gradient-based optimization algorithms are discussed here: simulated annealing and genetic algorithms. These algorithms can be used for unconstrained or constrained optimization.

14.3.3.1 Simulated Annealing

Simulated annealing is based on a phenomenon found in nature to optimize a complex problem. The phenomenon is the annealing of solids as found in the process of heating a solid to its liquid state and cooling it at a given rate such that thermal equilibrium is maintained. The atoms find a global minimum energy state. An algorithm developed by Metropolis[8] simulates this annealing process. Simulated annealing can be applied to discrete and continuous problems. The algorithm is defined in the following steps (minimizing the objective function):

1. A small random displacement of an atom is performed, resulting in a change in energy.
2. If the energy change is negative, the new configuration is desirable; therefore, it is kept.
3. If the energy change is positive, the new configuration represents a higher energy state, but may be accepted based on the Boltzmann probability function,

$$P = e^{\frac{-\Delta E}{k_b T}}$$
(14.19)

   where k_b is the Boltzmann constant and T is the current temperature.
4. The probability is compared to a random number drawn from a uniform distribution between 0 and 1. If smaller, the configuration is accepted.

For engineering design the "energy" is the objective function. The displacements of atoms are the perturbations of the design variables. The design is started with a high "temperature" producing a high value of the objective function. Random perturbations are made to the design variables. If the objective function is lower, it is accepted. If it is higher, it is accepted according to the probability of the Boltzmann function. This allows for escape from local optima.

As the temperature is gradually lowered, the probability that a worse design will be accepted decreases. This method can be computationally expensive, compared to gradient-based methods, because it usually needs many more analysis evaluations to converge. However, it can converge to the global optimum.

As the Boltzmann probability function is implemented for engineering design, ΔE is the difference between the previous and current designs, k_b is calculated as the running average of ΔE, and T is a unitless value corresponding to the ability of the process to accept a higher objective function. One does not set values of T; rather, one defines probabilities for accepting a worse design at the beginning of the search (on the order of 0.5) and at the end of the search (on the order of 1×10^{-6}). One also specifies the number of cycles that should be run corresponding to the number of temperatures and how quickly the design is cooled. The slower the design is cooled, the greater the number of required analysis evaluations.

One needs to specify the number of cycles the algorithm will run. This sets the number of "temperatures" and defines how quickly (or slowly) the design is "cooled." This selection is somewhat a matter of trial-and-error. A large number of cycles simulates better the annealing process, but may be computationally expensive. Each cycle perturbs each design variable randomly. Each design is evaluated for acceptance or rejection after every perturbation.

To reduce some unnecessary analysis calls, some implementations filter out a percentage of perturbations that are predicted by a Taylor approximation to be worse than the current design.

If the optimization search is started at an infeasible point, the maximum constraint violation is reduced (rather than the objective function) until a feasible design is found. Once a feasible design is found, the algorithm proceeds to optimize the objective function. For further information on simulated annealing, see References 9 through 11.

14.3.3.2 Genetic Algorithms

Genetic algorithms (GA) are a set of algorithms that mimic Darwin's theory of natural selection. They provide simulated operations for reproduction, crossover, and mutations in nature. Quoting from the *Handbook of Genetic Algorithms by Davis*[12]:

> In nature species are searching for beneficial adaptations to a complicated and changing environment. The "knowledge" that each species gains with a new generation is embodied in the makeup of chromosomes. The operations that alter the chromosomal makeup are applied when parents reproduce; among them are random mutation, inversion of chromosomal material, and crossover — exchange of chromosomal material between two parents chromosomes. Random mutation provides background variation and occasionally introduces beneficial material into a species' chromosomes. Inversion alters the location of genes on a chromosome, allowing genes that are co-adapted to cluster on a chromosome, increasing their probability of moving together during crossover. Crossover exchanges corresponding genetic material from two parents chromosomes, allowing beneficial genes on different parents to be combined in their offspring.

Golberg[13] has suggested four ways that genetic algorithms are different from gradient-based algorithms.

1. GAs work with a coding of the variables, not the variables themselves.
2. GAs search from a population of points, not a single point at a time.
3. GAs use objective function information, not derivatives or auxiliary knowledge.
4. GAs use probabilistic transition rules, not deterministic rules.

The steps for a genetic algorithm:

1. Start with a population of designs. These are often generated randomly. A design is often coded as a string of 1's and 0's to form a "genotype." Usually, a population of 20 to 100 designs works well. Each member of the population is a complete analysis call.
2. Determine the mating pool. This can be done using a "weighting roulette wheel," where the slots in the wheel are weighted according to the "fitness" of the objective function values.
3. Perform "crossover." This requires that one randomly select parents to mate from the mating pool, and then randomly select a crossover site.
4. Perform "mutation." The check for mutation is done on a bit-by-bit basis. The mutation rate is usually kept low (0.0005). If a bit mutates, it changes from 1 to 0 or vice versa.
5. The new population has now been created. The populations' strings are decoded to get the new fitness values. This gives a new generation of designs.
6. Go to step 2 and repeat until the objective is not improved or until the number of generation iterations is reached.

Other operations may be added to enhance the performance, namely, using multiple crossover points, linear normalization, and elitism. Much of the above discussion is taken from Borup.[14]

14.4 DESIGN OPTIMIZATION FOR COMBUSTION PROCESSES USING CFD

While the application of design optimization techniques to industrial combustion processes can speed the determination of optimal furnace and process designs, it still requires careful thought and consideration on the part of the design engineer. Design optimization requires the posing of an optimization problem. This requires more that just producing a computer model of the design for analysis purposes. The engineer must also be able to identify all of the variables that define the design or process and decide which are going to be allowed to vary (the design variables). The engineer must carefully consider what constraints, if any, must be applied to the optimization problem, keeping in mind which constraints are "soft" (those whose exact specification may be negotiable). Finally, the engineer must carefully consider how to define analysis functions that will be needed to apply the constraints and to define the objective function.

Engineering design optimization can be applied to the design of combustion equipment as well as the design of the overall operation process. The optimization problem may include static design variables (shape of melting container), as well as variables that are controllable during the process (fuel flow rate). Variables that are functions of time can be parameterized by breaking them into pieces (fuel flow at level 1 for time t_1, then ramp to level 2 during time interval t_2, etc.). If the process is too complicated or would take a long time to perform a computer simulation, it can be broken down into smaller problems. Some of the design variables and functions that may be important in industrial combustion are discussed next.

14.4.1 DESIGN VARIABLES

The choice of design variables is very important to successful optimization. On the one hand, one wants to include all of the variables that might be important so that one has an opportunity to find the overall best design. However, the addition of each design variable increases the dimension of the design space, increasing the time it takes to find the optimum. It is certainly possible to identify different sets of design variables and run optimizations for each set, then pick the overall best design. Also, one may be able to eliminate some variables that do not seem to be important.

Some examples of design variables include:

- Fuel flow rate
- Air/oxygen flow rate
- Burner size/flame shape
- Burner orientation/position
- Number of burners (overall energy input rate)
- Shape and size of furnace enclosure
- Heat flux input to the furnace/melter
- Location and size of exhaust port
- Amount and location of additional air/oxygen input

14.4.2 ANALYSIS AND OBJECTIVE FUNCTIONS

As discussed previously, analysis functions are outputs from the computer model that quantify important results. The objective function, which embodies the objective of the optimization process, is formed from among the analysis functions. It may simply be one of the analysis functions, or it may be a weighted combination of two or more analysis functions. If it is desired to use multiple objectives, one must be aware that there may be competing objectives. That is, one objective may benefit at the expense of another. Thus, the design engineer must assign weights (importance) to each objective to obtain the overall objective function. The final objective function should produce a single value that is to be minimized or maximized. Other analysis functions may be necessary for the application of constraints. Examples of possible analysis/objective functions in industrial combustion include:

- Temperature at some point in the furnace
- Total mass of particular species
- Total length of time of a process
- Percent of complete combustion
- Overall variation of temperature or heat flux
- Material processing time
- Total amount of fuel/air/oxygen

An example of the employment of analysis functions in an optimization problem could be as follows. Minimize the overall process time (objective function) with the following constraints: the temperature at any point in the furnace must not exceed T_{max} (inequality constraint); the maximum rate of generation of CO is G_{max} (inequality constraint); and the fuel/air ratio must be maintained to be exactly stoichiometric (equality constraint).

Inasmuch as shape is an important design variable for components and systems that involve fluid flow and heat transfer, a discussion of how to optimize the geometric shape for engineering designs is given in Section 14.5.

14.5 SHAPE OPTIMIZATION

14.5.1 INTRODUCTION

Many times, the performance of a component or system is highly dependent on its geometry or shape (turbine blades, airfoils, combustion chambers). As a result, the use of engineering design optimization to determine optimal shape has become a major research area and is known as *shape optimization*. Shape optimization was first developed for the field of structural design. It began as sizing optimization where the **dimensions** of structural members were the **design variables**. It then progressed to three-dimensional shape optimization by the use of "feature-based" design variables. Most mechanical design systems are composed of a series of features such as bosses, holes, fillets, slots, dimensions, etc. Once CAD systems became feature based, it was obvious to use the feature dimensions as the design variables. Along with the evolution of CFD, there has been an increased interest in shape optimization for fluid flow systems. The performance of systems that involve fluid flow and/or heat transfer are especially sensitive to geometry. For two-dimensional shapes, shape optimization has become commonplace. However, the capability to perform general three-dimensional shape optimization for fluid flow systems is still an area of research with some recent successes.

There are two major challenges in the field of three-dimensional shape optimization for fluid flow systems: (1) the capability to describe, mathematically, arbitrary 3-D geometry; and (2) the ability to maintain or obtain a suitable grid for the fluid flow after the shape of the adjoining boundary has been deformed. To meet the first challenge, one seeks a general methodology to parameterize three-dimensional shapes for shape optimization tasks; that is, to be able to define 3-D shapes in terms of a set of parameters. Techniques that have proven very powerful for the shape optimization of mechanical systems, such as feature-based design, are inadequate for fluid flow systems, because the geometries of fluid flow systems are composed of free-form surfaces that usually lack any obvious or convenient parameterizations. To make matters worse, the particular shape must be carefully defined, because even subtle shape changes can radically alter performance.

There are two fundamentally different approaches in which the second challenge to shape optimization can be solved. One is to recreate the analysis grid from the CAD geometry every time the shape is changed (remeshing). Presently, this is impractical for all but simple problems. The creation of the analysis grid for CFD is, in most cases, a tedious, time-consuming task. Therefore, a great deal of research has been devoted to automating grid generation. However, automatic grid generation remains an area of intense research and, at present, cannot answer the needs of shape optimization. This has remained a major stumbling block to not only shape optimization using CFD, but also the practical use of CFD in the design process.

The second way to solve the problem of grid manipulation is to deform the analysis grid as the shape optimization process proceeds in such a way that the quality of the grid is maintained during the deformation process. A method to do this must prevent undesirable distortions of the grids so that the accuracy of the analysis is preserved. After discussing some of the approaches that have been developed to handle shape optimization, the authors discuss a recent method that has many desirable attributes for shape deformation, called arbitrary shape deformation (ASD).

14.5.2 Shape Parameterization for Structural Systems

The first area of research for shape optimization was in the field of structural analysis and design. The methods therein developed have been extensively used and modified for the shape optimization of fluid flow systems. The use of numerical optimization techniques began shortly after the development of finite-element analysis when it was recognized that the finite-element method permits the efficient analysis of arbitrary shapes. Early techniques directly modified the finite-element mesh, and various parameterization techniques were developed that associated the design variables with the finite-element model. An example of this approach used the element boundary nodal coordinates as the design variables.[15] It was discovered, however, that there were several problems associated with this choice for the design variables. First, this approach often produced an excessively large number of design variables that made the solution of the problem inefficient and difficult. Second, when modifying the mesh by this method, it was difficult to maintain continuity (smoothness) and compatibility on the boundaries of the model.

Several researchers were able to solve many of these initial problems with shape parameterization by developing methods that defined the boundary shape of the structure using piecewise polynomials (e.g., Bezier and B-spline functions).[16,17] These functions allow the coordinates of the boundary nodes to be determined by a smaller number of control points. The coordinates of these control points or nodes become the design variables. This greatly reduced the number of design variables and solved many of the continuity and compatibility problems.

Unfortunately, the extension of the use of Bezier or B-spline functions to three-dimensional surfaces is generally impractical. The difficulties arise because of the large number of design variables (in 3-D) and the difficulty in maintaining model integrity while modifying the model. This results, in part, from the difficulty of maintaining three-dimensional slope and curvature relationships among the geometric entities of the three-dimensional model as the shape of the structure is modified. However, these methods do have an important advantage in that the optimization information is easily implemented in the design process because it uses parameters of the CAD model as the design variables.

14.5.3 Shape Parameterization for Fluid Flow Systems

Shape parameterization methods for shape optimization of fluid flow systems have borrowed extensively from the methods developed for structural shape optimization. For two-dimensional problems, mapping a B-spline to the boundary of the design model and using the control point coordinates as the design variables is commonly used.

Jameson[18] has produced significant work in the area of aerodynamic shape optimization, having worked on the inverse problem of finding the shape of a wing that will produce a specified pressure distribution. The wing shape is defined by a series of profiles generated by a conformal mapping from the unit circle. This is a powerful method for wing shape optimization, but it is limited in scope.

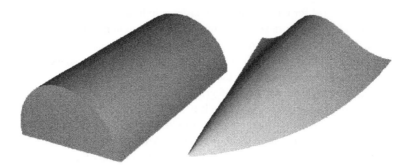

FIGURE 14.3   The initial and final shapes of a hypersonic vehicle optimized to maximize the max lift-to-drag ratio.

Two methods of shape parameterization have been used, almost exclusively, for three-dimensional problems. The first is similar to feature-based shape optimization, where the design variables are based on the basic dimensions of the design. The second is popular for shapes such as aircraft wings where, at selected cross-sections, a two-dimensional shape parameterization is performed with the resulting shape modification interpolated between the cross-sections.

A more powerful method of shape deformation is called free-form deformation (FFD). FFD was developed by Sederberg and Parry[19] and later employed for shape optimization by Landon et al.[20] for both shape parameterization and mesh modification. The use of FFD for shape optimization of a hypersonic vehicle is illustrated in Figure 14.3. This technique, which is used extensively in the field of computer graphics, embeds the model in a parametrically defined volume and then deforms the model by deforming the volume. The design variables are the coordinates of the control points of this volume. This method has several powerful features: It is independent of the model so that the shape parameters are not a function of the analysis grid, and it can be applied to either the CAD model or the analysis grid to produce the same shape. (The CAD model defines the space associated with a solid structure; the analysis grid is the numerical mesh that represents the region of fluid flow, adjacent to a solid structure.) FFD has found wide acceptance in the field of computer graphics, where it is used to produce many of the deformation and "morphing" effects now commonly seen in movies and on television.

FFD makes use of mappings from Cartesian space to parametric space and back again to modify a geometric entity. The first mapping, from Cartesian space to parametric space, must be done in such a way that a direct inverse mapping back to Cartesian space would reproduce the original geometry. To effect the deformation, the shape of the parametric space is modified before performing the inverse mapping to Cartesian space, thus changing the shape of the geometric entity.

These mappings are perhaps best understood by a physical analogy. Consider a flexible plastic ball as the object to be deformed. The first mapping is analogous to casting a cube of clear flexible plastic around the ball. The clear plastic forms the parametric space. Then, by deforming the plastic cube, the embedded ball is automatically deformed in the same general manner as the cube. Original and deformed balls are shown in Figure 14.4.

FFD has many powerful advantages for use in shape optimization. It provides a simple and intuitive method for two- and three-dimensional shape parameterization that produces a minimal set of design variables. It facilitates the manipulation of geometric models composed of complex free-form surfaces. Also, because FFD is independent of the mathematical definition of the model, it provides for the modification of both the analysis grid and the CAD model.

As powerful as it is, however, FFD suffers several limitations. One of these limitations is that by using a Bezier volume, the computational expense of the deformation dramatically increases as control points are added. Also, the movement of each control point produces a global deformation of the entire model, which is often undesirable. An even more severe limitation of FFD is related

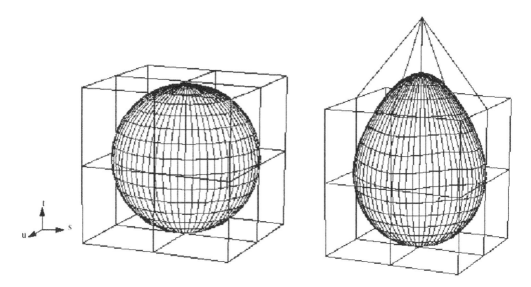

FIGURE 14.4   Ball before (left) and after (right) deformation with FFD.

to the topology of the volume. FFD can only produce a parallelepiped volume that limits its use dramatically by not being able to adequately handle complex topologies.

The problems related to the topology arise, for example, in the case of a Y-shaped object. To be able to deform independently each branch of the Y, a Y-shaped volume is needed. One approach might be to construct the required volume out of three FFD volumes, one for each branch. However, the problem of maintaining smoothness between the three FFD volumes where they meet becomes impossible for most situations. What is needed is a volume that, effectively, can be "shrink-wrapped" around any model of any complexity of topology.

14.5.4   ARBITRARY SHAPE DEFORMATION

Arbitrary shape deformation (ASD)[21,22] solves the limitations of the FFD for use in shape optimization. Figure 14.5 shows an ASD volume defined around a Y-shaped grid and the deformation of the grid using the ASD volume.

An ASD volume is created in a similar fashion to that of an FFD volume, that is, specifying control points and their connectivity. ASD is a generalization of a Bezier volume with the capabilities to model volumes of general topological structure. Therefore, ASD is able to produce both

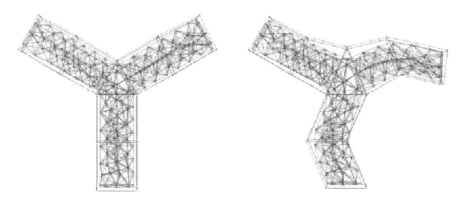

FIGURE 14.5   Y-shaped grid with ASD control points before (left) and after (right) deformation with ASD.

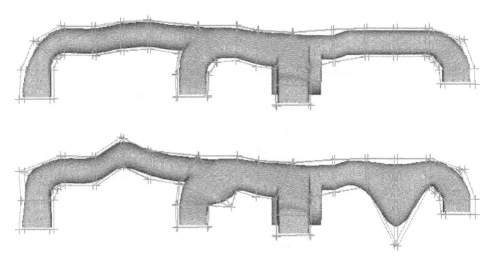

FIGURE 14.6 Undeformed (upper) and deformed (lower) automobile air-conditioning duct system enveloped in ASD volumes.

global and local deformations. The control mesh can also easily be refined to give the user more control of the volume. Finally, it is also possible to assign weights to the control points, thereby providing even more control to the deformation process.

For deformation, the ASD volume is manipulated by moving the control points. When the control points are moved, the underlying entities modify the individual Bezier volumes. Then the new Cartesian coordinates are calculated for each point in the analysis model. An undeformed automobile air-conditioning duct system enveloped by an ASD volume is illustrated in Figure 14.6. Also shown is a modified ASD volume and the resulting deformed duct system. (The duct has been deformed to demonstrate ASD and is not an optimal shape.)

ASD provides an answer to the problems of shape parameterization and analysis grid modification for three-dimensional fluid flow systems. ASD is able to produce a wide variety of shape modifications with a minimum number of shape parameters. Also, ASD is able to modify any type of analysis grid as well as the associated CAD geometry. This supplies the important link needed to tie shape optimization to the design process and the CAD database.

ASD guarantees that at least G^1 (first derivative) continuity is maintained throughout the ASD volume. This ensures that the analysis grid is smoothly modified and maintains the quality of the grid for even major deformations. This solves the problem of grid modification. In fact, based on this capability alone, ASD provides an important tool to the engineer for modifying the analysis grid for minor design changes without having to recreate the grid from scratch, even if he/she never does shape optimization. It has been shown that even with the substantial shape changes displayed, the analysis grid is still well-behaved.

The design variables used for shape deformation are usually the Cartesian coordinates of the ASD control points. Typically, one control point would produce three design variables, one for each direction x, y, and z. However, one of the key components to successful optimization is reducing the number of design variables while maintaining a rich design space. The grouping of control points together to form a single design variable can be implemented both as a method for reducing the number of design variables and for implementing geometric constraints. This grouping may enable all designated control points to move in the same manner. Alternatively, a functional relationship could be defined so that, for example, two control points may move the same distance but in opposite directions.

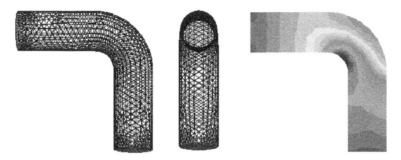

FIGURE 14.7   Initial elbow geometry and pressure contours.

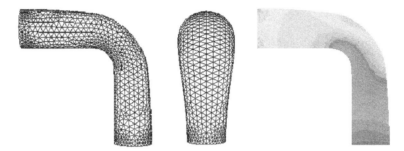

FIGURE 14.8   Optimized elbow shape and pressure contours.

In summary, the ASD provides for shape optimization using very few design variables to deform the geometry, deforms the shape of solid boundaries as well as the analysis grid at the same time, and can handle the deformation of very complex geometries.

14.6   ILLUSTRATIVE EXAMPLES OF SHAPE OPTIMIZATION

Three examples are given of engineering design optimization, and specifically of shape optimization employing arbitrary shape deformation methodology. These examples demonstrate the ability of engineering design optimization to optimize fluid flow components and systems, and include shape optimization for a 90° elbow, an automobile air-conditioning duct, and a combustor.

14.6.1   90° ELBOW

The shape of a 90° elbow was optimized to reduce the pressure drop. Figure 14.7 shows the initial shape and pressure contours of the elbow. The commercial CFD code FLUENT[23] was used to perform the analyses for this problem. The inlet is at the upper left and the outlet is at the lower right. An ASD volume was defined with seven sets of control points along the length of the elbow. *Two* design variables were defined by mapping control points together to allow transverse and radial deformations at the bend. Geometry constraints were applied to avoid interference with adjacent components. Figure 14.8 shows an improved (optimal) design in which a 25% reduction in pressure drop was achieved. The optimal shape actually eliminates a recirculation region at the bend.

14.6.2   AUTOMOBILE AIR-CONDITIONING DUCT

Figure 14.9 shows the initial and final shape of an automobile air-conditioning duct that had its shape optimized to minimize pressure drop. The larger duct toward the left was optimally deformed

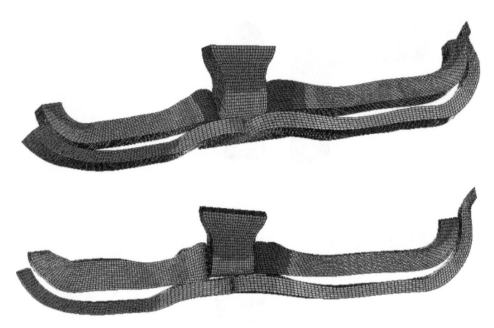

FIGURE 14.9   The initial (upper) and final (lower) shapes of an automobile air-conditioning duct.

by ASD such that the pressure drop was reduced by over 40%. The commercial CFD code StarCD performed the analysis calls during this optimization problem.

14.6.3 COMBUSTOR

This example is a combustion problem where the shape of a gas-fired combustor was optimized. Combustion problems are extremely complex because the analysis must simultaneously calculate the chemical reactions occurring, the thermal characteristics of the system, and the fluid dynamics of the air and chemicals produced from the combustion process.

The analysis of the burner is particularly critical because it determines the efficiency and performance of the combustor. The fluid dynamics near the burner determine the formation and concentrations of the pollutants such as CO and NOx.

Burners have been designed (low-NOx burners) that control the mixing and temperature in the near-burner region to minimize NOx formation. The optimal burner configuration depends on the combustor shape, inlet flow rates, fuel type, etc.

The objective of this optimization problem was to minimize the output of CO from the combustor. The analysis was performed with a two-dimensional axisymetric model using a code named PCGC-3.[24] The objective function was calculated by determining the amount of CO at each element at the outlet, and averaging those values to find the average CO output. An ASD volume was constructed around the model. Three control points along the outside edge were defined as design variables. These control points were allowed to move in the radial direction only.

The initial and optimized combustor shapes are displayed in Figure 14.10. Also shown are the corresponding CO levels. The optimization was able to significantly reduce the production of CO, with a very minor change in the shape of the combustor. The diameter of the combustor was increased. But more importantly, the angle of the boundary in the burner region was slightly changed. This had a dramatic effect on the flowfield in the burner region. The recirculation zone in the corner above the burner was nearly eliminated, and the flow coming out of the burner stayed attached to the wall of the combustor.

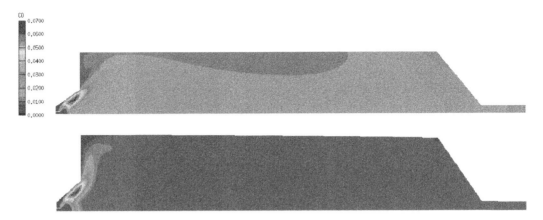

FIGURE 14.10   CO levels of the initial (upper) and optimized (lower) combustor shapes.

REFERENCES

1. Kreyszig, E., *Advanced Engineering Mathematics*, 6th ed., John Wiley & Sons, New York, 1988.
2. Powell, M.J.D., A fast algorithm for nonlinearly constrained optimization calculations, *Numerical Analysis, Dundee 1977,* Lecture Notes in Mathematics No. 630, Springer-Verlag, New York, 1978.
3. Powell, M.J.D., Variable metric methods for constrained optimization, *Math. Programming: the State of the Art*, A. Bachem, M. Grotschel, and B. Korte, Eds., Springer-Verlag, New York, 1983.
4. Powell, M.J.D., Ed., *Nonlinear Optimization 1981,* Academic Press, London, 1981.
5. Parkinson, A. et al., OptdesX Users Manual Release 2.0, Design Synthesis, Inc., Orem, UT.
6. Parkinson, A., Notes for ME 575, Design Optimization, Brigham Young University, Provo, UT, 1995.
7. Parkinson, A. and Wilson, M., Development of a hybrid GRG-SQP algorithm for constrained nonlinear programming, *J. of Mech. Trans. And Automation in Design, Trans. ASME,* 110, 308, 1988.
8. Metropolis, N. et al., *Journal of Chemical Physics,* 21, 1087-1092, 1953.
9. Kirkpatrick, S., Gelatt, C.D., and Vecchi, M.P., Optimization by simulated annealing, *Science*, 220(4598), 671, 1983.
10. Bochachevsky, I., Johnson, M., and Stein, M., Generalized simulated annealing for function optimization, *Technometrics*, 28(3), 209, 1986.
11. Aarts, E. and Korst, K., *Simulated Annealing and Boltzmann Machines: A Stochastic Approach to Combinatorial Optimization and Neural Computing*, Wiley, New York, 1989.
12. Davis, *Handbook of Genetic Algorithms*, Van Nostrand Reinhold, New York, 1991.
13. Golberg, D.E., *Genetic Algorithms in Search, Optimization, and Machine Learning*, Addison Wesley, 1989.
14. Borup, L., Optimization Algorithms for Engineering Models Combining Heuristic and Analysis Knowledge, Masters thesis, Brigham Young University, 1991.
15. Zienkiewicz, O.C. and Campbell, J.S., Shape optimization and sequential linear programming, *Optimum Structural Design*, R.H. Gallagher and O.C. Zienkiewicz, Ed., John Wiley, New York. 1973
16. Braibant, V. and Fleury, C., Shape optimal design using B-splines, *Comput. Meths. Appl. Mech. Engrg.*, 44(3), 247-267, 1984.
17. Schramm, U. and Pilkey, W.D., The coupling of geometric descriptions and finite elements using NURBs — A study in shape optimization, *Finite Elements in Analysis and Design*, 15, 11-34, 1993.
18. Jameson, A., Aerodynamic design via control theory, *J. Sci. Comput.*, 3, 233-260, 1988.
19. Sederberg, T.W. and Parry, S., Free-form deformation of solid geometric models, Proceedings of SIGGRAPH'86, *Computer Graphics,* 20, 151-160, 1986.
20. Landon, M.D., Hall, D.W., Udy, J.L., and Perry, E.C., Automatic supersonic/hypersonic aerodynamic shape optimization, *Proceedings of 12th AIAA Applied Aerodynamics Conference,* Colorado Springs, CO, June 20-23, 1994.

21. Perry, E.C., Balling, R.J., Landon, M.D., and Johnson, R.W., Aerodynamic shape optimization of internal fluid flow systems, *16th AIAA Applied Aerodynamics Conference,* Albuquerque, NM, June 15-18, 1998.
22. Perry, E.C., Balling, R.J., and Landon, M.D., A new morphing method for shape optimization, *7th AIAA/USAF/NASA/ISSMO Symposium on Multidisciplinary Analysis and Optimization,* St. Louis, MO, September 2-4, 1998.
23. Fluent Inc. (1999), FLUENT, URL: http://www.fluent.com/software/FLUENT/main.htm.
24. Hill, S.C. and Smoot, L.D., A comprehensive three-dimensional model for simulation of combustion systems: PCGC-3, *Energy and Fuel,* 7(6), 874, 1993.

ADDITIONAL RESOURCES

There is a wealth of information for those interested in learning more about optimization algorithms and their application to engineering design optimization and, in particular, CFD-based design optimization. This section lists references for related books, journals, and World Wide Web sites.

Books

Belegundu, A.D. and Chandrupatia, *Optimization Concepts and Applications in Engineering,* Prentice-Hall, Englewood Cliffs, NJ, 1998.

Avriel, M. and Golany, B., Eds., *Mathematical Programming for Industrial Engineers,* Marcel Dekker, 1996, (ISBN 0824796209).

Nocedal, J., Stephen J., and Wright, S.J., *Numerical Optimization (Springer Series in Operations Research),* Springer Verlag, New York, 1999 (ISBN 0387987932).

More, J.J. and Wright, St., *Optimization Software Guide,* SIAM, 1993.

Fletcher, R., *Practical Methods of Optimization,* John Wiley & Sons, New York, 1980.

Papalambros, P.Y. and Wilde, D.J., *Principles of Optimal Design,* Cambridge University Press, Cambridge, 1988.

Pike, R.W., *Optimization for Engineering Systems,* Van Nostrand Reinhold, New York, 1986.

Polak, E. and Polak, E., *Algorithms and Consistent Approximations,* Springer-Verlag, New York, 1997.

Press, W.H., Teukolsky, S.A., Vetterling, W.T., and Flannery, B.P., *Numerical Recipes in Fortran 77, The Art of Scientific Computing,* 2nd ed., Cambridge University Press, Cambridge, 1992.

Rao, S.S., *Engineering Optimization: Theory and Practice,* 3rd ed., John Wiley & Sons, New York, 1996.

Reklaitis, G.V., Ravindran, A., and Ragsdell, K.M., *Engineering Optimization: Methods and Applications,* John Wiley & Sons, New York, 1983.

Journals

American Institute of Aeronautics and Astronautics (AIAA) journals (See http://www.aiaa.org/):
1. *AIAA Journal*
2. *Journal of Aircraft*
3. *Journal of Guidance, Control, and Dynamics*
4. *Journal of Propulsion and Power*
5. *Journal of Spacecraft and Rockets*
6. *Journal of Thermophysics and Heat Transfer*

American Society of Mechanical Engineers (ASME) journals (See http://www.asme.org/):
1. *Journal of Fluids Engineering*
2. *Journal of Heat Transfer*
3. *Journal of Engineering for Gas Turbines & Power*
4. *Journal of Mechanical Design*
5. *Journal of Turbomachinery*

Computers & Fluids
Applied Mathematical Modeling

Journal of Computational Physics
Computer Methods in Applied Mechanics and Engineering
Mathematics of Computation
Applied Mathematics and Computation
International Journal for Numerical Methods in Fluids
Journal of Computational and Applied Mathematics
Theoretical & Computational Fluid Dynamics
International Journal of Heat & Mass Transfer

WEB SITES

The Optimization Technology Center, http://www-fp.mcs.anl.gov/otc/

Argonne National Laboratory and Northwestern University joined to form this Web site. The Center's mission is to make potential users in industry, government, and academia aware of how optimization techniques can aid their work, and to make the latest techniques widely available. Their site has NEOS: Network-Enabled Optimization System consisting of three elements:

   The NEOS Server for solving optimization problems remotely over the Internet
   The NEOS Guide to optimization technology, a Web-resident source of information about
      optimization algorithms and applications, including optimization software packages, opti-
      mization algorithms, and case studies
   The NEOS Tools, a library of advanced optimization software

Decision Tree for Optimization Software, http://plato.la.asu.edu/guide.html

Provided by H.D. Mittelmann and P. Spellucci, The Decision Tree for Optimization Software "aims at helping you identify ready to use solutions for your optimization problem, or at least to find some way to build such a solution using work done by others." Where possible, public domain software is listed. This site has software sorted by problem type to be solved, a collection of test results and performance tests (benchmarks), a short list of introductory books and tutorials, a list of software that aids in the formulation of an optimization problem or simplifying its solution, etc.

Operations Research Resources at Michael Trick's Operations Research Page, http://mat.gsia.cmu.edu/resource.html

This site contains a list of resource collections, specialized topic guides, a list of software packages and their descriptions, etc. Specifically, one can find much information on operations research-related topics, including societies, newsgroups, e-mail lists, resources, research groups, educational programs, companies, journals, etc.

The Guide to Available Mathematical Software (GAMS), http://gams.nist.gov/

The GAMS project of the National Institute of Standards and Technology (NIST) studies techniques to provide scientists and engineers with improved access to reusable computer software that is available to them for use in mathematical modeling and statistical analysis. One of the products of this work is an online cross-index of available mathematical software. This system also operates as a virtual software repository. It provides centralized access to such items as abstracts, documentation, and source code of software modules that it catalogs (source code for nonproprietary software only). Four software repositories are indexed here: three maintained for use by NIST staff (but accessible to public), and *netlib*, a publicly accessible software collection maintained by Oak Ridge

National Laboratory and the University of Tennessee at Knoxville. The vast majority of this software represents FORTRAN subprograms for mathematical problems that commonly occur in computational science and engineering. Among the packages cataloged are the IMSL, NAG, PORT, and SLATEC libraries; the BLAS, EISPACK, FISHPAK, FNLIB, FFTPACK, LAPACK, LINPACK, and STARPAC packages; and the DATAPLOT and SAS statistical analysis systems.

Opt-Net, http://optnet.itwm.uni-kl.de/opt-net/

Opt-Net is an electronic forum for the optimization community organized by SIGOPT under the auspices of the Deutsche Mathematiker Vereinigung (DMV). The basic services provided by Opt-Net are a weekly digest containing a moderated list of contributions, a unique e-mail address and White Pages for each member, and the Opt-Net archive.

15 Virtual Reality Simulations

Lori Freitag

CONTENTS

15.1 INTRODUCTION

Computer-based virtual reality simulations have been an active area of research since 1965 when Ivan Sutherland published "The Ultimate Display," which described the potential of computers to provide a window into virtual worlds.* However, only in the last 10 years has the technology necessary for creating interactive virtual reality simulations: high-speed computers, three-dimensional display

---

* Non-computer-based virtual environments were first introduced in 1940 for flight simulation and training.

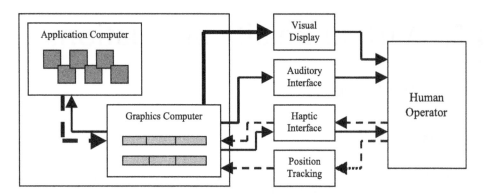

FIGURE 15.1 A schematic of a virtual environment.

devices, and various human computer interfaces — become widely available to both the research and industrial communities. Thus, their use has become increasingly common in a variety of application areas, including medicine and health care, education, hazardous operations, and, of most relevance to the computational combustion community, information visualization.[2,3] Virtual environments have helped to decrease prototype design time and cost, allowed engineers to intuitively evaluate three-dimensional data sets, and provided effective tools for training and education.

All the virtual environment systems described in this chapter conform to the definition given by the Committee on Virtual Reality Research for the National Research Council in its 1995 report:

Definition 1. *"A virtual environment system consists of a human operator, a human-machine interface, and a computer. The computer and the displays and controls in the interface are configured to immerse the operator in an environment containing three-dimensional objects with three-dimensional locations and orientations in three-dimensional space. Each virtual object has a location and orientation in the surrounding space that is independent of the operator's viewpoint, and the operator can interact with these objects in real time using a variety of motor output channels to manipulate them."*[3]

A schematic of a typical virtual environment system is shown in Figure 15.1. This system includes a graphics computer that runs the simulation code controlling input and output to the virtual world. Additional application-specific computations may be running on a separate, possibly remote, computer, and the virtual world may be updated using these computational results. Either or both of these computers can have multiple processors to speed computations and thereby enhance the immersive experience. The arrows in the figure indicate the direction of data flow; solid lines indicate output from the virtual environment; dashed lines indicate input to the virtual world; and the thickness of the arrows indicates the potential amount of data for each transaction.

The visual display device is the primary mechanism for immersing the user in the virtual world and can be a conventional computer monitor, a head-mounted display (HMD), or an elaborate room-sized, theater-based system. The auditory interface augments the visual channel by adding sound to the virtual world, which can create a significantly more realistic experience. For example, sound can be used to alert a user to an object or event that is outside his field of view. Haptic interfaces, such as wands and force-reflecting joysticks, provide additional input and output mechanisms for interacting with objects in the virtual environment. Position tracking devices are used to monitor the user's position and orientation in the real world, and this information is used to update the scene so that objects in the virtual world appear to be independent of the user's viewpoint. Other interface devices not illustrated in Figure 15.1 include the use of voice recognition interfaces to provide input to the virtual environment and motion and balance interfaces for applications such as flight training. For each component of the system, reconfigurable software is required so that the virtual environment can be easily extended to new applications or problem domains.[3]

Virtual reality simulations have the promise to fundamentally change the way we see and interact with scientific and engineering simulation data sets. Again we quote the 1995 report from the Committee on Virtual Reality Research and Development for the National Research Council.

"Virtual environment (VE) technology is a natural match for the analysis of complex, time-varying data sets… Furthermore, as the phenomena being represented are abstract, a researcher can perform investigations in VE that are impossible or meaningless in the real world. The real-time interactive capabilities promised by VE can be expected to make a significant difference in these investigations, with the potential to provide: the ability to quickly sample a data set's volume without cluttering the visualization; no penalty for investigating regions that are not expected to be of interest; and the ability to see the relationship between the data nearby in space or time without cluttering up the visualization. In short, real-time interaction should encourage exploration."[3]

Because a comprehensive treatment of all aspects of virtual reality simulations (which would require a book in and of itself!) is unfeasible, the goal of this chapter is to give the reader an overview of existing virtual reality technologies and describe their use in concrete applications. Toward this end, this overview focuses on commonly used visual display devices, auditory and haptic interfaces, and on the tracking technologies that allow the user to navigate through the virtual world. Then described are two virtual reality simulations that have been developed for commercial combustion applications. These simulations allow engineers to interactively visualize and analyze their application data sets in a virtual furnace or boiler, and in each case, there is a description of the custom toolkits that were developed to extend the virtual reality simulations beyond standard visualization techniques.

15.2   VIRTUAL REALITY TECHNOLOGIES AND HARDWARE

The technological advances for virtual reality simulations in the past 10 years have been remarkable. Users can immerse themselves in room-sized display devices and see, hear, and "touch" the virtual world around them. Additional functionalities that allow users to interact directly with an application code, make changes, and watch the results of those changes in real time have increased the levels of understanding and insight that can be obtained from application simulation codes.[4] When developing the hardware that makes these simulations possible, significant research challenges must be faced, including overcoming limitations that affect accuracy, resolution, and update rates for each subsystem; designing the user interfaces in such a way that navigation and interaction feel natural and intuitive; and finally, integrating all subsystems together to provide a high-performance, realistic, immersive experience.

The most useful experiences in virtual reality are derived from the appropriate stimulation of the visual, auditory, and haptic senses. Critical to the creation of useful human-computer interfaces for these senses is the understanding of how humans perceive and interact with the real world. Some findings within human factors research are discussed to explain how humans resolve three-dimensional images, how individual sound sources are located in three dimensions, and how the human sense of touch provides information about the objects around us. A general description of available human-computer interface technologies for these three senses is given, as well as specific examples to illustrate current state-of-the-art implementations.

15.2.1   Visual Display Devices

Visual display systems comprise four primary components: the display surface, the system that generates images, the tracking system that senses the position and orientation of the user's head and/or eyes, and, for some technologies, the system that positions the display relative to the user's eyes. Ideally, the virtual world created by these systems would be indistinguishable from the real world in terms of image quality, image extent, and perception update rate. Thus, it is critical to understand the aspects of the human vision system that affect our perception of the world around us. These include:

- *Resolution*: The human eye is capable of resolving 80 million polygons per picture[3] and roughly 4800 × 3800 pixels.[5] Note that for computer graphics, the number of polygons that can be resolved is bounded above by the number of pixels in the display device. Most visual displays do not support the 80 million pixels required to resolve an image to the limitations of human perception, but this limitation has not severely impeded useful work with computer graphics.
- *Depth perception*: Humans sense depth in a number of different ways, including the physiological cues of accommodation (tension of the muscle that changes the focal length of the eye) and convergence (the difference in direction of the eyes), which are both weak depth cues that work for short distances. For medium distances, the most important depth cue is binocular parallax, which is a result of the fact that each eye views the same scene differently due to their different locations. For far-field depth perception, psychological cues such as retinal image size (the difference between the sensed image size and its "known" real size), linear perspective, texture gradients (smooth objects are perceived as being further away), and overlap become increasingly critical in both monocular and binocular environments. See Haber and Hershenson[6] or Okoshi[7] for further information on depth perception.
- *Smooth motion perception:* The human eye perceives smooth motion at a rate of 20 frames/second and can perceive flicker at 30 frames/second.[3] The minimum acceptable frame update rate is approximately 8 to 12 frames/second; below 8 frames/second, the sense of being in an animated three-dimensional environment begins to fail.
- *Field of view (FOV):* The human field of view is 180° horizontal (H) by 120° vertical (V).[8]
- *Additional factors:* Color, lumination, and shading all affect the realism of computer-generated images.

Thus, to obtain a realistic visual experience, display devices must produce highly resolved, three-dimensional, and frequently updated images, and variances in these factors from the real world (especially in depth perception and update rate) can dramatically affect the user's experience. Update rate is primarily a function of the graphics computer and, therefore, largely independent of the particular display technology used. Other aspects of human vision such as depth perception, field of view, and resolution *are* affected by the choice of display technology, and three commonly used devices for generating three-dimensional images — head-mounted displays, head-coupled displays, and theater-style displays — are discussed.

15.2.1.1 Head-mounted Displays

The first head-mounted displays were patented in the mid to late 1950s and consisted primarily of binocular designs that displayed images from remotely located cameras to the user.[2] In 1968, Sutherland published "A Head Mounted Three-Dimensional Display" and was the first to use computer-generated images displayed to the user's eyes to produce a simple three-dimensional world.[9] Research using these displays continued over the next 20 years; and in 1989, VPL Research and Autodesk developed the first commercially available head-mounted displays and coined the term "virtual reality."[2] Head-mounted displays are now widely produced, and they differ primarily by resolution, field of view, and display technology.

As indicated by the name, these display devices are mounted directly on the user's head and consist of either a color or monochrome screen with images projected by one or two display elements. Only units containing two display elements can generate three-dimensional stereoscopic

FIGURE 15.2 The Datavisor® HiRes (left) and V8® (right) head-mounted displays. (Courtesy of n-vision and Virtual Research Systems, Santa Clara, CA, respectively.)

images. The display elements are either cathode-ray tubes (CRTs) or liquid crystal displays (LCDs), and special optics and mirror configurations can be used to focus and stretch the perceived field of view, which is typically 60° H × 45° V.[8] CRTs can be made quite small, with high resolution, brightness, and good contrast ratios, but their usability is affected because they are heavier, more expensive, and require more power than LCDs. In contrast, the LCDs are lightweight and use less power than CRTs, but they cannot easily be made with high resolution, and they have lower contrast and update rates. Because the display environment is attached to the user's head, navigation by position tracking is easily accomplished by directly measuring the display device position and orientation. The left image in Figure 15.2 shows the Datavisor® HiRes head-mounted display sold by n-vision, it provides a 78° horizontal field of view using 1-in. CRTs with 1280 × 1024-pixel resolution. On the right is Virtual Research System's V8® head mounted display. This system uses 1.3-in. active matrix LCDs with a 640 × 480-pixel resolution and a 60° diagonal field of view.

For both CRT and LCD units, the potential drawbacks include their weight, which can encumber the user and cause neck fatigue, eye strain resulting from the difference between the image focal depth and the perceived depth, and the limited field of view associated with many models.[3] In addition, because head-mounted displays are designed for single-person use, collaboration in the three-dimensional world is difficult but can be accomplished through networked systems of displays and avatars that represent other users.

15.2.1.2 Head-coupled Displays

Head-coupled displays are single-user displays that are mounted on the floor or desktop using a counterbalanced infrastructure that alleviates the weight problems associated with head-mounted displays. Typically, the user holds the display device by two handles and navigates through the space by pointing and turning the display head as if it were a pair of binoculars. Forward and backward navigation through the environment is handled by simple haptic interfaces, such as thumb buttons on the side of the display.

Fakespace BOOM® devices are the most well-known examples of head-coupled displays, and their BOOM 3C and BOOM HF products are shown in Figure 15.3. The display systems use CRT technology and have a resolution of 1280 × 1024 pixels, with field of views ranging from 30° to 140° H. Like-head mounted displays, tracking is provided by a mechanical system mounted on the arm that monitors the position and orientation of the visual display.

15.2.1.3 Theater-based Display Devices

Division demonstrated the first family of theater-based displays in 1992, and later that year, the CAVE® (CAVE Automatic Virtual Environment) was demonstrated by the University of Illinois at Chicago.[10] The goal in developing these systems was to obtain higher-resolution color images and

FIGURE 15.3 The BOOM® 3C and BOOM® HF sold by Fakespace Labs, Inc. (Courtesy of Fakespace Labs, Inc., Mountain View, CA.)

a larger field of view than had been previously achieved using head mounted or window-on-the-world type systems. In addition, developers hoped to reduce geometric distortion, the errors induced by head rotations, and the amount of hardware that encumbered the user.

These systems immerse one or more users in a room-sized display environment with images projected on two or more walls and the floor. Stereoscopic views are obtained by projecting alternating left and right eye images on each display surface at a rate of 96 Hz, and these images are resolved into a single, three-dimensional image using LCD shutter glasses (see Figure 15.4). The lenses of the shutter glasses are alternately opaque and clear and are synchronized with the projected images through the use of infrared emitters. The projector for each wall has a resolution of 1024 × 768 pixels, providing a total resolution of approximately 3000 × 1500 linear pixels for systems supporting three walls and a floor image. These displays are large enough to allow several users to be immersed simultaneously in the same virtual environment and interact with the same computational model, thereby greatly enhancing their utility for team-based design and engineering. One user is tracked by an electromagnetic tracking system, and the image orientation is calculated with respect to the head position of that user.

FIGURE 15.4 LCD shutter glasses.

Smaller, drafting table format displays are also available to provide a wide field of view window into the virtual world. The same shutter glasses and tracking devices are used to provide the user with a semi-immsersive display on a 4 × 5-foot rear-projected screen at a resolution of 1024 × 768 pixels. These devices are generally designed to be portable and are significantly less expensive than the CAVE®, as they require only one graphics pipe for image rendering. Figure 15.5 shows schematics of the CAVE® and ImmersaDesk® systems sold by Pyramid Systems.

Similar systems include the theater-style C2 developed at Iowa State University, which debuted in 1996, and the WorkBench Series of drafting table display devices developed by the German National Research Institute in 1993.[11]

15.2.2 AUDITORY INTERFACES

The goal of the auditory interface for a virtual environment is to produce high-fidelity sounds that can be altered, depending on the orientation and position of the user, and should, ideally, exclude all sounds not generated by the virtual environment.[3] To accomplish this goal, two aspects of human hearing must be replicated. First, because human ears are approximately 20 cm apart, the differences in the arrival time of incoming signals to each ear determine the sound's origin. Second, incoming signals are shaped by each individual's head, ear, and torso in ways unique to that individual, and this affects one's ability to locate objects in three dimensions by sound.[12]

The two auditory interface technologies readily available in current virtual environments are single-user earphones and loudspeaker systems. Earphones vary in size, weight, and coupling to the user. At one end of the spectrum, the user dons a large pair of earphones that completely encompasses the ear; at the other end, the earphones are inserted directly into the ear canal and reproduce sound by using vibrations applied directly to the ear canal wall. Earphones have the advantage that they can completely exclude external sounds from the virtual environment, but suffer from a tendency for sounds to appear as if they are internal to the user's head. That is, the sounds are not externalized, which limits their usefulness in the virtual environment. To overcome this difficulty, many systems employ head-related transfer functions (HRTFs), which take into account the shape of the ear, head, and torso to simulate the spectral shaping of sound waves as they encounter the user (see, for example, Wenzel et al.[13]). However, measuring each individual for personalized HRTF parameters requires a great deal of time, skill, and equipment, so that approximations for the "generic" user are often employed that can increase the distortion and location errors for individual users.[3,13] Loudspeaker technology has developed over the years from the first "mono" systems, which send one signal to all speakers simulating sound emanating from a point to the development of room-based, surround-sound systems containing eight or more loudspeakers. The simulation of sound origin on multiple speakers is accomplished by offsetting the signals sent to different speakers by a few microseconds. However, no loudspeaker system has been developed that tracks the user and changes the sound characteristics based on the user's position or orientation. Thus, if the user moves out of the intended working area or "sweet spot" of the system, the three-dimensional effect degrades. In addition, external noises are difficult to exclude, and the acoustics of the room are difficult to change to reflect changing environments in the virtual world. For more information on auditory display technologies, see the overview provided by Kramer.[14]

15.2.3 HAPTIC INTERFACES

The word "haptic" is defined as "relating to or based on the sense of touch,"[3] which for the human haptic system comprises two classes of information. First, mechanical, thermal, and chemical stimuli resulting from contact with an object are referred to as *tactile* information. Such information includes the texture of an object, its temperature, or even the force of weight associated with that object. The second class of information is *kinesthetic*, which is defined to be the sense of position or motion of the limbs. All interactions involve both classes of information, and again it is instructive

FIGURE 15.5 Schematic of the CAVE® (left) and ImmersaDesk® (right) display devices. (Courtesy of the Electronic Visualization Laboratory, University of Illinois at Chicago.)

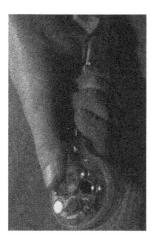 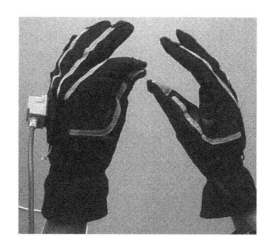

FIGURE 15.6 Passive haptic interfaces a wand interface (left) and pinch gloves (right). (Courtesy of the Electronic Visualization Laboratory, University of Illinois at Chicago.)

to consider the complexity of the human haptic system. For example, consider the human hand. It consists of 19 bones, connected to each other by frictionless joints and connected to 20 intrinsic and extrinsic muscles, which serve to activate 22 degrees of freedom.[3] The tactile system in the human finger tip can distinguish vibrations up to 1 kHz, and the spatial resolution of the fingerpad is 0.15 mm. Even without looking beyond the human hand, it is clear that developing a comprehensive haptic feedback device is an extremely challenging task.

In general, haptic interactions fall into four basic categories: (1) free motion of the hand or body in which there is no contact; (2) contact involving unbalanced forces such as pressing on an object; (3) contact involving self-equilibrating forces, such as squeezing an object; and (4) tactile interaction.[3] Haptic devices are therefore classified by whether or not they are able to reflect forces and by the number of degrees of freedom and types of contact forces they are capable of simulating.

The simplest haptic interfaces are passive devices such as three-dimensional mice, wands, or non-force reflecting joysticks (the left image in Figure 15.6 shows a wand device). These interfaces provide input mechanisms from the user to the computer, but do not provide output in the form of forces or tactile sensations from the computer to the user. Software that monitors button presses or the joystick position can be used to navigate through the three-dimensional world and to build user interfaces, such as menuing systems, for additional application-specific functionality. Bend-sensing and pinch gloves are interesting alternatives to wand and joystick interfaces. Bend-sensing gloves measure the flexure of the hand and relative positions of the fingers using one of three technologies: optical fiber, mechanical measurement, and strain gauges. Pinch gloves (shown on the right in Figure 15.6) are constructed with electrical sensors located in each fingertip. Contact between the sensors completes the conductive path, which is recognized by the simulation code and used to manipulate objects in the virtual environment. The main strength of glove interfaces compared to wands and joysticks is that they provide more intuitive gesture-based commands for manipulating the virtual environment. In all three cases (wands, joysticks, and gloves), the haptic device is usually augmented with tracking devices to provide additional location and orientation information to the computer program.

Force-reflecting and -displaying interfaces measure the position and contact force of the user's hand as input to the simulation code, and they also display contact forces and positions generated by the virtual world as output to the user. The first force feedback devices were developed in 1967 as part of the GROPE project at UNC Chapel Hill.[15] Since that time, many sophisticated interfaces that provide force-reflecting and force-displaying capabilities have been built using both ground-based joysticks and hand controllers or body-based, flexible exoskeletons for the hands, arms, or body. Most devices are designed either to replicate a specific tool or set of tools (such as surgical instruments) with which objects in the virtual environment are manipulated, or they are designed for direct human interaction with the virtual environment through the use of exoskeletons. Force-reflecting interfaces involving unbalanced forces require that the device be attached to the ground or desktop to provide a mechanism for "pushing back," whereas force-displaying mechanisms involve self-equilibrating forces and can be either ground- or body-based. Because they are physically attached, ground-based, force-reflecting devices tend to have a restricted workspace, a less intuitive mechanism for interacting with the virtual world, and fewer degrees of freedom than body-based systems. In contrast, exoskeletons are characterized by the fact that they are designed to fit over and move with the limbs or fingers of the user. They have the widest range of unrestricted motion and can incorporate good, inexpensive position tracking technology. However, these devices can be cumbersome for long-term use in a virtual environment. Figure 15.7 shows the PHANTOM® series of force-reflecting haptic interfaces sold by SensAble Technologies. These systems allow up to six degrees of freedom input and three degrees of freedom force feedback output, with workspaces ranging from 5 × 7 × 10 in. to 16 × 23 × 33 in.

Finally, haptic interfaces that provide tactile feedback attempt to simulate sensation applied to the skin typically in response to contact or other actions in a virtual world. Currently, the primary mechanisms for providing tactile feedback are *pneumatic stimulation,* which uses air jets, air pockets, or air rings to stimulate skin reactions; *vibrotactile stimulation,* which uses blunt pins, voice coils, or piezoelectric crystals; *electrotactile stimulation,* which attaches small electrodes to the users' fingers and simulates tactile feedback using electrical pulses; and *shape changers,* which consist of alloys that flex and change shape when electrical current is applied to them.[3]

15.2.4 TRACKING DEVICES

To enable navigation through the virtual world, the computer simulation program must be aware of the orientation and position of the user's head or eyes. This is accomplished by reporting the location of objects in the real world to the simulation program controlling the virtual world using position trackers. Given the position of real objects such as the user's head, the simulation code can update the scene accordingly, so that it appears as if objects are independent of the user's point of view. It is critical that the delay associated with the change in position and orientation of the user's head and the update of the scene be as small as possible. If the update takes longer than 50 ms, the immersion experience can be choppy or disjointed, causing the user to experience nausea or discomfort.[3] Thus, the goal of research in this area is to develop inexpensive devices that combine a high update rate and large working volume with high accuracy, precision, and ease of use.

The four primary technologies currently used for position tracking are mechanical, magnetic, optical, and acoustic.[8] *Mechanical trackers* operate by attaching a headband or jointed exoskeleton to the user or display device and determining the location and orientation through direct displacement measurements. These devices are commonly used with head-mounted or head-coupled displays and directly connect the user to the computer. They are fastest and most accurate of the four techniques, but can suffer difficulties of fit and measurement for different individuals. For example, exoskeleton gloves will not be properly aligned for every user's joints. In addition, the direct coupling of the user and computer limits movement and restricts the area of operation. Finally, tracking multiple body parts (e.g., the head and two arms) is challenging and encumbering. *Magnetic trackers* use electrical currents through perpendicular coils to produce electromagnetic fields that can monitor many different body parts and/or users at once. These systems have a low cost and modest but reasonable accuracy. However, latencies in these systems are often high, which affects the update rate; and magnetic effects caused by external electromagnetic fields can affect their accuracy. Finally, these system work best within a limited working volume of a few cubic meters of the magnetic sensor so that work space is restricted. *Optical tracking* uses infrared-light marker strategies to create bright spots in the images sensed by two or more cameras. Triangulation using

FIGURE 15.7 The PHANTOM® series of force feedback devices. (Courtesy of SensAble Technologies, Cambridge, MA.)

the known positions of the cameras is used to compute the position and orientation of the user. This is one of the most convenient methods for position tracking but suffers drawbacks due to expense, direct line-of-sight requirements, and the ability of other sources of light or glare to affect correctness. Finally, *acoustic trackers* use three high-frequency sound wave emitters in a rigid formation on the user that form the source for three receivers, also in rigid formation. Position and orientation are detected by phase differences or by time-of-flight mechanisms, and these methods achieve good accuracy at modest cost. However, the accuracy of these trackers can be affected by changes in the speed of sound in air due to temperature, pressure, or humidity changes in the environment. Like optical trackers, these devices suffer from the need for a direct line of sight and have a somewhat restricted working volume. More information on position tracking can be found in Vince.[8]

15.3   VIRTUAL REALITY SIMULATIONS FOR INDUSTRIAL COMBUSTION APPLICATIONS

Virtual reality simulations built with the hardware described in the previous section have been used to solve problems in a wide variety of scientific and engineering disciplines. To demonstrate the potential of virtual reality technology in the analysis of computational data, two industrial combustion applications are highlighted. These projects were developed as collaborations between Argonne National Laboratory and two industrial partners, Air Products and Chemicals, Inc. and Fuel Tech, Inc. (formerly Nalco Fuel Tech). This discussion starts by describing the case studies and the virtual environment used for these simulations. Then discussed are the interactive visualization techniques used to explore the simulation data. Also described are the custom toolkits that were developed for each application and used to enable engineers to interactively design or analyze the data sets in the virtual reality simulation.

15.3.1   INDUSTRIAL COMBUSTION APPLICATIONS

For each case study, the author presents a descriptive overview of the application and highlights the goal of the virtual reality simulation. Further information about these applications can be found in Diachin et al.[4] and Freitag and Urness.[16]

15.3.1.1   Case Study 1: Analyzing Aluminum Smelting Furnace Efficiency

In 1997 and 1998, the U.S. Department of Energy and Air Products and Chemicals, Inc., entered into a joint project to design, build, and test a new burner nozzle for industrial furnaces. The overall goal of this project was to develop a nozzle and fuel combination that achieves high efficiency and low nitrogen oxide (NOx) emissions at a low cost to the furnace operator. To determine the best fuel choice, a number of factors were examined, including the initial cost of the fuel, the level of NOx emissions, the amount and kind of contaminants in the molten aluminum alloy caused by the fuel, and the efficiency of the furnace.

To quantify these trade-offs before installation in a production aluminum smelting furnace, numerical simulations that coupled the combustion process to the molten aluminum were performed with the FLUENT computational fluid dynamics (CFD) package.[17] Numerous vector and scalar field values were computed at each cell center, and the following simulation data were of most interest:

- the velocity fields and temperature distributions in both the combustion gases and molten aluminum; in particular, the mixing and uniformity of the temperature distributions of combustion gases and the circulation characteristics and stagnation points in the molten aluminum, and
- the distribution of O_2 throughout the computational domain.

For the virtual reality simulation, a particular nozzle design is chosen and three different fuel choices are considered: air, pure oxygen (O_2), and air enriched with O_2. Determination of the best fuel choice requires evaluation and comparison of the three data sets to understand the differences in temperature distribution, O_2 levels in furnace, and mixing characteristics of each fuel type. Unfortunately, most visualization tools are limited to importing and displaying a single data set; multiple data sets must be loaded in successive order and displayed individually. To compare simulation results in this paradigm, the user must either remember the characteristics of interest from one data set to the next, or initiate multiple visualization sessions and display the windows side by side. In both cases, it is difficult — if not impossible — to isolate the differences among the data sets, particularly if the differences are significantly smaller than the physical features of the numerical simulation.

Project Goal: *To create a virtual reality simulation that allows multiple data sets to be displayed simultaneously in a virtual furnace; to develop paradigms for interactively analyzing the differences between the data sets, both directly and through image differencing; and to clarify for both the engineers and potential customers the advantages and disadvantages of each fuel choice.*

15.3.1.2   Case Study 2: Designing Pollution Control Systems

Between 1994 and 1996, Argonne National Laboratory and Fuel Tech, Inc., collaborated to develop an interactive virtual reality tool for designing injection-based pollution control systems in commercial boilers and incinerators. The pollution control system uses noncatalytic reagents that react with nitrogen oxides (NOx) in a certain temperature range to form nitrogen, water, and carbon dioxide. Optimal performance of the system is obtained by careful placement of the injectors in the boiler with respect to the flue gas temperatures and velocity fields.

Numerical simulations to obtain a steady-state solution of the flue gas flows in the boilers and incinerators were performed using the PHOENICS CFD package.[18] Again, numerous vector and scalar field values were computed at each cell center, and, in this case, the simulation data of most interest were the velocity fields and temperature distributions in the combustion chamber; in particular, recirculation zones in the flue gas flows and temperature zones that are optimal for the chemical reactions.

To design the pollution control system, sprays from candidate injector locations are computed and evaluated for time residence and coverage. Promising configurations are then coupled with the computational fluid dynamics model to compute the chemical reactions and estimated NOx reductions. Prior to the Argonne-Fuel Tech collaboration, this process was performed using a text-based input mechanism that was both non-intuitive and time-consuming.

Project Goal: *To allow engineers to visualize the flue gas flow velocities and temperature distributions; to eliminate non-intuitive, text-based mechanisms used to place and test injector configurations for time residence and coverage with an immersive, interactive, three-dimensional environment; to simulate injector sprays in real time as the injector position changes; and to provide intuitive tools that allow the engineer to quickly analyze each design configuration.*

15.3.2   THE VIRTUAL ENVIRONMENT

Both virtual reality simulations use the CAVE family of display devices with CrystalEyes® LCD shutter glasses. Ascension's Flock of Birds® magnetic tracker is used to follow the viewpoint of one user and also the position and orientation of a passive 3-D wand. In both applications, navigation through the virtual world is accomplished using the joystick functionality of the wand; users point the wand in the desired direction and use the up/down degrees of freedom to move forward or backward. The graphics computer controlling the virtual reality simulation is an SGI Origin2000

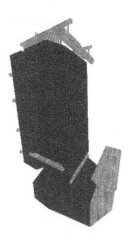

FIGURE 15.8 The exteriors of the virtual aluminum smelting furnace and tire incinerator.

with a RealityMonster graphics pipe for each display surface. For the Fuel Tech application, an additional application computer is used to calculate injector sprays and the resulting particle trajectories are communicated to the graphics computer using the CAVEcomm library developed at Argonne National Laboratory.[19]

For both applications, a realistic frame of reference is critical both for engineers developing and analyzing computational results, and later, for customers considering purchase of the company product. In each case, a representative virtual furnace or boiler is constructed using texture-mapped polygons. For the Air Products application, the texture images consist of digital pictures of an operating furnace and include, for example, the melt surface, the charging bins, and the exhaust stack. The left image in Figure 15.8 is the virtual aluminum smelting furnace that corresponds to the computational domain used in the FLUENT simulations. The right image is the exterior of the virtual tire incinerator used in the Fuel Tech application. In this case, digital pictures of an operating boiler are not readily available, and virtual pipes, waterwalls, headers, and ash hoppers are all constructed using polygons with simple metal or brick textures. To obtain an unimpeded view of the simulation data, the texture-mapped polygons are replaced with a wireframe representation of the furnace or boiler where key reference points, such as the burners and exhaust stacks, are retained.

The user interface to access the data and visualization toolkits is different for each application. For the Air Products application, a 3-D control panel with button and slider widgets is used. For the Fuel Tech project, a menuing system is used and additional three-dimensional scrollbar widgets are made available for certain interactions. These two user interfaces are shown in Figure 15.9. After experimenting with both systems, it was determined that the control panel approach was more flexible and less cumbersome than the menu-based approach for complex interactions.

15.3.3 DATA VISUALIZATION

In each case, visualization of the computational data is critical to the success of the virtual reality simulation. This section describes commonly used techniques to analyze both scalar and vector data and illustrates the techniques with snapshots from the case studies. In addition, each application required the development of customized interactive software. For the Air Products and Chemicals application, tools were developed that allowed users to directly compare different data sets; and for the Fuel Tech project, an interactive tool was designed to place injectors in the virtual boiler and to compute and analyze in real time the corresponding sprays for coverage and time residence.

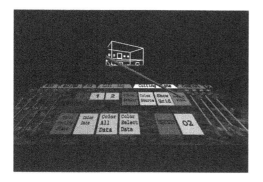

FIGURE 15.9 The mechanisms used for interacting with the virtual reality simulations: on the left, the control panel used with the Air Products and Chemicals application; on the right, the menuing system used with the Fuel Tech application.

15.3.3.1   Scalar Data

Scalar data can be represented in a number of different ways: for example, through the use of color maps, contouring algorithms, or volume rendering techniques. This section describes the use of these techniques in the virtual reality simulation and also discusses the methods developed to visualize scalar field differences among two or more data sets.

15.3.3.1.1 Color mapping

This visualization technique maps the scalar data values to colors through color lookup tables or general transfer functions. For example, suppose that the minimum and maximum scalar values are s_{min} and s_{max}, respectively. If a linear interpolation is desired, the color index, C, corresponding to the scalar value s is computed by

$$C = \left(s_{max} - s\right) / \left(s_{max} - s_{min}\right) \tag{15.1}$$

C is then indexed into the lookup table for the desired color mapping and assigned an RGB value. More general expressions, called transfer functions, for assigning C to s can be used to exaggerate small variations in the data using logarithmic schemes, to focus the color map on a specified subrange of scalar values, or to limit the number of colors used in the visualization. For the examples presented in this section, a linear interpolation function is used, and values of s corresponding to s_{min} are colored blue and values of s corresponding to s_{max} are colored red.

   The results of color mapping are typically applied to other derived visualization entities such as streamlines, cutting planes, isosurfaces, or vector glyphs to increase the amount of information available to the user. For example, Figure 15.10 depicts a cutting plane at the level of the aluminum smelting furnace burners in the air data set. Each figure corresponds to using a different temperature range to compute the color maps. The leftmost image shows the temperature range corresponding to the minimum in all the data sets to the maximum in all the data sets. The colors are primarily blue, indicating that, as expected, the combustion temperatures corresponding to the air data set are cool compared to the air/O_2 and O_2 data sets. The next two images show the color map corresponding to the minimum and maximum temperatures in air data set only and in the cutting plane itself. The rightmost color map shows a user-selected range to highlight temperatures from 1250° to 1500°F; areas of the cutting plane that fall outside this temperature range are culled.

15.3.3.1.2 Contouring

Contouring techniques construct two-dimensional lines or three-dimensional surfaces (isosurfaces) corresponding to constant scalar values. This discussion of contouring algorithms follows the

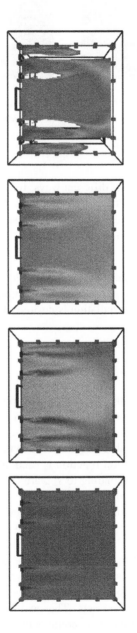

FIGURE 15.10 Four color maps using different ranges that show temperature at the level of the burners for the air fuel data set. The ranges from right to left are: over all data sets within the air data set, within the cutting plane, within a restricted, and user-defined range.

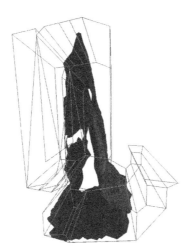

FIGURE 15.11 The isosurfaces corresponding to the optimal temperature zone in the tire incinerator.

presentation given in The Visualization Toolkit, and further details of these algorithms can be found there.[20]

The two primary methods for constructing contours of scalar data are *contour tracking* and *marching cube*[21] algorithms. The *contour tracking* algorithm starts by interpolating the discrete scalar values given at cell vertices onto edges in two dimensions and faces in three dimensions to determine where the contour intersects cell boundaries. It then follows or "tracks" the contour by computing where the contour leaves the current cell and enters a neighboring cell. The contour is followed until it closes back upon itself or exits the data set boundary. In contrast, the *marching cubes* algorithm computes the contour in each cell independently by using the fact that a limited number of topological relationships between the cell and the contour exist. For example, there are 16 possible topological states for a square cell and 256 possible cases for a cube, which can be reduced to 15 by symmetry. The state of each cell is determined by comparing the scalar values at the cell vertices to the contour value and marking them as inside or outside the contour. The state of the cell vertices are compared to the canonical cases to determine the contour topology, and the location of the contour on cell edges or faces is then computed using interpolation.

Both techniques for contour generation are prone to ambiguity in that there can be more than one way to construct the contour for a given topological state; and in three dimensions, this can lead to holes in the isosurfaces. This ambiguity can be resolved in several different ways, including decomposing the cubes into tetrahedra that exhibit no ambiguous cases but can lead to significantly more polygons in the isosurface representation, or by using the asymptotic behavior of the surface and choosing to join or break the isosurface based on the variation of the scalar variable across an ambiguous face.[22] Another approach involves extending the original 15 topological states in the marching cubes algorithm by six more cases that both resolve ambiguities and are compatible with the neighboring cases so that no holes are generated in the isosurface.

Figure 15.11 depicts the use of isosurfaces to highlight the optimal temperature zone for the Fuel Tech application. The inner (red) isosurface corresponds to a temperature of 1850°F, the outer (blue) isosurface corresponds to a temperature of 1650°F. The volume between the two isosurfaces shows the region of primary interest for injector particulate evaporation.

15.3.3.1.3 Volume rendering

Volume rendering techniques operate directly on volumetric data to produce an image without the use of intermediate surface geometry representations. Sophisticated transfer functions that map the value at a voxel into color and opacity values can be used to produce very realistic images. For example, four independent transfer functions can be defined to separately map scalar values into red, blue, green, and opacity, or gradient information could be used to highlight regions of rapid change in the data set, for example, a material property changing from air to muscle and then from muscle to bone. There are several commonly used methods for volume rendering, and this subsection briefly describes two of them: ray casting algorithms and two- and three-dimensional texture mapping. Again, the authors follow the discussion given in *The Visualization Toolkit*, and readers are referred to that text for further information.[20]

Ray casting techniques determine the color of a pixel by sending a ray through the pixel into the scene according to the current camera position. The two main steps of the algorithm are to determine the scalar values encountered along the ray and to process those values according to a ray function such as maximum or average value. Rays can be sampled either uniformly, which may lead to difficulty in choosing an appropriate sampling rate (large intervals may miss features of interest, small intervals are computationally expensive), or by voxels, which may be more efficient for linear interpolation and more accurate for certain ray functions. Interpolation of the scalar values is typically done using zero-order, or constant, interpolants or by trilinear interpolation. Once the scalar values are determined, they are assigned color and opacity values using the ray transfer function and added to the previously computed values along the ray. In general, ray casting is too computationally expensive to use in real-time interactive visualization environments, although more efficient variants exist, such as shear-warp algorithms that can determine early ray termination and have a pre-processing step to remove empty voxels from the data set.[23]

Two- and three-dimensional texture mapping techniques extract data samples along slices of the volumetric data. For two-dimensional texture mapping, the volume is decomposed into a set of orthographic slices along the axis most parallel to the viewing direction. As with ray casting, data values are interpolated to the slices using zero- or first-order techniques, and the color and opacity values determined by the transfer functions are used to create a texture image that is mapped to each slice. The planes are then sorted and displayed to the screen in a back-to-front manner so that incoming planes are blended with the on-screen data using transparency values. For three-dimensional texture mapping, the same technique is used, with the important distinction that the slice direction is not constrained to an axis direction. Thus, it is superior in its ability to sample the volume which yields higher quality images with fewer artifacts. Three-dimensional texture mapping and ray casting using trilinear interpolation are functionally equivalent and produce identical images, but the texture mapping techniques are generally faster due to the utilization of specialized graphics hardware.

15.3.3.1.4 Scalar data comparison techniques

Researchers have recently begun developing comparative visualization techniques for analyzing multiple data sets simultaneously. These methods fall into two primary categories: image comparison and data comparison. Image comparisons are performed either by displaying data set images side by side in the same visualization coordinate system or, more directly, by computing the difference in the *image* produced by the visualization tools.*[24] Data comparison is performed by creating an intermediate data set by combining information from two or more data sources and visualizing the resulting "differenced" data set.[26,27]

For the Air Products application, a comparative analysis tool was developed, using both image and data differencing techniques.[16] Here, an image differencing tool is described that plots simulation data from different runs on the same three-dimensional graph. The user can select any number of data sets and display the scalar fields from an interactively defined cutting plane using height profiles. Each data set is identified by a uniquely colored outline that corresponds to the control panel button color. In addition, the differences in the heights of the cutting plane are more clearly

---

* Non-computer-based virtual environments were first introduced in 1940 for flight simulation and training.

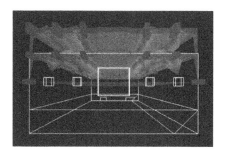 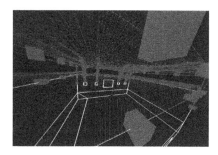

FIGURE 15.12 Image comparison of the temperature field for the three data sets in the aluminum furnace. The air data set is the coolest of the three; and the air/O_2 and O_2 data sets are nearly indistinguishable.

delineated if the user selects the option to color by data source. In this case, each data set is assigned a unique color, and subtle differences between the height profiles are easier to discern.

Figure 15.12 shows the use of the image comparison technique for the aluminum smelting furnace data. The cutting plane from which the data is displayed is shown by the flat, gray surface; the height of the data surfaces above the cutting plane corresponds to the temperature at that point. The air data set is roughly half the temperature of the air/O_2 and O_2 cases, and the air/O_2 and O_2 cases are nearly indistinguishable. The two images show the different options for displaying the height profiles: in the left image, the color of the data surface corresponds to temperature; in the right image, the color of the surface corresponds to data source and the height field is displayed with the computational grid.

15.3.3.2 Vector Data

In both case studies, understanding the velocity fields associated with the combustion gases is necessary for the design and analysis phase of the project. Toward this end, visualization tools were developed using vector glyph representations and interactively requested, animated streamlines to facilitate the analysis process.

15.3.3.2.1 Vector glyphs

Standard visualization techniques such as oriented glyphs can be used to represent vector and scalar fields simultaneously. For example, the velocity field can be represented using tetrahedral darts. The direction and length of each dart corresponds to the direction and magnitude of the vector field at that data point, and the color can correspond to any scalar quantity of interest such as temperature or chemical species concentration. Vector field glyphs can be manipulated in several ways to gain additional insight. For example, to isolate the flow fields in regions corresponding to some scalar field range, the vectors lying outside that region can be culled, or a subset of the field can be magnified for easier viewing and analysis. The leftmost image in Figure 15.13 shows the use of tetrahedral vector glyphs to visualize a swirling flow in a tangentially fired-boiler developed as part of the Fuel Tech project.

15.3.3.2.2 Streamlines

To animate a velocity field, streamlines are calculated from one or more initial positions by numerically integrating the differential equation

$$\partial x / \partial t = V_p, \text{ with } V_p = V_g \tag{15.2}$$

where x and V_p are the particle position and velocity vectors, respectively, and V_g is the fluid velocity vector. A number of numerical integration techniques can be used to solve this differential

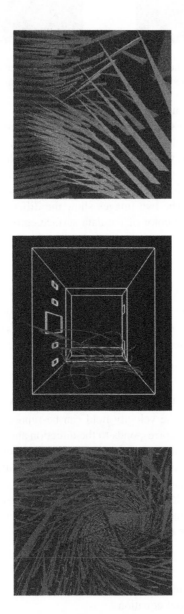

FIGURE 15.13 The use of vector field glyphs, streamlines, and animated streamlines to interactively explore the velocity fields in both combustion applications.

equation (see Darmofal and Haimes[28] for an overview); the most commonly used techniques are briefly reviewed here. In general, given any system of ODEs,

$$\partial x/\partial t = f\big(x(t)\big) \tag{15.3}$$

and initial condition x_0, numerical integration techniques provide an approximate solution in the form of a sequence of values x_i for $i = 0,\dots, N$, with $x_i \approx x(t_i)$ for some discrete time t_i. Two commonly used integration techniques are the forward Euler method given by:

$$x_i = x_{i-1} + h_i f\big(x_{i-1}\big), \quad i = 1, \dots, N, \tag{15.4}$$

where the ith time step, h_i, is given by $t_i - t_{i-1}$, and fourth-order Runge-Kutta (RK4) method defined by

$$x_i = x_{i-1} + 1/6 \left(F_1 + 2F_2 + 2F_3 + F_4\right)$$

$$F_1 = h_i f\big(x_{i-1}\big)$$

$$F_2 = h_i f\big(x_{i-1} + F_1/2\big)$$

$$F_3 = h_i f\big(x_{i-1} + F_2/2\big)$$

$$F_4 = h_i f\big(x_{i-1} + F_3\big)$$

The accuracy and stability of these methods critically depend on the time step, h_i. For massless particle trajectories in linearized flowfields, these stability requirements are well understood,[28] and one can use a variation of the criteria given in Darmofal and Haimes to ensure both accuracy and stability, and that the maximum error per unit time is bounded.[25]

Note that the forward Euler technique must be used with caution. In general, this first-order technique is more prone to error and stability problems than RK4. In fact, the author's experiments have shown that maintaining small global errors in a massless particle trajectory required excessively small time steps for the forward Euler technique, which led to large round-off error.[30] In practice, however, it was found that for the application problems considered here, only slight differences exist in the trajectories computed by forward Euler and RK4. Thus, the method is useful for depicting trends in large-scale flow features in near real time and can be used for initial analysis and product design.

15.3.3.2.3 Interactive velocity field analysis

To facilitate the study of the flue gas flows, we developed an interactive system that allows the user to initiate a streamline from any position within the virtual boiler or furnace. The starting point of the streamline is given by the location of the wand at the time of initiation. The streamline is calculated using the forward Euler technique and rendered in approximately 0.2 s.[4] By requesting a number of streamlines initiated from a two-dimensional cross-section of the boiler or furnace and displaying them simultaneously, large-scale structures in the flow can be easily identified and examined. For example, the middle image of Figure 15.13, shows the use of this technique in the Air Products application to examine the mixing characteristics in the combustion chamber. The rightmost image in Figure 15.13, shows a snapshot of multiple streamlines represented by animated vector glyphs to highlight the compression zone in the Fuel Tech tire incinerator. In each case, each point along the streamline is colored according to the scalar field of interest, in this case temperature, to provide the user with additional information about the simulation data set.

15.3.3.3 Particle Data

Many combustion applications contain massed or massed and evaporating particles as part of the simulation. One can either visualize the results of the simulation directly or include the computation of massed or massed and evaporating particles as part of the visualization environment for design and configuration of burners, nozzle injection systems, etc. This subsection describes the equations for massed and evaporating particles used to provide a simple modeling capability in the virtual reality simulation.

15.3.3.3.1 Massed and evaporating particles

The system of equations governing particles with mass is given by Equation (15.2) and

$$\partial V_p / \partial t = 18 \, \mu_g \left(V_g - V_p \right) / \left(\rho_p \, D^2 \right) + g \left(\rho_p - \rho_g \right) / \rho_p. \tag{15.5}$$

Here, μ_g is the viscosity of the fluid, ρ is the density, D is the particle diameter, and g is the gravitational acceleration vector. The first term on the right-hand side of Equation (15.5) is the fluid resistance imposed on the particle. A more general form of the fluid resistance includes a coefficient of drag, C_D, and a Reynolds number, N_{Re}. The term $C_D \, N_{Re}$ is absent from Equation (15.5) because it is assumed that the particle has a low Reynolds number and, hence, this term is equal to 24. The second term on the right-hand side of Equation (15.5) is the acceleration due to gravity. Note that by using the difference between the densities of the particle and fluid, buoyancy forces are included in this term as well.[4]

To include the effect of evaporation in the model, one again uses Equations (15.2) and (15.5). To efficiently account for the processes of heat and mass transfer, some simplifying assumptions are made: assume that the evaporation is heat transfer limited and that the droplet heating time is short compared with the droplet evaporation time. Thus, the temperature of the particle rises to near its boiling point and then begins to evaporate. This process is described by the equation

$$\partial T_p / \partial t = \left(N_{Nu} \pi k D \left(T_g - T_p \right) \right) / \left(m_p c_p^p \right), \tag{15.6}$$

where N_{Nu} is the Nusselt number, k is the thermal conductivity of the fluid, T_p and T_g are, respectively, the temperature of the particle and the gas, and c_p^p is the specific heat of the particle. Once a particle reaches its boiling temperature, all further heat gains from the fluid cause mass loss from evaporation without further changes in temperature. The evaporation is described by the equation

$$\partial m_p / \partial t = \left(Nu \pi k D \left(T_g - T_p \right) \right) / \left(H_v \right), \tag{15.7}$$

where H_v is the heat of vaporization of the particle.[4]

To obtain the particle's trajectory in the flow field, numerically integrate the systems of ordinary differential equations described above. Depending on the accuracy required, one can again use the forward Euler scheme or a fourth-order Runge-Kutta scheme. To maintain accuracy and stability for the nonlinear ODEs, use heuristic constraints to ensure that relative changes in particle characteristics are bounded. In particular, choose the time step that results in a change of less than 10% in particle velocity, temperature, and mass. Also enforce that a minimum number of time steps are taken in each cell.

15.3.3.3.2 Interactively modeling injector sprays

The massed, evaporating particle model was used to develop an interactive virtual reality simulation for the Fuel Tech application that allowed engineers to place injectors and quickly analyze the resulting spray coverage.[4] For each injector, compute a spray representation consisting of a large number of evaporating particles emanating from a nozzle. The orientation of the initial velocity vector is randomly distributed with respect to the orientation of the injector. In this case, sample distributions are chosen using a Gaussian distribution so that 95% of initial directions are contained in a 30° cone centered on the injector nozzle. The diameter of each particle is randomly sampled from a lognormal distribution around a distribution mean that is a linearly decreasing function from the cone center. The initial speed of every injected particle is assumed to be 30 m/s.

Up to 25 injectors can be placed on exterior boiler walls, and for each injector, we calculate 500 evaporating particle trajectories, which requires roughly 5 s per injector.[4] To highlight the areas of ideal injector placement with respect to temperature, we use a restricted color mapping with a range of 1650° to 1850°F. Fluid temperatures outside this range are indicated with lavender or dark purple for regions that are too hot or too cold, respectively. Thus, for maximum injector effectiveness, the evaporation of the particles (and hence the release of the chemical reagents) would take place primarily where the sprays are brightly colored. To optimize the injection system within the virtual environment, the user can interactively define and change the injector configuration. Each injector can be relocated by using the wand to select the injector and drag it to the new location. As it is relocated, a subset containing 20 of the 500 particle trajectories are computed and displayed as quickly as possible (roughly 0.5 s). Thus, the user can quickly identify potentially interesting locations by observing how the color map associated with the injector sprays changes as the injector location is changed. In addition, the specific spray configuration for each injector can be modified to study the effects of changes in the initial particle size, speed, and distribution.

Figure 15.14 shows the same injector location with three different spray configurations that vary in the initial mass of the injected particles and the shape of the injected spray. From right to left, the particles in the frames of Figure 15.14 are an average of 50 µm, 250 µm, and 750 µm in initial diameter. In the case where the particle size is very small, evaporation is almost immediate, and the injected spray covers very little of the flue gas flow in the boiler. As the particles increase in mass, they are no longer subject to immediate evaporation, and one clearly sees that their initial momentum carries them to the interior of the boiler domain. Finally, the third frame shows an injector configuration consisting of large particles in a flat fan. In this case, the momentum of the particles carries them across to the opposite wall and very little chemical reagent is released in the optimal zone.

15.4 CONCLUSIONS

The use of virtual reality simulations in engineering and scientific applications is more common now than at any time in the past. The affordability of the necessary computer and human-computer interface hardware, the development of software tools to enable application usage, and the research that demonstrates its utility for exploring three-dimensional simulation data sets have created an environment in which virtual reality applications are increasingly useful for engineering design and analysis. However, further work is required to reduce application development time and costs for these technologies to become widely accepted as industrial design and development tools.

FIGURE 15.14 Typical injector sprays for varying particle sizes and injector configurations. As the particle size increases, the momentum of the particles carries them further into the boiler interior and their time residence in the flue gases increases.

REFERENCES

1. I. Sutherland, The ultimate display, *Information Processing 1965, Proceedings IFIPS Congress*, 2, 506-508, 1965.
2. J. Vince, *Virtual Reality Systems,* ACM Press, 1995.
3. N. Durlach and A. Mavor, Eds., *Virtual Reality Scientific and Technological Challenges,* National Academy Press, Washington, D.C., 1995.
4. D. Diachin, L. Freitag, D. Heath, J. Herzog, B. Michels, and P. Plassmann, Collaborative virtual environments used in the design of pollution control systems, *The International Journal for Supercomputing Applications and High Performance Computing*, 10.2, 223-235, 1996.
5. M. McKenna and D. Zeltzer. Three-dimensional visual display systems for virtual environments, *Presence*, 1(4), 421-458, 1992.
6. R. Haber and M. Hershenson, *The Psychology of Visual Perception,* Holt, Rinehart, and Winston, New York, 1980.
7. T. Okoshi, *Three-dimensional Imaging Techniques,* Academic Press, New York, 1976.
8. J. Vince, *Essential Virtual Reality,* Springer-Verlag, London, 1998.
9. I. Sutherland, A head mounted three-dimensional display, *AFIPS Conference Proceedings*, 33, Part 1:757-764, 1968.
10. C. Cruz-Neira, D.J. Sandin, and T.A. DeFanti. Surround-screen projection-based virtual reality: the design and implementation of the CAVE, in *ACM SIGGRAPH 93 Proceedings*, ACM, 1993, 135-142.
11. W. Kruger and B. Frohlich, The responsive workbench, *IEEE Computer Graphics and Applications,* 14(3), 12-15, 1994.
12. E.G. Shaw, The external ear, in *Handbook of Sensory Physiology*, Springer-Verlag, New York, 1974.
13. E.M. Wenzel, M. Arruda, D.J. Kistler, and F.L. Wightman. Localization using non-individualized head-related transfer functions, *Journal of the Acoustical Society of America,* 94, 111-123, 1993.
14. G. Kramer, Ed., Auditory display: sonification, audification, and auditory interfaces, in *Proceedings of the Santa Fe Institute Studies in the Sciences of Complexity*, Vol. 18, Reading, MA, 1994, Addison-Wesley.
15. F.P. Brooks, M. Ouh-Young, and J. Batter, Project GROPE: haptic displays for scientific computation, *Computer Graphics*, 24(4), 177-185, 1990.
16. L. Freitag and T. Urness, Comparative visualization techniques to analyze aluminum smelting furnace efficiency in a virtual reality environment, P. Banerjee and T. Kesavadas, Eds., in *Industrial Virtual Reality: Manufacturing and Design Tool for the Next Millennium. Proceedings of the 1999 ASME International Mechanical Engineering Congress and Exposition*, 1999, 191-199.
17. FLUENT Corporation, http://www.fluent.com, 1999.
18. PHOENICS Reference Manual (CHAM/TR200), Phoenics, London, 1991.
19. T. L. Disz, M. E. Papka, M. Pellegrino, and R. L. Stevens. CAVEcomm User's Manual Document, World Wide Web Document, 1995, http://www.mcs.anl.gov/FUTURES_LAB/VR/APPS/C2C/.
20. W. Schroeder, K. Martin, and B. Lorensen, *The Visualization Toolkit, An Object-Oriented Approach to 3D Graphics,* Prentice-Hall PTR, Upper Saddle River, NJ, 1998.
21. W.E. Lorensen and H.E. Cline. Marching cubes: a high resolution 3D surface construction algorithm, *Computer Graphics*, 21(3), 163-169, 1987.
22. G. Nielson and B. Hamann, The asymptotic decider: resolving the ambiguity in marching cubes, in *Proceedings of Visualization '91*, IEEE Computer Society Press, Los Alamitos, CA, 1991, 83-91.
23. P. Lacroute and M. Levoy, Fast volume rendering using a shear-warp factorization of the viewing transformation, in *Proceedings of SIGGRAPH '94*, Vol. 25, Addison-Wesley, Reading, MA, 1994, 451-458.
24. K. Kim and A. Pang, Projection-based Data Level Comparisons of Direct Volume Rendering Algorithms, Technical Report UCSC-CRL-97-16, University of California at Santa Cruz, 1997.
25. H. Pagendarm and F. Post, Comparative visualization — Approaches and examples, in *Visualization in Scientific Computing*, Springer, 1995, 95-108.
26. H. Pagendarm and F. Post, Comparative visualization of flow features, in *Scientific Visualization: Overviews, Methodologies, and Techniques*, Nielson, Hagen, and Muller, Eds., CS Press, 1997, 211-227.
27. J. Trapp and H. Pagendarm, Data level comparative visualization in aircraft design, *Proceedings of Visualization '96*, 1996, 393-396.

28. D.L. Darmofal and R. Haimes. An analysis of 3D particle path integration algorithms, *Journal of Computational Physics*, 123, 182-195, 1996.

29. D. Diachin, L. Freitag, D. Heath, J. Herzog, and W. Michels, Interactive simulation and visualization of massless, massed, and evaporating particles, *Institute of Industrial Engineers Transactions,* Kluwer Academic Press, 30(7), 621-628, 1998.

30. D. Diachin and J. Herzog, Analytic streamline calculations for linear tetrahedra, in *13th AIAA Computational Fluid Dynamics Conference*, 1997, 733-742.

TRADEMARK SYMBOLS

Index

Index

Index

A

G

Printed and bound by CPI Group (UK) Ltd, Croydon, CR0 4YY

23/10/2024

01778254-0008